Chemistry

Chemistry

6th Edition

Rhobert Lewis
PhD FRSC
Chair of the Board NPTC Group of Colleges, Honorary Fellow and former Dean of Engineering and Science, the University of South Wales

Wynne Evans
PhD FRSC
Formerly Head of Chemistry at the University of South Wales and Associate Lecturer in Chemistry for the Open University

BLOOMSBURY ACADEMIC
LONDON • NEW YORK • OXFORD • NEW DELHI • SYDNEY

BLOOMSBURY ACADEMIC
Bloomsbury Publishing Plc, 50 Bedford Square, London, WC1B 3DP, UK
Bloomsbury Publishing Inc, 1359 Broadway, New York, NY 10018, USA
Bloomsbury Publishing Ireland, 29 Earlsfort Terrace, Dublin 2, D02 AY28, Ireland

BLOOMSBURY, BLOOMSBURY ACADEMIC and the Diana logo are trademarks of Bloomsbury Publishing Plc

First published in Great Britain 1997
Second edition published 2001
Third edition published 2006
Fourth edition published 2011
Fifth edition published 2018
This edition published 2026

Cover design: Jade Barnett
Cover image © ggw/Adobe Stock

A catalogue record for this book is available from the British Library.

ISBN: PB: 978-1-350-56661-3
ePDF: 978-1-350-56662-0
eBook: 978-1-350-56663-7

Typeset by Integra Software Services Pvt. Ltd.
Printed and bound in India

For product safety related questions contact productsafety@bloomsbury.com.

To find out more about our authors and books visit www.bloomsbury.com and sign up for our newsletters.

The sixth edition is dedicated to the late Dr Wynne Evans, able scientist, co-author, colleague, friend and a proud member of the 'Chemistry Set Generation'.

Periodic Table of Elements

Proton (atomic) number → 1
H
Hydrogen
Atomic mass/u → 1.008

Key
- Non-metals
- Main group metals
- d-block elements or transition metals, lanthanoids and actinoids

Groups	s 1	2	d 3	4	5	6	7	8	9	10	11	12	p 13	14	15	16	17	18
	1*	2											3	4	5	6	7	0
Periods 1							1 H Hydrogen 1.008											2 He Helium 4.003
2	3 Li Lithium 6.941	4 Be Beryllium 9.012											5 B Boron 10.811	6 C Carbon 12.011	7 N Nitrogen 14.007	8 O Oxygen 15.999	9 F Fluorine 18.998	10 Ne Neon 20.180
3	11 Na Sodium 22.990	12 Mg Magnesium 24.305											13 Al Aluminium 26.982	14 Si Silicon 28.086	15 P Phosphorus 30.974	16 S Sulfur 32.066	17 Cl Chlorine 35.453	18 Ar Argon 39.948
4	19 K Potassium 39.098	20 Ca Calcium 40.078	21 Sc Scandium 44.956	22 Ti Titanium 47.88	23 V Vanadium 50.942	24 Cr Chromium 51.996	25 Mn Manganese 54.938	26 Fe Iron 55.847	27 Co Cobalt 58.933	28 Ni Nickel 58.69	29 Cu Copper 63.546	30 Zn Zinc 65.39	31 Ga Gallium 69.723	32 Ge Germanium 72.61	33 As Arsenic 74.922	34 Se Selenium 78.96	35 Br Bromine 79.904	36 Kr Krypton 83.80
5	37 Rb Rubidium 85.468	38 Sr Strontium 87.62	39 Y Yttrium 88.906	40 Zr Zirconium 91.224	41 Nb Niobium 92.906	42 Mo Molybdenum 95.94	43 Tc Technetium (98)	44 Ru Ruthenium 101.07	45 Rh Rhodium 102.906	46 Pd Palladium 106.42	47 Ag Silver 107.868	48 Cd Cadmium 112.411	49 In Indium 114.82	50 Sn Tin 118.71	51 Sb Antimony 121.76	52 Te Tellurium 127.60	53 I Iodine 126.904	54 Xe Xenon 131.29
6	55 Cs Caesium 132.905	56 Ba Barium 137.327	57 La Lanthanum 138.906	72 Hf Hafnium 178.49	73 Ta Tantalium 180.948	74 W Tungsten 183.85	75 Re Rhenium 186.207	76 Os Osmium 190.2	77 Ir Iridium 192.22	78 Pt Platinum 195.08	79 Au Gold 196.967	80 Hg Mercury 200.59	81 Tl Thallium 204.383	82 Pb Lead 207.2	83 Bi Bismuth 208.980	84 Po Polonium (209)	85 At Astatine (210)	86 Rn Radon (222)
7	87 Fr Francium (223)	88 Ra Radium 226.025	89 Ac Actinium (227)	104 Rf Rutherfordium (267)	105 Db Dubnium (268)	106 Sg Seaborgium (269)	107 Bh Bohrium (270)	108 Hs Hassium (269)	109 Mt Meitnerium (278)	110 Ds Darmstadtium (281)	111 Rg Roentgenium (280)	112 Cp Copernicium (285)	113 Nh Nihonium (286)	114 Fl Flerovium (289)	115 Mc Moscovium (289)	116 Lv Livermorium (293)	117 Ts Tennessine (294)	118 Og Oganesson (294)

f

Lanthanoid series	58 Ce Cerium 140.115	59 Pr Praseodymium 140.908	60 Nd Neodymium 144.24	61 Pm Promethium (145)	62 Sm Samarium 150.36	63 Eu Europium 151.965	64 Gd Gadolinium 157.25	65 Tb Terbium 158.925	66 Dy Dysprosium 162.50	67 Ho Holmium 164.93	68 Er Erbium 167.26	69 Tm Thulium 168.934	70 Yb Ytterium 173.04	71 Lu Lutetium 174.967
Actinoid series	90 Th Thorium 232.038	91 Pa Protactinium 231.036	92 U Uranium 238.029	93 Np Neptunium 237.048	94 Pu Plutonium (244)	95 Am Americium (243)	96 Cm Curium (247)	97 Bk Berkelium (247)	98 Cf Californium (251)	99 Es Einsteinium (252)	100 Fm Fermium (257)	101 Md Mendelevium (258)	102 No Nobelium (259)	103 Lr Lawrencium (262)

*Older numbering system of Groups

An interactive periodic table can be found at https://periodic-table.rsc.org/

Contents

Preface

This book is designed to provide a course in 'chemical literacy' for those studying access and foundation courses at university or college, and for non-specialists on courses such as Forensic Science, Engineering or Health Sciences, for whom chemistry is a subsidiary subject in their first year at university. Much of the material will be suitable also for chemistry students at school or college. Teachers will wish to tailor their course to the needs of their students by selecting units and parts of units according to the background and the qualification being studied. Of the first twenty-two units, two (Gibbs Energy Changes and Organic Mechanisms) represent more advanced work while the special topics (Units 23–29) serve as case studies, allowing students to apply their knowledge to topics of practical importance.

Science involves abstraction and generalization. It is about applying the same principles to different cases, but its history and industrial and economic importance are rooted in practical work. Michael Faraday wrote that 'Chemistry is necessarily an experimental science; its conclusions are drawn from data, and its principles supported by evidence from facts'. Ideally, the work covered in this book should be supported by a judicious choice of laboratory work in which students take their own measurements, make their own observations and learn from their own mistakes.

This book starts 'from scratch', assuming only a minimal experience of mathematics and science. A key feature of each unit is the exercises which consolidate topics before the student proceeds further. The questions at the end of units are designed to extend and reinforce learning. To help students working mainly alone, the answers to all exercises and problems are provided at the back of the book.

The sixth edition has been revised throughout. The units on Medicinal Chemistry and Pollution have been largely re-written and there are two new units on Climate Change and Decarbonization and on Nanochemistry.

The website that accompanies the sixth edition may be found at https://www.bloomsburyonlineresources.com/chemistry and contains extensions to the units in the book.

Although this is a book for students, I hope that it will help and not hinder the teacher's task of coping with large groups, often with widely ranging ability and with differing subject interests. Accordingly, I welcome suggestions for improvements to the book and website from students and teachers alike.

RhL.
Brecon
April 2026

To the Student: How to Use this Book

- This book is designed so that you can work with the minimum of help.
- It has a glossary of key definitions, which have been highlighted in the units as **bold blue** text.
- Obtain advice on the topics (and within units, the sections) that are relevant to your studies at this stage. If you possess a sound knowledge of chemistry, you will find that much of Units 2, 4 and 6 is already known to you.
- There are exercises embedded in the text. Complete these as you go along.
- The boxes provide additional material (including some applications) that are not essential to the main body of the text and you may return to them after you have mastered the rest of the unit.
- Extension material is included in a website at https://www.bloomsburyonlineresources.com/chemistry. You may be tempted to skip Unit 1 (Numbers and Measurements) but if you are unfamiliar with this material or lack confidence with numbers and calculations, any time spent on this unit will prove invaluable in later units.

DRAWING CHEMICAL STRUCTURES

There are many software packages that allow chemical structures to be displayed on screen. Free ones include the RSC molecular database ChemSpider (https://www.chemspider.com) and ADC/ChemSketch (Freeware at https://www.acdlabs.com/resources/free-chemistry-software-apps/chemsketch-freeware). To see how these two programs work together, enter the ChemSpider site and click on 'Simple'. Enter the word 'Benzene' in the search box and press 'return'. Next press 'Download.mol'. This creates a file (e.g. '236 (1).mol') that is downloaded on your PC or laptop. Click on the file and the ChemSketch window appears along with the molecular structure of benzene. Clicking the ChemSketch icons '3D Optimization', followed by '3D view', generates a '3D ball-and-stick' model of benzene.

Acknowledgements

The authors and publishers wish to thank the following for permission to use copyright material: Hutchinson Education for Figure 11.3 from Surface Phenomena (1972) by S.R. Rao, Figure 27, p. 111. American Chemical Society for Figure 23.2 (b) from J. Chem. Educ. 2021, 98, 12, 3831–40. C. Zeng et al. for Figure 29.5 from Science Advances.

Photographs:
Sartorius UK Limited, p. 4; LGC Group, pp. 8, 149; Erwin Rossen, p. 15; United States Patent and Trademark Office, p. 20; Chip Clark, p. 18; Chemical Heritage Foundation Collections, p. 23; Juhele_CZ, p. 25; UKAEA, p. 27; Acoma, p. 32; Angharad and Catrin Lewis, p. 33; Reliance Precision Ltd, p. 37; Science Photo Library, pp. 50, 478, 494; The College of Chemistry, University of California, Berkeley, p. 61; Dover Publications, p. 61; Wellcome Library, pp. 62, 363, 466, 471, 459; Dr J.A. Stroscio, p. 74; Clint__Budd, p. 77; D-Stanley, p. 78; Zephyris, p. 80; Corbis, p. 79; Kilrock, p. 88; Flora Townsend, p. 88; Ted Bradley, p. 88; Gordon Roberts Photography, p. 89; Rufus Townsend, p. 91; A Meyers 91, p. 95; Betterware, p. 97; Patounis Soap Factory, p. 97; Wilco Oelen, pp. 101, 203; National Museum of Health and Medicine, p. 100; Shutterstock, pp. 103, 464; Ben Miles, p. 110; Lion Laboratories Ltd, pp. 108, 453; Edgar Fahs Smith Collection, Kislak Center for Special Collections, Rare Books and Manuscripts, University of Pennsylvania, p. 136; LGC Group, p. 151; Wolf Blass Wines International, p. 152; Keystone/Stringer/GettyImages, p. 168; Sven Horstmann, p. 174; Caspar Benson/GettyImages, p. 175; Ken Graham/Greenpeace, p. 179; Virojt Changyencham/GettyImages, p. 187; Mariana Ruiz Villarreal, p. 189; NASA, pp. 193, 402; Dennis S.K. Collection, p. 195; Dr.T, p. 197; Pslawinski, p. 205; Benjah-bmm27, p. 208; Hulton Archive / Stringer/GettyImages, p. 213; shank_ali, p. 218; IKA, p. 223; simonkr/GettyImages, p. 229; NASA, p. 233; IKEA, p. 244; Creativity103, p. 246; J. Allan Cash Ltd, p. 278; APS, p. 286; Jenway, p. 300; Camlab, p. 314; Horia Varlan, p. 326; Science & Society Picture Library, p. 341; Alamy, p. 341; Hetty Simpson, pp. 343, 487; Brecon Estate California, p. 347; Science and Society Picture Library/GettyImages, p. 365; Imperial College Archives, p. 377; Bettmann/GettyImages, p. 381; Agilent Technologies, Inc., p. 391; Analytic Jena AG, p. 400; A Krüss Optronic, p. 402; georgeclerk/GettyImages, p. 405; Jon Sullivan/GettyImages, p. 415; Bruker Corporation, p. 416; raclro/GettyImages, p. 427; Cary Bass, p. 248; R. G. Haire, US Department of Energy, p. 428; Press Association, pp. 434, 435; National Archives at Chicago, p. 435; NASA Earth Observatory, p. 439; kasio69, p. 441; Ralf Roletschek, p. 443; US Fish and Wildlife Service, p. 444; Matilda Simpson, p.477; Dr John Gray, p. 454; John Leech, p. 456; Bengt Nyman, p. 461; Ann Shaw and Carole Reeves, p. 462; Robert Kato, p. 464; Deutsche Bundesbank, Frankfurt am Main, p. 466; Kuni Takahashi/GettyImages, p. 469, Tim Ereneta, p. 469, MirageC/GettyImages, p. 470; Sebastian Schlüter; p. 472, Shutterstock, p. 475; Helen Lewis, p. 479; Keystone / Stringer/GettyImages, p. 479; Stfan H. Doerr, p. 480; Hulton Archive/Getty Images, p. 480; HejRonja, p. 484; Rolf Schulten/Bloomberg/GettyImages, p. 489; M J Richardson, p. 489; Thomas Wyke, p. 490; Sureshbup, p. 492; Aleksandar Kondinski, p. 495; Eric Wieser, p. 497; AlexanderAIUS, p. 497; OSRAM, p. 497; Guillaume Pelletier, p. 498; Tetiana Zhabska/Alamy, p. 498; Michael D. Buschmann/Manuel J. Carrasco/Suman Alishetty/Mikell Paige/Mohamad G/Alameh/Drew Weissman, p. 499; Drew Weissman, p. 499; Katalin Karikó, p. 499

It is a pleasure to acknowledge two of my mentors. The first is my former head in the Department of Physical Chemistry at Cambridge, the late Professor Sir John Thomas FRS, who has provided so much inspiration over the years. The second is Professor Bill George FRSC, whose infectious enthusiasm for chemistry kick-started both my and Wynne's academic careers.

Useful Physical Constants

Name	Value
Avogadro's constant (N_A)	6.022×10^{23} mol^{-1}
Charge on the electron	1.602×10^{-19} C
Atomic mass unit (u)	1.66054×10^{-24} g
Mass of an electron	9.10939×10^{-28} g
Mass of a proton	1.67262×10^{-24} g
Mass of a neutron	1.67493×10^{-24} g
Planck's constant (h)	6.626×10^{-37} kJ s
Planck's constant per mol	3.990×10^{-13} kJ s mol^{-1}
Speed of light in a vacuum (c)	2.998×10^{8} m s^{-1}
Universal gas constant (R)	8.3145 J mol^{-1} K^{-1}
Ionic product constant (K_w) at 25°C	1.008×10^{-14} mol^2 dm^{-6}
Specific heat capacity of water	4.18 J g^{-1} K^{-1}
Molar volume of an ideal gas at 20°C and 101 kPa	24.06 dm^3 (≈ 24 dm^3)
Molar volume of an ideal gas at 25°C and 100 kPa	24.79 dm^3

General Data

Standard ambient temperature and pressure (SATP) are 25°C and 100 kPa.

Conversion Factors

Volume	Pressure	Temperature
1 m^3 = 1000 dm^3 1 dm^3 = 10^{-3} m^3	1 atm = 760 torr = 101 325 Pa = 101.325 kPa 1 bar = 100 kPa	0 K = −273.15°C T(K) = T(°C) + 273.15

Using the NIST Website

http://webbook.nist.gov/chemistry

The National Institute of Standards and Technology (NIST) website is a huge free data library which gives information on many thousands of compounds. It is worthwhile becoming familiar with this site. Upon entering the website you are faced with a choice of search options. We will illustrate a search for the compound *benzene*, C_6H_6. Choose the search option 'Name'. Type in the name 'benzene' and click the boxes under 'other data': 'IR spectrum' and 'mass spectrum'. Then press 'Search'. The results of your search provide you with the formula of benzene, its IR spectrum (actually, a choice of spectra in gas and liquid phases) and its mass spectrum. A 3-D structure of your chosen molecule is also available. The CAS Registry Number is unique to that molecule. Try searching for benzene through its formula (C6H6) without specifying its name and without using subscripts. Lastly, try searching for a bigger molecule, such as the narcotic drug *cocaine*. The NIST database is so useful that it might be worthwhile setting up a shortcut icon on your tablet, PC or phone.

Units and Measurement

1

OBJECTIVES

This unit:

- Introduces standard notation
- Tests you on the use of your calculator
- Describes how to work out the units of a quantity
- Defines accuracy and precision
- Explains the different types of errors

CONTENTS

1.1 Very small and very big numbers

Science often involves very large and very small numbers. Such numbers may be cumbersome to write down, and an abbreviated notation (known as 'standard' or 'scientific' notation) is often used.

Think of the number 100. This is the same as 1×100. In standard notation we write this as 1×10^2. Similarly,

2300 becomes $2.3 \times 1{,}000$ or 2.3×10^3

6,749,008 becomes 6.749008×10^6

0.0012450 becomes 1.2450×10^{-3}

EXERCISE **1A**

Standard notation

Express the following in standard notation

(i)	0.0000345	**(iv)**	3.5
(ii)	300000000	**(v)**	602200000000000000000000
(iii)	0.0820575	**(vi)**	17

Sometimes, we are only interested in approximate values. The symbol ≈ means 'approximately', e.g. $6.023 \times 10^{23} \approx 6 \times 10^{23}$.

1.2 Logarithms

Logarithms to the base 10

The logarithm (or 'log') of a number to the base ten is the power to which the number 10 has to be raised in order to equal that number. For example, $100 = 10^2$. Therefore, the log of 100 is 2. Similarly, since $0.0001 = 1 \times 10^{-4}$, the log of 0.0001 is –4.

What is the log of 150? The log of 150 is the value of x in the expression

$$150 = 10^x$$

We carry out this operation on a calculator. In many types of calculator, this is done by entering *log*, the *number*, and then pressing the = button. The log of 150 is 2.176. We write this as

$$\log(150) = 2.176$$

What if you are provided with the log of x and asked to find x? Using the above example, how do we get back to 150 from 2.176? To do this we would need to evaluate $10^{2.176}$. (To carry out this operation on a scientific calculator we use the 10^x key, a common sequence of operations being *shift*, *10^x^*, *number*, and =.) We then write

$$10^{+2.176} = 150$$

Similarly,

$$10^{-0.9104} = 0.1229$$

Logarithms to the base e (natural logs)

The symbol 'e' is a mathematical constant (like π) where

$$e = 2.718\ldots$$

The logarithm of a number to the base e is the power to which the number e has to be raised in order to equal that number. For example, $e^{3.912} = 50$ so that the natural log of 50 is 3.912. In this book we symbolize natural logs as 'ln', e.g.

$$\ln(50) = 3.912$$

It follows from the definition of natural logs that $\ln(e^x) = x$.

Manipulating logarithms

The following general formulae are useful and apply to logs of any base:

$$\log(ab) = \log a + \log b$$

$$\log\left(\frac{a}{b}\right) = \log a - \log b$$

For example,

$$\log\left(\frac{yz}{km}\right) = \log y + \log z - \log k - \log m$$

Using your calculator

We are now in a position to summarize the type of calculations you need to be able to do on your calculator in preparation for later units. You will need to be able to

1 enter numbers in standard notation form;
2 add, subtract, divide and multiply numbers;
3 square numbers and find their square roots;
4 use the calculator memory;
5 calculate log x, ln x, e^x and 10^x.

The way you carry out such calculations varies slightly according to the make of your calculator. Refer to the calculator instructions for further information – or ask a knowledgeable friend! Now try Exercise 1B.

EXERCISE **1B**

Quick test on calculator use

Use your calculator to evaluate the following:

(i) $\left(\frac{45.6}{2.34}\right)^2$

(ii) $\sqrt{300.7}$

(iii) log (1.2×10^{-2})

(iv) $10^{-4.56}$

(v) ln (0.178×8.456)

(vi) $e^{-5.20}$

(vii) $e^{-E/RT}$, where E = 30 000, R = 8.3145 and T = 298

1.3 Units

International system of units

The international system of units (usually known as SI units, from the French *Système International*) consists of several *base units* from which all other units (such as those of volume or energy) are derived. Some of the base units are shown in Table 1.1.

Because the base units are sometimes too large or too small for use, SI prefixes (Table 1.2) are used to produce smaller or bigger units. For example, the milligram (= 0.001 g, and symbolized mg) is used if we are reporting small masses.

The cubic metre (written m^3) is too large for most purposes in chemistry, and the cubic decimetre, dm^3 (or litre) is commonly used. There are 1,000 dm^3 in 1 m^3. Also, there are 1,000 cubic centimetres (cm^3) in a cubic decimetre (see Figure 1.1). Summarizing,

$$1\ m^3 = 1000\ dm^3 = 1000000\ cm^3$$

Table 1.1 Base units

Property	Base unit	Symbol for unit
Mass	kilogram	kg
Length	metre	m
Time	second	s
Temperature	kelvin	K
Amount of substance	mole	mol

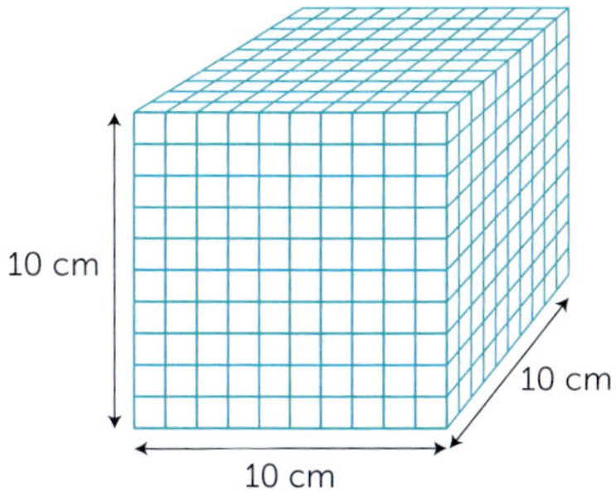

Figure 1.1 There are 1,000 cm^3 in 1 dm^3.

Table 1.2 SI Prefixes

Value	Prefix	Symbol
10^9, billion	giga-	G
10^6, million	mega-	M
10^3, thousand	kilo-	k
10^{-1}	deci-	d
10^{-2}	centi-	c

Value	Prefix	Symbol
10^{-3}	milli-	M
10^{-6}	micro-	μ
10^{-9}	nano-	N
10^{-12}	pico-	P
10^{-15}	femto-	F

You will also meet units raised to negative powers, such as m^{-3} (read as 'per metre cubed'). Remember,

$$m^{-3} = \frac{1}{m^3}$$

Now try Exercise 1C.

EXERCISE **1C**

Working with different units

(i) The radius of the hydrogen atom is approximately 40 pm (picometres). Convert this to nm (nanometres).

(ii) 1.0 cm^3 of a solution contains 0.010 g of salt. What is the mass of salt contained in 1 dm^3 of the same solution?

(iii) The wavelength of orange-yellow light is roughly 600 nm. Express the wavelength in metres.

Amount of substance

One of the most important physical quantities in chemistry is the *amount of substance*, which has the unit of **mole** (symbol mol). The greater the number of particles (atoms, ions or molecules) in a piece of material, the greater is the amount of substance. **Concentration** is another physical quantity which is particularly important to chemists. Concentration is a measure of the packing of particles per unit volume, and is commonly expressed in the units of mol per dm^3, usually written as mol dm^{-3}. Moles and concentration are dealt with in more detail in Units 8 and 9.

Temperature and energy

The hotness or coldness of a material is called its temperature. The units of temperature used in science are degrees **Celsius** (°C) or **kelvin** (K). For more details, see pages 157–158.

The SI unit of *energy* is the **joule** (J). 1 joule is the energy used up in pushing against a force of 1 newton (N) over a distance of 1 metre. This means that we can write

1 J = 1 Nm

Units of physical quantities

A physical quantity consists of a number and a unit. For example, suppose we measure the volume of a block and find it to be 4.5 cm^3:

4.5	cm^3
↑	↑
the number	the unit

Mathematically, the physical quantity consists of a number multiplied by a unit:

$$\text{physical quantity} = \text{number} \times \text{unit} \quad \textbf{(1.1)}$$

In our example,

$$\text{physical quantity (i.e. volume)} = 4.5 \times \text{cm}^3$$

For convenience, the physical quantity is usually written without the multiplication sign – here as 4.5 cm^3. This may be compared with the algebraic expression $4.5y$ (i.e. $4.5 \times y$).

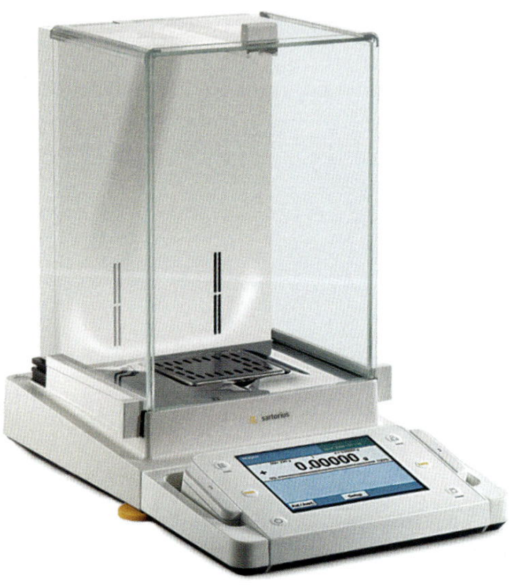

A laboratory balance. This balance measures masses as low as 0.0001 g.

Labelling axes on graphs

Suppose that we are plotting the volume of a gas (V, in dm^3) against the temperature of the gas (T, in kelvin, K). First, consider the y-axis. We might be tempted to label this axis as 'V (dm^3)'. However, we are not actually plotting the quantity *volume* but simply the *number* part of the quantity. Rearrangement of equation (1.1) shows that

$$\text{number} = \frac{\text{physical quantity}}{\text{unit}}$$

This shows that in plotting the number part we are really plotting

$$\frac{\textit{physical quantity}}{\text{unit}}$$

Hence the y-axis is labelled Volume/dm^3. This is usually written as V/dm^3. Similar reasoning leads to the x-axis being labelled Temperature/K or T/K.

Deriving the units of a quantity

To illustrate the derivation of the units of a quantity, consider the following question: with mass in kg and the volume in m^3, what are the units of density?

We start with the definition of **density**:

$$\text{density} = \frac{\text{mass}}{\text{volume}}$$

We find the units of density by substituting the units of mass and volume into the equation defining density:

$$\text{units of density} = \frac{\text{unit of mass}}{\text{units of volume}} = \frac{\text{kg}}{\text{m}^3} = \text{kg m}^{-3}$$

The *units of density* are therefore kilograms per cubic metre (see Example 1.1).

EXAMPLE **1.1**

The ratio $\frac{E}{RT}$

occurs frequently in chemistry. *E* is an energy per mole of substance (in units of joules per mole, symbolized J mol^{-1}), *T* is the temperature (in K) and *R* is a universal constant with the units of $\text{J mol}^{-1}\,\text{K}^{-1}$ (read as joule per mole per kelvin). What are the units of *E/RT*?

Answer

$$\text{units of } \frac{E}{RT} = \frac{\text{J mol}^{-1}}{\text{J mol}^{-1}\,\text{K}^{-1}\text{K}}$$

Cancelling like units and remembering that K is really K^{+1}

$$\frac{\cancel{\text{J}}\ \cancel{\text{mol}^{-1}}}{\cancel{\text{J}}\ \cancel{\text{mol}^{-1}}\ \cancel{\text{K}^{-1}}\ \cancel{\text{K}}}$$

so that *E/RT* is unitless, i.e. it is simply a number.

Comment

The most commonly encountered form of this expression is $e^{-E/RT}$ (e = 2.718). Note that as *E* increases, $e^{-E/RT}$ *decreases*. Now try Exercise 1D.

EXERCISE **1D**

Deriving units

(i) The energy (*q* joules) needed to raise the temperature of a material (of mass *m* grams) by ΔT kelvin may be calculated by the equation

$$q = m \times C \times \Delta T$$

where *C* is the specific heat capacity of the material. What are the units of *C*?

(ii) The mass *m* (in grams) of an amount of substance *n* (in moles) is related to the molar mass M of that substance by the expression

$$m = n \times M$$

What are the units of *M*?

1.4 Errors in experiments

Types of experimental error

Suppose we measure the temperature of a liquid. The difference between a *single* measurement of temperature and the *true* temperature is the *absolute error* of the measurement. The **absolute error** can have a positive or a negative value. Generalizing:

Absolute error = experimental value − true value

It is always good practice to repeat measurements. For example, if we are measuring the concentration of pesticide in a lake, we might fill a large bottle with lake water and later (back in the laboratory) withdraw 50 cm^3 portions of the lake water from the bottle and analyse each portion for pesticide using the same analytical technique. The *mean* pesticide concentration is then obtained by averaging the concentrations found in each 50 cm^3 portion. In this case, the absolute error in the pesticide measurements is the difference between the *mean* measurement and the true concentration.

There are two main kinds of error that we need to consider in experimental measurements.

1 **Random errors**, random because they cause repeat measurements on the same sample to go up and down. Random errors will cause successive measurements to be scattered, although averaging a large number of such measurements will produce a mean measurement which may not be greatly affected by random error. However, repeating a measurement many times may not be practical (for example, there may be too little sample available, or the measurements might be too time consuming or expensive to carry out). It is for this reason that precise measurements are highly desirable.

2 The second type are called **systematic errors**. A systematic error affects *all* repeat measurements, and makes all measurements either higher or lower than the 'true value'. Systematic errors *do not* average out, no matter how many repeat measurements are made.

Examples of random errors

Random errors are introduced whenever there is a *subjective* part to the experiment (such as estimating when a solution has reached the mark in a pipette, or recognizing the onset of a colour change during a titration), *or* where the experimental measurements are influenced by rapidly fluctuating conditions (e.g. air draughts).

Examples of systematic errors

A simple example of a systematic error is provided by a balance. Balances are often set to read a mass of zero before being used to weigh a sample. Suppose a speck of dust falls upon the pan of a balance *after* zeroing. This will cause the indicated mass of any object to be greater than the true mass. For example, if the speck has a mass of 0.0001 g, all objects will have an apparent mass which is 0.0001 g *too high*.

Another example of systematic error involves the analysis of chromium in blood. If the blood samples are stored in stainless steel vessels prior to analysis, then some chromium may dissolve out of the stainless steel into the sample. This introduces a systematic error which causes the measurement (here the chromium concentration) to be *overestimated*.

Systematic errors are often difficult to recognize, particularly in measurements of the concentration of substances (**quantitative analysis**) in which the concentrations of materials are being found in the presence of substances that *interfere* with the measurement (see Box 1.1).

EXERCISE **1E**

Systematic errors

In order to compare the alcohol content of several wines, a student poured samples of each wine into open test tubes and the next day analysed each for alcohol using a standard analytical technique. Comparison of the student's results with those obtained by other laboratories showed that her alcohol concentrations were consistently low. Suggest one reason for the systematic error. (Hint: What happens to wine when it is left open to the air?)

BOX **1.1**

Example of systematic errors – the analysis of aluminium ions (Al^{3+}) in tea

One way of determining the concentration of Al^{3+} in tea is to add a carefully selected complexing agent (usually a complicated organic compound) to the tea (Figure 1.2). In our example, this combines with the Al^{3+} ion to produce a red coloured substance (a 'coloured **complex**'):

$$Al^{3+} + \text{complexing agent} \rightarrow \underset{\text{(red colour)}}{Al^{3+}\text{ complex}}$$

The stronger the concentration of Al^{3+} in the tea, the stronger will be the intensity of the red colour.

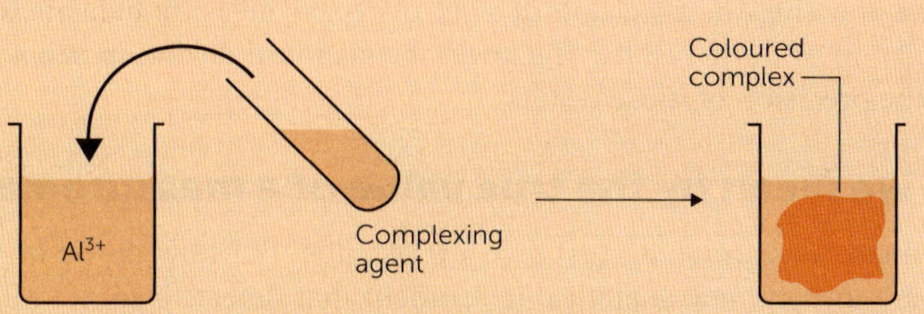

Figure 1.2 Determination of Al^{3+} ion by coloured complex formation.

However, if the tea contains traces of heavy metal ions (such as copper), it happens that these ions will also form red coloured compounds with the complexing agent. If we presumed that all the red colour was due to the Al^{3+} complex, this would produce a systematic error in which the Al^{3+} concentration in the tea is *overestimated*.

If the tea contains traces of fluoride ions (F^-), these ions react directly with the Al^{3+} producing stable aluminium fluoride complexes and so preventing the Al^{3+} ions from reacting with the complexing agent. This leads to a systematic error in which the Al^{3+} concentration in the tea is *underestimated*.

In an ideal measurement, we would separately measure the concentrations of ions (F^-, Cu^{2+} etc.) which interfere with the measurements of the Al^{3+} ion, and correct the measured Al^{3+} concentration accordingly.

Accuracy and precision

Repeat measurements on the same sample that are close together are said to be *precise*:

Precise measurements have a small random error.

A measurement which is close to the true value is said to be *accurate*:

Accurate measurements have a small systematic error.

The 'rifle shooting analogy' helps us to distinguish between **accuracy** and **precision** (Figure 1.3). In a rifle competition, the aim is to hit the bullseye. Competitor A is a precise shot (the shots are close together) but inaccurate (no bullseye); B is a precise and accurate shot (three bullseyes); C is neither precise nor accurate.

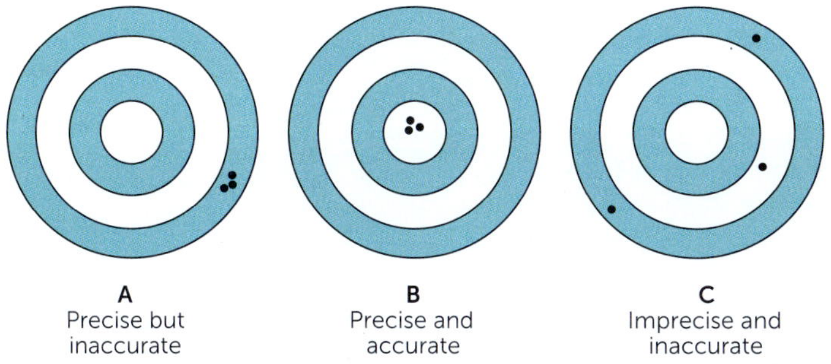

Figure 1.3 Accuracy and precision – the rifle shooting analogy.

We have already noted that measurements are usually repeated several times and this often means that the random errors will be nearly completely cancelled out in the mean measurement provided that enough repeat measures are made. This is why the precision of measurements is important: *the greater the precision, the fewer the number of repeat measurements that need to be made in order for the random errors to be eliminated*. On the other hand, the fewer the repeat measurements that are required, the quicker and cheaper are the measurements.

One way of showing the precision of a set of repeat measurements is to quote the standard deviation of those results. The lower the standard deviation of a set of results, the better is the precision of those results.

What do we mean by the true value of a measurement?

If a measurement is accurate, it must give a true value. But how do we know the true value? If we are using a new analytical instrument that detects lead, we might test the accuracy of the instrument by analysing solutions whose lead concentrations [Pb] are known (i.e. **standard solutions**). If the analysis is accurate, the difference

$$[\mathrm{Pb}]^{\text{in}}_{\text{standard}} - [\mathrm{Pb}]^{\text{found by}}_{\text{instrument}}$$

should be close to zero. If we are analysing mixtures where the true concentrations are unknown, the absence of systematic errors is essential if we are to have faith in the final result. In such cases, accuracy may be estimated by comparing the results of *different* analytical methods for the *same* sample.

Professional organizations (such as the American Society for Testing Materials) often publish the most reliable analytical methods in the form of **standard methods**, in which the likely sources of experimental error are highlighted.

Quality of analytical measurements in laboratories

Chemical analysis is a multi-billion dollar business in all industrialized countries, involving thousands of highly skilled scientists and technicians. Poor quality analysis is a barrier to international trade, technological advancement and the development of government policies such as health and safety. For example, unless the concentrations of pollutants can be reliably and accurately measured, regulations based on the highest permissible concentration of an airborne pollutant are unenforceable.

Analytical chemists distinguish between an *analytical technique* and an *analytical method*. The analytical technique used to take the measurement itself (such as gas

Quality control is very important in analytical laboratories, so that customers know that they can rely upon the results of any analysis.

chromatography or ultraviolet spectroscopy) is one part of a 'recipe' of operations that make up an analytical method. The analytical method may include details of sample preparation and how interference from other substances can be minimized. If the method is recognized by a national or international professional body it is called a **standard analytical method**.

See **Extension 1** on the website, https://www.bloomsburyonlineresources.com/chemistry. This gives details of one of the greatest scandals in analytical chemistry, that of the Theranos Company.

1.5 Reporting measurements

Significant figures and measurement uncertainty

The number of significant figures in a measurement is worked out by first expressing the measurement in standard notation and counting the number of figures (numbers, including zeroes) on the left-hand side. For example, a reported concentration 0.0234 parts per million is written as 2.34×10^{-2} in standard notation. Since 2.34 contains three figures, the concentration is being reported to *three* significant figures.[1]

If we asked someone to measure the length of a piece of wire with a standard ruler and they reported its length as 19.843 cm, we would have every right to be sceptical: 19.843 (or 1.9843×10^1) contains *five significant figures*, a number of figures that cannot be justified when we are using a ruler.

We might estimate the **uncertainty** in the length measurement as ±0.2 cm. This means that the length is estimated to be at worst 0.2 cm too high or 0.2 cm too low. It follows that we are justified in only including the first decimal place of the measurement and we then report the length as three significant figures as 19.8 ± 0.2 cm. Alternatively, we might report the measurement as simply 19.8 cm. Neglecting the ±0.2 cm is less informative, but because of an agreement between scientists about the meaning of significant figures, even simply writing 19.8 cm carries with it some information about the *minimum* uncertainties involved in the measurement.

To explain this further, suppose that you report the length of the wire to a friend as 19.8 cm but provide no further information. What could your friend say about the likely uncertainties in the measurement? By general agreement, it is assumed that the uncertainty in the measurement is *equal to at least one digit in the last significant figure*. In our example, reporting the length as 19.8 cm implies that the total uncertainty in the measurement is equal to at least one digit in the first decimal place. In other words, the *minimum* uncertainty is ±0.1 cm. As we have seen, the actual uncertainty is estimated to be greater, as ±0.2 cm.

In order to report the correct number of significant figures in a measurement, an estimate of the uncertainties is required. Sometimes this will be nothing more than an informed estimate of the likely effect of random errors. In more sophisticated measurements, further experiments may need to be carried out in order to assess the importance of both random and systematic errors.

What is the connection between uncertainty and absolute error? In the above examples, the uncertainty (the ± part) is an estimate of the *range of the absolute error* that is likely to apply to the reported measurement. For example, suppose the uncertainty is ±0.2 cm. This is equivalent to saying that the absolute error in the reported measurement is between 0.2 cm *above* the measurement and 0.2 cm *below* the measurement.

---

[1]Sometimes the number of significant figures is ambiguous. Consider, for example, the number 150. If we mean 1.50×10^2, then there are three significant figures. If we mean 1.5×10^2 then there are only two significant figures.

EXAMPLE **1.2**

The mass of a coin is displayed on a balance as 10.0078 g. The uncertainty of this measurement is estimated as ±0.002 g. How many significant figures are we justified in using when reporting the mass of the coin?

Answer

The uncertainty shows that three decimal places can be justified in the measurement. This means that we are justified in reporting the mass of the coin to five significant figures, i.e. as 10.008 g. This implies that the *minimum* uncertainty in the measurement is ±0.001 g.

Number of significant figures in a quantity calculated by multiplication or division

Suppose we carry out an experiment to find the density of a lump of metal. We require two measurements, namely: (i) the mass of the metal and (ii) its volume. Suppose the mass of the metal was reported as 10.0078 g whereas the volume of the metal (which is more difficult to determine accurately) was reported as 2.8 cm^3. The density is now calculated as

$$\text{density} = \frac{\text{mass}}{\text{volume}} = \frac{10.0078}{2.8} = 3.57421429 \text{ g cm}^{-3}$$

where the number 3.57421429 is the one that might be displayed on a calculator. It is absurd to report the density as 3.57421429 g cm^{-3}, since this would suggest an uncertainty of about ±0.00000001 g cm^{-3}! Again, there is a rule to guide us: the number of *significant figures* in the final calculated figure is *set equal to the number of significant figures in the most uncertain contributing measurement*. The density calculation depends upon measurements of volume and mass, but the volume measurement was reported to only two significant figures and is therefore the most uncertain of the two measurements. Accordingly, the density should also be reported to two significant figures:

density = 3.**6** g cm^{-3}

(where the **5** of the 3.**5**7421429 is rounded up to **6**).

EXERCISE **1F**

Significant figures and rounding up

(i) How many significant figures are present in the following numbers: (a) 0.02, (b) 20.02, (c) 890, (d) 0.00765?

(ii) The atomic mass of the oxygen-16 atom is 15.9949 atomic mass units, but a student uses an approximate value of 16.10. Is this justified?

REVISION QUESTIONS

Questions 1.1 to 1.4 are multiple choice questions. Select the correct answer.

1.1 The temperature of a liquid is reported as 25.0 ± 0.5°C. The 0.5 represents the estimated:

(a) Precision of the measurement
(b) Accuracy of the measurement
(c) Type of error in the measurement
(d) Uncertainty in the measurement

1.2 The concentration of calcium in water may be determined by atomic absorption spectrometry (AAS), in which the aqueous solution is sprayed into a flame which breaks down the calcium compounds into atoms. However, if phosphate ions (PO_4^{3-}) are present in the sample, they bind the calcium as calcium phosphate, which resists atomization in the flame. Therefore, the presence of phosphate in the sample causes the measured calcium concentration to be:

(a) Underestimated, causing a random error in the measurement
(b) Underestimated, causing a systematic error in the measurement
(c) Overestimated, causing a systematic error in the measurement
(d) Overestimated, causing a random error in the measurement

1.3 Measurements that are precise will always have a:

(a) Large systematic error
(b) Small systematic error
(c) Large random error
(d) Small random error

1.4 A solution containing exactly 10.0 ppm (parts per million) of manganese ion (Mn^{2+}) is sent to a laboratory to test the quality of its measurements. The lab analysed three samples of the solution, obtaining the following results:

Sample number	$[Mn^{2+}]$/ppm
1	10.0
2	10.5
3	11.0

The absolute error in the mean measurement is:

(a) 0.0 ppm
(b) 0.5 ppm
(c) 1.0 ppm
(d) 21.5 ppm

1.5 The radii of several atoms and ions (in different units) are as follows: Cr^{3+} 0.069 nm, F^- 1.36 × 10^{-6} cm, O 1.40 × 10^{-5} mm. Express the radii in metres and arrange the particles in order of increasing size.

1.6 The temperature of a water bath was reported as 27.1°C. What does the number of significant figures tell you about the minimum uncertainty involved in this measurement?

1.7 Report the following measurements to four significant figures:

(i) 0.12347 V
(ii) 12.45 m
(iii) 0.00355757 cm
(iv) 1200.5 K

1.8 Calculate $e^{-E/RT}$ (with E = 20,000 J mol^{-1} and R = 8.3145 J mol^{-1} K^{-1}) at

(i) T = 300 K and
(ii) T = 3000 K

In each case, express the result to three significant figures.

1.9 In spectroscopy, the **absorbance** (A) of a solution of a substance is defined by the expression

(a) $A = \varepsilon \times c \times b$

where ε (pronounced epsilon) is the molar absorption coefficient of the substance, c the concentration of the substance (in moles per metre cubed) and b the thickness of the sample (in metres); A is unitless. What are the units of ε?

1.10 Label the following as random or systematic errors:

(i) The variation of the mass recorded by a balance because of air draughts in the laboratory.
(ii) A thermostat in a water bath that registers 30°C when the real temperature of the water is 25°C.
(iii) The leaking of a gas cell that is being used to hold a sample of gas so that its pressure can be measured.

DID YOU KNOW?

Celsius and his temperature scale

The *Centigrade* temperature scale, as its name suggests, consists of 100 degrees between the two 'fixed points'. It was first proposed by the Swedish scientist Anders Celsius in 1742. Celsius suggested that 0°C should be the temperature of boiling water and 100°C should be the freezing point of water. The scale was reversed after his death to the current version with the boiling point of water being set at 100°C and the freezing point of water at 0°C. To avoid confusion with the SI prefix centi- and in honour of the discoverer, the temperature was renamed the *Celsius scale* in 1948.

2 Elements, Compounds and Reactions

CONTENTS

OBJECTIVES

This unit:

- Names and describes the various forms of matter
- Distinguishes between chemical and physical changes
- Shows how to construct formulae and equations
- Introduces the idea of competing reactions

2.1 Matter, mass and energy

The universe is composed of matter and energy. Matter is anything that occupies **space** and has **mass** – rocks, oceans, the air that we breathe and we, ourselves, are all composed of matter. On the other hand, *energy* has no shape or form – it is defined as the ability to do **work**.

Mass and weight are different. The *mass* of a body depends on the amount of matter (the number of atoms) it contains. Unlike *weight*, mass does not depend on location. A 10 g lump of copper still has a mass of 10 g on the Moon, even though the gravitational force on the copper is less on the Moon. Chemists use their balances to measure mass *not* weight.

Portland Bill lighthouse, built 1903. The original lens was rotated mechanically and to make this easier, the lens was floated in a 750 kg pool of the liquid element mercury.

Forms of matter

Matter is generally observed in three physical states: **solids, liquids** and **gases**. A fourth state of matter, **plasma**, is formed at extremely high temperatures and may be found in the atmospheres of stars. Matter is classified as either elements, compounds or **mixtures**.

All matter is made of **atoms**. For the moment we may think of atoms as extremely minute solid particles, but this is revisited in Unit 3. Atoms are the smallest part of an element that chemically reacts like a bulk sample of the element. If it were possible to take a piece of pure copper and divide it into smaller and smaller pieces, eventually we would have one atom of copper. Atoms are incredibly small – 35,000,000 copper atoms, laid end to end in a line, would be about 1 cm long. The existence of atoms has been proposed since the time of the early Greeks and it explains the behaviour of substances. The **scanning tunnelling microscope**, developed in 1981, enables us to see images of atoms (see Box 2.2).

1 Elements

Elements are basic forms of matter and cannot be broken down into simpler substances by chemical reactions. This is most easily visualized in terms of atoms. A sample of (pure) iron, for example, contains only iron atoms. At present, there are 118 named elements (see the Periodic Table, which is an arrangement of all known elements), of which 92 occur naturally and two (Tc and Pm) are made as byproducts in nuclear reactors. The others have been made in the laboratory, the most recent ones in vanishingly small amounts and with extremely expensive pieces of equipment. The atoms of the 118 elements are each different: they have different numbers of electrons and different arrangements of electrons within the atom. These differences cause the atoms to behave differently, and so bulk samples of elements have different chemical reactions.

Each of the 118 elements has a unique symbol. The symbol is not simply a shorthand for the element (for example, we might label a bottle of iron with the label Fe) it is also used to represent *one atom* of that element. Hence Na means one atom of sodium, whereas 2Na means two *separate* and unconnected atoms of sodium.

2 Compounds

Compounds consist of two or more elements joined together in *fixed proportions* to form a new substance. The term 'fixed proportions' means fixed percentages by mass. For example, no matter where sodium chloride (common salt) is found in the natural world, a sample of pure salt always contains 60.66 per cent of chlorine atoms and 39.34 per cent sodium atoms by mass.

Chemical reactions, also known as chemical changes, produce **new substances**: elements react together to form compounds, or compounds may be broken down into their constituent elements (or into other compounds).

Both elements and compounds are **substances**. Notice that the scientific meaning of the word 'substance' is very precise – a substance is a single pure form of matter, and not a mixture of different kinds of matter. Substances are relatively rare; and most naturally occurring materials on Earth are mixtures.

Pure elements and compounds, for example a pure sample of the element copper or of pure salt (the compound sodium chloride), have *reproducible* properties. For example, different samples of pure copper and pure salt behave in exactly the same way under identical conditions: they have the same melting points and undergo the same chemical reactions as any other sample of those substances. This is hardly surprising, since a sample of copper contains only copper atoms and salt always contains the same two atoms (Na and Cl) in the same proportion. Mixtures, because they can have varying compositions, often do not behave in the same way under identical conditions. Different samples of soil, for example, may have very different properties although they are all referred to as 'soil'.

The atomic arrangements within a sample of a compound are shown in the *chemical formula* for that compound. Experiments show that chlorine gas consists of connected pairs of chlorine atoms known as *chlorine molecules*, with the formula Cl_2. The formula for the compound sodium chloride is NaCl, showing that equal numbers of sodium and chlorine atoms are present, but this is a simplification as, unlike chlorine gas, the atoms are connected in a giant 3D lattice rather than in smaller clusters or pairs. There is a further complication that the atoms in NaCl are not electrically neutral but charged, so a better representation is Na^+Cl^-; this will be discussed further in Unit 4. One of the remarkable characteristics of compounds is that they have very different chemical properties (chemical reactions) from the elements used to make them.

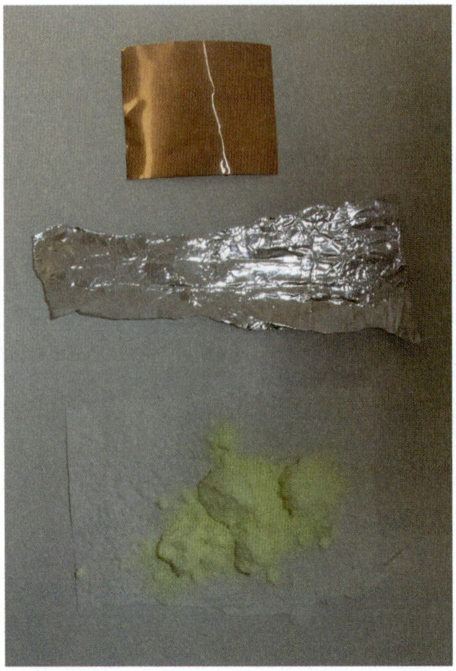

Figure. 2.1 The elements, copper, aluminium and sulfur.

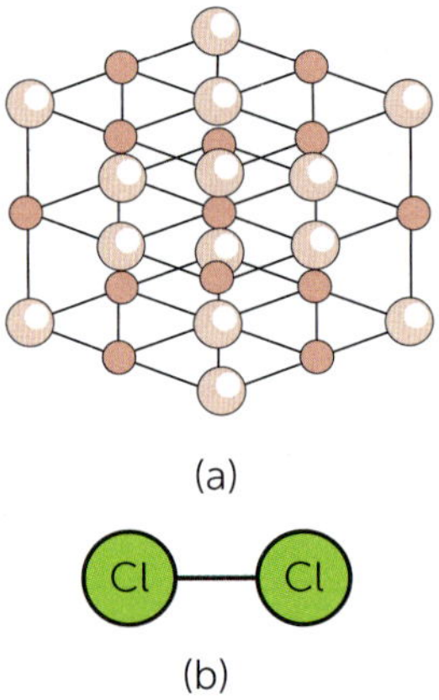

Figure. 2.2
(a) In solid sodium chloride, sodium ions and chlorine ions (chloride ions) are arranged in a giant lattice. (b) In chlorine gas the chlorine atoms are paired up. The pair of bonded atoms is known as a chlorine molecule.

3 Mixtures

Mixtures contain more than one substance (elements and/or compounds). The composition of a mixture can be varied and the chemical reactions of a mixture are simply the chemical reactions of its constituents. Air is a mixture of gases, but the precise percentage of its components varies with geography and the degree of local pollution.

BOX **2.1**

Purity of substances

Sometimes substances are referred to as 'pure substances' to emphasize, for example, that a sample of the substance sodium chloride (salt) contains only salt. In practice, even the purest of samples of substances contain traces of other substances. Purity is very important in chemistry. Guaranteed pure reagents were first advertised by the German chemical and pharmaceutical firm of Merck in 1888. Today, purity is expressed as a percentage, e.g. $\geq$ 95 per cent pure. Chemicals for experiments can normally be purchased in two grades. You will normally use **standard laboratory reagents** (SLR), which are quite pure and adequate for most general laboratory experiments. **Analytical reagents** (AR) are usually greater than 99.9 per cent pure. They are much more expensive and are used for specialist purposes such as instrumental analysis. Look at the labels of reagent bottles in your laboratory to check their purity!

Elements: metals and non-metals

Elements can be roughly divided into metals and non-metals.

1 Metals

Metals usually have the following properties:

- They have a lustre (a 'shiny' appearance).
- They conduct heat and electricity.

- They can be drawn into wire (are **ductile**).
- They can be hammered into thin sheets (are **malleable**).
- They make a 'clanging' sound when you hit them (are **sonorous**).
- In chemical reactions metals react with acids, form **basic oxides** and form **positively charged ions** (you may not understand these terms at this point, but will be able to refer back to these properties later on).

Examples of metals are gold, silver, copper and iron. Mercury is a metal that is liquid at room temperature.

DID YOU KNOW?

Caesium and gallium have melting points of 28.4 °C and 29.8 °C respectively and will be liquid on hot days.

2 Non-metals

Non-metals usually have the following physical properties:

- They are poor conductors of heat and electricity.
- They cannot be hammered into sheets or drawn into wire because they are brittle.
- They are not lustrous and not sonorous.
- In chemical reactions non-metals do not react with acids, they form **acidic oxides** and they either form **negatively charged ions** or else they form **covalent compounds**.

Examples of non-metals are iodine, oxygen, nitrogen and carbon.

As you will see later, some elements have intermediate properties between those of metals and non-metals and are called **metalloids** or **semimetals**. Examples include arsenic, silicon and germanium.

Atoms and molecules

Two or more atoms can link together to form new particles called **molecules**. The chemical behaviour of the new molecules is different from that of the atoms from which they are made. The molecules may be combinations of atoms of the *same* element. For example, under normal conditions the element oxygen exists as oxygen gas and consists

BOX 2.2

We know atoms exist because we can see them

We cannot look at atoms with a microscope that uses visible light because particles with diameters less than the wavelength of visible light (about 400 nm) cannot be distinguished.

Electron microscopes use electrons (small particles found inside the atom) to produce images of tiny objects. The **scanning tunnelling microscope** (STM) is the most sophisticated of this type of instrument, and it can produce images of the surfaces of elements which show the individual atoms. The atoms show up as blurred spheres.

An image of gold from a scanning tunnelling microscope. The small dots represent individual gold atoms. Image by Erwin Rossen.

of **diatomic** oxygen molecules (two atoms joined together) and is depicted as O_2. For this reason, the oxygen molecule is sometimes called 'dioxygen'. If the atoms joined are of *different* elements, the molecules are those of a compound. The compound known as water consists of molecules in which two hydrogen atoms are joined to one oxygen atom (see Box 2.3), so its formula is H_2O.

BOX **2.3**

Molecules

Examples of elements that normally exist as pairs of atoms (known as *diatomic molecules*) include: oxygen, hydrogen, nitrogen, chlorine, bromine and iodine (O_2, H_2, N_2, Cl_2, Br_2 and I_2). Phosphorus exists as a **tetratomic** molecule (P_4), whereas a common type of sulfur molecule contains eight atoms (S_8) (Figure 2.2). Molecules, made up of atoms, are also incredibly small. A single water molecule, consisting of two atoms of hydrogen joined to one atom of oxygen, has a mass of 3 x 10^{-23} g (Figure 2.4).

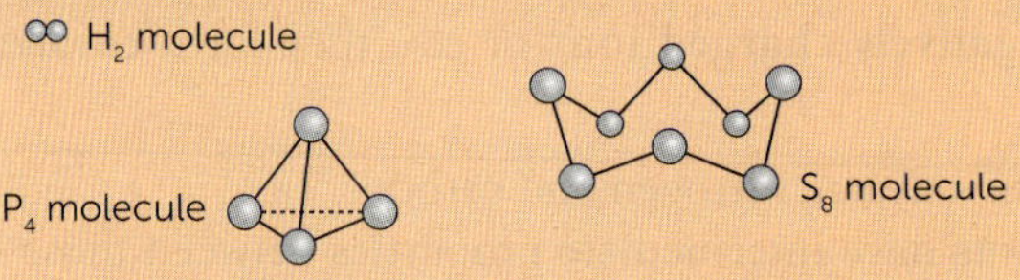

Figure. 2.3 Molecules of elements.

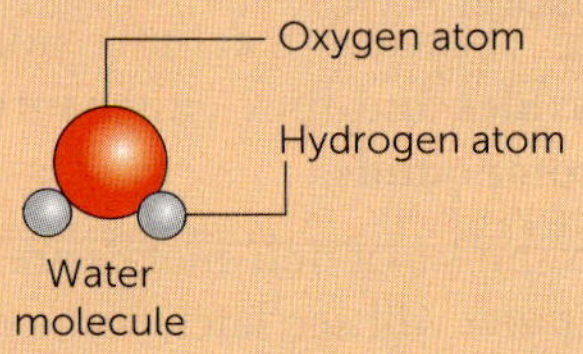

Figure. 2.4 Water molecule, H_2O.

2.2 Physical and chemical changes

Physical changes

The key feature of physical changes is that the ions or molecules do not change during that change. Examples of physical changes include boiling and dissolution. When water is boiled, the molecules of water *persist* and so steam has the same chemical reactions as liquid water. When solid salt dissolves in water, the giant lattice of Na^+ and Cl^- ions in the solid is broken up and the ions are free to move around in the water, but the ions have not changed and so the chemical reactions of the salty water are those of solid salt.

Chemical changes (chemical reactions)

A **chemical reaction** involves the conversion of one type of matter into another. New substances are formed. After a chemical reaction there **is no change in the total mass from that present initially**. This is the **law of conservation of mass** and it can be expressed in another form: in chemical reactions atoms are simply re-arranged by bonding together in different ways, but atoms are not created or destroyed. For example, charcoal, burning on your barbecue, may appear to do a vanishing act:

charcoal + oxygen gas (from air) → carbon dioxide gas

However, if you added the mass of oxygen, from the air, used up in the reaction and the mass of charcoal used up, you would find that the sum was exactly equal to the mass of (invisible) carbon dioxide produced.

EXERCISE **2A**

Physical and chemical changes

Classify the following changes as physical or chemical:

(i) the burning of coal

(ii) the freezing of water

(iii) the melting of solder

When a substance undergoes a chemical change, it changes into another type of matter, often with quite different chemical properties. For example, heated sodium reacts violently with chlorine. Sodium is a highly reactive metal, whereas chlorine is an extremely poisonous gas. After the reaction, sodium chloride (salt) is formed; this is not a particularly reactive substance and it is essential to life!

Physical and chemical properties

The physical properties of a particular type of matter are those properties of the substance that do not involve a chemical reaction. Examples are the state in which the substance exists (**solid**, **liquid**, **gas** or in **solution**), melting point, boiling point, colour and electrical conductivity.

The chemical properties of a substance describe the chemical reactions that a substance will undergo. For example, sodium metal reacts violently with water to form a solution of sodium hydroxide and hydrogen gas. The reactivity shown by sodium towards water is a chemical property (a chemical reaction) of sodium metal.

EXERCISE **2B**

Physical and chemical properties

Classify the following as physical or chemical properties:

(i) the **hardness** of a diamond

(ii) the low **density** of aluminium

(iii) the high toxicity of arsenic

Extensive and intensive properties

Any property of matter that can be measured can be classified as **extensive** or **intensive**.

Extensive properties depend on the amount of substance present; intensive properties do not.

Volume is an extensive property. If you have 100 cm^3 of water and add 50 cm^3 more, the total volume is 150 cm^3 – the volumes can be added to give a total.

Temperature is an intensive property. If you have water at a temperature of 50°C and add more water at 50°C the temperature of the whole is still 50°C and *not* 100°C.

EXERCISE **2C**

Extensive and intensive properties

Classify the following as extensive or intensive properties:

(i) length

(ii) colour

(iii) mass

(iv) density.

2.3 Chemical formulae

Symbols

We have already introduced the symbols of a few elements. These symbols sometimes consist of the first letter(s) of the name of the element; for example, hydrogen is given the symbol H, whereas the chemical symbol for bromine is Br. Some symbols, however, are not derived from the modern or English name of the element: gold has the chemical symbol Au (from the Latin *Aurum*), whereas tungsten is denoted by W (from *Wolfram*, a German word). Note that the first letter only of a chemical symbol is written as a capital letter. Box 2.4 gives the symbols of some elements.

EXERCISE **2D**

Using symbols and formulae

Name the elements in the following compounds and state how many atoms of each element are present:

(i) CsBr **(ii)** $CuCl_2$ **(iii)** Fe_3O_4 **(iv)** CH_4 **(v)** $PbSO_4$ **(vi)** $K_2Cr_2O_7$ **(vii)** Na_3PO_4 **(viii)** MnO_2.

EXERCISE **2E**

Main elements in the human body

Can you identify the main elements in the human body from their symbols?

C. HOPKINS
CaFe
NaCl – free on every table

Figure 2.5 Main elements (in bold) in the human body.

BOX **2.4**

Some symbols for elements (you are strongly advised to learn these!)

Element	Symbol	Element	Symbol	Element	Symbol	Element	Symbol
Aluminium	Al	Cobalt	Co	Manganese	Mn	Sodium	Na
Argon	Ar	Copper	Cu	Mercury	Hg	Sulfur	S
Arsenic	As	Fluorine	F	Neon	Ne	Tin	Sn
Barium	Ba	Gold	Au	Nitrogen	N	Titanium	Ti
Boron	B	Helium	He	Oxygen	O	Tungsten	W
Bromine	Br	Hydrogen	H	Phosphorus	P	Uranium	U
Caesium	Cs	Iodine	I	Platinum	Pt	Vanadium	V
Calcium	Ca	Iron	Fe	Potassium	K	Xenon	Xe
Carbon	C	Lead	Pb	Radon	Rn	Zinc	Zn
Chlorine	Cl	Lithium	Li	Silicon	Si		
Chromium	Cr	Magnesium	Mg	Silver	Ag		

EXERCISE **2F**

Symbols of elements

1 Using library books or the internet, find out the names from which the symbols of the following elements were derived:

(i) sodium (Na) **(ii)** iron (Fe) **(iii)** lead (Pb) **(iv)** tin (Sn) **(v)** potassium (K) **(vi)** silver (Ag).

2 Some of the latest elements discovered derive their names from famous scientists. See if you can find out the main contributions to our knowledge made by the following scientists:

(i) meitnerium (Mt) named after Lise Meitner
(ii) seaborgium (Sg) named after Glenn Seaborg
(iii) bohrium (Bh) named after Niels Bohr
(iv) rutherfordium (Rf) named after Ernest Rutherford
(v) oganesson (Og) named after Yuri Oganessian

Formulae

Chemical formulae give us two important pieces of information:

1 the elements that are present in a substance;
2 the relative number of atoms that make up that substance.

Perhaps the most well-known molecular formula is that of water, H_2O. Note that '$2H_2O$' does not refer to a new molecule with six atoms: it refers to *two separate molecules* of water, i.e. $2(H_2O)$.

Writing chemical formulae

Although it is necessary for a chemist to learn the symbols of the common elements, it is possible to work out the chemical formulae of common substances without having to learn them parrot fashion. Different elements have atoms with different combining powers. The combining power of an atom is known as its **valency**, and these valencies can be used to find out how many atoms of one element will combine with another. A table of the valencies of common elements and groups is shown on page 539. Again, these need to be *learned*.

To obtain a clearer picture of how a chemical formula is constructed, imagine that the valencies of atoms are 'arms', which link up with the arms of other elements' atoms.

Some elements have more than one valency, but the name of the compound should help you work out the formula. For example, in the compound iron(III) chloride, the iron atoms have a valency of three and its formula is $FeCl_3$, whereas in iron(II) chloride, the iron combines with a valency of two and the formula is $FeCl_2$.

Models of molecules alongside their molecular formulae

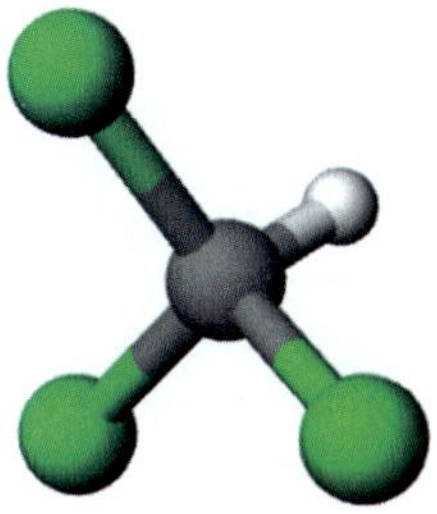

Trichloromethane (Chloroform) $CHCl_3$

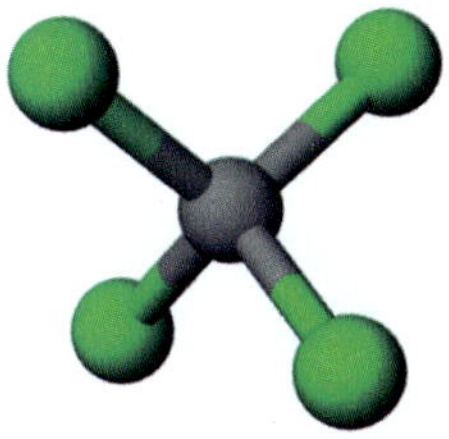

Tetrachloromethane CCl_4

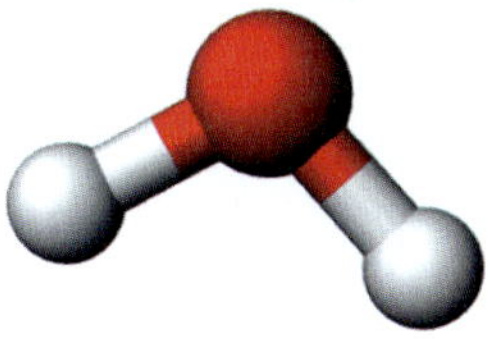

Water H_2O

EXAMPLE **2.1**

Construct the formula of magnesium oxide (a compound of magnesium and oxygen).

Answer

The symbol for magnesium is Mg and it has a valency of two. Imagine it as:

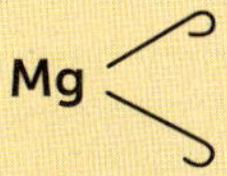

Similarly, oxygen could be thought of as:

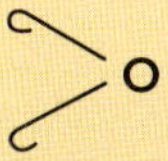

When magnesium combines with oxygen, the arms link up with one another and *there are no free arms left over*. Therefore, one atom of magnesium combines with one atom of oxygen to form magnesium oxide:

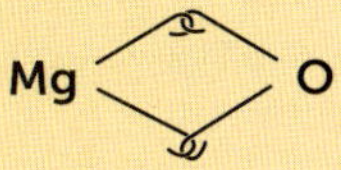

The formula is written as MgO. (Leave out the arms, they are only used to help you to construct the formula!)

There is another quick way to get the same answer. Write down the atoms with their valency underneath and use arrows to point to the *other* atom. The arrows then point to the ratio of the atoms in the formula. Here:

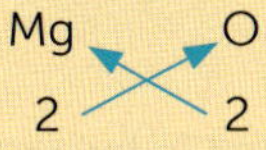

The ratio of Mg to O is 2:2 or 1:1, and the simplest formula is therefore Mg_1O_1 which we write as MgO.

EXERCISE **2G**

Simple formulae

Write formulae for the following compounds, using the valency table on page xxx.

(i) calcium oxide

(ii) aluminium fluoride

(iii) aluminium oxide

EXERCISE **2H**

More simple formulae

Write formulae for the following compounds:

(i) copper(I) oxide

(ii) copper(II) oxide

(iii) lead(IV) oxide

(iv) chromium(III) oxide

(v) vanadium(V) oxide

(vi) titanium(III) fluoride

(vii) disulfur dichloride

(viii) dioxygen difluoride

DID YOU KNOW?

Samuel Hopkins obtained a patent in 1790 for the purification of potassium containing ores ('potash'). This was the very first US patent and was signed by President George Washington.

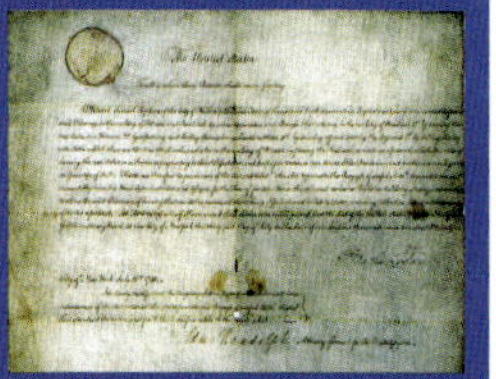

(Image provided by United States Patent and Trademark Office)

BOX **2.5**

Naming compounds

1 When a compound consists of a metal and a non-metal, the metal is written and named first. The Periodic Table will help you to distinguish whether an element is a metal or non-metal. Metals are shown to the left of the table, whereas non-metals are found to the right. The border between the two is shown as a zig-zag line. The non-metal part is named by putting 'ide' at the end, thus chlorine becomes chloride and oxygen becomes oxide.

Examples:
MgO magnesium oxide
Na_2O sodium oxide
$AlCl_3$ aluminium chloride
CaS calcium sulfide

2 If a compound is composed of two non-metals, the non-metal that is the closest to being a metal (closer to the metal – non-metal border in the table) is written first. If more than one compound of the same two non-metals exists then the prefixes mono(1), di(2), tri(3), tetra(4), penta(5), hexa(6) and so on are put in front of the non-metal to indicate the number of its atoms present in the compound.
Some examples:

CO carbon monoxide
CO_2 carbon dioxide
N_2O_3 dinitrogen trioxide
N_2O dinitrogen monoxide
P_4O_6 tetraphosphorus hexaoxide*
P_4O_{10} tetraphosphorus decaoxide*

* Often written hexoxide and decoxide (for ease of pronunciation).

Groups of atoms

For the purposes of writing formulae, sometimes groups of atoms act as if they have a valency of their own. Their valencies are also shown on page xxx.

EXAMPLE **2.2**

Write the formula of ammonium sulfate

Answer

The ammonium group is a combination of one nitrogen atom and four hydrogen atoms, but together they act as if the group has a valency of one:

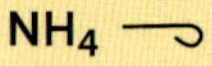

Similarly, sulfate has a valency of two:

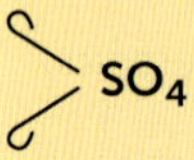

Two ammonium groups therefore combine with one sulfate group:

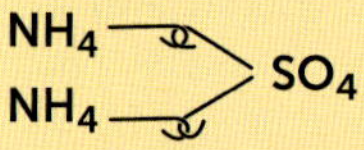

The formula is written as $(NH_4)_2SO_4$.

Alternatively, we could get the formula by writing the valencies under the two groups NH_4 and SO_4, with the arrows pointing to the opposite group:

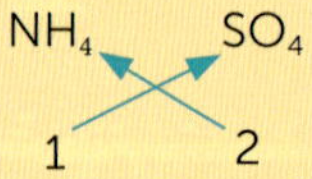

The ratio of NH_4 to SO_4 is therefore 2 to 1 and the simplest formula is $(NH_4)_2SO_4$.

Comment

Note that *if more than one group of atoms* are to be represented, each group is enclosed in brackets and the number of groups present is written as a subscript outside the brackets.

EXERCISE **2I**

Formulae involving groups of atoms

Write formulae for the following compounds:

(i) copper(II) sulfate

(ii) copper(II) nitrate

(iii) ammonium chloride

(iv) sulfuric acid (dihydrogen sulfate)

(v) nitric acid (hydrogen nitrate)

(vi) aluminium nitrate

(vii) calcium hydroxide

(viii) potassium hydrogencarbonate

(ix) calcium hydrogencarbonate

2.4 Writing chemical equations

Equations for chemical changes can be constructed by setting out the formulae of the reactants and products of a reaction. The same numbers of atoms of each element should appear on each side of the equation – we do not create or destroy atoms in chemical reactions – and the equation is then **balanced**. Balancing is particularly important when the reaction in question 'goes to completion' and the reaction is used in quantitative analysis (see Unit 8).

Chemical equations are best written in three stages:

1 Write the names of the reactants and products on each side of the equation.
2 Replace the names with formulae.
3 Balance the equation.

EXAMPLE **2.3**

Hydrogen gas (dihydrogen) reacts with oxygen gas (dioxygen) to form water. Write a balanced equation for the reaction.

Answer

1 hydrogen gas + oxygen gas → water

2 $H_2 + O_2 \rightarrow H_2O$

3 Count the number of atoms of each element on each side of the equation:

left-hand side – two atoms of H and two atoms of O
right-hand side – two atoms of H and one atom of O

We need to increase the number of atoms of O on the right-hand side. Numbers *in front* of the formulae must be added, thus $2H_2O$ means two molecules of water (containing four atoms of H and two atoms of O), so we can increase the number of O atoms on the right-hand side:

$$H_2 + O_2 \rightarrow 2H_2O$$

Now there are more hydrogens on the right-hand side (four) than on the left-hand side (two), but writing 2 in front of H_2 remedies the situation:

$$2H_2 + O_2 \rightarrow 2H_2O$$

The equation is now balanced – there are the same number of atoms of hydrogen and oxygen on each side. This equation tells us that two molecules of H_2 react

EXERCISE **2J**

Balancing equations

Balance the following equations:

(i) $N_2 + H_2 \rightarrow NH_3$

(ii) $Na + Cl_2 \rightarrow NaCl$

(iii) $Al + O_2 \rightarrow Al_2O_3$

(iv) $Fe_2O_3 + CO \rightarrow Fe + CO_2$

EXERCISE **2K**

Writing equations

Write balanced equations for the following reactions:

(i) The reaction of potassium with oxygen gas (dioxygen) to form potassium oxide.

(ii) The reaction of iron with chlorine gas (dichlorine) to form iron(III) chloride.

(iii) The reaction of lithium with water to form lithium hydroxide and hydrogen gas (dihydrogen).

(iv) The reaction of copper with nitric acid to produce copper(II) nitrate, nitrogen dioxide and water.

EXAMPLE **2.3** *(continued)*

with one molecule of O_2 to form two molecules of H_2O. The balanced equation is represented by Figure 2.6.

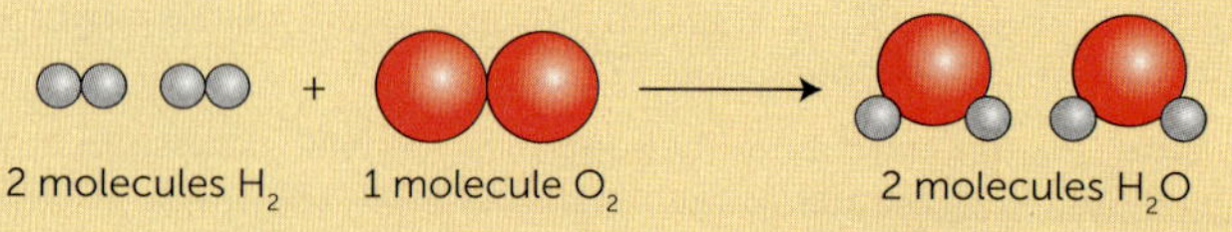

Figure. 2.6 Balanced equation for the reaction of hydrogen with oxygen.

Comment

Note that although chemists often use formulae instead of names (a bottle of water might be labelled H_2O) the formula H_2O, strictly speaking, refers to *one molecule* of water.

Additional comment

Before you try to balance an equation it is very important that you write correct formulae for the products and reactants. For example, you will come across a chemical with the formula H_2O_2. This compound is commonly called hydrogen peroxide and it is very different from water in its chemical behaviour. The extra oxygen in the formula makes a big difference to the nature of the substance!

Although equations are usually balanced using the simplest ratio of numbers, an equation remains balanced if higher numbers are used, as long as the correct ratio (and correct formulae) is maintained. For example

$$2Na + 2H_2O \rightarrow 2NaOH + H_2$$
$$4Na + 4H_2O \rightarrow 4NaOH + 2H_2$$

or

$$Na + H_2O \rightarrow NaOH + {}^1/_2H_2$$

are all ways of writing a balanced equation for the reaction of sodium with water because in each case the ratio of the species in the equation is 2:2:2:1 and the formulae of the species is unchanged.

State symbols

Substances in balanced equations often have **state symbols** written in brackets after them. The state symbols are (s) solid, (l) liquid, (g) gas and (aq) aqueous, i.e. dissolved in water. Symbols give a more detailed description of the reaction. For example, the equation

$$AgNO_3(aq) + NaCl(aq) \rightarrow AgCl(s) + NaNO_3(aq)$$

shows that silver nitrate and sodium chloride solutions react together to form solid silver chloride and sodium nitrate solution. It is important to include the state symbols in this equation, because solid silver nitrate and solid sodium chloride do not react.

2.5 Competing reactions

Sometimes the same starting substance(s) can produce two or more products. This means that there are *competing reactions* for the same reactant(s). This is illustrated by the reaction of aluminium and iodine. If powdered aluminium and iodine are mixed in a container, and two drops of water are added to speed up the reaction, the reaction starts after a few minutes and much heat is generated. The end result is a mixture of white solid aluminium oxide (Al_2O_3) with traces of yellow solid aluminium iodide (Al_2I_6). The products are produced by two different reactions of aluminium:

$$2Al(s) + 3I_2(g) \rightarrow Al_2I_6(s) \quad (1)$$

$$4Al(s) + 3O_2(g) \rightarrow 2Al_2O_3(s) \quad (2)$$

The oxygen comes from the surrounding air. If it was our intention to convert as much aluminium to aluminium iodide as possible, competing reaction (2) is a nuisance. If the experiment is repeated in an oxygen-free atmosphere (e.g. by flooding the reaction vessel with argon gas), only Al_2I_6 is produced. Competing reactions are commonplace in organic chemistry, where they may considerably reduce the yield of a desired product.

The reaction was carried out in an old saucepan in the open air (Fig 2.7). The reactants (Al) and iodine (I_2) are shown in (a) and (b). (c) Two drops of water were added to the mixed reactants to bring the reactants in contact, producing white hot aluminium and lots of I_2 vapour. (d) shows the final products, mostly white Al_2O_3 with some yellow Al_2I_6.

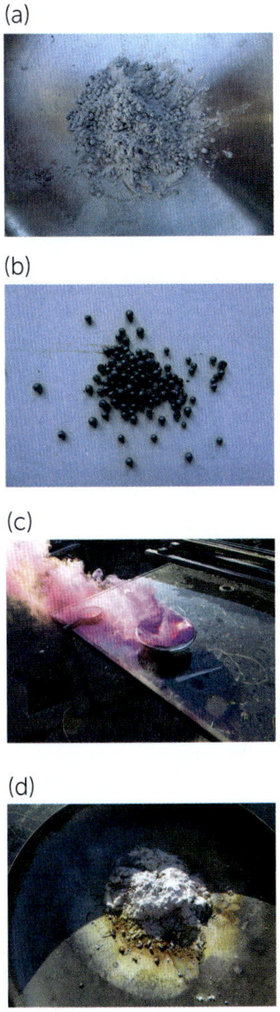

Figure 2.7 Reaction of aluminium powder with iodine (image: Rhobert Lewis).

BOX **2.6**

John Dalton (1766–1844) was the son of a poor Quaker couple. He spent most of his life in Manchester teaching and researching. Although the Greeks, as early as 400 BCE, believed that matter consisted of small particles which they called atoms, the theory was revived and elaborated on by Dalton. In 1808 he presented his **atomic theory** in which he stated:

- Elements are made up of particles called atoms, which are indestructible.
- The atoms of an element are all the same, but different from those of other elements. (We now know this is not quite true because of the existence of isotopes (Unit 3).)
- 'Compound atoms' (molecules as we know them today) are formed by small, whole-number combinations of atoms.
- A chemical change involves a new arrangement of atoms.

The theory, with a few important modifications, still holds true today.

Dalton proposed a cumbersome set of symbols for the elements (Figure 2.8) and the symbols we use today were suggested by Baron Jons Jacob Berzelius (1779–1848) of Sweden.

Figure 2.8 Early symbols for the elements as proposed by Dalton.

REVISION QUESTIONS

2.1 Classify each of the following as an element, compound or mixture:

(i) milk
(ii) white sand, or silicon dioxide
(iii) rubidium
(iv) soil
(v) carbon dioxide
(vi) brass
(vii) nickel
(viii) ink
(ix) barium
(x) chalk

2.2 The element lithium has a melting point of 181°C and a boiling point of 1,342°C. It is a soft metal that corrodes rapidly in air. Lithium also conducts electricity and reacts violently with water. Classify all of these properties of lithium as physical or chemical.

2.3 Ethanol (common alcohol) is a compound with a melting point of –114°C and a boiling point of 78°C. What is its physical state at room temperature (room temperature is about 20°C)?

2.4 Butane is the chemical name for the compound used in lighter fuel. It melts at –138°C and boils at 0°C. What is its physical state at room temperature?

2.5 Look up the words from which the *symbols* for the following elements are derived:

(i) copper
(ii) antimony
(iii) mercury

2.6 Write the correct symbols or formulae for each of the following:

(i) one atom of oxygen
(ii) two atoms of oxygen
(iii) two molecules of oxygen gas
(iv) three molecules of sulfur trioxide
(v) four atoms of neon

2.7 How many atoms of each element are present in the following?

(i) Li_2CO_3
(ii) $Ca(NO_3)_2$
(iii) $(NH_4)_3PO_4$

2.8 Write formulae for the following compounds:

(i) barium fluoride
(ii) tin(II) bromide
(iii) tin(IV) chloride
(iv) sodium ethanoate
(v) aluminium sulfide
(vi) potassium hydroxide
(vii) sodium bromate
(viii) potassium dichromate
(ix) ammonium phosphate
(x) lead(II) cyanide

2.9 Find out a common use for hydrogen peroxide (H_2O_2).

2.10 Liquid hydrogen peroxide slowly decomposes into water and oxygen gas. Write a balanced equation for the reaction.

2.11 Balance the following equations:

(i) $KClO_4 \rightarrow KCl + O_2$
(ii) $S_8 + AsF_5 \rightarrow S_{16}(AsF_6)_2 + AsF_3$
(iii) $Hg + NH_4I \rightarrow HgI_2 + H_2 + NH_3$
(iv) $Al_4C_3 + H_2O \rightarrow Al(OH)_3 + CH_4$
(v) $Zn + H_3PO_4 \rightarrow Zn_3(PO_4)_2 + H_2$
(vi) $Fe_2O_3 + Na_2CO_3 \rightarrow NaFeO_2 + CO_2$

2.12 Construct balanced equations (including state symbols) to describe the following reactions:

(i) Zinc metal burns in oxygen gas (dioxygen) to form solid zinc oxide.
(ii) Potassium metal reacts with water to form potassium hydroxide solution and hydrogen gas (dihydrogen).
(iii) Carbon dioxide gas reacts with magnesium metal to give solid magnesium oxide and carbon.

2.13 Suggest why, when an aluminium food container is placed in a dishwater, it turns grey.

DID YOU KNOW?

When iron was worth more gold

In ancient Egypt, bronze (an alloy of copper and tin) dominated, but the metal workers of those days did know of iron, even though they could not extract it from iron ore. The iron they were familiar with came from meteorites and it was extremely rare. One fine example of iron working was found in Tutankhamun's tomb, where an iron dagger was discovered with the other more famous priceless artifacts. The high nickel content of the iron confirms its meteoritic origin.

'Iron from the sky'. Tutankhamun's iron dagger. Worth more than gold at the time, it was a gift fit for a king.

Extension material is available on the website at https://www.bloomsburyonlineresources.com/chemistry.

3 Inside the Atom

CONTENTS

OBJECTIVES

This unit:

- Describes atomic structure and the evidence for quantization
- Defines isotopic mass and the atomic mass of elements
- Looks at the types of mass spectrometers
- Explains the use of mass spectrometry in molecular identification
- Discusses the spdf configuration of atoms
- Introduces the idea of an electron acting as a wave

In Unit 2 we saw that atoms are not created or destroyed in chemical reactions, but that they are simply reshuffled to produce new molecules. Do these observations mean that atoms are completely indestructible? This unit shows how atoms have been forced to reveal their inner secrets. Atoms are not indestructible – they do indeed have an inner structure, and although they often behave as if they were solid, the particle model of atoms suggests that they are mostly empty space!

3.1 Atomic structure

Pieces of history

In 1897, Sir J. J. Thomson (1856–1940) investigated the way that gases at low pressure conduct electricity. When high voltages were used, streams of minute particles were detected on a fluorescent screen. These particles possessed a negative electrical charge and were even lighter than hydrogen atoms. The particles were called **electrons**.

Thomson realized that the electrons must have been squeezed out of the atoms of gas by the high voltages used in his experiments. (This is one piece of evidence that atoms are not indestructible. Another piece of evidence is provided by radioactivity, discussed in Unit 23.) In other words, electrons are **subatomic particles**. Thomson realized that since atoms are usually observed to be electrically neutral, there must be positive charges in the atom too. He proposed that atoms consisted of a positively charged 'cement' in which rings of electrons were embedded. Thomson's model has been compared with currants in a bun, with the electrons as the currants and the dough as the positive charge.

Joseph John ('JJ') Thomson (left) and Ernest Rutherford. Both were awarded Nobel Prizes, Thomson in 1906 (for his work on atomic structure) and Rutherford in 1908 (for his work on radioactivity). Much to Thomson's delight, his son, George, also won the Nobel Prize in 1937.

In 1909 a group led by Lord Rutherford (1871–1937) carried out an experiment in which alpha particles from a radioactive material were fired at a very thin layer of gold metal in an evacuated bulb. (Alpha particles are much smaller than gold atoms.)

Rutherford's team repeated the experiment many times. They found that although most of the alpha particles passed through the foil undisturbed, very occasionally an alpha particle underwent a massive deflection – like a table tennis ball bouncing off a wall. Only about 1 in every 20,000 alpha particles bounced back in this way, and the deflections were so rare that many experimenters might have simply ignored them. But Rutherford and his workers realized that such deflections were very significant indeed. Rutherford proposed that the mass of an atom was concentrated in its centre (the **nucleus**). The chances of an alpha particle striking the nucleus were very small but, when such a collision did occur, the considerable mass of the nucleus ensured that the alpha particle was greatly deflected (Figure 3.1).

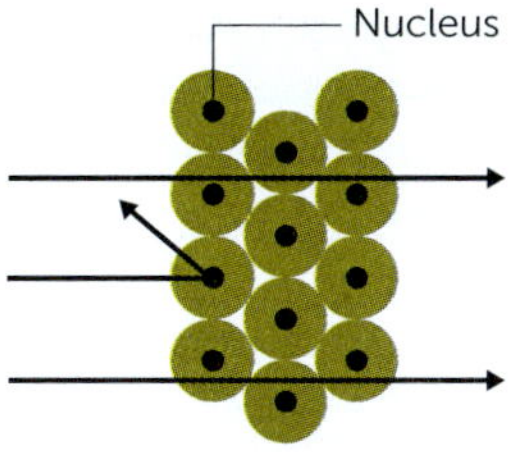

Figure 3.1 Explanation of the Rutherford experiment. The path of the alpha particles is shown by arrows. Only if an alpha particle makes a direct hit on a nucleus does it suffer a large deflection.

The 'insides' of atoms: a summary

The Rutherford model may be summarized as follows:

1 **Atoms consist of electrons circling the nucleus of the atom.** Electrons are negatively charged. By agreement, electrons are assigned a relative charge of −1. The circular paths of the electrons are called **shells** or **orbits**. The radius of an atom is equal to the radius of the outermost electron shell.

2 **The nucleus consists of two kinds of particles,** protons (which possess a relative charge of +1) and neutrons (with a mass approximately equal to that of the proton but with no electrical charge). Protons and neutrons are collectively known as nucleons.

3 **Protons and neutrons are very much heavier than electrons.** This means that most of the mass of an atom lies in its nucleus. The number of protons added to the number of neutrons is termed the **mass number** (symbolized A) of the atom.

4 **Most of the atom is empty space.** The picture we should keep in our mind is of minute 'specks' (the electrons) circling huge distances from a small (but relatively heavy) nucleus.

5 **Atoms are electrically neutral.** This is because the number of protons and electrons are equal. The number of protons in an atom is called its **atomic number** (symbolized Z).

6 **The atoms of each element are different because they have different numbers of protons** (i.e. different atomic numbers). The helium atom consists of a nucleus of two protons and two neutrons surrounded by two electrons, so we can write

$Z = 2 \qquad A = 4$

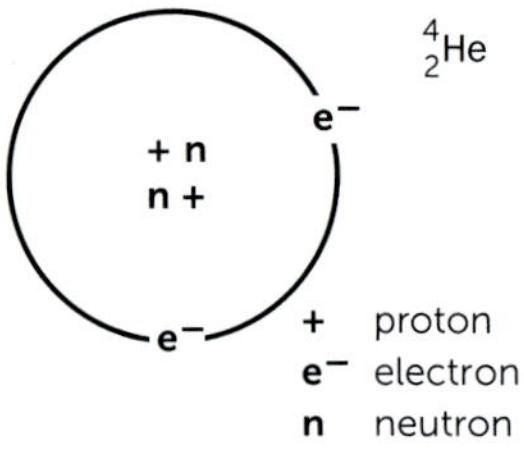

Figure 3.2 Helium atom.

Such an atom is shown diagrammatically in Figure 3.2. The position of the electrons is indicated by 'e^-'. This diagram is not to scale – if it were, the electrons would be several hundred metres from the symbols we have used to represent the nucleus.

The helium atom is represented as ^4_2He. This symbol is built up as follows:

- He is the chemical symbol for helium.
- The superscript '4' is the mass number.
- The subscript '2' is the atomic number.

Generalizing:

The number of neutrons N is the number of protons Z subtracted from the number of protons and neutrons A:

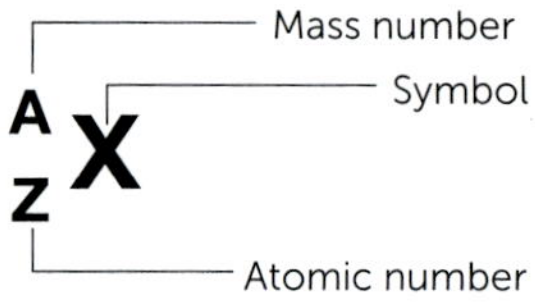

$N = A - Z$

EXERCISE **3A**

Structure of the atom

Atoms of carbon are represented as $^{12}_6C$. Sketch the atom, following the pattern of Figure 3.2.

Atomic mass scale

Table 3.1 gives more information about subatomic particles. In the fifth column, the mass of the particles is given in **atomic mass units**, symbolized as u, where

$1\,u = 1.66054 \times 10^{-24}\,g$

This tiny mass is *exactly equal to one-twelfth of the mass of one atom of carbon (isotope mass number 12)*. This means that the mass of one atom of $^{12}_6C$ is exactly 12 on the atomic mass scale. This is written as $m(^{12}_6C) = 12\,u$.

The alpha particle is also included in Table 3.1. This particle is a helium nucleus and is symbolized as $^4_2He^{2+}$. Now try Exercise 3B.

EXERCISE **3B**

Masses of particles

Answer these questions using Table 3.1.

(i) How much more massive is the proton than the electron?

(ii) How much more massive is the alpha particle than the proton?

Table 3.1 The mass and charge of subatomic particles

Particle	Symbol	Charge*	Mass/g	Mass/u	Approx. mass/u
Proton	p	+1	1.67262×10^{-24}	1.00727	1
Electron	e^-	−1	9.10939×10^{-28}	0.00054	0
Neutron	n	0	1.67493×10^{-24}	1.00867	1
Alpha	α	+2	6.64466×10^{-24}	4.00151	4

* Relative to the electron, which has an absolute charge of 1.602×10^{-19} coulombs.

3.2 Isotopes

Introduction

Atoms with the *same number of protons* but with *different numbers of neutrons* are called isotopes. In other words, **isotopes** of an element possess the same atomic number but different mass numbers. For example, consider two isotopes of carbon, $^{12}_{6}C$ and $^{13}_{6}C$. Carrying out a 'particle count' for atoms of both isotopes, we have

Count	$^{13}_{6}C$	$^{12}_{6}C$
number of electrons	6	6
number of protons	6	6
number of neutrons	7	6

In summary, atoms of both isotopes of carbon contain six protons and six electrons, but atoms of carbon-13 are heavier because they each contain an extra neutron. (Try Exercise 3C.)

EXERCISE **3C**

Numbers of protons, neutrons and electrons

How many protons, neutrons and electrons do neutral atoms of each of the following isotopes contain?

(i) The two main naturally occurring isotopes of uranium:

$^{235}_{92}U$ $^{238}_{92}U$

(ii) The three isotopes of hydrogen:

$^{1}_{1}H$ (normal hydrogen or protium)
$^{2}_{1}H$ (heavy hydrogen or deuterium)
$^{3}_{1}H$ (super heavy hydrogen or tritium)

Key points about isotopes

1 Isotopes have identical chemical reactions, although their compounds may have slightly differing physical properties (such as boiling point, or density). For example, hydrogen gas composed of normal hydrogen atoms ($^{1}_{1}H_2$) and hydrogen gas composed of deuterium atoms ($^{2}_{1}H_2$) display identical chemical reactions and both burn in oxygen, making normal water ($^{1}_{1}H_2O$) and 'heavy water' ($^{2}_{1}H_2O$), respectively. However, the density and boiling point of $^{2}_{1}H_2$ is greater than that of $^{1}_{1}H_2$, and that of $^{2}_{1}H_2O$ is greater than that of $^{1}_{1}H_2O$.

2 Some isotopes are **radioactive** (discussed further in Unit 23).

3 Atoms in a naturally occurring sample of the pure element (or of its compounds) may consist of one or more isotopes. The **abundance** of an isotope is defined as the percentage of atoms of that isotope in a sample of the element. The abundance of isotopes in nature (their **natural abundance**) is fixed. For example, about 75 per cent of all chlorine atoms in nature are atoms of $^{35}_{17}Cl$ (Table 3.2). This percentage does not change appreciably whether the chlorine occurs as chlorine gas, as sodium chloride or as hydrochloric acid, or (in the case of chlorine compounds) whether they were mined in Norway, Australia or Canada (see Figure 3.3).

EXERCISE **3D**

Isotopes

(i) Use Table 3.2 to answer the following:

(a) Which isotope of uranium is the most abundant in nature?

(b) Which element exists only as one isotope in nature?

(c) Which isotope of hydrogen is radioactive?

(d) Which isotope of hydrogen does not occur in nature?

(ii) Bromine consists of two naturally occurring isotopes, $^{81}_{35}Br$ and $^{79}_{35}Br$, and 50.69 per cent of all bromine atoms are of mass number 79. What percentage of bromine atoms has a mass number of 81?

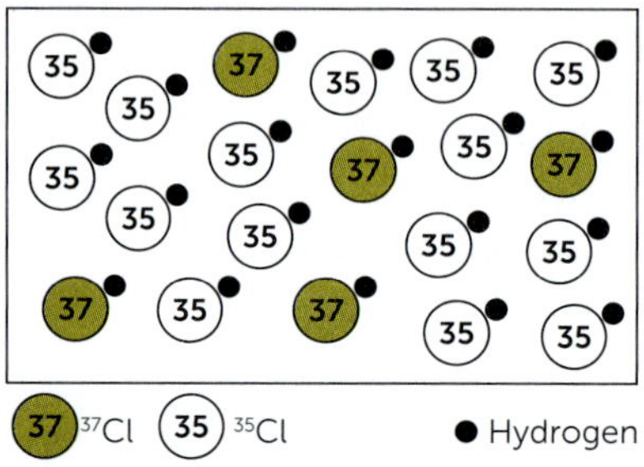

Figure 3.3 A collection of hydrogen chloride molecules (HCl) as found in nature. About 75 per cent of the HCl molecules contain the ^{35}Cl isotope.

Table 3.2 Isotopic and atomic data for selected elements

Element	Isotope	Natural abundance/%	Isotopic mass/u	Atomic mass of element/u
Hydrogen	$^{1}_{1}H$	99.985	1.0078	1.008
	$^{2}_{1}H$	0.01492	2.0140	
	$^{3}_{1}H$*	0	3.01605	
Chlorine	$^{35}_{17}Cl$	75.770	34.9689	35.45
	$^{37}_{17}Cl$	24.229	36.9659	
Uranium	$^{235}_{92}U$*	0.7205	235.0439	238.0
	$^{238}_{92}U$*	99.274	238.0508	
Carbon	$^{12}_{6}C$	98.893	12.0000	12.01
	$^{13}_{6}C$	1.107	13.0034	
Oxygen	$^{16}_{8}O$	99.759	15.9949	16.00
	$^{17}_{8}O$	0.0374	16.9991	
	$^{18}_{8}O$	0.2039	17.9992	
Fluorine	$^{19}_{9}F$	100	18.9984	18.9984

* Shows that the isotope is radioactive.

The atomic mass of an element is sometimes called its atomic *weight*. With the exception of fluorine, the atomic masses of the elements are expressed to four significant figures. Abundances and isotopic masses are taken from the *American Handbook of Physics*. The atomic masses of all elements are shown in the Periodic Table.

Atomic mass of elements and of isotopes

The mass of one atom of an isotope on the atomic mass scale is called its **isotopic mass** m. Some isotopic masses are listed in Table 3.2. Remember that the mass of atoms is defined with reference to a 'standard atom', i.e. one atom of carbon-12 for which $m(^{12}_{6}C) = 12.0000$ u.

Notice that isotopic masses (in atomic mass units) are very nearly equal in size to the mass number of an isotope. This is because the masses of the electrons, protons and neutrons are approximately 0, 1 and 1, respectively, on the atomic mass scale. Note also that the units of m and A are not the same – A is unitless.

If more than one isotope of an element exists in nature, the **average mass** of one atom of that element will be weighted towards the mass of the most abundant isotope. The average mass of one atom of an element in nature is called the **atomic mass** of that element, symbolized ***m*(element)**. The atomic mass of an element is worked out using the equation

$$m(element) = \frac{(m_1P_1) + (m_2P_2) + (m_3P_3) + \ldots}{100} \quad \textbf{(3.1)}$$

where m_1, m_2 and m_3 are the isotopic masses of isotopes 1, 2 and 3, and P_1, P_2 and P_3 are the percentage natural abundances of those isotopes.

An example: chlorine

Chlorine atoms in nature consist of those with mass number 35 and those with mass number 37. We start by approximating the isotopic masses as equal to the mass numbers:

$m(^{35}_{17}Cl) = 35$ u

$m(^{37}_{17}Cl) = 37$ u

If there were equal numbers of both isotopes in nature, the atomic mass of chlorine would be the average of 35 u and 37 u, i.e. 36 u. In doing this averaging, we are really using equation (3.1) with $P_1 = 50$ per cent and $P_2 = 50$ per cent:

$$\text{atomic mass of Cl} = \frac{(35 \times 50) + (37 \times 50)}{100} = 36 \text{ u}$$

The actual abundances (Table 3.2) are 75.770 per cent and 24.229 per cent. Substituting these values in equation (3.1) gives

$$m(Cl) = \frac{(35 \times 75.770) + (37 \times 25.229)}{100} = 35.48 \text{ u}$$

Use of *exact* isotopic masses (Table 3.2) gives the accepted value for the atomic mass of chlorine:

$$m(Cl) = \frac{(34.9689 \times 75.770) + (36.9659 \times 24.22)}{100} = 35.45 \text{ u}$$

EXERCISE **3E**

Working out atomic mass of the oxygen atom

Using the isotopic masses and abundances shown in Table 3.2, confirm that the atomic mass of oxygen is 16.00 u.

Table 3.2 shows that there is only one isotope of fluorine that occurs naturally. Accordingly, for this element:

$m(^{19}_{9}F) = m(F) = 18.9984$ u

Although we will use atomic masses (m) in units of u, they may also be defined so as to be unitless. The **relative atomic mass** (A_r) of one atom of an element is the average mass of an atom of that element relative to one atom of carbon-12. A_r and *m(element)* are numerically equal but A_r is unitless.

BOX **3.1**

The Saxon princess who became Queen Eadgyth of Magdeburg

Eadgyth (Edith) was the granddaughter of Alfred the Great and the half-sister of Athelstan, King of England. Athelstan sent Eadgyth to marry Otto, the king of Saxony, in 929 CE. She died when she was about thirty-six years old. An impressive tomb to Eadgyth was built in Magdeburg Cathedral around 1510, and it was thought that this was empty of human remains. However, when the tomb was opened in 2008, archaeologists found a lead box, with the inscription 'EDIT REGINE CINERES HIC SARCOPHAGVS HABET ... ' (The remains of Queen Eadgyth are in this sarcophagus ...). Inside the box were some bones of a woman.

The wear on her femur bones suggested that bones were those of a frequent horse rider, consistent with the nobility. But were they the bones of Eadgyth or someone else? Researchers at the University of Bristol's Department of Archaeology and the Institute of Anthropology at Mainz University studied the relative abundances of strontium and oxygen isotopes that were deposited in the teeth of the skeleton, with the isotope levels known to reflect the minerals in local waters where the individual grew up. Isotopic levels in the bones in the box were different from those found in bones of people born in Saxony, and were matched to people in Wessex. This led to the conclusion that the bones belonged to a person who spent their youth in Wessex, an area where ancient records record Eadgyth was born. The remains of Eadgyth, now known to be the oldest bones connected to British royalty, were reburied in a titanium coffin in 2010.

Statue of Queen Eadgyth in Magdeburg Cathedral, Germany. (Photo by Acoma).

Masses of molecules

The masses of molecules may also be expressed in atomic mass units. For example, use of the isotopic masses in Table 3.2 allows us to calculate the exact molecular mass of $^{1}H^{35}Cl$ as follows:

$$m(^{1}H^{35}Cl) = (1.0078\,u) + (34.9689\,u) = 35.9767\,u$$

Commercially available compounds usually contain atoms of isotopes of elements in the percentages of their natural abundances, i.e. such compounds are *not isotopically pure*. Therefore, in calculations of the yield of a chemical reaction, it is necessary to use the average atomic masses of each element; we should not use the isotopic masses. For example, for HCl we would use $m(H) = 1.008$ u and $m(Cl) = 35.45$ u. The average mass of the HCl molecule in nature is then

$$m(HCl) = (1.008\,u) + (35.45\,u) = 36.46\,u \approx 36.5\,u$$

Although we will use molecular masses (m) in units of u, they may also be defined so as to be unitless. The **relative molecular mass** (M_r) of a molecule is the mass of that molecule relative to an atom of carbon-12. M_r and m are numerically equal but M_r is unitless.

3.3 Mass spectrometer

How a mass spectrometer works and what it is used for

The isotopic masses and abundance of isotopes in a sample of an element are found using an instrument called a **mass spectrometer** (Figure 3.4).

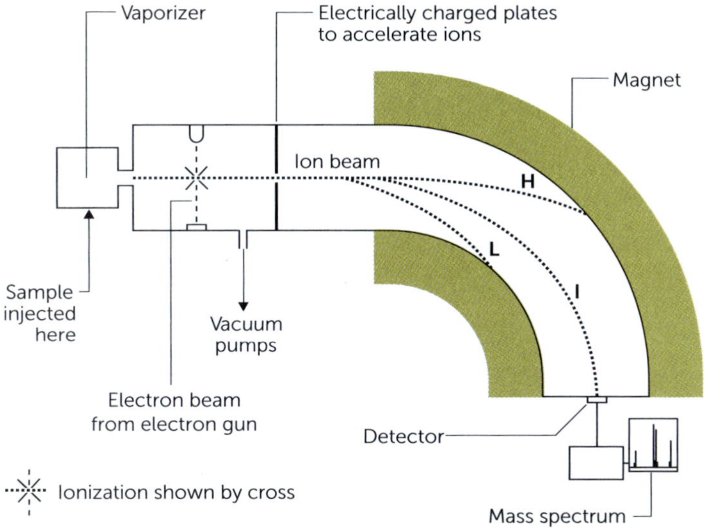

Figure 3.4 Diagram of a mass spectrometer. This type of spectrometer is said to be 'magnetic scanning' because it uses a varying magnetic field to sort the ions according to their charge and mass. The paths of ions of light (L), intermediate (I) and heavy (H) mass are shown. Powerful vacuum pumps keep gas pressures very low so that molecules or ions do not bump into each other.

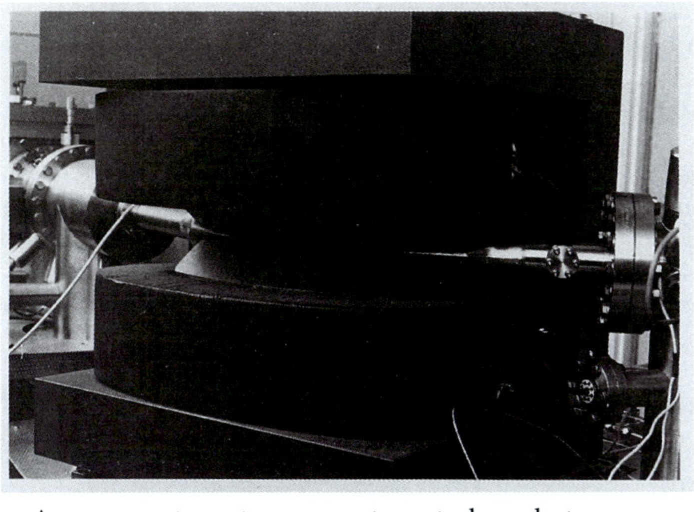

A magnetic scanning mass spectrometer. The thin steel spectrometer chamber is wedged between the poles of a powerful electromagnet (black) whose magnetic strength may be varied, so bending the beam of ions. Because the curvature of the chamber is fixed, only ions of one m/z value reach the detector for a particular magnetic field strength.

A mass spectrometer causes atoms to lose electrons, a process called **ionization**. Loss of an electron means that the positive charge of the protons in the nucleus is not balanced by the remaining electrons, and the atom is now positively charged (a **positive ion**). The positive ions are separated out in a mass spectrometer according to the ratio of the mass to charge *ratio* m/z where z is the number part of the relative charge of the ion (e.g. $z = 1$ for O^+ and $z = 2$ for Cl^{2+}) and m is the mass of the ion in atomic mass units. In the following discussion we will use the mass number of ions instead of the exact ion masses. For example,

$$\frac{m}{e}\left({}^{35}Cl^{+}\right) = \frac{35}{1} = 35$$

$$\frac{m}{e}\left({}^{37}Cl^{2+}\right) = \frac{37}{2} = 18.5$$

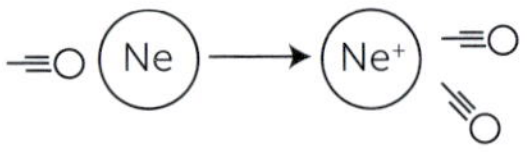

Figure 3.5 Ionization of a neon atom by an electron (O).

Figure 3.6 A straight beam of ions (····) is deflected by a magnet. For the same magnetic field strength, lighter ions (small m) bend from their path more than heavier ones.

Stages involved in a magnetic scanning mass spectrometer

As an example, suppose that neon gas (consisting of atoms of the isotopes neon-22, neon-21 and neon-20) has been injected into the mass spectrometer. The purposes of the main parts of the mass spectrometer are as follows:

1. The vaporizer is an oven which ensures that any sample is turned into a gas. (This is unnecessary with neon, which is a gas at room temperature.)
2. Electrons are fired at the gaseous sample. If one of these electrons collides with an atom, it may push out an electron from the atom, making a positive ion:

   atom + e^- (from gun) $\rightarrow$ ion + e^- (from gun) + e^- (from atom)

   See Figure 3.5. For neon:

   $^{22}Ne(g) + e^- \rightarrow {}^{22}Ne^+(g) + e^- + e^-$
   $^{21}Ne(g) + e^- \rightarrow {}^{21}Ne^+(g) + e^- + e^-$
   $^{20}Ne(g) + e^- \rightarrow {}^{20}Ne^+(g) + e^- + e^-$

   (Some doubly charged ions, such as $^{20}Ne^{2+}$, will also be produced but these will be less common than singly charged ions because more energy is required to make them.)
3. The ions are accelerated using charged plates. A narrow beam of ions is allowed to enter the rest of the spectrometer 'chamber'. The chamber is evacuated using powerful vacuum pumps which remove un-ionized atoms, and which ensure that there are very few molecules of air present that might collide with the ion beam.
4. The beam of ions would continue to travel in a straight line but, because a powerful electromagnet surrounds the chamber, the ions are bent downwards (Figure 3.6). The degree of bend experienced by an ion depends upon its m/z value and upon the strength of the magnetic field. For ions of a particular m/z ratio, there will be only *one* value of magnetic field strength which will bend those ions so that they strike the detector.
5. By slowly adjusting the strength of a magnetic field (a procedure called **scanning**), ions of a particular mass charge ratio (m/z ratio) are brought down to the ion detector one by one. The size of the electronic signal produced by the detector increases with the number of ions arriving per second. Since the atoms of all the neon isotopes are equally easy to ionize, the signal at each m/z value is proportional to the number of atoms of each isotope present in the original injected sample. A plot of signal intensity against m/z (called a **mass spectrum**) shows the relative abundance of isotopes for that element (Figure 3.7).

Application of mass spectrometry to molecules

If molecules are injected into a mass spectrometer, the molecules become ionized. For hydrogen chloride ($^1H^{35}Cl$):

$^1H^{35}Cl + e^- \rightarrow (^1H^{35}Cl)^+ + e^- + e^-$

$(^1H^{35}Cl)^+$ has a mass of 36 u and a charge of +1 so $m/z = 36/1 = 36$.

The ion obtained by ionizing the starting (i.e. parent) molecule is called the **parent ion**. Here, the parent ion is $(^1H^{35}Cl)^+$.

However, it often happens that the molecular ions produced in mass spectrometers have so much energy that they break up (i.e. **fragment**). For example,

$(^1H^{35}Cl)^+ \rightarrow {}^1H + {}^{35}Cl^+$

$m/z = 36 \quad m/z = 35$

Because HCl gas consists of both $^1H^{35}Cl$ and $^1H^{37}Cl$ molecules its mass spectrum will include peaks at $m/z = 36$ (due to $(^1H^{35}Cl)^+$), at 38 ($(^1H^{37}Cl)^+$), at 37 ($^{37}Cl^+$) and 35 ($^{35}Cl^+$). Peaks derived from species containing ^{35}Cl, the most abundant chlorine isotope, will be the most intense. Generalizing, the intensity and m/z value of peaks in a mass spectrum are very useful in helping chemists to decide which molecules were present in the original sample.

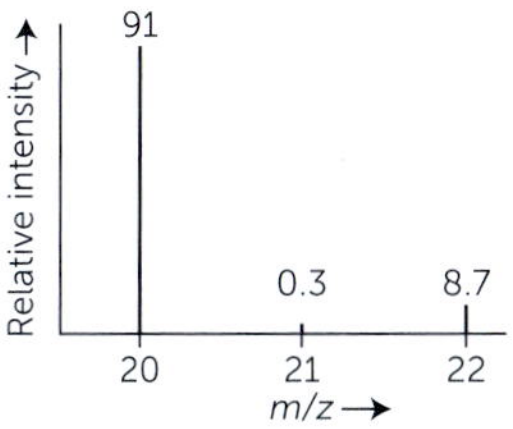

Figure 3.7 Only positively charged ions show up on a mass spectrum. This is the mass spectrum of neon gas. The three peaks are caused by the singly charged ions of the three isotopes of neon.

EXERCISE **3F**

Calculation of the atomic mass of neon from its mass spectrum

Use Figure 3.7 to estimate the atomic mass of the element neon.

EXERCISE **3G**

Mass spectra

(i) Calculate the m/z values of the following ions: $(^{12}C^{16}O_2)^+$, $(^{12}C^1H_3)^+$ and $(^{14}N_2)^{2+}$.

(ii) The mass spectrum of HF contains peaks at m/z = 20 and 19; explain this.

(iii) The mass spectrum of the organic compounds benzene (C_6H_6), nitrobenzene ($C_6H_5NO_2$) and phenol (C_6H_5OH) all contain an intense peak at m/z = 77 – why?

The molecular formula of water

We take it for granted that the molecular formula for water is H_2O. The historical difficulty was that chemists could not prove the formula of water without knowing the atomic masses of H and O, but they could not find the atomic masses of H and O without knowing the formula of water! Neither does the fact that the **electrolysis** of water gives hydrogen and oxygen in the ratio 2:1 prove that the formula of water is H_2O (see p. 119 in Unit 7). The way that the formula of water was established in the 1850s need not bother us here (see website for Unit 7), but nowadays we can easily confirm the formula of water from its mass spectrum (Figure 3.8), since the ions detected arise directly from the parent molecule H_2O. Specifically, the peak at m/z = 18 is due to the ion $(H_2O)^+$, which then fragments to $(OH)^+$ with m/z = 17.

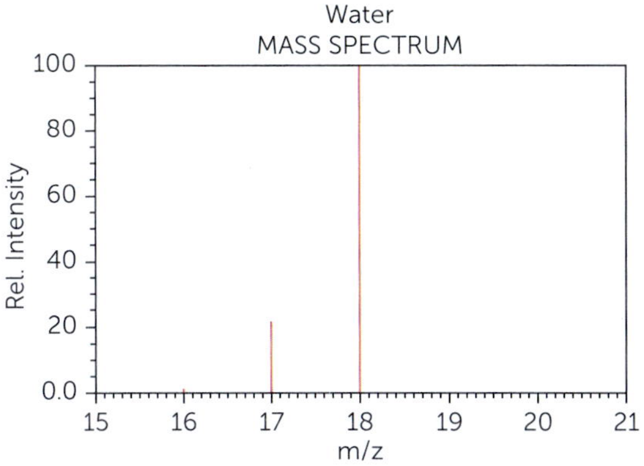

Figure 3.8

3.4 Types of mass spectrometers

Mass spectrometry is a technique that ionizes and selects (separates) ions, so allowing the identification of atoms or molecules present in the injected sample. Most mass spectrometers are used in tandem with gas or liquid chromatographs and for this purpose magnetic scanning has been superseded by **mass quadrupole spectrometers** ('quads') and also by **time-of-flight (TOF) mass spectrometers**. Users also have the option of 'hard' or 'soft' ionization (Figure 3.9).

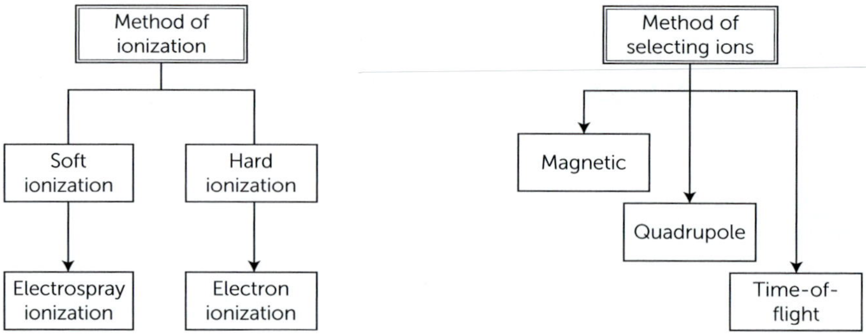

Figure 3.9 Classification of mass spectrometers.

Soft and hard ionization in mass spectrometers

In the magnetic scanning mass spectrometer instrument already discussed (Figure 3.4), ionization is achieved by firing electrons at the sample. This is referred to as '**electron ionization**' **(EI)**. Electron ionization is categorized as '**hard ionization**' because the energy absorbed by the molecules in the sample causes considerable fragmentation of the ions, producing many peaks in the mass spectrum. When the sample contains many large molecules, fragmentation is so extensive that there is a 'forest' of peaks which are difficult to interpret. For this reason, 'soft' ionization, which produces less fragmentation, is preferred.

One way in which 'soft ionization' may be achieved is to replace **EI** with **electrospray ionization** (**ESI**). In electrospray ionization, the sample is mixed with a solvent and sprayed into the instrument. The droplets of solution become intensely charged so the molecules present in the droplet become ionized and are then separated and detected. Whereas **EI** requires samples to be gaseous or volatile (and so easily vaporized), **ESI** can deal with mixtures containing substances with high boiling points, including biological molecules which would decompose if attempts were made to vaporize them by raising their temperature. In recognition of his role in developing ESI, John Bennett Fenn (1917–2010) was awarded a share of the 2002 Nobel Prize for chemistry.

Mass quadrupole and 'time-of-flight' spectrometers

Apart from magnetic scanning, ions may be separated (and hence selected and analysed) by mass quadrupole or 'time-of-flight' (TOF) technologies.

In a **mass quadrupole** scanning spectrometer, often simply known as a 'quad', the ions are separated electrically using four charged rods known as a quadrupole. The ions are first produced by an appropriate ionization method (EI or ESI) and then pass through the quadrupole in which a voltage is maintained between the rods (see Figure 3.10a). The trajectory of the ions in the electric field produced by the rods is complicated, but only ions of a certain mass to charge ratio (m/z value) may pass between the rods to reach the ion detector (Figure 3.10b): other ions strike the rods themselves and are pumped away. 'Quad mass spectrometers' are relatively cheap and compact and are the mass spectrometer of choice for routine analysis where the molecules have masses below 1000 u. As in all mass spectrometers, the visible outputs are in the form of ion abundance (or ion intensity) versus m/z ratio.

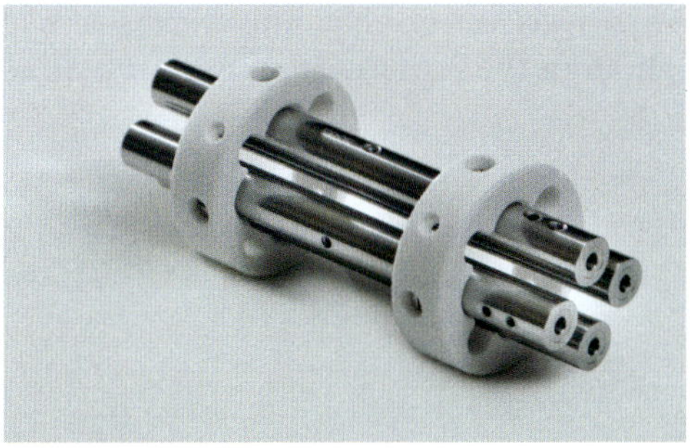

(a)

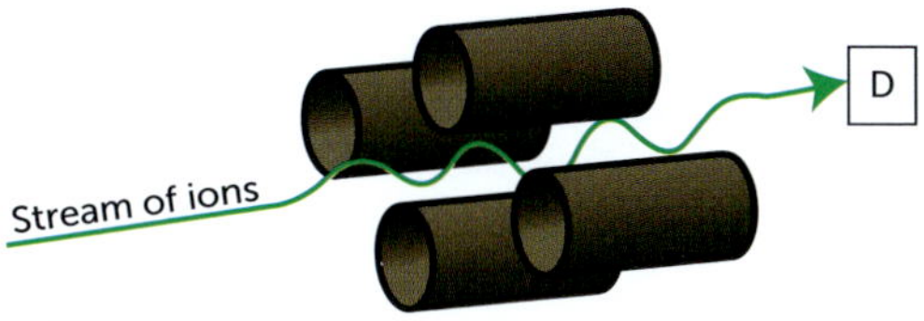

(b)

Figure 3.10 **(a)** An image of a quadrupole.

(b) Principle of mass quadrupole scanning. In a mass quadrupole scanning spectrometer the ions attempt to pass through the charged rods towards the detector, but only some (those with the *m*/*z* values selected by the operator) get through. In this diagram, the arrow shows the complicated trajectory of an ion beam (consisting of ions of a particular *m*/*z* value) that successfully makes it to the detector (D). Image courtesy of Reliance Precision Ltd.

TOF mass spectrometers use a different method in order to separate ions. TOF instruments measure the *time* it takes for the ions to complete their flights to the detector – hence the name. Once ions are formed, TOF mass spectrometers use an electric field to accelerate the ions. The time taken for ions to reach the detector depends on their mass, *m*. The lightest ions reach the detector first and the ions with the highest mass arrive last. Once again, the visible outputs are in the form of ion abundance versus *m*/*z* ratio. TOF mass spectrometers acquire *m*/*z* data much more rapidly than quadrupole mass spectrometers and this allows resolution of overlapping *m*/*z* peaks which are a feature of the mass spectra of complicated mixtures. TOF mass spectrometers can also be used to analyse much larger molecules ($m > 1000$ u), but they are more expensive than quad instruments.

3.5 Electronic structure of atoms

Bohr model of the atom

In 1913 Niels Bohr (1885–1962) suggested that the electrons circling the nucleus could only possess certain energies. In other words, the energy of the electron is **quantized**. The key features of the Bohr model (Figure 3.11) are:

1. Each shell of electrons lies at a definite radius (called the **Bohr radius**) from the nucleus. The electrons in each shell are at the *same energy*.
2. Only some values of Bohr radius are allowed. This means that only certain energy levels are allowed.
3. Each shell is labelled with a **principal quantum number** *n*. For example, the shell closest to the nucleus is labelled $n = 1$ and in the hydrogen atom its Bohr radius, r_0, is 52.9 pm from the nucleus. The higher the value of *n*, the higher the energy of the electron.
4. In order to change from one energy level E_1 to another E_2, the electron must gain (or lose) an amount of energy which is *exactly equal* to the difference in energy (symbolized ΔE) between the two levels:

   $\Delta E = E_2 - E_1$
5. Each shell can only hold a fixed number of electrons. For example, the first shell can hold up to two electrons and the second up to eight electrons. (The total number of electrons allowed in a shell is $2n^2$.)

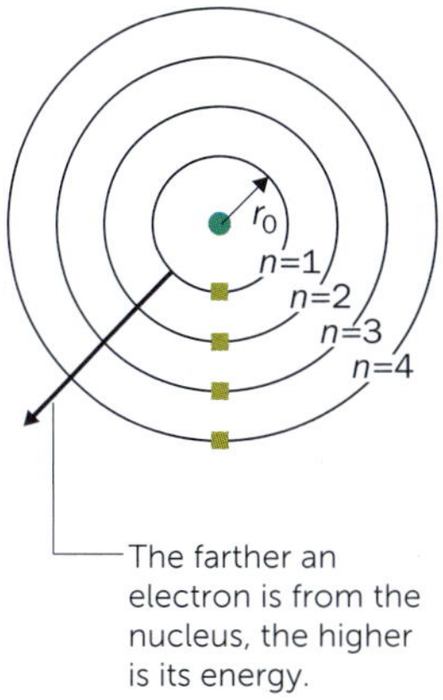

Figure 3.11 The Bohr model of the atom.
The electrons are shown as ■. The nucleus is represented as ●.

6 Electrons will occupy as low an energy level as possible. This means that if electrons are fed into the shells of an atom the *lowest energy shells are filled first.*

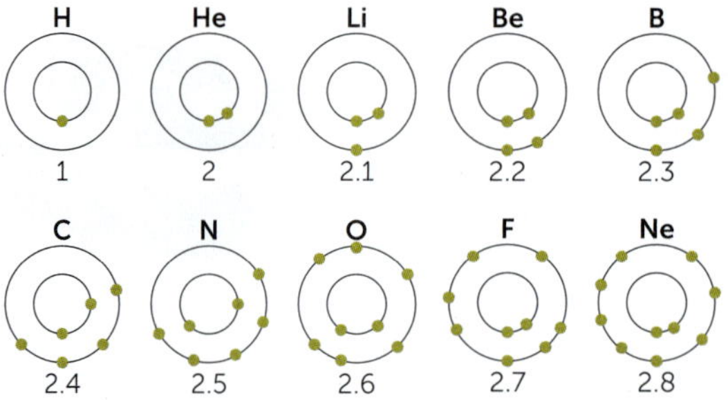

Figure 3.12 The electronic structure of atoms for elements with atomic numbers 1–10.

Figure 3.12 shows the arrangement of electrons in shells (the so-called **electronic structure** or **electronic configuration**) for atoms of elements with atomic numbers from 1 to 10 in their lowest energy levels. Also shown is the shorthand for such electronic structures; for example, 2.6 (read as 'two dot six') means that there are two electrons in the first shell (for which $n = 1$) and six electrons in the second shell (for which $n = 2$). The configurations of atoms containing the same shells of electrons as in helium, neon (or other 'noble gases' such as argon) are sometimes abbreviated, as in the following examples:

Li 2.1 becomes [He].1
Mg 2.8.2 becomes [Ne].2
Cl 2.8.7 becomes [Ne].7

Table 3.3 gives electronic structures for elements with atomic numbers 1–20.

3.6 Evidence for the existence of energy levels in atoms

You may have wondered what experimental evidence is available to support the idea that atoms contain electrons which lie in shells at definite energy levels. The truth is that the quantum theory of atoms was not arrived at quickly or easily, and a short account of such evidence will inevitably appear sketchy and incomplete. That said, there are two pieces of evidence which are difficult to explain without quantization, and we shall briefly review them because they introduce important ideas that will be useful later. They are **successive ionization energies** and **emission spectra**.

Successive ionization energies

Evidence for the existence of energy levels is provided by studies of successive ionization energies (ionization enthalpies) which may be measured electrically or spectroscopically. The **first ionization energy** (symbolized I_1 or $\Delta_i H^\ominus$) of an atom in the gaseous state is the minimum energy needed to completely remove the most easily removed electron from the attraction of the atomic nucleus. The outermost electrons are least tightly held by the nucleus and require less energy to completely break away from the attraction of the positively charged nucleus. It is for this reason that if energy is continually added to an atom, the *outer electrons are ejected first.*

Taking the potassium atom as an example, the first ionization energy is the energy needed for the change:

$$\begin{array}{lcl} K(g) & \rightarrow & K^+(g) + e^- \\ 2.8.8.1 & & 2.8.8 \end{array}$$

BOX **3.2**

Quantum theory

Most people have heard of the 'quantum theory' and the mere mention of it is guaranteed to stop most conversations in their tracks! Yet, many of the ideas behind the quantum theory are beautifully simple.

A quantity is quantized when it is allowed to take on only certain values. Time can take any value (2.5 seconds, 123.67877 seconds, 0.0143 seconds etc.) and is not quantized. On the other hand, the position of a painter on a ladder is quantized since he can only stand where there are rungs (Figure 3.13).

The idea of quantization was first put forward by Max Planck in 1900. One feature of the mathematics of the quantum theory is that it shows that energy depends upon 'magic numbers', which can only take on certain values: such numbers are known as **quantum numbers**. For example, in the case of the hydrogen atom, the energy *E* of an electron is proportional to the **principal quantum number** *n*, of the electron's shell. This quantum number can only take the values 1, 2, 3, 4 and so on. The fact that the value of *n* is restricted to certain values means that the value of *E* is also limited to certain values. Similar relationships apply to all atoms and in every case **the quantization of the energy of an electron is a natural consequence of the existence of quantum numbers.**

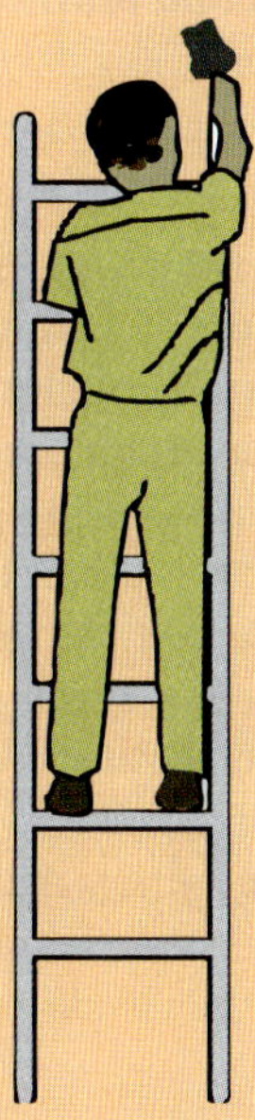

Figure 3.13 Whether they know it or not, the position of the painter on the ladder is **quantized**.

Table 3.3 The electronic structure for elements with atomic numbers 1–20. The last column will be filled in later on in this unit

Element and symbol	Atomic number	Bohr structure	s,p,d,f structure
Hydrogen H	1	1.	$1s^1$
Helium He	2	2. or [He]	$1s^2$
Lithium Li	3	2.1 or [He].1	$1s^2\ 2s^1$ or [He] $2s^1$
Beryllium Be	4	2.2 or [He].2	
Boron B	5	2.3 or [He].3	
Carbon C	6	2.4 or [He].4	
Nitrogen N	7	2.5 or [He].5	
Oxygen O	8	2.6 or [He].6	
Fluorine F	9	2.7 or [He].7	
Neon Ne	10	2.8 or [Ne]	
Sodium Na	11	2.8.1 or [Ne].1	
Magnesium Mg	12	2.8.2 or [Ne].2	
Aluminium Al	13	2.8.3 or [Ne].3	
Silicon Si	14	2.8.4 or [Ne].4	
Phosphorus P	15	2.8.5 or [Ne].5	
Sulfur S	16	2.8.6 or [Ne].6	
Chlorine Cl	17	2.8.7 or [Ne].7	
Argon Ar	18	2.8.8 or [Ar]	
Potassium K	19	2.8.8.1 or [Ar].1	
Calcium Ca	20	2.8.8.2 or [Ar].2	

EXERCISE **3H**

Ionization energy

Write down an equation representing the fourth ionization energy of beryllium (Be). Is there a fifth ionization energy?

The **second ionization energy** (symbolized I_2 or $\Delta_{i(2)}H^{\circ}$) is the energy required for the change:

$$\underset{2.8.8}{K^{+}(g)} \rightarrow \underset{2.8.7}{K^{2+}(g)} + e^{-}$$

The third and successive ionization energies are defined in a similar way.

Figure 3.14 shows how the ionization energies for potassium change as we increase the number of electrons removed. (The log of the ionization energies has been plotted in order to compress the ionization energies so as to make the *y*-axis a reasonable size.)

Each point in Figure 3.14 represents one electron. The obvious feature is that the points are clumped together into four groups. *This provides direct evidence that the nineteen electrons in the potassium atom belong to four groups. These groups are the electron shells.* Moving from left to right in Figure 3.14, the groups contain one, eight, eight and two electrons, respectively. This is the reverse of the familiar 2.8.8.1 configuration for potassium (*reverse* because in ionization the outer electrons are removed *first*). Although we have only plotted the ionization energies for potassium, similar patterns are seen for atoms of the other elements.

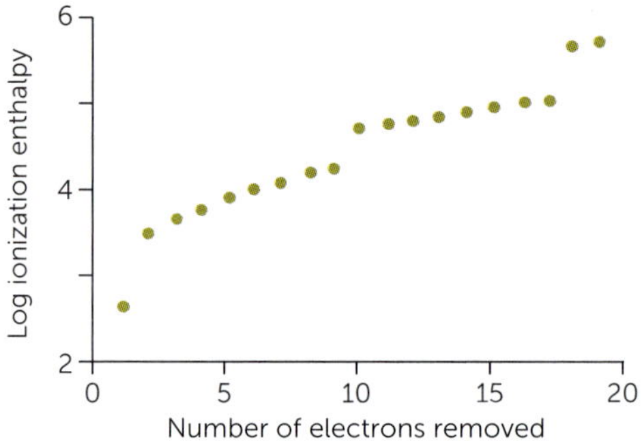

Figure 3.14 Successive ionization energies (kJ mol^{-1}) for the potassium atom (2.8.8.1).

Emission spectra of atoms

Evidence for the quantization of electrons in atoms also comes from **emission spectra**. If a sample of atoms is heated, or exposed to light of a suitable wavelength, the outer electrons of some atoms may gain energy, and the atom is then said to be in an **excited state**. One way in which this energy may be lost is by **emission**, in which the electron loses energy as **light**. The electron then occupies a lower energy level. The yellow colour seen when common salt is sprinkled into a flame is an example of such an emission.

The frequency (ν, pronounced 'nu') of the emitted light depends upon the amount of energy lost by the electron, ΔE. The exact relationship between the two quantities is described by the **Planck equation**:

$$\Delta E = h\nu$$

where h is a universal constant known as the **Planck constant**.

If the emission of light from an element is studied using an instrument that allows the measurement of the frequency of light (a **spectrometer**), it is noticed that several frequencies of light are usually emitted: these frequencies make up the emission spectrum of the element. Just as importantly, however, there are some frequencies that are not emitted. This means that only some values of energy change ΔE are allowed. *The simplest way of explaining this is to assume that the energy of the electron in the atom may only possess certain values.* In other words, the energy of electrons in atoms is quantized. (More details about light emission, and calculations using the Planck equation, are given in Unit 22.)

3.7 More advanced ideas about electronic structure[1]

Sublevels of energy

According to the Bohr picture of the atom, all electrons in the same shell possess the same energy. More detailed experiments (also involving ionization energies and spectra) show that with the exception of electrons in the first shell this is not the case, and the main energy levels of atoms possess **sublevels** of energy. The sublevels include the four types **s, p, d** and **f**:

s sublevels contain up to **2** electrons; the electrons are called 's electrons'.
p sublevels contain up to **6** electrons; the electrons are called 'p electrons'.
d sublevels contain up to **10** electrons; the electrons are called 'd electrons'.
f sublevels contain up to **14** electrons; the electrons are called 'f electrons'.

A full description of a sublevel also requires the principal quantum number n. For example, the third shell of electrons ($n = 3$) contains s, p and d sublevels of energy, and these are labelled 3s, 3p and 3d. The symbol $3s^1$ would show that there is one electron in the 3s sublevel, and $3p^6$ would show that the 3p sublevel is full.

The order in which the sublevels are filled is:

1s 2s 2p 3s 3p 4s 3d ...
Lower energy Higher energy

Note that (rather oddly) the 4s sublevel is filled before the 3d. Using superscripts to show the maximum number of electrons in these levels, the sublevels may be written as follows:

$\mathbf{1s^2\ \ 2s^2\ \ 2p^6\ \ 3s^2\ \ 3p^6\ \ 3d^{10}\ \ 4s^2}$ ...

You are strongly advised to learn this before carrying on!

Note also that despite the order of energies of the sublevels, when *writing* the electronic structure of an atom, sublevels with the same value of n are grouped together (i.e. we *write* 3d before 4s).

Think of the aluminium atom, with an electronic configuration of 2.8.3. A more detailed description of its energy levels is obtained by feeding all thirteen electrons into the sublevels, giving

$1s^2\ \ 2s^2\ \ 2p^6\ \ 3s^2\ \ 3p^1$

or

$[Ne]\ \ 3s^2\ \ 3p^1$

Electronic structure of the 'first transition series' of elements

The elements titanium to copper are known as the **first transition series**. The electronic structures of these atoms (and of zinc and scandium) in their lowest energy states are shown in Table 3.4.

Notice that the 4s and 3d levels do not fill up in a completely regular fashion.

> EXERCISE **3I**
>
> **Electron configuration of atoms of the elements: s,p,d notation**
>
> Use the ideas outlined in section 3.7 to complete the last column of Table 3.3 by entering the detailed electronic structures of the first twenty elements using the s,p,d notation.

[1] These ideas are not essential for an understanding of Unit 4.

This is because the $3d^5$ and $3d^{10}$ configurations are particularly stable, and chromium and copper would 'prefer' to possess these structures, even if the 4s sublevel remains unfilled.

The electronic structure of elements is given on pages 503–504.

Wave nature of electrons

Experiments have shown that electrons sometimes behave as if they were waves. The idea that the electron can behave as a particle or as a wave is called **wave–particle duality**.

Think of a water wave produced by dropping a stone in a pond. Freeze this picture in your mind and ask yourself the question: where is the water wave? The answer, of course, is that the water wave consists of many ripples that are spread out upon the surface of the water. Some ripples are more pronounced than others, but the wave does not belong to a single spot.

This idea of an 'uncertainty' in the position of a wave is carried over into the atom. If electrons behave like waves we cannot pinpoint their exact position. All we can do is to consider the **probability** of finding the electron at a point or in a region of space.

Table 3.4 Electronic structure of the first transition series of metals

Element and symbol	Atomic number	Bohr structure	s,p,d structure
Scandium	21	2.8.9.2 or [Ar].3	$1s^2\ 2s^2\ 2p^6\ 3s^2\ 3p^6\ 3d^1\ 4s^2$ or [Ar] $3d^1\ 4s^2$
Titanium	22	[Ar].4	[Ar] $3d^2\ 4s^2$
Vanadium	23	[Ar].5	[Ar] $3d^3\ 4s^2$
Chromium	24	[Ar].6	[Ar] $3d^5\ 4s^1$
Manganese	25	[Ar].7	[Ar] $3d^5\ 4s^2$
Iron	26	[Ar].8	[Ar] $3d^6\ 4s^2$
Cobalt	27	[Ar].9	[Ar] $3d^7\ 4s^2$
Nickel	28	[Ar].10	[Ar] $3d^8\ 4s^2$
Copper	29	[Ar].11	[Ar] $3d^{10}\ 4s^1$
Zinc	30	[Ar].12	[Ar] $3d^{10}\ 4s^2$

Figure 3.15 The orbital for a 1s electron in the hydrogen atom. The radius of the sphere in which it is 90 per cent likely that the electron will be found is about 100 pm. The single radius at which the 1s electron is *most likely* to be found is a distance 52.9 pm from the nucleus. This may be compared with the Bohr theory of the atom, where it was assumed that the electron was *certain* to be found at a radius of 52.9 pm.

Orbitals

Suppose we were able to take a snapshot photograph of the 1s electron in the hydrogen atom at an instant. A fraction of a second later we take another snapshot. By taking thousands of such snapshots, and superimposing all of them on a single frame, we would end up with a single picture showing thousands of points where the electron had been located. The greater the number of points in a region, the greater is the probability of finding the electron there in the future.

Such snapshots would show that the **volume** in which it is 90 per cent probable that the s electron would be found is spherical. The volume is given the name **atomic orbital**, usually abbreviated to **orbital**. The orbital of a 1s electron is called a 1s orbital, and it possesses a radius of about 100 pm (100×10^{-12} m). However, the *single* radius at which the 1s electron is most likely to be found is at 52.9 pm (the Bohr radius of the $n = 1$ shell). But – and this is the crucial difference between the wave model and the Bohr model – the wave model states there is always a chance that the electron will be somewhere outside this radius (Figure 3.15).

All s orbitals are spherical; 2s orbitals have larger radii than 1s orbitals.

Similarly, orbitals containing p electrons are termed p orbitals. There are three types of **p orbital** (labelled p_x, p_y and p_z), which are normally of equal energy but which have different directions in space. The shape of a p orbital is often described as a 'dumb-bell' (Figure 3.16). Orbitals containing d electrons are termed **d orbitals.** There are five types of d orbitals, and each is normally of equal energy (Fig 3.17).

Each orbital (whether $2p_z$, 1s, $3d_{xy}$ or 2s etc.) in an atom holds **a maximum of two electrons**. If there are two electrons present, the electrons must be spinning in opposite ways (one anticlockwise, one clockwise) and we say that the electrons are **paired**. This is shown using a box for the orbital and arrows for the electrons. Up and down arrows confirm that their spins are opposite, as follows: ↑↓

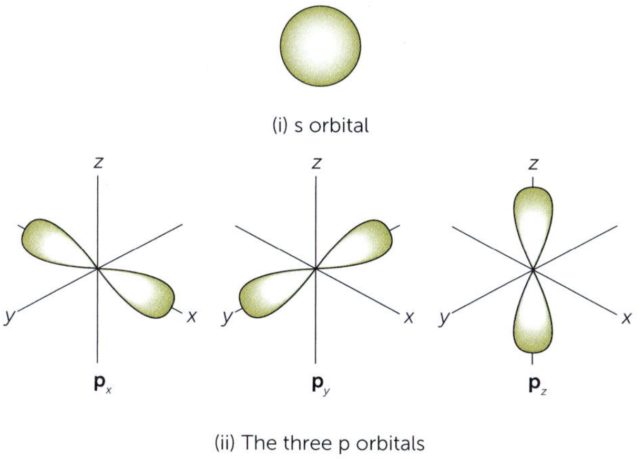

Figure 3.16 The shape of an s orbital and of the three types of p orbitals. Each s orbital and each p orbital holds up to two electrons.

d orbitals

The five d orbitals normally have the same energy. They have the shapes shown in Figure 3.17.

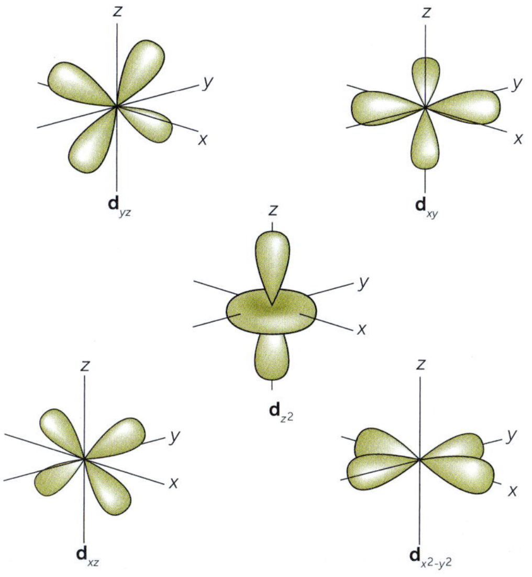

Figure 3.17 The shape of five types of d orbitals. Each of the five d orbitals holds up to two electrons.

Orbitals and sublevels

We now return to the order in which the sublevels of electrons in atoms are filled,

$1s^2$ $2s^2$ $2p^6$ $3s^2$ $3p^6$ $4s^2$ $3d^{10}$ ...

where the superscripts show the maximum number of electrons in each sublevel. Electrons in a 1s orbital occupy the 1s energy sublevel, electrons in a 2s orbital occupy the 2s sublevel and electrons in a 3s orbital occupy the 3s sublevel. Electrons in any of the three 2p orbitals are normally equal in energy and occupy the 2p sublevel. There are five 3d orbitals: electrons in these orbitals will occupy the 3d sublevel.

The use of 'boxes' may help make this clearer. In the aluminium atom (electronic configuration $1s^2$ $2s^2$ $2p^6$ $3s^2$ $3p^1$) the thirteen electrons occupy the following orbitals:

Electrons in orbitals:	↑↓	↑↓	↑↓	↑↓	↑↓	↑↓	↑
			p_x	p_y	p_z		p_y, p_x or p_z
Type of orbitals:	1s	2s		2p		3s	3p

where the $3p^1$ electron occupies either the $3p_y$, $3p_x$ or $3p_z$ orbital, all of which are of equal energy.

The 1s, 2s, 2p and 3s orbitals in the phosphorus atom ($1s^2$ $2s^2$ $2p^6$ $3s^2$ $3p^3$) are filled in the same way as for aluminium, with the complication that, without additional knowledge, it is unclear whether the 3p electrons feed into two of the 3p orbitals, e.g.:

↑↓	↑
$3p_x$	$3p_y$

or whether they enter all three p orbitals:

↑	↑	↑
$3p_x$	$3p_y$	$3p_z$

It turns out that the second distribution is the correct one – to minimize repulsion, electrons 'prefer' to occupy orbitals singly. This is known as **Hund's rule.**

EXERCISE **3J**

Detailed electron configuration of atoms of the elements

By following the pattern for Al, write out the electronic 'box structures' (i) for atoms with atomic numbers 1–10, and (ii) for the transition metals Mn, Fe and Cu.

REVISION QUESTIONS

3.1 Which of the following is false?

(i) The atoms of an element all have the same atomic number.

(ii) 1 u = one-twelfth of the mass of one atom of $^{13}_{6}C$.

(iii) Mass number = atomic number + number of neutrons.

(iv) The atoms of an element are identical, but different from atoms of other elements.

(v) A mass spectrometer detects both ions and neutral atoms.

3.2 The following table gives information on the four naturally occurring isotopes of sulfur:

Isotope	Number of neutrons	Number of electrons	Number of protons	Natural abundance /%
$^{32}_{16}S$				95.0
$^{33}_{16}S$				0.76
$^{34}_{16}S$				4.2
$^{36}_{16}S$				0.021

(i) Fill in the second, third and fourth columns of the table.

(ii) Estimate the atomic mass of sulfur to two significant figures. Why is your answer an estimate?

3.3 Write equations showing the formation of Mg^+ and Mg^{2+} in the chamber of a mass spectrometer. What are (i) the m/z values, and (ii) the 2.8.1-type electronic configurations of these ions?

3.4 Explain the following, writing equations for the formation of any ions in the mass spectrometer chamber:

(i) The mass spectrum of fluorine gas (F_2) contains two peaks, at m/z = 19 and 38.

(ii) The mass spectrum of hydrogen cyanide (HCN) contains peaks at *m*/*z* = 27, 26 and 13.5.

(iii) The mass spectrum of chlorobenzene (C_6H_5Cl) contains peaks at *m*/*z* = 77, 112 and 114.

(iv) An intense line at *m*/*z* = 28 is observed when an attempt was made to run the mass spectrum of a compound. (**Hint:** It is suspected that the spectrometer is leaking.)

(v) A mixture of methane (CH_4) and chlorine gas (Cl_2) yielded a mass spectrum with peaks at 70, 72, 74, 16, 52 and 50.

3.5 Write equations for the burning of normal hydrogen gas ($^{1}_{1}H_2$) and deuterium gas ($^{2}_{1}H_2$) in oxygen. Would you expect the density of normal and heavy water to be the same?

3.6 Use Table 3.2 to calculate the exact mass in (i) atomic mass units, (ii) grams, of (a) a molecule of heavy water ($^{2}_{1}H_2{}^{16}_{8}O$) and (b) an average water molecule in nature (work to four significant figures).

3.7 Briefly explain why atomic emission spectra provide experimental evidence for the existence of electron energy levels (shells) in atoms.

3.8 The successive ionization energies (in kJ mol^{-1}) for argon are 1,521, 2,666, 3,931, 5,771, 7,328, 8,781, 11,996, 13,842, 40,761, 46,188, 52,003, 59,654, 66,201, 72,920, 82,570, 88,620, 397,710 and 426,910. Plot the log of the ionization energy against the number of electrons removed (as in Figure 3.14). Why does your graph support the electronic structure of argon as being 2.8.8? What is the total amount of energy (in kJ mol^{-1}) required to remove all eighteen electrons from the argon atom?

3.9 What is an orbital? Comment upon the size of an orbital in which it is *certain* that an electron will be found.

3.10 Write down the electronic configuration of (i) Cl, (ii) Cl^- (a chlorine atom which has gained an electron), (iii) Cl^+, (iv) He, (v) Ne and (vi) Cu^{2+}, using the 2.8.1 notation. What is special about the electronic configurations (ii), (iv) and (v)?

3.11 Repeat question 3.10 using the s,p,d,f notation. Write out the structure for the ion Cl^+ using the electrons in boxes notation.

3.12 Are atoms mostly empty space?

Further information about models of atoms and the application of mass spectrometry to organic molecules will be found in Extension 3 on the website at https://www.bloomsburyonlineresources.com/chemistry

4 Bonding Between Atoms

CONTENTS

OBJECTIVES

This unit:

- Explains why elements react
- Describes the extreme types of bonding between atoms
- Links the properties of compounds to the bonding within those compounds
- Extends the description of bonding in compounds to include 'intermediate' bonding types

4.1 Why atoms combine

You have already seen that atoms of different elements react together to form compounds. The microscopic hooks introduced in Unit 2 are not real and are only useful tools to predict formulae. The forces that hold atoms together in compounds are called **chemical bonds**. There are three main classes of chemical bonds: ionic, covalent and metallic and all three are electrostatic in nature, meaning that they derive from the charges (and in particular, the electrons) in atoms. In order to understand the nature of these bonds, perhaps we should first ask the question: Why do atoms bother to combine in the first place?

Noble gases

Clues to the answer to this question may be found by considering a group of elements known as the noble (or **inert**) gases: helium (He), neon (Ne), argon (Ar), krypton (Kr), xenon (Xe) and radon (Rn). Inert means 'inactive' or 'idle', and describes the chemical reactivity of these elements. They are unreactive and for this reason are said to be **chemically stable**. He, Ne and Ar do not react with any other elements, whereas Kr and Xe react only with highly reactive fluorine. What is it that makes them so unwilling to react?

The electronic structures of the atoms of these elements are listed in Table 4.1. With the exception of helium, the atoms have eight (or $ns^2\,np^6$, where n is a whole number greater than one) electrons in the outer shells of their atoms. The arrangement is called a **stable octet** of electrons. There seems to be a stability associated with this arrangement, and with the unique electron configuration of helium, which makes the gases unreactive.

Lewis symbols

In 1916, Kossel and Lewis proposed that atoms of elements react with each other in order to achieve the stable electron arrangements of the noble gases. In order to obtain these stable arrangements, atoms may gain, lose or share electrons. These processes are generally represented by using Lewis symbols – an atom of an element is represented by the symbol for that element and the electrons in *the outer shell* of the atom are shown as dots, small circles or crosses. For example, the Lewis symbol for the sodium atom is

Na°

The symbol for sodium is Na and the electronic structure of the sodium atom is 2.8.1, so there is **one** electron in the last electron shell, which we represent by one dot (or cross – the choice is yours!).

Table 4.1 Electronic structures of the noble gases*

Inert gas	Electronic structure (Bohr model)	Electronic structure (s,p,d,f notation)
He	2	$1s^2$
Ne	2.8	[He] $2s^2\ 2p^6$
Ar	2.8.8	[Ne] $3s^2\ 3p^6$
Kr	2.8.18.8	[Ar] $3d^{10}\ 4s^2\ 4p^6$
Xe	2.8.18.18.8	[Kr] $4d^{10}\ 5s^2\ 5p^6$
Rn	2.8.18.32.18.8	[Xe] $4f^{14}\ 5d^{10}\ 6s^2\ 6p^6$

* To save space, the convention [noble gas] is used and refers to the electronic configuration of that gas. So, for example, [He] replaces $1s^2$.

4.2 Ionic bonding

When an atom of sodium reacts with an atom of chlorine, sodium chloride is formed. The Lewis symbols for sodium and chlorine are

Although the complete electron arrangements of each atom are written underneath each Lewis symbol, *only the outer electrons* are shown by dots and crosses because they are the only ones that take part in bonding.

When a sodium atom reacts with a chlorine atom, the sodium atom loses one electron to chlorine (to obtain the stable electron arrangement of neon), while the chlorine atom, in gaining an extra electron, attains the stable electron arrangement of argon. This can be represented in an equation:

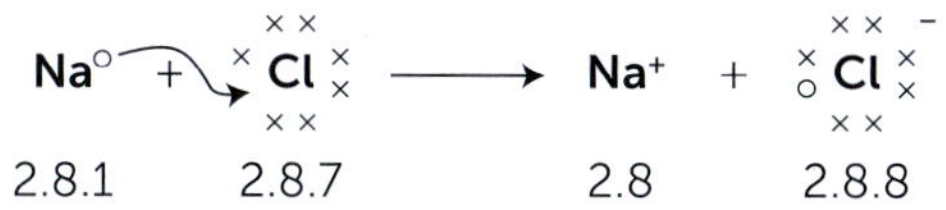

Before the reaction the neutral sodium atom had eleven electrons and eleven protons. After giving an electron to chlorine, the sodium atom has ten electrons and eleven protons, so it is not electrically neutral – it carries an overall positive charge equal to the charge of one proton (+1). The new species is written Na^+ and is called a sodium **ion** (see Box 4.1). The chlorine atom that has accepted an electron now has seventeen protons and eighteen electrons, so it has an overall charge of –1, is written as Cl^- and is called a chloride ion.

EXERCISE **4A**

Lewis symbols

Write Lewis symbols (using dots or crosses) for the following neutral atoms. The number of electrons in each atom is shown in brackets:

(i) F (9)

(ii) K (19)

(iii) Be (4)

(iv) S (16)

(v) P (15)

(vi) Ca (20)

(vii) C (6)

(viii) Al (13)

(ix) Ar (18)

(x) H (1)

BOX **4.1**

Ions

Ions are atoms (or groups of atoms) that carry electrical charges. Examples of ions are K^+, F^-, CO_3^{2-}, N^{3-}, Ca^{2+} and Al^{3+}.

Positively charged ions (e.g. K^+, Ca^{2+} and Al^{3+}) are called **cations**, whereas negatively charged ions (e.g. F^-, CO_3^{2-} and N^{3-}) are called **anions**.

Note that once a sodium atom has given an electron to a chlorine atom, that electron is indistinguishable from the other electrons present in the outer shell of the chloride ion. All eight electrons now 'belong' to the chloride ion. Crosses and dots are used to represent the electrons, simply to show which atoms they originated from. The use of a cross, or a dot, does not imply that the electrons are different.

These changes are shown below:

Before

Na atom	Cl atom
electrons = 11 × (–1)	electrons = 17 × (–1)
protons = 11 × (+1)	protons = 17 × (+1)
Neutral	Neutral

After

Na^+ ion	Cl^- ion
electrons = 10 × (–1)	electrons = 18 × (–1)
protons = 11 × (+1)	protons = 17 × (+1)
Charge = +1	Charge = – 1

Sodium chloride is therefore composed of Na^+ and Cl^- ions in a 1:1 ratio. The compound is called an **ionic** compound (because it is composed of ions), and the bonding in the compound is termed **ionic**. Ionic bonding arises because the oppositely charged Na^+ and Cl^- ions attract each other. Ionic compounds are formed whenever one atom (or group of atoms) transfers electrons to another. Note that when sodium and chlorine react, they both end up with the electron configuration of a noble gas *but they do not change into noble gases*; they are still the same elements. This is because the identity of an atom depends on the number of protons in its nucleus and these remain the same for sodium and chlorine both before and after the reaction.

The 'ordinary' chemical equation for this reaction tells us that the product of the reaction is sodium chloride, but gives no information about the bonding in the compound formed:

$$2Na(s) + Cl_2(g) \rightarrow 2NaCl(s)$$

For this reason, it is better written as producing pairs of ions

$$2Na(s) + Cl_2(g) \rightarrow 2Na^+, 2Cl^-(s)$$ (i.e. two pairs of Na^+ and Cl^- ions)

'The comma between the ions represents the ionic bond. Solid sodium chloride consists of equal numbers of sodium and chloride ions arranged in 3-D. Because the ions are so small, there are trillions of ions in even the smallest grain of NaCl'

EXAMPLE 4.1

Magnesium reacts with fluorine to form magnesium fluoride. Write an equation to describe the bonding using Lewis symbols.

Answer

The bonding in the compound can be shown by writing an equation using Lewis symbols, as follows:

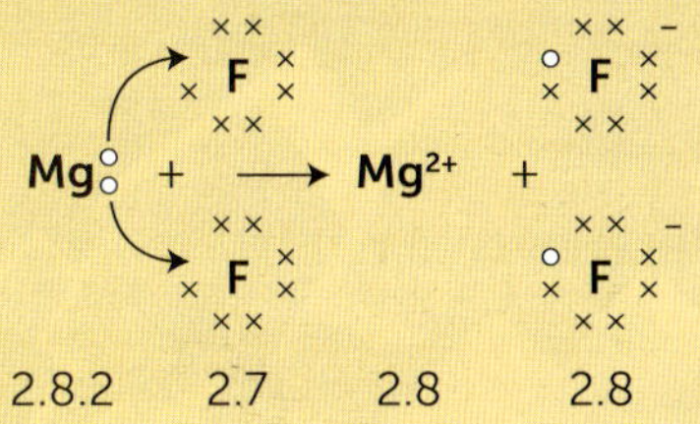

Comment

Magnesium needs to lose two electrons to attain the stable electronic configuration of neon, so two fluorine atoms accept one electron each to also gain the same electron arrangement as neon. Magnesium ions (ten electrons, twelve protons) with a charge of 2+ are formed, while fluorine forms F^- ions. Magnesium fluoride is an ionic compound in which the Mg^{2+} ions and F^- ions are in a ratio of 1:2.
The best equation that can be written for this reaction is therefore

$$Mg(s) + F_2(g) \rightarrow Mg^{2+}, 2F^-(s)$$

EXERCISE 4B

Reaction of potassium with chlorine

The equations using Lewis symbols provide us with more information than the usual equations for chemical reactions. The reaction of potassium metal with chlorine gas is shown in Figure 4.1.

(i) Write an equation, using Lewis symbols, that shows the changes in electronic structure that occur when potassium atoms react with chlorine atoms.

(ii) Write the balanced equation for the reaction of potassium with chlorine gas (Cl_2).

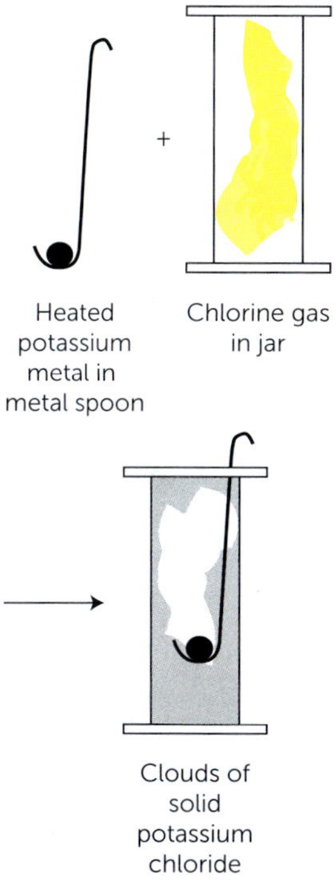

Figure 4.1 The violent reaction of silvery potassium metal with yellow chlorine gas. The yellow colour of the chlorine disappears, and crystals of white potassium chloride are formed.

EXERCISE 4C

Writing equations for the formation of ionic compounds using Lewis symbols

Write equations for the formation of the following compounds, using Lewis symbols. Use Table 3.3 for electronic arrangements:

(i) magnesium oxide (MgO) from Mg and O atoms

(ii) calcium fluoride (CaF_2) from Ca and F atoms

(iii) aluminium chloride ($AlCl_3$) from Al and Cl atoms

(iv) lithium oxide (Li_2O) from Li and O atoms

(v) aluminium oxide (Al_2O_3) from Al and O atoms.

Arrangement of ions in sodium chloride

In a sodium chloride crystal, the Na^+ and Cl^- ions are arranged in a *giant lattice* structure. The 'building brick' of this structure, a **unit cell**, is as shown in Figure 4.2. The larger spheres represent Cl^- ions, whereas the smaller spheres represent Na^+ ions. A crystal of sodium chloride consists of many trillions of these unit cells stacked together in the lattice.

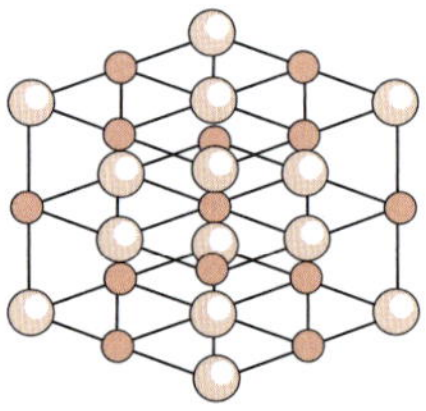

(a)

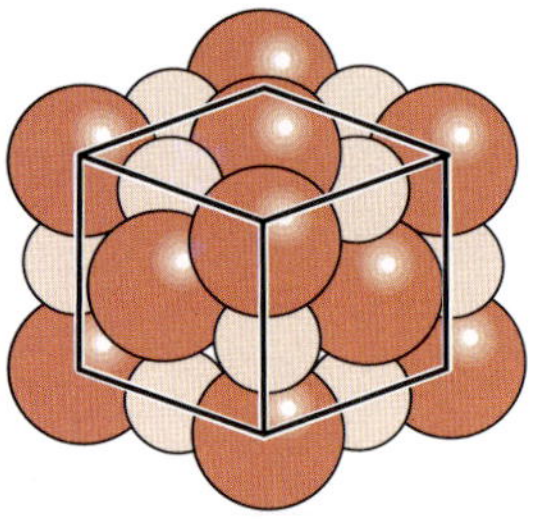

(b)

Figure 4.2 The unit cell of sodium chloride: (a) ball-and-stick model; (b) space-filling model.

Cubic salt crystals. An orderly arrangement of Na^+ and Cl^- ions gives rise to this shape.

In the unit cell, each sodium ion is surrounded by six chloride ions and each chloride ion is surrounded by six sodium ions, so preserving the 1:1 ratio overall. The ions are held together by the attraction of their opposite charges (electrostatic forces). These forces are very strong. X-ray diffraction allows us to 'see' ions in an ionic compound, as described in Box 4.2.

All ionic substances consist of ions arranged in a giant lattice. Different ionic structures have their ions arranged in different patterns. The sodium chloride structure is one of the simplest arrangements.

BOX **4.2**

Electron density map for sodium chloride

How do the ideas of Kossel and Lewis fit in with the modern view of the arrangement of electrons in an atom? The modern theory teaches that electrons have a wave-like nature. In regions where there is a high probability of finding the electron, there is a high concentration of electrical charge and we say that the electron density is large. By similar reasoning, electron density is small where there is a low probability of finding the electron. Through using a technique known as X-ray diffraction, it is possible to obtain **electron density** contour maps of ionic substances. Figure 4.3 shows such a map for sodium chloride – each contour line connects points with the same electron density, and the nucleus of each atom is represented by a dot at the centre of each ion. Notice that there is a space between the nuclei where there are no contours; this is a characteristic feature of the electron density map for an ionic substance.

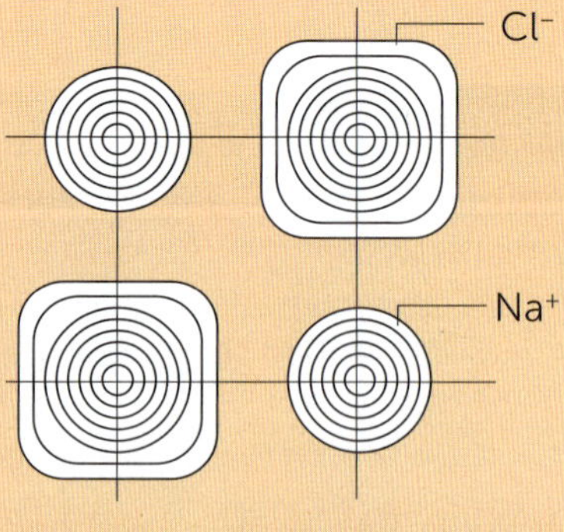

Figure 4.3 Contours to show the electron density distribution around sodium and chloride ions.

Properties of ionic compounds

Ionic compounds have certain properties that are common to all of them:

1 High melting and boiling points (they are non-volatile)

They are solids at room temperature. When a substance melts, then boils, its particles separate. A great deal of energy is needed to break down the lattice structure because the ions are attracted together by strong forces. The structure must therefore be heated to a high temperature before it melts.

2 Soluble in water

The hydrogen atoms in water molecules have a slight positive charge, shown as δ^+, and the oxygen atom has a slight negative charged, δ^-. (The charge separation in a molecule is called a **dipole**.) When ionic compounds are put into water, its fractional positive or negative charges attract the ions in the lattice, pulling them into the solution (*dissolution*). The ions in the solution are tightly surrounded by water molecules, a process called *hydration*. Because of its ability to dissolve ionic compounds, water is said to be a **polar solvent**. This is shown in Figure 4.4.

3 Conduct electricity when molten or in aqueous solution

In order for a substance to conduct electricity, it must possess **charged particles** that can *move* (in a metal, mobile electrons can carry the electric current). Solid ionic compounds *do not conduct electricity* – the ions are held firmly by strong electrostatic forces and cannot move. When the substance is melted or dissolved in water, however, the ions move freely and can carry an electric current. An aqueous solution, or melt, of an ionic substance that behaves in this way, is called an **electrolyte**.

4 Ionic crystals shatter easily

When a force is applied to the crystal, the layers of ions in the crystal structure can 'slip' so that similarly charged ions are next to one another. The like charges of the ions repel one another and the crystal structure shatters.

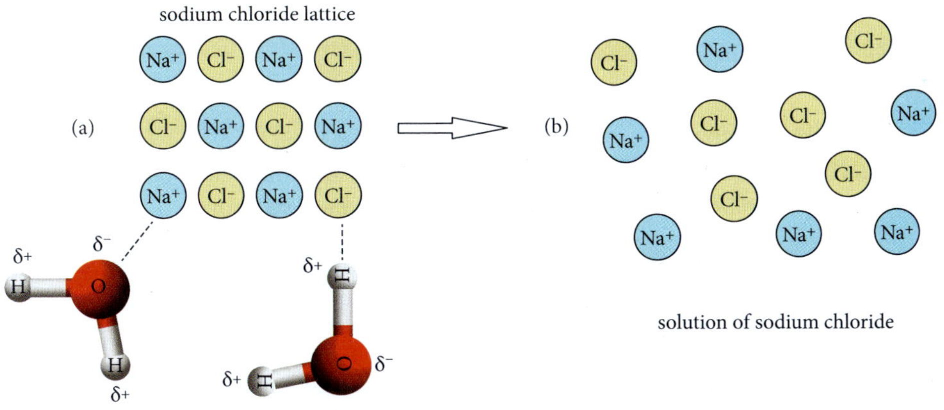

Figure 4.4 An ionic compound (sodium chloride) dissolving in water. The hydrogen and oxygen atoms in water molecules have partial positive and negative charges which attract (– – –) the negative (Cl^-) and positive (Na^+) ions in the crystal lattice. This breaks up the lattice (**dissolution**). The ions in the solution are surrounded by water molecules but this is not shown in (b).

When ionic bonding is likely to occur

Ionic bonding occurs between elements when the energy required to remove the outer shell electrons (the **ionization energy**) of one of those elements is relatively low.

Elements with such low ionization energies are *metals*; they generally contain *no more than three electrons* in the outer shell of their atoms. Therefore, compounds containing a metal tend to be ionic. Non-metals generally contain more than three electrons in the outer shells of their atoms. Note that although hydrogen contains one electron in the first shell of its atoms, the ionization energy for that atom is very high because it is close to, and highly attracted to, the positive nucleus. Hydrogen is therefore classified as a non-metal and most of its compounds are not ionic.

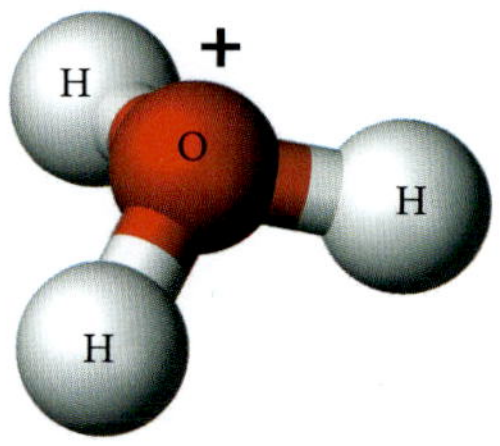

The hydronium ion, H_3O^+. The +1 charge sits on the oxygen atom.

BOX **4.3**

Hydrogen ion

The energy required for the process

$$H(g) \rightarrow H^+(g) + e^-$$

(i.e. the ionization energy) is very high at 1,310 kJ mol^{-1}. (This is higher than the first ionization energy for xenon at 1,170 kJ mol^{-1}!). The hydrogen ion can be formed only when it is **solvated**. Solvation occurs when the hydrogen ion attaches itself to a solvent molecule; the energy released when the hydrogen ion is solvated makes it 'worthwhile' ionizing hydrogen to H^+. The most common solvent with which to solvate H^+ is water itself and the hydrogen ion in water should be written as H_3O^+ or $H^+(aq)$ after the process

$$H^+(aq) + H_2O(l) \rightarrow H_3O^+(aq)$$

Solvation using water is given the special name of **hydration** and the ion H_3O^+ is commonly called the **hydronium ion** or oxonium ion. Remember that although you may see the symbol H^+, $H^+(aq)$ or a reference to the hydrogen ion, in water all of these should be taken to mean $H_3O^+(aq)$.

4.3 Covalent bonding

It may seem unlikely that simply sharing electrons would allow atoms to achieve stability, but experiments show that this is indeed the case!

What type of bonding occurs if it takes too much energy to remove electrons from one of the atoms taking part in bonding? This situation occurs when two or more non-metals combine – their atoms have to *share* electrons to achieve the stable electronic configuration of a noble gas. This type of bonding is called **covalent** bonding.

Consider chlorine gas; the Lewis symbol for chlorine is:

2.8.7

If two chlorine atoms share one electron, they achieve the stable electron arrangement of argon. We can represent this on a diagram called a **Lewis structure**:

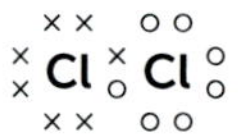

2.8.8 2.8.8

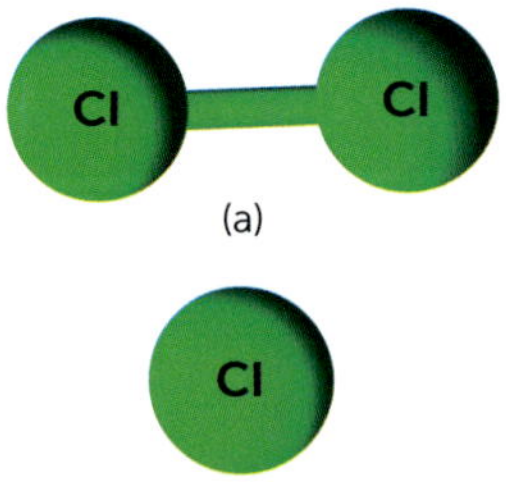

(a)

(b)

Chlorine molecule: (a) side on, (b) viewed end on (along the bond axis) with the second Cl atom obscured.

A **molecule** of chlorine is formed – there are no charged particles. The electrons of one chlorine atom are usually represented by dots and those of the other by crosses, to enable you to see where they come from. *Remember, though, that once the electrons are involved in the covalent bond there is no way to distinguish between them.*

Chlorine gas therefore exists as a **diatomic molecule**, chemical formula Cl_2, because this allows its atoms to achieve the stable electronic arrangement of a noble gas. The **structural formula** of the molecule is written as

Cl——Cl

where '——' represents a pair of shared electrons, i.e. a **covalent bond**.

EXAMPLE **4.2**

Ammonia (NH_3) is a covalent substance. Draw a Lewis structure to represent the bonding in ammonia.

Answer

Nitrogen, which has the electronic structure 2.5, shares its outer electrons with three hydrogens (electron arrangement 1). In this way nitrogen, and the three hydrogens, achieve the electronic structure of noble gases:

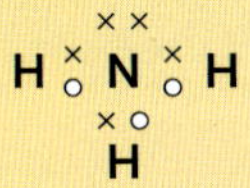

N electron arrangement 2.8
H electron arrangement 2

Comment

The structural formula of ammonia is written:

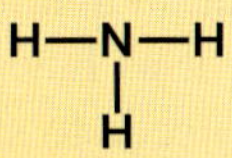

Remember, —— means 'a pair of shared electrons'. Also note that the three N–H bonds are identical, i.e. the bond lengths and bond strengths are the same.

'Ball-and-stick' model of the ammonia molecule. The nitrogen atom is coloured blue.

EXAMPLE **4.3**

Molecules of tetrachloromethane (often called carbon tetrachloride) have the formula CCl_4. What is the Lewis structure of the molecule?

Answer

Carbon has the electronic structure 2.4 and chlorine the electronic structure 2.8.7. One carbon atom shares electrons with four chlorine atoms so that all atoms achieve the electronic structure of a noble gas:

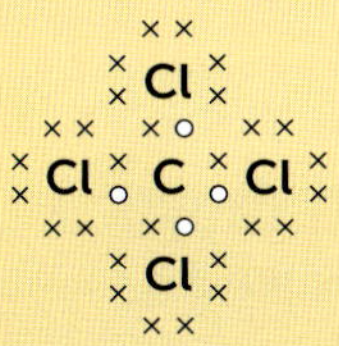

C electron arrangement 2.8
Cl electron arrangement 2.8.8

Comment

The four C–Cl bonds are identical and the structural formula of CCl_4 is written as

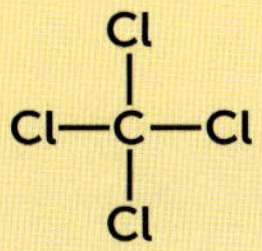

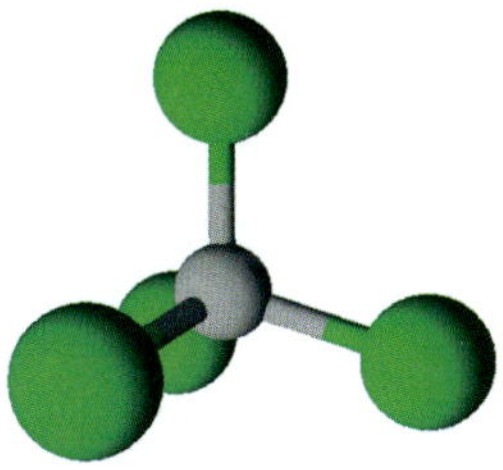

Model of the tetrachloromethane molecule. The chlorine atom are coloured green.

EXERCISE **4D**

Writing Lewis structures

Write Lewis structures for the following covalent molecules, then write their structural formulae using — to represent a shared pair of electrons:

(i) H_2
(ii) CH_4
(iii) $SiCl_4$
(iv) F_2
(v) H_2O
(vi) OCl_2
(vii) PCl_3
(viii) ICl (iodine has seven electrons in the outer shell of its atoms)
(ix) HBr
(x) $CHCl_3$ (C is the central atom).

Valency of an atom

Can you now see where we got the idea of 'arms' to describe combining power, or valency, in Unit 2? *The number of 'arms' that an atom or group possesses is actually a simple way of describing how many electrons that an atom or group needs to give away, accept (ionic bonding) or share (covalent bonding) in order to obtain the stable configuration of a noble gas.* Carbon, for example, has the electronic structure 2.4. We describe it as having a valency of four, or four 'arms'; this is just another way of saying that it needs to share four more electrons to achieve the stable electronic structure of neon, namely 2.8.

Multiple bonds

Sometimes, more than one pair of electrons is shared. Consider oxygen gas (O_2):

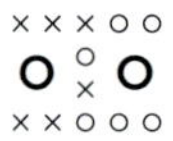

Here, each oxygen shares *two pairs* of electrons to achieve the electron configuration of neon. The structural formula of this oxygen molecule is written as

O═O

A **double bond** exists between the oxygen atoms. In a structure where three pairs of electrons are shared, the bond is called a **triple bond**.

EXERCISE **4E**

Covalent molecules with multiple bonds

Write Lewis structures for the following, then write their structural formulae:

(i) CO_2
(ii) N_2
(iii) $COCl_2$ (C is the central atom)
(iv) CS_2
(v) H_2CO (C is the central atom).

BOX **4.4**

Families of carbon compounds

Carbon forms many, many covalent compounds in which the carbon atoms are covalently bonded to each other and to other atoms, such as hydrogen. Carbon has four outer shell electrons to share with other atoms and, in these examples, achieves the stable electronic structure of neon by sharing electrons with hydrogen or another carbon atom. Carbon atoms can share one, two or even three electrons with each other. Some examples are:

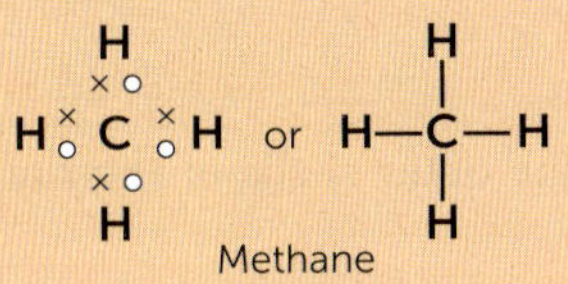

Methane
CH_4

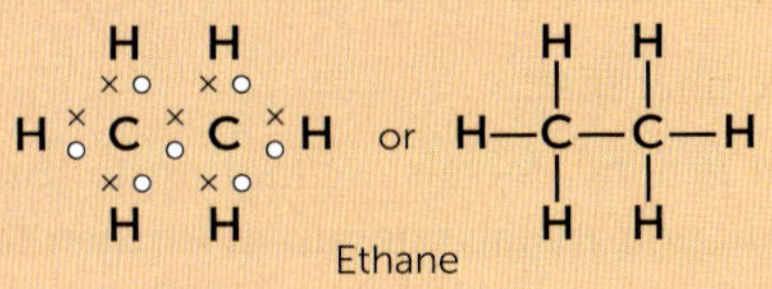

Ethane
C_2H_6

Multiple bonds may also be formed:

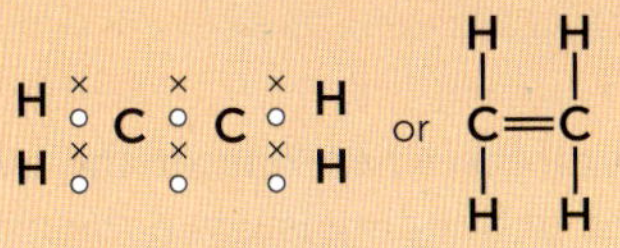

Ethene
C_2H_4

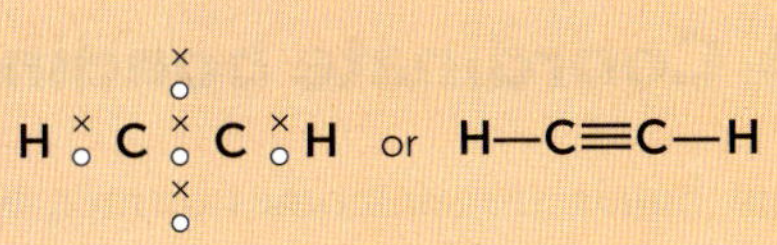

Ethyne (acetylene)
C_2H_2

BOX **4.5**

Electron density map for hydrogen (H_2)

The results of quantum mechanical calculations can be used to plot the electron density distribution in a hydrogen molecule. The contour lines join regions of the same electron density. Notice that there is a region of high electron density between the nuclei, which corresponds to the covalent bond in the Lewis structure. This electron density attracts the positive nuclei and prevents repulsion between the two positive nuclei from driving the atoms apart – **it 'glues' the atoms together**. A region of high electron density, holding atoms together in this way, is a characteristic feature of covalent molecules (Figure 4.5).

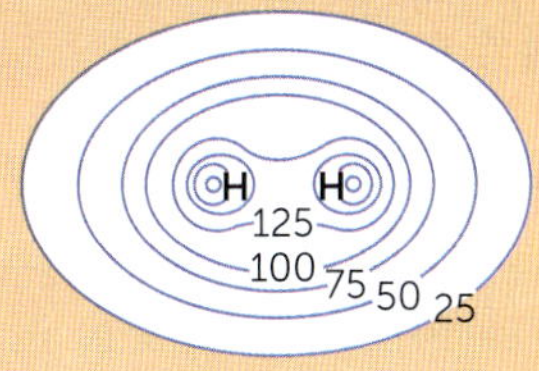

Figure 4.5 An electron density map for hydrogen. The numbers give an indication of the electron densities.

Properties of covalent compounds

Covalent compounds have the following characteristic properties:

1 Low melting points and boiling points (they are volatile)

They are often liquids or gases at room temperature (think of O_2, H_2O, N_2 and CO_2). The molecules are not attracted towards each other by strong electrostatic forces – so they have enough energy to move apart even at relatively low temperatures. The forces that attract molecules together (the **intermolecular forces**) are called **van der Waals forces** and are relatively weak.

2 Low solubility in water

Covalent compounds do not contain ions and therefore do not mix readily with water molecules. Covalent compounds tend to dissolve readily in **non-polar organic solvents**, such as benzene, which also contain covalent molecules.

3 Do not conduct electricity

Covalent compounds cannot conduct electricity because they do not contain ions.

How to predict whether compounds are ionic or covalent

Some general rules:

metal + non-metal → ionic bonding
non-metal + non-metal → covalent bonding

Remember that:

- In general, metals contain up to three electrons in the outer shells of their atoms and non-metals contain more than three electrons.
- Hydrogen should be classified as a non-metal.

EXERCISE **4F**

Formation of ionic or covalent compounds

What type of bonding do you think would be present in compounds formed by the combination of the following atoms?

(i) hydrogen and chlorine, which form hydrogen chloride

(ii) potassium and fluoride, which form potassium fluoride

(iii) silicon and hydrogen, which form silane, SiH_4.

4.4 Coordinate bonding

Covalent bonds are sometimes formed between two atoms when one of the atoms provides *both* of the shared electrons. This type of covalent bonding is given the names **coordinate** or **dative** bonding. The two electrons that one atom donates for sharing are called a **lone** (or non-bonding) **pair** of electrons.

Boron atoms have the electron arrangement 2.3 and can share only three electrons with other atoms. In the compound NH_3BF_3, the nitrogen atom provides both of the shared electrons to boron, so that the boron atom is surrounded by an octet of electrons:

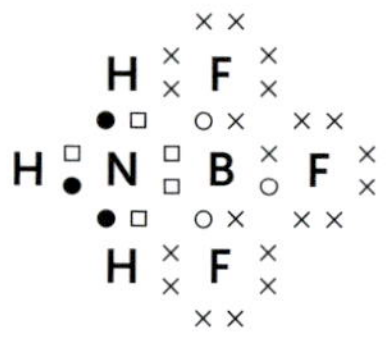

To help you distinguish where the electrons come from:

● = electrons from the H atom
□ = electrons from the N atom
× = electrons from the F atom
○ = electrons from the B atom.

The structural formula of the molecule may be written

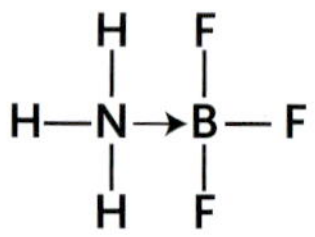

where ' — ' represents a covalent bond, where one electron has been donated from each atom, and '→' represents a coordinate covalent bond, where both electrons arise from one atom (in this case the nitrogen). In this way, all the atoms in the molecule end up with the electronic configuration of an inert gas. There is no difference between an 'ordinary covalent bond' and a coordinate bond once they are formed (*the electrons no longer 'belong' to a particular atom taking part in the covalent bond*) and the above structure is often represented as

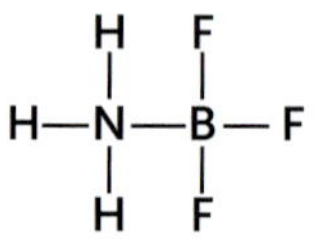

EXERCISE **4G**

Coordinate bonding

(i) Draw a Lewis structure for carbon monoxide (CO), bearing in mind it contains a multiple bond and a coordinate bond. Draw the structural formula of the molecule.

(ii) Draw a Lewis structure for the hydrated hydrogen ion, or hydronium ion H_3O^+ (see p. xx).

(iii) Draw a Lewis structure for the ammonium ion, NH_4^+.

4.5 Ionic and covalent compounds – two extremes

Although sodium chloride is regarded by chemists as an ionic compound and the hydrogen molecule is considered to be 'completely covalent', the vast majority of chemical compounds fall between these two extremes. Water, for example, is regarded as a covalent substance, and its molecules are neutral overall, but the molecules do have a slight positive charge at one end and an equal negative charge at the other end, i.e. there is a slight degree of ionic character. By the same token, some compounds that are generally regarded as ionic, such as calcium iodide (CaI_2), have some covalent character.

Polar covalent molecules

The electrons in a bond between two identical atoms, such as that present in H_2, can be considered to be equally shared between both of them. If the atoms are not identical, however, the electrons may be more attracted to one atom than the other. For example, in hydrogen fluoride, HF, the Lewis structure might be more accurately written as

in which the bonding pair of electrons is slightly closer to the F atom. Fluorine attracts electrons in a covalent bond much more than hydrogen – it has a high **electronegativity**. Because the shared electrons are more associated with the fluorine atom than with the hydrogen atom, the structural formula of the molecule is sometimes written as

$$H^{\delta+}—F^{\delta-}$$

Here, δ means 'slightly': the hydrogen end of the bond is slightly positively charged and the other end of the bond is slightly negatively charged, and this known as a **polar molecule**. The molecule is *neutral overall* because the charges at either end balance out. A covalent bond which has a positive end and a negative end is called a **polar covalent bond** and a molecule which has a positive end and a negative end is said to possess a **dipole** (two poles).

BOX **4.6**

Pauling's electronegativity values

The ability of an atom to attract a pair of shared electrons in a covalent bond is called the **electronegativity** of an atom.

Linus Pauling calculated relative electronegativity values for a number of elements. He assigned a value of 4 to fluorine, the most electronegative element. Electronegativity values are shown below:

BOX **4.6** *(continued)*

H 2.1						
Li 1.0	Be 1.5	B 2.0	C 2.5	N 3.0	O 3.5	F 4.0
Na 0.9	Mg 1.2	Al 1.5	Si 1.8	P 2.1	S 2.5	Cl 3.0
K 0.8	Ca 1.0				Se 2.5	Br 2.8

If two elements combine, the difference in their electronegativity values (always expressed as a positive number) can help decide whether the bonds holding them together are ionic or covalent:

1 The difference between the electronegativity values for sodium and chlorine is

$3.0 - 0.9 = 2.1.$

This is a large difference and, in general, differences greater than 1.7 indicate that compounds are ionic.

2 The difference in electronegativity values between the two atoms in a hydrogen molecule is

$2.1 - 2.1 = 0.0$

A value of 0 shows that the compound is completely covalent.

These examples are two extremes. By calculating these differences for covalent bonds between atoms, we can compare the **polarities** of covalent bonds. For example, the H–Cl bond (difference in electronegativities of H and Cl = 0.9) is much more polar than the H–Br bond (difference in electronegativities of the elements = 0.7).

Electronegativities can also show which ends of a polar covalent bond can be marked δ^+ and δ^-. The atom that has the highest electronegativity, or has the greatest attraction for the shared electrons, is marked δ^-. In the HBr molecule, Br has the largest electronegativity value (2.8) and so the polarity in the molecule is shown as

$H^{\delta+} - Br^{\delta-}$

EXERCISE **4H**

Comparing the polarity of bonds

Work out the differences in electronegativities between the elements in the following covalent bonds, rank them in order of increasing polarity, and show the direction of each dipole:

(i) Be – H

(ii) O – H

(iii) C – H

(iv) C – Cl

(v) C – O.

Polarization in ionic compounds

A small, positively charged ion (a **cation**) in an ionic compound can attract the electrons of a neighbouring negatively charged ion (an **anion**) towards it and distort the anion. When this happens, the anion is said to be **polarized**. This distortion can be better represented by using the electron density model of an ionic compound, rather than using Lewis symbols. The outer electron density contours of a purely ionic bond, and an ionic bond which is polarized, are shown in Figure 4.6.

Notice that, in the polarized case, more electron density is concentrated between the ions. A region of high electron density between atoms is characteristic of covalent compounds, so that the polarized compound in Figure 4.6(b), although still mainly ionic, has some covalent character. When is significant polarization likely to occur? We can answer this question by applying the following rules.

The polarization of an ionic bond (i.e. its covalent character) is high if:

1 The cation is small and carries a high positive charge.
2 The anion is large and carries a high negative charge.

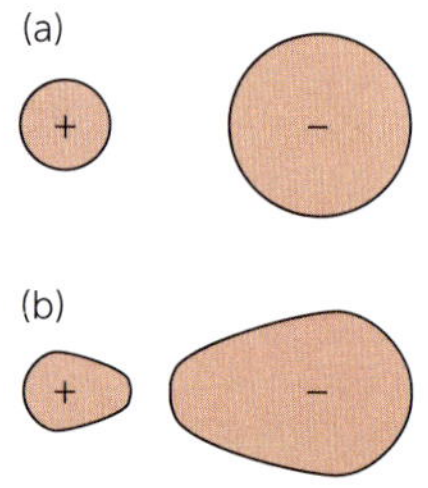

Figure 4.6 Polarization: (a) purely ionic; (b) polarized.

Compounds in which both conditions apply will have a great deal of covalent character. An example of a strongly polarizing cation is Al^{3+}, which is small and carries a large positive charge. The anion I^- is large; the degree of polarization in AlI_3 is so large that it is considered to be a covalent compound. However, the compound AlF_3 has far more ionic character, because F^- is a much smaller anion and is not polarized to the same degree as I^-.

Ionic and covalent bonding: a summary

The overall picture is summed up in Figure 4.7. Again, the outer contours of electron density maps have been used instead of Lewis symbols.

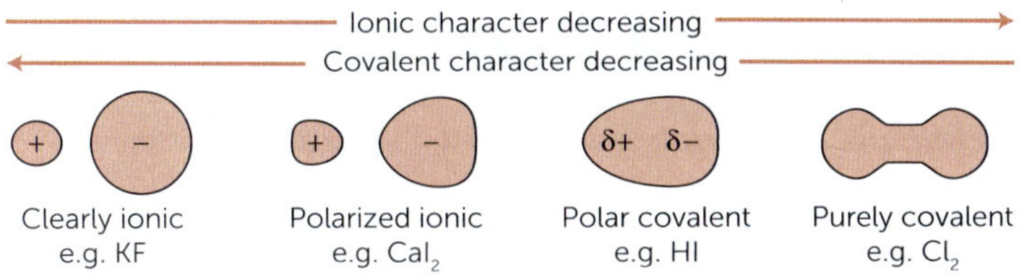

Figure 4.7 An overview of bonding types.

Ionic compounds with polyatomic ions

Both ionic and covalent bonding exist within some compounds. Consider the deadly poisonous compound potassium cyanide (KCN). This is an ionic compound; it contains the ions K^+ and CN^-. A Lewis structure for the cyanide anion can be written as

× C electrons
○ N electrons
● electron donated by potassium

The whole is written in square brackets with the negative charge outside to indicate that the negative charge is spread across the ion and does not belong to one particular atom. This 'spreading' of the negative charge or **delocalization** of the charge makes the ion more stable (see s.4.6). The atoms C and N in the cyanide ion achieve the stable electronic arrangement of inert gases. They do this by covalent bonding *within* the ion. The structural formula of the ion is written:

$[C{\equiv}N]^-$

Ions such as CN^- that contain more than one atom are called **polyatomic ions**. Other examples of polyatomic anions that form ionic compounds, but have covalent bonding *within* the anion, are hydroxide (OH^-), nitrate (NO_3^-), sulfate (SO_4^{2-}), carbonate (CO_3^{2-}), hydrogencarbonate (HCO_3^-) and phosphate (PO_4^{3-}).

EXERCISE 4I

Polarization

The sizes of some cations and anions are listed below:

Cation	Radius/pm	Anion	Radius/pm
Li^+	60	F^-	136
Na^+	95	Cl^-	181
K^+	133	S^{2-}	184
Mg^{2+}	65		
Al^{3+}	53		

(i) In the series of ionic chlorides: LiCl, NaCl and KCl, which compound would have the most covalent character?

(ii) From the selection of anions and cations above, which compound would you expect to be the most ionic?

(iii) Which compound would you expect to have the most covalent character?

EXERCISE **4J**

Bonding in polyatomic ions

Write Lewis structures for:

(i) the hydroxide ion, OH^-

(ii) the phosphonium ion, PH_4^+ (P is the central atom).

4.6 Resonance and delocalization

The structural formula for the carbonate ion, CO_3^{2-}, might be written as:

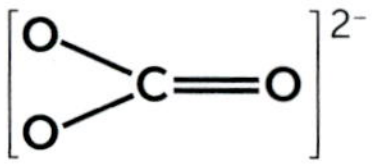

but there are two other alternative structures

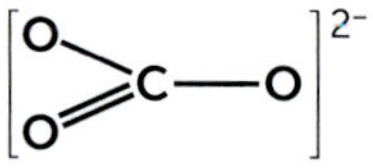

or

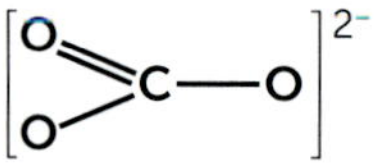

These are 'resonance structures'. Which one is correct? Actually, none of them is exactly correct. In cases like this the Lewis model of bonding is not completely adequate.

The lengths of the bonds in compounds can be measured. Double bonds between atoms are shorter than single bonds between the same atoms, so we might expect that measurement of the bond lengths in the carbonate ion would reveal two long bonds and one short one. In fact, all the C–O bonds are *of the same length*; this bond length is shorter than would be expected for a 'normal' C–O bond, but longer than that for C=O. The 'real' structure of the carbonate ion is a combination of the three structures that we have already written and can be represented by:

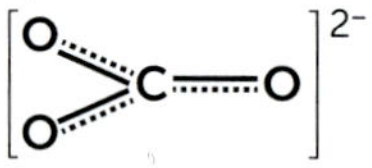

This 'mixing' of structures is called resonance and the resulting structure, such as the one written above, is called a **resonance hybrid**. Resonance does not mean that the molecule is 'flipping' between structures. This idea is similar to the relationship between a donkey, a mule and a horse. The mule is the 'resonance hybrid' – it is neither a donkey nor a horse but has features of both of them.

The structure of the carbonate ion can *also* be described by the idea of **delocalization**, whereby the electrical charge (here, of two electrons) is spread out among all four atoms. The ideas of resonance hybrids and delocalization are *equivalent* and both lead to a thermodynamically preferred ('more stable') molecule. In both ideas, electrons are not localized on any single atom. We may think of resonance hybrids as showing the delocalization through the contributing Lewis structures for the molecule in question. *In general, the more the electrons can be spread out, the more stable is the molecule.*

Box 4.7 gives some historical information on scientists who have contributed to bonding theory.

EXERCISE **4K**

Resonance structures

(i) There are two possible structural formulae that might represent NO_2^-. Draw both of these.

(ii) Draw the best single structural formula of the NO_2^- ion; it is a resonance hybrid.

BOX **4.7**

G. N. Lewis and Linus Pauling

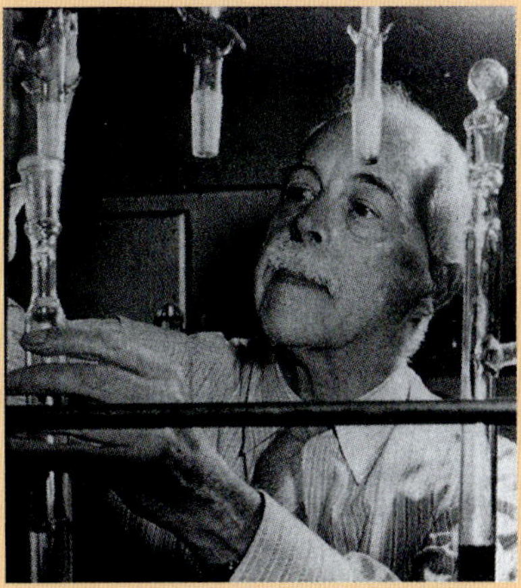

Gilbert Newton Lewis was an American chemist. His theories led to a clearer understanding of the concept of valency.

Courtesy of the College of Chemistry, University of California, Berkeley

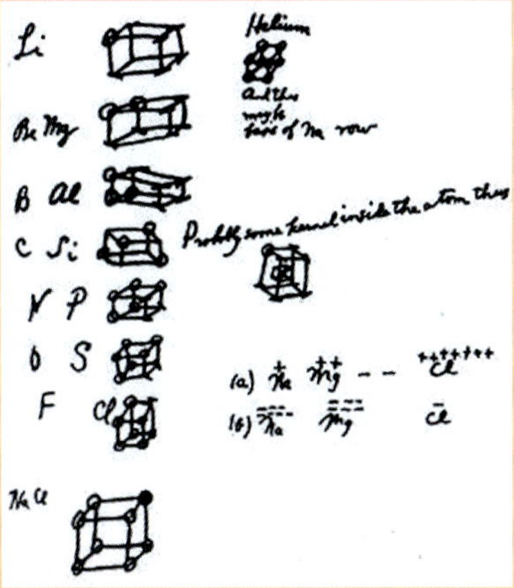

Early sketches of his ideas, by G. N. Lewis.

Courtesy of the College of Chemistry, University of California, Berkeley

Linus Pauling, an American chemist and twice Nobel Prize winner. His book *The Nature of the Chemical Bond* is regarded as a classic by chemists.

Gilbert Lewis (1875–1946) was the first American to win the Nobel Prize. This was awarded in 1914. Referring to covalent bonding, he wrote 'an electron may form part of the shell of two different atoms and cannot be said to belong to either one exclusively'. He was not without his critics, however. It was argued that this statement was the same as saying that if a couple have two dollars in a joint bank account and six dollars each besides (total fourteen dollars), they both have eight dollars apiece (total sixteen dollars). Lewis structures, however, still explain the properties of a great deal of covalent compounds today – the bank account analogy does not hold true in the world of chemistry!

Linus Pauling (1901–1994), who proposed electronegativity values, wrote *The Nature of the Chemical Bond* in 1939. It is still regarded as a classic chemistry textbook. He received not one, but **two** Nobel Prizes: one for chemistry in 1954 and one for peace in 1962.

REVISION QUESTIONS

4.1 Write Lewis symbols for the following atoms:

(i) Li
(ii) B
(iii) N
(iv) O
(v) Si

4.2 Write Lewis symbols for the following ions:

(i) S^{2-}
(ii) H^-
(iii) P^{3-}
(iv) Ca^{2+}
(v) Br^-

4.3 Use Lewis symbols to describe the reactions between the following elements:

(i) Mg and S
(ii) Ca and Br
(iii) K and S
(iv) Mg and N

4.4 Why does sodium not form the Na^{2+} ion in its compounds?

4.5 Draw Lewis structures for the following molecules:

(i) H_2S
(ii) CF_4
(iii) I_2
(iv) PH_3

4.6 Write a Lewis structure for F_2CO (C is the central atom).

4.7 The ion BF_4^- contains a coordinate bond. Write:

(i) the Lewis structure
(ii) the structural formula of the ion

4.8 Draw Lewis structures for the following ions:

(i) ClO^-
(ii) NO_2^+
(iii) NH_2^-

4.9 Use the electronegativity values in Box 4.6 to decide which of the following bonds is the most polar:

(i) N–O
(ii) P–Cl
(iii) Si–Br

4.10 What is the electronegativity difference between the atoms Mg and N? Is the bond between these atoms likely to be ionic or covalent?

4.11 Two Lewis structures exist for ozone (O_3), each of which contains one double and one single bond.

(i) Write the structural formulae of both of the Lewis structures.
(ii) The actual structure of ozone is a resonance hybrid of both structures. Draw a diagram to show the structure of the resonance hybrid.

4.12 So far we have not used the s,p,d,f .... notation to describe bonding. Use this notation to show the reaction between sodium and chlorine atoms to produce sodium chloride.

4.13 When demonstrating the reaction of potassium (in an iron spoon) with chlorine in a gas jar (Figure 4.1), a brown solid is produced as well as white potassium chloride. What might be causing this?

DID YOU KNOW?

Sir **Humphry Davy** (1778–1829) was an English scientist, inventor and poet, who discovered metals such as potassium, sodium, barium, strontium, calcium and magnesium. His most famous invention was a safety lamp for use in coal mines where the highly inflammable gas methane is often found. In 1831, Davy appointed a young Michael Faraday as his assistant. Faraday would become even more famous than Davy.

Davy first proposed the name for the element with the symbol Al and called it *alumium*, but he changed his mind a few years later, and called it *aluminum*. Many British users disliked this name, however, because it didn't conform with the 'ium' ending given in the names of metals, and adopted the name *aluminium*. Americans still prefer *aluminum*, which is why you will see this spelling in their textbooks.

More information on models of bonding, including resonance and delocalization, will be found in Extension 4 at https://www.bloomsburyonlineresources.com/chemistry.

More About Bonding

5

OBJECTIVES

This unit:

- Discusses exceptions to the octet rule
- Explains how we can predict the shapes of molecules
- Distinguishes between polar and non-polar molecules
- Describes the different types of intermolecular attractive forces

CONTENTS

After reading this unit you should realize how useful Lewis's bonding theory is! With small additions, the basic ideas can be used to explain a number of phenomena: how molecules get their shape; why metals conduct electricity; what makes diamond such a hard substance; and why water is a liquid under normal conditions. All these questions are answered in the following pages. First, however, we need to discuss how some exceptions to the ideas developed in Unit 4 can be explained.

5.1 Exceptions to the octet rule

Compounds in which the central atom is surrounded by fewer than eight electrons

The idea that there is a stability associated with the electron arrangements of the atoms of noble gases is sometimes called the **octet rule**. This rule applies to many compounds, but it is not always obeyed. (Unfortunately scientific theories are often not as 'neat' as we would like them to be!) When does the octet rule break down?

Two important exceptions to remember are beryllium (Be, electronic structure 2.2) and boron (B, electronic structure 2.3). According to the reasoning in Unit 4, they should react by losing two and three electrons, respectively. They do not react in this way, however, because their ionization energies (*the energies required to remove their outer electrons*) are very high. Because the atoms are small, the outer electrons in their atoms are close to the positive nucleus and are strongly attracted to it. Other atoms with two and three electrons in their outer shells, such as magnesium and aluminium, can lose these electrons when they form compounds because the electrons are farther away from the nucleus and more shielded from its attractive charge by full shells of electrons. Instead, beryllium and boron choose to share electrons and form covalent

> EXERCISE 5A
>
> **Lithium**
>
> Beryllium and boron do not react by losing electrons, whereas lithium (Li, 2.1) does. Can you suggest a reason for this?

compounds. Those compounds are stable enough to exist at room temperature, but the central atoms do not have a stable octet of electrons surrounding them.

For example, BF_3 is a covalent substance and its formula may be written:

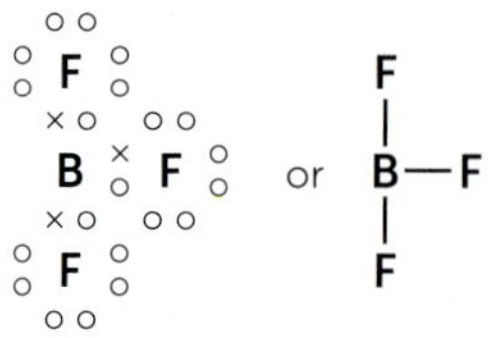

In this compound, boron is surrounded by only six electrons. The compound readily reacts with any substance that will give it the extra electrons it needs to obtain a stable octet. Two examples of such reactions are shown in the following equations:

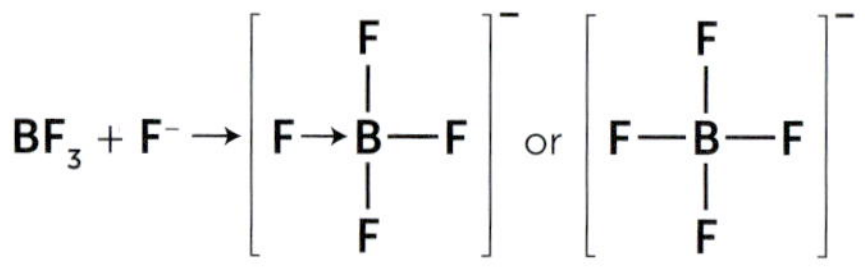

and

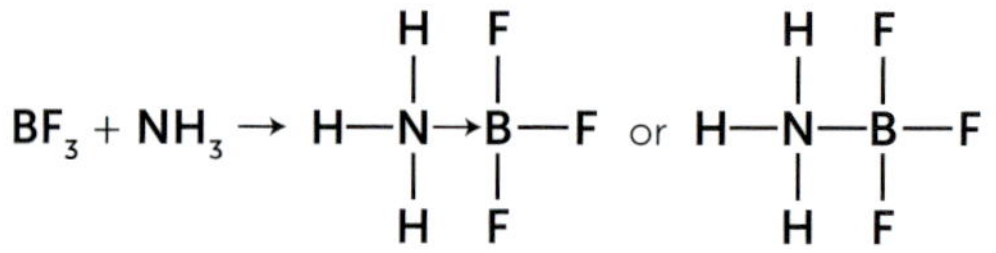

> EXERCISE **5B**
>
> **More about beryllium compounds**
>
> **(i)** Draw a Lewis structure and also a structural formula for covalent $BeCl_2$.
>
> **(ii)** In $BeCl_2$, how many electrons are in the outer shell of the Be atom?
>
> **(iii)** $BeCl_2$ reacts with Cl^- ions to form the **polyatomic** ion $[BeCl_4]^{2-}$. Draw the structure of this ion.
>
> **(iv)** How many electrons are in the outer shell of the Be atom in $[BeCl_4]^{2-}$?
>
> **(v)** Which do you think is the more stable compound, $BeCl_2$ or $[BeCl_4]^{2-}$? Why?

Compounds that contain an atom with an expanded octet

The other class of compounds in which the octet rule is not obeyed is covalent compounds in which the central atom can accommodate more than eight electrons in its outer shell.

Remember that the first electron shell of an atom can hold up to two electrons ($1s^2$), the second shell can hold eight electrons ($2s^2\ 2p^6$) and the third shell can hold eighteen electrons ($3s^2\ 3p^6\ 3d^{10}$). If the central atom in a covalent compound has electrons in the third (or a higher) shell, then it can use its empty d orbitals to hold extra electrons. In theory, such elements can share up to nine pairs of electrons (making a total of eighteen), but the number of atoms around the central atom is restricted by the size of the central atom; it gets too 'crowded'. It is rare to find the central atom sharing more than six pairs of electrons.

EXAMPLE **5.1**

Phosphorus forms two fluorides PF_3 and PF_5:

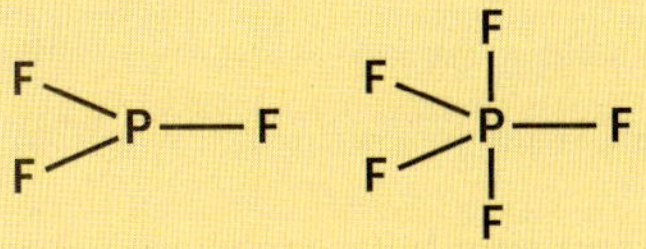

In PF_3, the octet rule is obeyed:

But in PF_5, the central phosphorus atom has 10 electrons surrounding it:

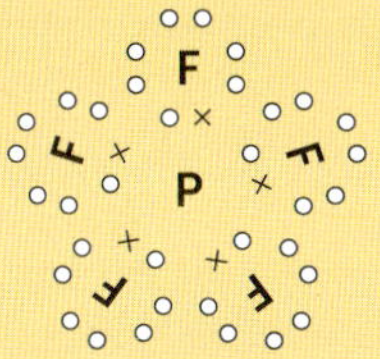

Here, phosphorus uses its empty 3d orbitals to accommodate the extra electrons.

EXERCISE **5C**

Covalently bonded atoms with an expanded octet

Draw Lewis structures for the following molecules or ions:

(i) **(a)** SF_6

(b) XeF_4

(c) BrF_3

(d) IF_5

(e) PCl_6^-

(f) ClF_4^+

(ii) NF_3 exists, but NF_5 does not – suggest why.

(iii) IF_7 exists, but ClF_7 does not – suggest a reason for this.

5.2 Shapes of molecules

If we can draw Lewis structures for covalent molecules, we can predict their shapes by applying a few simple rules. The theory that accounts for the shapes of molecules is called **valence shell electron pair repulsion theory** (VSEPR theory) and is based on the reasoning that electron pairs attempt to get as far away from other electron pairs as possible because their negative charges repel each other.

Molecules with bonding electron pairs only

1 Consider the molecule $BeCl_2$

This can be represented as

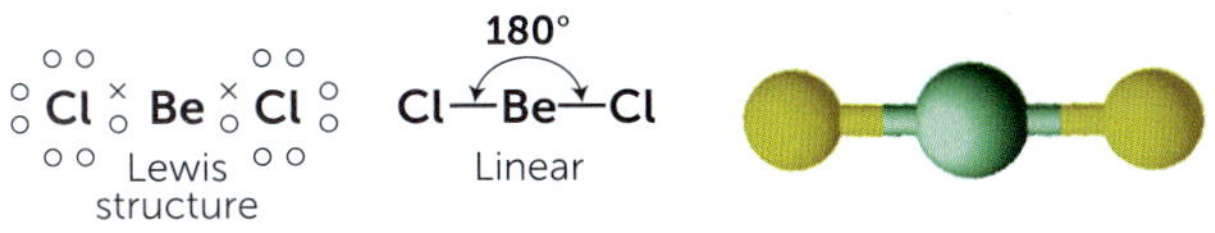

There are two shared electron pairs (**bonding pairs**) around the central beryllium atom. These two electron pairs arrange themselves as far apart from each other as they can, so that the angle between the two Be–Cl bonds is 180° and the molecule is **linear.**

2 The BF_3 molecule

Now let us apply this theory to a situation where there are three pairs of bonding electrons around the central atom, as in BF_3:

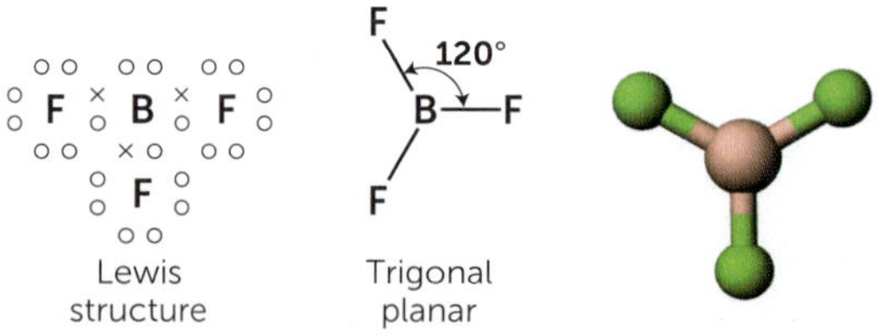

The three electron pairs around boron minimize the repulsion between them when the molecule adopts the **trigonal planar** shape. This molecule is flat. The fluorine atoms are in the same plane, at the three corners of a triangle drawn around the central atom. The three angles between the B–F bonds are all 120°.

3 Methane, CH_4

To minimize repulsion, the four electron pairs adopt a **tetrahedral** shape, with a bond angle of 109° 28′:

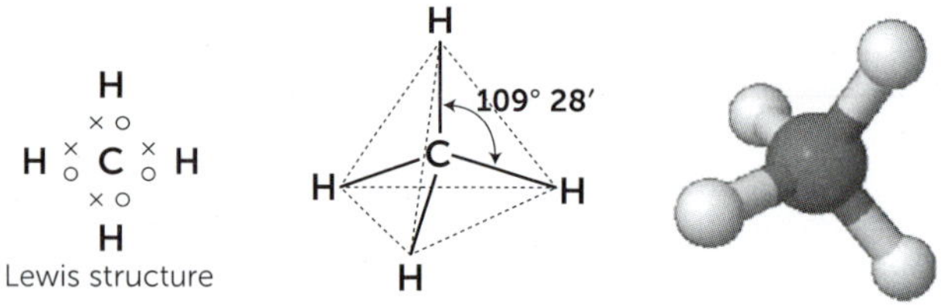

4 The PF_5 molecule

The five electron pairs adopt a **trigonal bipyramidal** shape, with bond angles of 120° and 90°:

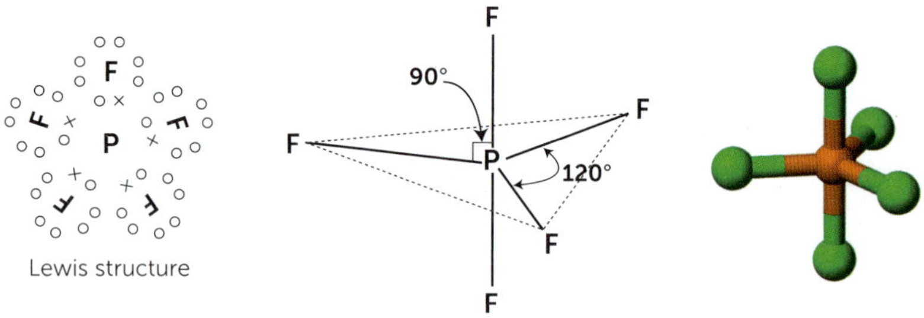

5 The SF_6 molecule

In this molecule the six electron pairs arrange themselves in an **octahedral** shape. All bond angles are 90°. Four of the F atoms are in the same plane (flatness).

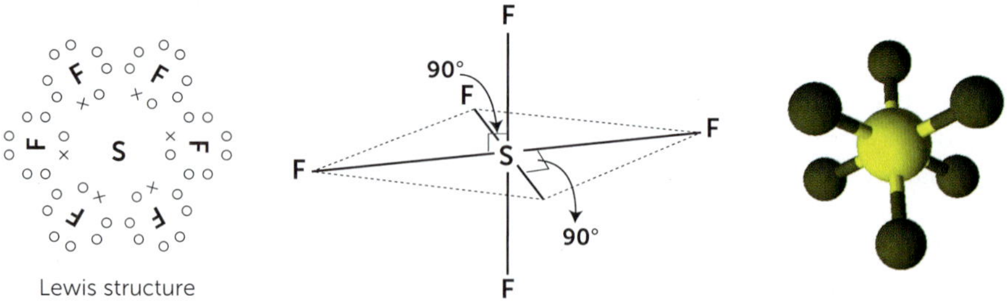

To summarize, Table 5.1 shows the shapes adopted by covalent compounds with different numbers of electron pairs around the central atom.

Table 5.1 Shapes of covalent molecules

Number of electron pairs in outer shell of central atom	Shape of molecule with respect to the electron pairs		Bond angles	Examples
2	Linear		180°	$BeCl_2$
3	Trigonal planar	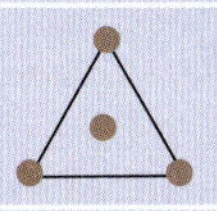	120°	BF_3
4	Tetrahedral	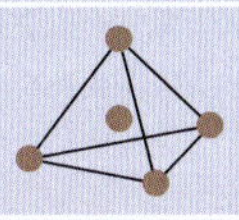	109° 28′	CH_4
5	Trigonal bipyramidal	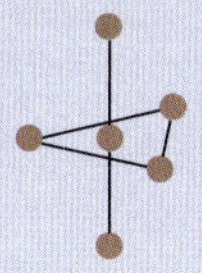	120° and 90°	PF_5
6	Octahedral	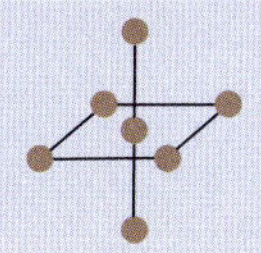	90°	SF_6

EXERCISE **5D**

Shapes of molecules

Using the Lewis structure for the following molecules or polyatomic ions, consider the number of bonding pairs around the central atom, then determine the shapes of the molecules:

(i) BCl_3

(ii) SiF_4

(iii) CCl_2F_2 (C is the central atom)

(iv) $[BF_4]^-$

(v) $[PF_6]^-$.

Remember that in all the shapes, the electron pairs are *as far apart from each other as possible.*

Molecules with lone pairs

In all the examples in Table 5.1, the electron pairs are shared between atoms – they are **bonding pairs**. Some molecules have pairs of electrons that are not shared by two atoms. These electrons are called **lone pairs**.

The order of repulsion between electron pairs is:

lone pairs repel lone pairs more than **lone pairs repel bonding pairs** more than **bonding pairs repel bonding pairs**

EXAMPLE **5.2**

What is the shape of the ammonia molecule?

Answer

The Lewis structure of the ammonia molecule is

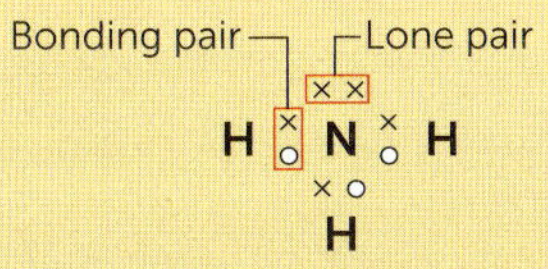

Because lone pairs are not shared between two positive nuclei, they are closer to the central nucleus than bonding pairs and repel nearby electron pairs more than bonding pairs. In the ammonia example, there are four electron pairs around the central nitrogen atom but the angle between the N–H bonds is not 109° 28′ but 107°. This lower angle between the N–H bonds is a result of the lone pair pushing

EXAMPLE **5.2** *(continued)*

the bonding pairs closer together; the lone pair repels the bonding pairs more than they repel each other. The shape of the ammonia molecule *with respect to the electron pairs* is a distorted tetrahedron. The structure is shown in Figure 5.1.

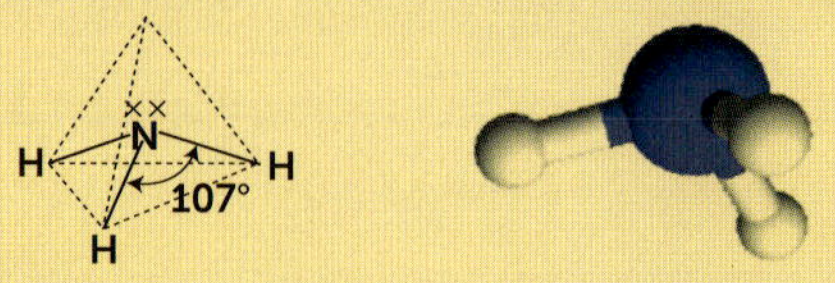

Figure 5.1 The structure of the ammonia molecule.

Comment

The ammonia molecule is also described as *trigonal pyramidal with respect to its N–H bonds* because it looks like a three-cornered pyramid.

EXAMPLE **5.3**

What is the shape of the water molecule?

Answer

The Lewis structure for water is

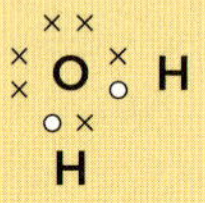

There are four electron pairs around the central oxygen atom: two bonding pairs and two lone pairs. Again, the shape of the water molecule *with respect to the electron pairs* is **distorted tetrahedral**, but the repulsive effect of two lone pairs is to push the O–H bonds together so that the bond angle between them is 105°. The structure is shown in Figure 5.2.

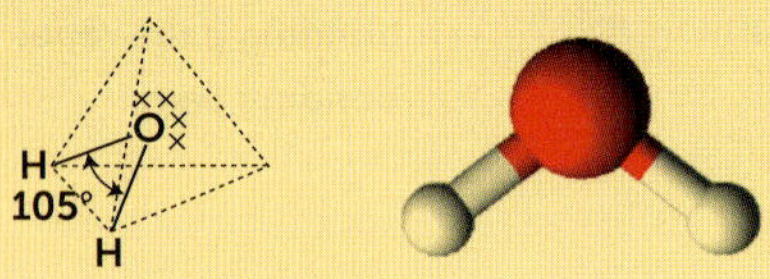

Figure 5.2 The structure of the water molecule.

Comment

The water molecule is described as '**bent**' or '**V-shaped**' *with respect to its O–H bonds.*

EXERCISE **5E**

Shapes of molecules with lone pairs

By first drawing Lewis structures of the following species, then considering the number and nature of the electron pairs around the central atom, predict the shapes of the following molecules:

(i) H_2S

(ii) PH_3

(iii) IF_5

(iv) OF_2

(v) SCl_2.

5.3 Shapes of molecules with multiple bonds

How do we work out the shape of a molecule when it contains double, or even triple bonds? We can still use VSEPR theory, but a double or triple bond is treated as though it were a single bond, that is, one bonding pair of electrons. Carbon dioxide has the structural formula:

O=C=O

Although each double bond contains two shared pairs of electrons, they are held together in the same place that a bonding pair of electrons would be held in a single bond. Therefore, if each double bond is treated as a single bond, the carbon atom has the equivalent of two bonding pairs of electrons around it and the shape of the molecule is **linear.**

> What about molecules consisting of only two atoms (diatomic molecules)? Because two atoms connected by a bond will always be seen as extending in a straight line, diatomic molecules like O_2, CO and HCl, are said to be *perforcely* (i.e. necessarily) linear.

EXAMPLE **5.4**

What is the shape of ethyne (common name acetylene)?

Answer

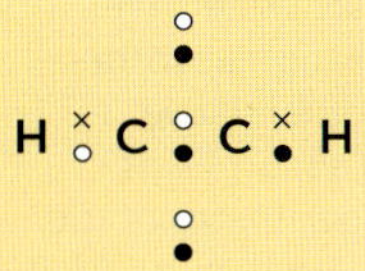

Lewis structure

H—C≡C—H

Structural formula

or

Here the triple carbon to carbon bond and single C–H bond on either side of the carbon atoms are regarded as single bonds for the purposes of working out the overall shape of the molecule. The molecule is **linear**.

EXAMPLE **5.5**

What is the shape of sulfur dioxide, SO_2?

Answer

For the purposes of working out the shape, the central sulfur atom acts as if it has three electron pairs around it, one of which is a lone pair. The molecule is therefore a distorted trigonal planar shape with respect to the electron pairs, or V-shaped with respect to the S=O bonds:

Comment

The angle between the S=O bond, at 119.5°, is less than the usual trigonal planar angle of 120°. The S=O bonds are 'pushed' together by the repulsion of the lone pair, but the double bonds have greater electron density than single bonds so the effect is not so pronounced.

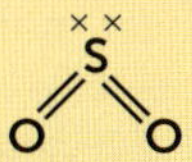

or

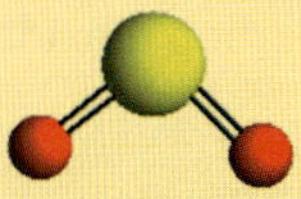

EXERCISE **5F**

Shapes of molecules with multiple bonds

What are the shapes of the following?

(i) the carbonate ion, CO_3^{2-}

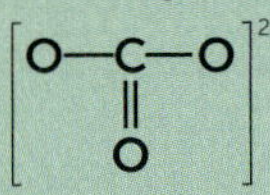

(ii) sulfur trioxide, SO_3

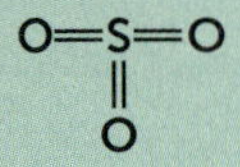

(iii) carbon disulfide, CS_2 (draw the Lewis structure, then the structural formula first!)

(iv) the sulfate ion, SO_4^{2-}

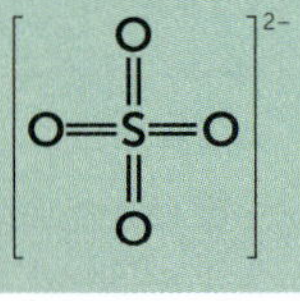

(v) the nitrate ion, NO_3^-

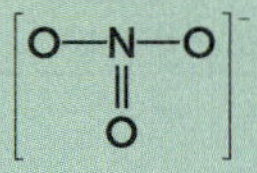

BOX **5.1**

Finding out molecular shapes

How do we know that VSEPR theory is good for predicting the shapes of molecules? The predictions have been shown to be true by a number of techniques that can be used to gain information about the shapes of molecules.

1 X-ray diffraction X-ray diffraction can be used to determine the arrangement of atoms in a crystal. Simply put, this involves using X-rays (which are electromagnetic radiation with wavelengths about the same as the distances between neighbouring atoms in a crystal) to take 'pictures' of the crystal structure.

2 Infrared spectroscopy A pair of covalently bonded atoms vibrate when they absorb electromagnetic radiation. The frequencies at which molecules absorb infrared (IR) radiation depend upon the type and number of atoms in the molecule, the types of bonds present (single, double etc.) and the shape of the molecule. For simple molecules we can work backwards and deduce molecular shape from the experimentally determined IR spectrum.

3 Dipole moment measurements We have already discussed polar covalent bonds. Molecules that contain these bonds *may* have a positive end and a negative end (a dipole), depending on their geometry. Measurement of the magnitude of dipoles (**the dipole moment**), therefore, could give valuable information about the shape of a molecule. Dipole moments are discussed in the next section.

4 Microwave spectroscopy Molecules rotate in the gaseous state. Upon absorbing microwave radiation, the molecules rotate faster. The frequencies of radiation absorbed by molecules and the relative intensities of such absorptions depend upon the geometry of the molecules. A mathematical analysis of the microwave spectrum of a molecule can provide chemists with bond angles and bond lengths. This information is much more accurate than that obtained by infrared spectroscopy. Studies of the effect of a magnetic field on a microwave spectrum allows the dipole moment of the molecule to be calculated.

5.4 Molecules with and without dipoles

In this section we consider **bond dipoles** *within* a molecule and whether the molecule *overall* has a **molecular dipole**. To work out whether a molecule has a dipole or not (i.e. whether or not the value of molecular dipole is non-zero or zero), we need to consider

1 The number and nature of polar covalent bonds (i.e. the magnitude of the bond dipoles);
2 The shape of the molecule.

Molecules without dipoles

Carbon dioxide is a linear molecule, containing carbon and oxygen atoms:

O=C=O

The electronegativity difference between carbon (electronegativity 2.5) and oxygen (electronegativity 3.5) is 1, so the C=O bond is **polar** and its bond dipole is > 0:

$C^{\delta+}=O^{\delta-}$

The carbon dioxide molecule, however, *does not have an overall molecular dipole* because it is linear:

$O^{\delta-} = C^{\delta+} = O^{\delta-}$

The dipoles of each polar C=O bond are equal in size but point in opposite directions and cancel each other out. So, *although it contains polar bonds, the molecule is non-polar overall.*

Carbon dioxide is an example of those molecules which, although they contain polar bonds, do not have an overall dipole because the geometry of the molecules allows the bond dipoles to cancel each other out. These geometries include trigonal planar, tetrahedral, trigonal bipyramidal and octahedral: these types of molecule are highly symmetrical. Note that in order for the bond dipoles to cancel out, *all the covalent bonds must be the same.* Some examples with highly polar bonds but with zero molecular dipoles overall are

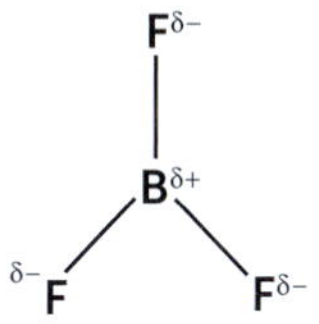

Trigonal planar

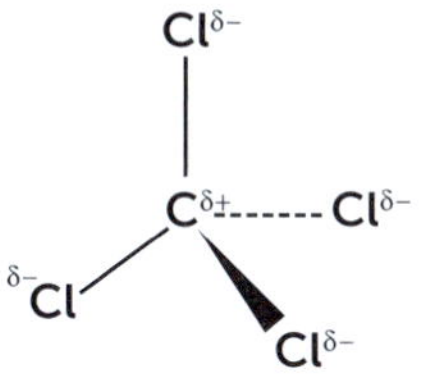

Tetrahedral

Molecules with dipoles

A water molecule is V-shaped and, because of the difference in electronegativities between oxygen and hydrogen, it contains polar covalent bonds:

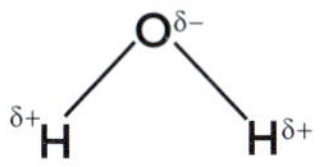

Because the molecule is not linear, it has a positive end and a negative end – it has a molecular dipole and *is described as polar overall*, as shown in Figure 5.3.

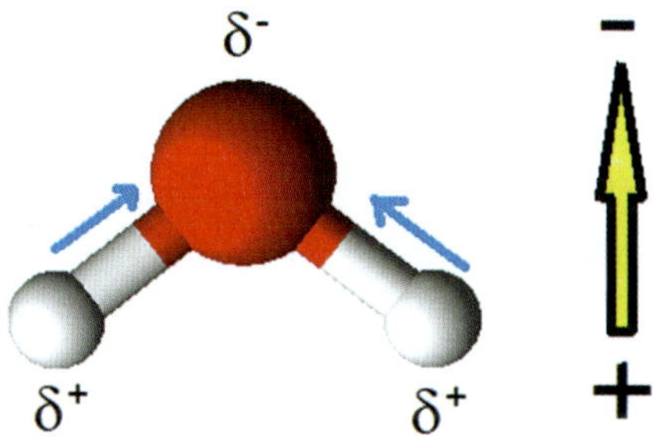

Figure 5.3 Water, H_2O, a polar molecule. The blue arrows show the directions of the two bond dipoles from the equally polar and equally angled H–O bonds. The overall *molecular* dipole is the 'vector sum' of the two bond dipoles. The *molecular* dipole of the water molecule is NOT zero and the yellow arrow shows its direction. By convention, arrows are always shown pointing to the more negative part of the molecule.

EXERCISE **5G**

Polar and non-polar molecules

1 Which of the following molecules have an overall dipole moment?

(i) CS_2 (iii) $SiCl_4$ (v) $CHCl_3$.
(ii) NH_3 (iv) H_2S

2 An electron density map for hydrogen fluoride, HF, is shown in Figure 5.4.

(i) Does the molecule have a dipole?
(ii) In which direction is the dipole?

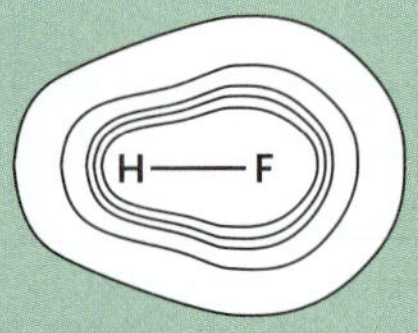

Figure 5.4 Electron density map for hydrogen fluoride. (Note that the electron density increases going from the outer to the inner contours.)

BOX **5.2**

Demonstrating the polarity of molecules

You can easily demonstrate that water molecules are polar by rubbing a plastic ruler with a dry duster, then bringing it close to a narrow stream of water flowing from a burette. The stream of water will bend because the polar water molecules are attracted to the charged rod. Do the same experiment with hexane, instead of water. Hexane molecules are non-polar and the stream of hexane is not deflected (Figure 5.5). (Liquids are coloured for clarity).

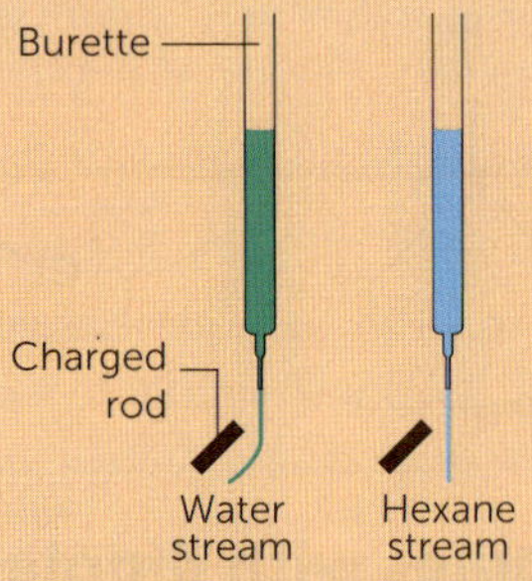

Figure 5.5 Demonstration of the polarity of water and hexane molecules.

5.5 Metallic bonding

As we have seen, metals have very distinctive properties. Any successful model of the bonding in metals must account for these properties. Solid metals have a **crystalline** structure – a crystal can be regarded as a rigid substance in which particles are arranged in a repeating pattern. We can regard a metallic crystal as consisting of a lattice of positive ions, surrounded by a cloud of electrons that are free to move. The electrostatic attraction between the positive ions and the negatively charged cloud of electron density 'glues' the whole structure together. Each atom of the metal has given up its outer (i.e. **valence**) electrons to become a positive ion. These valence electrons form part of the electron cloud and, because they are free to move, they no longer 'belong' to the particular atom from which they have been released. This is illustrated in Figure 5.6.

The model explains the properties of metals:

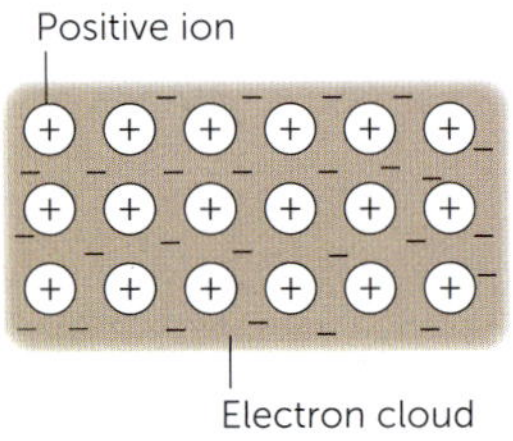

Figure 5.6 Metallic bonding.

1 Metals are shiny because the free electrons absorb light that falls on the metal and then re-emit the light back. Most of the light that falls on the metal is re-emitted, so the metal appears shiny. This is why metals are used in mirrors. Most metals are a silvery colour and this is because of the particular combination of visible light reflected (not absorbed) by the metal.
2 The free electrons in the structure allow metals to conduct electricity *in the solid state* (in contrast to ionic compounds). There is no chemical change to the metal when electricity is passed through it. If a potential difference is applied to the metal,

the negative cloud of electrons will be attracted to the positive potential. These same electrons can transfer kinetic energy through the solid quickly, explaining why metals are good conductors of heat.

3 The structure can be distorted without breaking into pieces, because it is held together as a whole by the electron cloud. This is why metals are malleable and ductile. The strength of the metallic bond depends upon the ratio of the number of free electrons in the electron cloud per ion. The more valence electrons an atom of a metal releases into the electron cloud, the stronger the metallic bond.

EXERCISE **5H**

Metallic bonding

(i) How many electrons are there in the valence shell of magnesium?

(ii) Draw a diagram, similar to Figure 5.6, to illustrate metallic bonding in magnesium.

(iii) How many electrons are there in the valence shell of aluminium?

(iv) Arrange the following metals in order of increasing strength of their metallic bonds: magnesium, aluminium and sodium.

(v) Which metal in (iv) would you expect to have the highest melting point and why?

5.6 Giant molecules

Allotropes of carbon

The element carbon exists in different forms, called **allotropes**, which differ in the way that the carbon atoms are bonded together. Allotropes are different forms of the same element; in each form the atoms are arranged in different ways. Three allotropes of carbon are: diamond, graphite and buckminsterfullerene.

BOX **5.3**

Allotropes

Some other elements which exist as different allotropes are the following:

1 Oxygen consists of diatomic molecules in *oxygen gas* (O_2) and triatomic molecules in *ozone* (O_3). These allotropes have very different chemical reactions.

2 There are two common allotropes of phosphorus: *white phosphorus* (P_4, its structure is shown on page 16), which spontaneously burns in air and is very poisonous; and *red phosphorus*, stable in air and non-poisonous, which consists of chains of P_4 units (Figure 5.7). Red P is deposited on the rough side of match boxes. When struck by a match, small amounts of Red P turn into white P, which immediately ignites, setting light to the matchhead.

3 Sulfur can exist in many forms including *rhombic sulfur* and *monoclinic sulfur*. They both contain S_8 molecules (page 16), but crystallize in different forms. *Plastic sulfur*, which consists of chains of sulfur atoms, is a form of sulfur that can be 'stretched'.

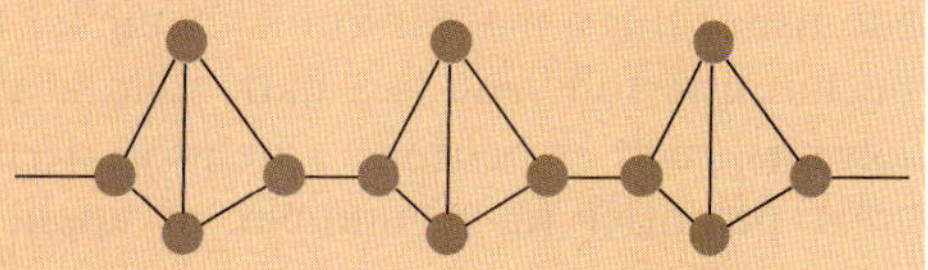

Figure 5.7 The structure of red phosphorus.

Diamond

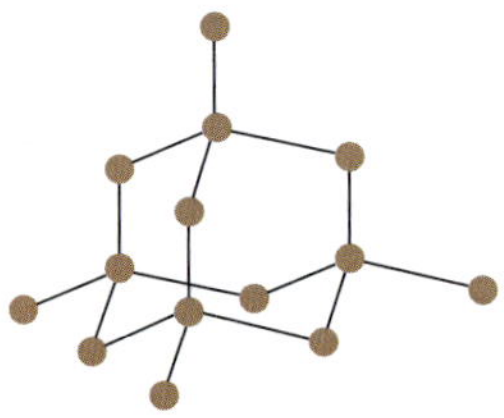

Figure 5.8 The arrangement of carbon atoms in diamond.

Diamond is a transparent solid that does not conduct electricity. In diamond, each carbon atom forms single covalent bonds to four other carbon atoms, which are at the corners of a tetrahedron. The four carbon atoms are, in turn, covalently bonded to four other carbon atoms and so on. This bonding exists throughout the entire crystal. A part of this arrangement is shown in Figure 5.8. When all the atoms in a crystal are covalently bonded to one another throughout the whole crystal, the solid is termed a **network solid**.

Diamond is a very hard substance because the carbon–carbon single bonds are very strong and extend throughout the whole structure. In order to melt diamond, many of these bonds need to be broken and this requires a high temperature. This explains why diamond has a very high melting point (3823 K).

Graphite

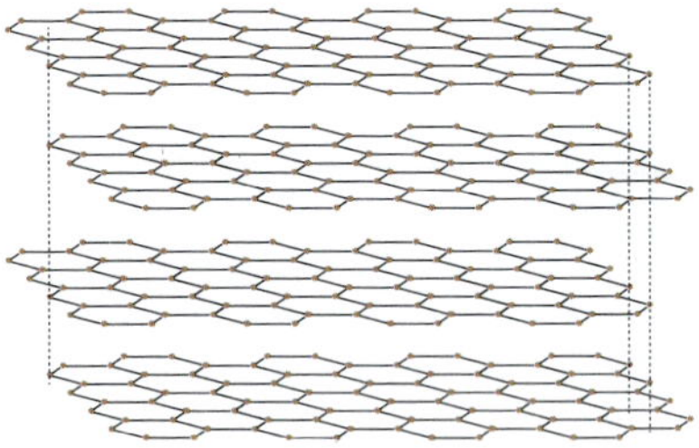

(a)

The arrangements of carbon atoms in graphite.

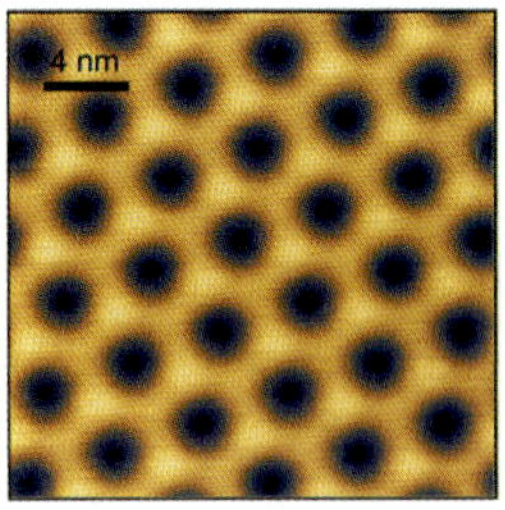

(b)

Atomic Force Microscope image of single layer of graphite ('graphene') from above the ring plane. This confirms the presence of the hexagonal carbon ring. (Courtesy of Dr J. A. Stroscio, https://doi.org/10.1038/s41467-021-22886-7).

Figure 5.9 Structure of graphite.

Graphite is another allotrope of carbon and another example of a network solid. Graphite is a soft, blackish solid that conducts electricity (unusual for a covalently bonded substance!). In graphite, there are layers of carbon atoms. Within each layer, each carbon atom is covalently bonded to three others in a trigonal planar arrangement (Figure 5.9).

Although the covalent bonds between the carbon atoms within the layers are strong, the layers are held together by weak forces (**London dispersion forces**) and can slide over one another easily. Graphite therefore has a 'slippery' feel, and is used as a lubricant and as the so-called 'lead' in pencils. The pencil marks paper because the layers of graphite are easily rubbed off.

Each carbon atom uses only three of its valence electrons in bonding to three other carbon atoms and therefore has one valence electron left over. These electrons form a cloud above and below the layers, rather like that in the metallic bond. The free electrons allow graphite to conduct electricity and account for its shiny appearance. Soot consists of small crystals of graphite.

Buckminsterfullerene

Unlike diamond and graphite, which have been known for many centuries, this third allotrope of carbon was discovered only in 1985. The allotrope has the formula C_{60} and, although it cannot be classified as a giant molecule in the same way as diamond and graphite are giant molecules, it is still much larger than other molecular solids such as S_8 or P_4. The shape of C_{60} resembles domes designed by the American architect Buckminster Fuller, hence its name. The arrangement of carbon atoms also resembles a football (Figure 5.10), with twenty hexagons and twelve pentagons; because of this a more informal name for the allotrope is sometimes used – **buckyball**. Buckminsterfullerene is a dark solid at room temperature and dissolves in covalent solvents, such as benzene. Chemical derivatives of the allotrope are believed to have great potential for a number of applications, including lubricants, batteries and semiconductors. Three scientists – Harry Kroto, Richard Smalley and Robert Curl – were jointly awarded the 1996 Nobel Prize for Chemistry for the discovery of C_{60}. For more about carbon allotropes, see Unit 29.

EXERCISE 5I

Another allotrope of carbon?

Chemists in the USA have reported the discovery of yet another allotrope of carbon, called LAC. LAC consists of long chains of carbon atoms, where the alternate carbon–carbon bonds are of different lengths. Remembering that carbon atoms have four valence electrons, can you suggest a structure for LAC? (**Hint:** LAC stands for linear acetylenic carbon.)

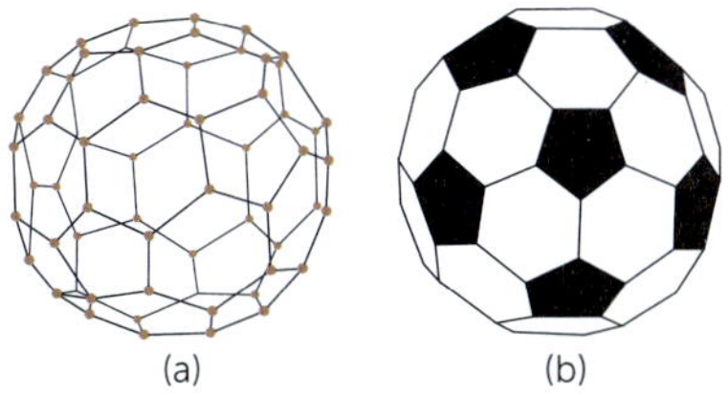

Figure 5.10 The arrangement of carbon atoms in buckminsterfullerene: (a) is the arrangement of the carbon atoms; (b) is a football for comparison.

5.7 Forces between covalent molecules

So far, we have discussed the bonding within covalent molecules (***intra*molecular bonding**). In this section, we will describe the forces that hold covalent molecules (and neutral atoms) together in the liquid and solid states (***inter*molecular bonding**). Intermolecular forces have the general name **van der Waals forces**, after the Dutch scientist Johannes van der Waals (1837–1923). The different types of intermolecular forces discussed are shown in Table 5.2.

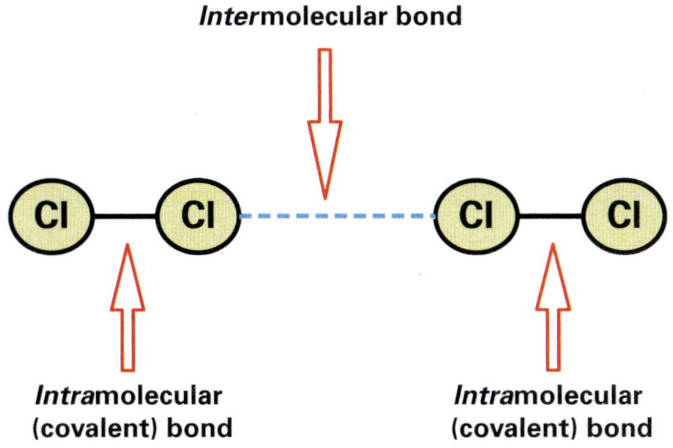

Figure 5.11 Bonds within and between chlorine molecules, Cl_2. *Intra*molecular Bonds are 'inside' molecules. The attractive forces between molecules, known as *Inter*molecular Bonds, are generally much weaker.

Table 5.2 Types of intermolecular forces

Types of molecules	Types of intermolecular forces present between molecules
Molecules with dipoles	(i) dipole–dipole attractive forces or hydrogen bonding attractive forces (ii) London dispersion attractive forces
Molecules (or atoms) without dipoles	London dispersion attractive forces only

Dipole–dipole interactions

Molecules which have dipoles, $X^{\delta+}$–$Y^{\delta-}$, tend to be attracted towards one another, as shown in Figure 5.12. The negative end of one molecule is attracted to the positive end of another and vice versa. Because these attractions are between *partial* charges (δ^+ and δ^-), they are much weaker electrostatic interactions than exist between ions in ionic substances. The melting and boiling points of substances that consist of molecules with dipoles, therefore, tend to be much lower than those of ionic solids, but higher than those substances that consist of non-polar molecules of similar size. Hydrogen bonding (p. 77) is a special case of dipole–dipole attraction.

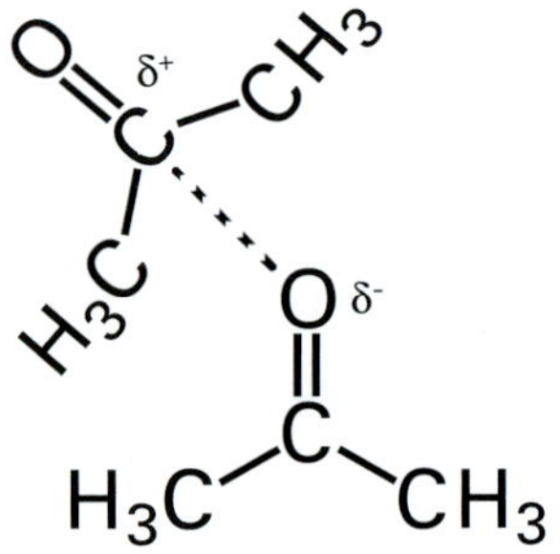

Fig 5.12 The intermolecular dipole-dipole attraction (- - - -) between propanone (acetone) molecules. The attraction is caused by the carbonyl group C=O, which is highly polar, with a partial negative charge on the oxygen atom and a partial positive charge on the carbon atom.

London dispersion forces

London dispersion forces are another type of van der Waals force. They are attractive forces between *all* atoms and therefore between all molecules even if additional attractive forces (e.g. dipole–dipole attractions) also exist. Without London dispersion forces argon atoms (for example) could not be solidified.

How do these attractive London forces arise? Although non-polar molecules and inert gas atoms have no dipole moment, the electron cloud in these particles, at one instant, may be denser on one side of the particle than the other. This causes a **temporary dipole**. The particle has, for a brief moment, one end with a small positive charge and the other end with an equally small negative charge. This temporary dipole causes (**induces**) another temporary dipole in a nearby particle – the positive end of the first particle attracts electron density from a neighbouring particle, making one end of the second particle slightly electron deficient and the other end electron rich. This process of particles inducing dipoles in nearby particles continues throughout. Weak attractive forces between the particles in the substance exist, at any particular instant. As the electron density in molecules or atoms shifts around, these dipoles change direction (fluctuate) but they are always present (Figure 5.13).

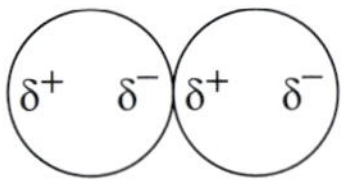

Figure 5.13 Temporary dipole-induced dipole attractions.

The attractions due to London dispersion forces increase as the molecules or atoms get larger. The number of electrons increases and the fluctuations in the electron density produce bigger temporary dipoles. Molecular shape also influences the strength of the London interactions.

London dispersion forces are responsible for attractions between **all types of molecules and between all atoms**, both polar and non-polar, and except for very large molecules are generally much weaker than other intermolecular forces.

BOX **5.4**

Arrangement of molecules in solid iodine

Covalent compounds can form crystals, just like ionic substances. In black, shiny crystals of iodine, the iodine molecules (I_2) are arranged in a lattice structure but they are held together by London dispersion forces only because iodine molecules (I–I) have no dipole. Because these forces are weak the iodine is a volatile substance; the crystal structure can be pulled apart very easily. The arrangement is shown in Figure 5.14

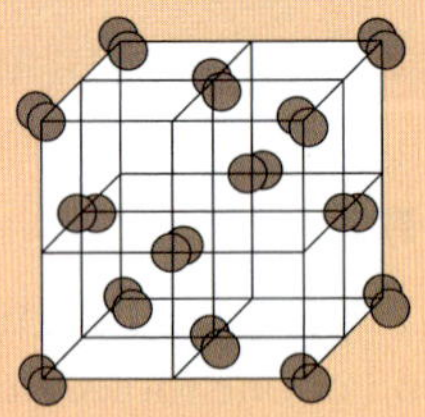

Figure 5.14 The unit cell in solid iodine. The pairs of overlapping circles indicate the orientations of the I_2 molecules.

Solid iodine

BOX **5.5**

London dispersion forces and the shapes of molecules

Molecules of pentane (C_5H_{12}) exist with the same number of electrons, but with very different shapes. One molecular form (**isomer**) of C_5H_{12} is cylindrical, whereas the another molecular form is spherical. The cylindrical form has a boiling point of 36°C, while the spherical form boils at 10°C, suggesting that weaker London interactions exist between the spherical molecules. The valence electrons in the molecules with a spherical shape are 'hidden' deeper within the molecule than those in a cylindrical molecule, and since attraction falls off with distance the molecule cannot interact as strongly with neighbouring molecules (Figure 5.15).

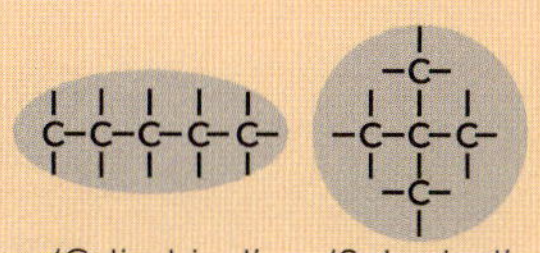

'Cylindrical' pentane 'Spherical' pentane

Figure 5.15 Two different shapes of pentane molecules.

EXERCISE **5J**

London dispersion forces

(i) Arrange the following in order of decreasing boiling points. Explain your answer.

Inert gas	Diameter of atom/pm
He	100
Ne	130
Ar	190
Kr	220

(ii) At room temperature, F_2 and Cl_2 are gases, whereas Br_2 is a liquid and I_2 a solid. Explain these observations in terms of the strength of intermolecular forces within these substances.

A drop of water falling through the air. The water molecules are held within the drop by hydrogen bonding, discussed in the next section

Hydrogen bonding

Hydrogen bonding is an extra strong type of dipole–dipole attraction *in which the tiny hydrogen atom acts as a bridge between two electronegative atoms A and B*. A **hydrogen bond** follows the general formula

$A^{\delta-}\text{–}H^{\delta+} \cdots\cdots B^{\delta-}$

Examples of hydrogen bonding are $F^{\delta-}\text{–}H^{\delta+} \cdots O^{\delta-}$, where A is F and B is O, and $F^{\delta-}\text{–}H^{\delta+} \cdots F^{\delta-}$, where A and B are F (see Example 5.6). A and B are the electronegative atoms, such as F, O or N. Such atoms possess one or more lone pairs of electrons.

EXAMPLE **5.6**

A molecule of hydrogen fluoride (HF) has a large dipole moment. The partial positive charge on the hydrogen is attracted to the lone pair of electrons on the partially negatively charged fluorine atom of a neighbouring molecule – it forms a **hydrogen bond** with the fluorine. Liquid hydrogen fluoride contains zigzag chains of molecules of HF joined together by hydrogen bonds (Figure 5.16).

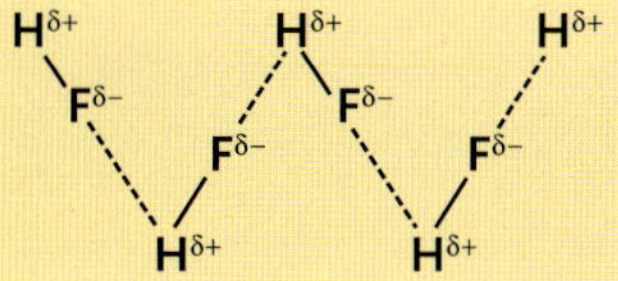

Figure 5.16 Hydrogen bonding between molecules of hydrogen fluoride (hydrogen bonds are usually represented by dotted lines).

EXAMPLE 5.7

Hydrogen bonds also exist between water molecules in liquid water and in ice. Because each oxygen atom in the water molecule has two lone pairs of electrons, it can attract two hydrogen atoms from other water molecules (Figure 5.17).

Icebergs float on water. The open arrangement of water molecules in ice partially collapses when the substance melts, so that the liquid has a higher density up to about 4°C.

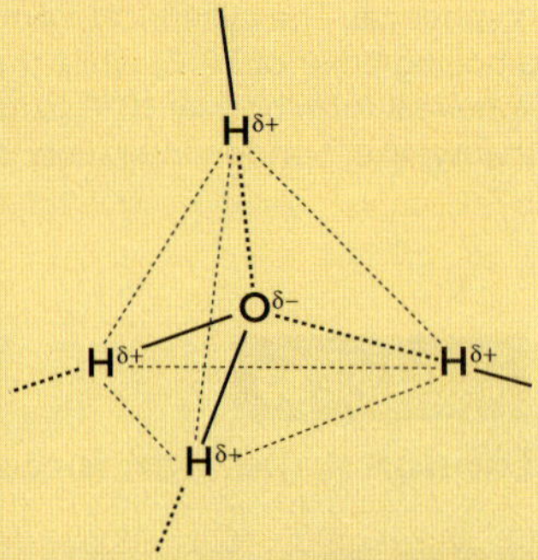

Figure 5.17 Hydrogen bonding in water. Note that the O···H bonds are longer and weaker than the O–H bonds.

Comment

Hydrogen bonding has a pronounced effect on the properties of water:

1. Water has a much higher melting or boiling point than compounds of similar formula containing hydrogen. For example, H_2S, H_2Se and H_2Te have the same general formulae but are all gases at room temperature. Water, of course, is a liquid at room temperature because more energy is needed to separate H_2O molecules. Water molecules have much stronger attractive forces between them, due to hydrogen bonding, than do H_2S, H_2Se and H_2Te molecules.

2. Water has a high surface tension. The surface of water forms a kind of 'skin' on which insects can walk, or a needle can float. The high surface tension of water is due to hydrogen bonding – the intermolecular forces within the water pull surface molecules inwards.

3. Liquid water is a much more polar solvent than is expected even after its highly polar O–H bonds are considered. This is because the extensive hydrogen bonding in liquid water amplifies the polarity of the whole network of molecules.

4 A remarkable consequence of hydrogen bonding is the low density of ice. A substance is generally more dense in the solid state than the liquid state, but ice is *less dense* than water at 0°C (ice floats on water). The tetrahedral arrangement of water molecules in ice creates a very open structure (Figure 5.18). When ice melts this structure breaks up and the water molecules approach each other more closely, so that the liquid has a higher density.

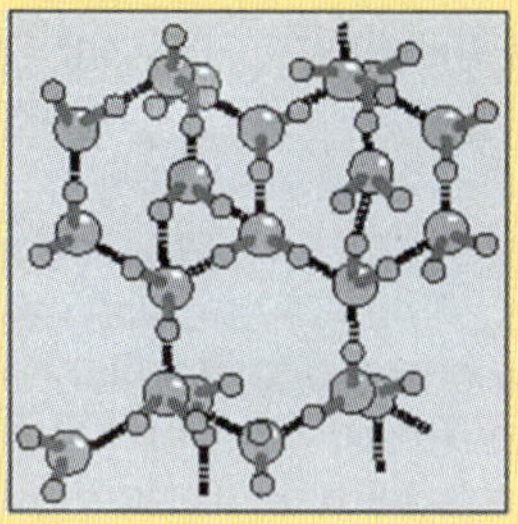

Figure 5.18 Arrangement of water molecules in ice (dotted lines show hydrogen bonds).

BOX **5.6**

The boiling points of hydrides

The graph in Figure 5.19 shows the boiling points of four sets of hydrides with similar formulae. Note that for the set CH_4, SiH_4, GeH_4 and SnH_4 the boiling points increase with increasing mass of the molecules. The hydrides NH_3, H_2O and HF, however, have boiling points much higher than would be expected from the trends shown by the other members of their set because the process of boiling requires the breaking of hydrogen bonds. These compounds also have abnormally high melting points, and latent heats of melting and vaporization because of hydrogen bonding between their molecules.

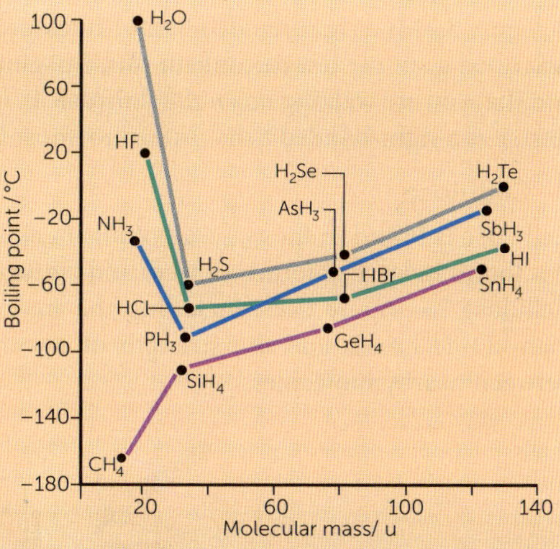

Figure 5.19 Plot of the boiling points of hydrides.

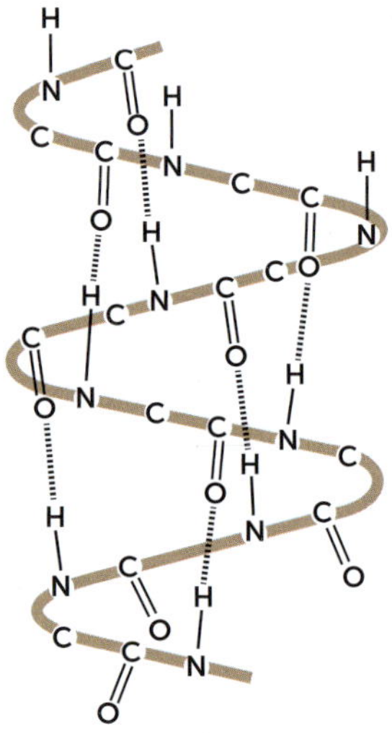

Figure 5.20 Structure of a protein.

Hydrogen bonding in biology

1 Proteins

Proteins are the basic 'building blocks' of living tissue. Protein molecules consist of long chains of atoms, containing polar >C=O and H–N< bonds. Hydrogen bonding can occur between these two groups:

$$>C^{+}=O^{-}\text{------}H^{+}—N^{-}<$$

Many protein molecules have the chains of atoms twisted or coiled, and these coils are held together by many hydrogen bonds. These hydrogen bonds help to give the protein its shape (Figure 5.20).

Because the hydrogen bonds holding the protein molecule together are weak, relative to the covalent bonds in the long chain of the molecule, they can be broken by heating. *Clear* egg white contains proteins suspended in water. When the egg white is heated, the increased vibrations of the protein molecules break the hydrogen bonds giving them their shape and they unravel. The protein turns the familiar colour of cooked egg white. When a protein loses its shape (and function) in this way it is said to be **denatured**.

2 DNA

Deoxyribonucleic acid (DNA) stores the genetic information present in living cells. It allows the cell to make proteins with a definite sequence of atoms. The DNA molecule consists of *two* long chains of molecules twisted around each other (i.e. a *double* helix Figure 5.21) with the two chains held together by hydrogen bonds. For this discovery, F. Crick, M. Wilkins and J. Watson gained the Nobel Prize for Medicine in 1962. The DNA in Figure 5.21 is a right-handed helix. Professor Larry Moran (University of Toronto) has described a simple test to confirm this. Imagine that the DNA helix is a spiral staircase and that you are walking *down* the staircase. If, in doing so, you are turning right at every turn, it is a right-handed helix. In nature, most DNA double helices are right-handed.

3 Cellulose

Cellulose is found in the cell wall of plant cells and helps to give plants their structure. It is found in large amounts in trees and cotton fibres (which are nearly pure cellulose). Cellulose has many uses, including the making of paper. Cellulose molecules have many O–H bonds and the strength of wood is due, in part, to hydrogen bonding between nearby molecules.

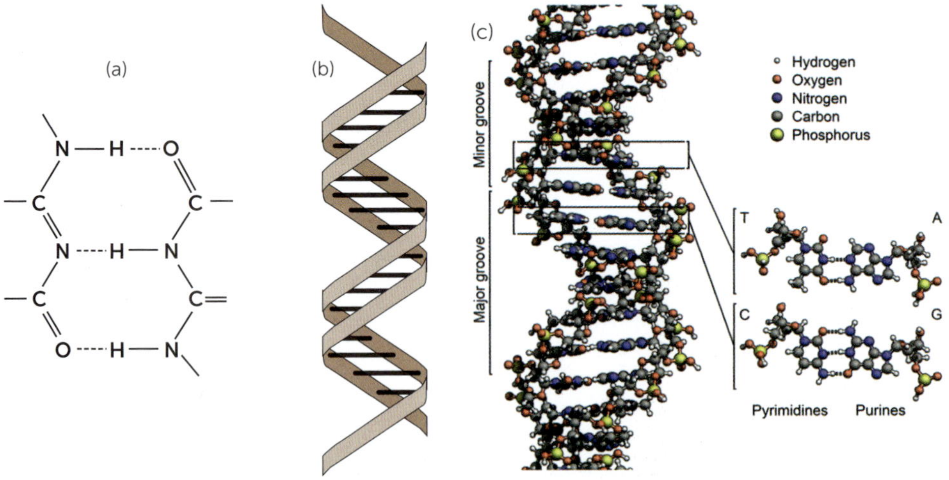

Figure 5.21 The double helix of DNA. (a) Typical hydrogen bonding interaction (– – –) between strands. (b) Two strands of DNA – the black horizontal bars represent hydrogen bonds. (c) Molecular picture of DNA, showing the complexity and size of the molecule. The symbols T, A, C and G represent different types of amino acids. (Image by Zephyris.)

Strength of bonds: an overview

Units 4 and 5 have discussed the different ways in which atoms bond together and the various attractions between molecules. Table 5.3 gives an idea of the relative strengths of some of these interactions.

Table 5.3 Examples of the approximate relative strengths of interparticle attractions

Interacting species	Description	Strength of attractive forces
Intramolecular bonding		
Ions e.g. Na^+ and Cl^-	Ionic bonding – attraction between oppositely charged ions	Strong
Atoms sharing electron pairs e.g. Cl–Cl	A covalent bond – two nuclei share an electron pair	Strong
Metal atoms e.g. Cu	Metallic bonding – metal cations surrounded by an electron cloud	Strong
Intermolecular bonding		
Polar molecules e.g. H–Br · · · H–Br	Dipole–dipole – attraction between dipoles	Medium
Non-polar and polar molecules and/or atoms e.g. I_2 · · · Ar	London forces between molecules and/or atoms caused by temporary dipoles in electron clouds	Weak (for small molecules)
Covalently bonded H and covalently bonded O, F or N e.g. O–H · · · F–H	A hydrogen bond – a 'special' dipole–dipole attraction	Medium (but sometimes stronger than dipole–dipole)

BOX **5.7**

Relative strengths of London forces and hydrogen bonds

Are hydrogen bonds stronger than the sum of the London dispersion forces? *Not necessarily.* The dispersion forces between many very *large* molecules, because they contain a large number of electrons, cause those large molecules to have a greater total *inter*molecular attraction than that between small molecules such as water, even though hydrogen bonds are present between water molecules.

Cooking oil contains molecules with high molecular masses and therefore many electrons. The London forces *between* the molecules are so large that the oil has a higher boiling point than water.

REVISION QUESTIONS

5.1 Draw Lewis structures for the following:

(i) ICl_3

(ii) BeF_2

(iii) $BFCl_2$

(iv) XeF_2

(v) CH_2Cl_2

5.2 Determine the shapes of the molecules in Question 5.1.

5.3 Draw a Lewis structure for covalent HCN. Determine the shape of the molecule.

5.4 Which of the following molecules have a dipole moment?

(i) SF_6

(ii) XeF_2

(iii) SO_3

(iv) SO_2

(v) CH_2Cl_2

5.5 What physical evidence is there for the existence of London dispersion forces?

5.6 Which metal would you expect to have the highest melting point K or Ca? Explain your reasoning.

5.7 Explain why diamond does not conduct electricity, but graphite does.

5.8 Explain the type of intermolecular bonding that exists in solid carbon dioxide (dry ice).

5.9 Water molecules have stronger hydrogen bonding between them than ammonia molecules. Why?

5.10 Consider the following boiling points: argon (−186°C); propanone (56°C) and water (100°C. Discuss the intermolecular forces present in each case and so justify the order of boiling points.

5.11 Since the discovery of buckminsterfullerene, a number of carbon molecules with a similar spherical structure (called fullerenes) have been made. It was once believed that any fullerene smaller than C_{60} would be too unstable to isolate in bulk, but in 1998 the Zettl Research Group announced the preparation of C_{36}. As with other fullerenes, the spherical molecule of C_{36} is composed of pentagons and hexagons of carbon atoms. Using a model kit (or modelling clay and straws) make a model of the molecule and decide how many hexagons and pentagons it contains.

5.12 Outline the role of intermolecular forces when ice is heated producing liquid water.

5.13 Dipoles (also called dipole moments) are measured in units called debyes (D). The molecular dipole (moment) of beryllium difluoride is zero. What does this tell us about the shape of the molecule BeF_2?

DID YOU KNOW?

So Near and Yet So Far ...

Scientists are confident that deep within the Earth, where the pressure and temperature are both extremely high, huge diamonds exist, some the size of footballs. Unfortunately, because of the lower temperature and pressure at the Earth's surface, by the time the diamond is expelled from volcanoes most of the diamond has been converted to graphite!

See the website at **https://www.bloomsburyonlineresources.com/chemistry** for extension material to support this unit, as well as for more advanced theories of bonding.

Reactions of Ions in Solution

6

OBJECTIVES

This unit:

- Explains how to construct ionic equations
- Discusses the reactions of HCl(g), H_2SO_4(l) and NH_3(g) with water
- Lists the reactions of acids and bases
- Outlines how simple tests may be used to identify common gases and ions

CONTENTS

In Unit 4 we saw that ions are atoms (or groups of atoms bonded together) which have lost or gained electrons. Loss of electrons produces positively charged ions (**cations**), while the gain of electrons produces negatively charged ions (**anions**). This unit includes some of the reactions of common ions. Such reactions are responsible for most of the coloured products and bubbling gases often associated with the chemistry laboratory.

6.1 Dissolution of salts in water

Ions produced when salts dissolve

Water is a covalently bonded substance, and if we want substantial numbers of ions in water we have to introduce them ourselves. The easiest way is to add a soluble salt such as sodium chloride.

Dissolving sodium chloride destroys its crystal structure and frees the Na^+ and Cl^- ions, which can then disperse through the body of the water. This process is called **ionic dissociation**. The ions are attracted to the surrounding water molecules and are said to be **hydrated**. We represent this by 'aq', an abbreviation for 'aqueous' which means watery. In our example we write $Na^+(aq)$ and $Cl^-(aq)$. The sum of the dissociation and hydration steps is called **dissolution**. The dissolution may be summarized by the equation:

$$Na^+,Cl^-(s) \xrightarrow{H_2O} Na^+(aq) + Cl^-(aq)$$

The table of valencies for common ions on page 539 lists many of the ions met in the college laboratory. We can use the table to work out the ions produced when soluble salts are dissolved in water.

EXAMPLE **6.1**

What ions are produced when magnesium nitrate dissolves in water?

Answer

The formula of magnesium nitrate is $Mg(NO_3)_2$. Its ionic formula is $Mg^{2+},2NO_3^-$. This generates magnesium ($Mg^{2+}(aq)$) and nitrate ($NO_3^-(aq)$) ions in water:

$$Mg^{2+},2NO_3^-(s) \xrightarrow{H_2O} Mg^{2+}(aq) + 2NO_3^-(aq)$$

BOX **6.1**

Composition of natural waters

It has been estimated that there is 1.4×10^{21} dm^3 of water on the Earth, with 97 per cent of the water being present in the oceans. About 8×10^{18} dm^3 is tied up as fresh water in rivers and 3×10^{19} dm^3 as ice in the ice-caps and glaciers. Table 6.1 shows the average concentrations of ions in rain, river water and ocean water although concentrations do vary with locality.

The main ions to be found in the oceans are sodium and chloride, i.e. sea water contains a relatively high concentration of common salt. Note that fluoride ions are not found in fresh water.

Table 6.1 Average composition ($mg\ dm^{-3}$) of rain, river water and sea water

Ion	Rain	River water	Sea water
Na^+	2.0	6.3	10,770
K^+	0.3	2.3	398
Mg^{2+}	0.3	4.1	1,290
Ca^{2+}	0.6	15.0	412
Cl^-	3.8	7.8	19,500
SO_4^{2-}	2.0	11.2	900
HCO_3^-	0.1	58.4	28
F^-	0.0	0.0	1.3

Source: *Encyclopaedia of Physical Science and Technology*, Academic Press (1987) Vol 5, p. 260.

EXERCISE **6A**

Dissolution of salts in water

Write equations to show the dissolution in water of:

(i) calcium nitrate

(ii) potassium sulfate

(iii) sodium carbonate decahydrate ($Na_2CO_3 \cdot 10H_2O$)

6.2 Ionic equations

Reaction of zinc metal with copper ions

The reaction between copper(II) sulfate solution and zinc metal produces zinc(II) sulfate solution and copper metal:

$$CuSO_4(aq) + Zn(s) \rightarrow ZnSO_4(aq) + Cu(s)$$

During the reaction, the blue copper(II) sulfate solution changes to colourless zinc(II) sulfate solution and the reddish coloured copper metal settles out.

The equation suggests that zinc atoms react directly with copper(II) sulfate. However, this is misleading – copper(II) sulfate is a soluble salt and is therefore fully dissociated into ions in solution. We can get a better picture of the reaction (more precisely, the reaction mechanism) if we write out the ions that are present as reactants and as products. In doing so, we will be writing out an **ionic equation**.

How to assemble the ionic equation for the reaction between copper(II) sulfate and zinc metal

1 Look at the table of valencies and symbols of ions (page 539). This shows that copper(II) ions are symbolized Cu^{2+} and that sulfate ions are symbolized SO_4^{2-}. In solution, copper(II) sulfate consists of separate copper(II) and sulfate ions. In all ionic equations, solids (whether metals or insoluble compounds) are represented by their chemical formulae. Here, we represent the zinc metal as Zn(s). The left-hand side of the chemical equation may now be written as

$$Cu^{2+}(aq) + SO_4^{2-}(aq) + Zn(s) \rightarrow$$

2 Zinc sulfate is water soluble and we write $ZnSO_4(aq)$ as $Zn^{2+}(aq)$ and $SO_4^{2-}(aq)$, respectively. The copper metal is symbolized as Cu(s). The right-hand side of the equation becomes:

$$\rightarrow Zn^{2+}(aq) + SO_4^{2-}(aq) + Cu(s)$$

3 The two halves are now written together:

$$Cu^{2+}(aq) + SO_4^{2-}(aq) + Zn(s) \rightarrow Zn^{2+}(aq) + SO_4^{2-}(aq) + Cu(s)$$

4 The sulfate ion appears unchanged on both sides of the equation. This shows that the sulfate ion is not involved in the chemical reaction. The sulfate ion is termed a **spectator ion** and may be eliminated in the same way as identical terms may be cancelled on either side of a mathematical equation:

$$Cu^{2+}(aq) + \cancel{SO_4^{2-}}(aq) + Zn(s) \rightarrow Zn^{2+}(aq) + \cancel{SO_4^{2-}}(aq) + Cu(s)$$

giving

$$Cu^{2+}(aq) + Zn(s) \rightarrow Zn^{2+}(aq) + Cu(s)$$

The last equation shows that the reaction does not involve sulfate ions, but that the products are made as a result of reaction between copper(II) ions and zinc atoms.

The reaction of $Cu^{2+}(aq)$ and Zn(s) is also an example of a **redox reaction** (see page 105).

(a) (b) (c)

Figure 6.1 The reaction of copper(II) sulfate solution with zinc metal.
(a) Copper sulfate solution and zinc
(b) After dipping the zinc into the copper sulfate, red/brown copper metal appears on the surface of the zinc.
(c) After 30 minutes, most of the copper ions have precipitated as copper. Nearly all the copper sulfate has been used up. Zinc sulfate solution is colourless.

EXERCISE **6B**

Ionic equations

Assemble the ionic equation for the reaction of copper(II) nitrate with zinc metal:

$$Cu(NO_3)_2(aq) + Zn(s) \rightarrow Zn(NO_3)_2(aq) + Cu(s)$$

Solubilities of ionic compounds

Table 6.2 shows the solubilities of common ionic compounds. If an ionic compound is insoluble, then that substance will fall out of (or **precipitate** from) solution when the appropriate ions are present. For example, if copper(II) sulfate solution is added to sodium carbonate solution, as Table 6.2 shows, the combination of copper(II) and carbonate ions forms a precipitate of copper(II) carbonate:

$$Cu^{2+}(aq) + CO_3^{2-}(aq) \rightarrow \underset{\text{copper carbonate}}{CuCO_3(s)}$$

Where the compound is listed in Table 6.2 as slightly soluble, the precipitate may appear only as a milkiness in the solution.

Table 6.2 The solubilities of common salts. With three exceptions, all the cations are colourless in solution. All the anions are colourless in solution

Cation	Cl^-, Br^-, I^-	SO_4^{2-}	CO_3^{2-}	S^{2-}	OH^-	NO_3^-	CH_3COO^-
Na^+, K^+, NH_4^+							
Mg^{2+}			●	d	●		
Ca^{2+}		●	●	d	○		
Ba^{2+}		●	●	d	○		
Al^{3+}			–	d	●		
Zn^{2+}			●	●	●		
Cu^{2+} (blue)			●	●	●		
Fe^{2+} (pale green)			●	○	●		
Fe^{3+} (yellow brown)			–	d	●		
Pb^{2+}	●	●	●	●	●		
Ag^+	●	○	●	●	–		

A blank indicates that the salt is soluble, ● = salt is insoluble, ○ = salt is slightly soluble, d = decomposes in water, – = does not exist

EXERCISE **6C**

Solubilities

(i) Which zinc salts are insoluble in water?

(ii) Name two compounds which when dissolved in water will produce a precipitate of barium sulfate.

(iii) Name two compounds which when dissolved in water will produce a precipitate of iron(II) hydroxide.

6.3 Producing ions in water by chemical reaction

The dissolution of an ionic salt is an example of physical change, since the ions are present both in solid salt and in solution. By definition, covalent substances contain no ions, but some covalent substances may *produce* ions as a result of their chemical reaction with water. The presence of such ions may be confirmed by testing the solution to see if it conducts electricity, but it is the new reactions of the solution – the reactions of the ions produced – that conclusively prove that chemical change has taken place. The following sections include three examples: the reactions with water of hydrogen chloride gas, pure liquid sulfuric acid and ammonia gas.

Reaction of hydrogen chloride with water

Hydrogen chloride (HCl) is a gas at room temperature. The HCl molecule is covalent but it reacts with water producing a mixture of **hydronium** and chloride ions that is commonly referred to as hydrochloric acid (Figure 6.2):

$$HCl(g) + H_2O(l) \rightarrow \underset{\text{hydronium ion}}{H_3O^+(aq)} + Cl^-(aq)$$

The hydronium ion is a **hydrated hydrogen ion**, and is often abbreviated as $H^+(aq)$. The reaction of HCl(g) with water may alternatively be written as

$$HCl(g) \xrightarrow{H_2O} H^+(aq) + Cl^-(aq)$$

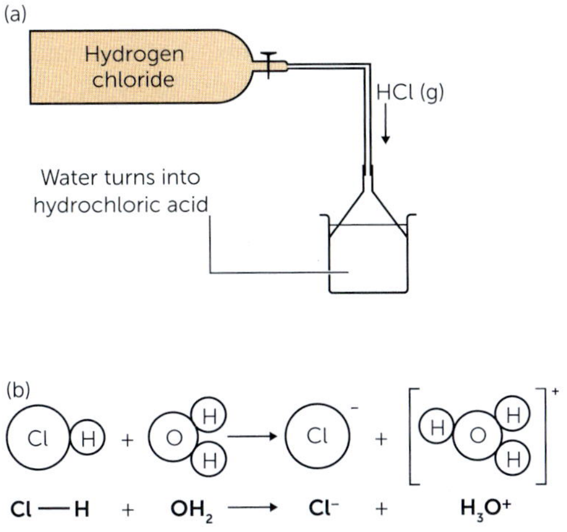

Figure 6.2 Making hydrochloric acid. (a) Hydrogen chloride gas from a cylinder is brought into contact with water. A funnel maximizes the contact between gas and liquid. At room temperature, 100g of water will hold up to 37g of HCl(g). (b) Shows the reaction of water and HCl(g) on a molecular scale.

The reaction between HCl gas and water makes hydrogen chloride extremely soluble in water (see the fountain experiment, page 188). A considerable amount of heat is also given out, although this is quickly absorbed by the water. There are two old names for hydrochloric acid: *spirits of salt* and *muriatic acid*, and the acid is still sold under these names.

Reactions of hydrochloric acid

The reactions of hydrochloric acid are simply the reactions of the hydronium and chloride ions. Concentrated and dilute hydrochloric acid display similar acidic behaviour, but the concentrated acid reacts at a faster rate.

Hydrochloric acid is highly corrosive and is used as a powerful cleaning agent.

EXERCISE **6D**

Hydronium ion

In the reaction between HCl(g) and H_2O(l), a proton (H^+) bonds to the water molecule by sharing a pair of electrons from the oxygen atom in the water molecule. Do you remember the name of this type of covalent bond?

EXERCISE **6E**

Reaction of hydrogen bromide and water

Hydrogen bromide, HBr(g), dissolves in water in a similar way to HCl(g) to make hydrobromic acid, HBr(aq). Represent this by an ionic equation.

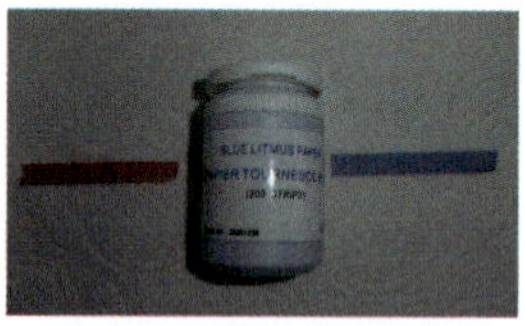

Litmus (from the Old Norse *litmose*) is a mixture of dyes extracted from lichens, first extracted in 1300. The image shows blue litmus paper. One strip has turned red after being dipped in acid. (Image by Flora Townsend.)

The H^+(aq) ion (or H_3O^+(aq) ion) is the ion that is responsible for the *reactions of all acids*. Such reactions are known as acidic properties (see page 92). One of the best known properties of H^+(aq) ions and, therefore, of acids in general, is their ability to cause **acid–base indicators** to change colour. The commonest indicator is litmus (litmus is usually absorbed in paper), which turns red in an acidic solution and blue in a basic (alkaline) solution. It follows that dry HCl gas does not affect dry blue litmus paper, but if the HCl gas or the litmus is damp, then the litmus will turn red.

The presence of the chloride ion in hydrochloric acid may be confirmed by adding silver ions, when a white precipitate of silver chloride (AgCl) is seen:

$$Ag^+(aq) + Cl^-(aq) \rightarrow AgCl(s)$$

Silver nitrate solution is usually used as the source of silver ions. A precipitate is obtained by mixing silver ions with *any* chloride and this is used as the basis for a test to detect the presence of a chloride. In testing unknown solutions using silver nitrate, it is usual to add dilute nitric acid to prevent the precipitation of other insoluble silver compounds, such as silver carbonate and silver sulfite.

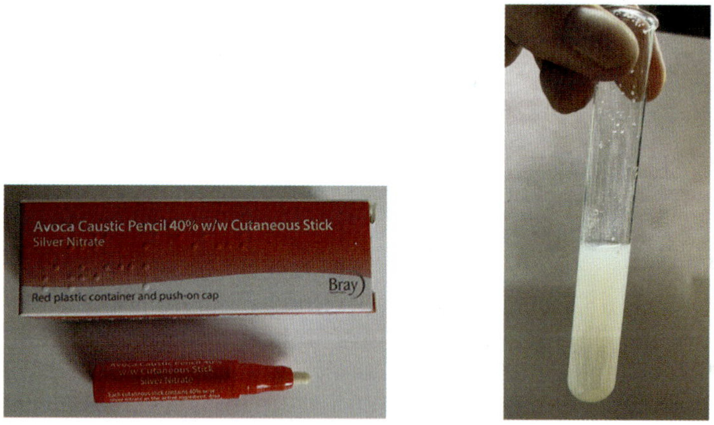

(a) (b)

Confirming the presence of Ag^+(aq).

(a) Although silver ions are usually used to detect chloride ions, we can also use chloride ions to detect the presence of silver ions. For example, silver nitrate is an ingredient in a solid used to treat warts (see photo (a)). How do we confirm the presence of Ag^+(aq)?

(b) This is how to test for silver ions. If the solid is added to warm water, the silver ions enter solution. The solution is then acidified with a few drops of nitric acid and a few crystals of sodium chloride added. The white precipitate of silver chloride (AgCl) confirms the presence of Ag^+(aq). (Photo by Ted Bradley.)

Before moving on, it is interesting to note that hydrogen chloride gas dissolves in benzene (a non-polar solvent) without producing any ions. Such a solution contains hydrogen chloride molecules that are weakly attracted (solvated) to the surrounding benzene molecules. As we might expect, a solution of HCl in benzene does not conduct electricity and it does not show any of the reactions of hydrochloric acid.

Bags of fertilizer. Fertilizers have greatly increased the productivity of crop farmers, and may have prevented large-scale starvation. Many fertilizers are made from ammonia (NH_3), sulfuric acid (H_2SO_4) and phosphoric acid (H_3PO_4).

EXERCISE **6F**

Reactions of silver nitrate

(i) Which of the following would give a white precipitate when added to a mixture of silver nitrate and nitric acid?

- **(a)** potassium sulfate
- **(b)** magnesium nitrate
- **(c)** calcium chloride.

(ii) If a solution containing bromide ions (Br^-) is added to silver nitrate solution, a cream-coloured precipitate is produced. If a solution containing iodide ions (I^-) is added to silver nitrate solution, a pale yellow precipitate is produced. What do you think these precipitates are? Write an ionic equation for each precipitation reaction.

Reaction of pure sulfuric acid with water

Sulfuric acid (H_2SO_4), also known as sul*ph*uric acid, is usually sold as concentrated sulfuric acid, which contains 98 per cent sulfuric acid and 2 per cent water by mass: this is sufficiently pure to justify the name pure sulfuric acid.

Pure sulfuric acid (like hydrogen chloride) is covalent. Unlike HCl(g), however, it is a liquid at room temperature. Pure sulfuric acid is a colourless oily liquid that looks and pours like glycerine, but there the similarity ends. If pure sulfuric acid is added to water, a violent reaction occurs. Much heat is produced. The resulting colourless solution is called dilute sulfuric acid. The overall equation for the reaction is:

$$H_2SO_4(l) \xrightarrow{H_2O} 2H^+(aq) + SO_4^{2-}(aq)$$

The reaction of pure sulfuric acid with water is so violent that dilute sulfuric acid should always be prepared by slowly adding the pure acid to water. Under no circumstances should water be added to the pure acid. So much heat would be generated where the two liquids mix that some of the mixture might spit out into your face (Figure 6.3).

Pure sulfuric acid is so greedy for water that it is commonly used as a drying agent. As most gases (apart from ammonia) are only slightly soluble in the pure acid, they may be dried by bubbling the gas through the acid contained in a glass bottle. However, the greed of the acid for water also means that it is extremely corrosive to the skin and eyes.

Did you know?

The acid murderer

Between 1944–1949, sulfuric acid was headline news for the most gruesome of reasons. John Haigh murdered people for money, trying to get rid of their bodies by dissolving them in pure sulfuric acid. Investigators found fragments of undissolved bones and dentures. He was convicted of murdering six people but he claimed he had killed nine. He was executed on the 10 August 1949.

Figure 6.3 Always add acid to water, never water to acid.

Reactions of concentrated sulfuric acid

Dilute sulfuric acid does not react with copper metal even when hot. Concentrated sulfuric acid does react with copper metal but *only* when *hot*, producing blue copper sulfate and pungent smelling sulfur dioxide gas:

$$Cu(s) + 2H_2SO_4(l) \rightarrow \underbrace{Cu^{2+}(aq) + SO_4^{2+}(aq)}_{\text{blue copper sulfate solution}} + SO_2(g) + 2H_2O(l)$$

This is *not* a reaction that is typical of acids because it does not involve $H^+(aq)$. (Remember, pure sulfuric acid is a covalent liquid – it does not contain ions.) In this reaction sulfuric acid is acting as an **oxidizing agent** (see Unit 7), bringing about the oxidation of copper atoms to copper(II) ions:

$$Cu \rightarrow Cu^{2+}$$

while the sulfuric acid is reduced to SO_2.

Reactions of dilute sulfuric acid

The reactions of dilute sulfuric acid are really the reactions of the $H^+(aq)$ and $SO_4^{2-}(aq)$ ions. The $H^+(aq)$ ions provide the acidic properties. The presence of a sulfate ion is confirmed by adding barium ($Ba^{2+}(aq)$) ions, when a dense white precipitate of barium sulfate is observed:

$$Ba^{2+}(aq) + SO_4^{2-}(aq) \rightarrow BaSO_4(s)$$

The usual source of barium ions is barium chloride solution. In order to make this test specific for $SO_4^{2-}(aq)$ ions, hydrochloric acid (but not sulfuric acid!) is added to prevent barium sulfite or barium carbonate from precipitating. (See Exercise 6G.)

EXERCISE **6G**

Reactions of barium chloride

Which of the following would give a white precipitate when added to a mixture of barium chloride and hydrochloric acid?

(i) potassium sulfate

(ii) magnesium carbonate

(iii) sodium chloride

Reactions of nitric acid

Hydrochloric and sulfuric acids are two of the so-called 'mineral acids'. The third is nitric acid, HNO_3 (old name: *aqua fortis*, meaning strong water), which is used to make fertilizers and explosives. Concentrated nitric acid is a highly corrosive, colourless liquid (sometimes it appears pale yellow due to impurities) usually sold with a concentration of about 70 per cent of acid by mass; the rest is water. 'Fuming nitric acid', used in nitration (see Unit 20), contains 86 per cent HNO_3. When *very dilute*, nitric acid behaves as an acid (like hydrochloric acid) due to the presence of H^+(aq) ions. However, at moderate to high concentrations, nitric acid acts as an oxidizing agent even when cold. For example, nitric acid converts copper to green copper(II) nitrate, the acid being reduced to nitrogen dioxide (NO_2) and nitric oxide (NO) gases.

Nitric acid slowly decomposes in sunlight and is sometimes stored in dark bottles. (Image by Rufus Townsend.)

Concentrated sulfuric acid is used to identify the nitrate ion (NO_3^-), in a test called the **brown ring test**. This involves the careful addition of concentrated sulfuric acid to a solution of the suspected nitrate. A solution of iron(II) sulfate is then slowly added to the mixture. If a nitrate is present, the mixture produces nitric oxide gas (NO), which reacts with Fe^{2+}(aq) to produce a brown ring of $Fe(NO)^{2+}$ at the junction of the acid mixture and the $FeSO_4$ solution:

$$Fe^{2+}(aq) + NO(aq) \rightarrow Fe(NO)^{2+}(aq)$$

Reaction of ammonia gas with water

Ammonia gas (NH_3(g)) is extremely soluble in water. The reaction is

$$NH_3(g) + H_2O(l) \rightleftharpoons \underset{\text{ammonium ion}}{NH_4^+(aq)} + OH^-(aq)$$

Ammonia solution is sometimes (incorrectly) called ammonium hydroxide. A saturated solution at room temperature contains about 28 per cent of ammonia by mass. The density of such a solution is $0.880\,g\,cm^{-3}$. For this reason a saturated ammonia solution is referred to as '880 ammonia'.

The OH^-(aq) ion is the **hydroxide** ion. It is this ion that gives solutions of bases their characteristic reactions; for example, it causes red litmus to turn blue.

The $\rightleftharpoons$ symbol reflects the fact that the reaction is a **reversible** or **equilibrium** reaction, and the reaction can proceed in either direction depending on the temperature.

Heating a mixture of ammonium ions with hydroxide ions causes them to recombine, making ammonia gas and water. For example, heating ammonium chloride with sodium hydroxide solution produces ammonia. This is used as a laboratory test to confirm the presence of the ammonium ion. The ammonia gas is detected by its very characteristic smell (usually described as pungent) and by the fact that it turns damp red litmus blue (Figure 6.4 and Exercise 6H).

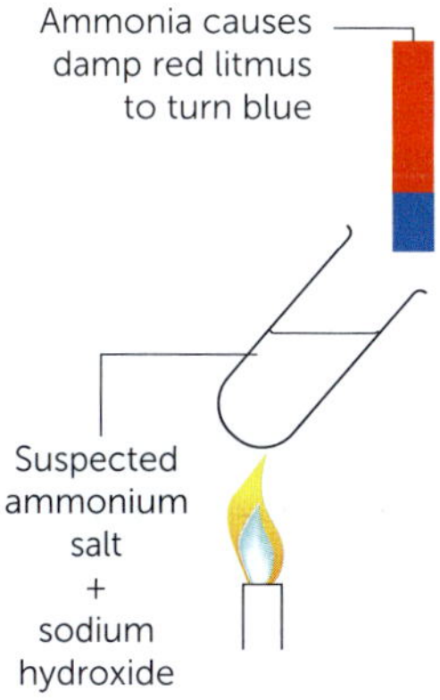

Figure 6.4 Testing a solution to see if it contains ammonium (NH_4^+) ions. If ammonia is detected, the presence of the ammonium ions is confirmed.

EXERCISE **6H**

Reactions that produce ammonia gas

Which of the following would be expected to produce ammonia gas when heated with sodium hydroxide solution?

(i) lead(II) nitrate **(ii)** ammonium carbonate **(iii)** calcium carbonate.

Write an ionic equation for the reaction(s) that are involved.

6.4 Acids and bases

Acids

An ***acid*** *is a substance that produces $H^+(aq)$ ions when dissolved in water.*

The commonest acids in the laboratory are the 'mineral acids': sulfuric acid (H_2SO_4), hydrochloric acid (HCl) and nitric acid (HNO_3). Another common acid is ethanoic acid (acetic acid, CH_3COOH, also written as CH_3CO_2H), a smelly liquid that boils at 118°C. Ethanoic acid is the chemical that gives vinegar its sour taste. Its full structure is as follows:

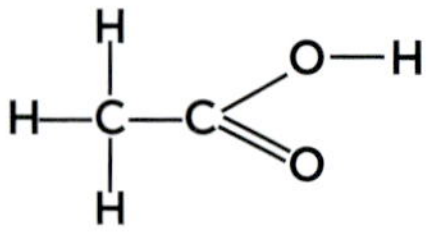

Vinegar contains about 6 per cent v/v ethanoic acid.

Only one of the hydrogen atoms in this molecule makes a hydrogen ion in solution, and for this reason this hydrogen atom is called an **acidic hydrogen**. In ethanoic acid, the acidic hydrogen is the one bonded to the oxygen atom. The ionization of ethanoic acid may then be represented by the equation

$$CH_3COOH(l) \xrightleftharpoons{H_2O} CH_3COO^-(aq) + H^+(aq)$$

ethanoate (or acetate) ion

or

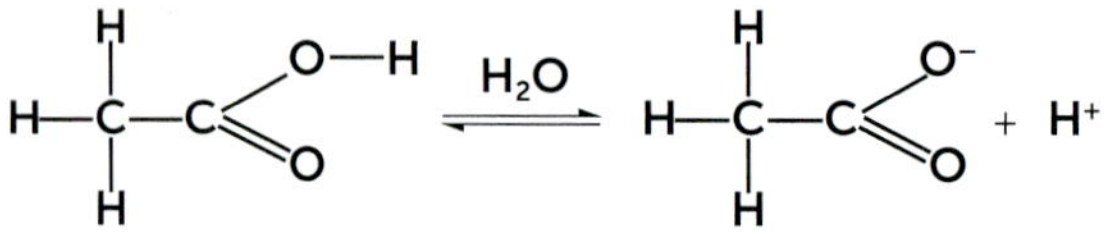

Lemon juices contain about 6 per cent v/v citric acid.

The $\rightleftharpoons$ sign shows that only some of the ethanoic acid molecules are ionized in solution.

EXERCISE **6I**

Number of acidic hydrogen atoms

(i) How many acidic hydrogen atoms do molecules of (a) sulfuric acid and (b) hydrogen chloride possess?

(ii) Hydrochloric acid forms salts known as chlorides. What are the names of the types of salts produced from (a) ethanoic acid and (b) nitric acid?

Bases and alkalis

*A **base** is a substance that reacts with an acid in solution producing a **salt** and water only.*

Generalizing:

acid + base → salt + water

This reaction is known as **neutralization**.

*An **alkali** is a base that dissolves in water.*

See Figure 6.5. A solution of an alkali contains the hydroxide ion, $OH^-(aq)$.

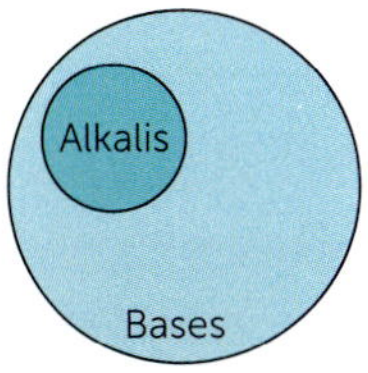

Figure 6.5 Bases and alkalis – although all alkalis are bases, not all bases are alkalis.

Alkalis are usually hydroxides of metals. The common alkalis are the hydroxides of calcium, potassium and sodium. They are all ionic solids which completely dissociate into ions in water, for instance:

$$\underset{\text{sodium hydroxide}}{Na^+,OH^-(s)} \xrightarrow{H_2O} Na^+(aq) + \underset{\text{hydroxide ion}}{OH^-(aq)}$$

Ammonia solution is also regarded as an alkali because it contains the hydroxide ion.

Bases which are insoluble in water include the oxides of metals such as magnesium oxide (Mg^{2+},O^{2-}) and copper(II) oxide (Cu^{2+},O^{2-}) and organic compounds (compounds based on carbon) which contain nitrogen atoms such as propylamine ($C_3H_7NH_2$).

Ammonia solution is used as a household cleaner.

EXERCISE **6J**

Dissolution of potassium and calcium hydroxides

Write an equation showing the dissolution of potassium and calcium hydroxides in water.

6.5 Reactions of acids

The reactions of acids are the reactions of the $H^+(aq)$ ion.

1 Effect on acid–base indicators

Taking litmus as our example:

$$\text{red coloured litmus} \underset{H^+(aq)}{\overset{OH^-(aq)}{\rightleftharpoons}} \text{blue coloured litmus}$$

2 Reaction with bases (neutralization)

Acids neutralize bases, producing a salt and water only. Examples of reactions of alkalis with acids include

$$NaOH(aq) + HCl(aq) \rightarrow NaCl(aq) + H_2O(l)$$

$$2KOH(aq) + H_2SO_4(aq) \rightarrow K_2SO_4(aq) + 2H_2O(l)$$

Both neutralization reactions reduce to the ionic equation

$$H^+(aq) + OH^-(aq) \rightarrow H_2O(l)$$

or, in terms of the hydronium ion:

$$H_3O^+(aq) + OH^-(aq) \rightarrow 2H_2O(l)$$

Two examples of the reactions of **water** insoluble bases with acids are

$$CuO(s) + H_2SO_4(aq) \rightarrow CuSO_4(aq) + H_2O(l)$$

$$MgO(s) + 2HNO_3(aq) \rightarrow Mg(NO_3)_2(aq) + H_2O(l)$$

Both reactions reduce to the ionic equation

$$O^{2-}(s) + 2H^+(aq) \rightarrow H_2O(l)$$

where $O^{2-}(s)$ is the oxide ion in the solid state.

The reaction of organic bases such as propylamine $C_3H_7NH_2$ with acids also produces salts:

$$C_3H_7NH_2(l) + HCl(aq) \rightarrow C_3H_7NH_3{\cdot}Cl(aq)$$

($C_3H_7NH_3{\cdot}Cl$ is an ionic salt similar to NaCl, and is better represented as $C_3H_7NH_3^+,Cl^-$. It dissociates in aqueous solution, forming $C_3H_5NH_3^+(aq)$ and $Cl^-(aq)$ ions.)

Organic acids containing long carbon chains are known as 'fatty acids'. The neutralization of fatty acids by alkalis produces salts that are soapy to the feel. Fatty acids are present on our skin, and this explains why alkalis feel soapy. Sodium stearate (common soap) is produced by neutralizing stearic acid with sodium hydroxide. (For more about soap, see page 182.)

3 Reaction with metals

Solutions of acids react with some metals, making hydrogen gas and a salt. For example, magnesium ribbon reacts with hydrochloric acid, producing magnesium chloride and hydrogen gas. The hydrogen gas is produced so quickly that the solution fizzes:

$$Mg(s) + 2HCl(aq) \rightarrow MgCl_2(aq) + H_2(g)$$

The ionic equation for this reaction is:

$$Mg(s) + 2H^+(aq) \rightarrow Mg^{2+}(aq) + H_2(g)$$

We may test for hydrogen on a **small** scale by exploding a hydrogen–air mixture in a flame (Figure 6.6), with the violence of the explosion depending on the proportions involved:

$$2H_2(g) + O_2(g) \rightarrow 2H_2O(l)$$

Other metals that react with $H^+(aq)$ include calcium, zinc, aluminium and iron. Potassium and sodium react with explosive violence. Copper, silver and gold do not react.

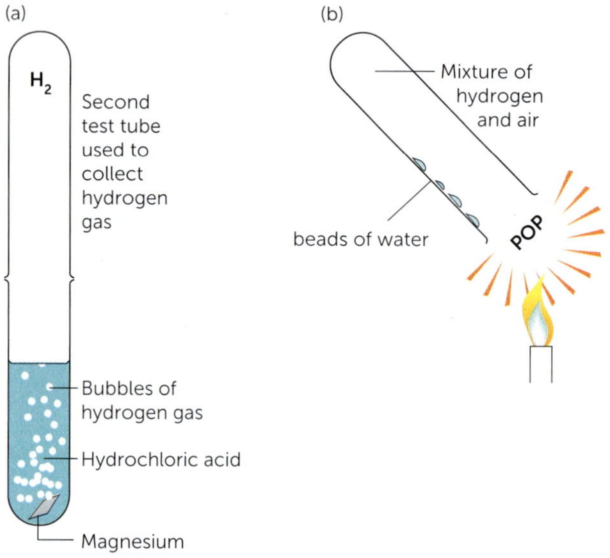

Figure 6.6 Testing for hydrogen gas. An inverted test tube is used to collect the hydrogen (which rises because it is less dense than air). After about twenty seconds, the open end of the test tube is exposed to a flame. A 'pop' (small explosion) confirms the presence of hydrogen.

Death of the Zeppelin Hindenburg (1937). The airship *Hindenburg* contained 200,000 m^3 of hydrogen in lacquered cotton fabric gas bags. A stray spark of unknown origin caused the fabric to burn, releasing hydrogen. Because the hydrogen gas didn't get premixed with oxygen in the air, it burned but did not explode. The gas *helium* is also less dense than air, and since it does not burn is much safer to use in airships than hydrogen. However, helium is much more expensive.

EXERCISE **6K**

Reaction of iron with dilute sulfuric acid

Write (i) a chemical equation and (ii) an ionic equation for the reaction of iron with dilute sulfuric acid to produce iron(II) sulfate and hydrogen gas.

4 Reaction with carbonates and hydrogencarbonates

All acids react with carbonates to produce carbon dioxide gas. For example, the ethanoic acid in vinegar will fizz with sodium carbonate. The reaction is

$$\underset{\text{ethanoic acid in vinegar}}{2CH_3COOH(aq)} + Na_2CO_3(aq) \rightarrow \underset{\text{sodium ethanoate}}{2CH_3COONa(aq)} + CO_2(g) + H_2O(l)$$

The ionic equation for this reaction is

$$\underset{\text{any acid}}{2H^+(aq)} + \underset{\text{carbonate ion}}{CO_3^{2-}(aq)} \rightarrow CO_2(g) + H_2O(l)$$

Hydrogencarbonates (also called bicarbonates) also fizz with acid:

$$H^+(aq) + \underset{\text{hydrogencarbonate ion}}{HCO_3^-(aq)} \rightarrow CO_2(g) + H_2O(l)$$

(Note that hydrogencarbonates decompose upon *heating* giving $CO_2(g)$:

$$2HCO_3^-(aq) \rightarrow CO_2(g) + H_2O(l)$$

The decomposition of solid sodium hydrogencarbonate ('kitchen bicarb') is utilized in cooking, with the CO_2 gas causing cake mixtures to rise.)

5 Reaction with sulfites

Sulfites contain the ion SO_3^{2-}. (Don't confuse this ion with the sulfate ion (SO_4^{2-}), which does not react with acids.) Acids react with sulfites making the sharp-smelling gas sulfur dioxide (SO_2). The ionic equation is

$$\underset{\text{sulphite ion}}{SO_3^{2-}(aq)} + 2H^+ \rightarrow SO_2(g) + H_2O(l)$$

Hydrogensulfites (also called bisulfites) also react with acid:

$$H^+(aq) + \underset{\text{hydrogensulfite ion}}{HSO_3^-(aq)} \rightarrow SO_2(g) + H_2O(l)$$

6 Reaction with sulfides

The reaction of acids with sulfides (compounds containing the sulfide ion, S^{2-}) produces the foul-smelling poisonous gas hydrogen sulfide – one of the products of rotten eggs. The ionic equation for the reaction is

$$\underset{\text{sulfide ion}}{S^{2-}(s)} + 2H^+(aq) \rightarrow H_2S(g)$$

Metal ions often react with sulfide ions to produce insoluble sulfides. For example,

$$Cu^{2+}(aq) + S^{2-}(aq) \rightarrow \underset{\text{black}}{CuS(s)}$$

Sodium sulfide solution or hydrogen sulfide gas are used as the source of sulfide ions. Solutions of Na^+, Ca^{2+} and K^+ will not produce precipitates because their sulfides are water soluble. Zinc sulfide (ZnS) is white, while lead(II) and silver sulfides (PbS and Ag_2S) are black.

6.6 Acids produced when gases CO_2, SO_2 and NO_2 dissolve in water

Carbon dioxide and sulfur dioxide gases are both water soluble. Some of the dissolved gas molecules then react with water producing hydrogencarbonate and hydrogensulfite ions, respectively. The reactions are:

$$CO_2(aq) + H_2O(l) \rightleftharpoons \underset{\text{hydrogencarbonate ion}}{HCO_3^-(aq)} + H^+(aq)$$

$$SO_2(aq) + H_2O(l) \rightleftharpoons \underset{\text{hydrogensulfite ion}}{HSO_3^-(aq)} + H^+(aq)$$

The $\rightleftharpoons$ sign shows that these reactions do not go to completion, i.e. attempts to react the gases completely with water will always produce an equilibrium mixture of unreacted gas and hydrogencarbonate or hydrogensulfite and hydrogen ions. Under the same conditions, more SO_2 molecules react with water than do CO_2 molecules. This causes $SO_2(g)$ to be more soluble in water than $CO_2(g)$, so that saturated solutions of SO_2 are more acidic (contain a higher H^+ concentration) than solutions of CO_2. This explains why $CO_2(g)$ only turns damp blue litmus paper a purple/red colour, whereas $SO_2(g)$ produces a much more definite red colour.

In the presence of oxygen, a solution of sulfur dioxide is very slowly converted to dilute sulfuric acid at room temperature:

$$SO_2(aq) + {}^1/_2O_2(aq) + H_2O(l) \rightarrow SO_4^{2-}(aq) + 2H^+(aq)$$

In the atmosphere, the conversion of $SO_2(aq)$ to sulfuric acid within clouds occurs faster, and is the main source of acid rain (see Unit 24).

If CO_2 or SO_2 are dissolved in *alkaline* solutions, all the hydrogencarbonate and hydrogensulfite ions are converted to carbonates and sulfites, respectively:

$$HCO_3^-(aq) + OH^-(aq) \rightarrow CO_3^{2-}(aq) + H_2O(l)$$

$$HSO_3^-(aq) + OH^-(aq) \rightarrow SO_3^{2-}(aq) + H_2O(l)$$

Summarizing, when CO_2 and SO_2 gases are bubbled into a beaker of water, a mixture of undissolved gas and hydrogencarbonate or hydrogensulfite ions are present. However, in the presence of OH^-(aq), the only species present are carbonate or sulfite ions.

Nitrogen dioxide (NO_2(g)) is a toxic brown gas that dissolves in water producing an acidic cocktail of hydrogen ions, nitrate ions (NO_3^-(aq)) and nitric oxide gas (NO):

$$3NO_2(g) + H_2O(l) \rightarrow 2H^+(aq) + NO(g) + 2NO_3^-(aq)$$

6.7 Reactions of the hydroxide ion

The hydroxide ion in solution is normally written as OH^-(aq). A dilute solution of sodium hydroxide, Na^+, OH^- (old name 'caustic soda') is a convenient source of OH^-(aq) in the laboratory.

1 Effect on litmus

Litmus turns blue in basic solution.

2 Neutralization of acids

The OH^-(aq) ion neutralizes acids, giving a salt and water only.

3 Reaction with NH_4^+ ion

Ammonia gas is produced:

$$NH_4^+(aq) + OH^-(aq) \xrightarrow{\text{heat}} H_2O(l) + NH_3(g)$$

(a) (b) (c)

Sodium hydroxide (NaOH, *caustic soda* or *natrium hydroxide*) is a corrosive alkali used to make soap and to clear drains. Soap is produced by the reaction of NaOH and oil or fat. (c) shows the cutting of soap into slabs. (Images (b) and (c) by courtesy of Patounis Soap Factory, Corfu.)

4 Reaction with some metal ions in solution: the use of precipitation in analysis

The hydroxide ion causes insoluble metal hydroxides to precipitate out of solution. For example, if we add a few drops of sodium hydroxide solution to (blue) copper(II) sulfate solution, a beautiful blue precipitate of copper(II) hydroxide is observed:

$$Cu^{2+}(aq) + 2OH^-(aq) \rightarrow \underset{\text{blue precipitate}}{Cu(OH)_2(s)}$$

Examples of coloured hydroxides: iron(II) hydroxide (dirty green), iron (III) hydroxide (red/brown), copper (II) hydroxide (blue) and nickel (II) hydroxide (light green)

The $Cu^{2+}(aq)$ comes from the copper(II) sulfate, and the $OH^{-}(aq)$ comes from the sodium hydroxide. Similar reactions take place with other metal ions:

$$Zn^{2+}(aq) + 2OH^{-}(aq) \rightarrow Zn(OH)_2(s)$$

zinc ion (colourless) — white precipitate

$$Al^{3+}(aq) + 3OH^{-}(aq) \rightarrow Al(OH)_3(s)$$

aluminium ion (colourless) — white precipitate

$$Fe^{2+}(aq) + 2OH^{-}(aq) \rightarrow Fe(OH)_2(s)$$

iron(II) ion (pale - green colour) — white precipitate which almost instantaneously turns dirty green

$$Fe^{3+}(aq) + 3OH^{-}(aq) \rightarrow Fe(OH)_3(s)$$

iron(III) ion (yellow/brown colour) — reddish-brown precipitate

$$Pb^{2+}(aq) + 2OH^{-}(aq) \rightarrow Pb(OH)_2(s)$$

lead(II) ion (colourless) — white precipitate

$$Mg^{2+}(aq) + 2OH^{-}(aq) \rightarrow Mg(OH)_2(s)$$

magnesium ion (colourless) — white precipitate

Zinc, aluminium and lead(II) hydroxides will dissolve to make clear solutions if we add excess sodium hydroxide solution. This happens because all three hydroxides are **amphoteric** (see Box 6.2). This effect is useful because it helps to distinguish between magnesium hydroxide (whose white precipitate is insoluble in excess hydroxide) and zinc, aluminium and lead(II) hydroxides (whose white precipitates are soluble in excess hydroxide).

Since zinc, aluminium and lead(II) hydroxides are white, additional tests are needed to distinguish between them. A quick identification test for lead(II) ions depends upon the fact that lead(II) sulfate is insoluble in water, whereas aluminium and zinc sulfates are soluble. Addition of dilute sulfuric acid to lead(II) ions will therefore precipitate the lead(II) sulfate:

$$Pb^{2+}(aq) + SO_4^{2-}(aq) \rightarrow PbSO_4(s)$$

white precipitate

Zinc and aluminium ions may be distinguished by heating a small amount of their hydroxides on a metal spatula. Aluminium hydroxide decomposes to form white aluminium oxide. Zinc hydroxide also decomposes into the oxide, but zinc oxide (although white when cold) is yellow when hot.

BOX **6.2**

Amphoteric hydroxides

Hydroxides (like all bases) react with acids. For example, aluminium hydroxide dissolves in acid giving $Al^{3+}(aq)$:

$$Al(OH)_3(s) + 3H^{+}(aq) \rightarrow Al^{3+}(aq) + 3H_2O(l)$$

colourless

However, some hydroxides will also react with the hydroxide ion if it is present in high enough concentration. For example, with aluminium hydroxide the reaction is:

$$Al(OH)_3(s) + OH^{-}(aq) \rightarrow [Al(OH)_4]^{-}(aq)$$

colourless

In this reaction, aluminium hydroxide is acting as an acid. Aluminium hydroxide is said to be **amphoteric**, meaning that it acts like a base or an acid depending on whether excess $H^{+}(aq)$ or $OH^{-}(aq)$ is present. Zinc and lead(II) hydroxides are also amphoteric and both react in a similar way to aluminium hydroxide.

Silver hydroxide does not exist. When hydroxide ions are added to silver ions, a dark brown precipitate of silver oxide is formed:

$$2Ag^+(aq) + 2OH^-(aq) \rightarrow Ag_2O(s) + H_2O(l)$$

Because potassium and sodium hydroxides are very soluble, they do not precipitate when OH^- ions are added to sodium or potassium salts. Calcium hydroxide is not very soluble in water, and appears as a white cloudiness when a concentrated solution of OH^- is mixed with a concentrated solution of calcium salts. $Na^+(aq)$, $K^+(aq)$ and $Ca^{2+}(aq)$ may be identified using **flame tests** (see Unit 12, Table 12.1).

6.8 Identification of common gases

Table 6.3 shows some of the properties of common gases at room temperature and pressure, arranged under the headings of **C.O.L.D.S**:

Colour of gas

Odour (smell) of gas (pungent means very strongly smelling)

Litmus – effect of gas on damp litmus (red or blue)

Density of gas compared with air (heavier than air depicted as H, lighter as L, roughly the same as =)

Solubility of gas in water at room temperature (categorized as virtually insoluble, very soluble, slightly soluble etc.)

Many gases *extinguish* a flaming splint of wood on a test-tube scale. Gases that burn (or explode) on a test tube scale include three odourless gases, methane, hydrogen and carbon monoxide. Methane is difficult to distinguish from other hydrocarbons: this is where a gas chromatography–mass spectrometry instrument comes in handy! If you want to read more about the reactions of common gases consult a trusted old chemistry textbook, such as *A New Certificate Chemistry* by Holderness and Lambert (any edition, Heinemann). It makes good bedtime reading.

Table 6.3 Properties of common gases

Name	Formula	Colour (blank means colourless)	Odour (blank means odourless)	Litmus paper (blank means no effect)	Density compared with air	Solubility in water	Notes
Ammonia	NH_3		Pungent	red → blue	L	very[3]	The only common alkaline gas
Carbon dioxide	CO_2			blue → claret[1]	H	slightly	Gives positive 'limewater test'[2]
Carbon monoxide	CO				L	insoluble	Burns with a blue flame
Chlorine	Cl_2	Pale yellow	Pungent	bleaches it	H	soluble	
Hydrogen	H_2				L	insoluble	See 'pop test' Figure 6.6
Hydrogen chloride	HCl		Pungent	blue → red	H	very[3]	
Hydrogen sulfide	H_2S		Rotten eggs	blue → claret[1]	H	slightly	

Methane	CH_4				L	insoluble	Difficult to distinguish from other hydrocarbon gases
Nitric oxide	NO		Unknown		=	insoluble	Reacts instantly in air (oxygen) producing red-brown NO_2
Nitrogen	N_2				=	insoluble	Does not burn
Nitrogen dioxide	NO_2	Red brown[5]	Pungent	blue → red	H	soluble	
Nitrous oxide	N_2O		Sickly smell		H	soluble	Also known as 'laughing gas'. Will relight a glowing splint[4]
Oxygen	O_2				=	slightly	Relights a glowing splint[4]
Sulfur dioxide	SO_2		Pungent	blue → red	H	soluble	

[1] Forms a weak acid in solution, so the red colour is better described as claret.

[2] See Revision question 6.5(ii).

[3] These gases are so soluble that they can be used in the fountain experiment (see Unit 19, Fig 11.9).

[4] N_2O also relights a glowing splint although less dramatically than O_2. N_2O is easily distinguished from O_2 by its smell and its greater solubility in water.

[5] NO_2 becomes lighter coloured when cooled. See Unit 15, Exercise 15I.

BOX **6.3**

Chlorine and phosgene as war gases

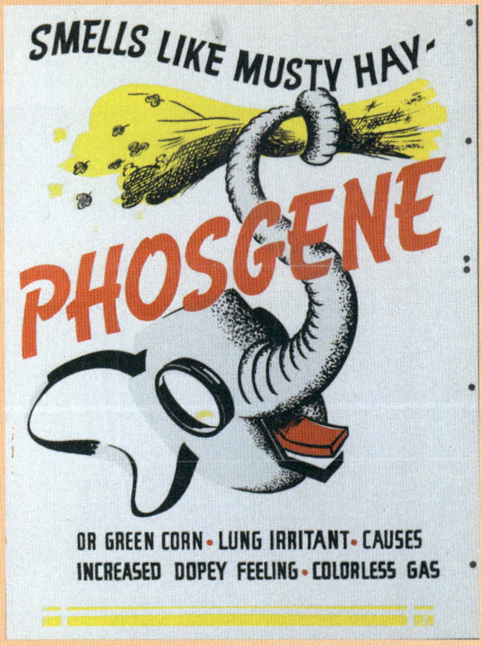

Image courtesy of the National Museum of Health and Medicine.

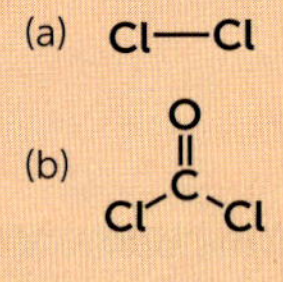

Top left: Warning poster about phosgene gas from the Second World War.

Above: molecular structures of (a) chlorine (Cl_2) and (b) phosgene ($COCl_2$).

Chlorine and phosgene were used as poison gases in the First World War (the Great War), 1914–1918. The first poison gas attack was led by chemist Fritz Haber who later earned the Nobel Prize for the conversion of nitrogen and hydrogen into ammonia. Phosgene is even more irritating than chlorine, but unlike chlorine it is colourless and less smelly, so it is less easily detected. In practice, phosgene was often used in the First World War in combination with chlorine, with the chlorine helping to spread the denser phosgene along the ground. It is estimated that there were hundreds of thousands of casualties due to poison gas (including chlorine, phosgene and another poison gas called mustard gas) in the First World War.

Chlorine gas is yellow.

6.9 Use of reactions in the identification of ions in solution

The identification of simple ions using test-tube experiments relies on the chemistry described earlier in this unit.

Table 6.4 Common ions whose reactions are discussed in this book

Cations	Anions
H^+, NH_4^+, K^+, Na^+, Ca^{2+}, Cu^{2+}, Fe^{2+}, Fe^{3+}, Ni^{2+}, Pb^{2+}, Al^{3+}, Zn^{2+}, Mg^{2+}, Ba^{2+}, Ag^+	OH^-, Cl^-, SO_4^{2-}, Br^-, I^-, SO_3^{2-}, S^{2-}, CO_3^{2-}, NO_3^-

EXAMPLE **6.2**

A solution of a simple salt (AB, where A is the cation and B is the anion) is pale green. Addition of a mixture of barium chloride solution and dilute hydrochloric acid produces a white precipitate. Addition of sodium hydroxide solution to AB produces a dirty green precipitate. Identify A and B.

Answer

The colour of the solution suggests that the iron(II) ion is present. This is confirmed by the green precipitate with hydroxide ion. The precipitate with $BaCl_2$ + HCl confirms the presence of a sulfate.

A = Fe^{2+}(aq); B = SO_4^{2-}(aq); AB = $FeSO_4$

EXAMPLE **6.3**

A solution, containing a cation Z and an anion Y, fizzed with sodium carbonate solution and gave a cream-coloured precipitate when added to a mixture of silver nitrate solution and dilute nitric acid. Identify Z and Y.

Answer

The fizzing suggests that Z is probably H^+(aq). The cream-coloured precipitate is probably silver bromide. Therefore, the unknown is a solution of hydrobromic acid.

Z = H^+(aq); Y = Br^-(aq); ZY = HBr(aq)

EXERCISE **6L**

Identification of unknowns

(i) A colourless solution CD contains an anion C and a cation D. Addition of sodium hydroxide solution to CD gave a white precipitate H that dissolved in excess hydroxide to produce a clear solution. H turns yellow when strongly heated in a flame. Addition of CD to a mixture of silver nitrate solution and dilute nitric acid gave a white precipitate. Identify C and D.

(ii) A colourless solution is either magnesium carbonate, sodium nitrate or dilute sulfuric acid. Addition of sodium sulfite solution to the unknown (with warming) produces a sharp-smelling gas. What is the identity of the colourless solution? Suggest a test which would confirm your answer.

(iii) The labels have fallen off four bottles. One of the bottles contains copper(II) sulfate solution, another sodium hydroxide solution, the third sodium carbonate solution and the fourth dilute hydrochloric acid. Using only these bottles (and clean test tubes) plan tests which would enable you to identify the contents of each bottle.

REVISION QUESTIONS

6.1

(i) Cold pure sulfuric acid does not react with zinc metal, whereas dilute sulfuric acid reacts very rapidly. Explain this.

(ii) Explain why ammonia gas turns damp red litmus paper blue.

(iii) Why should concentrated sulfuric acid not be used as a drying agent for ammonia gas?

(iv) Under what condition is pure sulfuric acid a powerful oxidizing agent?

6.2 Write equations showing the ions formed when the following dissolve in water:

(i) copper(II) sulfate crystals ($CuSO_4 \cdot 5H_2O$)

(ii) lithium hydroxide (LiOH, i.e. Li^+,OH^-).

6.3 Oxalic acid is a toxic acid found in rhubarb leaves. Its structure may be written as $(COOH)_2$. Both hydrogen atoms are acidic. Write down a chemical and ionic equation for the reaction of oxalic acid with: (i) potassium hydroxide solution, (ii) zinc metal.

6.4 Write chemical *and* ionic equations for the reaction of:

(i) magnesium carbonate with dilute sulfuric acid

(ii) ammonium sulfate and hot sodium hydroxide

(iii) sodium sulfide and dilute nitric acid

(iv) sodium sulfite (a preservative) and vinegar.

6.5

(i) Much more CO_2(g) is produced annually than SO_2(g), yet SO_2(g) causes greater damage to limestone buildings than CO_2. Why?

(ii) The reaction of clear calcium hydroxide solution ('limewater') with carbon dioxide is the basis of the limewater test for CO_2. During the reaction, a white precipitate is produced. Name the precipitate and write an ionic equation for the reaction.

6.6 An insoluble blue solid, containing an anion J and a cation K, fizzed violently with dilute sulfuric acid, producing a gas and a blue solution B. The gas produced had no smell and solution B gave a blue precipitate with sodium hydroxide solution. Identify J and K. Explain your reasoning.

6.7 A solution of a simple salt GH contains a cation G and an anion H. GH gave a red-brown precipitate when mixed with sodium hydroxide solution. GH gave a brown ring with concentrated H_2SO_4 and $FeSO_4$. Identify the cation and anion present.

6.8 A simple salt ST (containing an anion S and a cation T) is insoluble in water. ST produced hydrogen sulfide when warmed with hydrochloric acid. The resulting solution was divided into two. One half gave a white precipitate with sodium sulfate solution. The other half produced a white precipitate with sodium hydroxide solution which was soluble in excess NaOH. Identify S and T. Explain your reasoning and include ionic equations for any reactions that occur.

6.9 The detrimental effect of acid rain upon fish is partly due to the release of toxic aluminium ions (Al^{3+}(aq)) under acidic conditions. Explain this by writing ionic equations which show what happens to Al^{3+}(aq) ions (i) in an alkaline environment, (ii) in a previously alkaline environment which has been overwhelmed with acid rain.

6.10

(i) Why is the smell of nitric oxide gas (NO(g)) unknown?

(ii) Identify the three gases X, Y and Z. X relights a glowing splint and has a sickly smell. Y has a pungent smell and turns damp red litmus blue. Z is colourless, extinguishes a burning splint and turns a clear solution of calcium hydroxide milky.

DID YOU KNOW?

The 'liquid gold' that the Germans missed

The Danish laboratory of the famous physicist Niels Bohr was a safe haven for Jewish scientists during the early part of the Second World War. Bohr had donated his own Nobel Prize (200 g of 23 carat gold) to charity, but two other Nobel Prize winners left their inscribed medals with him in 1940, just before Denmark was invaded by Germany.

Fearful that the medals would be discovered by the invading Germans, one of Bohr's colleagues came up with the idea of dissolving the medals in acid and storing the gold solution ('liquid gold') on the laboratory shelf with other chemicals. Despite a thorough search in Bohr's laboratory, the gold survived undetected and new medals were cast after the war.

Gold is resistant to most acids but a fuming mixture of concentrated hydrochloric and nitric acids, known as *Aqua Regia* (meaning 'Royal Water'), does dissolve gold and this was the acid used in Bohr's laboratory.

Information on the history of chemistry will be found in Extension 6 at https://www.bloomsburyonlineresources.com/chemistry

7 Oxidation and Reduction

CONTENTS

OBJECTIVES

This unit:

- Explains how to work out oxidation numbers and write redox equations
- Shows you how to use standard reduction potentials to decide if reactions can occur
- Discusses corrosion and redox reactions in nature
- Describes the electrolysis of water

7.1 Redox reactions

A **redox** reaction is an important type of chemical reaction. In such a reaction, one reactant is **oxidized**, so causing the **simultaneous reduction** of a second reactant.

Definitions of oxidation and reduction

Oxidation was once regarded as a chemical reaction in which oxygen was added to a substance, whereas reduction occurred when oxygen was lost.

For example, in the reaction

$$CO_2(g) + C(s) \rightarrow 2CO(g)$$

carbon is oxidized to carbon monoxide (an atom of carbon gains an oxygen atom), while carbon dioxide is reduced to carbon monoxide (a molecule of carbon dioxide loses an atom of oxygen). The definition was later broadened to include hydrogen: oxidation was removal of hydrogen from a substance, while reduction took place when hydrogen was added.

In this section we introduce a definition of oxidation and reduction for reactions where ions are involved:

The oxidation of a chemical species occurs when electrons are lost and the reduction of a chemical species occurs when electrons are gained.

This is easy to remember as **OIL RIG** (**o**xidation **i**s electron **l**oss; **r**eduction **i**s **g**ain of electrons).

For example, when copper reacts with oxygen (**is oxidized**) to form copper(II) oxide, the chemical equation may be written as

$$2Cu(s) + O_2(g) \rightarrow 2CuO(s)$$

In this reaction, a copper atom loses two electrons:

$$Cu \rightarrow Cu^{2+} + 2e^- \qquad (A)$$

and according to our OIL definition this is an oxidation. Each oxygen atom in an oxygen molecule gains two electrons from the copper atom. This is reduction:

$$O + 2e^- \rightarrow O^{2-}$$

Or, since oxygen gas (O_2) reacts with copper the reduction is better shown as

$$O_2 + 4e^- \rightarrow 2O^{2-} \qquad (B)$$

and we can also state that the oxygen *molecule* is reduced. The overall equation for the reaction between copper and oxygen can be obtained by adding the two 'half-reactions' (A and B):

1 First multiply the equation for the oxidation of copper by two, so that this half-reaction contains the same number of electrons as the half-equation for the reduction of O_2:

$$2Cu \rightarrow 2Cu^{2+} + 4e^-$$
$$O_2 + 4e^- \rightarrow 2O^{2-}$$

2 Then add the two half-reactions (think of the arrow as an equals sign in maths), and the number of electrons on each side cancels out:

$$2Cu + O_2 + \cancel{4e^-} \rightarrow 2Cu^{2+} + 2O^{2-} + \cancel{4e^-}$$

3 The overall equation is

$$2Cu(s) + O_2(g) \rightarrow 2Cu^{2+}(s) + 2O^{2-}(s)$$

Remember that it is impossible to oxidize a species without another species being reduced at the same time. Just as a bank loan must involve a borrower and a lender, reduction and oxidation occur together in every REDOX reaction.

EXERCISE **7A**

The reaction of zinc with aqueous copper(II) sulfate

When solid zinc is added to blue copper(II) sulfate solution, the solution becomes paler and eventually colourless. A red-brown solid (copper) is deposited at the bottom of the test tube. This is shown in Figure 7.1: (For photos, see p. 85).

In this reaction, the zinc metal supplies two electrons to the copper ions and copper is precipitated.

The two half-reactions are:

$Zn(s) \rightarrow Zn^{2+}(aq) + 2e^-$ ($Zn^{2+}(aq)$ is colourless, Zn(s) is grey)
$Cu^{2+}(aq) + 2e^- \rightarrow Cu(s)$ ($Cu^{2+}(aq)$ is blue, Cu(s) is red-brown)

(i) Which half-reaction represents reduction?

(ii) Which half-reaction represents oxidation?

(iii) Write the overall redox equation for the reaction.

(iv) The sulfate ions are not included in the redox equation. Why?

EXERCISE **7B**

Oxidation and reduction

In the following reactions, which substances are oxidized and which are reduced?

(i) $2Zn(s) + O_2(g) \rightarrow 2ZnO(s)$

(ii) $Mg(s) + Cu^{2+}(aq) \rightarrow Mg^{2+}(aq) + Cu(s)$

(iii) $Cl_2(g) + 2I^-(aq) \rightarrow 2Cl^-(aq) + I_2(aq)$

(iv) $2Na(s) + Cl_2(g) \rightarrow 2NaCl(s)$

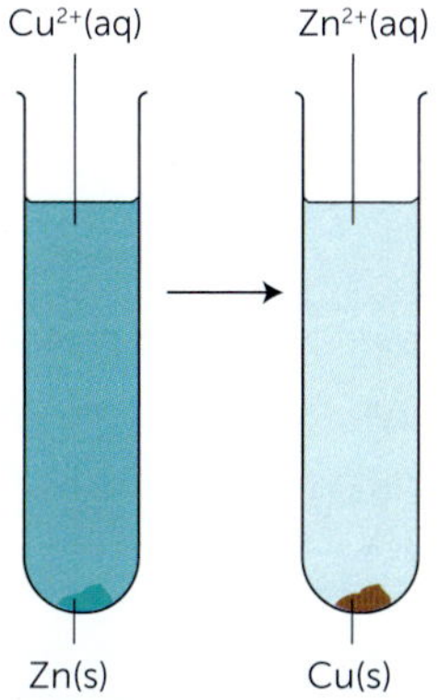

Figure 7.1 Reaction of zinc with copper(II) sulfate solution.

7.2 Oxidation numbers

Oxidation numbers enable us to work out whether oxidation or reduction has occurred in a chemical reaction. This system works for **ALL** reactions, not just the reactions involving ions described in section 7.1, and it reaches the same conclusions. The **oxidation number** or **oxidation state** of an atom is a positive or negative number that is decided using agreed rules (see Box 7.1). Oxidation numbers are used to widen the definition of oxidation and reduction, with **oxidation being defined as an increase in oxidation number and reduction as a decrease in oxidation number.**

The oxidation number system has been agreed by scientists as a standardized way of defining oxidation or reduction more broadly. *It does not tell you anything about the bonding in a compound.* For example, the fact that carbon has an oxidation number of +4 in some of its compounds does *not* mean that it exists as +4 ions in these compounds.

BOX **7.1**

Oxidation numbers

The following elements nearly always have the same oxidation number in their compounds. They are used as 'standards' to help assign the oxidation numbers of other elements.

There are important exceptions to the elements marked with an asterisk (*), but they will not be dealt with here. The different oxidation numbers of Cl and O will be dealt with at a later stage in the unit.

Element(s)	Oxidation number
K^+ Na^+	+1
Mg^{2+} Ca^{2+}	+2
Al^{3+}	+3
H^+ (or covalent H)*	+1
F^- (or covalent F)	−1
Cl^- (or covalent Cl)*	−1
O^{2-} (or covalent O)*	−2

EXERCISE **7C**

Assigning oxidation numbers

Assign oxidation numbers to the underlined atoms in the following examples:

(i) $\underline{S}O_2$
(ii) $\underline{N}H_3$
(iii) $\underline{S}_8$
(iv) $\underline{Cr}Cl_5$
(v) $\underline{N}O_3^-$
(vi) $\underline{Mn}O_4^-$
(vii) $\underline{Xe}F_2$
(viii) $Ca(\underline{V}O_3)_2$
(ix) $\underline{Bi}O_3^-$
(x) $\underline{Cr}_2O_7^{2-}$

Rules for working out oxidation numbers

1 The oxidation number of an atom of an *uncombined element* is *zero*. For example, the oxidation number of nitrogen in $N_2(g)$ or $N_2(s)$ or calcium in Ca(s), is zero.

2 The oxidation number of the ion of an element is equal to its charge. For example, the oxidation number of copper in Cu^{2+} is +2 and the oxidation number of oxygen in O^{2-} is – 2.

3 The *algebraic sum* of the oxidation numbers of the atoms in the formula of an electrically neutral compound is zero. For $Ca(OH)_2$, for example, the sum of the oxidation numbers is

[+2 for Ca] + [2(−2) for O] + [2(+1) for H] = 0

4 For convenience, shared electrons in covalent compounds are assigned to the element having the greater electronegativity. For example, the oxidation number for phosphorus in PCl_3 is +3 (chlorine is – 1), since phosphorus is less electronegative than chlorine.

5 The *algebraic sum* of the oxidation numbers of all the atoms of an ion is equal to the charge on the ion. In SO_4^{2-}, for example, the sum of the oxidation numbers is

[+6 for S] + [4(−2) for O] = – 2

Remember that during chemical reactions an increase in oxidation number signifies OXIDATION; a decrease, REDUCTION.

EXERCISE **7D**

Oxidation ladders

Some elements have many oxidation states. You can construct redox 'ladders' like the one for sulfur.

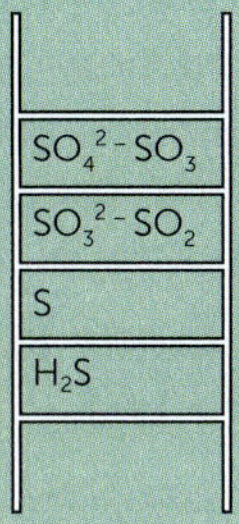

(i) What is the oxidation state of sulfur on each rung of the 'ladder'?

(ii) SO_2 reacts with water to form the *sulfite* ion SO_3^{2-}. Is this a redox reaction?

(iii) H_2S reacts with SO_2 to precipitate sulfur:

$$2H_2S(g) + SO_2(g) \rightarrow 2H_2O(l) + 3S(s)$$

Which species is reduced? Which species is oxidized?

EXERCISE **7E**

Redox ladder for chlorine

Chlorine can have different oxidation states. Using the following species, construct a redox ladder for chlorine:

Cl_2, HCl, ClO_3^-, ClO^-, Cl^-, $HClO_4$

Note that those species with the lowest oxidation number for chlorine should be at the bottom of the ladder, whereas those with the highest oxidation number for chlorine are at the top.

EXERCISE **7F**

Oxidizing and reducing agents

Look back at the redox ladder for sulfur in Exercise 7D.

(i) Which sulfur species can act as both reducing and oxidizing agents in different chemical reactions?

(ii) Which sulfur species can only act as oxidizing agents and why?

(iii) Which sulfur species can only act as reducing agents and why?

7.3 Oxidizing and reducing agents

An **oxidizing agent** is a substance that causes oxidation in another chemical species. A **reducing agent** causes the reduction of another chemical species. For ionic reactions, **oxidizing agents** accepts electrons and, in doing so, becomes **reduced**, whereas reducing agents supply the electrons needed for reduction and so itself becomes **oxidized**.

EXAMPLE **7.1**

When iron is heated in chlorine gas, iron(III) chloride is produced. The 'ordinary' equation is:

$$2Fe(s) + 3Cl_2(g) \rightarrow 2FeCl_3(s)$$

The redox half-equations are

$$Fe(s) \rightarrow Fe^{3+}(s) + 3e^-$$
$$Cl_2(g) + 2e^- \rightarrow 2Cl^-(s)$$

Here, iron is acting as a *reducing agent* because it supplies electrons to chlorine and is *oxidized*.

Chlorine is acting as an *oxidizing agent* because it accepts electrons from iron and is *reduced*.

7.4 Writing and balancing redox equations

In order to write a redox equation, it is first necessary to write the two half-reactions that identify the oxidation and reduction processes taking place. The overall redox equation is then obtained by adding these two half-reactions together, so that the electrons in each half-reaction cancel out.

To write the half-reactions, follow these simple rules:

1 **Identify the atoms that are oxidized and reduced**, using the oxidation number method.

Most of the reactions that you will come across at this stage, will occur in neutral or acid solution, and step 2 applies.

2 **Balance the half-reactions:**

(i) Make sure that there are the same number of atoms of the element that is oxidized (or reduced) on each side of the half-reaction.

(ii) If there are any oxygen atoms present, balance them by adding water molecules to the other side of the half-reaction.

(iii) If there are hydrogen atoms present, balance them by adding hydrogen ions on the other side of the half-reaction.

(iv) Make sure that the half-reactions have the same overall charge on each side by adding electrons.

Note that the rules are slightly different if the reaction occurs in basic (alkaline) solution: hydrogen atoms are balanced using H_2O molecules and then the same number of OH^- ions are added to the opposite side of the equation to balance the oxygens. Carry on as before, adding electrons to balance the charges.

EXAMPLE **7.2**

Write the overall redox equation for the oxidation of iron(II) ions to iron(III) ions by the manganate(VII) or permanganate ion, MnO_4^-, in acid solution. The manganate(VII) ion reacts to form the manganese(II) ion, Mn^{2+}.

Answer

The oxidation half-reaction

The question tells you that iron(II) is oxidized, but you can check on this by working out the oxidation numbers of iron before and after the reaction:

$Fe^{2+} \rightarrow Fe^{3+}$; by oxidation numbers: $+2 \rightarrow +3$

Therefore, an increase in oxidation number, i.e. oxidation, has occurred.

There are the same number of atoms on each side of this half-reaction, and no oxygen or hydrogen atoms, so all that remains to be done is to make sure that the overall charge is the same on each side of the half-equation. By adding one electron to the right-hand side, the overall charge on each side of the equation becomes +2 (*think of an electron as a unit negative charge, which will cancel out a positive charge*):

$Fe^{2+} \rightarrow Fe^{3+} + e^-$

The half-equation above represents the oxidation reaction.

The reduction half-reaction

The manganate(VII) ion reacts to form manganese(II). Check that this is a reduction by working out the oxidation numbers of manganese before and after the reaction:

$MnO_4^- \rightarrow Mn^{2+}$;   by oxidation numbers: $+7 \rightarrow +2$

Therefore, a decrease in oxidation number, i.e. reduction, has occurred.
There are the same number of manganese atoms on each side of the half-reaction, but the oxygen atoms need to be balanced:

1 Balance the oxygens with water molecules:

$$MnO_4^- \rightarrow Mn^{2+} + 4H_2O$$

2 Now H atoms have been introduced, so they need to be balanced with H^+ ions:

$$MnO_4^- + 8H^+ \rightarrow Mn^{2+} + 4H_2O$$

3 Balance the charges on each side of the half-reaction by adding electrons:

$$MnO_4^- + 8H^+ + 5e^- \rightarrow Mn^{2+} + 4H_2O$$

Each side of the equation now has an overall charge of +2. The above equation represents the reduction reaction. To construct the overall redox equation, the two half-reactions must be added together so that the electrons cancel out:

$Fe^{2+} \rightarrow Fe^{3+} + e^-$ (oxidation)
$MnO_4^- + 8H^+ + 5e^- \rightarrow Mn^{2+} + 4H_2O$ (reduction)

The first half-reaction must be multiplied by five, then the equations added together:

$5Fe^{2+} \rightarrow 5Fe^{3+} + 5e^-$ (oxidation)
$MnO_4^- + 8H^+ + 5e^- \rightarrow Mn^{2+} + 4H_2O$ (reduction)

$$MnO_4^- + 8H^+ + \cancel{5e^-} + 5Fe^{2+} \rightarrow Mn^{2+} + 4H_2O + 5Fe^{3+} + \cancel{5e^-}$$

The overall redox equation is therefore

$$MnO_4^-(aq) + 8H^+(aq) + 5Fe^{2+}(aq) \rightarrow Mn^{2+}(aq) + 4H_2O(l) + 5Fe^{3+}(aq)$$

EXERCISE **7G**

Writing and balancing redox equations

1 Write half-equations for:

(i) the oxidation of I^- to IO_3^-
(ii) the reduction of NO_3^- to NO in acidic solution
(iii) the reduction of $Cr_2O_7^{2-}$ to Cr^{3+} in acidic solution
(iv) the reduction of H_2O_2 to H_2O in acidic solution
(v) the oxidation of S^{2-} to SO_4^{2-} in *basic* solution.

2 Write an overall redox equation for:

(i) The reaction of copper with silver nitrate, in aqueous solution, to produce copper(II) nitrate and silver.
(ii) The oxidation of copper metal to copper(II) nitrate by concentrated nitric acid (*regard the reduction of nitric acid as the ion NO_3^- being reduced to brown NO_2 gas*).
(iii) The reduction of I_2 by thiosulfate, $S_2O_3^{2-}$, to I^- in aqueous solution. Thiosulfate is oxidized to tetrathionate, $S_4O_6^{2-}$.
(iv) The oxidation of I^- to I_2 by H_2O_2 in acid solution (H_2O_2 is reduced to H_2O). (Note that in H_2O_2, the oxidation number of oxygen is – 1.)
(v) The oxidation of H_2O_2 to O_2 by ClO_2^- (ClO_2^- is reduced to Cl^-).

BOX 7.2

$2K^+,Cr_2O_7^{2-}$

K^+,MnO_4^-

Dichromate ($Cr_2O_7^{2-}$) ions and manganate (IV) ions (MnO_4^-) are very important oxidizing agents in acidic solution ($H^+(aq)$) (See Unit 9, p146). The photographs are of samples of the potassium salts of these ions. The half-reactions for these oxidizing agents are:

$$Cr_2O_7^{2-} + 14H^+ + 6e^- \rightarrow 2Cr^{3+} + 7H_2O$$

orange green

$$MnO_4^- + 8H^+ + 5e^- \rightarrow Mn^{2+} + 4H_2O$$

violet/pink colourless

Note that Manganate (IV) is also called Permanganate. $Cr^2O_7^{2-}$ is orange in solution. MnO_4^- is violet or pink in solution, depending on its concentration. Photos by Ben Miles.

7.5 Redox couples

Think of a piece of zinc metal, Zn(s), dipped in a solution of silver ions, $Ag^+(aq)$. A reaction is seen to occur, and solid silver is produced:

$$Zn(s) + 2Ag^+(aq) \rightarrow Zn^{2+}(aq) + 2Ag(s)$$

The half-reactions that make up this overall reaction are:

$$Zn(s) \rightarrow Zn^{2+}(aq) + 2e^-$$

and

$$2Ag^+(aq) + 2e^- \rightarrow 2Ag(s)$$

The fact that there is a reaction between Zn(s) and $Ag^+(aq)$ suggests that the silver ions want to accept electrons more than the zinc atoms want to keep them.

No reaction takes place when zinc metal is added to magnesium ions, $Mg^{2+}(aq)$. This suggests that magnesium ions in solution have a smaller tendency to accept electrons, and form Mg(s), than do silver ions, forming Ag(s).

The pairs of species involved here, $Ag^+(aq)/Ag(s)$ and $Mg^{2+}(aq)/Mg(s)$ are known as **redox couples**. Therefore, our experiments suggest that the $Ag^+(aq)/Ag(s)$ redox couple is a *stronger oxidizing agent* than the $Mg^{2+}(aq)/Mg(s)$ redox couple. Since a strong oxidizing agent may be described as a weak reducing agent, we could also say that the $Mg^{2+}(aq)/Mg(s)$ redox couple is a *stronger reducing agent* than the $Ag^+(aq)/Ag(s)$ redox couple.

The relative oxidizing or reducing strengths of redox couples are expressed in terms of their **standard electrode potentials**, $E^{\ominus}$, which have the units of volts. It is impossible to measure the standard electrode potentials of redox couples in isolation, without introducing other metals into the electrical circuit. In practice, the redox couple under study is connected to a **reference redox couple** in an **electrochemical cell**. Voltage measurements then give a relative potential difference. The reference redox couple used is $H^+(aq)/H_2(g)$, which (under agreed standard conditions) is given an $E^{\ominus}$ value of zero.

The standard hydrogen electrode (SHE)

Experimentally, the $H^+(aq)/H_2(g)$ couple is arranged in the form of a hydrogen **electrode**. A hydrogen electrode is made up of a piece of platinum foil, coated with fine particles of platinum, dipped into a solution of hydrogen ions, with hydrogen gas bubbling over the surface of the platinum (Figure 7.3).

The platinum provides a surface upon which either of the following reactions may occur:

$$2H^+(aq) + 2e^- \rightarrow H_2(g)$$

or

$$H_2(g) \rightarrow 2H^+(aq) + 2e^-$$

These two possibilities are summarized by the equation

$$2H^+(aq) + 2e^- \rightleftharpoons H_2(g)$$

or it could also be written as

$$2H_3O^+(aq) + 2e^- \rightleftharpoons H_2(g) + 2H_2O(l)$$

The standard hydrogen electrode (SHE; Figure 7.3) is a hydrogen electrode where $H^+(aq)$ and $H_2(g)$ are in their **standard states** (see page 218). This means that the concentration of $H^+(aq)$ is 1 mol dm^{-3}, and the hydrogen gas is at a pressure of 1 bar. By agreement, the potential difference between the $H^+(aq)$ solution and the Pt electrode in the SHE is set to zero at all temperatures. This is written as:

$$E^\ominus(H^+(aq),H_2(g)) = 0$$

where the use of the symbol $\ominus$ shows that the substances are in their standard states.

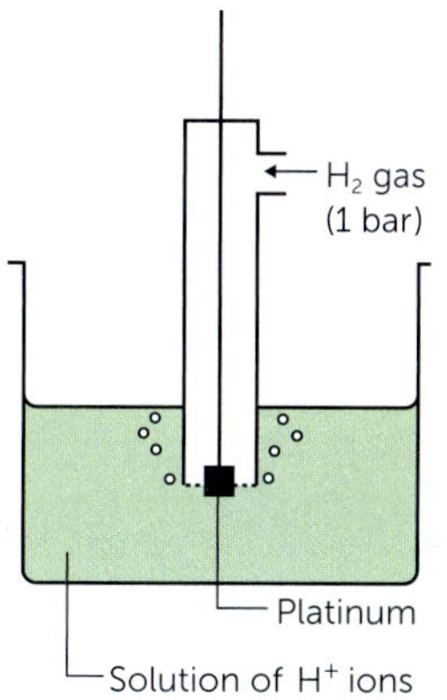

Figure 7.3 The standard hydrogen electrode.

Other standard electrodes

The electrode of the $Ag^+(aq)/Ag(s)$ redox couple consists of a rod of silver dipped in a solution of silver ions. The $Ag^+(aq)/Ag(s)$ electrode becomes the **standard** $Ag^+(aq)/Ag(s)$ electrode if the silver is pure, and the silver ion concentration is 1 mol dm^{-3}.

The electrode of the $Fe^{3+}(aq)/Fe^{2+}(aq)$ redox couple has a different construction. It consists of a piece of platinum (Pt) dipped in a mixture of both ions. In an electrochemical cell, the Pt connects the electrode to the rest of the cell. The Pt also provides a surface upon which electrode reactions may occur. The **standard** $Fe^{3+}(aq)/Fe^{2+}(aq)$ electrode contains both ions at a concentration of 1 mol dm^{-3}.

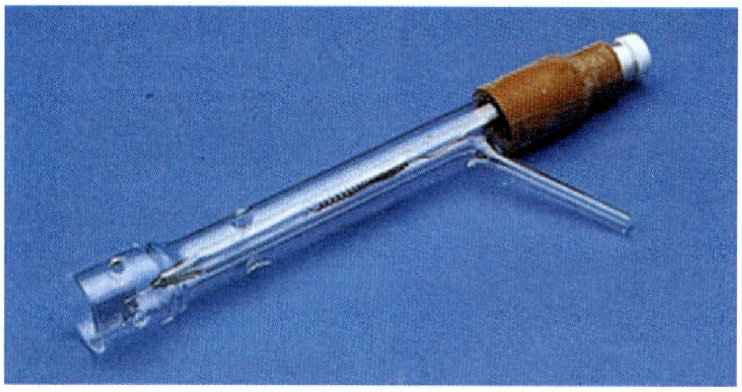

A standard hydrogen electrode.

Measuring $E^\ominus$ values by making an electrochemical cell

Measuring $E^\ominus(Ag^+(aq)/Ag(s))$

We will now see how the standard electrode potential of a redox couple may be found by experiment. Figure 7.4 shows the electrochemical cell composed of standard hydrogen and standard silver ion/silver electrodes. All measurements are made at 25°C. The two electrodes are connected by:

1. a wire through which electrons can pass;
2. a **salt bridge** (a glass tube that contains a saturated solution of an ionic salt, such as KNO_3) which completes the electrical circuit.

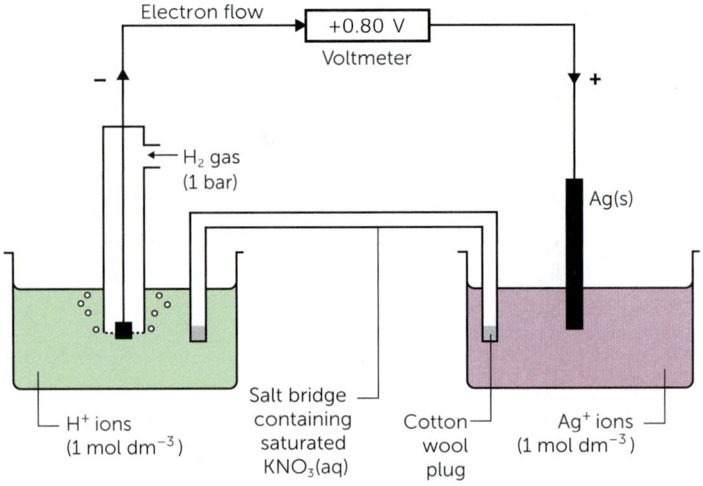

Figure 7.4 Electrochemical cell used in the determination of $E^\ominus$ (Ag^+(aq)/Ag(s)).

If we measure the voltage of the cell (the **cell potential**) using a high-resistance voltmeter, we will not be using up any electrical current produced by the cell, and very little chemical reaction takes place. Under these conditions, the measured voltage does not fall during the measurement.

Which electrode is connected to the positive terminal of the voltmeter? By trial and error, it is found that only if the Ag^+(aq)/Ag(s) electrode is connected to the positive terminal is the cell potential **positive.** (A voltage of +0.80 V is actually obtained; if the connections were reversed, the voltage reading would be – 0.80 V.) Because of this, the Ag^+(aq)/Ag(s) electrode is said to be the cell **cathode** (+) and the hydrogen electrode the cell **anode** (–).

Electrons flow from the anode to the cathode. This means that the hydrogen electrode is losing electrons (oxidation); this is only possible if the H^+(aq)/H_2(g) redox couple is undergoing the reaction:

$H_2(g) \rightarrow 2H^+(aq) + 2e^-$

The electrons then pass through the circuit and move to the cathode where they cause the Ag^+(aq)/Ag(s) redox couple to be reduced within the Ag^+(aq)/Ag(s) electrode:

$Ag^+(aq) + e^- \rightarrow Ag(s)$

The overall cell reaction is:

$H_2(g) + 2Ag^+(aq) \rightarrow 2H^+(aq) + 2Ag(s)$

(The same reaction occurs outside a cell if hydrogen gas is bubbled into a solution containing silver ions. However, in that case heat – not electricity – is made.)

The **cell diagram** for any electrochemical cell follows the pattern:

reactant → product	‖	reactant → product
(–) **anode**		(+) **cathode**
(in which oxidation takes place)		(in which reduction takes place)

where || symbolizes the salt bridge. In our case:

Pt | H_2(g) | H^+(aq) || Ag^+(aq) | Ag(s)
goes to goes to

which shows that H_2(g) is oxidized to H^+(aq) (the product at the anode), while Ag^+(aq) is reduced to Ag(s) (the product at the cathode).

The standard potential of the cell, $E^\ominus = +0.80$ V, is related to the standard electrode potentials for each redox couple by the following equation:

$$E^\ominus \text{(cell)} = E^\ominus_R - E^\ominus_L$$

which applies to all cells, and in which $E^\ominus_R$ is the standard electrode potential of the right-hand electrode (cathode) and $E^\ominus_L$ the standard electrode potential of the left-hand electrode (anode) as they appear in the cell diagram. Here, $E^\ominus_R = E^\ominus(Ag^+(aq)/Ag(s))$ and $E^\ominus_L = E^\ominus(H^+(aq),H_2(g)) = 0$ (by definition). Thus:

$$+0.80 \text{ V} = E^\ominus(Ag^+(aq)/Ag(s)) - 0$$

And $E^\circ(Ag^+(aq)/Ag(s)) = +\ 0.80$ V.

Measuring $E^\ominus(Zn^{2+}(aq)/Zn(s))$

Suppose we wanted to measure $E^\ominus(Zn^{2+}(aq)/Zn(s))$. Our first step is to set up the electrochemical cell, comprising the SHE and standard $Zn^{2+}(aq)/Zn(s)$ electrode (consisting of pure Zn metal dipped into 1 mol dm^{-3} $Zn^{2+}(aq)$) connected by a salt bridge.

If the standard $Zn^{2+}(aq)/Zn(s)$ electrode is connected to the negative pole of the voltmeter, the voltage reading is found to be +0.76 V. The positive voltage shows that the $Zn^{2+}(aq)/Zn(s)$ electrode is the anode. At the anode, the $Zn^{2+}(aq)/Zn(s)$ redox couple undergoes oxidation:

$$Zn(s) \rightarrow Zn^{2+}(aq) + 2e^-$$

The electrons cause the $H^+(aq)/H_2(g)$ redox couple to be reduced within the cathode of the SHE:

$$2H^+(aq) + 2e^- \rightarrow H_2(g)$$

The overall cell reaction is

$$Zn(s) + 2H^+(aq) \rightarrow Zn^{2+}(aq) + H_2(g)$$

The cell diagram is

$$Zn(s)\ |\ Zn^{2+}(aq)\ ||\ Pt\ |\ H^+(aq)\ |\ H_2(g)$$

The cell potential is +0.76 V:

$$E^\ominus\text{(cell)} = E^\ominus_R - E^\ominus_L$$
$$0.76 = E^\ominus(H^+(aq),H_2(g)) - E^\ominus(Zn^{2+}(aq)/Zn(s))$$
$$0.76 = 0 - E^\ominus(Zn^{2+}(aq)/Zn(s))$$

or,

$$E^\ominus(Zn^{2+}(aq)/Zn(s)) = -\ 0.76 \text{ V}$$

We have now looked at two electrochemical cells. In the first, where the $E^\ominus$ of the redox couple under investigation proved to be positive, the $H^+(aq)/H_2(g)$ couple undergoes oxidation (Figure 7.5(a)). In the second, where the $E^\ominus$ of the redox couple under investigation proved to be negative, the $H^+(aq)/H_2(g)$ couple undergoes reduction (Figure 7.5(b)). Generalizing:

1 **A negative $E^\ominus$ means that a redox couple is a stronger reducing agent than the $H^+(aq)/H_2(g)$ couple.**
2 **A positive $E^\ominus$ means that a redox couple is a weaker reducing agent than the $H^+(aq)/H_2(g)$ couple.**

The reverse statements apply to the oxidizing power of these redox couples.

EXERCISE 7H

Electrochemical cells and $E^\ominus$

A cell, consisting of the SHE (as anode) and the standard Cu^{2+}(aq)/Cu(s) electrode, gives a standard cell potential of +0.34V at 25°C.

(i) Write down the cell diagram.

(ii) Calculate $E^\ominus(Cu^{2+}(aq)/Cu(s))$.

(iii) Write down the overall reaction that occurs in the cell.

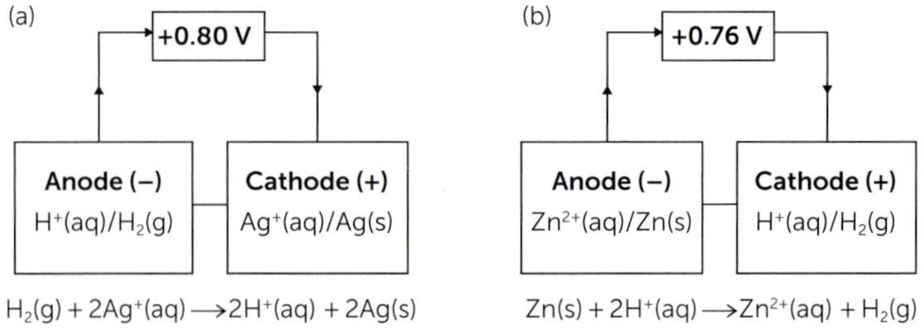

Figure 7.5 The flow of electrons in cells involving the SHE in which (a) Zn^{2+}(aq)/Zn(s) and (b) Ag^+(aq)/Ag(s) form the other electrode.

Table 7.1 Standard electrode potentials

	Redox couple	Reaction equation when redox couple is reduced	$E^\ominus$ at 298 K/V
Strong Oxidizing Agents	$F_2(g)/F^-(aq)$	$F_2(g) + 2e^- \rightarrow 2F^-(aq)$	+2.87
	$Cl_2(g)/Cl^-(aq)$	$Cl_2(g) + 2e^- \rightarrow 2Cl^-(aq)$	+1.36
	$Br_2(l)/Br^-(aq)$	$Br_2(l) + 2e^- \rightarrow 2Br^-(aq)$	+1.09
	$Ag^+(aq)/Ag(s)$	$Ag^+(aq) + e^- \rightarrow Ag(s)$	+0.80
	$Fe^{3+}(aq)/Fe^{2+}(aq)$	$Fe^{3+}(aq) + e^- \rightarrow Fe^{2+}(aq)$	+0.77
	$I_2(s)/I^-(aq)$	$I_2(s) + 2e^- \rightarrow 2I^-(aq)$	+0.54
	$Cu^{2+}(aq)/Cu(s)$	$Cu^{2+}(aq) + 2e^- \rightarrow Cu(s)$	+0.34
	$H^+(aq)/H_2(g)$	$2H^+(aq) + 2e^- \rightarrow H_2(g)$	0 (by definition)
	$Fe^{2+}(aq)/Fe(s)$	$Fe^{2+}(aq) + 2e^- \rightarrow Fe(s)$	−0.44
	$Cr^{3+}(aq)/Cr(s)$	$Cr^{3+}(aq) + 3e^- \rightarrow Cr(s)$	−0.74
Strong Reducing Agents	$Zn^{2+}(aq)/Zn(s)$	$Zn^{2+}(aq) + 2e^- \rightarrow Zn(s)$	−0.76
	$Mg^{2+}(aq)/Mg(s)$	$Mg^{2+}(aq) + 2e^- \rightarrow Mg(s)$	−2.37
	$Na^+(aq)/Na(s)$	$Na^+(aq) + e^- \rightarrow Na(s)$	−2.71

Key points about the standard electrode potentials of redox couples

1 $E^\ominus$ values are also known as **standard reduction potentials**. Table 7.1 shows selected $E^\ominus$ values that were measured at 25°C. Note that the redox couples are listed with the oxidized species *first* (e.g. Na^+(aq)/Na(s), not Na(s)/Na^+(aq)).

2 Measurements of $E^\ominus$ may be carried out at other temperatures. $E^\ominus$ values vary with temperature, apart from $E^\ominus(H^+(aq)/H_2)$, which by agreement is set to zero at all temperatures.

3 Some $E^\ominus$ values, such as $E^\ominus(Na^+(aq)/Na(s))$, cannot be found using simple electrochemical cells, and are calculated from other data.

4 Strictly speaking, $E^\ominus$ values apply to couples, and *not* to single species. For example, we cannot speak of the standard electrode potential of sodium, $E^\ominus$(Na). The fact that $E^\ominus(Na^+(aq)/Na(s)) = -2.71$V, shows specifically that sodium has a strong tendency to form Na^+ ions in solution (and not to form, say, Na^-).

5 $E^\ominus$ values may be used to predict whether redox reactions, starting from pure reactants in solution, are 'allowed' to occur. ('Allowed to occur' means that the reaction has a large value for the equilibrium constant K.) See p. 115

6 The greater the difference in $E^\ominus$ values between the couples making up the between the two half-cells occurring in an electrochemical cell, the greater the equilibrium constant K and greater the overall voltage of the cell and the greater the energy transferred per mole of electrons.

BOX **7.3**

Dry cells

The trouble with 'wet' cells is that the solutions (the *electrolyte*) might leak out. In the dry cell, the electrolyte is a damp paste. This type of cell is used in torches, calculators and clocks. Many dry cells use the reactions of zinc and manganese(IV) oxide:

$Zn(s) \rightarrow Zn^{2+}(aq) + 2e^-$ (oxidation)

$MnO_2(s) + H_2O(l) + e^- \rightarrow MnO(OH)(s) + OH^-(aq)$ (reduction)

Other reactions occur, but we will not discuss them here. A diagram of a dry cell is shown in Figure 7.6.

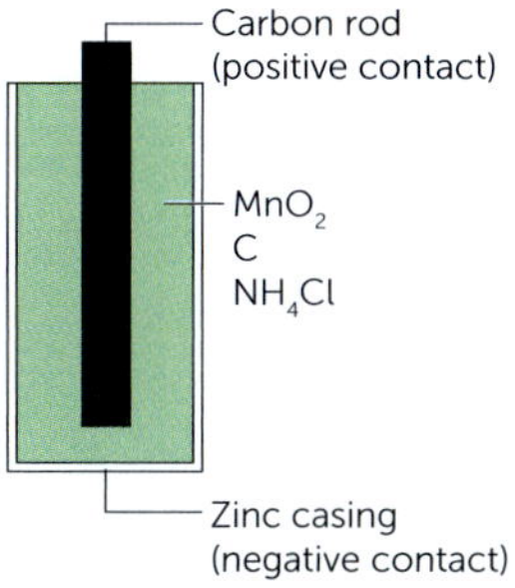

Figure 7.6 A dry cell.

Predicting whether ionic redox reactions are allowed to happen

Will a possible redox reaction between two ions be allowed to happen? To find out, we look at the standard electrode potentials ($E^\ominus$ values) of the two half-reactions that make up that redox reaction. The key point to remember here is that the half-reaction having the more positive electrode potential will always undergo reduction, while the half-reaction having the more negative electrode potential undergoes oxidation.

To help you apply this easily, follow the following procedure:

1 Write down both half-reactions, with the half-reaction possessing the *more negative* (or *least positive*) $E^\ominus$ value at the top.

2 Then draw *anticlockwise arrows* to predict whether the overall reaction can occur.

EXAMPLE **7.3**

Will hydrogen gas reduce $Fe^{3+}(aq)$ to $Fe^{2+}(aq)$?

Answer

Write down both half-reactions, with the half-reaction possessing the more negative (or least positive) potential at the top:

$2H^+(aq) + 2e^- \rightarrow H_2(g)$ $E^\ominus = 0.00V$

$Fe^{3+}(aq) + e^- \rightarrow Fe^{2+}(aq)$ $E^\ominus = +0.77V$

Draw anticlockwise arrows:

$2H^+(aq) + 2e^- \rightarrow H_2(g)$ $E^\ominus = 0.00V$

$Fe^{3+}(aq) + e^- \rightarrow Fe^{2+}(aq)$ $E^\ominus = +0.77V$

The direction in which the arrows point shows the reaction that is allowed is the one in which H_2 changes to H^+ and Fe^{3+} changes to Fe^{2+}. The predicted overall reaction is therefore:

$H_2(g) + 2Fe^{3+}(aq) \rightarrow 2Fe^{2+}(aq) + 2H^+(aq)$

EXAMPLE **7.3** *(continued)*

Comments

The conclusion is that if we start from the pure reactants in solution, H_2(g) and Fe^{3+}(aq), the reaction can occur. In other words, hydrogen gas is capable of reducing Fe^{3+}(aq) to Fe^{2+}(aq) at room temperature. (Another way of expressing this is that the reaction is 'thermodynamically allowed' or that the redox reaction equilibrium constant K is large. See Unit 16.)

Note, however, that *we cannot predict how fast (or slow!) the reaction will be*. Although the reactions of ions are often fast, the reaction may be so slow that nothing will appear to happen.

In addition, our predictions only apply to solutions (since $E^\ominus$ values cannot be obtained for dry gases or solids) and at room temperature.

EXAMPLE **7.4**

Will bromine water (Br_2(aq)) react with an aqueous solution of potassium chloride (Cl^-(aq))?

Answer

Following the same procedure as above:

$$Br_2(aq) + 2e^- \rightarrow 2Br^-(aq) \quad E^\ominus = +1.09V$$
$$Cl_2(aq) + 2e^- \rightarrow 2Cl^-(aq) \quad E^\ominus = +1.36V$$

In this case, the direction of the arrows shows that Br_2(aq) and Cl^-(aq) *stay as they are*. The conclusion is that *there is no reaction* at room temperature.

EXERCISE **7I**

Predicting reactions

Using the standard reduction potentials in Table 7.1, determine whether the following reactions may occur:

(i) $Br_2(aq) + I^-(aq) \rightarrow I_2(aq) + 2Br^-(aq)$

(ii) $Cu(s) + 2H^+(aq) \rightarrow Cu^{2+}(aq) + H_2(g)$

(iii) $Zn(s) + 2Fe^{3+}(aq) \rightarrow Zn^{2+}(aq) + 2Fe^{2+}(aq)$

(iv) $2Fe^{3+}(aq) + 2I^-(aq) \rightarrow 2Fe^{2+}(aq) + I_2(aq)$.

7.6 Activity series of metals

The more negative the standard electrode potential, $E^\ominus$, of a redox couple M^{n+}(aq)/M(s), the more powerful a reducing agent is that metal when changing to its ion. This means it is a more reactive metal because it loses electrons more easily. We can arrange the metals in order of reducing power, producing an **activity series** of metals (Table 7.2). The metals at the top of the series are more reactive than those below.

A metal in solution can react with a metal ion that appears below it in the table. Hydrogen is included so that the reactivity of metals with aqueous acids may be worked out; when a metal above hydrogen is added to an acidic solution, a reaction will occur and hydrogen gas is given off.

Never mix strongly reducing metals, such as potassium, with acid solutions – the reactions are extremely violent.

Table 7.2 The activity series of metals

Reduction half-reaction	$E^\ominus$ (M^{n+}(aq)/M(s))/V (at 298 K)
$K^+(aq) + e^- \rightarrow K(s)$	−2.92
$Ca^{2+}(aq) + 2e^- \rightarrow Ca(s)$	−2.87
$Na^+(aq) + e^- \rightarrow Na(s)$	−2.71
$Mg^{2+}(aq) + 2e^- \rightarrow Mg(s)$	−2.37
$Al^{3+}(aq) + 3e^- \rightarrow Al(s)$	−1.67
$Zn^{2+}(aq) + 2e^- \rightarrow Zn(s)$	−0.76
$Fe^{2+}(aq) + 2e^- \rightarrow Fe(s)$	−0.44
$Ni^{2+}(aq) + 2e^- \rightarrow Ni(s)$	−0.25
$Sn^{2+}(aq) + 2e^- \rightarrow Sn(s)$	−0.14
$Pb^{2+}(aq) + 2e^- \rightarrow Pb(s)$	−0.13
$2H^+(aq) + 2e^- \rightarrow H_2(g)$	0
$Cu^{2+}(aq) + 2e^- \rightarrow Cu(s)$	+0.34
$Ag^+(aq) + e^- \rightarrow Ag(s)$	+0.80
$Au^+(aq) + e^- \rightarrow Au(s)$	+1.68

EXERCISE **7J**

Activity series of metals

Write redox equations for any reactions that might occur when the following substances are mixed:

(i) magnesium and silver nitrate

(ii) copper and lead(II) nitrate solution

(iii) nickel and copper(II) sulfate solution

(iv) zinc and dilute hydrochloric acid (H^+(aq))

(v) copper and dilute sulfuric acid (H^+(aq)).

7.7 Corrosion of iron

Rusting

Many metals, including iron, react with air and/or water (**corrode**). The reactivity of metals with air or water can be predicted using reduction potentials. For a neutral solution:

$$2H_2O(l) + 2e^- \rightarrow H_2(g) + 2OH^-(aq) \qquad E^\ominus = -\ 0.42V$$

In the case of iron:

$$Fe^{2+}(aq) + 2e^- \rightarrow Fe(s) \qquad E^\ominus = -\ 0.44V$$
$$Fe^{3+}(aq) + e^- \rightarrow Fe^{2+}(aq) \qquad E^\ominus = +0.77V$$

Therefore pure water has only a slight tendency to oxidize Fe(s) to Fe^{2+}(aq).

However, if oxygen is present, the following half-reaction can occur in neutral solution (pH = 7):

$$O_2(g) + 4H^+(aq) + 4e^- \rightarrow 2H_2O(l) \qquad E^\ominus = +0.82V$$

Check, by using anticlockwise arrows, that this couple can oxidize Fe to Fe^{2+} and then from Fe^{2+} to Fe^{3+}.

Both oxygen and water can therefore oxidize Fe to Fe^{2+}, and oxygen is further capable of oxidizing Fe^{2+} to Fe^{3+}. These reactions are known as **rusting**.

Rust is a brown, insoluble compound of formula $Fe_2O_3{\cdot}xH_2O$ (the x in the formula indicates that it has a variable composition) that is formed when iron reacts with air and water. Rusting requires the presence of oxygen, water and ionic substances dissolved in the water (**electrolytes**). In the absence of any one of these, little rusting will occur.

EXERCISE **7K**

Rusting

(i) Explain why iron nails do not rust if they are placed in a sealed tube containing boiled water.

(ii) Why do cars that are kept by the sea rust more rapidly?

(iii) Rusting occurs much more quickly in areas that are polluted. Can you give a reason for this?

EXERCISE **7L**

Tinning

Iron (steel) cans are sometimes coated with tin.

(i) Explain why the tin does *not* act as a sacrificial anode here. You are given the following $E_{\ominus}$ (298K) values:

$Sn_{2+}(aq) + 2e- \rightarrow Sn(s)$ $E^{\ominus} = -0.14V$

$Fe_{2+}(aq) + 2e- \rightarrow Fe(s)$ $E^{\ominus} = -0.44V$

(ii) Scratching the tin layer is observed to cause greater corrosion of the exposed iron than would be the case if the iron had not originally been coated with tin. (In other words, once scratched, the presence of tin accelerates the corrosion of iron.) Explain this observation.

7.8 Redox reactions in nature

Nitrogen fixation

Living organisms need nitrogen to make proteins, the 'building bricks' of plants and animals. Although nitrogen gas is abundant, making up approximately four-fifths of the atmosphere, most living organisms cannot obtain it directly from the air because it is an unreactive gas. Nitrogen becomes available to plants and animals via the **nitrogen cycle** (Figure 7.7).

Atmospheric nitrogen is particularly unreactive; however, it is oxidized to NO when lightning flashes. Oxygen in the atmosphere oxidizes NO to NO_2, which reacts with rain water to form the acids HNO_2 and HNO_3. These acids react with metal oxides and carbonates in the soil, to form nitrate and nitrite salts.

Plants obtain nitrogen in the form of nitrate and ammonium ions. Nitrates are very soluble in water and reach the roots of plants easily; the nitrate ion is then taken up

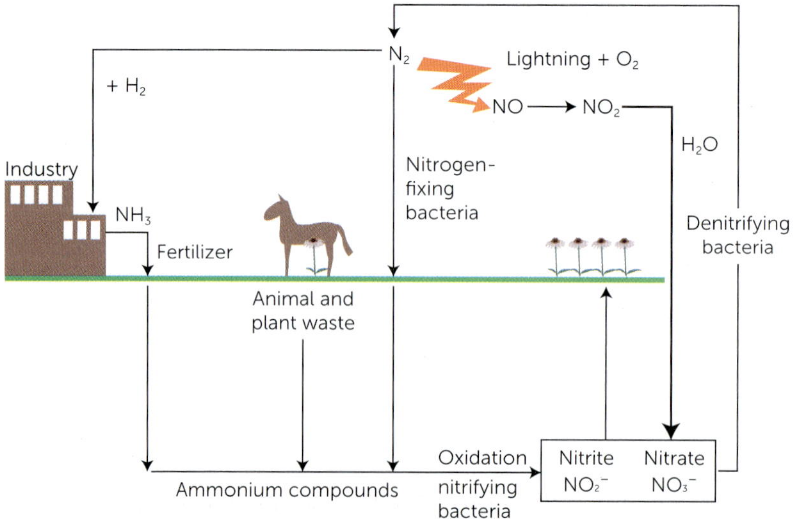

Figure 7.7 The nitrogen cycle.

and reduced to ammonia by the plant. Nitrogen-fixing bacteria that live in the soil, or in **nodules** on the roots of plants, convert (i.e. **biologically fix**) atmospheric nitrogen into ammonium salts. Animals eat plants and, when both of these organisms die, their organic matter eventually decays into ammonium compounds. Nitrifying and denitrifying bacteria convert ammonium compounds into NO_3^- and NO_2^-, then N_2O and N_2. In this way nitrogen is returned to the atmosphere and the cycle is complete.

The availability of nitrogen to plants (and therefore to animals) is vital for producing food. This is why *fertilizers*, ammonium and nitrate salts, are produced industrially to help supply nitrogen to crops.

The **Haber–Bosch process** is an industrial process whereby nitrogen gas is reduced to ammonia using an iron catalyst, and is the first step in the production of fertilizers:

$$3H_2(g) + N_2(g) \rightleftharpoons 2NH_3(g)$$

This is a very expensive process, requiring high temperatures and pressures, and the expense is one of the reasons why developing countries find difficulty in feeding their populations. (For more about the process, see Unit 15.) Biological nitrogen fixation, however, requires only solar energy as its power source and chemists are currently trying to find catalysts that can act in the same way as nitrogen-fixing bacteria.

Electron transport in living systems

Both **photosynthesis** and **respiration** are energy conversion processes and involve redox reactions.

Photosynthesis is a process that occurs when plants take in carbon dioxide and water to make sugar. Photosynthesis needs solar energy and is catalysed by **chlorophyll**, the green substance in leaves; oxygen is also produced. The overall reaction is:

$$6CO_2(g) + 6H_2O(l) \rightarrow C_6H_{12}O_6(s) + 6O_2(g)$$

For more about photosynthesis, see page 414.

Respiration is a reaction that supplies living organisms with the energy they need. It is the opposite reaction to photosynthesis. The overall reaction is:

$$C_6H_{12}O_6(s) + 6O_2(g) \rightarrow 6CO_2(g) + 6H_2O(l)$$

Note that energy is stored in glucose during photosynthesis and released again during respiration.

Respiration involves a series of redox reactions, some of which involve the participation of iron-containing substances called **cytochromes**. Cytochromes are electron carriers. They accept electrons from better reducing agents and give electrons to better oxidizing agents. The iron 'flips' back and forth between Fe^{3+} (the oxidized form) and Fe^{2+} (the reduced form). A similar set of cytochromes are used for the transport of electrons when green plants undergo photosynthesis.

> EXERCISE **7M**
>
> **Nitrogen cycle**
>
> Work out the oxidation numbers of all the nitrogen-containing species in the previous section. Construct an 'oxidation state' ladder for nitrogen, using these species.

7.9 Electrolysis

Electrolysis is the decomposition of a substance in a conducting medium using electricity. Water, for example, shows no tendency to decompose except at extremely high temperatures, yet it can be forced to decompose at room temperature using the external energy of an electrical current.

Electrolysis of water

The electrolysis of liquids is often carried out using a Hofmann voltameter. It consists of two outer tubes each connected electrically to the poles of a direct current (DC)

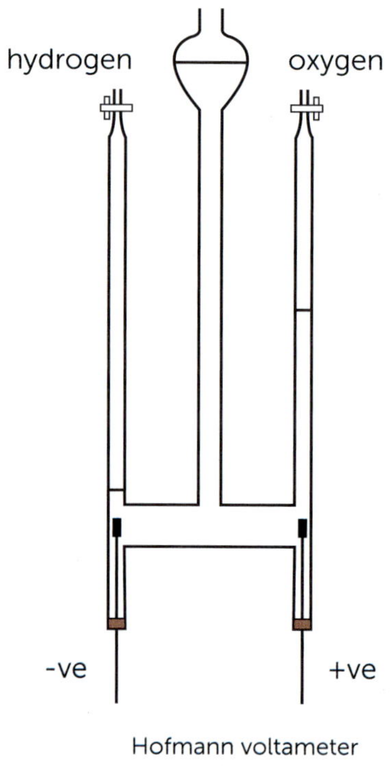

Figure 7.8 Hofmann voltameter containing acidified water. Hydrogen gas collects above the cathode (-) and oxygen gas above the anode (+).

generator or battery. The decomposition of water occurs at the surface of two charged electrodes made of a small piece of platinum foil (coloured black in Figure 7.8). Since water is a poor conductor of electricity, some dilute sulfuric acid is added to provide ions that carry current. The reduction of the hydrogen ions from water occurs at the (negatively charged) cathode, producing hydrogen gas:

$2H^+(aq) + 2e^- \rightarrow H_2(g)$ (electron gain)

The oxidation of water occurs at the (positively charged) anode, producing oxygen gas:

$2H_2O(l) - 4e^- \rightarrow 4H^+(aq) + O_2(aq)$ (electron loss)

The volume of hydrogen to oxygen gas is observed to be in the ratio of 2:1 (Figure 7.8). Knowing that the formula of water, hydrogen and oxygen is H_2O, H_2 and O_2, the electrolysis of water over both electrodes may be summarized as:

$2H_2O(l) \rightarrow 2H_2(g) + O_2(g)$

Hydrogen and oxygen may be tested for by attaching rubber tubing to the taps at the top of the Hofmann voltameter, opening the taps and collecting the gases in an inverted test tube. Hydrogen gas 'pops' in a flame. If a *glowing* splint is put into the tube of oxygen, it re-ignites the splint.

The formula of water

What is the proof that 'water is H_2O?' This is not nearly as straightforward as it might at first appear! The electrolysis of water is sometimes thought to be the proof needed. According to Avogadro's Law (see Unit 10), volumes of gases at the same temperature and pressure may be equated with numbers of molecules. Therefore, the fact that electrolysis produces hydrogen and oxygen gases in the volume ratio 2:1 means that twice as many hydrogen molecules are produced from electrolysis as oxygen molecules. Does this prove that the formula of the water molecule is H_2O? The answer is 'no'. The formula of water is indeed H_2O, but we cannot jump to this conclusion without more information.

To explain further, imagine that the formula of water was HO. The gas volume ratio would still be 2:1 if the formula of hydrogen was H (and not H_2):

$2HO(l) \rightarrow 2H(g) + O_2(g)$

The 2:1 ratio would also be obtained with hydrogen and oxygen as H_2 and O_2 respectively but with the formula of water as H_4O_2:

$H_4O_2(l) \rightarrow 2H_2(g) + O_2(g)$

A summary of the position is as follows. Only if we take for granted that both hydrogen and oxygen molecules are diatomic does the two to one volume observed from the electrolysis of water confirm the formula of water is in the simple ratio H_2O. Even then, its exact molecular formula might be a multiple of the 2:1 ratio e.g. H_4O_2. Historically of course, the molecular formula of hydrogen and oxygen gases (H_2 and O_2) were unknown as was the molecular formula of water itself. Electrolysis alone does not prove that the formula of water is H_2O.The modern proof of the formula of water is to be found in the mass spectrum of water (see p. 35 Unit 3).

For more about the historical determination of the formula of water, see Extension 6 of the website. For more about electrolysis and its uses, see Extension 7 on the website at https://www.bloomsburyonlineresources.com/chemistry.

REVISION QUESTIONS

7.1 What are the oxidation numbers of the named element in the following compounds?

(i) C in CF_4
(ii) Mn in MnO_2
(iii) S in SO_4^{2-}
(iv) Sn in $SnCl_4$
(v) N in Mg_3N_2.

7.2 Use standard electrode potentials to predict whether copper metal could reduce aqueous iron(III) ions to iron(II) at room temperature.

7.3 Using standard electrode potentials, predict whether reactions could occur in the following situations. Write balanced redox equations for those reactions that could occur at room temperature.

(i) An iron nail is placed in an aqueous solution of copper(II) sulfate ($CuSO_4$(aq)).
(ii) A silver ring is placed in an aqueous solution of zinc nitrate ($Zn(NO_3)_2$(aq)).

7.4 In the equation

$$MnO_4^-(aq) + 8H_3O^+(aq) + 5Au(s) \rightarrow Mn^{2+}(aq) + 12H_2O(l) + 5Au^+(aq)$$

(i) Which species is oxidized and what is its change in oxidation number?
(ii) Which species is reduced?
(iii) Which species is acting as an oxidizing agent?

7.5 When an aqueous solution of iron(III) ions is reacted with sulfur dioxide gas (this forms an acid solution), aqueous iron(II) ions and the sulfate ion (SO_4^{2-}) are produced. Write a balanced overall redox equation for the reaction.

7.6 In the following cell, all solutions are of concentration 1 mol dm^{-3} and at 25°C:

(i) Mark the direction of electron flow.
(ii) Give equations for the processes occurring at each of the electrodes.
(iii) Calculate $E^\ominus$ (cell).
(iv) Draw a cell diagram for the system.

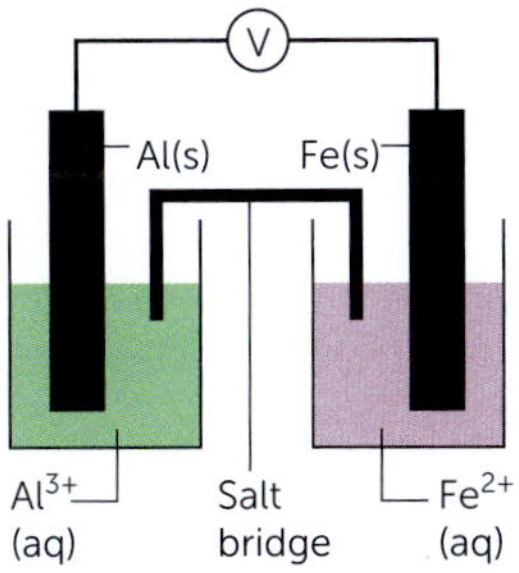

7.7 Tarnishing of silver jewellery occurs because the silver metal reacts with hydrogen sulfide in the presence of air to form silver sulfide (Ag_2S). The silver object can be 'cleaned' by dipping it into an aqueous solution of sodium chloride, contained in an aluminium metal container.

(i) Write down symbols for the ions contained in silver sulfide.
(ii) Using the standard electrode potentials for silver and aluminium, work out whether a reaction occurs between the species present in the container.
(iii) Why is this a more desirable way of cleaning the silver than using silver polish?

7.8 Write a balanced redox equation for the oxidation of concentrated hydrochloric acid (H^+, Cl^-) to chlorine (Cl_2) by solid manganese(IV) oxide (MnO_2). The manganese(IV) oxide is reduced to Mn^{2+}(aq).

7.9 In basic solution, zinc metal is oxidized by the nitrate ion (NO_3^-(aq)) to $[Zn(OH)_4]^{2-}$(aq). The nitrate ion is reduced to ammonia. Write a balanced ionic equation for this reaction.

7.10 In the reaction

$$Ni^{2+}(aq) + Zn(s) \rightarrow Ni(s) + Zn^{2+}(aq)$$

which species is (i) the oxidizing agent, (ii) the reducing agent?

7.11
(i) How would you confirm that hydrogen and oxygen are produced in the electrolysis of water?
(ii) A teacher reports that he has obtained equal volumes of gases in the electrolysis of water. What might have happened to cause this?

7.12 Lithium metal is used in portable batteries. Sodium is not. Why? You have the following data:

Standard Electrode potentials at 298 K
$E^\ominus$ (Na^+(aq)/Na(s)) = −2.71 V
$E^\ominus$ (Li^+(aq)/Li(s)) = −3.04 V

Densities of the metal at 298 K
Na 0.534 g cm^{-3} Li 0.971 g cm^{-3}

More information about the extraction of metals is found in Extension 7 at https://www.bloomsburyonlineresources.com/chemistry.

8 The Mole

CONTENTS

OBJECTIVES

This unit:

- Explains why chemists work in moles
- Describes how to perform chemical calculations, involving masses and gas volumes using moles
- Defines 'percentage yield'
- Explains how to work out which substance is the limiting reagent in a chemical reaction

8.1 Molecular mass

The idea of **molecular mass** was introduced in Unit 3. This is the mass of one molecule of substance on the atomic mass scale. Molecular masses are calculated using the atomic masses of the constituent atoms. A list of approximate atomic masses for atoms of elements is shown in Table 8.1. You should use these for calculations, unless you are instructed otherwise.

Examples of calculations of molecular mass

$$m(H_2O) = 1 + 1 + 16 = 18\,u$$

$$m(C_6H_5Cl) = (6 \times 12) + (5 \times 1) + (35.5) = 112.5\,u$$

$$m(H_2SO_4) = (2 \times 1) + (32) + (4 \times 16) = 98\,u$$

Substances such as sodium chloride (Na^+, Cl^-) and copper(II) nitrate (Cu^{2+}, $2NO_3^-$) consist of ions (not molecules) and the term 'molecular mass' is not strictly appropriate. Nevertheless the 'formula mass' of these substances is calculated in a similar way to neutral molecules. Examples are

$$m(Na^+, Cl^-) = 23 + 35.5 = 58.5\,u$$

$$m(Cu^{2+}, 2NO_3^-) = 63.5 + 2(14 + 16 + 16 + 16) = 187.5\,u$$

Table 8.1 Approximate atomic masses of selected elements

Element	Symbol	Approximate atomic mass/u
Hydrogen	H	1
Helium	He	4
Carbon	C	12
Nitrogen	N	14
Oxygen	O	16
Fluorine	F	19
Neon	Ne	20
Sodium	Na	23
Magnesium	Mg	24
Aluminium	Al	27
Phosphorus	P	31
Sulfur	S	32
Chlorine	Cl	35.5
Potassium	K	39
Calcium	Ca	40
Iron	Fe	56
Nickel	Ni	59
Copper	Cu	63.5
Zinc	Zn	65
Bromine	Br	80
Silver	Ag	108
Tin	Sn	119
Iodine	I	127
Barium	Ba	137
Gold	Au	197
Mercury	Hg	201
Lead	Pb	207
Uranium	U	238

EXERCISE **8A**

Calculating the molecular and formula mass

Write down the mass (u) of the following:

(i) nitric acid, HNO_3

(ii) magnesium sulfate, $MgSO_4$

(iii) ethyne, C_2H_2

(iv) sulfur molecules, S_8

(v) ethyl ethanoate, $CH_3COOC_2H_5$

(vi) hydrated iron(III) nitrate, $Fe(NO_3)_3 \cdot 9H_2O$

(vii) one atom of neon.

8.2 Moles

The mole

How do you identify a chemist? One simple way is to ask the question: 'What is a mole?' A keen gardener will answer 'A small furry animal that digs holes in the lawn'; a doctor or nurse will answer 'a dark spot on skin'; and the head of a company may answer 'a spy'. A chemist will **always** answer 'a standard number of particles' or, being more specific, 'just over six hundred thousand million, million, million particles'. When chemists calculate amounts of reacting substances they constantly work (and think!) in moles.

To explain why chemists think in moles, consider a reaction

$$A + B \rightarrow AB$$

You have already learned that the equation tells us that one particle of A reacts with one particle of B to form one particle of the compound AB. If a chemist wishes to get an exact amount of A to react with an exact amount of B, and not have excess A or B left over, then equal numbers of the particles of A and B must be reacted together.

The particles might be atoms, molecules or ions. Such particles are very small, so in order that the amounts involved are practical to work with, a great many particles of A must be added to the same number of particles of B.

Balances in the laboratory are used to weigh amounts of substance in grams and not atoms, molecules or ions, so it is *not* useful to weigh out, say, 1 g of A to react with 1 g of B – A and B are different substances so their particles *do not have the same mass.* Equal masses of A and B *will not contain the same number of particles.* Chemists work in moles, because *one mole of any substance contains the same number of particles.* Note that the *name* of the amount is **mole** and the symbol for the *unit* is **mol**.

Avogadro's constant

The number of particles (atoms, molecules or ions) in one mole of a substance is defined as being

the number of atoms contained in exactly 12 grams of carbon-12.

This number is extremely large and has been found by experiment to be

602,200,000,000,000,000,000,000 or 6.022×10^{23},
often approximated to 6×10^{23} mol^{-1}

This value is called **Avogadro's constant** and is symbolized N_A.

From Table 8.1,

$m(C) = 12$ u
$m(He) = 4$ u
$m(H_2O) = 18$ u
$m(H_2) = 2$ u

and we can reason as follows:

- One atom of carbon is three times as heavy as one atom of helium. Therefore, one thousand atoms of carbon are three times as heavy as one thousand atoms of helium. Therefore, if a sample of pure carbon has three times the mass of a sample of pure helium the two samples must have the same number of atoms, so that 12 g of carbon and 4 g of helium must each contain the same number of atoms – this number is numerically equal to N_A.

Similar reasoning produces the following conclusions:

- One molecule of water is 4.5 times as heavy as one atom of helium, so 18 g of water contains N_A *molecules* of water, whereas 4 g of helium contains N_A *atoms* of helium.
- One atom of helium is twice as heavy as one molecule of hydrogen, so 4 g of helium contains N_A *atoms* of helium and 2 g of hydrogen contains N_A *molecules of hydrogen.*

Generalizing, the mass of a substance in grams that contains N_A particles is numerically equal to the atomic or molecular mass of that substance in atomic mass

units (u). This mass is called the **molar mass** of the substance because it is the mass of one mole of that substance.

The symbol for molar mass is M. Its units are normally grams per mole (g mol^{-1}). For example,

$M(H_2O) = 18\ g\ mol^{-1}$

See Table 8.2 and Figure 8.1.

Table 8.2 The relationship between mass, molar mass and the number of moles for selected substances. (for simplicity, N_a has been taken as $6 \times 10^{23}\ mol^{-1}$)

Substance	Mass of substance/g	Formula of substance	Molar mass /g mol^{-1}	Number of moles of substance in the stated mass	Number of particles in the stated mass
Carbon	12	C	12	1	6×10^{23}
Carbon	120	C	12	10	6×10^{24}
Oxygen gas	32	O_2	32	1	6×10^{23}
Oxygen gas	3.2	O_2	32	0.1	6×10^{22}
Sodium chloride	58.5	NaCl	58.5	1	6×10^{23}*
Sodium chloride	0.0585	NaCl	58.5	0.001	6×10^{20}*
Benzene	78	C_6H_6	78	1	6×10^{23}
Benzene	7800	C_6H_6	78	100	6×10^{25}

* This is the number of Na^+,Cl^- ion pairs.

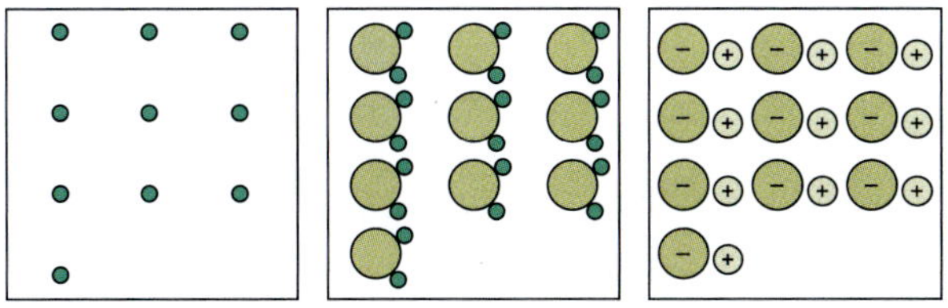

Figure 8.1 An equal number of moles of neon, water and sodium chloride.

EXERCISE **8B**

Moles, molar mass and Avogadro's constant

(i) Neon gas consists of single atoms. What mass of neon contains 6×10^{23} atoms?

(ii) How many potassium ions are there in 94 g of potassium oxide, $2K^+,O^{2-}$?

(iii) Magnesium metal consists of magnesium atoms. What mass of magnesium contains 6×10^{22} atoms?

(iv) What is the molar mass of copper(II) sulfate, $CuSO_4$?

Relationship between moles, molar mass and the mass of substance

To convert a mass of substance into moles, the following equation can be used:

$$\text{amount of substance in moles} = \frac{\text{mass}}{\text{molar mass}}$$

Notice that units of molar mass are grams per mol, so the units of mass must be grams.

The rearranged forms of this equation are also useful:

$$\text{molar mass} = \frac{\text{mass}}{\text{amount of substance in moles}}$$

and

mass = amount of substance in moles × molar mass

EXAMPLE **8.1**

How many moles of hydrated copper(II) sulfate ($CuSO_4 \cdot 5H_2O$) are contained in 5.00 g of the substance?

Answer

$m((CuSO_4 \cdot 5H_2O) = (63.5 + 32 + 4(16) + 5(18)) = 249.5\ u$

Therefore,

$M(CuSO_4 \cdot 5H_2O) = 249.5\ g\ mol^{-1}$.

$$\text{The amount of copper sulfate in moles} = \frac{\text{mass}}{\text{molar mass}}$$

$$= \frac{5.00}{249.5} = 0.0200\ \text{mol}$$

EXAMPLE **8.2**

0.059 mol of calcium carbonate ($CaCO_3$) is required in an experiment. What mass of $CaCO_3$ needs to be weighed out?

Answer

$M(CaCO_3) = 100\ g\ mol^{-1}$

Therefore, mass of calcium carbonate

= amount of calcium carbonate in moles × molar mass
= 0.059 × 100 = 5.9 g

EXAMPLE **8.3**

0.20 mol of a compound containing carbon and hydrogen has a mass of 3.2 g. What is the molecular mass of the compound? What is its likely formula?

Answer

$$\text{molar mass} = \frac{\text{mass}}{\text{amount of substance in moles}}$$

$$= \frac{3.2}{0.20}$$

$$= 16\ g\ mol^{-1}$$

Since M(compound) = 16 g mol^{-1}, m(compound) = 16 u.

Carbon has an atomic mass of 12. The molecule is likely to be methane (CH_4) with

$m(CH_4) = 12 + 1 + 1 + 1 + 1 = 16\,u$

Comment

Where there could be confusion, you should always be careful to specify *exactly* the substance you are referring to. For example, the statement '1 mol of hydrogen' could mean 1 mol of H atoms *or* 1 mol of H_2 molecules. These have different molar masses, 1 g mol^{-1} and 2 g mol^{-1}, respectively.

EXERCISE **8C**

More calculations involving moles, mass and molar mass

1 What is the mass of:

(i) 1 mol of N atoms

(ii) 4 mol of Fe atoms

(iii) 1.50 mol of Cl^- ions

(iv) 20 mol of Na_2SO_3?

2 How many moles of substance are present in:

(i) 40 g of calcium metal (Ca)

(ii) 123.2 g of CCl_4

(iii) 0.49 g of SO_4^{2-}

(iv) 14 g of N_2?

3 A pure solid consists of either (i) sodium chloride (NaCl), (ii) copper carbonate ($CuCO_3$), or (iii) sodium carbonate (Na_2CO_3). 1.06 g of the solid contains 0.01 mol of compound. What is the chemical identity of the solid?

4 Calculate:

(i) The number of molecules of sulfur (S_8) in 16 g of solid sulfur.

(ii) The number of aluminium ions in 0.056 g of aluminium oxide ($2Al^{3+}$, $3O^{2-}$) ($N_A = 6 \times 10^{23}\,mol^{-1}$).

Determining formulae by experiment

To work out a formula by experiment, it is necessary to find out how many moles of each element combine together to make the compound. Magnesium metal, in the form of a strip (*magnesium ribbon*), is heated in air until all the magnesium is combined with oxygen. The diagrams in Figure 8.3 show different steps in the experiment, in order.

Some sample results are:

Mass of crucible = 20.40 g
Mass of crucible + Mg = 22.26 g
Mass of crucible + magnesium oxide = 23.52 g

The formula of magnesium oxide can be worked out by answering the following questions (answers in brackets):

1 Find the mass of magnesium used in the experiment. (22.26 – 20.40 = 1.86 g)

2 Calculate how many moles of magnesium atoms the answer to **1** represents. (1.86/24 = 0.078)

3 Find the mass of magnesium oxide formed. (23.52 – 20.40 = 3.12 g)
4 Find the mass of oxygen that combines with magnesium. (3.12 – 1.86 = 1.26 g)
5 How many moles of oxygen atoms does the answer to **4** represent? (1.26/16 = 0.079)
6 What is the ratio of the number of moles of magnesium atoms to number of moles of oxygen atoms (0.078:0.079) as a whole number? (almost exactly 1:1)
7 The simplest formula of magnesium oxide is therefore MgO.

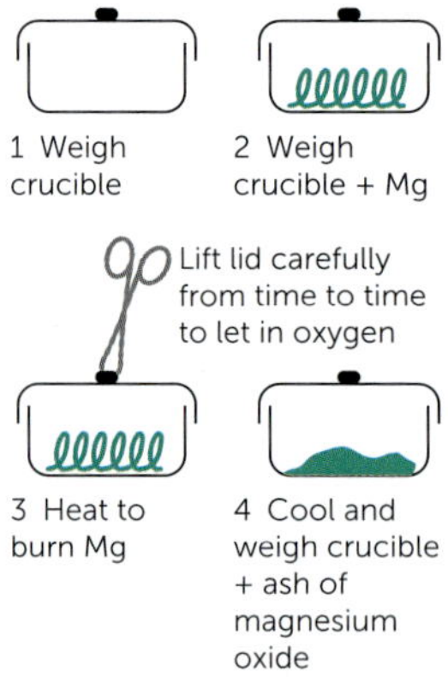

Figure 8.3 Experiment to determine the formula of magnesium oxide.

8.3 Percentage composition by mass

The percentage of an element in a compound is found by using the formula of the compound and the atomic masses of the elements in the compound. It is more correctly called the **percentage composition by mass** of the element in that compound.

The percentage composition by mass of an element is given by the formula

$$\frac{\text{molar mass of element} \times \text{number of atoms of the element in the formula of the compound}}{\text{molar mass of compound}} \times 100$$

EXERCISE 8D

Calculating percentage composition by mass

Find the percentage composition by mass of:

(i) sulfur in Na_2SO_4

(ii) sodium in $Na_2CO_3 \cdot 10H_2O$

(iii) oxygen in $Al(NO_3)_3$

(iv) bromine in $CaBr_2$

(v) sulfate in $(NH_4)_2SO_4$

EXAMPLE 8.4

Calculate the percentage composition by mass of nitrogen in ammonium nitrate, NH_4NO_3.

Answer

$m(N) = 14\,u$

$m(NH_4NO_3) = 14 + (4 \times 1) + 14 + (3 \times 16) = 80\,u$

number of atoms of N = 2

$$\text{percentage composition by mass} = \frac{14 \times 2 \times 100}{80} = 35\%$$

Simplest formula

Just as the percentage composition by mass can be calculated from the formula of a compound, the simplest formula of a compound can be found from the percentage composition by mass of each element. The simplest formula is also known as the **empirical formula**. Here, 'empirical' means 'obtained from experimental results'. To find the simplest formula:

1 Write down the symbols of the elements involved in the compound.
2 Write the percentage composition of each element under its symbol.
3 Divide each percentage by the molar mass of that element ('the atomic mass in grams') to obtain a molar ratio.
4 Find the simplest, whole number, molar ratio to work out the formula of the compound.

EXAMPLE **8.5**

Calculate the simplest, or empirical, formula for a compound that contains 27.3 per cent C and 72.7 per cent O by mass.

Answer

Symbols of elements involved	C	O
Percentage composition	27.3	72.7
Divide by molar masses	$\frac{27.3}{12}$	$\frac{72.7}{16}$
Molar ratio	2.3	4.5
Simplest whole number ratio (this can often be obtained by dividing through by the smallest number)	$\frac{2.3}{2.3}$	$\frac{4.5}{2.3}$
This is approximately	1	2

Therefore, the simplest formula of the compound is CO_2.

Comment

Sometimes the composition of elements in the compound is given as mass of elements in a given mass of compound. In this case, the procedure is the same, but the mass of each element is substituted for the percentage composition by mass.

EXERCISE **8E**

Calculating simplest (or empirical) formulae

(i) When methane is analysed it is found to contain 75 per cent C and 25 per cent H. What is its empirical formula?

(ii) 15.3 g of aluminium atoms combine with 13.6 g of oxygen atoms to form an oxide. What is the simplest formula of the oxide?

(iii) 0.4764 g of an oxide of iron contained 0.3450 g of iron. What is the empirical formula of the oxide?

(iv) A compound contains sodium (32.3 per cent), sulfur (22.5 per cent) and oxygen only. What is its empirical formula?

Finding molecular formulae

The simplest formula is not always the 'true' or molecular formula: it is only the simplest ratio of the atoms contained in the substance. If, for example, you calculated the simplest formula of hydrazine (N_2H_4) from its percentage composition by mass of nitrogen and hydrogen, you would obtain NH_2. In order to establish the molecular formula, you need the molecular mass as well as the empirical formula.

EXERCISE **8F**

Determining molecular formulae

(i) Work out the empirical and molecular formulae for a compound that has a molar mass of 30 g mol^{-1} and is of percentage composition 80 per cent carbon and 20 per cent hydrogen.

(ii) Nicotine contains 74.9 per cent carbon, 8.7 per cent hydrogen and 17.3 per cent nitrogen. The compound contains two nitrogen atoms per molecule. What are the empirical and molecular formulae of nicotine?

EXAMPLE **8.6**

Dichloroethane has a molecular mass of 99 u. Analysis of a sample shows that it contains 24.3 per cent carbon, 4.1 per cent hydrogen and 71.6 per cent chlorine. What is its molecular formula?

Answer

Calculate the simplest formula:

Element	C	H	Cl
Percentage composition	24.3	4.1	71.6
Divide by molar masses	$\frac{24.3}{12}$	$\frac{4.1}{1}$	$\frac{71.6}{35.5}$
Molar ratio	1	2	1

Therefore, the simplest formula is CH_2Cl.
The molecular mass of $CH_2Cl = 12 + (2 \times 1) + 35.5 = 49.5$ u, so it is not the molecular formula of dichloroethane.
Multiply the number of each atom by two, to maintain the same ratio, and you get $C_2H_4Cl_2$ which has a molecular mass of

$$(2 \times 12) + (4 \times 1) + (2 \times 35.5) = 99 \text{ u}$$

So, the molecular formula of dichloroethane is $C_2H_4Cl_2$.

8.4 Water of crystallization

The formulae of some ionic compounds may be written as $AB \cdot nH_2O$, where '$\cdot\ nH_2O$' indicates that n molecules of water are associated with each AB unit within the crystal lattice. Because the water is within the lattice, the substance is perfectly dry to the touch but it contains **water of crystallization** and is said to be **hydrated**. The water may be driven off by heating (**dehydration**) and the residue, that now does not contain water of crystallization, is termed **anhydrous**.

For example, if blue copper(II) sulfate pentahydrate crystals are heated, it passes through several stages, eventually becoming white, anhydrous copper(II) sulfate:

$$\underset{\text{blue}}{CuSO_4 \cdot 5H_2O(s)} \rightarrow \underset{\text{white}}{CuSO_4(s)} + 5H_2O(l)$$

If water is added to anhydrous copper(II) sulfate, it turns blue again.

EXERCISE **8G**

Changes that occur when copper(II) sulfate pentahydrate is heated

The dehydration of copper(II) sulfate pentahydrate by heating is a 'stepwise' process:

(i) First, the pentahydrate loses water to become the trihydrate.

(ii) Then the monohydrate is formed.

(iii) Finally, the anhydrous salt remains.

(iv) If the anhydrous salt is heated very strongly, it decomposes to copper(II) oxide and sulfur trioxide.

Write balanced equations for all of these reactions.

8.5 Calculating amounts from equations

Calculating masses

You have already seen that chemical equations give the following information:

1 The chemical composition of the reactants and products in a reaction.
2 The simplest whole number relationship of the molecules involved in the reaction.

Using the mole concept, together with the atomic masses of the elements involved, and assuming the reaction 'goes to completion' (i.e. that the maximum amount of product possible is made, see section 8.8) we are now in a position to work out the masses of substances involved in chemical reactions.

EXAMPLE **8.7**

What mass of iodine will react completely with 10 g of aluminium to form aluminium iodide (Al_2I_6)? (Assume the reaction goes to completion.)

Answer

Write the balanced equation:

$$2Al(s) + 3I_2(s) \rightarrow Al_2I_6(s)$$

Work out the relationship between the molecules involved:

$$2 \text{ atoms Al} + 3 \text{ molecules } I_2 \rightarrow 1 \text{ molecule } Al_2I_6$$

You could also write:

$$4 \text{ atoms Al} + 6 \text{ molecules } I_2 \rightarrow 2 \text{ molecules } Al_2I_6$$

or

$$200 \text{ atoms Al} + 300 \text{ molecules } I_2 \rightarrow 100 \text{ molecules } Al_2I_6$$

or even

$$2 \times 6.022 \times 10^{23} \text{ atoms Al} + 3 \times 6.022 \times 10^{23} \text{ molecules } I_2 \rightarrow 1 \times 6.022 \times 10^{23} \text{ molecules } Al_2I_6$$

The last line can also be written as

$$2 \text{ mol Al} + 3 \text{ mol } I_2 \rightarrow 1 \text{ mol } Al_2I_6$$

Convert the numbers of moles to grams using molar masses, $M(\text{Al}) = 27 \text{ g mol}^{-1}$ and $M(\text{I}) = 127 \text{ g mol}^{-1}$:

$$2 \times 27 \text{ g Al} + 3 \times 254 \text{ g } I_2 \rightarrow 1 \times 816 \text{ g } Al_2I_6$$

or

$$54 \text{ g Al} + 762 \text{ g } I_2 \rightarrow 816 \text{ g } Al_2I_6$$

We now have the ratios between the quantities of reactants and products in grams.

In the question, you start with 10 g of aluminium, so what masses of iodine and aluminium iodide are involved if the above ratio is to apply?

EXAMPLE **8.7** *(continued)*

For 1 g of aluminium:

$$1\text{g Al} + \frac{762}{54}\text{g}I_2 \rightarrow \frac{816}{54}\text{g } Al_2I_6$$

Therefore, for 10g aluminium:

$$10\text{ g Al} + \frac{762 \times 10}{54}\text{g } I_2 \rightarrow \frac{816 \times 10}{54}\text{g } Al_2I_6$$

or

$$10\text{ g Al} + 141\text{ g } I_2 \rightarrow 151\text{ g } Al_2I_6$$

So, 10 g of aluminium reacts with 141 g of iodine.

Comment

The calculation is written in simple steps to help you understand. In practice, you can leave out some of the more obvious steps and ignore the calculations involving Al_2I_6, since the question does not ask for them – it only asks for the relationship between aluminium and iodine.

EXERCISE **8H**

Calculations based on equations

(i) What mass of magnesium oxide is formed by burning 1.2 g of magnesium metal in oxygen?

$$2Mg(s) + O_2(g) \rightarrow 2MgO(s)$$

(ii) What mass of oxygen is formed when 4.34 g of mercury(II) oxide is completely decomposed by heating?

$$2HgO(s) \rightarrow 2Hg(l) + O_2(g)$$

(iii) What mass of copper(II) nitrate is needed to produce 16 g of copper(II) oxide upon decomposition?

$$2Cu(NO_3)_2(s) \rightarrow 2CuO(s) + 4NO_2(g) + O_2(g)$$

(iv) How much graphite (carbon, C) must be burned in oxygen gas to produce 2.2 g of carbon dioxide gas? (You provide the balanced equation.)

(v) Solid copper(II) carbonate decomposes on heating to solid copper(II) oxide and carbon dioxide gas. How much copper(II) oxide is obtained when 1.28 g of the carbonate is heated? (Again, first provide the balanced equation.)

8.6 Calculating gas volumes

When gases are involved in reactions, it is easier to measure their volumes than to weigh them and find their masses. Although we can calculate the masses of gases used up or produced in chemical reactions using the methods described in the previous section, it is often more convenient to calculate the volumes involved.

At this point, it is worth remembering that:

One mole of ANY gas occupies a volume of 24 dm^3 at room temperature and pressure.

'Room temperature' is taken as 20°C or 293 K. 'Normal' **atmospheric pressure** is 1 bar (101 325 Pa) – see page 161.

EXAMPLE **8.8**

What volume of carbon dioxide can be obtained by decomposing 50 g of calcium carbonate at room temperature and pressure?

Answer

Write the balanced equation:

$CaCO_3(s) \rightarrow CaO(s) + CO_2(g)$

Work out the mole relationship between the substances you are considering:

1 molecule $CaCO_3 \rightarrow$ 1 molecule CO_2

Therefore,

1 mol $CaCO_3 \equiv$ 1 mol CO_2

Note that '$\rightarrow$' can be replaced by '$\equiv$' which can be taken to mean 'reacts with' or 'reacts to form' according to the amounts shown by the equation.

Work out the mole relationship between the substances you are considering: Convert moles to grams in the case of calcium carbonate and moles to cubic decimetres in the case of carbon dioxide, because the question asks about the volume of the gas:

100 g $CaCO_3 \equiv$ 24 dm^3 CO_2

Therefore,

50 g $CaCO_3 \equiv$ 12 dm^3 CO_2

So, 12 dm^3 CO_2 are produced.

EXERCISE **8I**

Calculating gas volumes from equations

All gas volumes are measured at room temperature and pressure.

(i) What is the volume occupied by 0.25 mol of chlorine gas?

(ii) What is the mass of 600 cm^3 of sulfur dioxide gas?

(iii) What volume of hydrogen is obtained from the reaction of 4.8 g of magnesium metal with excess hydrochloric acid?

$Mg(s) + 2HCl(aq) \rightarrow MgCl_2(aq) + H_2(g)$

(iv) What volume of oxygen is needed for complete combustion of 2 dm^3 of propane (C_3H_8)?

$C_3H_8(g) + 5O_2(g) \rightarrow 3CO_2(g) + 4H_2O(l)$

(v) What volume of hydrogen chloride gas is produced by the reaction of 40 cm^3 of chlorine with hydrogen? (First write a balanced equation.)

(vi) What mass of water is made when 10 dm^3 of hydrogen gas burns completely in excess air? (First write a balanced equation.)

8.7 Limiting reagents

We have very often specified that a particular reagent is **in excess**; this means that there are more than enough moles of it to react with the other reactant. Under these circumstances it is the reagent that is *not* in excess that controls the amount of product. This reagent is called the **limiting reagent**.

EXAMPLE **8.9**

In order to understand the idea of a limiting reagent, imagine that you are packing biscuits. If each packet can hold ten biscuits, the 'reaction' is

1 empty packet + 10 loose biscuits → 1 packet of biscuits

You have twenty-five empty packets and 200 loose biscuits. How many full packets of biscuits can you make?

Answer

Twenty with 5 empty packets left over. Here, you do not have enough biscuits to fill all the packets. The number of loose biscuits (the *limiting reagent*) controls the number of packets of biscuits you can make.

Comment

If you have twelve empty packets and 150 biscuits, you can make twelve packets of biscuits and thirty biscuits remain. This time it is the number of empty packets that is the limiting reagent.
In a chemical reaction, unless the exact amounts of each reagent (as specified by the balanced equation for the reaction) are reacted together, then one reagent is always limiting. The limiting reagent can be spotted by comparing the ratio of the amounts of reactants available with the ratio of the amounts of reactants obtained from the balanced equation for the reaction.

EXAMPLE **8.10**

Copper reacts with silver nitrate solution according to the equation

$$Cu(s) + 2AgNO_3(aq) \rightarrow Cu(NO_3)_2(aq) + 2Ag(s)$$

If 0.50 mol of copper is added to 1.5 mol of silver nitrate, which is the limiting reagent and how many moles of silver are formed?

Answer

Decide which is the limiting reagent.

According to the equation, 1 mol Cu ≡ 2 mol $AgNO_3$, so

$$0.50 \text{ mol Cu} \equiv 2 \times 0.50 = 1.0 \text{ mol } AgNO_3$$

but there are 0.50 mol Cu and 1.5 mol $AgNO_3$. Therefore, $AgNO_3$ is present in excess and Cu is the limiting reagent.

Calculate how many moles of silver are formed.

Use the amount of the limiting reagent to find the amount of product. According to the equation, 1 mol Cu ≡ 2 mol Ag. Therefore,

$$0.50 \text{ mol Cu} \equiv 2 \times 0.50 = 1.0 \text{ mol Ag}$$

So 1.0 mol Ag is formed.

EXERCISE **8J**

Limiting reagents

(i) What is the limiting reagent when 0.5 mol of solid phosphorus and 0.5 mol of oxygen react according to the equation

$$P_4(s) + 5O_2(g) \rightarrow P_4O_{10}(s)$$

and how many moles of P_4O_{10} are formed?

(ii) What is the limiting reagent when 1.5 g of magnesium and 1.5 g of nitrogen combine according to the equation

$$3Mg(s) + N_2(g) \rightarrow Mg_3N_2(s)$$

and how many grams of Mg_3N_2 are formed?

(iii) Zinc reacts with heated copper(II) oxide to form zinc oxide and copper metal. If 3.0 g of zinc are reacted with 3.0 g copper(II) oxide, which is the limiting reagent? What is the mass of copper metal formed?

8.8 Reactions that do not go to completion

The previous calculations of amounts of reactants consumed or produced have assumed that the reactions '**go to completion**' so that 100 per cent of the limited reagent is converted to product. Where a reaction involves the relative numbers of moles of reactant as in the chemical equation (1:1, 3:1 etc), going to completion means that all the reactants are completely converted to products, but where one reactant is in excess, going to completion means that at the end of the reaction only the *limiting reactant* is completely consumed.

There are two phenomena that prevent a reaction going to completion:

(a) Competing side reactions in which other products are formed.

(b) Equilibrium reactions. In equilibrium reactions, there may be a significant amount of limiting reactants present in the final equilibrium mixture. This is considered in detail in Unit 15.

Even if a reaction does go to completion, and despite the best of experimental techniques, there may be some losses of product during the transference of material within the experimental routine.

Percentage yield of a product

Whatever the reason, the yield of the product obtained by experiment, may be lower than expected; sometimes it is a great deal lower. It is usual to report your **percentage yield** when you are doing preparative chemistry:

$$\text{percentage yield} = \frac{\text{experimental yield}}{\text{theoretical yield}} \times 100\%$$

where the experimental yield is the mass of product obtained and the theoretical yield is the mass of product you calculated you would get from the balanced equation.

EXAMPLE **8.11**

Methane is converted to carbon dioxide and water when burned in a plentiful supply of oxygen (*complete combustion*):

$$CH_4(g) + 2O_2(g) \rightarrow CO_2(g) + 2H_2O(l)$$

If 10 g of CO_2 were obtained when 16 g of CH_4 were burned in a limited supply of oxygen gas, what would be the percentage yield of carbon dioxide?

Answer

If 16 g of methane is burned in oxygen and 10 g of carbon dioxide is formed, then the theoretical yield of carbon dioxide formed can be found from the equation:

$$CH_4(g) + 2O_2(g) \rightarrow CO_2(g) + 2H_2O(l)$$

1 mol $CH_4(g)$ produces 1 mol $CO_2(g)$

16 g $CH_4(g)$ produces 44 g $CO_2(g)$

$$\text{percentage yield} = \frac{\text{experimental yield}}{\text{theoretical yield}} \times 100\%$$

$$= \frac{10}{44} \times 100$$

$$= 23\%$$

Comment

In a limited supply of oxygen, not all of the methane would be converted to carbon dioxide – carbon monoxide and soot (carbon) would also be formed in competing reactions. The reaction does not go to completion and there is *incomplete combustion* of the methane.

EXERCISE **8K**

Percentage yields

A student prepared solid magnesium oxide by burning 4.8 g of magnesium metal in excess air. Unfortunately, much of the white magnesium oxide produced in this way formed a white smoke that dispersed into the air of the laboratory. The student recovered 4.5 g of magnesium oxide. What was the percentage yield of magnesium oxide?

BOX **8.1**

Avogadro

Lorenzo Romano Amedeo Carlo Avogadro de Quaregnae di Cerreto (1776–1856) had a very impressive name – and an equally impressive intellect. He graduated in law at 16 but became interested in science and mathematics and eventually gave up law to teach these subjects.

He had read that one volume of oxygen gas always reacted with twice the volume of hydrogen gas to form water. If the temperature was high enough, then the water was formed as steam and exactly two volumes of steam were produced. He concluded that two *particles* of hydrogen gas must react with one *particle* of oxygen gas to produce two *particles* of water. After looking at the results of the experiments of another famous chemist, Gay-Lussac, in which reacting gas volumes had been measured, Avogadro came up with the following law (see Unit 10):

> *Equal volumes of different gases (at the same temperature and pressure) contain the same number of particles..*

He called the gas particles 'molecules'. If Avogadro's law is applied to volumes of reacting gases, it is possible to deduce their formulae. The law was discounted by the scientific community until very late in his lifetime.

REVISION QUESTIONS

8.1 Calculate the molecular mass of:
(i) nitrogen gas (N_2)
(ii) copper(II) chloride
(iii) phosphoric acid (H_3PO_4)
(iv) calcium nitrate
(v) sodium sulfate decahydrate.

8.2 What is the mass of one mole of:
(i) chlorine atoms (Cl)
(ii) chlorine molecules (Cl_2)
(iii) phosphorus molecules (P_4)
(iv) iodide ions (I^-)
(v) buckminsterfullerene molecules (C_{60}).

8.3 Calculate the number of moles in:
(i) 30 g of oxygen molecules (O_2)
(ii) 31 g of phosphorus molecules (P_4)
(iii) 25 g of calcium carbonate
(iv) 3.28 g of sulfur dioxide
(v) 980 g of sulfuric acid.

8.4 Calculate the mass of each of the following:
(i) 0.500 mol of NaCl
(ii) 0.250 mol of CO_2
(iii) 2 mol of H_2
(iv) 2.50 mol of $CuSO_4 \cdot 5H_2O$
(v) 100 mol of HNO_3.

8.5 A sample of ammonia weighs 2.00 g. What mass of sulfur dioxide contains the same number of molecules as are in 2.00 g of ammonia?

8.6 What mass of magnesium metal would react completely with 8 g of sulfur?

$$Mg(s) + S(s) \rightarrow MgS(s)$$

8.7
(i) What mass of oxygen would be produced by heating 4.25 g of sodium nitrate?

$$2NaNO_3(s) \rightarrow 2NaNO_2(s) + O_2(g)$$

(ii) What *volume* of oxygen, at room temperature and pressure, does the answer to (i) represent?

8.8 When sodium hydrogen carbonate (NaHCO3) is heated, the following change occurs:

$$NaHCO_3(s) \rightarrow Na_2CO_3(s) + H_2O(l) + CO_2(g)$$

(i) Balance the equation written above.
(ii) What is the mass of two moles of $NaHCO_3$?
(iii) What is the mass of CO_2 produced when 2.0 mol of $NaHCO_3$ are decomposed? What is the volume of this mass at room temperature and pressure?
(iv) What volume of carbon dioxide, at room temperature and pressure, would be obtained if 4.2 g of $NaHCO_3$ were completely decomposed?

8.9 What percentage by mass is water in the salt $MgSO_4 \cdot 7H_2O$?

8.10 When the gas ethene is analysed, it is found to contain 85.72 per cent carbon and 14.28 per cent hydrogen by mass. The molar mass of ethene is 28 g mol^{-1}. What is its molecular formula?

8.11 Aluminium was burned in a stream of chlorine so that it was all converted to aluminium chloride. The following results were obtained:
Mass of aluminium at the start of the experiment = 13.50 g.
Mass of aluminium chloride at the end of the experiment = 66.75 g.
From these results, determine the formula of aluminium chloride.

8.12 Chlorhexidine (containing C, H, Cl and N) is used in mouthwashes. It contains 52.3 per cent C, 5.94 per cent H and 27.7 per cent N, and has a molar mass of 505 g mol^{-1}. What is its molecular formula?

8.13 A 50.1 mg sample of vitamin C (chemical name ascorbic acid) was found to contain 20.5 mg C, 2.27 mg H, 27.3 mg O. Its molar mass is 176 g mol^{-1}. What is its molecular formula?

8.14 Washing soda crystals consist of the compound sodium carbonate decahydrate.
(i) Write the chemical formula of this compound.
(ii) How many moles of water molecules are to be found in a packet of crystals containing 1 kg of washing soda?
(iii) What is the percentage of water in the crystals?

8.15 Vitamin A (chemical name retinol) has the chemical formula $C_{20}H_{30}O$.
(i) What is its molar mass?
(ii) How many molecules of vitamin A are present in a 15 mg tablet, assuming the tablet contains vitamin A only?

DID YOU KNOW?

Although the amount 'mole' usually refers to small particles, it can be used to describe any objects. The reason it isn't used more widely is because it is such a huge number.

If every person on Earth (population around 8 billion) were to have a mole of dollars ($) equally divided between them at birth, and they all lived to eighty, each person would have to spend about $30,000 per *second* of their lives in order to die penniless. And that excludes any interest on their savings!

Information on how Avogadro's constant is determined is found in Extension 8 at https://www.bloomsburyonlineresources.com/chemistry

Calculating Concentrations

9

OBJECTIVES

This unit:

- Describes how the concentrations of solutions can be expressed in appropriate units
- Gives examples of how volumetric analysis can be used to find the concentration of solutions
- Explains what is meant by 'pH'

CONTENTS

9.1 Concentration of solutions

We have already discussed how to calculate masses of solids or volumes of gases for the reactants or products in balanced equations. What if solutions are involved in the reaction? An aqueous solution may be concentrated or dilute. That is, it may contain a large or small amount of solid dissolved in a given amount of water, as shown in Figure 9.1.

How can we express the concentrations of the solutions?

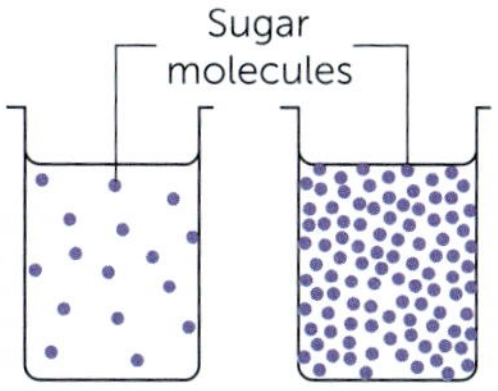

Figure 9.1 Sugar solutions at two different concentrations.

Molar concentration

It is common to measure the concentration of a solution in *moles per cubic decimetre* (**mol dm^{-3}**). For example, a solution of sodium chloride, of concentration 1 mol dm^{-3}, contains 1 mol (or a mass of 23 + 35.5 = 58.5 g) of NaCl dissolved in 1 dm^3 of *solution*. *Always use the term 'solution' and not water.* A mixture of 58.5 g of NaCl dissolved in 1 dm^3 of water is *not* quite the same thing.

Different textbooks may express concentration as M (molar) or mol L^{-1} (moles per litre). Both these units may be taken to mean the same as mol dm^{-3}.

Some examples of molar concentration are:

1 A solution of sodium chloride of concentration 2.00 mol dm^{-3} contains 2.00 mol (or 2.00 × 58.5 = 117 g) of NaCl dissolved in 1 dm^3 of solution.

2 Sodium chloride solution of concentration 0.500 mol dm^{-3} contains 0.500 mol (or 0.500 × 58.5 = 29.3 g) of NaCl dissolved in 1 dm^3 of solution.

Being able to prepare solutions of known concentration is one of the most important skills needed in the chemistry laboratory. Confidence in calculations involving solutions comes with hard work and practice. Take your time in working through this Unit!

In general for solutions,

amount of substance in mol = volume × molar concentration

When using this equation, volume is expressed in dm^3 and concentration is in mol dm^{-3}.

Rearranged forms of this equation may be needed:

$$\text{volume} = \frac{\text{amount in mol}}{\text{molar concentration}}$$

or

$$\text{molar concentration} = \frac{\text{amount in mol}}{\text{volume}}$$

EXAMPLE **9.1**

What is the concentration of a solution of sodium hydroxide (NaOH) that contains 0.10 mol of NaOH dissolved in 250 cm^3 of solution?

Answer

amount of NaOH in mol = volume × molar concentration

Substituting, and remembering that 1 dm^3 = 1,000 cm^3, we must divide 250 cm^3 by 1,000 to convert it to dm^3,

$$0.10 = \frac{250}{1000} \times \text{molar concentration}$$

Therefore the concentration of the NaOH solution is

$$\frac{1000 \times 0.10}{250} = 0.40 \text{ mol dm}$$

EXAMPLE **9.2**

How many grams of sodium chloride are there in 500 cm^3 of a solution that has a concentration of 0.500 mol dm^{-3}?

Answer

amount of NaCl in mol = volume × molar concentration

$$= \frac{500}{1000} \times 0.500$$
$$= 0.250 \text{ mol}$$

To convert moles into grams:

$$\text{amount of NaCl in mol} = \frac{\text{mass NaCl}}{\text{molar mass NaCl}}$$

$$0.250 = \frac{\text{mass NaCl}}{58.5}$$

which gives

mass of NaCl = 0.250 × 58.5 = 14.6 g

Therefore, 14.6 g of NaCl are dissolved in 500 cm^3 of solution.

BOX **9.1**

Concentrations of pure solids and pure liquids

Concentration is measured in $mol\,dm^{-3}$. This is a measure of the number of particles per unit volume of the substance. In pure solids or pure liquids the number of particles in a given volume can be calculated from the density and this is constant, at constant temperature. Since the density of a given pure solid or pure liquid is constant, at constant temperature, **so is its concentration.**

1 The density of copper metal at 298 K is $8.92\,g\,cm^{-3}$. This is equivalent to $8{,}920\,g\,dm^{-3}$, or a concentration of $8{,}920/63.5\,mol\,dm^{-3}$, which approximates to $140\,mol\,dm^{-3}$.

2 The density of pure water at 298 K is $1\,g\,cm^{-3}$, equivalent to $1{,}000\,g\,dm^{-3}$ or a concentration of $1{,}000/18\,mol\,dm^{-3}$. This approximates to $56\,mol\,dm^{-3}$.

EXERCISE **9A**

Concentration of solutions

(i) Sodium chloride (58.5 g) is dissolved in water. More water is then added so that the total volume is $100\,cm^3$. What is the concentration of the solution in $mol\,dm^{-3}$?

(ii) Calculate the concentration (in $mol\,dm^{-3}$) of hydrated copper(II) sulfate ($CuSO_4 \cdot 5H_2O$) in a solution originally prepared by dissolving 0.1 g of the solid in water and making up to $50\,cm^3$ with water.

(iii) What volume of sodium chloride solution of concentration $0.2\,mol\,dm^{-3}$ contains 10 mol of salt?

(iv) What mass of hydrated sodium carbonate ($Na_2CO_3 \cdot 10H_2O$) should be dissolved in water and made up to exactly $1\,dm^3$ in order to make the solution a concentration of $0.100\,mol\,dm^{-3}$?

Ion concentrations

When ionic solids are dissolved in water, they are regarded as completely ionized in solution. It is sometimes necessary to express the solution in terms of the concentrations of the ions present, as shown in the examples below:

1 An aqueous NaCl solution, of concentration $1\,mol\,dm^{-3}$, contains $Na^+(aq)$ and $Cl^-(aq)$ ions. Each NaCl unit is dissociated in solution as follows:

$$NaCl(aq) \rightarrow Na^+(aq) + Cl^-(aq)$$

As 1 mol NaCl produces 1 mol Na^+ and 1 mol Cl^-, a solution of NaCl of concentration $1\,mol\,dm^{-3}$ contains a concentration of $1\,mol\,dm^{-3}$ $Na^+(aq)$ ions and $1\,mol\,dm^{-3}$ $Cl^-(aq)$ ions.

2 An aqueous $CaCl_2$ solution, of concentration $1\,mol\,dm^{-3}$, contains $Ca^{2+}(aq)$ and $Cl^-(aq)$ ions. Each $CaCl_2$ unit is dissociated in solution as follows:

$$CaCl_2(aq) \rightarrow Ca^{2+}(aq) + 2Cl^-(aq)$$

A solution of $CaCl_2$ of concentration $1\,mol\,dm^{-3}$ contains a concentration of $1\,mol\,dm^{-3}$ $Ca^{2+}(aq)$ and contains a concentration of $2\,mol\,dm^{-3}$ $Cl^-(aq)$.

3 Similarly, a $0.1\,mol\,dm^{-3}$ solution of $Na_3PO_4(aq)$ contains $Na^+(aq)$ at a concentration of $0.3\,mol\,dm^{-3}$ and $PO_4^{3-}(aq)$ at a concentration of $0.1\,mol\,dm^{-3}$.

EXERCISE **9B**

Assuming the compounds are completely ionized in solution, identify the ions that exist in the following aqueous solutions and state the concentration of each ion.

(i) Potassium chloride, KCl ($0.1\,mol\,dm^{-3}$).

(ii) Sodium carbonate, Na_2CO_3 ($0.125\,mol\,dm^{-3}$).

(iii) Ammonium sulfate, $(NH_4)_2SO_4$ ($0.25\,mol\,dm^{-3}$).

(iv) Copper nitrate, $Cu(NO_3)_2$ ($0.15\,mol\,dm^{-3}$).

9.2 Standard solutions

A **standard solution** is a solution of known concentration. **Primary standards** are substances that are used to make up standard solutions by accurately weighing out a given mass of the standard, then dissolving the standard in **deionized water** (i.e. pure water) to make up a known volume of solution. Primary standards should be:

1 chemically stable in the solid and aqueous states
2 pure
3 soluble in water.

Anhydrous sodium carbonate is an example of a primary standard. Standard solutions of primary standards can be used to standardize other solutions, as you will see in the next section.

Standard solutions are made up in special flasks, called **volumetric flasks**, which have been calibrated to contain a definite volume of liquid. A graduation line is marked on the neck of the flask. When the flask contains solution to this level it contains the volume of solution marked on the front of the flask. Some volumetric flasks are shown in Exercise 9C.

EXERCISE **9C**

Calculating the concentration of standard solutions

An aqueous solution containing exactly 0.1000 mol of anhydrous sodium carbonate (Na_2CO_3) is placed in each of the volumetric flasks, and carefully diluted with deionized water to each of the volumes shown. Calculate the concentration of each solution.

(i) V = 2,000 cm^3 **(ii)** V = 1,000 cm^3 **(iii)** V = 500 cm^3 **(iv)** V = 250 cm^3 **(v)** V = 100 cm^3

BOX **9.2**

Making a standard solution

A standard solution of, for example, sodium carbonate, is made by weighing out the required mass of sodium carbonate using a very sensitive balance called an **analytical balance**. The mass is generally measured to the nearest 0.0001 g. The solid is then *completely* dissolved in deionized water in a beaker and the resulting solution transferred to a volumetric flask. The beaker is rinsed with deionized water a few times and the washings are also transferred to the flask, so that *all* the sodium carbonate solution is transferred. The flask is then carefully topped up to the graduation mark on the flask with deionized water, so that the bottom of the solution meniscus is 'sitting on' the graduation line when viewed at eye level.

Finally, the flask is shaken well, so that the concentration of the solution it contains is homogeneous. The sequence of operations, (i)–(v), is illustrated below

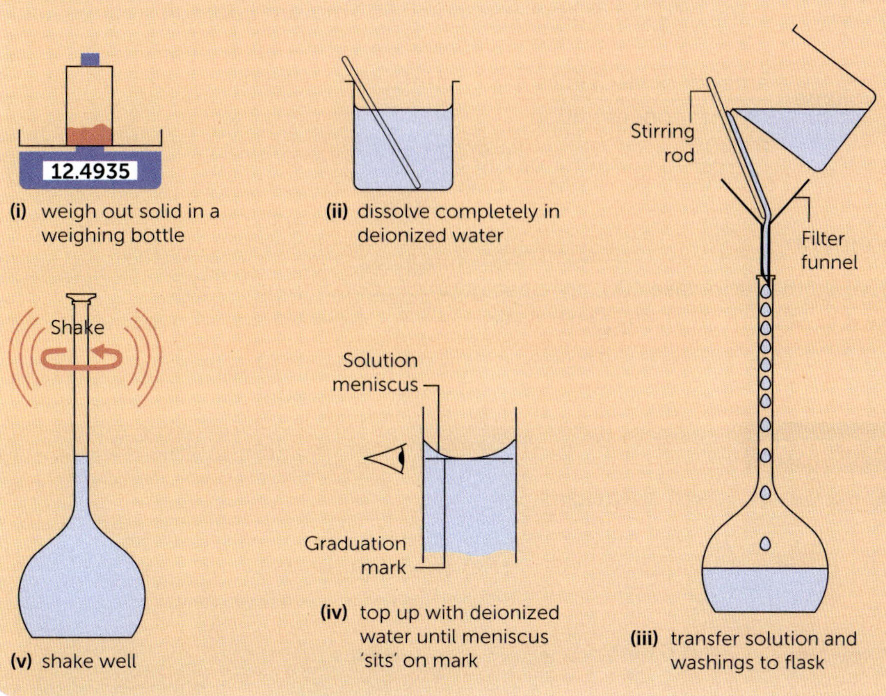

Diluting standard solutions

A dilute standard solution can also be made by carefully diluting a more concentrated standard solution to a known total volume. In this instance, a known volume of the concentrated standard solution is delivered into a standard volumetric flask, and the solution diluted ('made up') to the graduation line with deionized water. **Pipettes** are designed to deliver specific volumes of liquid. They can contain various volumes (from 0.50 to 200 cm^3) and the volume of the pipette is marked on its bulb or shown by graduation marks on the side of the pipette (Figure 9.2).

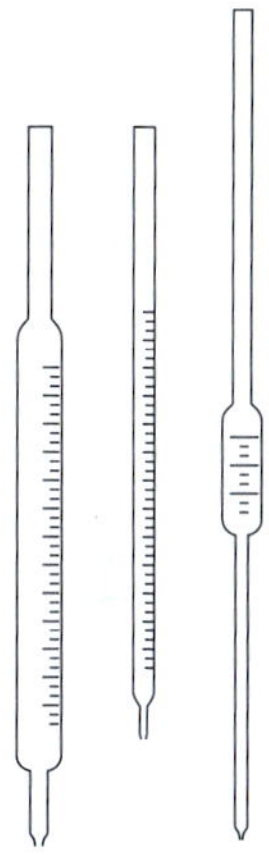

Figure 9.2 Different types of pipette.

EXAMPLE **9.3**

How much of a sodium chloride solution, of concentration 0.500 mol dm^{-3}, would you take and dilute in order to make exactly 500 cm^3 of a solution of concentration 0.100 mol dm^{-3}?

Answer

First, work out how many moles of sodium chloride you would end up with.

Since the amount of NaCl in moles = volume × the molar concentration, 500 cm^3 of a solution of NaCl of concentration 0.100 mol dm^{-3} contains

$$= \frac{500}{1000} \times 0.100 = 0.0500 \text{ mol}$$

Students often find calculations involving dilution difficult at first. It is worthwhile getting to grips with these calculations before you go to the lab!

EXAMPLE **9.3** *(continued)*

What volume of the solution of concentration 0.500 mol dm^{-3} does this correspond to?

$$\text{volume} = \frac{\text{amount of NaCl in mol}}{\text{molar concentration}}$$

$$= \frac{0.0500}{0.500} = 0.100\ dm^3 \text{ or } 100\ cm^3$$

Comment

You would therefore take 100 cm^3 of solution of concentration 0.500 mol dm^{-3} and dilute it to 500 cm^3 in a volumetric flask.

EXERCISE **9D**

Making up standard solutions

(i) What mass of anhydrous sodium carbonate would you have to weigh out to make 500 cm^3 of a standard solution of concentration 0.010 mol dm^{-3}?

(ii) If, when making the solution in (i), you did not add enough water to bring the volume of the solution completely up to the 500 cm^3 graduation mark on your volumetric flask, what effect would that have on the concentration of the standard solution?

(iii) Calculate the concentration of a solution that results when 50 cm^3 of a sodium hydroxide solution, of concentration 0.1 mol dm^{-3}, is pipetted into a 250 cm^3 volumetric flask and made up to the graduation mark with deionized water.

(iv) You have some acid of concentration 16.0 mol dm^{-3}. How much of the concentrated acid would you dilute in order to make 1,000 cm^3 of acid solution at a concentration of 2.00 mol dm^{-3}?

(v) How much of a copper(II) sulfate solution, of concentration 2 mol dm^{-3}, would you need to take and dilute ('make up') to 200 cm^3 in order to end up with a solution of concentration 0.5 mol dm^{-3}?

9.3 Volumetric analysis

The procedure involved in finding the concentration of a solution is called **volumetric analysis**. It involves reacting a solution of *known* concentration (a standard solution) with one of *unknown* concentration, in order to determine the **equivalence point**. The reactions involved are chosen so that they go to completion (see section 8.8, p. xx).

The equivalence point is reached when the reactants have reacted together in an exact ratio of quantities, as given in the balanced equation for the reaction. For example, in an acid–base titration, the equivalence point involves addition of exactly enough base to neutralize any acid present.

The technique whereby one reactant is slowly added to a second reagent until the equivalence point is reached is called a **titration**. **Chemical indicators** are substances that change colour at a particular ratio of reactant concentrations called the **endpoint**. In a titration, a chemical indicator is chosen so that the endpoint and equivalence point *are the same*.

By measuring the volumes of the solutions that have reacted together, and using the balanced equation for the reaction, the unknown solution concentration can be determined. If carried out carefully, volumetric analysis is quite an accurate technique and a skilled worker should determine an unknown concentration within an error of no more than 0.2 per cent.

An acid–base titration

Acid-base titrations are a type of volumetric analysis in which the (previously unknown) concentration of an acid can be determined by titrating it with a standard alkaline solution. The acid is neutralized by the alkali. The procedure is as follows:

1 A measured volume of the acid is delivered into a conical flask, using a pipette as shown in Figure 9.3.
2 A few drops of a suitable indicator are added to the acid solution.
3 The standard alkaline solution is added from a burette. A burette is a long glass tube with graduation marks on it, with a stopcock at the bottom end. It allows the user to measure the volume of solution delivered through the stopcock. The alkaline solution is added until the endpoint is reached and the indicator changes colour; the total volume of alkaline solution added is then noted. A diagram of this procedure is shown in Figure 9.4.

Some typical results from such a titration are shown in Example 9.4.

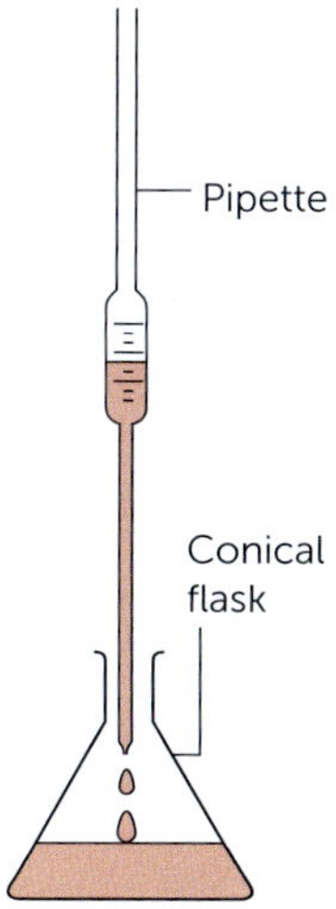

Figure 9.3 Delivering a measured volume of acid.

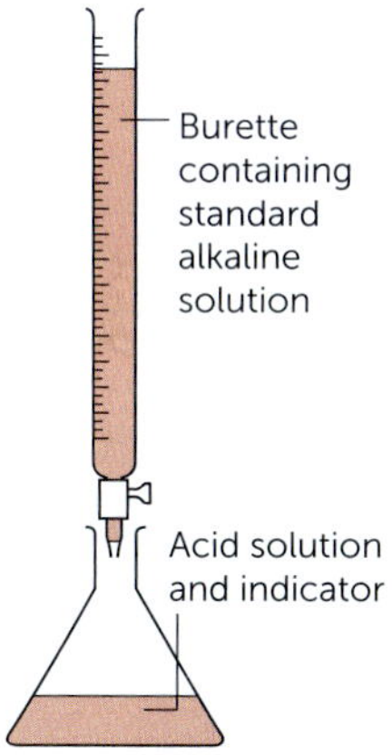

Figure 9.4 Addition of standard alkaline solution.

Methyl orange indicator is an acid–base indicator used in acid–base titrations. It is red at lower pH and yellow at higher pH.

EXAMPLE **9.4**

Hydrochloric acid, of unknown concentration, was titrated with a standard solution of potassium hydroxide of concentration 0.500 mol dm^{-3}. It was found that 25.0 cm^3 of the acid reacted with 37.5 cm^3 of the potassium hydroxide solution. Calculate the concentration of the hydrochloric acid.

Answer

Write the balanced equation for the reaction:

$$HCl(aq) + KOH(aq) \rightarrow KCl(aq) + H_2O(l)$$

Work out the molar relationship between the two reactants:

$$1\text{ mol HCl} \equiv 1\text{ mol KOH}$$

Express this ratio as a fraction:

$$\frac{\text{amount of HCl in moles}}{\text{amount of KOH in moles}} = \frac{1}{1}$$

Since the amount of substance in moles = volume × molar concentration, we can substitute:

$$\frac{\text{vol} \times \text{molar conc HCl}}{\text{vol} \times \text{molar conc KOH}} = \frac{1}{1}$$

Substitute the values given in the equation:

$$\frac{25.0/1000 \times \text{molar concentration of HC}}{37.5/1000 \times 0.500}$$

Therefore, the molar concentration of HCl is

$$\frac{37.5 \times 0.500}{25} = 0.750\text{ mol dm}^{-3}$$

EXERCISE **9E**

Calculation of unknown concentrations by titration

Use approximate atomic masses for these calculations.

(i) How much dilute sulfuric acid of concentration 1.0 mol dm^{-3} will react completely with 25 cm^3 of potassium hydroxide solution of concentration 4.0 mol dm^{-3}?

$$H_2SO_4(aq) + 2KOH(aq) \rightarrow K_2SO_4(aq) + 2H_2O(l)$$

(ii) A solution of sodium hydroxide contains 4.0 g dm^{-3}. An endpoint is reached when 25 cm^3 of this solution react with 35 cm^3 of dilute nitric acid. Calculate the concentration of the nitric acid.

$$NaOH(aq) + HNO_3(aq) \rightarrow NaNO_3(aq) + H_2O(l)$$

(iii) A standard solution was prepared by dissolving 2.6061 g of anhydrous sodium carbonate in deionized water and the solution diluted to 250 cm^3. A 25.0 cm^3 portion of this solution was titrated against hydrochloric acid, using a suitable indicator. The endpoint was reached after 18.7 cm^3 of acid had been added. Calculate the concentration of the acid.

$$Na_2CO_3(aq) + 2HCl(aq) \rightarrow 2NaCl(aq) + H_2O(l) + CO_2(g)$$

(iv) How many cubic centimetres of a solution of hydrochloric acid, of concentration 2.0 mol dm^{-3}, are required to react completely with 50 cm^3 of a solution of $Ba(OH)_2$ that contains 68.4 g dm^{-3}?

$$2HCl(aq) + Ba(OH)_2(aq) \rightarrow BaCl_2(aq) + 2H_2O(l)$$

Redox titrations

The technique of titration may also be used to determine the concentration of an unknown solution, when the solutions react together in a **redox** reaction. Potassium permanganate (or potassium manganate(VII), $KMnO_4$) is a powerful oxidizing agent and is used to estimate the concentration of many reducing agents. $KMnO_4$ is violet or pink in solution (depending on its concentration) but is virtually colourless when reduced to Mn^{2+} – it can therefore act as a **self-indicator**. As it is added to the reducing solution it turns colourless, but at the endpoint a very small amount of excess permanganate will be present. The excess is shown when the reacting mixture turns pale pink. The half-reaction for the reduction in *acid solution* is:

The permanganate ion, MnO_4^- is almost black as a solid salt (e.g. potassium permanganate), but it is violet (or, at lower concentration, pink) when dissolved in water.

$$\underset{\text{violet/pink}}{MnO_4^-(aq)} + 8H^+(aq) + 5e^- \rightarrow \underset{\text{colourless}}{Mn^{2+}(aq)} + 4H_2O(l)$$

Two common reactions involving oxidation by potassium permanganate are:

1 Oxidation of iron(II) to iron(III):
$Fe^{2+}(aq) \rightarrow Fe^{3+}(aq) + e^-$

2 Oxidation of oxalates to carbon dioxide. This reaction needs to be carried out at 70°C, because it is too slow at room temperature:
$C_2O_4^{2-}(aq) \rightarrow 2CO_2(g) + 2e^-$

Potassium dichromate (or potassium dichromate(VI), $K_2Cr_2O_7$) is also used as an oxidizing agent in similar reactions to those described for potassium permanganate. In acid solution, it reacts according to the following half-reaction:

$$\underset{\text{orange}}{Cr_2O_7^{2-}(aq)} + 14H^+(aq) + 6e^- \rightarrow \underset{\text{green}}{2Cr^{3+}(aq)} + 7H_2O(l)$$

The green colour of the chromium(III) ion 'masks' the orange colour of the dichromate ion, so the oxidizing agent cannot act as a self-indicator. The solution is used together with a few drops of a **redox indicator**, such as diphenylamine. This indicator is oxidized to an intense blue compound in the presence of a slight excess of dichromate.

The dichromate ion, $Cr_2O_7^{2-}$, is orange in a solid salt (e.g. sodium dichromate) and in solution.

EXAMPLE **9.5**

Iron(II) sulfate is oxidized by potassium permanganate in acid solution. The overall ionic equation is

$$5Fe^{2+}(aq) + MnO_4^-(aq) + 8H^+(aq) \rightarrow Mn^{2+}(aq) + 4H_2O(l) + 5Fe^{3+}(aq)$$

What volume of 0.010 mol dm^{-3} iron(II) sulfate will be oxidized by 25.00 cm^3 of 0.020 mol dm^{-3} permanganate solution?

Answer

Work out the molar relationship between the two reactants:

$$5 \text{ mol } Fe^{2+} \equiv 1 \text{ mol } MnO_4^-$$

Express this ratio as a fraction:

$$\frac{\text{amount of } Fe^{2+} \text{ in moles}}{\text{amount of } MnO_4^- \text{ in moles}} = \frac{5}{1}$$

Since the *amount of substance in moles* = *volume* × *molar concentration*, we can substitute:

$$\frac{\text{vol} \times \text{molar conc } Fe^{2+}}{\text{vol} \times \text{molar conc } MnO_4^-} = \frac{5}{1}$$

Substituting values into the equation:

$$\frac{\text{vol} \times 0.010}{25/1000 \times 0.020} = \frac{5}{1}$$

Therefore, the volume of Fe^{2+} = 0.250 dm^3 = 250 cm^3.

EXERCISE **9F**

Redox titrations

(i) 25 cm^3 of a solution of an iron(II) salt needed 13.5 cm^3 of a 0.020 mol dm^{-3} solution of potassium dichromate for complete oxidation to iron(III). Write the balanced overall redox equation for the reaction and calculate the concentration of Fe^{2+}(aq) in the original solution.

(ii) Cerium(IV) ions are reduced by oxalate ions in solution:

$$2Ce^{4+}(aq) + C_2O_4^{2-}(aq) \rightarrow 2Ce^{3+}(aq) + 2CO_2(g)$$

Ferroin indicator is used in the titration. Calculate the volume of 0.050 mol dm^{-3} cerium(IV) solution needed to exactly react with 25 cm^3 of 0.050 mol dm^{-3} oxalate ion solution.

(iii) Calculate the volume of a 0.020 mol dm^{-3} solution of potassium permanganate that is needed for the complete oxidation of 0.0070 mol of iron(II) oxalate solution. (*Be careful! – there are two oxidations occurring here – the oxidation of iron(II) to iron(III) and the oxidation of the oxalate ion to carbon dioxide.*)

Reactions of solids with solutions

Calculations involving the reaction of a solid with a solution can be worked out in a similar manner.

EXAMPLE **9.6**

What mass of zinc will react completely with 50.0 cm^3 of 0.100 mol dm^{-3} sulfuric acid?

Answer

Write the balanced equation for the reaction:

$$Zn(s) + H_2SO_4(aq) \rightarrow ZnSO_4(aq) + H_2(g)$$

Work out the molar relationship between the two reactants:

$$1\ mol\ Zn \equiv 1\ mol\ H_2SO_4$$

Express this ratio as a fraction:

$$\frac{\text{amount of Zn in moles}}{\text{amount of } H_2SO_4 \text{ in moles}} = \frac{1}{1}$$

Since, for a solution, the number of moles = volume × molar concentration, we can substitute this expression for H_2SO_4 **only** (zinc is a solid, not a solution):

$$\frac{\text{amount of Zn in moles}}{\text{vol} \times \text{molar concentration } H_2SO_4} = \frac{1}{1}$$

Substitute the values given in the equation:

$$\frac{\text{amount of Zn in moles}}{50.0/1000 \times 0.100} = \frac{1}{1}$$

Therefore, the number of moles of zinc needed is

$$\frac{50.0}{1000} \times 0.100 = 0.0050 \text{ mol}$$

and the mass of zinc required = 0.0050 × 65 = 0.325 g.

EXERCISE **9G**

Problems involving the reactions of solids with solutions

(i) What is the volume of sulfuric acid of concentration 1.50 mol dm^{-3} which would be needed to react with 7.95 g of copper(II) oxide?

$$H_2SO_4(aq) + CuO(s) \rightarrow CuSO_4(aq) + H_2O(l)$$

(ii) Magnesium hydroxide is a compound used to neutralize stomach acid (dilute hydrochloric acid) and so cure acid indigestion. The reaction is

$$Mg(OH)_2(s) + 2HCl(aq) \rightarrow MgCl_2(aq) + 2H_2O(l)$$

The magnesium hydroxide in an indigestion tablet was found to react with 20 cm^3 of 0.50 mol dm^{-3} HCl. How many grams of $Mg(OH)_2$ were in the tablet?

(iii) Carbon dioxide gas is prepared by the action of 2.00 mol dm^{-3} hydrochloric acid on marble:

$$CaCO_3(s) + 2HCl(aq) \rightarrow CaCl_2(aq) + H_2O(l) + CO_2(g)$$

What volume of acid is required to give 10.0 dm^3 of carbon dioxide at room temperature and pressure?

(iv) What mass of solid copper(II) carbonate is needed to react completely with 50 cm^3 of sulfuric acid of concentration 0.50 mol dm^{-3}?

Gravimetric analysis

When two solutions react to completion to form an insoluble product, the concentration of one of the solutions can be worked out from the mass of precipitate formed in the reaction. The process is called gravimetric analysis. The precipitate is filtered off and dried thoroughly. The precipitate is heated and weighed several times until the mass of the product does not change (heated to constant mass); this process ensures that the insoluble product is thoroughly dried.

EXAMPLE **9.7**

Barium chloride solution reacts with the sulfate ion in aqueous solution to form insoluble barium sulfate. The ionic equation for the reaction is as follows:

$$Ba^{2+}(aq) + SO_4^{2-}(aq) \rightarrow BaSO_4(s)$$

If excess barium chloride solution was added to a 250.0 cm^3 sample of an unknown solution to precipitate all the sulfate ions present as barium sulfate, and 1.2210 g of dry $BaSO_4$ precipitated, what was the original concentration of sulfate ion in the unknown sample?

Answer

$$m(BaSO_4) = 137.327 + 32.066 + (4 \times 15.999) = 233.389\ g\ mol^{-1}$$

(Do not use approximate atomic masses in calculations of gravimetric determinations; use the atomic masses in the Periodic Table on page x.)

amount of $BaSO_4$ = 1.2210/233.389 = 5.2316×10^{-3} mol

Since 1 mol of $BaSO_4$ contains 1 mol SO_4^{2-} ions:

amount of SO_4^{2-} present = 5.2316×10^{-3} mol

This amount of SO_4^{2-} was originally present in 250.0 cm^3 solution, therefore the concentration of SO_4^{2-} ions in original solution is

$$5.2316 \times 10^{-3} \times 4 = 0.02093\ mol\ dm^{-3}$$

EXERCISE **9H**

Gravimetric analysis

The concentration of nickel ions in a solution can be determined by precipitation of scarlet nickel dimethylglyoxime, $Ni(DMG)_2$. The ionic equation for the reaction is as follows:

$$Ni^{2+}(aq) + 2DMGH \rightarrow Ni(DMG)_2(s) + 2H^+(aq)$$

where DMG = $C_4H_7O_2N_2$.

A student used a 100.0 cm^3 sample of a 2,000 cm^3 solution and precipitated all of the nickel in the sample as $Ni(DMG)_2$. The precipitate was filtered off and heated to constant mass. The results were:

mass of container	30.8749 g
mass of container + precipitate (1st weighing)	30.9885 g
mass of container + precipitate (2nd weighing)	30.9877 g
mass of container + precipitate (3rd weighing)	30.9877 g

(i) What was the mass of the dry precipitate?

(ii) Why were the precipitates heated to constant mass?

(iii) Calculate the concentration of nickel ions in the solution.

(iv) What was the mass of nickel in the original 2,000 cm^3 of solution?

(use $m(Ni) = 58.69$ u, $m(C) = 12.011$ u, $m(H) = 1.008$ u, $m(O) = 15.999$ u, $m(N) = 14.007$ u)

9.4 Other units of concentration

Parts per million

Very low concentrations of substances are usually expressed in terms of parts per million (ppm) or even parts per billion (ppb) by mass, where:

$$\text{ppm} = \frac{\text{mass of trace substance}}{\text{mass of sample}} \times 10^6; \quad \text{ppb} = \frac{\text{mass of trace substance}}{\text{mass of sample}} \times 10^9$$

1 ppm of a species is often expressed in the units:

1 mg of the species per kg of a sample

or

1 ppm is equivalent to 1 mg kg^{-1}

See Box 9.3 for examples.

EXERCISE 91

Parts per million by mass

(i) What total mass of mercury is present in a fish of mass of 2.0 kg, which has been found to contain mercury at a concentration of 0.60 ppm (or 0.60 mg kg^{-1})?

(ii) A 5.00 kg sample of soil is found to have a total lead content of 2.00 g. What is the concentration of lead in the soil, in ppm (mg kg^{-1})?

Trace concentrations in gaseous mixtures

If we are dealing with gaseous mixtures, it is often more convenient to express ppm and ppb by *volume*, rather than by mass.

$$1 \text{ ppm by volume } (\text{ppm}((\text{v})) = \frac{\text{volume of trace substance}}{\text{total volume of sample}} \times 10^6$$

or

1 ppm by volume is equivalent to 1 cm^3 in 10^3 dm^3 of sample

These units are useful for describing the levels of pollutants in air. For example, the concentration of NO_2 in the atmosphere around a large city was 15 ppm by volume; this corresponds to 15 cm^3 of the gas in a sample of 10^3 dm^3 of air.

ppm in aqueous solution

For dilute aqueous solutions at room temperature, the density of the solution is close to the density of water (1 g cm^{-3}), so the volume of the solution in cubic centimetres may be used instead of the mass of the solution in grams. For dilute aqueous solutions, therefore, we can use:

1 ppm = 1 mg of substance dissolved in 1 dm^3 of water

and

1 ppb = 1 μg of substance dissolved in 1 dm^3 of water

These units are more convenient for low concentrations because they avoid using the very low numbers that would be involved if the concentrations were to be expressed in mol dm^{-3}; water pollutants are often given in ppm units. Also, we may use these units to describe trace concentrations of metals when we are not sure of the exact nature of the species involved. For example, drinking water may be found to have a total aluminium content at a concentration of 3 ppm. The nature and concentration of the different aluminium compounds present in a sample of drinking water is dependent on pH and can be a very complex problem to solve; the ions Al^{3+}, $AlOH^{2+}$ and $Al(OH)_2^+$ are but a few species that may be found. Without exact knowledge of the species present, calculations of molar concentration cannot be made, but it is always possible to express concentration in terms of the total mass of metal present.

BOX **9.3**

How little is 'little'?

One lump of sugar (6 g) dissolved in:

Tea pot 0.6 dm^3		1 % = 1 per cent	10 g/kg	10 mg/g
Bucket 6 dm^3		0.1 % = 1 per thousand	1 g/kg	1 mg/g 1 milligram = 0.001 g
Tank lorry 6000 dm^3		1 ppm = 1 per million	1 mg/kg	1 μg/g 1 microgram = 0.000 001 g
Super-tanker 6 million dm^3		1 ppb = 1 per billion	1 μg/kg	1 ng/g 1 nanogram = 0.000 000 001 g
Reservoir 6 billion dm^3		1 ppt = 1 per trillion	1 ng/kg	1 pg/g 1 picogram = 0.000 000 000 001 g
Bay 6 trillion dm^3		1 ppq = 1 per quadrillion	1 pg/kg	1 fg/g 1 femtogram = 0.000 000 000 000 001 g

Reproduced courtesy of LGC Group, www.lgcgroup.com

EXAMPLE **9.8**

If a sample of drinking water contains aluminium ions at a concentration of 3.0 ppm, how many mol dm^{-3} of aluminium ions does this represent?

Answer

The sample contains 3.0 mg of Al^{3+} in 1 dm^3, or

0.0030 g of Al^{3+} in 1 dm^3

or

$$\frac{0.0030}{27} \text{ mol of } Al^{3+} \text{ in } 1\,dm^3 = 1.1 \times 10^{-4}\ mol\,dm^{-3} \text{ of } Al^{3+}$$

EXAMPLE **9.9**

If 20 cm^3 of a standard solution of magnesium ions (Mg^{2+}) of concentration 100 ppm was diluted to 200 cm^3, what would be the concentration of the new solution?

Answer

First, work out the factor by which you have diluted the solution by dividing the new volume by the old volume. The 100 ppm solution of Mg^{2+} has been diluted from 20 cm^3 to 200 cm^3, i.e. by a factor of

200/20 = 10

The new concentration is worked out by dividing the old concentration by the dilution factor:

$$\text{concentration of the new diluted solution} = \frac{100}{10} = 10 \text{ ppm of } Mg^{2+}$$

EXAMPLE **9.10**

A solution of NaCl has a concentration of 0.100 mol dm^{-3}. What is its concentration in ppm (mg dm^{-3})?

Answer

1.00 mol of NaCl has a mass of (23 + 35.5) = 58.5 g. Therefore a 0.100 mol dm^{-3} solution contains 5.85 g dm^{-3} or 5,850 mg dm^{-3} of NaCl. This is also 5,850 ppm.

Comment

When the salt dissolves in water 1 mol of NaCl produces 1 mol of Na^+ ions. A 0.1 mol dm^{-3} solution contains 2.3 g dm^{-3} of Na^+, or 2,300 ppm (mg dm^{-3}) Na^+, and 3.55 g dm^{-3} or 3,550 ppm (mg dm^{-3}) Cl^-.

BOX **9.4**

Measuring the alcohol content in beverages

The most common measure of the strength of an alcoholic drink expresses alcoholic strength as a percentage of alcohol by volume, or ABV, at a temperature of 20°C. For instance, a whisky of 40 per cent **ABV** will contain 40 per cent of pure alcohol, with the rest made up of water plus various additives.

Units of alcohol are used as a guideline for the consumption of alcoholic beverages. A unit is defined as 10 cm^3 of pure alcohol (the chemical name is ethanol). The number of units in a drink can be determined by multiplying the volume of the drink (in cm^3) by its ABV, and dividing by 1,000.

Label from a wine bottle clearly showing the alcohol content. Wines generally contain between 8 per cent and 14 per cent ethanol.

EXERCISE **9J**

Aqueous concentrations in ppm and ppb

(i) Drinking water is considered acceptable if it contains less than 500 ppm of chloride ions (Cl^-). What is this concentration in mol dm^{-3}?

(ii) A standard solution contains 1,000 ppm of chloride ions. If a 50 cm^3 sample of this standard solution were diluted to 250 cm^3, what would be the concentration of the new solution?

(iii) An aqueous solution of silver ions (Ag^+) is of concentration 0.0005 ppm. Express this concentration in ppb.

EXERCISE **9K**

Percentage compositions

(i) Glucose sugar has the formula $C_6H_{12}O_6$. If you dissolved 15.5 g of glucose in 100 g of water, what would be the concentration of glucose, expressed as a percentage composition by mass?

(ii) A concentrated ammonia solution is labelled as 28 per cent (w/w). How many grams of ammonia are in 150 g of the solution?

(iii) A wine bottle has 12 per cent (v/v) written on the label. This describes the ethanol content of the wine. What volume of ethanol is in a 1.5 L bottle of wine?

(iv) How many units of alcohol are in one pint (568 ml) of beer at 5 per cent ABV?

Concentration as percentage composition

Many solutions in industry are expressed in one of two ways:

1 Percent by mass (or 'weight'), often written as percentage (w/w):

$$= \frac{\text{mass of solute}}{\text{mass of solution}} \times 100$$

2 Percent by volume, often written as percentage (v/v):

percentage composition by volume

$$= \frac{\text{volume of liquid solute}}{\text{volume of solution}} \times 100$$

Concentration expressed as mole fraction

Sometimes the physical properties of solutions, such as **vapour pressure** (a measure of the number of molecules escaping from the surface of the liquid, see Unit 10), depend upon the relative numbers of particles of the components of the solution. In these cases, the composition of a solution can be described in terms of the **mole fraction** of each of the components. The mole fraction of any component is calculated by dividing the number of moles of that component by the total number of moles present in the mixture.

For example, in a mixture of A and B, the mole fraction of A (symbol X_A) is defined as follows:

$$X_A = \frac{n_A}{n_A + n_B}$$

where n_A is number of moles of A present and n_B is number of moles of B present. Similarly, for B,

$$X_B = \frac{n_B}{n_A + n_B}$$

(Note that $X_A + X_B = 1$.)

Also note that:

1 A mole fraction has no units, it is a ratio.
2 A mole fraction has a value between 0 and 1.

EXERCISE **9L**

Mole fractions in a four-component mixture

(i) In a four-component mixture of P, Q, R and S, write an expression for the mole fraction of R, X_R.

(ii) Show that $X_P + X_Q + X_R + X_S = 1$.

(iii) What is the composition of the mixture if $X_P = 1$?

EXERCISE **9M**

Calculating mole fractions

(i) A mixture consists of 1 mol of ethanol (C_2H_5OH), 3 mol of methanol (CH_3OH) and 6 mol of water. What is the mole fraction of methanol present?

(ii) A mixture consists of 37.25 g of potassium chloride dissolved in 1,000 g of water. What is the mole fraction of potassium chloride present?

9.5 pH scale

The **pH** scale is a way of expressing $[H_3O^+]$, or $[H^+(aq)]$, as a *small* number. Note that the square brackets, [ ], stand for 'the concentration of' (in mol dm^{-3}). The pH of a solution can be calculated using the equation

$$pH = -\log [H^+(aq)]$$

Alternatively, the hydrogen ion concentration can be calculated using the equation

$$[H^+(aq)] = 10^{-pH}$$

The pH is a measure of the *acidity* of a solution. A low pH corresponds to a high value of $[H_3O^+]$ or $[H^+(aq)]$, showing the solution to be strongly acidic (Table 9.1).

The values of $[H^+(aq)]$, and thus pH, can vary depending on the acidity or basicity of the solution. The pH scale below shows the range of common pH values.

0 1 2 3 4 5 6 | 7 | 8 9 10 11 12 13 14

←increasingly acidic   increasingly basic →

Acidic solutions have pH values below 7. Solutions with pH values above 7 are basic. A pH of 7 at 25°C (the pH of pure water) is neutral. A pH meter can be used to measure the pH of a solution. It consists of an electrode that is dipped into the solution and connected to a device that gives a digital readout. Alternatively, universal indicator paper may be used. It contains a mixture of dyes that changes to different colours depending on the pH of the solution (see s.17.5, Unit 17).

Table 9.1 Approximate pH values of common substances

Substance	pH
Gastric juice	1.2–3.0
Vinegar	3.0
Lemonade	3.0–3.5
Tomato juice	4.2
Coffee	5.0
Saliva	6.6
Milk	6.7
Pure water	7.0
Blood	7.4
Eggs	7.8

EXAMPLE **9.11**

What is the pH of a solution of 0.020 mol dm^{-3} nitric acid, HNO_3, at 25°C?

Answer

As the acid may be assumed to be completely dissociated (the concentration of the acid equals the concentration of hydrogen ions),

$[H^+(aq)] = 0.020$

$pH = -\log[H^+(aq)] = -\log[0.020] = 1.70$

Since pH is a logarithmic quantity, it has no units.

EXAMPLE **9.12**

The pH of blood is 7.4. What is the hydrogen ion concentration in blood?

Answer

$pH = -\log[H^+(aq)] = 7.4$

$[H^+(aq)] = 10^{-7.4}$

$[H^+(aq)] = 4 \times 10^{-8}$ mol dm^{-3}

EXERCISE **9N**

The pH scale

Classify the following as strongly acidic, weakly acidic, strongly basic or weakly basic:

(i) lemon juice (pH 4)

(ii) rain water (pH 6 because it contains dissolved carbon dioxide)

(iii) bicarbonate of soda (or sodium hydrogencarbonate, pH 12)

(iv) sodium hydroxide of concentration 0.1 mol dm^{-3}

(v) soap solution (pH 9)

(vi) wine (pH 3.5)

(vii) household ammonia (pH 12).

REVISION QUESTIONS

(assume that all reactions go to completion)

9.1 What mass of solute must be used in order to prepare the following solutions?

(i) 250 cm^3 of 0.200 mol dm^{-3} sodium hydrogencarbonate ($NaHCO_3(aq)$)

(ii) 2.50 dm^3 of 0.500 mol dm^{-3} potassium nitrate ($KNO_3(aq)$)

(iii) 750 cm^3 of 0.050 mol dm^{-3} oxalic acid ($H_2C_2O_4 \cdot 2H_2O(aq)$).

9.2 Standard solutions of sodium hydroxide are usually obtained in the laboratory by dilution of a commercial standard solution, or they are standardized by acid–base titrations.

Why do you think that solid sodium hydroxide is not very suitable to use as a primary standard? (**Hint:** What happens to solid sodium hydroxide in air?)

9.3 What volumes of the following solutions will neutralize 50 cm^3 of 0.10 mol dm^{-3} sulfuric acid?

(i) 0.20 mol dm^{-3} potassium hydroxide solution

(ii) 0.05 mol dm^{-3} of sodium carbonate solution.

9.4 Lead(II) nitrate solution ($Pb(NO_3)_2$) reacts with potassium iodide solution (KI) according to the following ionic equation:

$$Pb^{2+}(aq) + 2I^-(aq) \rightarrow PbI_2(s)$$

(i) What volume of 1.00 mol dm^{-3} KI(aq) is needed to react exactly with 25.0 cm^3 of $Pb(NO_3)_2$(aq) of concentration 2.00 mol dm^{-3}?

(ii) This reaction can be used for the gravimetric determination of lead(II) ions in solution. What mass of PbI_2(s) would be precipitated from the lead(II) nitrate solution in (i)? (Use accurate atomic masses in this calculation.)

9.5 What mass of the following will react with 25 cm^3 of 0.250 mol dm^{-3} nitric acid?

(i) solid potassium hydroxide (KOH)
(ii) solid copper carbonate ($CuCO_3$).

9.6 Sodium thiosulfate ($Na_2S_2O_3 \cdot 5H_2O$) solution reacts with aqueous iodine in a redox reaction according to the equation

$$2S_2O_3^{2-}(aq) + I_2(aq) \rightarrow S_4O_6^{2-}(aq) + 2I^-(aq)$$

A solution, of volume 500 cm^3, contains 12.400 g of dissolved hydrated sodium thiosulfate.

(i) What is the concentration of the solution in mol dm^{-3}?
(ii) 11.8 cm^3 of this solution reacted exactly with 25 cm^3 of a solution of aqueous iodine. What is the concentration of the iodine solution in g dm^{-3}?

9.7 How much of a standard solution of Ca^{2+}(aq) of concentration 1,000 ppm should you take and dilute in order to form 2 dm^3 of a solution of concentration 50 ppm?

9.8 A bottle of aqueous hydrochloric acid (HCl) has 35 per cent (w/w) on the label. What mass of solution contains 7.0 g of HCl?

9.9 Complete this table:

Substance	Molar concentration / mol dm^{-3}	Metal ion concentration/ppm or mg dm^{-3}
KBr	0.050	
Na_2SO_4	0.010	
$Pb(NO_3)_2$	0.0060	

9.10 Calculate the mass (in grams) of lead in 200 cm^3 of a solution of lead nitrate that contains 100 mg of lead nitrate per dm^3 of solution.

9.11 What mass of $Na_2CO_3 \cdot 10H_2O$ is required to make 1 dm^3 of a 20 ppm solution of Na^+(aq)?

9.12

(i) Calculate the pH of a solution whose [H^+(aq)] = 4.48 × 10^{-9} mol dm^{-3}.
(ii) Calculate the hydrogen ion concentration (in mol dm^{-3}) for a solution with a pH value of 5.93.

9.13 Normal levels of cholesterol in blood serum are around 150 mg in 100 cm^3. Convert this to (i) ppm, (ii) per cent composition by mass (assume the cholesterol is in aqueous solution).

9.14 The agreed limit for the concentration of aluminium in drinking water is 200 ppb. A 50 cm^3 sample of the drinking water from your home was found to contain 5 μg of aluminium. Does the aluminium content of your drinking water fall below this limit?

For more about ionic equilibria and pH see Unit 17.
Extension material to support this unit is available on our website at www.bloomsburyonlineresources.com/chemistry.

Gases, Liquids and Solids 10

OBJECTIVES

This unit:

- Defines heat and temperature
- Looks at changes of state
- Applies the kinetic molecular theory of matter
- Discusses the gas laws and the idea of vapour pressure
- Shows how to use the ideal gas equation in calculations
- Introduces phase diagrams and supercritical fluids

CONTENTS

10.1 Energy and temperature

The hotness or coldness of an object is measured by its **temperature** (Box 10.1). If we make an object hot, we have transferred energy to that object. This energy is called thermal energy or '**heat**'. Another way of thinking of temperature is that it is the *condition* that determines the direction of heat flow: heat always travels from hot to cold.

10.2 Changes in the state of matter

Everyday observations show that solids have a definite shape and volume. Liquids possess the shape of their container but their volume is fixed and does not depend upon the volume of their container. Gases take up the shape and volume of their container (Figure 10.2(a)).

Kinetic molecular theory of matter

The idea that matter consists of moving particles (molecules, atoms or ions) is the basis of the **kinetic molecular theory** (or simply, **kinetic theory**). Its two main assumptions are:

1. The higher the temperature of a collection of particles, the higher the *average* kinetic energy of those particles. On average, particles move *faster* at higher temperature.
2. There is a stickiness (attractive force) between all particles.

The attractive forces between particles in a material pull the particles together, whereas the movement of the particles (the kinetic energy of the particles) pushes them apart. Changes of state occur when *one of these factors dominates*. For example, if we raise the temperature of a solid, the particles will eventually possess enough energy to partially break away from each other and melting takes place, producing a liquid state. Further heating causes the particles to move so fast that the attractive forces in the liquid are unable to hold them even loosely together and they become well separated from each other. This process is called evaporation, and the resulting gas consists of particles which (at low gas pressure) move entirely independently of each other.

BOX **10.1**

Temperature scales

The units of temperature are degrees Celsius (°C) or kelvin (K) (Figure 10.1). The size of one kelvin and one degree Celsius is the same, and the two are interconverted using the expression

$$T(\text{K}) = T(°\text{C}) + 273.15$$

Although there is no theoretical limit to the upper temperature of matter, the lowest temperature that is permitted in nature is − 273.15°C (0 K): this temperature is known as **absolute zero**. Absolute zero is often approximated to − 273°C.

Even at 0 K all motion does not cease. Particles in a solid at 0 K would vibrate very slightly and so matter still possesses some energy, even at absolute zero. This energy is called the **zero point vibrational energy**.

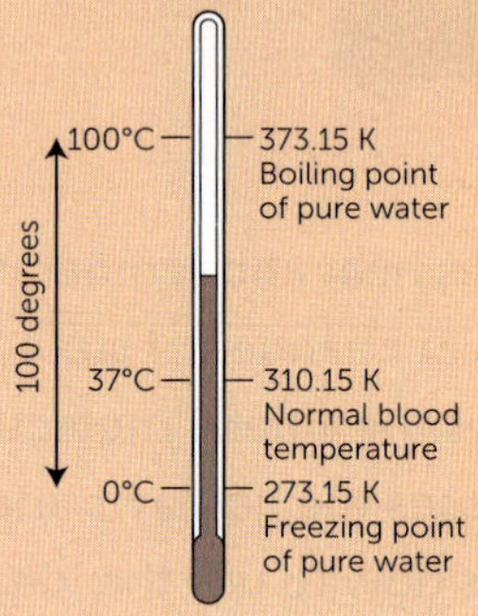

Figure 10.1 Temperature scales.

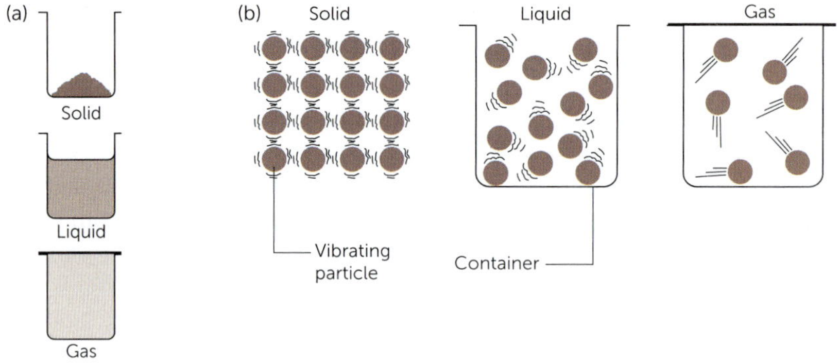

Figure 10.2 Gases, liquids and solids.

(a) The bulk appearance of the three states of matter.

(b) The molecular picture.

Similar explanations apply (but in reverse) to the other changes of state. These 'phase changes' are shown in Figure 10.3.

The 'molecular picture' of the three states of matter is summarized in Figure 10.2(b), with the particles in a solid vibrating, the particles in a liquid sliding over each other (explaining why liquids take up the shape of their container) and the particles in a gas moving so rapidly that they take up the shape *and* volume of their container.

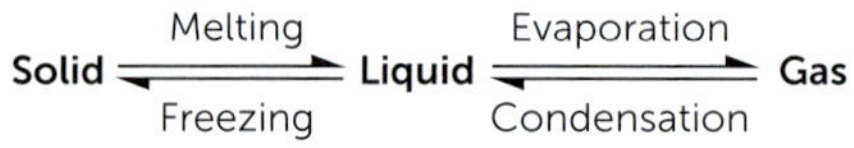

Figure 10.3 Changes of phase.

The position of atoms (and ions) in solids and liquids may be studied using **X-ray diffraction**. This technique confirms that atoms in a liquid are in a state of orderliness which is intermediate between those of gases and solids.

EXERCISE **10A**

Changes of phase

What is the significance of the double pointed arrows (⇌) in Figure 10.3?

Explanations based upon the kinetic molecular theory

The kinetic theory of matter explains many everyday observations including:

1 The relative density of solids, liquids and gases

The density of a material tells us how closely packed the particles are. Experiments show that solids usually possess the greatest density, and gases (in which the molecules are moving fastest) the least. Gases have low densities because the gaps between the particles are relatively great. For this reason, gases have a volume that may easily be reduced by applying pressure. This is what we mean when we say that gases are easily compressed.

2 The variation of gas pressure with temperature

Gas pressure results from the bombardment of the container walls by molecules of the gas (Figure 10.2(b)). At high temperatures, the gas molecules are moving faster and both the number of collisions per second and the energy of the colliding molecules increase. This causes an increase in gas pressure.

3 The evaporation of liquids

A drop of alcohol in a dish slowly vanishes. This 'vanishing act' is no magic, because the alcohol has simply become alcohol gas (alcohol *vapour*). Evaporation of liquid molecules from the liquid surface occurs at all temperatures. Only fast-moving molecules are able to escape from the attractive forces of the other molecules in the liquid. Since the number of fast-moving molecules increases as the average kinetic energy of the sample of molecules increases, the rate of evaporation of a liquid increases with temperature.

4 Diffusion

The spreading out of particles is called **diffusion**. For example, a gas will fill up its container because the gas molecules are in constant and random movement, and a crystal of copper(II) sulfate will slowly dissolve in water (even without agitation) because the water molecules are continually striking the crystal and the individual copper and sulfate ions disperse into the water.

EXERCISE **10B**

Kinetic theory

Use the kinetic theory to explain the following:

(i) Two beakers are heated with identical Bunsen flames for the same length of time. Beaker A contains less water than beaker B. Why does the temperature of the water in beaker A rise faster than that of beaker B?

(ii) The drying of clothes on a line is an example of evaporation. Explain why clothes dry faster on hot, windy days.

(iii) The smell of a stink bomb released in a corner of a room soon spreads to the rest of the room.

Heating and cooling curves

A heating curve is a plot of the temperature of a substance as it is being slowly heated. Phase changes are taking place on the flat parts of the curve (Figure 10.4).

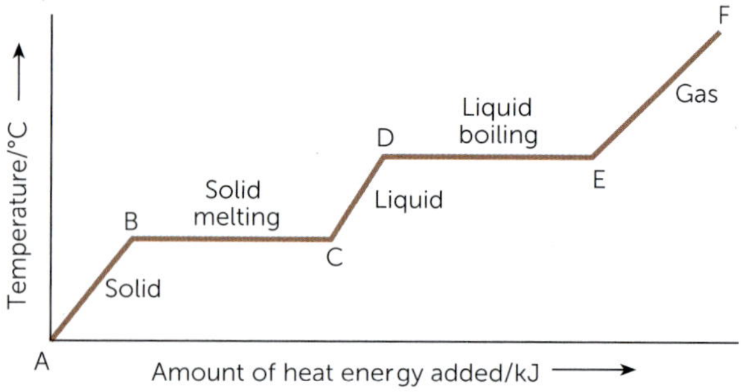

Figure 10.4 Heating curve for a pure substance.

Melting

As we start to heat the solid substance, its temperature rises (from A to B). The slope of the line AB depends upon the **specific heat capacity** of the solid. As soon as the first drop of liquid is produced (at B), the temperature of the liquid–solid mixture remains constant (from B to C). This constant temperature is the **melting point** or **freezing point** (or **melting temperature**) of the substance. The constant temperature arises because all the added energy is used to melt the solid: no energy is used to raise the temperature of the substance. At point C, all of the solid has melted. The **normal melting point** of water is its melting point at a surrounding air pressure of 101 kPa (1 atm) which is 0°C.

Boiling

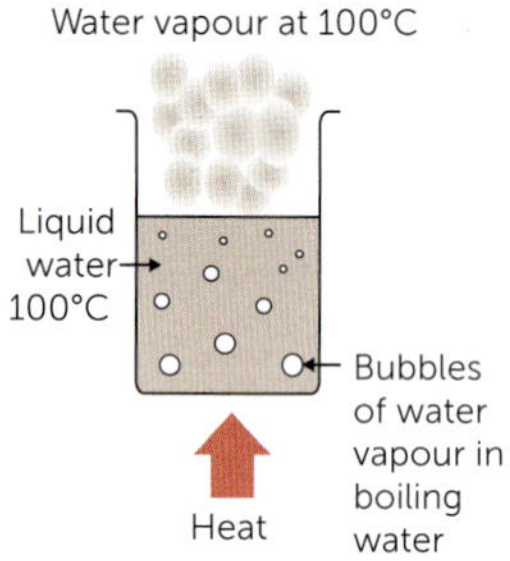

Figure 10.5 Boiling water.

As we continue heating, the liquid becomes hotter (C to D). The slope of the line CD depends upon the specific heat capacity of the liquid. At D, the liquid begins to boil. When the liquid boils, its temperature does not rise – all the added heat is used to boil the liquid. Both the liquid and gas (vapour) are at the same temperature. This constant temperature is called the **boiling point** (or **boiling temperature**) of the substance. The boiling point of water at an air pressure of 1 atmosphere (101 kPa) is 100°C. (At this temperature, the vapour pressure of water equals one atmosphere, see s10.7.)

At E, all the substance is present as a gas. If we contained the gas in a sealed vessel and then applied more heat, its temperature would continue to rise. The slope of EF depends upon the specific heat capacity of the gas.

If a substance is cooled (e.g. by putting it in a freezer) the resulting **cooling curve** is the mirror image of Figure 10.4. The energy changes of Figure 10.4 are also reversed. For example, steam condensing to liquid water and liquid water that is freezing both **give out** heat. Now try Exercise 10C.

EXERCISE **10C**

Boiling

When cooking vegetables by boiling them in water on a cooker, does the speed at which the vegetables cook depend upon the rate at which the water is boiled?

10.3 Gas laws

Think of a gas in a sealed container. We can measure the number of moles n of gas inside, the pressure P of the gas (Box 10.2), the volume V of the gas (which equals the volume of the container) and the temperature T of the gas. The observed relationships between P, V, n and T are summarized by four laws: Charles' law, Boyle's law, Avogadro's law and Dalton's law. Gases that obey these laws perfectly are known as **ideal gases**. Real gases behave approximately as ideal gases at room temperature and at atmospheric pressure. Gases become less ideal at higher pressures (> 1 atm) and at low temperatures (especially close to their boiling point).

If a gas is at 25°C and at a pressure of exactly 100 kPa (1 bar), the gas is said to be at **standard ambient temperature and pressure** (symbolized SATP). An earlier set of standard conditions, known as **standard temperature and pressure** (STP), refers to a gas at 0°C and exactly 1 atm pressure.

BOX **10.2**

Gas pressure

$$\text{Pressure} = \frac{\text{force}}{\text{area}}$$

The SI unit of force is the newton (N), and the SI unit of area is the square metre (m^2). So, the SI unit of pressure is newton per square metre ($N\,m^{-2}$) also known as the **pascal** (Pa).
The pressure of air at sea level is said to be one 'standard atmosphere' (often simply called 'one atmosphere'). The symbol for (standard) **atmosphere** is **atm**:

$$1\text{ atm} = 101325\text{ N m}^{-2} = 101.325\text{ kPa}$$
$$1\text{ atm} \approx 1 \times 10^5\text{ N m}^{-2} \approx 101\text{ kPa}$$

Exactly 100 kPa is called 1 **bar**:

$$100\text{ kPa} = 1\text{ bar} = 0.987\text{ atm}$$

Note that 1 bar is very nearly equal to 1 atm. The 'torr' is still used as a unit of pressure in some laboratories. One torr is the pressure of air that supports 1 mm of mercury in a barometer at sea level. There are 760 torr in one standard atmosphere:

$$1\text{ atm} = 760\text{ torr} = 101.325\text{ kPa}$$

EXERCISE **10D**

Pressure

Use the information in Box 10.2 to convert the following to pascals:

(i) 20 torr

(ii) 10 atm

(iii) 107,000 $N\,m^{-2}$

(iv) 1 kPa

Charles' law

The expansion of a fixed mass of gas at constant pressure as its temperature is raised is predicted by **Charles' law:**

The volume of a gas is directly proportional to its temperature in kelvin.

One way of allowing a sealed mass of gas to expand is to use a piston (Figure 10.6). Since the piston handle is free to move, the pressure of the gas inside the piston will adjust itself so that it equals the pressure on the piston handle (the 'external pressure'). For example, if the external pressure (due to the air) is 1 bar, the gas inside the piston will also take up a pressure of 1 bar.

Mathematically, Charles' law may be expressed as

$$V \propto T$$

where V is the volume of gas and T is its temperature in kelvin. The symbol $\propto$ means 'proportional to' and shows (for example) that if T doubles, V also doubles.

Charles' law can be used only where the volume of a gas is *allowed* to change. For example, heating a sample of gas in a sealed metal tube will not alter the volume of the gas because the rigid walls of the container prevent expansion.

Suppose a piston contains 100 cm^3 of hydrogen gas at 300 K (the temperature of the laboratory) and is under an external air pressure of 1 bar. Imagine that the piston is heated so that the gas eventually reaches a constant temperature of 600 K. Heating causes the temperature of the gas to rise. This causes the gas molecules to collide more often and more violently with the inner wall of the piston handle, resulting in a momentary increase in hydrogen gas pressure above 1 bar. Since the pressure of the hydrogen gas is greater than the pressure on the piston handle, the gas forces the piston handle back. In this way the gas increases its volume, but since the same number of molecules of hydrogen gas are now spread out in a larger space, the number of collisions per second on the inner wall of the piston handle falls and the pressure of the hydrogen gas begins to fall back towards 1 bar. The piston handle stops moving when the volume of the gas in the piston has been increased so that the pressure of the hydrogen equals 1 bar. Charles' law predicts that a doubling of the hydrogen gas temperature doubles the volume of the gas – see Figure 10.6.

If the piston is allowed to cool, the reverse effect occurs and the piston handle moves inwards, so reducing the volume of gas inside.

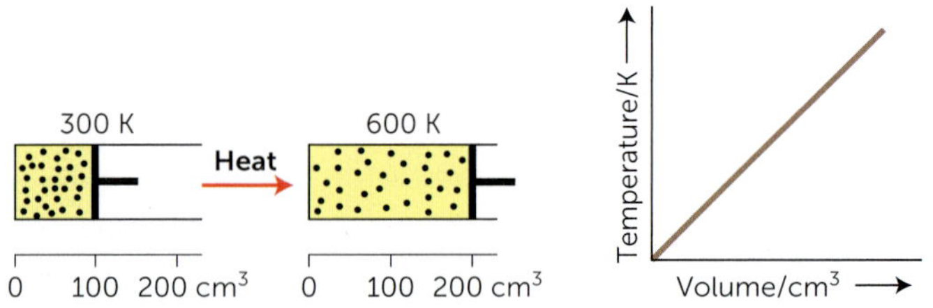

Figure 10.6 Charles' law.

EXERCISE **10E**

Charles' law

A sample of nitrogen at 500 K occupies 73.0 cm^3. Predict the volume of the gas at 2,500 K. (Assume that its pressure has not changed.)

Boyle's law

Think of a sample of gas of volume *V* contained in a piston (Figure 10.7). If we push the piston handle inwards, the pressure on the gas exceeds atmospheric pressure and the volume of the gas drops. Boyle's law states that for a fixed mass of gas at a particular temperature

The volume (V) of a gas is inversely proportional to the pressure (P) of the gas.

Mathematically, Boyle's law is expressed as

$$V \propto \frac{1}{P}$$

Figure 10.7 illustrates the case where the applied pressure increases from 1 bar to 2 bar. This halves the gas volume, as predicted by the above equation.

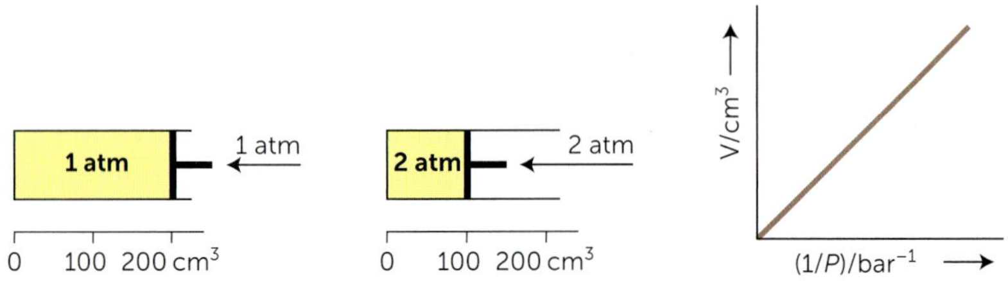

Figure 10.7 Boyle's law.

EXERCISE **10F**

Boyle's law

A gas contained in a piston at a pressure of 30.0 kPa occupies a volume of 200.0 cm^3 at 300 K. The pressure of the gas is slowly increased to 90.0 kPa without a change in temperature. What is the new volume of the gas?

Avogadro's law

Avogadro's law states that at a fixed pressure and temperature

The volume of a gas is proportional to the number of moles (or molecules) of gas present.

Mathematically we can write

$$V \propto n$$

where n is the number of moles of gas.

The following statement follows from Avogadro's law:

Equal volumes of all gases at the same temperature and pressure contain the same number of molecules.

It also follows that one mole of *any* gas occupies the same volume at the same temperature and pressure. This volume is known as the **molar volume** of the gas. Experiments show that the molar volume of an ideal gas at SATP is 24.79 dm^3. The molar volume of an ideal gas at 1 bar and 20°C is about 24 dm^3 (Figure 10.8), and it is this volume that we will use in routine calculations.

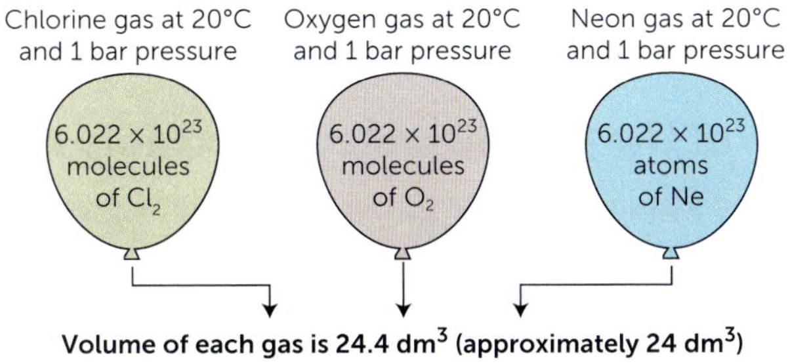

Figure 10.8 Molar volume.

EXERCISE **10G**

Avogadro's law

0.50 g of zinc reacts completely with excess dilute sulfuric acid at 20°C. Calculate the volume of hydrogen produced at this temperature. (The molar volume of a gas at 20°C and 1 atm pressure is 24 dm^3; m(Zn) = 65 u.)

Dalton's law of partial pressures

This law applies to gases that do not react chemically. It states that

The total pressure of a gas mixture is the sum of the partial pressures of the individual gases in the mixture.

The **partial pressure** of a gas in a mixture is defined as the **pressure that gas would exert if it alone occupied the container.**

Mathematically, Dalton's law may be expressed as

$$P_T = p_A + p_B + p_C \ldots$$

where P_T is the total pressure of the mixture and p_A, p_B, p_C ... are the partial pressures of the gases A, B, C ... in the mixture.

An example will show how easy it is to use Dalton's law. Suppose we add nitrogen to an initially empty gas cell so that the pressure of the nitrogen in the cell becomes 1.0 bar. If we now add an amount of oxygen (such that the pressure of oxygen if it alone occupied the gas cell would be 0.5 bar), a measurement would show that the total pressure of the mixture is 1.5 bar. This is predicted by Dalton's law:

$P_T = pN_2 + pO_2 = 1.0 + 0.5 = 1.5$ bar

Dalton's law 'works' because the gaps between gas molecules are so large that the molecules of the constituent gases do not interfere with each other. The total pressure of the mixture is then the sum of the pressures exerted by the individual gases.

The most important gas mixture is, of course, the air around us (Table 10.1).

EXERCISE 10H

Air

Use the data in Table 10.1 to calculate the volume of argon in 1.0 dm^3 of air. (Assume the air is dry and at sea level.)

Table 10.1 Composition of dry air at sea level

Gas	Average percentage (by volume)
Nitrogen N_2	78.08
Oxygen O_2	20.95
Argon Ar	0.93
Carbon dioxide CO_2	0.04
Neon Ne	0.018

10.4 Kinetic molecular theory of gases

Assumptions of the kinetic molecular theory of gases

The kinetic theory of gases is a physical model that may be expressed in mathematical form. It assumes:

1 Gases consist of tiny particles (molecules or atoms) that are in chaotic and random motion.
2 The particles *do not* attract or repel each other.
3 The collision of two molecules does not alter the overall energy of both molecules.
4 The volume of the particles is negligible compared with the volume of the container.

As we have already noted, assumption 1 immediately provides us with an explanation of gas pressure. The particles frequently collide with the molecules which make up the container walls: it is the force of these collisions which (on a gigantic scale) is responsible for the pressure of a gas.

Assumption 2 is a simplifying feature which applies *only* to gases at low pressures where the particles are so far apart from each other that they behave as independent 'mathematical points'. Intermolecular forces *cannot* be ignored in the solid and liquid state, nor in explaining why gases condense to form liquids. This is one reason why solids and liquids are theoretically more complicated to deal with than gases, and why there is no simple equivalent of the ideal gas equation (discussed below) for these states of matter.

Assumption 3 is another way of stating that the average kinetic energy of particles in a gas remains fixed at a particular temperature. If the energy of particles were lost in collisions (either wall–gas particle or gas particle–gas particle collisions) the total energy of the gas would be continuously draining away, and (contrary to experimental observation) the pressure of a gas would fall with time.

Kinetic energy of gas molecules

Since the molecules of a gas are in constant motion, they all possess **kinetic energy**. The greater the kinetic energy of a gas molecule, the greater its speed. Experiments show that molecules in a sample of gas are not all travelling at the same speed. Figure 10.9

shows the spread of molecular speeds of nitrogen at 300 K and 3,000 K, and of chlorine gas at 300 K.

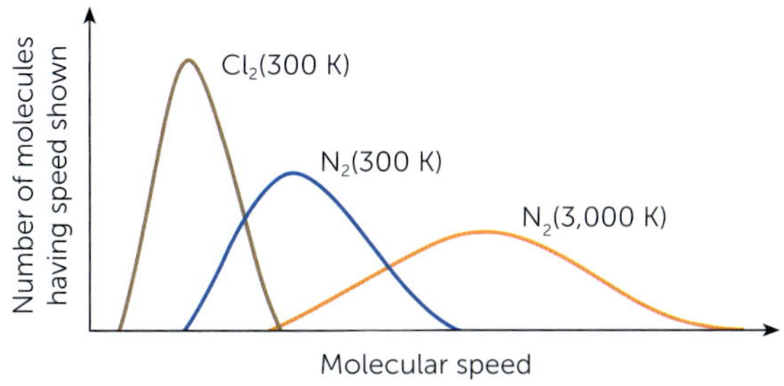

Figure 10.9 The spread of molecular speeds in chlorine and nitrogen gases.

1 The **average speed** of gas particles at a particular temperature depends upon the mass of the gas molecules, with smaller molecules (such as N_2) moving faster than heavier ones (such as Cl_2). Even relatively heavy molecules travel at average speeds of several hundred metres per second at room temperature!
2 The average speed of molecules in a gas also *increases* with the temperature of the gas, but the spread of speeds is greater at higher temperatures.

10.5 Ideal gas equation

The kinetic theory of gases reduces to a remarkably simple mathematical form, known as the **ideal gas equation**.

$$PV = nRT$$

where P is the pressure of a gas (Pa), V is the volume of gas (m^3), T is the gas temperature (K) and n is the number of moles of gas. R is the 'universal gas constant' with a value of 8.3145 J mol^{-1} K^{-1}.

Key points about the ideal gas equation

1 The ideal gas equation shows that P, V, n and T are not independent. These variables cannot take on any values we like – nature has dictated that if three of them are known, then the fourth is fixed.
2 Remarkably, the equation contains no factor which is *specific* to any gas. The type of gas (helium, chlorine, oxygen, water vapour, etc.) is irrelevant.
3 The rearranged forms of the ideal gas equation are often used in calculations:

$$V = \frac{nRT}{P}; T = \frac{PV}{nR}; P = \frac{nRT}{V}; n = \frac{PV}{RT}$$

4 The interrelationships between P, V, n and T, summarized in the gas laws of Charles, Boyle and Avogadro, are 'inbuilt' into the ideal gas equation, and the equation makes the separate use of these laws in calculations redundant. For example, the ideal gas equation shows that if the pressure and the number of moles of gas do not change, then V is proportional to T. In other words, it predicts Charles' law.

The mathematics of the ideal gas equation is examined further in **Extension 10** on the website, https://www.bloomsburyonlineresources.com/chemistry

Examples in the use of the ideal gas equation

Note that, to make calculations easier, you always express *T* in kelvin, *V* in cubic metres, *P* in pascals and *n* in moles.

EXAMPLE **10.1**

Calculate the volume of 1.0000 mol of gas at exactly 20°C at a pressure of 101.325 kPa.

Answer

$T = 20 + 273.15 = 293.15\ \text{K}$

Rearranging the ideal gas equation,

$$V = \frac{nRT}{P} = \frac{1.0000 \times 8.3145 \times 293.15}{101\,325} = 0.024\,055\ \text{m}^3 \quad \text{(to five significant figures)}$$

Remembering that $1\ \text{m}^3 = 1000\ \text{dm}^3$,

volume = $0.024055 \times 1{,}000 = 24.055\ \text{dm}^3$

EXAMPLE **10.2**

Calculate the number of molecules of methane in $0.50\ \text{m}^3$ of the gas at a pressure of 2.0×10^2 kPa and a temperature of exactly 300 K.

Answer

$2.0 \times 10^2\ \text{kPa} = 2.0 \times 10^5\ \text{Pa}$

$$n = \frac{PV}{RT} = \frac{2.0 \times 10^5 \times 0.50}{8.3145 \times 300} = 40 \quad \text{(to two significant figures)}$$

To convert moles to molecules, we multiply by Avogadro's constant N_A:

number of molecules = $6.022 \times 10^{23} \times 40 = 2.4 \times 10^{25}$

EXERCISE **10I**

Ideal gas law

Calculate the pressure of 1.0 mol of helium in a $2.0\ \text{dm}^3$ container at 20.0°C.

Experimental measurements of the deviations of real gases from ideal behaviour

An ideal gas is one that exactly follows the ideal gas equation. Rearranging the ideal gas equation for *n*, we obtain:

$$n = \frac{PV}{RT}$$

If we use one mole of a gas (*any* gas) then the ratio *PV*/*RT* is predicted to have a numerical value of 1 *at all pressures*. If a gas does not obey the ideal gas law, the ratio will be either greater than 1 or less than 1.

Experiments show that at pressures below 1 bar (roughly atmospheric pressure), the ratio *PV*/*RT* does not deviate appreciably from one for most gases. This gives us a rough guide – **most real gases behave nearly ideally if their pressure is below 1 bar**. Even at a pressure of several bars, many gases behave sufficiently ideally that the use of the ideal gas equation remains justified.

A plot of PV/RT against P for three gases is shown in Figure 10.10. Deviations from ideal behaviour for CO_2 occur because, above 1 bar pressure, the attractive forces between the gas molecules can no longer be neglected. A single particle striking the wall of the containing vessel will be slightly attracted (and therefore 'held back') by other molecules in the bulk of the gas. As a consequence, the force with which the molecule strikes the container wall is *reduced*. On a bulk scale this causes the pressure of the gas to be *lower* than that of an ideal gas. Because P is lower than that for an ideal gas, the ratio PV/RT is smaller than it would be for an ideal gas. For example, PV/RT for CO_2 is less than 1 between 10 and 600 bar.

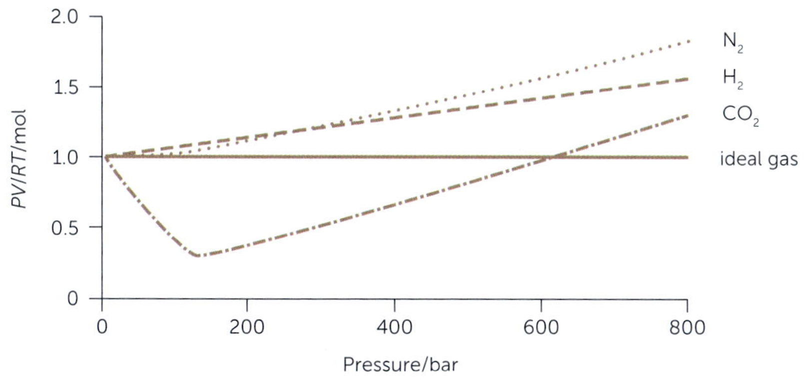

Figure 10.10 Deviations from ideal gas behaviour. The behaviour of a gas following the ideal gas equation at all pressures is shown by a continuous line. The true behaviour of hydrogen (H_2, at 273 K), nitrogen (N_2 at 273 K) and carbon dioxide (CO_2, at 313 K) is shown by the discontinuous lines.

EXERCISE **10J**

Deviation from ideal gas behaviour

Why would you expect carbon dioxide to deviate from ideal behaviour more than hydrogen or nitrogen?

At yet higher pressures, it is no longer reasonable to assume that the volume of the gas molecules is negligible compared with the container volume. This makes PV/RT greater than one because the volume V is greater than that for an ideal gas. For CO_2, this is observed at pressures above 600 atm and the effect dominates the H_2 and N_2 plots throughout the pressure range shown.

10.6 Adsorption of gases on solids

Charcoal is used in gas masks to remove toxic gases. This is one example of a general phenomenon known as **adsorption**.

In adsorption, a substance (the **adsorbate**) sticks to the *surface* of another material (the **adsorbent**). This is distinct from **absorption**, in which a substance penetrates another material. The difference between the two phenomena may be remembered by thinking of a bath sponge. If dust sticks to the surface of the dry sponge then adsorption has occurred, but if water is taken into the sponge, absorption has occurred.

It is often impossible to decide whether adsorption, absorption or a mixture of both is occurring. For this reason, the noncommittal term 'sorption' is used for both processes. An example of this difficulty is found in chromatography, where a combination of solute absorption and adsorption may bring about the separation of a mixture on a chromatographic column (Unit 21).

Gas masks contain activated charcoal that adsorbs toxic gases in the surrounding air. This photograph of a family out shopping in London was taken during a tear gas test in 1941 (Second World War).

The forces that bind the adsorber to adsorbent may be physical (i.e. intermolecular forces) or chemical (when chemical bonds are formed). The adsorption of gases by charcoal is physical, whereas the adsorption of gases on some catalysts is chemical. Toxic gases adsorb to charcoal better than the oxygen and nitrogen of the air because they are often relatively large molecules and are frequently polar. This increases the intermolecular forces between adsorbate and adsorbent.

The charcoal used in gas masks is called 'activated charcoal', and is made by heating wood or coal in carbon dioxide, water vapour or in a limited supply of air. Activated charcoal has an enormous surface area – crucial to adsorption – as high as 1,000 m^2 of gas per gram of charcoal! At a given temperature and pressure, a sample of activated charcoal will adsorb only a fixed mass of adsorbate, but the mass increases as the pressure of adsorbate gas is increased or if the temperature is lowered. The charcoal may be regenerated by heating, which drives off the adsorbate.

Adsorption will also occur in solution. For example, charcoal beds are used to adsorb trace organic compounds (such as pesticides and dyes) from water.

10.7 Vapour pressure

Meaning of vapour pressure

A **vapour** is a gas in contact with a liquid of the same substance. For example, the water 'gas' above the surface of liquid water is described as water vapour. The gas pressure of the water vapour is known as its **vapour pressure**.

Suppose that liquid water is placed into a container and that the gas (mainly air) above the water is pumped away using a powerful pump, and the container sealed without allowing air to re-enter. Some of the liquid water then evaporates so that the only gas in the space above the liquid water is water vapour. Figure 10.11(a) shows how a pressure measuring device might be used to measure the vapour pressure of the water.

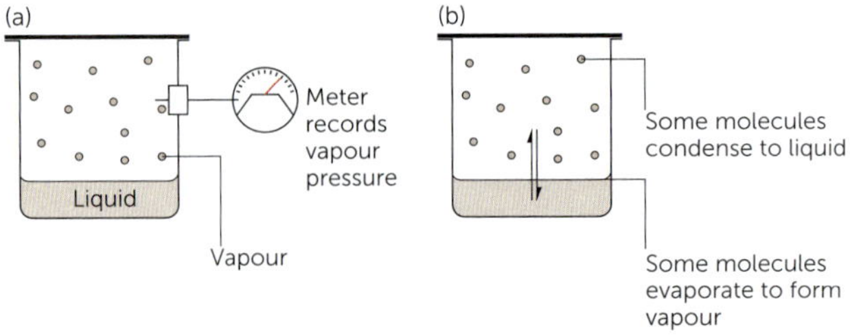

Figure 10.11 Vapour pressure: (a) experimental measurement of saturated vapour pressure; (b) the dynamic equilibrium that gives rise to the saturated vapour pressure.

The vapour pressure increases as soon as the container is sealed, but the rate of increase slows down after a few seconds because, although water molecules continue to evaporate, some water molecules also condense back to form liquid water. After a few more seconds the vapour pressure levels off and reaches a maximum called the **saturated vapour pressure** – so-called because the air is saturated with water vapour at that temperature. (For pure water at 20°C, the saturated vapour pressure is 2.33 kPa.) Even at this stage, molecules continue to evaporate and condense, but a **dynamic equilibrium** has been achieved in which the number of molecules evaporating per second equals the number of molecules that are condensing per second (Figure 10.11(b)). For this reason, saturated vapour pressure is often called **equilibrium vapour pressure**. We represent the equilibrium by the equation

$$\text{liquid} \rightleftharpoons \text{vapour}$$

If the container was open to the air, the water vapour would continually diffuse away and the vapour pressure would not reach its equilibrium value. More and more water would slowly evaporate into the air until (after several days) no more water is left in the container.

Figure 10.12 shows the saturated vapour pressure of water at various temperatures. The figure also shows the way that the saturated vapour pressure of ethanol varies with temperature. The vapour pressure of water and ethanol (like that of all liquids) increases with temperature. This is expected, since at the higher temperature more liquid molecules have sufficient energy to escape the attractive forces holding molecules in the liquid.

At the boiling point of a liquid, the saturated vapour pressure *equals* the external air pressure. At sea level, the air pressure is 1 atm (101 kPa). The boiling point of a liquid at 1 atm external pressure is called the **normal boiling point** of the liquid. The normal boiling points of water and ethanol are 100°C and 78°C, respectively.

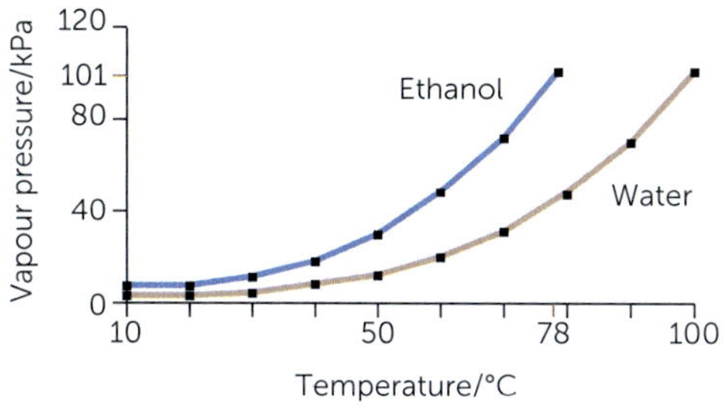

Figure 10.12 The vapour pressure of ethanol and water between 10°C and 100°C.

EXERCISE 10K

Boiling point

Explain the following:

(i) The boiling point of water drops with increasing altitude.

(ii) Salty water boils at a higher temperature than pure water.

Vapour pressure and total gas pressure

If a *sealed vessel* contains a liquid X (such as ethanol or water) and a gas Y (such as nitrogen), it is usually safe to assume that the gas Y is saturated with the vapour of X. As expected from Dalton's law, the total gas pressure above the liquid is the sum of the gas and vapour pressures. This is illustrated by Example 10.3.

EXAMPLE 10.3

A sealed bottle contains carbon monoxide gas (CO) and liquid water at 20°C. A pressure gauge measured the total gas pressure in the bottle as 81.2 kPa. What is the partial pressure of carbon monoxide in the bottle? The saturated vapour pressure of water at 20°C is 2.33 kPa.

EXAMPLE **10.3** *(continued)*

Answer

The presence of liquid water in the bottle shows that there is enough water present to saturate the CO with water vapour (Figure 10.13). Since

$$\text{total gas pressure} = 81.2\,\text{kPa} = p_{CO} + p_{H_2O}$$

where p_{CO} and p_{H_2O} represent the vapour pressure contributions from the two substances. Since $p_{H_2O} = 2.33\,\text{kPa}$,

$$p_{CO} = 81.2 - 2.33 = 78.9\,\text{kPa}$$

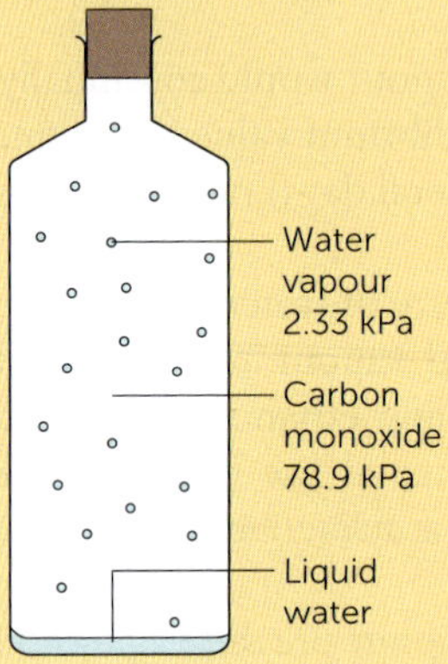

Figure 10.13 Vapour pressure and total gas pressure.

EXERCISE **10L**

Vapour pressure calculations

A corked test tube at 25°C contains air and a few drops of liquid ethanol. The total gas pressure in the tube is 101 kPa. What is the partial pressure of air in the tube? (Saturated vapour pressure of ethanol at 25°C is 7.9 kPa.)

Volatility of solvents

Ethanol evaporates much more readily than water; therefore, we say that ethanol is more **volatile**. This is reflected in Figure 10.12, where the vapour pressure of ethanol exceeds the vapour pressure of water at every temperature.

One way to estimate the volatility of solvents at room temperature is to saturate pieces of filter paper with solvent, and time how long it takes for all the solvent to evaporate. Alternatively, the boiling point of a solvent may be taken as a rough indication of its volatility. One of the most volatile solvents available in the laboratory is ethoxyethane (diethyl ether), with a normal boiling point of 34.5°C (see Box 10.3).

BOX **10.3**

Flash point of solvents

Many solvents are good fuels, and will readily burn in air. This can be a safety hazard in laboratories and factories. The temperature at which a solvent catches fire in the *absence* of a flame is called its **autoignition temperature**. This is surprisingly high for most solvents (Table 10.2).

The lowest temperature at which the vapour pressure of a pure solvent in air is sufficient to be ignited by a naked flame is called the **flash point** of the solvent (Figure 10.14). This can be very low (Table 10.2). For

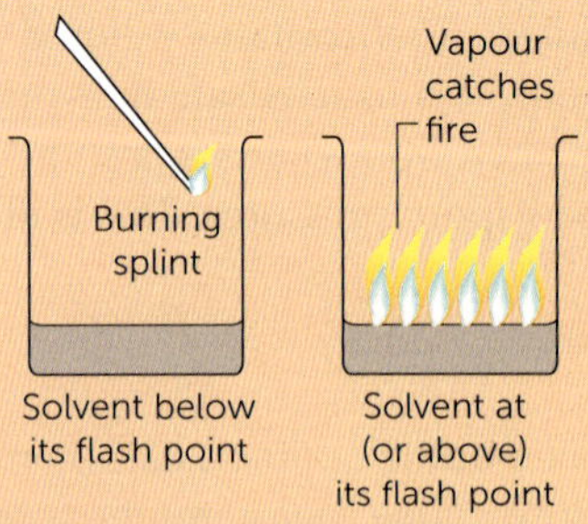

Figure 10.14 Solvent flash point.

example, the flash point of petroleum ether (a fraction obtained from the distillation of crude oil and which boils between 30 and 40°C) is – 51°C. This means that petroleum ether stored in an open (or leaking) container will still represent a serious fire hazard even if stored in a freezer operating at – 51°C. Flash points are used to classify the **flammability** of solvents.

A **flammable** liquid is one with a flash point less than 55°C, a **highly flammable** liquid is one with a flash point less than 21°C and an **extremely flammable** liquid is one with a flash point less than 0°C and a boiling point of 35°C or less.

Table 10.2 Flash points and ignition temperatures of common solvents (see also page xxx)

Solvent	Flash point/°C	Boiling point/°C	Autoignition temperature/°C
Propanone (acetone)	−20	56	465
Benzene	−11	80	498
Ethoxyethane (diethyl ether)	−41	34.5	180
Ethanol	−13	78	363
30–40 petroleum spirit	−51	30–40	230
Butan-1-ol	−29	118	343

10.8 Critical temperature and pressure

If the volume of a gas (or vapour) is reduced by compressing it (as in a piston), or cooling it, or both, the gas may condense to a liquid. Experiments show that a gas *cannot be liquefied* by pressure alone unless it is at (or below) a temperature known as the **critical temperature**. Put another way, *the critical temperature is the highest temperature at which a liquid can exist.*

At a molecular level, molecules in a sample at a temperature above the critical temperature simply have too much energy to 'stick together' – no matter what the applied pressure.

The pressure needed to liquefy a gas at its critical temperature is called its **critical pressure**. The critical temperatures and pressures for some gases are given in Table 10.3. For example, if ethanol vapour is *above* 240.7°C, no amount of pressure will convert the vapour to liquid. If the temperature of the ethanol vapour was exactly 240.7°C, a pressure of 61 atm would need to be applied to the vapour in order to force it to condense to a liquid. We discuss this further in section 10.10.

Table 10.3 Critical temperatures and pressures for common gases

Gas	Normal boiling point/°C	Critical temperature/°C	Critical pressure/atm
Ammonia	−33	132.4	114
Carbon dioxide	−78 (sublimes)	31.0	74
Ethanol	78	240.7	61
Hydrogen	−252	−240.0	13
Oxygen	−183	−118.6	53
Nitrogen	−196	−146.9	34
Water	100	373.9	218

EXERCISE **10M**

Critical temperature and pressure

(i) A mixture of hydrogen and oxygen at – 119°C is subjected to a very high pressure. Would you expect liquid hydrogen and liquid oxygen to condense?

(ii) Historically, hydrogen was one of the last gases to be liquefied in the laboratory – why?

10.9 Phase diagrams

Phase diagrams show the most stable (equilibrium) arrangement of the phases of a material at a particular pressure and temperature. For example, the most stable molecular arrangement might be that the material exists as all solid, or as a mixture of solid and liquid. Although they vary from substance to substance, a generalized phase diagram for a pure substance is shown in Figure 10.15. The y-axis is pressure (P, in bar or atm) and the x-axis is temperature (T, in °C or K). The curves are produced by joining a large number of experimentally determined points, each point being a measurement of the pressure and temperature of the system in whatever phase or phases it exists.

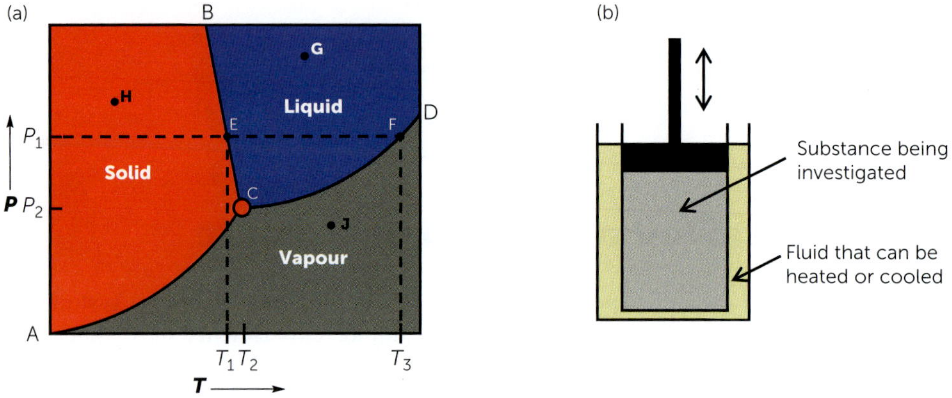

Figure 10.15 (a) Generalized phase diagram for a pure substance. (b) Equipment used to investigate phase changes.

The principles of the experimental investigation of phase changes are illustrated in Figure 10.15(b). A metal cylinder with a tight fitting piston is completely filled with the substance under investigation (e.g. water or carbon dioxide). The pressure on the substance can be increased (by mechanically pushing down the piston) or reduced (by drawing up the piston). The cylinder is immersed in a drum containing a fluid that can be cooled or heated to reach the desired temperature.

The key to understanding phase diagrams is to be able to read the P and T coordinates correctly by going straight across for pressure and straight up for temperature. Two examples are made explicit in Figure 10.15 by using dashed lines. Point E represents a pressure P_1 and a temperature T_1. Point F represents a pressure P_1 and a temperature T_3.

The curves CB, AC and CD are very important. Any P and T coordinates lying on these curves represent conditions where two phases are in equilibrium, i.e. where two phases can co-exist without the system changing over time. If a system has pressure and temperature coordinates lying in the coloured areas (e.g. H, G and J), but not precisely on the curves CB, AC and CD, only one phase is present at equilibrium.

In what follows, we will refer to the 'most stable' molecular arrangements. In the language of thermodynamics (Unit 16), this represents the lowest Gibbs energy for that system.

We will apply the phase diagram to the liquid water/ice/water vapour system:

- Suppose we alter the temperature and pressure of the system to P_1 and T_1, corresponding to point E on the curve CB. If we could see inside the metal cylinder (perhaps through a toughened glass window) we would see only ice and water. This is because E represents a pressure and temperature at which ice and liquid water (and not water vapour) can co-exist in dynamic equilibrium.The mixture of ice and liquid water is completely stable and does not alter provided that the pressure and temperature stay the same. If we carefully controlled any changes in T and P so that the system moved along curve CB, the system would continue to consist of ice and liquid water in equilibrium with each other, although the proportions of ice and water would be different.

- If the pressure of the system at E is suddenly increased and the temperature suddenly decreased to correspond to point H, which lies in the pure solid zone, the stable system is now solid ice and any existing water begins to freeze. If we change T and P to correspond to point G, the stable arrangement of the water molecules is then pure liquid only. At lower pressure and higher temperature (e.g. point J) the stable system is water vapour only and any liquid present will begin to vaporize.
- If P_1 = 101 kPa (1 atm), then it is found that for the water system, T_1 = 0°C, the **normal freezing point** of water.
- Point F, also at a pressure of 101 kPa, lies on the liquid–vapour equilibrium line. T_3 is 100°C, the normal boiling point of water.
- Curve CD is of practical importance. The temperature at which a liquid and vapour are at equilibrium is the boiling point of the system. Therefore, CD represents the variation of the boiling point of water with external pressure. The boiling point increases with increasing pressure and vice versa.
- Point C is a unique point. At this temperature (T_2) and pressure (P_2) only, all three phases (ice, liquid water and water vapour) co-exist at equilibrium. For this reason, point C is known as the **triple point**. For water, the corresponding pressure (labelled P_2) and temperature (T_2) are 0.6 kPa (0.006 atm) and 0.01°C (293.16 K), respectively. The slightest deviation from these conditions means that the three phases are no longer the thermodynamically preferred molecular arrangement for the system.

10.10 Supercritical fluids

In section 10.8 we defined critical pressure and critical temperature. For CO_2:

- Critical temperature (T_c) is 31°C (304.1 K): this is the highest temperature at which liquid CO_2 can exist. Above this temperature, even a pressure of 100,000 atm would not produce liquid CO_2.
- Critical pressure (P_c) is 74 atm: this is the pressure needed to liquefy CO_2 at its critical temperature of 31°C.

We have already introduced the generalized phase diagram (Figure 10.15) for the solid, liquid and gaseous phases of a pure substance. Figure 10.16 shows the diagram again (we will apply it to CO_2) but with the pressure and temperature extended to higher values. At temperature/pressure coordinates within the grey zone, the most stable phase of CO_2 is gas vapour; within the blue region it is CO_2 liquid that is the most stable; in the red region it is solid CO_2 that is the most stable.

The coordinate T_c, P_c is called the **critical point** of the substance. At temperatures above the critical temperature and above the critical pressure (a region of pressures and temperatures shown in Figure 10.16 as a yellow box) the substance is said to be a **supercritical fluid**: here, the word 'super' means 'above', i.e. the supercritical fluid only exists above its critical pressure and critical temperature. Note that at the critical point there is no change of state when pressure is increased or if the temperature is increased: the system remains in a supercritical state or phase.

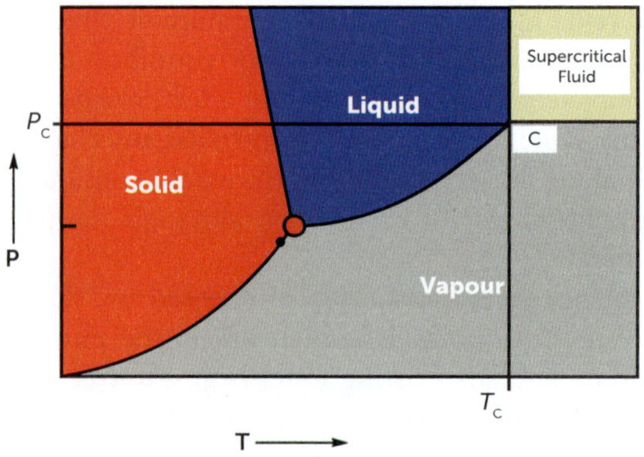

Figure 10.16 Generalized phase diagram showing the region in which the substance exists in a new phase, the supercritical state. C represents the critical point, which for CO_2 corresponds to the critical temperature (T_c) of 31°C and a critical pressure (P_c) of 74 atm.

Ethane, C_2H_6, is a colourless gas at room temperature and pressure. Figure 10.17 shows some remarkable photographs of ethane taken by Dr Sven Horstmann, Laboratory for Thermophysical Properties, Germany. At or close to the critical point, fluids become opaque and white (a phenomena given the grand name 'critical opalescence') and this is seen in Figure 10.17(b).

(a) (b) (c)

Figure 10.17 Ethane in a pressure container. In (a), the ethane gas is pressurized below its critical temperature and both (clear) liquid and gas co-exist; (b) is the same ethane sample but exactly at its critical temperature of 32.17°C and at its critical pressure of 48.08 atm (48.72 bar). The system is now uniformly opaque and since there is only one phase there is no meniscus; (c) shows the sample of ethane above its critical point. The fluid is clear. The ethane is neither gas nor liquid: it is in the supercritical state. Photographs kindly provided by Dr Sven Horstmann.

Substances in the supercritical region are neither gases nor liquids and their density is intermediate between that of the liquid and gas states. They therefore spread out (diffuse) in a similar way to (but less effectively than) a gas and they act as solvents in a similar way to (but less effectively than) a liquid. High diffusion rates and solvation (dissolving) power are both highly desirable in solvent extraction and so the use of supercritical fluids in solvent extraction effectively incorporates both these properties in the solvent. Increasing the pressure of the supercritical fluid increases its solvating power but reduces its rate of diffusion. The use of supercritical fluids to extract solutes from organic material is called **supercritical fluid extraction** (SFE). This technique is used in the decaffeination of coffee. Special equipment is needed to maintain the solvent in a supercritical state. Since SFE requires that the solvent is above its critical temperature some heating is often required (although in the case of CO_2 this is minimal since its critical temperature is only 31°C). The requirement that the solvent be above its critical pressure is usually more demanding as pressure-resistant vessels are required (for CO_2 P_c = 74 atm, i.e. 74 times atmospheric pressure). We do not usually require an exact temperature or pressure to be maintained, provided that both P_c and T_c are exceeded during the extraction.

Why supercritical carbon dioxide is a useful solvent

Supercritical carbon dioxide ($scCO_2$) is a particularly useful solvent because:

1. It is easily removed (leaving the extracted material behind) because it will evaporate in the open lab.
2. $scCO_2$ is a better solvent than gaseous CO_2 because it is denser and dissolves a wider range of substances (including hydrocarbons and slightly polar compounds) and penetrate solids more effectively. In the case of caffeine (in coffee beans), $scCO_2$ is able to extract (i.e. dissolve) caffeine without extracting the bigger (and less soluble) molecules that give coffee (and roasted coffee) its characteristic taste.
3. Experiments show that $scCO_2$ is a good solvent even in the presence of water. This is important as water is often present in mixtures containing the solute that we wish to extract. (The temperature required to maintain a mixture of CO_2 and water in its supercritical state is higher than 31°C, but is easily achieved experimentally.)
4. The low critical temperature of CO_2 means that organic compounds are less likely to be decomposed.
5. Carbon dioxide is non-toxic. For example, if halogen-containing solvents were used to extract caffeine from coffee beans, there is the danger than traces of toxic solvent might be present in packaged coffee beans.

DID YOU KNOW?

Liquid water is a highly polar molecule but supercritical water is non-polar! While the individual water molecules always have polar O–H bonds, whatever its phase, experiments show that supercritical water has fewer hydrogen bonds, indicating that the network found in liquid water is largely absent. (See also Example 5.7, p. 78.)

DID YOU KNOW?

Airbags in cars save lives by protecting passengers with a bag of gas. The gas bag provides a large surface area over which the force of a collision is absorbed, so reducing injury. In one design of airbag, the pressurized gas is produced when an ionic solid called guanidinium nitrate, ($CH_6N_4O_3^+.NO_3^-$), is detonated upon collision, decomposing into nitrogen gas, which inflates the airbag in less than 0.1 s:

$$(CH_6N_4O_3)^+.NO_3^- \longrightarrow 3H_2O(g) + 2N_2(g)$$

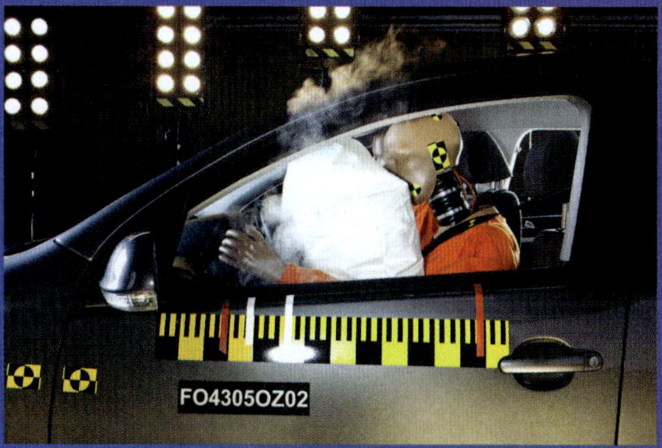

REVISION QUESTIONS

10.1

(i) Convert:

(a) 0.124 torr to standard atmospheres

(b) Exactly 1.5 atm to pascals

(c) 89 kPa to newtons per square metre ($N\,m^{-2}$)

(d) 101.325 kPa to bar

(e) −100°C to kelvin

(ii) The annual mean atmospheric pressure in Teddington is 101.5 kPa. The range is 104.0 kPa (high) to 97.0 kPa (low). Calculate the volume of one mole of ideal gas at 20°C at all three pressures.

10.2 In an experiment (in which all gases were at the same temperature and pressure before and after reaction) it was found that 1.21 dm^3 of hydrogen gas reacted completely with 1.21 dm^3 of chlorine gas to produce 2.42 dm^3 of hydrogen chloride. Using Avogadro's law, explain whether or not these volumes confirm the equation

$$H_2(g) + Cl_2(g) \rightarrow 2HCl(g)$$

10.3 Potassium chlorate decomposes upon heating in the presence of a catalyst to give oxygen gas:

$$2KClO_3(s) \rightarrow 2KCl(s) + 3O_2(g)$$

Calculate the volume of oxygen gas made at room temperature and pressure (20°C and 1 atm) when 10.0 g of chlorate is completely decomposed.

10.4 25.0 kPa of chlorine gas was fed into a previously evacuated cell with a volume of 0.0100 m^3. Oxygen gas was then added to the cell so that its partial pressure was 50.0 kPa. The cell temperature is exactly 20°C. Calculate:

(i) the total pressure of gas in the cell in pascals

(ii) the mole fraction of oxygen gas in the gas mixture, where

$$\text{number of moles of } O_2 = \frac{\text{number of moles of } O_2}{\text{total number of moles of gas}}$$

10.5 A metal can of volume 2000.0 cm^3 contains 2.0 g of nitrogen gas and 6.0 g of hydrogen gas at exactly 200°C. What is the total pressure of gas in the can?

10.6 A flexible leak-proof balloon of volume 1.00 m^3 was filled with helium to a pressure of 1.00×10^5 Pa ($N\,m^{-2}$) at 293 K. When allowed to ascend to an altitude where the helium pressure was 5.00×10^4 Pa, the balloon volume had increased to 1.66 m^3. Use the ideal gas equation to calculate the number of moles of He in the balloon at 293 K. Then calculate the atmospheric temperature at the new altitude.

10.7 Experiments show that the molar volumes of hydrogen, carbon dioxide and ammonia at 1 atm pressure (101,325 Pa) and 273 K are H_2 22.4 dm^3, CO_2 22.3 dm^3, NH_3 22.1 dm^3. What do these values suggest about the degree to which these gases deviate from the ideal gas equation? Explain the trend. (Start this problem by calculating the molar volume of an ideal gas under these conditions.)

10.8 A stoppered 2.0 dm^3 flask was half-filled with water at exactly 40°C. Calculate the number of water molecules in the water vapour. (The saturated vapour pressure of water at 40°C is 7370 Pa.)

10.9 Why do solvents with high vapour pressures at room temperature usually possess low flash points?

10.10

(i) Sodium azide has also been used in car airbags. Upon detonation, it produces nitrogen gas: $2NaN_3(s) \rightarrow 2Na(s) + 3N_2(g)$. Calculate the volume of N_2 from the ignition of 100 g of azide at room temperature and pressure.

(ii) Old cars are often crushed before disposal. What are the dangers in crushing cars with unexploded airbags?

10.11

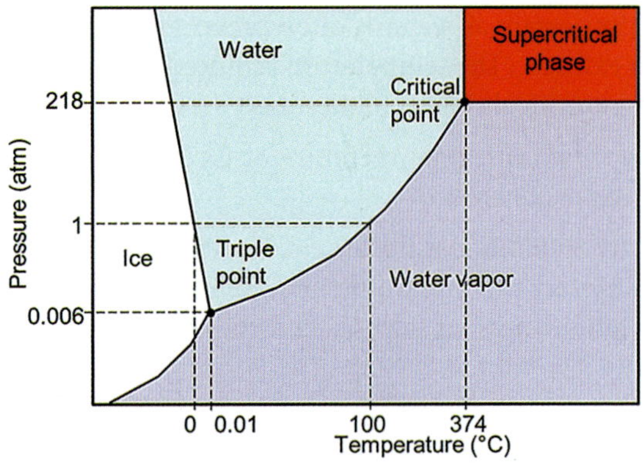

The diagram shows the phase diagram of water (Courtesy of Dr Marc Deshusses, Duke University). It is extended into the supercritical phase.

(i) What do you notice about the scale of both pressure and temperature on the diagram?

(ii) Explain the significance of the Triple Point and the Critical Point.

(iii) What state is water in at (a) −0.10°C and 0.8 atm; (b) 110°C and 200 atm; (c) 100°C and 1 atm; (d) 375°C and 219 atm?

Further information to this unit can be found in Extension 10 at https://www.bloomsburyonlineresources.com/chemistry

Solutions and Solubility

11

OBJECTIVES

This unit:

- Examines solvent miscibility and immiscibility
- Explains the basis of 'like dissolves like'
- Looks at distribution ratios and gas solubility
- Discusses the principles of osmosis
- Introduces colloids

CONTENTS

11.1 Solubility

A **solution** is a mixture consisting of a **solvent** (the 'dissolver') and the **solute** (the substance that is being dissolved). For example, if we dissolve sugar in water, the water is the solvent, the sugar the solute and the sugary water is the solution. If we keep adding sugar to some water, a point will be reached when the water will not be able to hold any more sugar. The solution is now said to be **saturated**. Adding more sugar simply results in sugar settling on the bottom of the container. Raising the temperature of the solution allows the water to hold more sugar before it becomes saturated. Many solids, like sugar, are more soluble at higher temperatures, although the reverse applies to gases, which are less soluble in hot water than in cold water.

Rules of solubility

The word 'polar' was introduced in Unit 5. A polar substance is a substance that contains ions or consists of **polar molecules**. A polar solvent is a solvent that consists of polar molecules.

We start by stating the following 'rules of thumb':

1 If a *polar substance* dissolves, it dissolves only in *polar solvents*.
2 If a *non-polar substance* dissolves, it dissolves only in *non-polar solvents*.

These generalizations are summarized in the rule *like dissolves like*. Solvents may be placed in order of polarity by testing their solubility in each other. The order of solvents in Table 11.1 was obtained in this way. Of the common solvents, water is the most polar and the hydrocarbons heptane and hexane the least polar.

Table 11.1 Polarity of common solvents – in order of increasing polarity with heptane the least polar and water the most polar

Solvent	Formula	Density at 25°C/g cm^{-3}
Heptane	$CH_3(CH_2)_5CH_3$	0.68
Hexane	$CH_3(CH_2)_4CH_3$	0.66
Cyclohexane	C_6H_{12}	0.77
Tetrachloromethane[1]	CCl_4	1.58
Methyl benzene[2]	$C_6H_5CH_3$	0.86
Ethoxyethane[3]	$C_2H_5OC_2H_5$	0.71
Dichloromethane	CH_2Cl_2	1.32
Propan-2-ol	$CH_3CH(OH)CH_3$	0.78
Tetrahydrofuran	C_4H_8O	0.89
Trichloromethane[4]	$CHCl_3$	1.48
Ethanol[5] (absolute)	CH_3CH_2OH	0.79
Ethyl ethanoate[6]	$CH_3COOC_2H_5$	0.90
Propanone[7]	CH_3COCH_3	0.79
Methanol[8]	CH_3OH	0.79
Ethanenitrile[9]	CH_3CN	0.78
Dimethyl sulfoxide	CH_3SOCH_3	1.10
Water	H_2O	1.00

Alternative names: [1]Carbon tetrachloride, [2]Toluene, [3]Diethyl ether, [4]Chloroform, [5]Ethyl alcohol, [6]Ethyl acetate, [7]Acetone, [8]Methyl alcohol, [9]Acetonitrile.

Miscibility

If, when two solvents are mixed, a single layer (consisting of a solution of the two solvents) is produced, the solvents are said to be **miscible**. If two layers are produced and both layers consist of pure solvent, the liquids are said to be **immiscible** (Figure 11.1). If two layers are produced, the solvent with the lowest density floats on the top.

The word 'layer' is often replaced by the word **phase**. Thus, a mixture of hexane and water produces *two phases*.

Table 11.2 shows which pairs of common solvents are immiscible, with • denoting immiscibility. For example, the table shows that water is immiscible with trichloromethane, ethyl ethanoate and hexane.

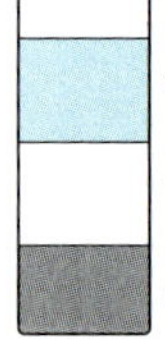

Figure 11.1 Three immiscible liquids – tetrachloromethane, mercury and water: mercury (density 13.6 g cm^{-3} at 25°C) sinks to the bottom; tetrachloromethane (density 1.6 g cm^{-3}) occupies the middle position; and water (density 1.0 g cm^{-3}) floats on top.

Partially miscible solvents

Few solvents are truly immiscible, and even though two liquids may not appear to mix, there will still be a tiny amount of each solvent present in the other layer. Table 11.3 shows the solubilities of organic solvents in water, and of water in organic solvents. The units of the solubilities are grams of organic solvent per 100 g of saturated water, and grams of water per 100 g of saturated organic solvent.

Suppose, for example, that 100 g of hexane is mixed with 100 g of water at room temperature. The liquids are almost immiscible, but the hexane layer would contain about 0.01 g of water, and the water layer would contain about 0.001 g of hexane.

Table 11.2 Miscibility of common solvents. a circle denotes immiscibility. (Adapted, with permission of fisher scientific UK, from their 1996 chemicals catalogue)

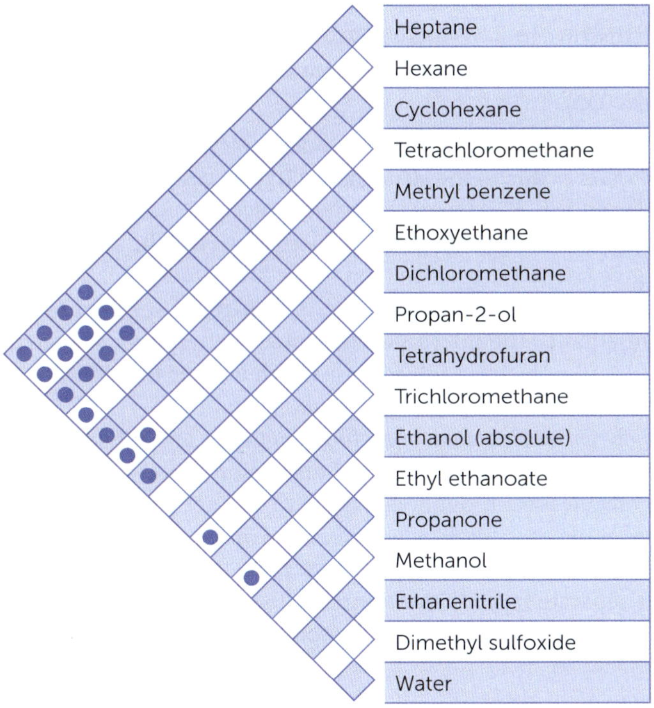

Table 11.3 Approximate solubilities of organic liquids in water, and of water in organic liquids, at 25°C

	Concentration/g per 100g	
Solvent	**Organic liquid in water phase**	**Water in organic phase**
Hexane	0.001	0.01
Cyclohexane	0.01	0.01
Tetrachloromethane	0.08	0.008
Methyl benzene	0.074	0.03
Ethoxyethane	6.9	1.3
Trichloromethane	0.82	0.056
Ethyl ethanoate	8.7	3.3

A seabird affected by oil pollution. Oil destroys a bird's insulation by clogging up its feathers. The bird responds by attempting to preen its feathers so swallowing or inhaling toxic compounds in the oil. Attempts to clean birds with detergents are often unsuccessful.

EXERCISE **11A**

The captain's bet

The captain of a sailing ship challenged the leader of his mutinous crew to a bet, with the winner to take the ship as prize. The captain mixed oil and beer in a large earthenware jug and immediately poured the mixture into two pewter tankards. The captain quickly picked up a tankard and drank down its contents in one go, but the crewman was unable to drink more than a mouthful. Explain how the captain won his bet.

EXERCISE **11B**

Concentrations in two phases

Cyclohexane is mixed with pure water, producing two layers.

(i) Which liquid forms the top layer?

(ii) How much water is contained in 130 cm^3 of the cyclohexane layer?

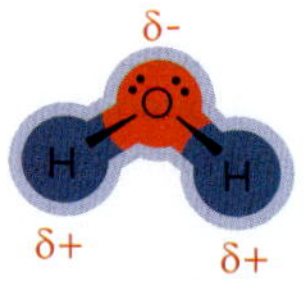

The water molecule is polar and it can attract ions or other polar molecules

Justification of the rule that 'like dissolves like'

As a generalization, polar substances will dissolve in polar solvents and non-polar substances will dissolve in non-polar solvents. This is called the 'like dissolves like' rule. In order to rationalize the rule, we rely upon a knowledge of the forces between ions, atoms and molecules:

1. Inorganic salts (such as Na^+, $Cl^-(s)$) are polar by definition. The solubility of sodium chloride in water is due to the strong attractive forces between the Na^+ and Cl^- ions, and the polar ends ($O^{\delta-}$–$H^{\delta+}$) of the water molecules. Salts such as sodium chloride are also fairly soluble in ethanol and methanol (which, although less polar than water, still give rise to fairly strong ion–molecule forces).
2. Salts (such as Na^+, Cl^-) do not dissolve in non-polar solvents such as hexane and tetrachloromethane. The intermolecular forces between ions (such as Na^+ and Cl^-) and molecules of non-polar solvents are much weaker than the attractions between the oppositely charged ions in the ionic lattice.
3. Solvents such as hexane and tetrachloromethane are soluble in similar solvents (such as heptane) because the London dispersion forces between the two *different* kinds of molecules are roughly equal to (or greater than) the London dispersion forces between the *same* molecules and so the mixing of both liquids is unhindered. However, if water is added to heptane or hexane, the water molecules 'stick together' and form a separate layer (*phase*) because the hydrogen bonds between the water molecules are much stronger than the forces between the water and hydrocarbon molecules.
4. Even large covalent molecules may be soluble in water if they contain sufficiently polar bonds. This is the case for sucrose (table sugar, $C_{12}H_{22}O_{11}$), which contains eight polar $O^{\delta-}$–$H^{\delta+}$ groups per molecule and so is able to link with the water molecules using hydrogen bonding. However, the general rule is that within a series of organic compounds, solubility in water *decreases* as the number of carbon atoms in the molecular chain *increases*. For example, consider ethanoic and hexanoic acids:

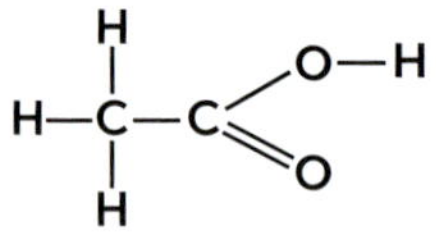

Ethanoic acid

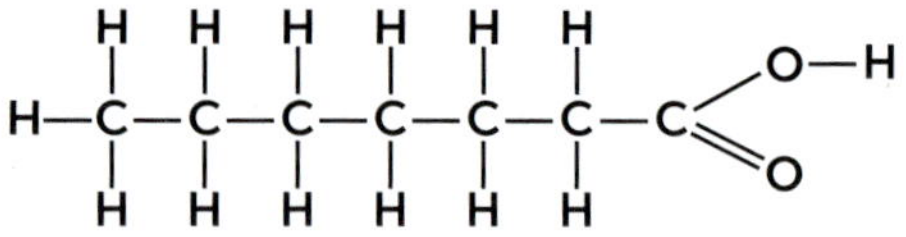

Hexanoic acid

Ethanoic acid is completely miscible with water in all proportions. This means that ethanoic acid will never form two layers with water, no matter how much acid is present. However, hexanoic acid is virtually insoluble in water. Although both acids contain the highly polar C=O and –O–H groups, the long carbon chain of the hexanoic acid molecule increases the size of the London dispersion forces between the acid molecules; this causes the acid molecules to stick together. Even the possibility of hydrogen bonding between the acid and water molecules is not enough to make hexanoic acid soluble in water.

Hydrophilic and hydrophobic ends of large organic molecules

We can look at large organic molecules in an even simpler way, by thinking of organic molecules such as hexanoic acid as consisting of a polar 'head' (the –COOH part) and a zigzag hydrocarbon 'tail' (the $-CH_2-$ chain), as shown in Figure 11.2. The polar heads are attracted to water and are said to be **hydrophilic** (water loving), whereas the tails attract each other by London dispersion forces (and so pull away from water molecules) and are said to be **hydrophobic** (water hating).

EXERCISE 11C

Effect of chain length upon solubility

Look at this list of alcohols and their formulae:

butan-1-ol	$CH_3CH_2CH_2CH_2OH$
hexan-1-ol	$CH_3CH_2CH_2CH_2CH_2CH_2OH$
nonan-1-ol	$CH_3CH_2CH_2CH_2CH_2CH_2CH_2CH_2CH_2OH$
octan-1-ol	$CH_3CH_2CH_2CH_2CH_2CH_2CH_2CH_2OH$

Which of these alcohols is the most soluble in water?

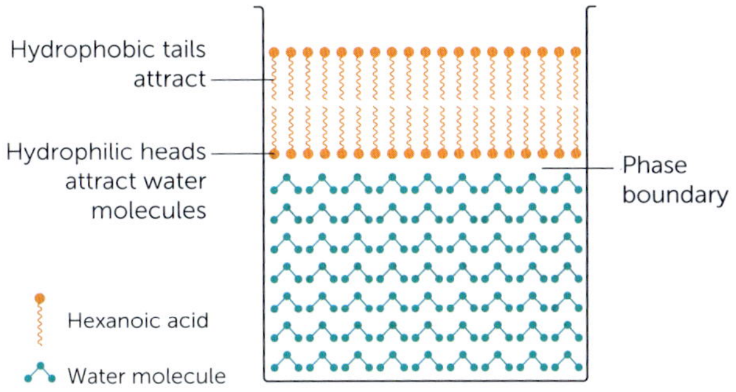

Figure 11.2 The arrangement of molecules in a mixture consisting of a small quantity of hexanoic acid mixed with water. Two distinct phases are produced, with the less dense acid floating on top.

The application of these ideas to soap action (**detergency**) is discussed in Box 11.1. Detergency is an example of **emulsification** (or **solubilization**) in which fats are prevented from forming two layers in water-based mixtures.

11.2 Making drugs more soluble

Even if a compound contains highly polar (partially charged) functional groups, the presence of large carbon chains or rings – a feature of large organic molecules, including many drugs – makes them virtually insoluble. This is an important issue in the pharmaceutical industry, since drug molecules will often only work if they are water-soluble, so allowing them to enter the bloodstream.

One way of making drugs soluble while retaining their medically useful properties is to convert them to an ionic salt. This makes the compound soluble because of the particularly strong attraction between the ion (with its +1 charge) and the water molecules. A common tactic for nitrogen-containing drugs is to convert the molecule to its hydrochloride salt by reaction with hydrochloric acid. The general equation is:

$$-N + HCl \longrightarrow -N^{+}-H \cdot Cl^{-}$$

In this equation we have not shown the molecular fragments attached to the nitrogen atom and the dot signifies the attraction (the ionic bond) between the positive ion and the chloride ion. As far as solubility is concerned, the hydrochloride behaves like common salt ($Na^+.Cl^-$), and so the whole molecule dissolves in water. There are more than 200 commercially available drugs sold in their hydrochloride form.

BOX **11.1**

Soaps and detergents

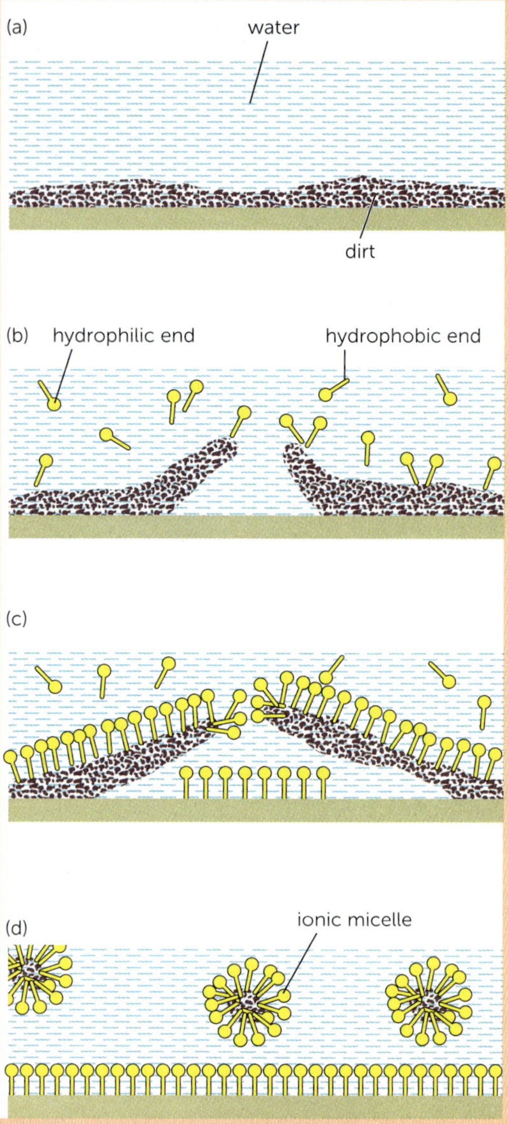

Figure 11.3 The cleaning action of soap or detergents. The stearate or detergent ion is symbolized as ○═.

What are soaps?

Soaps are salts of fatty acids. The commonest soap is sodium stearate, made by reacting sodium hydroxide (*caustic soda*) with stearic acid:

$$NaOH + C_{17}H_{35}COOH \rightarrow C_{17}H_{35}CO_2^-,Na^+ + H_2O$$

sodium hydroxide — stearic acid — sodium stearate

The sodium stearate dissociates into sodium and stearate ions in solution.

The $C_{17}H_{35}-$ part of the soap consists of a long chain of carbon atoms, with the COO^-Na^+ forming an ionic 'end' to the molecule. The full structure of soap may be written as

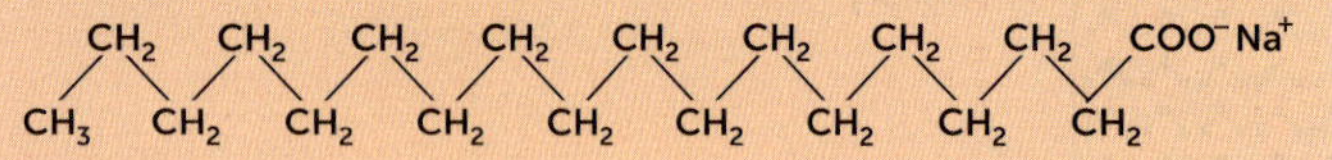

or as

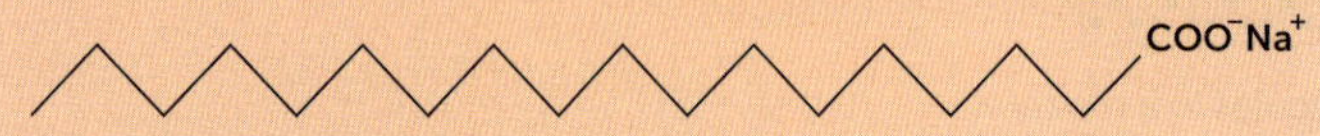

What are detergents?

Detergents are soap substitutes made in the oil industry. They possess similar molecules to soaps, with the sulfonate group (OSO_3^-) being the commonest ionic group. For example, the detergent sodium lauryl sulfate ($C_{17}H_{35}OSO_3^-,Na^+$) has the following structure:

The sodium lauryl sulfate dissociates into sodium and lauryl sulfate ions in solution.

Cleaning action of soaps and detergents

Dirt is a greasy material in which waste solid particles get stuck. Water cannot dislodge dirt (Figure 11.3(a)). The cleaning action of soap and detergents is identical and involves the loosening of the solid particles from surfaces (normally off clothes or plates or dishes). It can be thought of as taking place in three steps:

1 The hydrocarbon end (the hydrophobic end) of the stearate or detergent ion is attracted to the grease in the dirt. The ionic (hydrophilic end) of the soap or detergent is attracted to the water (Figure 11.3(b)).
2 The stearate or detergent ion reduces the stickiness (the so-called interfacial tension) between the dirt and the surfaces of the clothes or dishes. This is done by forming a layer of stearate or detergent ions on the clothes or dishes (Figures 11.3(c) and (d)). As a result, grease droplets are encouraged to enter the water phase.
3 The stearate or detergent ions surround small droplets of grease and form tiny clumps (containing typically 50–100 ions) called ionic micelles (Figure 11.3(d)). The overall result is a colloidal dispersion of oil droplets which, together with the dirt particles, can be physically washed away.

EXERCISE **11D**

Draw the structure of chloroprocaine hydrochloride (used as an anaesthetic for short operations). The parent drug, chloroprocaine, contains a benzene ring:

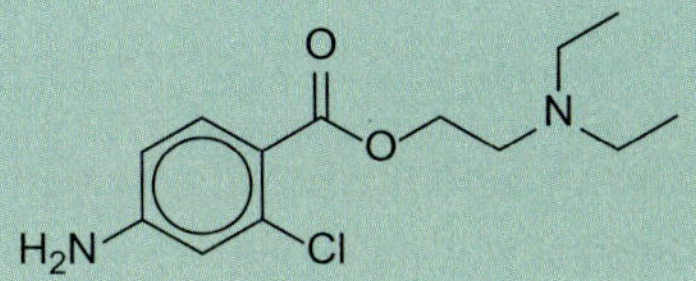

11.3 Dynamic nature of dissolution

Equilibrium involved in dissolving substances

Think of a beaker containing saturated aqueous potassium iodide solution, with undissolved (solid) K^+,I^- at the bottom of the beaker. Experiments show that although the concentration of dissolved K^+,I^- remains fixed at that temperature, the ions in the solution and in the solid are continuously exchanging places. Suppose, at an instant, a pair of K^+(aq) and I^-(aq) ions come together and fall to the bottom of the beaker. Simultaneously, a pair of K^+ and I^- ions in the solid break up into the solution. In other words, a **dynamic equilibrium** exists between the iodide in the solid and in solution. We represent this as:

$$K^+,I^-(s) \rightleftharpoons K^+(aq) + I^-(aq)$$

Examples of dynamic equilibria in solution are not restricted to the dissolving of ionic compounds. As we have seen, a mixture of hexane and water consists of two layers, one containing a small amount of water in hexane and the other a small amount of hexane in water. In fact, the dissolved hexane and water molecules are continuously exchanging places with the bulk solvents and we can write:

water in water layer $\rightleftharpoons$ water in hexane layer

also

hexane in hexane layer $\rightleftharpoons$ hexane in water layer

Other examples of dynamic equilibria we shall discuss in this unit include the dissolving of gases in water and the distribution of a solid between two solvents.

11.4 Solubility of ionic compounds: the common ion effect

Suppose we completely dissolve some sodium chloride in a test tube of water. This solution contains Na^+(aq) and Cl^-(aq). Now suppose we add some concentrated hydrochloric acid (H^+(aq) and Cl^-(aq)). A fine white precipitate of sodium chloride appears. This is called the **common ion effect** (Figure 11.4).

What is happening? The addition of the acid floods and saturates the solution with chloride ions. Excess chloride ions must be removed and this is achieved by precipitating some of the sodium ions alongside the chloride ions, i.e. by solid sodium chloride falling out of solution. This is discussed further on page 280.

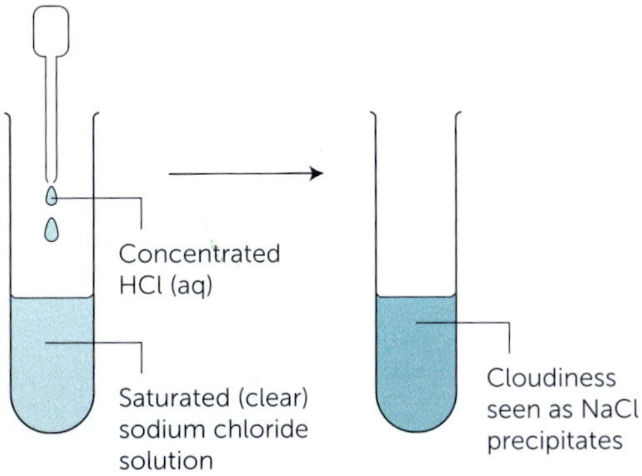

Figure 11.4 Common ion effect. (The precipitate of NaCl is actually colourless.)

11.5 Distribution of a solute between two solvents

Distribution ratio

If a solute is soluble in two solvents, and the solvents are immiscible, it is found that the concentration of solute in one layer is always bigger than the concentration of solute in the other layer by a fixed ratio. This ratio (symbolized K_d and known as the **distribution** or **partition ratio**) is a constant at a particular temperature:

$$K_d = \frac{\text{concentration of solute in solvent A}}{\text{concentration of solute in solvent B}}$$

The ratio is fixed only at that particular temperature. This equation does not hold if the solute dissociates, associates (such as forming pairs of hydrogen bonded molecules) or otherwise chemically reacts with the solvents present.

Our first example is the distribution of solid iodine (the *solute*) between water and tetrachloromethane (the *solvents*). If we add a few crystals of iodine to some water and tetrachloromethane (CCl_4) in a flask and shake the mixture well, the iodine colours the water a pale brown colour and the tetrachloromethane a purple colour. There is more iodine in the organic layer (the tetrachloromethane layer) than in the aqueous layer (the water layer). At 25°C experiments show that

$$K_d(I_2) = \frac{\text{concentration of } I_2 \text{ in } CCl_4}{\text{concentration of } I_2 \text{ in } H_2O} = 85$$

K_d is unitless, because the units of concentration cancel out.

It is important to realize that provided we do not saturate the solvent, the value of K_d at one temperature is the same whether we add 0.1 g, 1 g, or 2 g (and so on) of iodine to the mixture of solvents. This is because the partitioning of a solute between two solvents is another example of a dynamic equilibrium in which the relative concentration of the iodine in the two phases is constantly being maintained:

$$I_2(\text{in } H_2O) \rightleftharpoons I_2(\text{in } CCl_4)$$

If we add iodine to *one* layer and shake, the iodine concentration in both layers alters so as to maintain the value of K_d.

The distribution ratio allows us to calculate the concentration of solute in one layer, knowing the concentration of solute in the other. For example, if the concentration of iodine in the aqueous layer was determined by titration with thiosulfate ion and found to be 1.0×10^{-3} mol dm^{-3}, then the concentration of iodine in the tetrachloromethane layer must be 85 times as great:

concentration of I_2 in CCl_4 layer = $85 \times 1.0 \times 10^{-3} = 8.5 \times 10^{-2}$ mol dm^{-3}

EXERCISE **11E**

Distribution ratio

The distribution ratio of the gas sulfur dioxide (SO_2) between water and trichloromethane (chloroform) is about 0.90 at room temperature.

In an experiment, sulfur dioxide was bubbled through a mixture of chloroform and water. At equilibrium the concentration of SO_2 in the chloroform layer was 0.10 mol dm^{-3}. What is the concentration of sulfur dioxide in the water layer?

Solvent extraction

Frequently in organic chemistry, it is necessary to separate an organic compound from a watery mixture. This is done by adding an organic solvent (such as dichloromethane or ethoxyethane) and shaking – a process called **solvent extraction**. The organic solvent contains most of the desired organic compound (the solute) and the compound is obtained by evaporation of the organic layer. The arithmetic of solvent extraction is discussed in Unit 21.

Chromatography

One of the most important ways of separating mixtures, known as **chromatography**, is really a series of solvent extractions. More details are given in Unit 21.

11.6 Solubility of gases in water

The solubilities of common gases are given in Table 11.5. Some gases (e.g. H_2, O_2, N_2, CH_4) are only slightly soluble in water. The dissolution of these gases involves the molecules of gas being surrounded by the solvent and may be represented by the equation

$$\text{gas molecules} \xrightleftharpoons{\text{water}} \text{dissolved molecules}$$

This is a physical change. The dissolved gas may be driven out of solution by heating.

Other gases (e.g. NH_3, HCl, NO_2) **chemically react** with water so are more soluble than H_2, O_2, etc. The dissolution of such gases may be thought of as occurring in two stages:

$$\text{gas molecules} \xrightleftharpoons{\text{water}} \text{dissolved molecules} \xrightleftharpoons{\text{water}} \text{products}$$

For example,

$$NH_3(g) \xrightleftharpoons{\text{water}} NH_3(aq)$$

followed by

$$NH_3(aq) + H_2O(l) \rightleftharpoons NH_4^+(aq) + OH^-(aq)$$

Table 11.5 Solubilities of common gases at 0°C. The pressure of the gases in contact with the water is 101 kPa (1 atm). The bottom six gases chemically react with water

Gas	Solubility of gas/m^3 of gas per m^3 of water
Helium He	0.0094
Hydrogen H_2	0.021
Nitrogen N_2	0.024
Carbon monoxide CO	0.035
Oxygen O_2	0.049
Methane CH_4	0.054
Argon Ar	0.056
Carbon dioxide CO_2	1.7
Chlorine Cl_2	4.6
Hydrogen sulfide H_2S	4.7
Sulfur dioxide SO_2	80
Hydrogen chloride HCl	506
Ammonia NH_3	1,300

Variation of the solubility of gases with the partial pressure of the gas

How does the solubility of gases vary with the pressure of the dissolving gas? Because a gas at high pressure will be 'pushed' more into the water than a gas at low pressure, we predict that the greater the (partial) pressure of the dissolving gas, the greater the concentration of gas in solution (Figure 11.5). This turns out to be the case for all gases:

The solubility of a gas increases with its partial pressure.

The data in Table 11.5 refer to the case when the partial pressure of the dissolving gas is 1 atm (101 kPa). Notice that ammonia is the most soluble of all the gases listed.

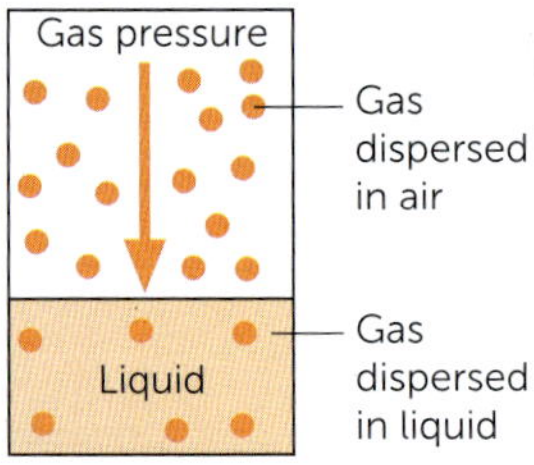

Figure 11.5 According to Henry's law, as the partial pressure of the *dissolving* gas pushing down on the solvent doubles, the concentration of *dissolved* gas also doubles.

Measurements of the partial pressure of a gas and its equilibrium concentration in solution may be plotted graphically. Some gases (Figure 11.6) give straight-line plots and obey the equation

$$c = K_H \times p$$

where c is the concentration of gas in solution (in mol dm^{-3}), p is the partial pressure of the dissolving gas (in atm) above the solution and K_H is a constant (with the units mol dm^{-3} atm^{-1}). This equation is the mathematical expression of **Henry's law,** which may be stated as follows:

The equilibrium concentration of a gas in a solvent at a particular temperature is proportional to the partial pressure of the dissolving gas.

K_H values are called Henry's law constants (Table 11.6). They vary with temperature.

Gases that chemically react with water follow Henry's law only at *very* low partial pressures (for SO_2, less than 0.001 atm). However, since calculations involving gas solubilities often involve gases (such as SO_2 and CO_2), which occur at trace levels in the atmosphere, Henry's law constants are still useful.

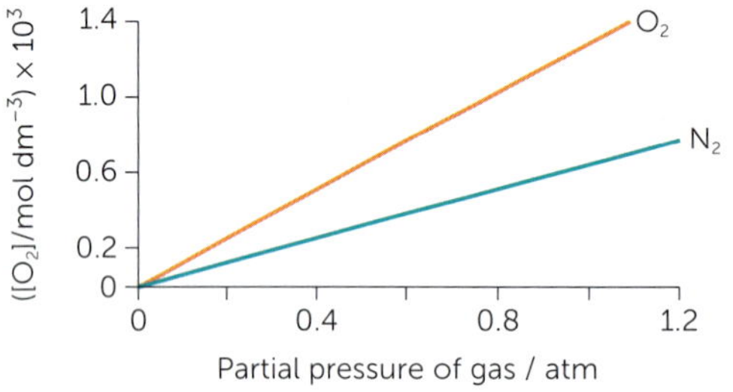

Figure 11.6 The solubilities of O_2 and N_2 at different partial pressures at 25°C. The slopes are equal to the Henry's law constants for these gases.

The solubility of a gas is unaffected by the presence of other gases. Above about 1 atm (101 kPa) *all* gases show significant deviations from Henry's law because the gas mixture behaves non-ideally.

Table 11.6 Selected henry's law constants for water at 25°C

Gas	K_H/mol dm^{-3} atm^{-1}
O_2	1.28×10^{-3}
CO_2	3.38×10^{-2}
H_2	7.90×10^{-4}
CH_4	1.34×10^{-3}
N_2	6.48×10^{-4}
NO	2.0×10^{-4}

Variation of the solubility of gases with the temperature of the solution

Figure 11.7 shows the concentration of dissolved oxygen in water (in equilibrium with dry air containing 20.9 per cent O_2 by volume at 1 atm total pressure) at different temperatures. As expected, the concentration of dissolved O_2 falls with increasing temperature. The concentrations vary from 14.6 to 7.6 mg dm^{-3} between 0 and 30°C. This means that over this temperature range, 1 dm^3 of water contains between 10 and 6 cm^3 of dissolved oxygen gas.

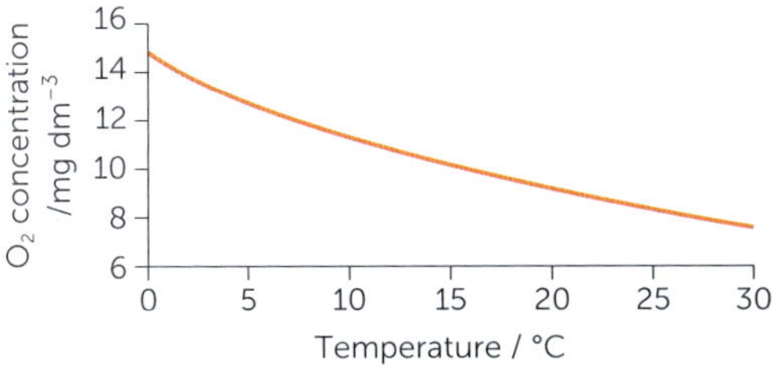

Figure 11.7 Variation of the concentration of dissolved oxygen in pure water with temperature.

The presence of dissolved solids in water further reduces the solubility of oxygen. Sea water contains roughly 10,000 mg dm^{-3} NaCl. At 25°C, this reduces the solubility of oxygen from 8.4 to 7.6 mg dm^{-3}. Typically, fish growth is inhibited below about 6 mg dm^{-3} of dissolved oxygen.

Demonstrations involving gas solubility

1 Lemonade bottle

A bottle of lemonade is essentially a solution of carbon dioxide in flavoured water and so Henry's law applies:

$$c(CO_2) = K_H \times p_{CO2}$$

where $c(CO_2)$ is the concentration of CO_2 dissolved in the water. When a new bottle of lemonade is opened, some of the carbon dioxide gas above the liquid escapes. Since p_{CO2} falls, so does the concentration of CO_2 in the water – the CO_2 is lost as bubbles, which rush out of the water.

2 Carbonating water

If we try and force CO_2 gas into a bottle of water that contains drink concentrate it will spray everywhere. This is because the presence of concentrate in the water reduces the solubility of the carbon dioxide – the excess CO_2 has nowhere to go and so it forces some of the liquid out of the bottle. A less dramatic effect is observed if you carbonate water first and add concentrate later – but note the release of CO_2 as the concentrate is added.

A fizzy drink. Dissolved CO_2 makes drinks fizzy and contributes to the tangy taste.

3 Diving

As divers descend in water, the pressure from the column of sea water above them increases. To counteract this, the divers' regulators provide them with air from cylinders at a pressure equal to the pressure from the surrounding sea. This means that scuba divers breathing from a cylinder of air (containing 21 per cent O_2 and the rest N_2) will be breathing higher partial pressures of oxygen under water than they would at sea level. This can cause CO_2 poisoning because the higher oxygen pressure triggers less frequent breathing which allows CO_2 levels in the divers' blood to build up. To match this, divers use 'air' mixtures that contain smaller percentages of oxygen, the exact percentage depending on the depth at which they operate.

Operating underwater (at higher pressure) causes more gas (notably nitrogen) to dissolve in the blood. If the diver returns to the surface too quickly, the nitrogen bubbles out from the blood, and the diver suffers 'the bends'. The bends can be reduced by replacing the nitrogen in the diver's cylinder with helium gas, which (as shown in Table 11.5) is less soluble in water. Even so, divers often have to allow their bodies to slowly equilibrate with the gases they breathe. This is accomplished by descending or ascending slowly.

4 Floating candle

If we float a candle on water and cover the candle with a jam jar, the candle eventually goes out. During this period, the water level rises (Figure 11.8).

The burning of the candle consumes oxygen. Therefore, we might expect the volume of air to be reduced by about 21 per cent, i.e. for the water level to rise by about one-fifth. The observed reduction in volume is much less than this because (i) the candle goes out before the oxygen gas contained in the jar is exhausted, and (ii) the carbon dioxide that is produced during burning is only slightly soluble in water. (If CO_2 were *completely insoluble*, the production of one mole of CO_2 for every mole of O_2 consumed would produce no change in the overall volume of gas in the jar.)

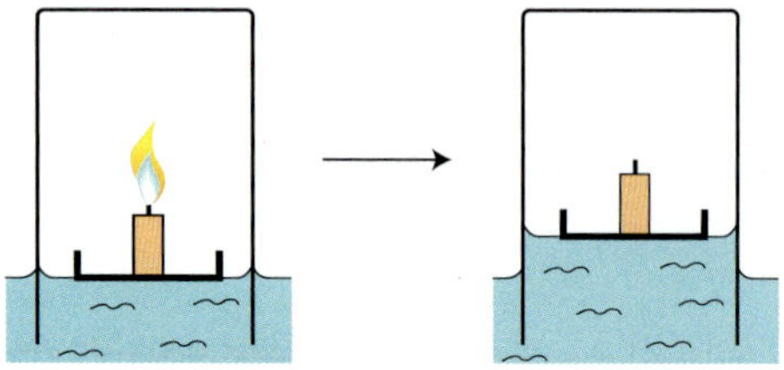

Figure 11.8 A burning candle consumes oxygen.

5 Fountain experiment

Very soluble gases (such as ammonia and hydrogen chloride) can be made to perform the 'fountain experiment'. Figure 11.9 shows the experiment carried out with HCl(g). A flask is fitted with a cork and tube and filled with dry hydrogen chloride. Without delay, the flask is lowered into a deep bucket of water. So much gas dissolves in the water rising up the tube that a partial vacuum is created in the flask. This draws up water into the flask so fast that it sprays like a fountain. If blue litmus is added to the bucket, the water entering the flask turns red because of the hydrochloric acid formed and a red fountain is observed. If the experiment is repeated using ammonia, red litmus is used and the colour change is red to blue.

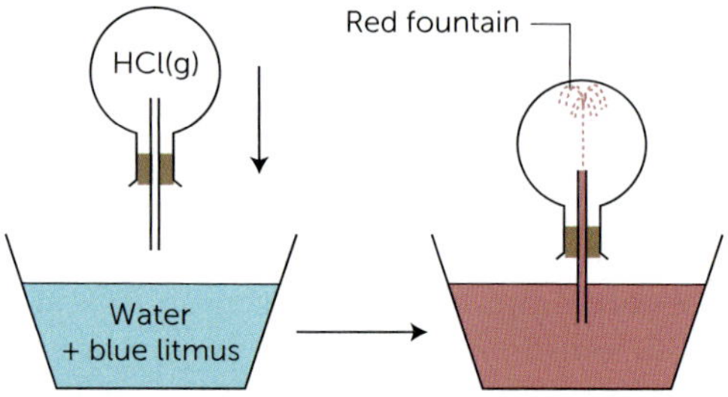

Figure 11.9 The fountain experiment using hydrogen chloride gas.

DID YOU KNOW?

One of the commonest causes of fish death in ponds is lack of dissolved oxygen. If summer temperatures rise, then the oxygen content of the pond falls. If extra plant material gets into the pond, this may slowly decompose, consuming even more of the dissolved oxygen.

11.7 Osmosis

Example of osmosis

When red blood cells are put into water they burst. What is happening? The cell membrane allows only water to pass through. The fluid in the red cells contains a mixture of organic compounds and inorganic ions, and is therefore more concentrated than the water outside. Water enters through the cell walls in an attempt to dilute the solution within the cell. Eventually, so much water enters that the cell wall bursts apart.

If fresh red blood cells are placed in a 2 per cent solution of salt, the cells shrink. In this case, the water within the cells enters the more concentrated solution outside the cell, and the cells dehydrate. See Figure 11.10.

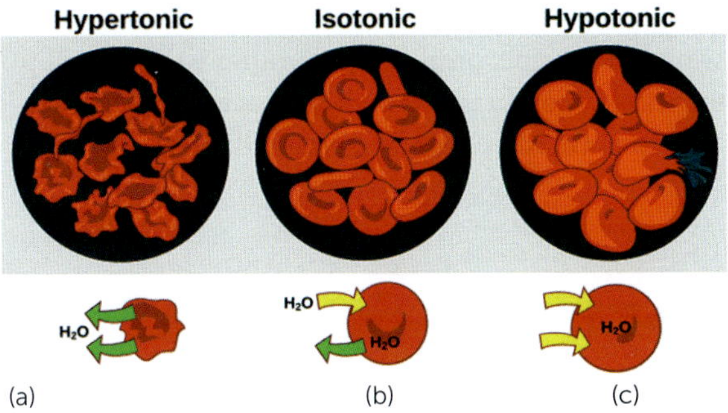

Figure 11.10

(a) If red cells are placed in salt solution, the cell loses water in an attempt to equalize salt concentrations inside and out. This causes the cell to shrivel.

(b) In blood, the concentration of solutes inside and outside the cell are equal, there is no net migration of water and the structure of the red cells is preserved.

(c) If red cells are placed in pure water, the cell attempts to equalize concentrations by allowing in water, causing the red cells to burst. (Drawing by Mariana Ruiz Villarreal)

Key points about osmosis

1. Osmosis is the passage of a *solvent* (usually water) from a zone of low concentration to one of high concentration. The solution of higher concentration is said to be **hypertonic** ('hyper-' means more), whereas the solution of lower concentration is said to be **hypotonic** ('hypo-' means less). After osmosis is complete, the solutions are equally concentrated and are said to be **isotonic** ('iso-' means the same). An example of osmosis is shown in Figure 11.10.
2. The membrane allows solvent to pass through but not solute; it is called a **semipermeable membrane.**
3. The movement of solvent from low to high concentration is similar to the spreading out of a gas (diffusion) from high gas pressure to low pressure. The idea of pressure is also applicable to osmosis, and the osmotic pressure may be thought of as the force acting per unit area of membrane that drives solvent molecules through the membrane.

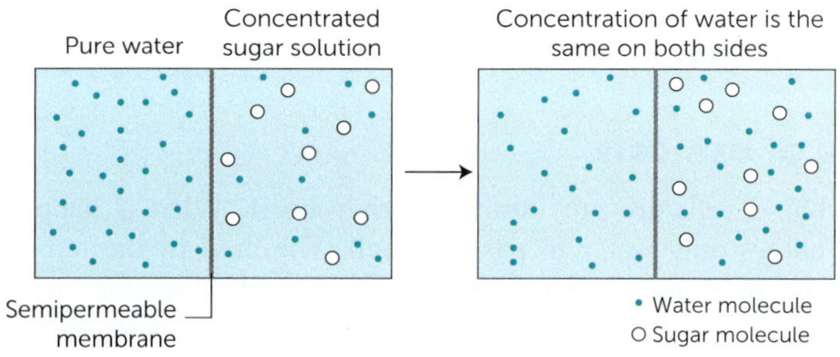

Figure 11.11 Osmosis involves the movement of a solvent across a semipermeable membrane.

Semipermeable membranes may be natural or artificial. No membrane is semipermeable to all solutes. Some membranes keep out ions (such as Na^+ and Cl^-) whereas others will keep out only larger molecules (such as metal-complex ions or sugars). The exact way that semipermeable membranes work is still poorly understood. A common semipermeable membrane is cellulose acetate, which will not allow sugar molecules or hydrated ions (made bulky by being attached to water molecules) to pass through.

11.8 Colloids

What is a colloid?

A **colloid** is a mixture of solute and solvent in which the particles of solute are intermediate in size between those found in true solutions (such as sugar in water) and those in suspensions (such as chalk or sand in water).

Examples of colloids include tea, starch and milk in water, gelatine (a protein) in water, aluminium hydroxide 'gel', and soapy water. Colloids are sometimes called **colloidal solutions, colloidal suspensions** or **colloidal dispersions**.

Key points about colloids

1 The particles in a colloid have diameters in the range 1–1000 nm. Such particles are too small to be seen by optical microscopes. The particles are either very large molecules (such as proteins) or else aggregates of smaller molecules or ions.

2 The particles of colloids do not settle at the bottom of the container in the same way that suspensions do, nor can they be filtered off using ordinary filter papers. However, colloids can be separated from true solutions using a special membrane – such a separation is called **dialysis**.

3 Colloids can be distinguished from true solutions by shining light through the solution. The relatively large particles of a colloid scatter the light. This is known as the **Tyndall effect** (Figure 11.12). A familiar example of this effect is car headlights in fog.

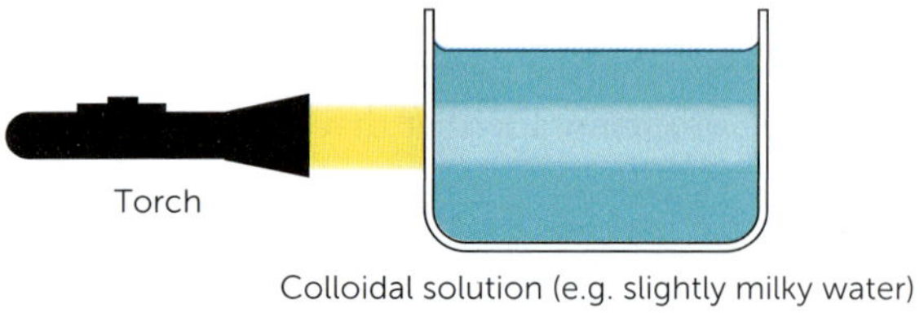

Figure 11.12 The Tyndall effect.

Classification of colloids

Colloids do not only involve solids and liquids. The classification of common types of colloid is shown in Table 11.7.

Emulsions

Dispersions of liquids in liquids, or **emulsions**, are important commercially, and occur in foods and in living organisms.

Suppose we violently shake 1 cm^3 of oil with 10 cm^3 of water for a few seconds. Upon stopping shaking, the oil remains in the form of tiny droplets for a fraction of a second before the mixture reverts to two layers again. The purpose of **emulsification** (also called **solubilization**) is to ensure that the solute (here the oil) remains dispersed permanently in the form of tiny droplets. To achieve this, we add an **emulsifying agent**.

Table 11.7 Classification of common types of colloid

Colloid type	Composition	Examples
Sol	Solid particles dispersed in a liquid	Most paints, starch and tea in water, inks, soapy water, many proteins in water
Gel (or solid emulsion)	Continuous network of solid particles dispersed in a liquid	Jellies, gelatine
Aerosol	Solid or liquid dispersed in a gas	Smoke in a room, fog
Foam	Gas dispersed in liquid	Beer foam, soapsuds, whipped cream and shaving foam
Emulsion	Liquid dispersed in liquid	Milk, mayonnaise, butter

Milk is an emulsion of butterfat droplets in an aqueous mixture of sugars, mineral salts and proteins. Ice cream is another emulsion with a similar composition. The emulsifying agents in both emulsions are proteins with smaller amounts of phospholipids, mainly lecithin. In egg yolk, the emulsifying agents are phospholipids and cholesterol. Dispersed nanoparticles also form colloids (see Unit 29).

Emulsions play an important role in the digestion of fats in the intestines. In the small intestine, fats are emulsified by the alkaline secretions of the pancreas and the liver. Why is it necessary to emulsify fats before they can be digested? The answer is that the rate at which fats are digested depends upon the surface area of fat 'solution' exposed to the intestines. For example, if 1 cm^3 of oil is dispersed to make oil droplets of 5 nm diameter, the total area of all the drops is about 1,200 m^2! Emulsification spreads out the oil so effectively that enzymes can break down fats more rapidly.

Soap

The colloidal particles of soap and other detergents are called **ionic micelles**. Such micelles typically contain 50–100 ions stuck together (see Box 11.1). Ionic micelles are important in explaining the emulsification of fats and oils by soaps and detergents. Apart from their use in washing clothes, soaps are used as emulsifying agents in many cosmetics.

REVISION QUESTIONS

11.1 Which of the following are true?

(i) Water is more soluble in hexane than hexane is soluble in water.

(ii) Long-chained organic molecules are usually more soluble in water than shorter-chained molecules.

(iii) The osmotic pressure of a solution increases during osmosis.

(iv) The hydrophobic end of a molecule is attracted to water molecules.

(v) An aerosol is defined as a dispersion of a solid in a gas.

11.2 Use Table 11.2 to find out whether the following pairs of liquids are immiscible or miscible, in each case explaining your answer by considering the intermolecular forces involved:

(i) Propan-2-ol and water.

(ii) Propanone and methyl benzene.

(The molecular formulae are listed in Table 11.1.)

11.3 The solubility of sodium chloride (in grams NaCl per 100 g solvent) at 25°C in three solvents (A, B and C) is (A) 36.1 g, (B) 1.3 g and (C) 0.1 g. Which solvent is likely to be the least polar? Explain your answer.

11.4 The distribution of a drug between the oil octanol and water is sometimes used to mimic the absorption of drugs from the aqueous solution in the stomach into the (lipid) stomach wall. If the K_d (octanol/water) for a drug is 1,000 at 25°C, and the concentration of drug in the stomach (i.e. in the aqueous layer) is 0.01 mg dm^{-3}, estimate the equilibrium concentration of drug in the stomach wall.

11.5 The solubility of chlorine gas at 0°C and an air pressure of 1 atm is about 4.6 m^3 of gas per cubic metre of water. Assuming no subsequent reaction, estimate the concentration of dissolved gas (in mol dm^{-3}) in a saturated solution under these conditions. (The molar volume of an ideal gas at 0°C and 1 atm pressure is 22.4 dm^3.)

11.6 The solubility of oxygen gas in water at a pressure of 1 atm is 14.6 and 7.6 mg dm^{-3} at 0°C and 30°C, respectively. Show that these concentrations correspond to about 10 and 6 cm^3 of dissolved oxygen gas per dm^3 of water. (Take the molar volume of oxygen as 22.4 dm^3 at 0°C and as 24.9 dm^3 at 30°C.)

11.7 Light from a torch is spread out when passed through soapy water – why? Could the particles causing this effect be filtered off with standard laboratory filter paper?

11.8 The shell of an egg is mainly calcium carbonate, and is easily dissolved by soaking the egg in vinegar. If a raw egg is left in a jar of vinegar for three days, it swells up. If the egg is now placed in syrup (concentrated sugar solution) it shrinks. Explain these observations.

11.9 The formula of solid soap is $RCOO^- Na^+$, where R is a long organic chain. To purify soap, salt is added to a soap solution, so precipitating out the pure soap. This is called 'salting out'. Explain how this works.

Further information on osmosis and dialysis can be found in Extension 11 at https://www.bloomsburyonlineresources.com/chemistry.

Chemical Families

12

OBJECTIVES

This unit:

- Takes you on a guided tour of the Periodic Table
- Describes the similarities and changes in the properties of the members of selected groups of the table
- Introduces the chemistry of the d-block elements (transition elements)
- Looks at trends in the properties of the elements across a period of the table

CONTENTS

12.1 Periodic Table

At first glance, the known elements present a bewildering array of different properties. Sodium, for example, is a very reactive solid that has a shiny appearance and conducts electricity, whereas neon is a non-conducting gas that does not react with anything. How do chemists organize the elements so that they can explain the properties of known elements and predict the properties of, as yet, undiscovered elements?

The Russian chemist Dmitri Mendeleev made a major contribution to chemistry when he presented his Periodic Table in 1869. The arrangement of the elements in that Periodic Table arose from observations that certain groups of elements show similar properties. For example F_2, Cl_2, Br_2 and I_2 are all coloured, volatile, poisonous and reactive non-metals, whereas Li, Na and K are all soft, light metals that tarnish in air.

The modern-day Periodic Table, has the atoms of elements listed in order of increasing atomic number. The numbered vertical columns are called **groups** (you could think of them as **families**) and elements in the same group often have similar properties. Some of the groups have special names:

1 Group 1 elements (Li, Na, K, Rb, Cs and Fr) are called the **alkali metals**.
2 Group 2 elements (Be, Mg, Ca, Sr, Ba and Ra) are the **alkaline earth metals**.
3 Group 15 elements (N, P, As, Sb, Bi and Mc) are (rarely) called the **pnicogens**.
4 Group 16 elements (O, S, Se, Te, Po and Lv) are sometimes referred to as the **chalcogens**.
5 Group 17 elements (F, Cl, Br, I, At and Ts) are commonly known as the **halogens**.
6 Group 18 elements (He, Ne, Ar, Kr, Xe, Rn and Og) are the **noble gases**.

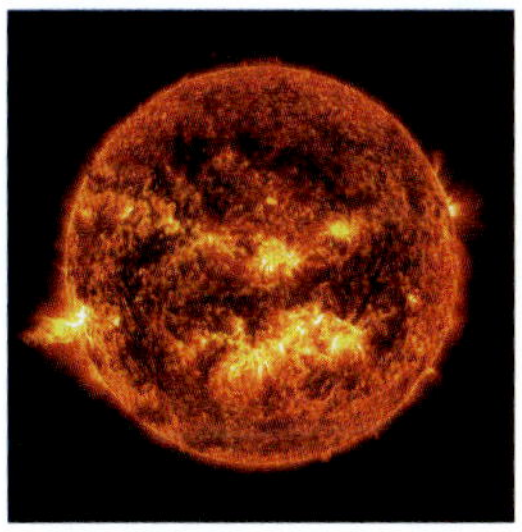

The Sun, our nearest star, is made up of 70 per cent of the element hydrogen (image by NASA).

EXERCISE **12A**

Electron arrangements of elements in a group

(i) Atoms of Li (atomic number 3) have the electronic structure 2.1. Write down the electron arrangements (Bohr model and s,p,d,f structures) for the other members of Group 1 – Na, K and Rb. Can you give a general expression for the electron arrangements of the elements in Group 1?

(ii) Write down the electronic structure (Bohr model and s,p,d,f arrangements) for the elements in Group 17 – F, Cl, Br, I. Give a general expression for the electron arrangements of the elements in this group.

For more about the chemistry of hydrogen, see Extension 12 at https://www.bloomsburyonlineresources.com/chemistry.

An older numbering system for the groups is also shown on the Periodic Table, whereby Groups 13, 14, 15, 16, 17 and 18 are shown as Groups 3, 4, 5, 6, 7 and 0, respectively. This system does not number the block of elements known as the d-block elements (transition elements). The recently discovered heavy elements are not discussed in this unit because, as yet, we know very little about their chemistry.

Other points to note about the Periodic Table are:

- The block of elements between Groups 2 and 13 is named the d-block elements or transition elements.
- The horizontal rows in the table are called **periods**. Period 1 consists only of the two elements H and He. Period 2 consists of the elements Li, Be, B, C, N, O, F and Ne. The bold zigzag line in the table separates the metals from the non-metals. The metals are on the left and you can see that there are a great many more metals in the table than non-metals.
- Hydrogen does not really belong in any group of the table, so it is usually placed at the top, on its own.

In order that the Periodic Table should fit on a single page, the lanthanoids (elements 58–71) and actinoids (elements 90–103) series are usually written below the table. We will not concern ourselves with the chemistry of these elements in this book.

After completing Exercise 12A, you should have gained a valuable clue as to why elements in the same group often have similar physical and chemical properties:

All the elements in a given group have the same number of electrons in the outermost shell of their atoms.

Since the type of bonding in compounds of an element is dependent on its ionization energy, which is related to the number of electrons in the outermost shell of its atoms, then we might expect those elements with similar electronic configurations to exhibit similar properties. Although this is generally true of the elements in a group, there are gradations (gradual changes) of the properties of the elements within a group.

If you write the electronic configuration of the elements in s,p,d,f notation, you should be able to see that:

- Group 1 and Group 2 elements are often called the s-block elements because they have s-orbital electrons in the outermost shell of their atoms.
- Groups 13 to 18 elements are called the p-block elements because they have p-orbital electrons in the outermost shell of their atoms.
- The elements in the block *between* Groups 2 and 13 are known as the d-block elements, because d electrons generally play an important part in determining their chemistry.

12.2 Group 1 elements – the alkali metals: Li, Na, K, Rb, Cs, Fr

Electronic configuration

These elements have one electron in the outermost shell of their atoms; their chemistry is dominated by the tendency to lose this electron and they are good **reducing agents**. The general ionic equation for this loss of an electron can be written as

$$M \rightarrow M^+ + e^-$$

Most of the compounds of the **alkali metals** are therefore ionic, and the alkali metals have an oxidation state of +1 in their compounds.

Alkali metals are extracted from their salts by electrolysis (see Extension 7 on the website).

EXERCISE **12B**

Ionization energies of the alkali metals

Whereas the first ionization energies (*the energy to remove one electron*) of the alkali metals are relatively low, the second ionization energies are very high.

Metal	First ionization energy/kJ mol^{-1}	Second ionization energy/kJ mol^{-1}
Li	520	7298
Na	494	4563
K	419	3051
Rb	403	2632
Cs	376	2420

(i) Would you expect Na to form a Na^{2+} ion?

(ii) Why do you think K^+ is much more stable than K^{2+}?

(iii) Suggest reasons why the second ionization energy of the metals is a great deal higher than the first.

Reactions of the alkali metals

1 They are all soft metals which tarnish in air, and for this reason they are generally stored under oil. The products obtained on burning the metals with oxygen, in the air, depend upon the metal:

$4Li(s) + O_2(g) \rightarrow 2Li_2O(s)$ the **normal oxide** is formed
$2Na(s) + O_2(g) \rightarrow Na_2O_2(s)$ a **peroxide** is formed
$K(s) + O_2(g) \rightarrow KO_2(s)$ for K, Rb and Cs a **superoxide** is formed, M^+,O_2^-

2 They react readily with water to form the corresponding alkali and hydrogen. The reaction becomes very violent as the group is descended. A general equation for this reaction is

$$2M(s) + 2H_2O(l) \rightarrow 2MOH(aq) + H_2(g)$$

The ionic equation is

$$2M(s) + 2H_2O(l) \rightarrow 2M^+(aq) + 2OH^-(aq) + H_2(g)$$

All the Group 1 hydroxides are very soluble in water.

Sodium is a soft metal and it is easily cut with a knife. (Image from the Dennis S.K. collection.)

3 The metals and their compounds display characteristic flame colours. If a moistened platinum wire is dipped into a substance containing the alkali metal and introduced into a Bunsen flame, the flame becomes coloured. The colours are shown in Table 12.1.

4 They form water-soluble, thermally stable carbonates (but note that Li_2CO_3 can be decomposed with strong heating in a Bunsen flame).

5 With the exception of $LiHCO_3$, which exists only in solution, they form solid hydrogencarbonates. Sodium hydrogencarbonate (sodium bicarbonate, used in baking powder) decomposes on heating:

$$2NaHCO_3(s) \rightarrow Na_2CO_3(s) + H_2O(l) + CO_2(g)$$

6 They form nitrates that decompose to *nitrites* and oxygen on strong heating (but note that $LiNO_3$ decomposes to the oxide). The general equation for decomposition of the nitrates to nitrites is

$$2MNO_3(s) \rightarrow 2MNO_2(s) + O_2(g)$$

7 Their salts (usually colourless or white) are ionic and not hydrolysed in solution (they do not react with water, they only dissolve in water).

Table 12.1 Flame colours of the alkali metals and their compounds

Alkali metals	Flame colour
Lithium	red
Sodium	yellow
Potassium	lilac (crimson through blue glass)
Rubidium	red
Caesium	blue

Note that the yellow emission of sodium is very intense. For example, even if a sample of potassium chloride contains only 1 per cent sodium chloride, the flame colour will appear mainly yellow.

EXERCISE **12C**

Trends in the physical properties of the alkali metals

The boiling points of the metals are as follows:

Metal	Boiling point/K
Li	1,615
Na	1,156
K	1,033
Rb	–
Cs	942

(i) Explain the trend shown by the boiling point values as the group is descended. (Hint: Consider the strength of the metallic bonding in the elements.)

(ii) Predict a boiling point value for Rb.

(iii) The alkali metals are so soft that they can all be cut with a penknife. They become softer as the group is descended. Can you think of an explanation for this trend?

(iv) The metals have low densities (Li, Na and K float on water). Why do you think this is?

The elements become *more reactive* as the group is descended. The atomic radius of the elements increases as the group is descended and the single electron in the outer shell of the atoms becomes farther away and more shielded from the pull of the nucleus by full shells of inner electrons. The outer electrons therefore become less strongly bound down the group and the ionization energy of the elements decreases. This is shown by the values given in Exercise 12B.

Trends in the chemical properties of the alkali metals and their compounds are summarized in Table 12.2.

The chemistry of francium (Fr) is not as well studied as that for the rest of the alkali metals, because it is radioactive and scarce. The longest-lived isotope of the element has a half-life of about 20 min.

EXERCISE **12D**

Chemistry of francium

From your knowledge of the alkali metals and the trends in their properties, predict:

(i) The reactivity of francium towards air and water. Write balanced equations for the reactions that occur.

(ii) The formula, solubility and thermal stability of francium carbonate.

(iii) The effect of heat on francium nitrate. Write an equation for the reaction.

Table 12.2 Trends in the chemical properties of the alkali metals and their compounds

Metal	Reactivity with air and water	Thermal stability of carbonates in a Bunsen flame	Thermal stability of hydroxides	Thermal stability of nitrates	Solubilities of carbonates
Li		decomposes		decomposes to oxide	
Na, K, Rb, Cs	more vigorous ↓	stable and increasing stability ↓	stable and increasing stability ↓	decompose to nitrites and increasing stability ↓	soluble and increasing solubility ↓

12.3 Group 2 elements – the alkaline earth metals: Be, Mg, Ca, Sr, Ba, Ra

The chemistry of the **alkaline earth metals** elements (except Be) is dominated by their tendency to lose two electrons, and their compounds are generally ionic. The general equation for this reaction is

$M \rightarrow M^{2+} + 2e^-$

The metals tend to be less reactive than their corresponding alkali metals because removal of two electrons requires more energy. The alkaline earth metals also have higher densities and melting points than the alkali metals. Reactivity of the Group 2 metals increases as the group is descended (Table 12.3).

Some chemical properties of Group 2 metals (with the exception of Be, which forms mostly covalent compounds) are:

1 They burn in air to form normal oxides (barium forms some peroxide). The general equation is

$2M(s) + O_2(g) \rightarrow 2MO(s)$

2 The alkaline earth metals, when heated, also react with nitrogen to form ionic nitrides:

$3M(s) + N_2(g) \rightarrow M_3N_2(s)$

3 Ca, Sr and Ba react with water to form the hydroxide and hydrogen gas:

$M(s) + 2H_2O(l) \rightarrow M(OH)_2(aq) + H_2(g)$

Magnesium reacts with cold water only slowly, but rapidly reacts to form the oxide when heated in steam:

$Mg(s) + H_2O(l) \rightarrow MgO(s) + H_2(g)$

4 The metals react with aqueous acids, liberating hydrogen:

$M(s) + 2HCl(aq) \rightarrow MgCl_2(aq) + H_2(g)$

5 Some alkaline earth metals have characteristic flame colours, shown in Table 12.4.

6 Their carbonates decompose on strong heating into carbon dioxide and the oxide:

$MCO_3(s) \rightarrow MO(s) + CO_2(g)$

7 The Group 2 nitrates decompose into the oxide, brown gaseous nitrogen dioxide and oxygen:

$2M(NO_3)_2(s) \rightarrow 2MO(s) + 4NO_2(g) + O_2(g)$

8 Their hydroxides are not as thermally stable as those of the alkali metals. Group 2 hydroxides decompose on heating:

$M(OH)_2(s) \rightarrow MO(s) + H_2O(l)$

EXERCISE 12E

Physical properties of the alkaline earth metals

Can you explain why the melting points of the alkaline earth metals are higher than those of the corresponding alkali metals?

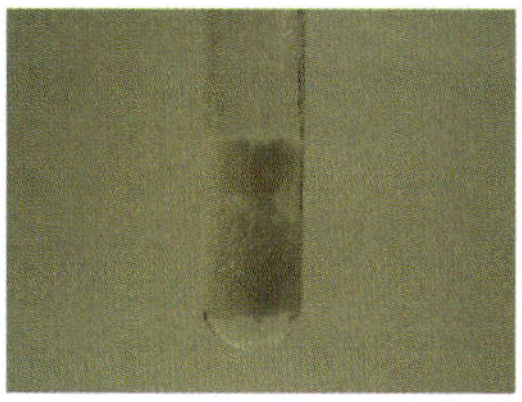

Reaction of magnesium and acid, producing a salt and bubbles of hydrogen gas.

Table 12.3 Trends in the chemical properties of the alkaline earth metals and their compounds

Metal	Reactivity with air, water and acids	Solubility of sulfates	Solubility of carbonates	Basic strength of hydroxides	Solubility of hydroxides
Be	poor		does not exist	amphoteric	insoluble
Mg		soluble		weak	insoluble
Ca	increases	insoluble	increasingly		
Sr		increasingly	insoluble	increasingly	increasingly
Ba	↓	insoluble ↓	↓	strong ↓	soluble ↓

Table 12.4 Flame colours of the alkaline earth metals and their compounds

Element	Flame colour
Calcium	brick red
Strontium	scarlet
Barium	apple green

BOX **12.1**

Hardness of water

Hardness in water is caused by the presence of dissolved calcium and magnesium compounds, such as calcium hydrogencarbonate or magnesium sulfate. The hardness of the water can be gauged by its ability to form a thick lather with soap. Water that lathers easily is called soft, whereas water that produces very little lather and a scum, is termed hard. Soap is sodium stearate, and scum forms as a result of the precipitation of an insoluble stearate:

$$Ca^{2+}(aq) + 2St^{-}(aq) \rightarrow CaSt_2(s)$$

where St^{-} = stearate ion ($C_{18}H_{35}O_2^{-}$)

Lather cannot form until all the calcium or magnesium ions have been precipitated. Note that soapless detergents, which are an ingredient of many washing-up liquids and detergents, do not form a scum with hard water.

Hard water is produced when rainwater, which is weakly acidic because of the presence of dissolved carbon dioxide, falls upon limestone or chalk, which are both forms of calcium carbonate. The acidic solution dissolves calcium carbonate producing a solution of calcium hydrogencarbonate, or hard water:

$$H_2O(l) + CO_2(g) + CaCO_3(s) \rightleftharpoons Ca(HCO_3)_2(aq)$$

This type of hard water is called temporary hard water, because the hardness can be removed by boiling. Permanently hard water contains calcium and/or magnesium salts which are not decomposed when the water is boiled. Calcium hydrogencarbonate solution decomposes on boiling to calcium carbonate (scale), which can block pipes or fur kettles in hard water districts:

$$Ca(HCO_3)_2(aq) \rightleftharpoons H_2O(l) + CO_2(g) + CaCO_3(s)$$

Stalactites and stalagmites are also found in hard water districts. When drops of calcium hydrogencarbonate solution collect on the roof of a cave, the solution decomposes, some of the water evaporates, carbon dioxide is lost and a tiny precipitate of calcium carbonate is left behind. Over many years these specks of calcium carbonate accumulate to form a **stalactite**. Solution that drips onto the floor undergoes the same decomposition, forming a stalagmite (Figure 12.1).

Temporary or permanent hard waters are commonly softened by **ion exchangers**, which remove the Group 2 metal ions and replace them with sodium ions. Although hard water has disadvantages, it does have a pleasant taste and the calcium compounds it contains are good for bones and teeth.

Figure 12.1 Formation of stalactites and stalagmites.

EXERCISE **12F**

Atomic and ionic radii of the s-block elements

The atomic and ionic radii of the elements of Groups 1 and 2 are shown below:

Element	Atomic radius /pm	Ionic radius /pm
Li	145	60
Na	180	95
K	220	133
Rb	235	148
Cs	266	169

Element	Atomic radius /pm	Ionic radius /pm
Be	105	31
Mg	150	65
Ca	180	99
Sr	200	113
Ba	215	135

(i) For each element, why is its ionic radius much smaller than its atomic radius?

(ii) Why do the atomic radii of the elements increase as the groups are descended?

(iii) Although they are in different groups, the chemistry of Li is quite similar to that of Mg. Why do you think this is?

12.4 Group 14 elements: C, Si, Ge, Sn, Pb

All these elements have four electrons in the outermost shell of their atoms (or the general outer electronic structure ns^2np^2, where n is a whole number greater than one). The elements form compounds with similar formulae, but they show marked changes in properties as the group is descended from carbon to lead.

About the elements

The element carbon exists mainly in two allotropic forms, diamond and graphite, and has a very large branch of chemistry (organic chemistry) concerned with the compounds that it forms because of its ability to form long chains by bonding with atoms of itself. The ability of the atoms of an element to covalently bond with themselves is called **catenation**.

Silicon and germanium are metalloids, whereas tin and lead are metals. Silicon and germanium have structures similar to that of diamond, tin exists in allotropic forms, but lead exists in only one metallic form.

There are two historical events often connected with the formation of grey tin at low temperatures. The first is that during the winter invasion of Russia, in 1812, Napoleon's troops tin coat buttons crumbled causing soldiers to freeze to death. This seems unlikely, since most soldiers did not use expensive tin buttons at the time. The second reference is to Captain Scott's ill-fated 1912 expedition to the South Pole. It has been suggested that fuel cans deposited for the return journey leaked and lost their contents because the steel cans had been soldered with tin which crumbled in the cold. This is plausible, but we may never know the truth.

BOX **12.2**

Allotropic forms of tin

Tin has three crystalline forms:

$$\alpha\text{-tin} \underset{}{\overset{13^\circ C}{\rightleftharpoons}} \beta\text{-tin} \overset{161^\circ C}{\rightleftharpoons} \gamma\text{-tin}$$

α-tin (grey tin) has a structure like diamond; β-tin (white tin) metallic; γ-tin metallic

'White tin' is the shiny silvery metal we are most familiar with. Below 13°C, powdery grey α-tin slowly forms on the white, metallic β-tin allotrope and the metal crumbles.

Catenation and multiple bonds

Apart from carbon, catenation does not occur to any great degree in the chemistry of the other elements. Si and Ge do form hydrides that might be compared with the lighter hydrocarbons, but they are not as stable. Catenation in these elements involves the formation of Si–Si and Ge–Ge covalent bonds, which are longer and therefore weaker than C–C bonds.

Carbon is also the only member of the group that is able to form multiple bonds with itself or other elements, such as oxygen. Although the dioxides of carbon and silicon have similar formulae (CO_2 and SiO_2), they have very different physical and chemical properties, as shown in Table 12.5.

Carbon dioxide molecules are simple, linear, non-polar molecules with the structure

$$O^{\delta-}{=}C^{\delta+}{=}O^{\delta-}$$

Silica does not have the same structure, even though it has the formula SiO_2. Silicon cannot form multiple bonds to oxygen and forms single covalent bonds instead. Silica occurs in several forms, but in all of them silicon has covalent single bonds to four oxygen atoms in a tetrahedral arrangement. Many units are bonded this way to form a network solid (not unlike the diamond structure), resulting in a high-melting-point solid. The formula of silica is SiO_2 because each silicon atom has a 'half-share' of four oxygen atoms, as shown in Figure 12.2.

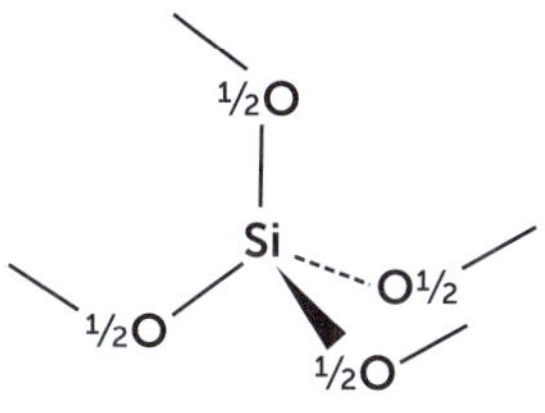

Figure 12.2 The structure of silica.

Table 12.5 Properties of carbon and silicon dioxides

Compound	Carbon dioxide (CO_2)	Silicon dioxide or silica (SiO_2)
Form under normal conditions	colourless gas	sand or quartz (both solids)
Chemical properties	weakly acidic gas, reacts with bases to form salts (carbonates)	unreactive
	slightly soluble in water: the weakly acid solution formed is carbonic acid	insoluble in water
	in soft drinks under pressure it carbonates them or makes them fizzy	
	used in fire extinguishers because it does not support combustion	

EXERCISE **12G**

Silicon and germanium hydrides

(i) Write a structural formula for the covalent compound Si_6H_{14}.

(ii) Describe the bonding and shape of the molecule Ge_2Cl_6.

BOX **12.3**

Dry ice

Solid carbon dioxide (dry ice) is obtained by releasing CO_2 from cylinders so that it expands and cools. A 'snow' is produced that is pressed into blocks. The solid carbon dioxide is used to refrigerate ice cream, meat and other foodstuffs. Solid carbon dioxide **sublimes** (changes directly from a solid to a gas) into carbon dioxide vapour at normal atmospheric pressure so that it is liquid-free when used as a refrigerant – hence the name **dry ice**.

Oxidation states

The elements of Group 14 combine with oxidation numbers +2 and +4. As the group is descended, there is an *increased tendency* for the elements to combine with other elements using the oxidation number +2 instead of +4. Carbon generally exhibits oxidation state four in its compounds, whereas the other elements in the group can form tetravalent and divalent compounds. The most important oxidation states for the elements are shown in Table 12.6. Lead(IV) compounds, however, are oxidizing agents, because lead(II) is the most stable oxidation state for the element. This gradual preference for the valency (or oxidation number/state) of 2 instead of 4 can be explained by the **inert pair effect**, described in Box 12.4.

Most of the compounds of Group 14 elements are covalent, with the exception of compounds of tin and lead in oxidation state +2, which are normally described as ionic.

Table 12.6 Oxidation states of group 14 elements. very unusual oxidation states are in brackets, preferred oxidation states are in boxes

Element	Oxidation states	
Carbon C	(2)	$\boxed{4}$ e.g. CCl_4
Silicon Si	(2)	$\boxed{4}$ e.g. SiO_2
Germanium Ge	2	$\boxed{4}$ e.g. $GeCl_4$
Tin Sn	2	$\boxed{4}$ e.g. SnO_2
Lead Pb	$\boxed{2}$ e.g. PbO	4

EXERCISE **12H**

Oxidation numbers (states) of Group 14 elements

(i) Using Lewis symbols, describe the bonding in $SnCl_4$ and $SnCl_2$.

(ii) PbO_2 is a strong oxidizing agent, whereas PbO is not. Can you offer an explanation?

(iii) Whereas CH_4 is a very stable compound, PbH_4 is not. Why do you think this is?

(iv) Can you think of a compound where carbon (very unusually) has oxidation state +2?

(v) Write the formulae for two oxides of germanium. Which would you expect to be the more stable?

BOX **12.4**

Inert pair effect

Carbon has the electronic structure 2.4 or $1s^2 2s^2 2p^2$. If we use the 'electrons in boxes' model the electronic structure can be written as

[↑↓] [↑↓] [↑|↑|]
1s 2s 2p

If an electron is promoted from a 2s orbital to a 2p orbital, the electronic structure will be

[↑↓] [↑] [↑|↑|↑]
1s 2s 2p

Now the carbon can combine with a covalency of four, because it has four unpaired electrons that can combine with the unpaired electrons of other elements, or groups, to make four covalent bonds. Remember that when covalent bonds are formed, energy is released, and the formation of four covalent bonds will release more energy than the formation of two covalent bonds. Energy is required to promote an electron from the 2s orbital to the 2p orbital, but this is more than 'paid back' in the case of carbon when the element forms four covalent bonds. Because carbon is a small atom, it forms short and strong covalent bonds and a great deal of energy is released when they are formed. As Group 14 is descended the covalent bonds formed by the elements become longer and weaker (because their atoms are larger) and the energy release gained by the formation of four covalent bonds does not become 'worth' the energy of promotion of *n*s electrons into *n*p orbitals. This is why the oxidation state of +2 becomes more favourable as compared with that of +4, and lead(IV) compounds are good oxidizing agents.

Hydrolysis of tetrachlorides

The tetrachlorides of Group 14 (XCl_4) are liquids with a tetrahedral structure. All, *except* CCl_4, react with water (or *hydrolyse*) to form HCl and the oxides (XO_2).

EXERCISE **12J**

Semiconductors

(i) Diamond is a very good insulator because the valence electrons around its carbon atoms tend to stay firmly in place. Why do the electrons in Si–Si covalent bonds delocalize much more easily?

(ii) In a sample of silicon, if electrons from Si–Si covalent bonds are set free, what stops the structure from collapsing?

EXERCISE **12I**

Hydrolysis of Group 14 chlorides

(i) Write a balanced equation for the hydrolysis of $SiCl_4$.

(ii) In the hydrolysis of $SiCl_4$, Si–Cl bonds must be broken. Why is the Si–Cl bond easier to break than a C–Cl bond?

Semiconductors

Silicon and germanium are used as **semiconductors**. Semiconductors have electrical resistances somewhere between those for conductors and insulators, and the electrical resistance of a semiconductor decreases as the temperature gets higher. In metals, which are good conductors, the valence electrons are able to break free from the individual atoms and move through the metal when a potential difference is applied across it. In insulators, the valence electrons cannot move freely around the material in the same way. The valence electrons in semiconductors, however, can break free of their atoms if a little energy is applied. As the temperature rises more valence electrons are able to break away (**become delocalized**) and the electrical conduction of the substance increases. The semiconducting properties of silicon are used in integrated circuits known as **silicon chips**. The element silicon contains many silicon atoms covalently bonded to each other in tetrahedral arrangements so that a giant molecule is built up. At very low temperatures, the electrons in the covalent bonds that hold the silicon atoms in place tend to stay associated with their 'parent' atoms. As silicon is heated, however, more of these electrons gain sufficient energy to delocalize and the resistance of the silicon decreases.

BOX **12.5**

Doping semiconductors

The electrical properties of semiconductors can be changed by adding small amounts of other substances. This process is known as **doping**. In ***n*-type** semiconductors, a small number of Group 15 element atoms, such as arsenic, are spread out in a sample of silicon. The arsenic atoms use four of their five valence electrons to bond with surrounding silicon atoms and the fifth electron is delocalized in the structure. This makes the substance a much better electrical conductor than pure silicon. The name *n*-type is derived from the fact that the doped material has an excess of negative charge. This is shown in Figure 12.3.

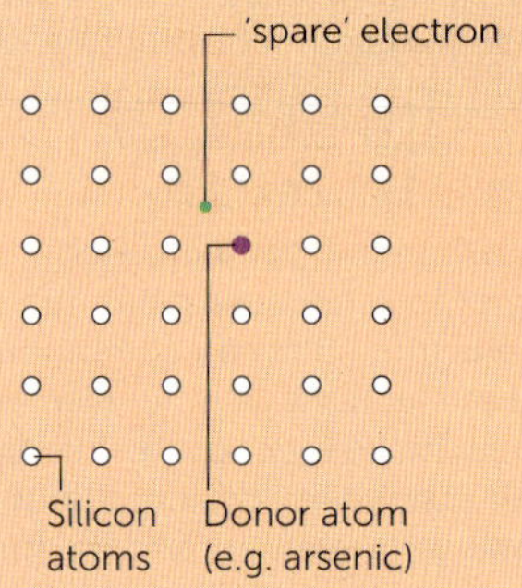

Figure 12.3 An *n*-type semiconductor.

If small traces of a Group 13 element, such as boron, are added to silicon then a deficiency of electrons is created. Boron has only three valence electrons to bond with four surrounding silicon atoms. The incomplete covalent bond created contains a 'hole' where an electron should be. A neighbouring valence electron from a silicon atom can jump into this hole, creating a new hole ready to be occupied by another electron, and so on. In this way electrons move throughout the sample and, again, the doped substance is a better conductor of electric current. These types of doped semiconductors are called ***p*-type**; the *p* refers to an excess of positive charge.

12.5 Group 17 elements – the halogens: F, Cl, Br, I, At

The **halogens** have seven electrons in the outermost shell of their atoms (with the general outer electronic structure ns^2np^5, where n is greater than one) and gain the stable electronic configuration of a noble gas either by accepting an electron to form a negatively charged ion, or by sharing an electron to form a covalent bond. Some of their common properties are:

1 The non-metallic elements exist as stable diatomic molecules: F_2, Cl_2, Br_2 and I_2.
2 They are volatile elements: fluorine and chlorine are gases at room temperature, whereas bromine and iodine are a liquid and a volatile solid, respectively. Solid iodine sublimes when warmed.
3 They are coloured; the colours of the elements are shown in Table 12.7.
4 Halogens combine directly with most metals to form ionic solids e.g.

$$2Na(s) + Cl_2(g) \rightarrow 2NaCl(s)$$

From left to right: chlorine, bromine and iodine at room temperature. Chlorine is a gas, bromine is a dark liquid with red/brown vapour and iodine is a dark solid with a violet vapour.

Some iodides have covalent character, however, because of the high polarizability of the iodide ion.

1 They form gaseous (with the exception of HF, bp 20°C), covalent hydrides with the general formula HX. The thermal stability of these hydrides decreases down the group (HF most stable). The hydrides form acidic aqueous solutions.
2 Their silver salts become increasingly covalent from AgF to AgI. A test for aqueous halides is the precipitation of the silver salt using silver nitrate solution. The general ionic equation is

$$Ag^+(aq) + X^-(aq) \rightarrow AgX(s)$$

The halogen can be identified from the colour of the precipitate formed: AgCl is white, AgBr is cream and AgI is pale yellow. See page 88.

When exposed to light, silver chloride and silver bromide decompose into silver and the halogen. The finely divided silver appears black. This is why silver halides are used in photographic films.

3 The reactivity of the halogens decreases as the group is descended. Chlorine will displace bromine and iodine from bromides and iodides, respectively, and bromine will displace iodine from iodides.

$$Cl_2(aq) + 2Br^-(aq) \rightarrow 2Cl^-(aq) + Br_2(aq)$$

Silver nitrate solution is used to test for Cl^-, Br^- or I^-. It decomposes in sunlight and is often stored in brown bottles.

In the reaction above, chlorine is acting as an oxidizing agent and the bromide ion is acting as the reducing agent.

The trends apparent within the group are shown in Table 12.8.

Table 12.7 Colours of the halogens at room temperature

Element	Colour
Fluorine	pale yellow gas
Chlorine	pale green gas
Bromine	red/brown liquid
Iodine	black solid, purple in organic solvents or vapour phase

Table 12.8 Trends in the properties of the halogens and their compounds

Element	Colour	Melting and boiling points	Stability of hydrides	Colour of silver salt	Oxidizing ability
F Cl Br I	getting darker ↓	increasing ↓	decreasing ↓	white cream pale yellow	decreases ↓

EXERCISE **12K**

Halogens

Astatine, the fifth halogen, is little studied because its isotopes are radioactive and short-lived. From your knowledge of the physical and chemical properties of the halogens, predict the physical and chemical properties of the element. Include:

(i) the colour of the element
(ii) its state at room temperature
(iii) the stability of hydrogen astatide
(iv) its strength as an oxidizing agent
(v) the nature of the bonding present in its silver salt
(vi) the colour of AgAt.

12.6 Group 18 elements – the noble gases: He, Ne, Ar, Kr, Xe, Rn

These elements, with the exception of helium, have the outer electronic configuration ns^2np^6. There is a stable octet of electrons in the outer shells of their atoms.

Group 18 elements have the following properties in common:

1 The elements are all colourless gases.
2 The lighter members of the group do not combine with any substance; xenon reacts directly only with highly reactive fluorine:

$$Xe(g) + F_2(g) \rightarrow XeF_2(s)$$

Although they are all gases, the elements become denser as the group is descended.

EXERCISE **12L**

Compounds of xenon

(i) Write a balanced equation for the reaction of xenon with fluorine to form xenon tetrafluoride.

(ii) Draw a Lewis structure for the product of the reaction in (i). What sort of shape would the molecule have?

(iii) Why are all the fluoride compounds of xenon strong oxidizing agents? What is the reduction product in these reactions?

BOX **12.6**

Discovery of the noble gases

In 1888 the brilliant polymath Lord Rayleigh (1842–1915) found that the density of nitrogen, obtained from the air, had a higher value than that obtained for pure nitrogen which had been prepared by chemical means. Together with William Ramsay (1852–1916), he sought an explanation. The two scientists eventually concluded that a new gas was present in samples of nitrogen obtained from the air. Because Ramsay, despite many chemical experiments, could not get this new gas to combine with another substance, the gas was named **argon** (from the Greek 'idle'). The two scientists realized that there was room for another group of elements in the Periodic Table. We now call this group of elements the noble gases, group 18 in the Periodic Table and by 1900 all members of the group had been identified. The last element in the group was called niton (Nt) until 1923 when its name was changed to radon (Ra).

Chemists were intrigued by the apparent unreactivity of the gases and attempts followed to react them with other substances. Argon was found not to react with the most reactive element of all, fluorine, yet H. G. Wells wrote that the Martians in the *War of the Worlds* attacked Earth with a poisonous argon compound! In the early twentieth century, when the electronic configurations of the elements were known, the fact that the atoms of the inert gases had completed shells of electrons was accepted as a reason for their lack of reactivity. However, the story was not finished: in the early 1960s xenon fluoride, XeF_4, was prepared. Although the ionization energies of the outer electrons in argon are too high for the atoms to act as electron donors in reaction with fluorine, xenon atoms are larger and the outer electrons are more shielded from their nuclei. In fact, the first ionization energy for xenon is very close to that obtained for oxygen. Xenon and the heavier inert gases can therefore form compounds under the right conditions.

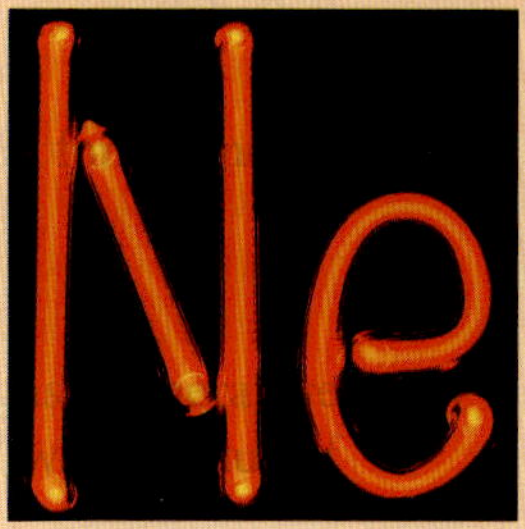

A neon discharge lamp.
(Image by Pslawinski.)

12.7 Elements of the first transition series

The block of elements between Group 2 and Group 13 of the Periodic Table are known as the **transition elements** or **d-block elements** (Sc to Zn and the elements below them). The **elements of the first transition series** are those elements that have partly filled d orbitals in any of their common oxidation states, and are the block of elements headed by Ti to Cu. Here, we will look mainly at the properties of the **first transition series**: Ti, V, Cr, Mn, Fe, Co, Ni and Cu. These elements are typical metals and are often referred to as the transition metals. They have very similar physical properties. The changes in the atomic radii and first ionization energies across the first transition series are small, because each increase in nuclear charge is well shielded by the inner 3d electrons and only a small increased attraction is noticed by the outer electrons in the 4s subshell. See Box 12.7.

Large samples of the *pure metals* in the first transition series are all silvery to varying degrees, apart from copper (shown here) which has a beautiful red-brown colour.

BOX 12.7

Changes across the first transition series

Element	Ti	V	Cr	Mn	Fe	Co	Ni	Cu
Electronic structure	$[Ar]3d^2 4s^2$	$[Ar]3d^3 4s^2$	$[Ar]3d^5 4s^1$	$[Ar]3d^5 4s^2$	$[Ar]3d^6 4s^2$	$[Ar]3d^7 4s^2$	$[Ar]3d^8 4s^2$	$[Ar]3d^{10} 4s^1$
Atomic radius/pm	146	131	125	129	126	125	124	128
First ionization energy/ $kJ\ mol^{-1}$	658	650	653	717	759	758	737	746
Melting point/K	1,930	2,160	2,130	1,520	1,800	1,770	1,730	1,360

EXERCISE 12M

Magnetic properties of the d-block elements

Write down the electronic configurations ('electrons in boxes') of the following metals and decide whether the metals are paramagnetic or diamagnetic:

(i) V **(ii)** Cr **(iii)** Zn.

Which metal do you think shows the strongest paramagnetism?

The transition metals have the following properties in common:

1. They are hard and have high melting points, both indications that strong metallic bonding exists within the metals.
2. They form alloys with one another and with other metals. Transition metals can form alloys with each other because their atoms are similar in size – the atoms of one metal can occupy positions in another metal's lattice. (Steel is an alloy of iron with other transition metals such as chromium.)
3. Metallic Fe, Co and Ni are **ferromagnetic** – they are able to become permanently magnetized. Other metals and their compounds can be **paramagnetic** – they are weakly attracted into a magnetic field. Paramagnetism is caused by the presence of **unpaired** spinning electrons. Two electrons paired up, in the same orbital, spin in opposite directions and their moments cancel each other out. If a substance has all its electrons paired up it is termed **diamagnetic**, and the substance is not attracted into a magnetic field.
4. Many of the metals can exist in a wide range of oxidation states. These are shown in Box 12.8. The 3d electrons are very close in energy to the 4s electrons and both types of electrons can be involved in bonding.

   Notice that the later metals in the series do not display anywhere near the maximum oxidation states possible. This is because the nuclear charge increases across the series and the outer electrons are held more tightly to the atoms.

BOX 12.8

Oxidation numbers of the d-block elements

Metal	Sc	Ti	V	Cr	Mn	Fe	Co	Ni	Cu	Zn
					7					
				6	6	6				
Oxidation states			**5**	5	5	5	5			
		4	**4**	4	**4**	4	4	4		
	3	**3**	**3**	**3**	3	**3**	**3**	3	3	
		2	2	**2**	**2**	**2**	**2**	**2**	**2**	**2**
		1	1	1	1	1	1	1	1	

(Important oxidation numbers are in bold type.)

1 Many transition metals, or their compounds, are important catalysts. For example, Fe is used as a catalyst in the Haber process to synthesize ammonia:

$$N_2(g) + 3H_2(g) \rightleftharpoons 2NH_3(g)$$

When the substance acts as a catalyst, bonds are formed between atoms of the reactants and the catalyst surface, so weakening the bonds in the reactants. A transition metal's ability to exist in different oxidation states allows it to easily make and break bonds with other species, so acting as an efficient catalyst.

2 The transition metals form **complex ions**. A complex ion consists of a central metal ion with molecules or ions attached to the metal ion by coordinate bonding. Some main group metals, such as Al, can form complex ions, but the transition metals form a much greater variety of complexes. Transition metal complexes are described in more detail in the next section.

3 Compounds of the transition metals tend to be coloured. Although the d electrons in the atoms of transition metals are **degenerate** (of the same energy), in complex ions of the transition metals, the d electrons become split in energy, some at a lower energy than others. The difference between the energies of the d electrons is of the same order as the energies of light photons in the visible region of the spectrum. If white light is passed through a solution of a transition metal complex, d electrons of lower energy may absorb some of the light and be promoted to a higher energy level, so the solution absorbs some of the visible light. The unabsorbed light is responsible for the colour observed.

Common colours of transition metal ions in aqueous solution are:

Cu^{2+}	sky blue	Cr^{3+}	violet
Fe^{2+}	pale green	Co^{2+}	pink
Fe^{3+}	yellow brown	Mn^{2+}	pale pink
Ni^{2+}	green		

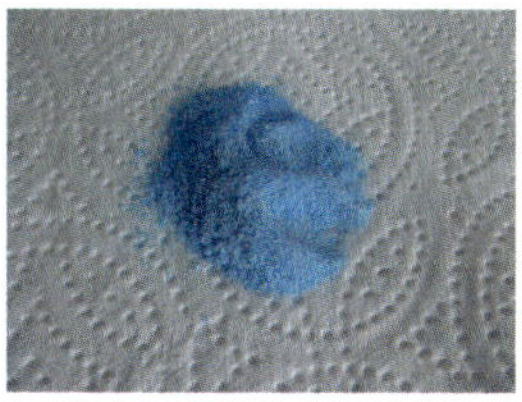

Copper sulfate pentahydrate, $CuSO_4{\cdot}5H_2O$, a coloured compound of a transition metal. Anhydrous copper sulfate, $CuSO_4$, without the water ligands, is colourless.

EXERCISE **12O**

d-block compounds and their colours

(i) Why is the complex ion $[Sc(H_2O)_6]^{3+}$ colourless?

(ii) Why are aqueous solutions of zinc compounds colourless?

EXERCISE **12N**

Oxidation states of the transition metals

Transition metals achieve their oxidation states *first* by the loss of the outer s electrons and *then* by the loss of the underlying d electrons, which are very close in energy.

(i) What is the electronic structure of atoms of Mn in oxidation state +2?

(ii) What is the electronic structure of atoms of Mn in oxidation state +3?

(iii) What do you think is the maximum oxidation state of Mn? Can you think of a common ion that has Mn in this oxidation state?

Transition metal complexes

In transition metal complexes the ions or groups attached to the metal cation are called **ligands**. Copper(II) sulfate solution, for example, contains the complex ion $[Cu(H_2O)_6]^{2+}$ in which six water ligands are attached to the central copper(II) ion by coordinate bonds. This complex is often simply written as $Cu^{2+}(aq)$.

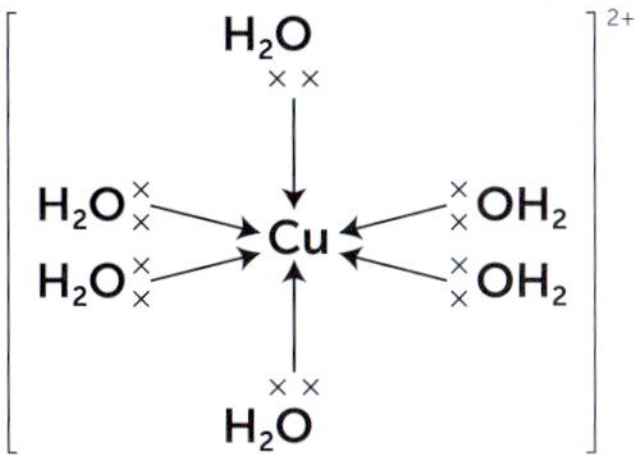

Notice that a transition metal complex ion is written in *square* brackets and the overall charge on the ion is written *outside* the brackets. In this example, the copper ion carries two positive charges and the water molecules are neutral, so the overall charge on the ion is 2+. The number of atoms directly attached to the central metal atom (in this example, six) is called the **coordination number**.

BOX **12.9**

Understanding the names of complex ions

Cationic (and neutral) complexes

Cationic complex ions are named by indicating the name and number of the ligands, followed by the name and oxidation state of the central metal ion:

$[Cu(H_2O)_6]^{2+}$	is the hexaaquacopper(II) ion
$[Ag(NH_3)_2]^+$	is the diamminesilver(I) ion
$[FeCl(H_2O)_5]^+$	is the pentaaquachloroiron(II) ion

Notice that water and ammonia produce 'aqua' and 'ammine' complexes. Negatively charged ligands, such as Cl^- (chloro), end in 'o'. The prefixes 'di', 'tri', 'tetra', 'penta' and 'hexa' are used to show the number of ligands present and ligands are named in alphabetical order.

Anionic complexes

If the complex ion is **anionic**, the rules are followed as for cationic complexes but 'ate' is added to the metal's name:

$[PtCl_4]^{2-}$	is the tetrachloroplatinate(II) ion
$[Ni(CN)_4]^{2-}$	is the tetracyanonickelate(II) ion
$[Zn(OH)_4]^{2-}$	is the tetrahydroxozincate(II) ion

Notice that 'ate' is added to the Latin name of a metal, if it exists. Anionic complexes of copper therefore end in 'cuprate' (from *cuprum*), whereas those of iron end in 'ferrate' (from *ferrum*).

EXERCISE **12P**

Complex ions

Four Cl^- ions can act as ligands to form a complex ion with Cu^{2+}:

(i) What is the overall charge of the complex ion?

(ii) Write the formula for the complex ion.

(iii) What is the coordination number of the complex ion?

(iv) Name the complex.

Some coloured ions of transition metals. From top to bottom and left to right, aqueous solutions of: $Co(NO_3)_2$ (red); $K_2Cr_2O_7$ (orange); K_2CrO_4 (yellow); $NiCl_2$ (turquoise); $CuSO_4$ (blue); $KMnO_4$ (purple). (Image courtesy of Benjah-bmm27.)

EXERCISE **12Q**

Names of complex ions

1 Name the following ions:

(i) $[Fe(CN)_6]^{3-}$

(ii) $[Zn(NH_3)_4]^{2+}$

(iii) $[Ti(H_2O)_6]^{3+}$

(iv) $[Cu(NH_3)_4(H_2O)_2]^{2+}$

(v) $[CoBr(NH_3)_5]^{2+}$

(vi) $[PtCl_4(NH_3)_2]$.

2 What are the formulae of the following ions?

(i) hexachloroplatinate(IV)

(ii) tetraaquadichlorochromium(III)

(iii) hexacyanopalladate(IV)

(iv) tetrabromocobaltate(II)

(v) hexaaquairon(III).

Coordination numbers two, four and six are found in transition metal complexes, with coordination number six commonly found.

Complexes with coordination number two tend to be **linear**, for example

$[Ag(NH_3)_2]^+$ $\quad$ $[H_3N \rightarrow Ag \leftarrow NH_3]^+$

Those with coordination number four tend to be **square planar** or **tetrahedral**, for example

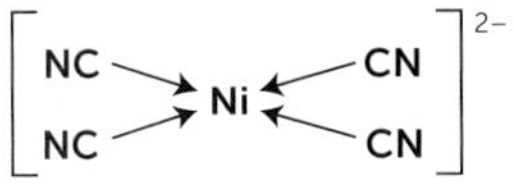

Square planar

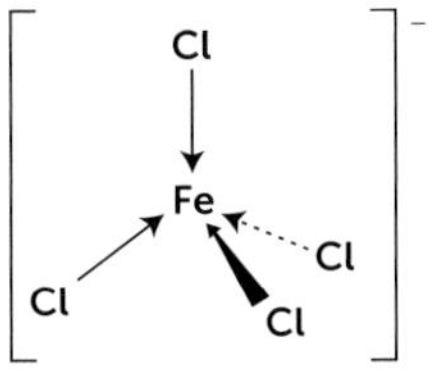

Tetrahedral

Complexes with coordination number six tend to be octahedral, for example

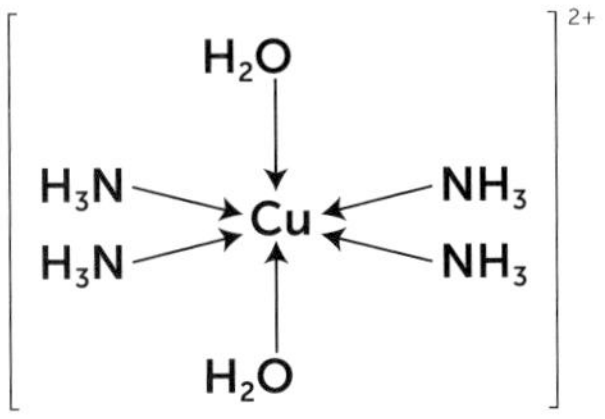

Octahedral

The ligands shown in the previous examples only donate one electron pair to the central ion. They are **monodentate** (or 'one tooth') ligands. Ligands that can donate more than one electron pair to the central metal ion are **polydentate** ('many toothed') ligands. Thus a ligand that donates two electron pairs is **bidentate**, whereas one that donates three electron pairs is **tridentate** and so on. A ligand which has the common name ethylenediamine (abbreviation 'en') has the structure

This has two electron pairs to donate to the metal and acts as a bidentate ligand, for example in the octahedral complex $[Co(en)_3]^{3+}$:

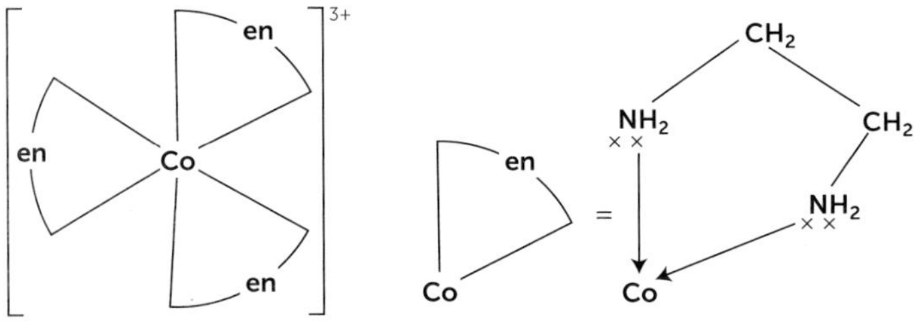

Because the ligand gives both electron pairs to the same metal, it is also known as a **chelate** ('claw-like') or **chelating ligand**. Notice that rings are formed that include the metal and the ligand, and the coordination number of the metal is six.

BOX **12.10**

Chelates in living systems

The stability of chelate complexes makes a very important contribution in the chemistry of living things. Porphyrins are a group of such compounds and they are derived from the molecule porphine:

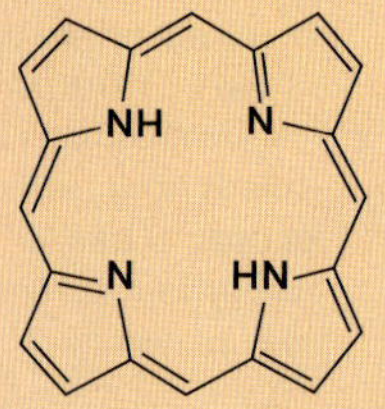

Porphine is a heterocyclic molecule – a cyclic compound with at least two different **elements** as ring members (in this case carbon and nitrogen). When the molecule loses the two protons attached to the nitrogen atoms it becomes a **tetradentate** ligand. The nitrogen atoms each have a lone pair of electrons and can complex with a metal by forming dative covalent bonds:

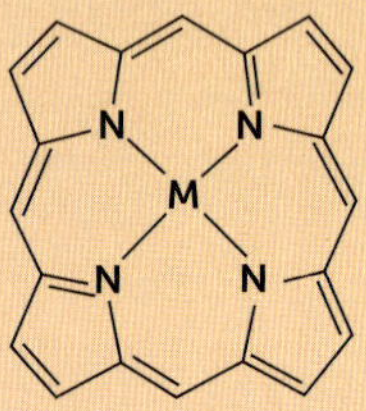

The metal sits in the middle of the molecule, in a square planar geometry, and four chelate rings are formed – this makes the complex very stable.

Haemoglobin contains iron porphyrins and is responsible for oxygen transfer within the blood. It picks up molecular oxygen in the lungs and delivers it to the rest of the body. Iron as Fe^{2+} binds to oxygen, when it then becomes Fe^{3+}. When oxygen is delivered to its destination, the iron reverts to Fe^{2+}. If you have a shortage, or deficiency, of iron in your body, your red blood cells will not contain enough haemoglobin. These red blood cells can't carry enough oxygen to your organs and tissues and this can lead to tiredness and breathlessness – symptoms of anaemia.

The binding of oxygen is affected by molecules such as carbon monoxide (CO) (for example, from **smoking** or faulty central heating systems). Haemoglobin's binding affinity for CO is many times greater than its affinity for oxygen, meaning that small amounts of CO dramatically reduce haemoglobin's ability to transport oxygen. When haemoglobin combines with CO, it forms a very bright red compound called **carboxyhaemoglobin**, which causes the skin of victims of CO poisoning to appear pink.

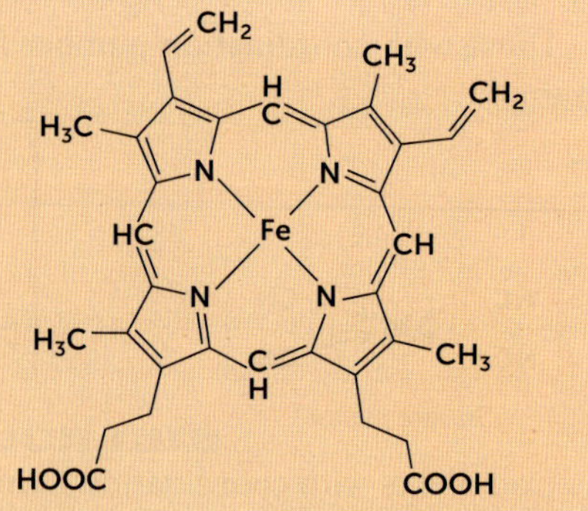

The iron porphyrin complex in haemoglobin

EXERCISE **12R**

Polydentate ligands

The ethylenediaminetetraacetate ion ($EDTA^{4-}$) is a polydentate ligand. Its structure is as follows:

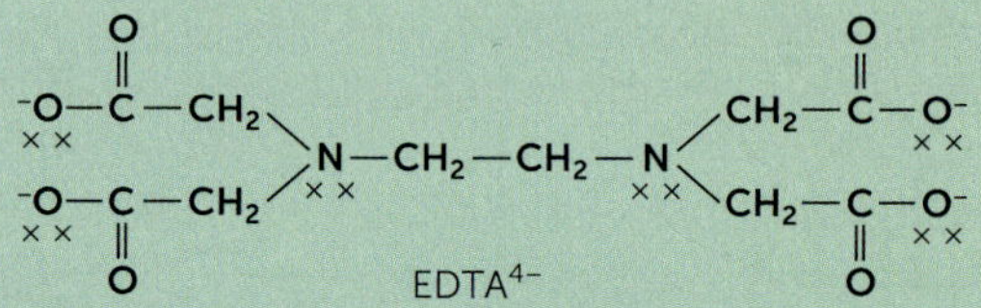

$EDTA^{4-}$

The lone pairs that are available for donation to a metal ion are shown on the diagram. $EDTA^{4-}$ also coordinates with Ca^{2+} or Mg^{2+} and is used to remove these ions from water (to 'soften' water) and is therefore often an ingredient of shampoos.

(i) How many lone pairs can $EDTA^{4-}$ coordinate to a metal ion?

(ii) One molecule of a hydrated metal ion reacts with $EDTA^{4-}$, so that all of the water molecules are replaced by one molecule of $EDTA^{4-}$:

$[M(H_2O)_6]^{2+} + EDTA^{4-}$

What is the overall charge of the new complex ion formed?

12.8 Variation of properties of elements within groups and periods

Changes in properties within groups

By now, you will have noticed that, although elements in the same group have similar properties, there are definite variations in the properties of the elements and their compounds as any group in the Periodic Table is descended.

As any group in the Periodic Table is descended, the general trends are as follows:

1 The elements become more metallic

On descending a group, the atoms become larger and the outer electrons become further away and more shielded from the positive charge of the nucleus that binds them to the atom. The ionization energies of the outer electrons therefore decrease. Metals react by losing electrons to form positive ions. The 'ease' of this reaction is related to the energy required to take away the outer shell electrons. Therefore, as the ionization energies of the elements decrease, the metallic character becomes more pronounced. This trend may be difficult to see in Groups 1 or 2, because the elements are all metallic, but the trend is very apparent in Group 14.

2 Atomic and ionic radii increase

Examples from Group 1 and 2 elements are shown in Exercise 12F. Atomic and ionic radii for the halogens are shown below:

Symbol	Atomic radius/pm	Symbol	Ionic radius/pm
F	50	F^-	136
Cl	99	Cl^-	181
Br	114	Br^-	195
I	133	I^-	215

3 The chlorides and oxides of the elements become more ionic

These properties are related to the metallic nature of the element – metallic chlorides tend to be *ionic*, whereas non-metallic chlorides are covalent. Also, metallic oxides tend to be *ionic and basic*. Highly basic oxides such as Na_2O form *hydroxides* when dissolved in water:

$$Na_2O(s) + H_2O(l) \rightarrow 2NaOH(aq)$$

Non-metallic oxides tend to form covalent, *acidic oxides*. These oxides produce acids when dissolved in water, for example:

$$SO_2(g) + H_2O(l) \rightarrow H_2SO_3(aq)$$

4 The elements become *less electronegative*

The ability of an atom in a molecule to attract electron density to itself is related to the size of the atom. The further away the electrons in the bond are from the nucleus, the more weakly they are attracted to the atom. In Group 17, therefore, fluorine is the most electronegative element and iodine is the least electronegative.

Changes in properties across a period

Consider the elements of the second period, excluding Ne:

EXERCISE **12S**

Atomic and ionic radii

(i) Explain why the radius of a sodium atom is *larger* than that of a sodium ion, whereas that of a chlorine atom is *smaller* than a chloride ion.

(ii) The ions Na^+ and Mg^{2+} are isoelectronic (contain the same number of electrons), but Mg^{2+} is a much smaller ion. Explain why this should be so.

(iii) N^{3-} and F^- are also isoelectronic, but N^{3-} is the larger ion – why?

Element	Li	Be	B	C	N	O	F
Electronic structure	2.1	2.2	2.3	2.4	2.5	2.6	2.7
s,p,d,f structure	$1s^22s^1$	$1s^22s^2$	$1s^22s^22p^1$	$1s^22s^22p^2$	$1s^22s^22p^3$	$1s^22s^22p^4$	$1s^22s^22p^5$

As we cross the period, from Li to F, the second shell of electrons is being filled. Each addition of one electron is accompanied by the addition of one proton into the nucleus, since the atoms in the elements are electrically neutral. Electrons in the *same shell* have little effect in shielding each other from nuclear charge, so that the *effective nuclear charge* felt by the outer electrons increases from left to right across the period. As a result of this increased 'pull' on the outer electrons by the nucleus:

1 The atomic radii of the elements *decrease*

In period 2, for example, the atomic radii decrease from Li at 145 pm to F with an atomic radius of 50 pm.

2 The first ionization energy *increases*

The energy required to remove one electron (the first ionization energy) increases as the effective nuclear charge felt by the electron increases. The variation of first ionization energy across the second period is shown in Figure 12.4.

Notice that there is a *general increase* in ionization energy from left to right in the period. The 'saw-tooth' appearance of the plot is due to the slight differences in energies involved in removing electrons from filled or exactly half-filled subshells. A similar pattern is seen if the first ionization energies are plotted for the third period (Na to Ar). Note that sodium atoms have the lowest first ionization energy and argon atoms the highest.

3 The elements become *less metallic*

The tendency of atoms of the elements to lose electrons decreases from left to right across a period, so the elements gradually become less metallic. On the extreme right-hand side the halogens react, as typical non-metals, by *gaining electrons*. For example, for chlorine in the reaction of sodium metal with chlorine gas,

$$Cl_2(g) + 2e^-(g) \rightarrow 2Cl^-(s)$$

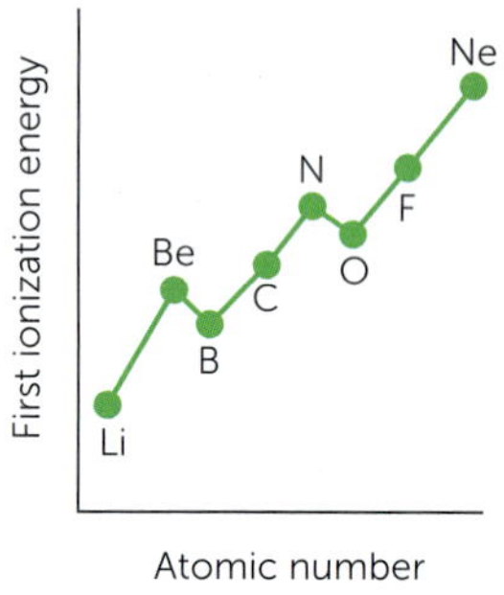

Figure 12.4 First ionization energies of the elements of period 2.

The halogens also react by sharing electrons to form covalent bonds. For example, in the reaction of hydrogen gas with chlorine gas,

$$H_2(g) + Cl_2(g) \rightarrow 2HCl(g)$$

4 Character of chlorides and oxides

The fact that the elements become less metallic, from left to right across a period, is shown by the characters of their chlorides and oxides. The chlorides of metals on the left-hand side of a period are *ionic* and *dissolve in water* to form *neutral solutions* whereas the chlorides of non-metals on the right-hand side of the table are *covalent and react with water*. There is a gradual change between these two extremes across the period. Consider the chlorides of period 3:

$NaCl(s)$	$MgCl_2(s)$	$AlCl_3(s)$	$SiCl_4(l)$	$PCl_3(l)$	$S_2Cl_2(l)$	$Cl_2(g)$
⟵ white ionic ⟶ solids, dissolve in water		⟵ covalent compounds, hydrolyse to ⟶ produce acidic solutions of HCl(aq), e.g. $SiCl_4(l) + 4H_2O(l) \rightarrow SiO_2 \cdot 2H_2O(s) + 4HCl(aq)$				

The oxides of period 3 change from *ionic to covalent* across the period. Oxides of the left-hand side metals are basic, whereas non-metallic oxides of the right-hand side elements are acidic. There is a gradual change from *basic to acidic* in the character of the oxides across the period:

$Na_2O(s)$ $MgO(s)$ $Al_2O_3(s)$ $SiO_2(s)$ $P_4O_6(s)$ $SO_2(g)$ $SO_3(l)$ $Cl_2O(g)$

⟵ ionic solids ⟶ ⟵ covalent ⟶
⟵ basic ⟶ amphoteric ⟵ acidic ⟶

For example, MgO(s) is basic, it *reacts with acids to give a salt and water*:

$$MgO(s) + 2HCl(aq) \rightarrow MgCl_2(aq) + H_2O(l)$$

The oxide $Al_2O_3(s)$ is **amphoteric**, it reacts with strong acids to produce salts:

$$Al_2O_3(s) + 3H_2SO_4(aq) \rightarrow Al_2(SO_4)_3(aq) + 3H_2O(l)$$

It will also react with a strong alkali to produce a salt:

$$Al_2O_3(s) + 2NaOH(aq) + 3H_2O(l) \rightarrow \underset{\text{sodium aluminate}}{2Na[Al(OH)_4](aq)}$$

The sulfur oxides are acidic; they react with water to form acids:

$$SO_2(g) + H_2O(l) \rightarrow H_2SO_3(aq)$$
$$SO_3(l) + H_2O(l) \rightarrow H_2SO_4(aq)$$

BOX **12.11**

Distribution of metals and non-metals within the Periodic Table

The trends in metallic and non-metallic character of the elements within groups and periods in the Periodic Table may be summarized in the following diagram:

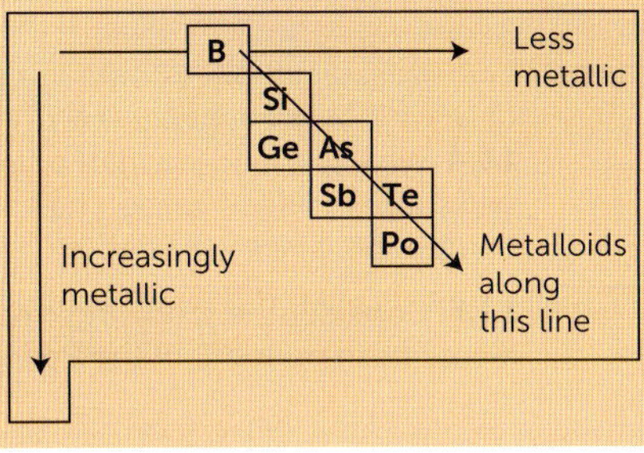

BOX **12.12**

Mendeleev's Periodic Table

Dmitri Mendeleev, a Russian chemist and professor at the University of St Petersburg. He was known to cut his hair only once a year, in the spring!

Dmitri Mendeleev (1834–1907), a Siberian chemist, was not the first scientist to notice similarities between some groups of elements, but he was the first to organize the elements into a Periodic Table and use the concept to successfully predict the properties of then undiscovered elements.

Convinced that the atomic weights (now atomic masses) of the elements were related to their properties and anxious to present some sort of 'ordering' system to the chemistry of the elements, he published his **Periodic Law:**

Elements placed according to the value of their atomic weights present a clear periodicity of their properties.

Periodic means 'reoccurring at regular intervals'. The order of atomic weights (now called atomic masses) did not always adhere to the law and Mendeleev had to deviate from a strict order where elements were obviously

BOX **12.12** *(continued)*

misplaced. Iodine, for example, clearly belonged with the halogens even though the order that Mendeleev worked with did not predict this. Mendeleev realized that some elements had yet to be discovered, so he left gaps for those elements and dared to forecast some of their properties. Later, when these elements were discovered, some of his predictions were found to be astonishingly accurate and led to acceptance of the law and great excitement among the scientific community. Consider his predictions, for example, for the element gallium, which he named 'Eka-aluminium':

Predicted in 1871	Discovered 1875
Eka-aluminium	Gallium
atomic weight 68	atomic weight 69.9
atomic volume 11.5	atomic volume 11.7
specific gravity 6.0	specific gravity 5.96

Period	Group I	II	III	IV	V	VI	VII	VIII
1	H							
2	Li	Be	B	C	N	O	F	
3	Na	Mg	Al	Si	P	S	Cl	
	K	Ca	*	Ti	V	Cr	Mn	Fe Co Ni
4	Cu	Zn	*	*	As	Se	Br	
	Rb	Sr	Y	Zr	Nb	Mo	*	Ru Rh Pd
5	Ag	Cd	In	Sn	Sb	Te	I	

The asterisks * marked spaces for elements which had yet to be discovered. They were scandium, gallium, germanium and technetium.

The classification of elements in the Periodic Table is one of the greatest contributions made to modern chemistry. Today's improved Periodic Table positions the elements in order of atomic number rather than atomic weight because scientists have realized that, and to some extent why, the properties of elements depend on their atomic number (which gives us the number and arrangements of the electrons in their atoms).

The hunt for new elements continues. To date, elements between 93 and 118 have been synthesized artificially by nuclear reactions. The heaviest elements are very unstable and cannot be of any practical use, but scientists predicted an 'island of stability' would exist around element 110. Recently, isotopes of elements in the range between 110 through 114 have been found to decay more slowly than isotopes of nuclei nearby in the Periodic Table. Attempts have been made to predict the properties of these new elements using periodic trends.

REVISION QUESTIONS

12.1 Using only the symbols shown in this outline of the Periodic Table

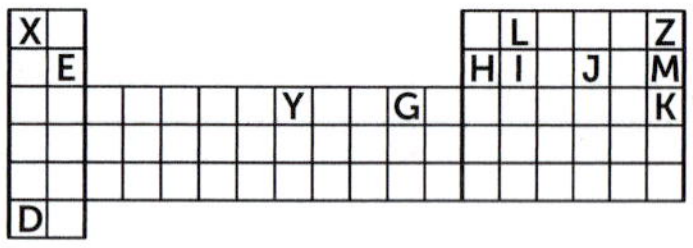

give the symbols for:

(i) an element in Group 13 of the Periodic Table
(ii) an element in period 2 of the table
(iii) a transition metal
(iv) an alkali metal
(v) a noble gas
(vi) an element with atoms which have the electron arrangement 2.8.4
(vii) an element which forms an ion with a 2– charge
(viii) an element which forms an ion with a 3+ charge
(ix) the most reactive metal
(x) an element that forms a great many compounds in which it exhibits catenation.

12.2 Radium is an alkaline earth metal.

(i) How would you expect it to behave when reacting with water? Write a balanced equation for this reaction.
(ii) Comment on the thermal stability and solubility of radium carbonate.

12.3 Tin-plated metal cans are generally not used in very cold countries – why?

12.4 What is the formula of a simple compound of carbon and sulfur in which carbon has oxidation state +2? (The compound is not very stable.)

12.5 Consider the following oxides formed by some elements in period 2 of the Periodic Table:

Li_2O, BeO, B_2O_3, CO_2, NO_2

(i) Which oxide would you expect to be the most basic? Write an equation for its reaction with water.
(ii) Which oxide would you expect to be the most acidic?
(iii) Give the formula of an amphoteric oxide.

12.6 What type of bonding exists in (i) RbCl (ii) ICl?

12.7 The graph shows the boiling points of the first 18 elements:

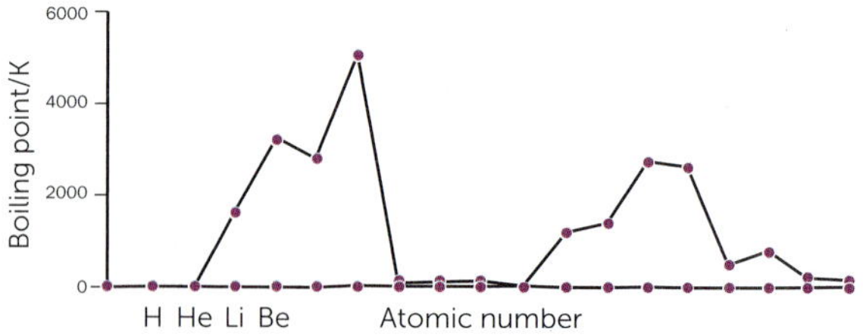

(i) Complete the graph by labelling the elements.
(ii) Which two elements have the highest boiling points? Describe the bonding in these elements.
(iii) Name two elements with very low boiling points. Describe the bonding in these elements.

(iv) Name two elements which exhibit metallic bonding. Describe metallic bonding.

(v) Is there a periodic variation in boiling point among the elements of the second and third periods? Explain.

12.8 Consider the complex compound $[NiCl_2(H_2O)_4]$.

(i) What is the coordination number of the central metal atom?

(ii) What is the oxidation state of the nickel?

(iii) Name the compound.

12.9 Write formulae for:

(i) the tetraamminediaquacopper(II) ion

(ii) the hexabromoplatinate(IV) ion.

12.10 Sketch and describe the structures of:

(i) $[Cu(NH_3)_2]^+$

(ii) $[Co(NH_3)_6]^{3+}$

DID YOU KNOW?

THE ELEMENT THAT NEVER WAS...

The element helium was first detected in the spectrum of the Sun, and scientists, eager to achieve the ensuing fame, were eager to see if they could identify other new elements in the same way.

During the solar eclipse of 17th August 1869, a green emission line at a wavelength of 530 nm was observed in the corona of the sun. This line had not been previously identified, and it did not appear in the spectrum of any known element. The 'new element' was given the provisional name coronium.

Unfortunately for all concerned, the emission line was later found to belong to the highly ionized form of iron, the Fe^{13+} ion. There was no new element after all!

A solar eclipse, with the solar corona visible.

13 Energy Changes in Chemical Reactions

CONTENTS

The study of enthalpy, entropy and Gibbs energy is called thermodynamics from the Greek words 'thermo' meaning heat and 'dynamics' meaning power or force.

OBJECTIVES

This unit:

- Defines different types of standard enthalpy change
- Explains Hess's law
- Introduces calculations involving enthalpy changes
- Discusses fuels, nutrition and explosives

13.1 Enthalpy and enthalpy changes

The energy of a substance under constant atmospheric pressure (such as in an open beaker) is called its **enthalpy** (symbol, H). The enthalpy of elements and compounds cannot be measured or calculated, but differences in enthalpy (**enthalpy changes**), symbolized ΔH or $\Delta_r H$, are easily measured in the laboratory. If a chemical or physical change is carried out at constant pressure, the Law of Conservation of Energy dictates that the amount of energy taken or given out by the reaction equals the *change* in enthalpy, ΔH, that has taken place.

More about enthalpy changes

When a reaction occurs, the mixture of reactants and products (but not the container, any inert solvent or the surroundings) is called the **reaction system** or 'the reaction'. The enthalpy change ΔH of a reaction that goes to completion

reactants $\rightarrow$ products

is defined by the equation

ΔH = (sum of enthalpies of products) − (sum of enthalpies of reactants)

Using the mathematical symbol Σ for sum, and the letters R and P for reactants and products:

$$\Delta H = \Sigma H_P - \Sigma H_R \qquad (13.1)$$

Suppose that energy ('heat') is given out during a chemical reaction. This is the case for the burning of fuels. Such a reaction is said to be **exothermic**, and the container holding the reaction mixture will be warm or hot to the touch. This loss of energy shows that the reactant molecules were originally at a higher level of enthalpy than the product molecules. Therefore, for exothermic reactions, ΔH (as defined by equation (13.1)) is **negative**. This is represented in Figure 13.1(a), which shows the total enthalpy of the reactants and products before the reaction starts and after it is complete. In this sense, the *x*-axis (labelled 'progress of reaction') may be thought of as a time axis.

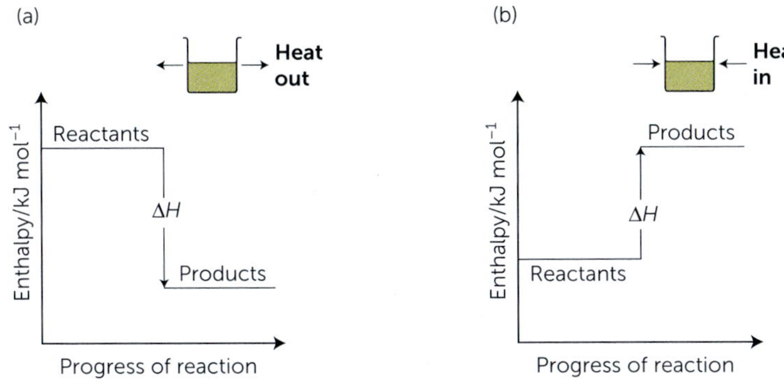

Figure 13.1 Enthalpy diagrams for (a) exothermic and (b) endothermic reactions.

Figure 13.1(b) shows the change of enthalpy for a reaction in which heat is taken in (an **endothermic reaction**). The container holding the reaction mixture will be cool or cold to the touch. In this case, the reactant molecules are at a lower level of enthalpy than the product molecules. Therefore, for endothermic reactions, ΔH is **positive**.

Explaining the sign and size of enthalpy changes using the idea of bond energies

Chemical reactions involve the breaking of old bonds and the formation of new bonds. *When bonds are formed, energy is released. When bonds are broken, energy is absorbed.* Strong bonds require more energy to break than weak bonds, but when strong bonds are created, more energy is released than is the case with weak bonds. If the bonds in the product molecules are (overall) stronger than those in the reactant molecules, the products lie at a lower enthalpy than the reactants and the reaction is exothermic (Equation (13.1)). The reverse applies to endothermic reactions, where the bonds in the reactant molecules are stronger than those in the product molecules and so the products lie at a higher enthalpy than the reactants. Whether a reaction is exothermic or endothermic is sometimes a close-run thing, but the size and sign of ΔH is simply a matter of 'bond strength arithmetic'. We will return to this in s13.6.

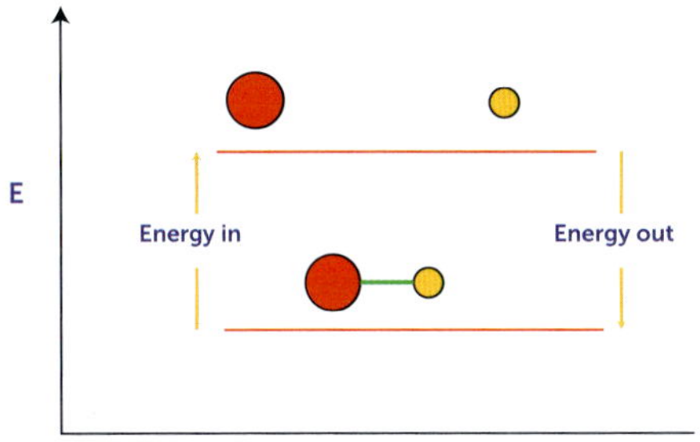

Energy changes in forming and breaking a chemical bond. The bond is coloured green. The reaction of two previously separated atoms, forming a bond, releases energy. The reverse process, that of breaking a chemical bond, requires the absorption of energy. Most chemical reactions involve the formation or breaking of several bonds, and whether the overall reaction is exothermic or endothermic is a matter of 'bond energy arithmetic'.

Once the enthalpy change of a reaction is known, we may write it down on the right-hand side of the chemical equation to which it applies. For example, the reaction between ethene and hydrogen is exothermic with $\Delta H = -137$ kJ per mol of reacted ethene at 25°C:

$$CH_2{=}CH_2(g) + H_2(g) \rightarrow CH_3CH_3(g) \qquad \Delta H = -137\ \text{kJ mol}^{-1}$$

The combination of a chemical equation and its ΔH value is known as a **thermochemical equation**.

The burning of acetylene (HC≡CH) in oxygen produces so much heat that part of the flame reaches 3000°C. This is high enough to melt most metals (iron melts at 1535°C) and is used to melt metals together, an operation known as gas welding.

BOX **13.1**

Cooling packs

Instant ice packs are used to cool food and also to numb sports injuries on the field of play. The ice packs do not require refrigeration and can be activated simply by striking them with a fist.

The ice packs consist of a thin plastic bag containing water, which is contained in a larger outer bag containing ammonium nitrate. When the pack is thumped, the bag containing the water breaks and the ammonium nitrate dissolves in the water. Ammonium nitrate is used because when it dissolves heat is taken in

$NH_4NO_3(s) \rightarrow NH_4^+(aq) + NO_3^-(aq) \quad \Delta H \approx +26\,kJ\,mol^{-1}$

A mixture containing 50 per cent (by mass) of water and ammonium nitrate reaches a temperature of about −15°C.

EXERCISE **13A**

Enthalpy change

The decomposition of silver chl*orite* upon heating,

$2AgClO_2(s) \rightarrow 2Ag(s) + Cl_2(g) + 2O_2(g) \quad \Delta H \approx 0$

involves changes in enthalpies which are approximately zero. Sketch an enthalpy diagram (following the pattern in Figure 13.1) for such a reaction.

Standard enthalpy changes of reaction ($\Delta H^\ominus$)

The enthalpy change of a reaction that goes to completion when the reactants and products are in their *standard states* is called the **standard enthalpy change of reaction**, $\Delta H^\ominus$ (with the superscript $^\ominus$ sign being read as 'standard'):

$$\Delta H^\ominus = \Sigma H_P^\ominus - \Sigma H_R^\ominus \qquad \textbf{(13.2)}$$

What is a standard state? **A substance is said to be in its standard state when it is in its pure form at an atmospheric pressure of 1 bar** (100 kPa). Temperature is not part of the definition of a standard state and (for example) the standard state of iron is pure iron at a pressure of 1 bar whatever its temperature, although it happens that at 25°C the iron would be a solid and at 1,600°C the iron would be a liquid.

The idea of a standard state is equally applicable to compounds. For instance, the standard state of sodium chloride would be pure sodium chloride at a pressure of 1 bar. The standard state of a solution is normally 1.0 mol dm^{-3} at 1 bar.

Where elements may exist in several allotropic forms, one form is chosen as the **reference state**. For example, graphite – and not diamond – has been chosen as the reference state for the element carbon. Similarly, dioxygen (O_2), – and not ozone (O_3) – has been chosen as the reference state for the element oxygen.

It is most common for standard enthalpy changes to be measured at 25°C (more precisely, at 298.15 K), or else for standard enthalpy changes obtained at other temperatures to be theoretically corrected to 25°C. Such enthalpy changes are represented by the notation $\Delta H^\ominus$ (298 K). Unless otherwise stated, we will assume that all standard enthalpy changes refer to a reaction where the temperature at the beginning and end of the reaction is 298 K, and we shall therefore take $\Delta H^\ominus$ to mean $\Delta H^\ominus$ (298 K). (In fact, provided there are no changes in phase, the variation of $\Delta H^\ominus$ with *small* changes in temperature is often negligible, and $\Delta H^\ominus$ may then be considered as independent of temperature.)

Key points about enthalpy changes

1 ΔH values are meaningless unless we state how many moles of product are made, or else include a chemical equation to which that enthalpy change applies.

The enthalpy change of a reaction is an **extensive property**. This means that the enthalpy change depends upon the amounts of reactants consumed, which, in turn, control the quantities of product made. In the absence of any other information, we always assume that the ΔH value for a reaction is produced when the number of moles of reactants that combine are those indicated by the chemical equation. So, when we write the thermochemical equation

$$N_2(g) + 3H_2(g) \rightarrow 2NH_3(g) \qquad \Delta H^{\ominus} = -\ 92.22\ \text{kJ mol}^{-1}$$

we immediately understand that the stated enthalpy change arises when *three* moles of hydrogen gas *completely* react with *one* mole of nitrogen gas to make *two* moles of ammonia. If we use up a different number of moles of reactants, the predicted enthalpy change is calculated by simple proportion, as in Example 13.1.

EXAMPLE **13.1**

Calculate the amount of heat produced when 0.32 g of methanol is completely burned in excess oxygen at 25°C.

$$2CH_3OH(l) + 3O_2(g) \rightarrow 2CO_2(g) + 4H_2O(g) \qquad \Delta H^{\ominus} = -726\ \text{kJ mol}^{-1}$$

Answer

amount of methanol (in mol) = $\dfrac{0.32\ \text{g}}{32\ \text{g mol}^{-1}}$ = 0.010 mol

The thermochemical equation shows that two moles of methanol burn to produce 726 kJ of heat. Therefore, the enthalpy change that accompanies the burning of 0.010 mol of methanol is

$$\frac{0.010}{2} \times -726 = -\ 3.6\text{kJ}$$, i.e. 3.6kJ of energy ('heat') is given out

2 The physical states of the reactants and products must be clearly shown.

The importance of specifying the physical states of the substances involved in a reaction may be illustrated by referring to the reaction of hydrogen and oxygen gases, with gaseous water as the product:

$$2H_2(g) + O_2(g) \rightarrow 2H_2O(g)$$

$\Delta H^{\ominus}$ for this reaction is -483.64 kJ mol^{-1}. Experiments show that $\Delta H^{\ominus}$ for the reaction

$$2H_2(g) + O_2(g) \rightarrow 2H_2O(l)$$

(in which 2 mol of *liquid* water is made) is -571.66 kJ mol^{-1}. The difference in the standard enthalpy changes for these reactions,

$$-571.66 - (-483.64) = -88.02\ \text{kJ mol}^{-1}$$

is the energy released when 2 mol of $H_2O(g)$ condense to form 2 mol of $H_2O(l)$.

EXERCISE **13B**

Thermochemical equations

(i) Explain why the statement 'for the reaction of sodium with chlorine, ΔH° = −411.15 kJ mol^{-1}' is ambiguous. What extra information would make sense of this statement?

(ii) Given that

$C(s) + O_2(g) \rightarrow CO_2(g) \quad \Delta H^\circ = -393.15\ kJ\ mol^{-1}$

calculate the amount of heat given out when 120 g of carbon is completely burned in excess air.

3 Reverse reactions possess ΔH values which are equal in size, but opposite in sign, to the forward reaction.

What is ΔH° for the following reaction?

$2H_2O(g) \rightarrow 2H_2(g) + O_2(g)$

The answer is simply that since the reaction is *exactly* the reverse of the equation

$2H_2(g) + O_2(g) \rightarrow 2H_2O(g)$

then the enthalpy change is equal in size but opposite in sign, i.e. +483.64 kJ mol^{-1}.

We summarize this in the following rule:

Reactions that are exothermic in the forward direction are endothermic in the reverse direction.

This rule applies to both physical and chemical changes. For example, heat is absorbed in vaporization because liquid molecules require energy to escape from their neighbours into the gas phase. Therefore, vaporization is always endothermic. The **standard enthalpy of vaporization** of acetone at 25°C is +29.1 kJ per mol of acetone:

$CH_3COCH_3(l) \rightarrow CH_3COCH_3(g) \qquad \Delta_{vap}H^\circ = +29.1\ kJ\ mol^{-1}$

It follows that 29.1 kJ of heat energy is given out when 1 mol of acetone *condenses* into acetone liquid:

$CH_3COCH_3(g) \rightarrow CH_3COCH_3(l) \qquad \Delta H^\circ = -29.1\ kJ\ mol^{-1}$

The principle that if reactions are reversed the energy changes are equal in size but opposite in sign is important because, if this were not so, it would be possible to destroy or create energy, and the Law of Conservation of Energy would be broken.

EXERCISE **13C**

Enthalpy changes of back reactions

The formation of magnesium oxide from its elements at 298 K obeys the thermochemical equation

$2Mg(s) + O_2(g) \rightarrow 2MgO(s)$

$\Delta H^\circ = -601.7\ kJ\ mol^{-1}$

How much energy would be required to decompose one mole of magnesium oxide into oxygen gas and magnesium metal?

4 If a reaction can occur by more than one route, the enthalpy change is the same, whichever route is followed.

This is a statement of **Hess's law**. As an example, we shall look at the production of hydrogen iodide gas from iodine and hydrogen. The reaction can occur in one stage in which hydrogen iodide gas is made directly from *solid* iodine:

$H_2(g) + I_2(s) \rightarrow 2HI(g) \qquad \Delta H^\circ = +53\ kJ\ mol^{-1}$ **(13.3)**

We call this route A.

Alternatively, we can carry out this reaction in *two* stages:

$I_2(s) \rightarrow I_2(g) \qquad \Delta H^\circ = +62\ kJ\ mol^{-1}$ **(13.4)**

in which solid iodine is first vaporised, followed by

$I_2(g) + H_2(g) \rightarrow 2HI(g) \qquad \Delta H^\circ = -9\ kJ\ mol^{-1}$ **(13.5)**

We shall call reactions (13.4) and (13.5) route B.

Both routes are illustrated in Figure 13.2(a), which shows the enthalpy changes for the reaction in which 1 mol of reactants is converted into 2 mol of HI.

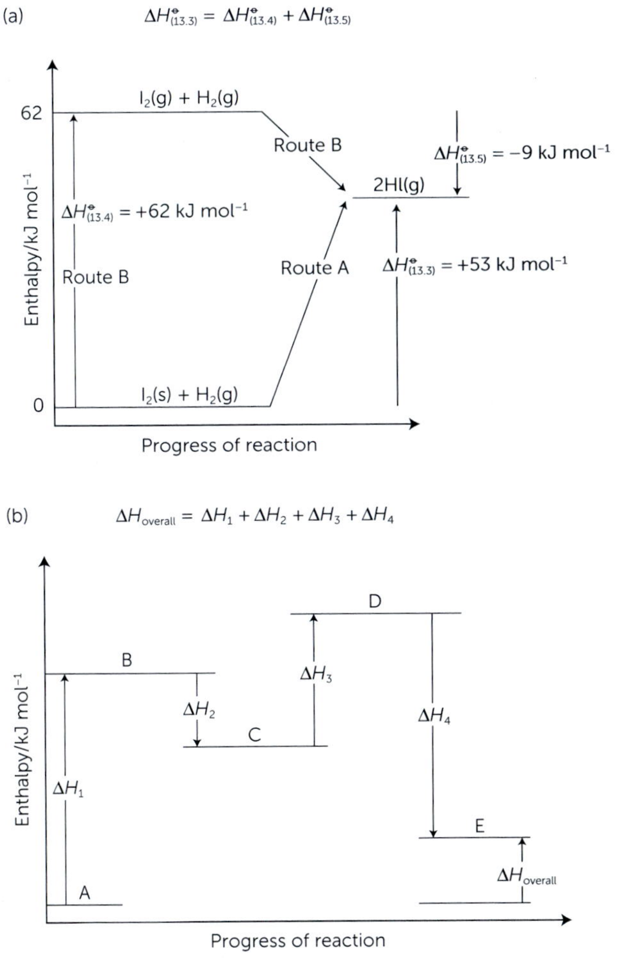

Figure 13.2 Hess's law: (a) the manufacture of hydrogen iodide by two routes; (b) the general case.

The diagram shows that standard enthalpy change for reaction (13.3) is the *sum* of the enthalpy changes for reactions (13.4) and (13.5):

$$\Delta H^{\ominus}_{(13.3)} = \Delta H^{\ominus}_{(13.4)} + \Delta H^{\ominus}_{(13.5)}$$
$$= (+62) + (-9) = +53 \text{ kJ mol}^{-1}$$

Generalizing, if a reaction A → E is carried out in several stages,

$$A \xrightarrow{1} B \xrightarrow{2} C \xrightarrow{3} D \xrightarrow{4} E$$

then the enthalpy change for the overall reaction,

$$\Delta H(A \rightarrow E)$$

is the sum of the individual enthalpy changes:

$$\Delta H(A \rightarrow E) = \Delta H(A \rightarrow B) + \Delta H(B \rightarrow C) + \Delta H(C \rightarrow D) + \Delta H(D \rightarrow E)$$

or,

$$\Delta H_{overall} = \Delta H_1 + \Delta H_2 + \Delta H_3 + \Delta H_4$$

as illustrated in Figure 13.2(b).

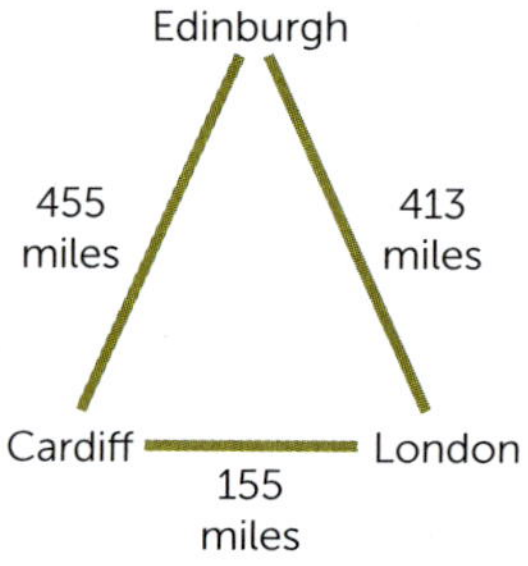

Figure 13.3 A geographical analogy of Hess's law.

Hess's law: a geographical analogy

Cardiff is 155 miles from London. Suppose that instead of travelling directly from Cardiff to London, you travelled first from Cardiff to Edinburgh (455 miles) and then from Edinburgh to London (413 miles) as in Figure 13.3. Whichever route you select, Cardiff is still only 155 miles from London. Similarly, whatever the chemical route taken in Figure 13.2(a), the *energy difference* between $I_2(s) + H_2(g)$ and $2HI(g)$ remains at 53 kJ mol^{-1}.

13.2 Use of Hess's law in calculations

Hess's law allows us to work out a ΔH value for a reaction without carrying out that reaction in the laboratory. This is achieved by algebraically manipulating (adding, subtracting or multiplying by a factor) the chemical equations of related reactions so that they produce the chemical equation for the reaction whose enthalpy change is required. The value for ΔH is then obtained by applying the same operations to the enthalpy changes of the related reactions. This procedure is illustrated in Example 13.2.

EXAMPLE **13.2**

From the thermochemical equations

$$H_2(g) + Cl_2(g) \rightarrow 2HCl(g) \qquad \Delta H^{\ominus}_{(13.6)} = -184.6\ \text{kJ mol}^{-1} \qquad \textbf{(13.6)}$$

$$2NH_3(g) \rightarrow 3H_2(g) + N_2(g) \qquad \Delta H^{\ominus}_{(13.7)} = +\ 92.2\ \text{kJ mol}^{-1} \qquad \textbf{(13.7)}$$

$${}^1/_2N_2(g) + 2H_2(g) + {}^1/_2Cl_2(g) \rightarrow NH_4Cl(s) \qquad \Delta H^{\ominus}_{(13.8)} = -314.4\ \text{kJ mol}^{-1} \qquad \textbf{(13.8)}$$

calculate the enthalpy change for the reaction

$$HCl(g) + NH_3(g) \rightarrow NH_4Cl(s) \qquad \Delta H^{\ominus}_{(13.9)} = ? \qquad \textbf{(13.9)}$$

Answer

We start by multiplying equation (13.6) by $-{}^1/_2$, giving

$$-{}^1/_2H_2(g) + -{}^1/_2Cl_2(g) \rightarrow -HCl(g)$$

The arrow in this equation may be likened to the equals sign (=) in a mathematical equation, and the chemical equation may be rearranged in a similar way to algebraic equations giving:

$$HCl(g) \rightarrow {}^1/_2H_2(g) + {}^1/_2Cl_2(g) \qquad \textbf{(13.10)}$$

(Compare this to the rearrangement of $-x + -z = -y$, to give $y = x + z$.) The rearrangement puts HCl(g) on the left-hand side of an equation, as in equation (13.9).

Dividing equation (13.7) throughout by 2 gives

$$NH_3(g) \rightarrow {}^3/_2H_2(g) + {}^1/_2N_2(g) \qquad \textbf{(13.11)}$$

which keeps $NH_3(g)$ on the left-hand side, as in equation (13.9).

Adding equations (13.10) and (13.11) to equation (13.8) gives

$$HCl(g) + NH_3(g) + {}^1/_2N_2(g) + 2H_2(g) + {}^1/_2Cl_2(g)$$
$$\rightarrow {}^1/_2H_2(g) + {}^1/_2Cl_2(g) + {}^3/_2H_2(g) + {}^1/_2N_2(g) + NH_4Cl(s)$$

Collection and cancellation of like terms gives:

$$HCl(g) + NH_3(g) \rightarrow NH_4Cl(s)$$

which we recognize as equation (13.9). The *enthalpy change* is now obtained by carrying out similar operations to the ΔH *values* as follows:

$$\Delta H^{\ominus}_{(13.9)} = -{}^1/_2\Delta H^{\ominus}_{(13.6)} + {}^1/_2\Delta H^{\ominus}_{(13.7)} + \Delta H^{\ominus}_{(13.8)}$$
$$= -{}^1/_2(-184.6) + {}^1/_2(92.2) + (-314.4) = -176.0\ \text{kJ mol}^{-1}$$

EXERCISE **13D**

Application of Hess's law

Given the following standard enthalpy changes at 298 K,

$S(s) + O_2(g) \rightarrow SO_2(g)$ $\Delta H^\ominus = -296.8\ kJ\ mol^{-1}$ **(13.12)**

$SO_2(g) + {}^1/_2O_2(g) \rightarrow SO_3(g)$ $\Delta H^\ominus = -98.9\ kJ\ mol^{-1}$ **(13.13)**

calculate the standard enthalpy change for the reaction:

$S(s) + {}^3/_2O_2(g) \rightarrow SO_3(g)$ $\Delta H^\ominus = ?$ **(13.14)**

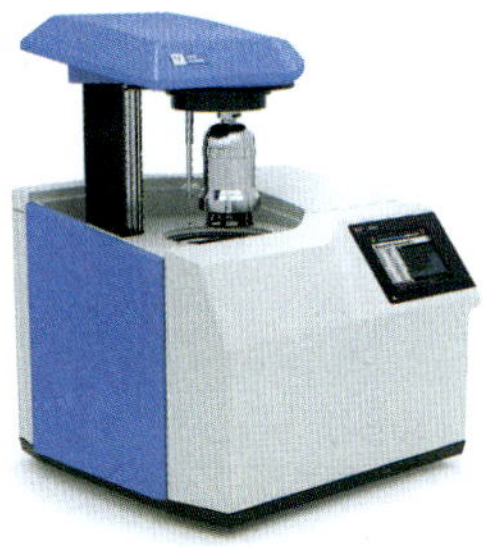

Figure 13.4 Oxygen bomb calorimeter. The fuel (e.g. diesel) is placed in a metal crucible inside the stainless steel 'bomb' (the small bell-like object in the photo). The bomb is lowered into the instrument and the fuel electrically ignited in the presence of oxygen gas. The temperature of the burning fuel is rapidly measured, allowing a calculation of the enthalpy change of combustion. See https://www.youtube.com/watch?v=vWA7dUuZOQU

(Copyright, IKA)

Measuring ΔH

The experimental determination of exothermic enthalpy changes is called **calorimetry**. It involves carrying out the chemical reaction under study in an insulated container to minimize the loss or gain of thermal energy ('heat'). For measurements of enthalpy of combustion, the container is a strong stainless steel vessel (a 'bomb calorimeter', Figure 13.4). The key measurement is that of temperature, and the temperature increase produced is used to calculate the enthalpy change. For an example where a simple coffee polystyrene cup may be used as a calorimeter to measure the enthalpy change of the reaction of Zn with lead ions see Extension 13 on the website.

Table 13.1 Examples of special standard enthalpy changes

Name	Symbol	Description	Example of process
Enthalpy of formation	$\Delta_f H^\ominus$	one mole of compound made from its elements	$K(s) + {}^1/_2Cl_2(g) \rightarrow KCl(s)$
Enthalpy of combustion	$\Delta_c H^\ominus$	one mole of a fuel burned in excess oxygen	$CH_4(g) + 2O_2(g) \rightarrow CO_2(g) + 2H_2O(l)$
Lattice enthalpy	$\Delta_{latt} H^\ominus$	one mole of a crystal completely separated into isolated particles in the gas phase	$K^+,Cl^-(s) \rightarrow K^+(g) + Cl^-(g)$
Enthalpy of neutralization	$\Delta_{neut} H^\ominus$	one mole of water formed by the neutralization of an acid by a base	$CH_3COOH(aq) + NaOH(aq) \rightarrow CH_3COONa(aq) + H_2O(l)$
Bond enthalpy of a specific bond A–B	$\Delta H^\ominus$(A–B) or $D^\ominus$(A–B) (the *D* stands for dissociation)	one mole of bonds broken: all species in the gas phase	$HCl(g) \rightarrow H(g) + Cl(g)$
Ionization energy (also known as ionization enthalpy)	$\Delta_i H^\ominus$ or simply *I*	one mole of atoms (or molecules) ionized: all species in the gas phase	$Na(g) \rightarrow Na^+(g) + e^-$ (this is the equation for the first ionization energy)
Enthalpy of electron gain (−$\Delta_{eg}H^\ominus$ is called the electron affinity)	$\Delta_{eg} H^\ominus$	one mole of anions being formed: all species in the gas phase	$F(g) + e^- \rightarrow F^-(g)$
Enthalpy of vaporization	$\Delta_{vap} H^\ominus$	one mole of vapour being formed from liquid without a change in temperature	$H_2O(l) \rightarrow H_2O(g)$
Enthalpy of fusion	$\Delta_{fus} H^\ominus$	one mole of liquid being formed from the solid without a change in temperature	$H_2O(s) \rightarrow H_2O(l)$
Enthalpy of atomization	$\Delta_{atm} H^\ominus$	one mole of substance broken up into its isolated atoms in the gas phase	$Ca(s) \rightarrow Ca(g)$

13.3 Special kinds of standard enthalpy change

Some examples of special standard enthalpy changes are listed in Table 13.1.

Standard enthalpy of formation

The standard enthalpy of formation of a compound $\Delta_f H^\ominus$ is the standard enthalpy change at a particular temperature when *one mole of compound is made from its elements.*

The standard enthalpies of formation used in this book are those which have been measured (or calculated) at 25°C (298 K), and we shall take $\Delta_f H^\ominus$ as meaning $\Delta_f H^\ominus$ (298 K). The units of $\Delta_f H^\ominus$ are kJ *per mole of compound* made.

Key points about standard enthalpies of formation

- $\Delta_f H^\ominus$ is a *standard* enthalpy change, and therefore applies to a reaction in which the reactants (here, the elements) react to completion and in which both the elements and compound involved are present in their pure forms at an atmospheric pressure of 1 bar.
- $\Delta_f H^\ominus$ at a particular temperature is a constant for a compound. Extensive compilations of $\Delta_f H^\ominus$ values are available. Selected values are shown in Table 13.2.
- If an element exists in more than one allotrope, $\Delta_f H^\ominus$ refers to the compound producing reaction in which that element is present in its reference state.

Table 13.2 Standard enthalpies of formation of selected elements and their compounds at 298K

Substance	Formula and state	$\Delta_f H^\ominus$ /kJ mol^{-1}	Substance	Formula and state	$\Delta_f H^\ominus$ /kJ mol^{-1}
Aluminium	Al(s)	0	Copper(II) oxide	CuO(s)	−157.3
Aluminium oxide	Al_2O_3(s)	−1675.7	Copper(II) sulfate	$CuSO_4$(s)	−771.36
Bromine	Br_2(l)	0	Copper(II) sulfate	$CuSO_4 \cdot 5H_2O$(s)	−2279.7
Bromine	Br_2(g)	+30.91	Fluorine	F_2(g)	0
Hydrogen bromide	HBr(g)	−36.40	Fluoride ion	F^-(aq)	−332.63
Calcium	Ca(s)	0	Hydrogen fluoride	HF(g)	−271.1
Calcium oxide	CaO(s)	−635.09	Hydrogen	H_2(g)	0
Calcium chloride	$CaCl_2$(s)	−795.8	Hydrogen atom	H(g)	+217.97
Carbon (graphite)	C(s) (graphite)	0	Deuterium	D_2(g)	0
Carbon (diamond)	C(s) (diamond)	+1.895	Water	H_2O(l)	−285.83
Carbon monoxide	CO(g)	−110.53	Water	H_2O(g)	−241.82
Carbon dioxide	CO_2(g)	−393.51	Deuterium oxide	D_2O(l)	−294.60
Tetrachloromethane	CCl_4(l)	−135.44	Hydrogen ion	H^+(aq)	0
Hydrogen cyanide	HCN(g)	+135.1	Iodine	I_2(s)	0
Methane	CH_4(g)	−74.81	Iodide ion	I^-(aq)	−55.19
Ethyne (acetylene)	C_2H_2(g)	+226.73	Hydrogen iodide	HI(g)	+26.48
Butane	C_4H_{10}(g)	−126.15	Iron	Fe(s)	0
Cyclohexane	C_6H_{12}(l)	−156.4	Iron(III) oxide	Fe_2O_3 (haematite)	−824.2
Benzene	C_6H_6(l)	+49.0	Nitrogen	N_2(g)	0
Ethanol	C_2H_5OH(l)	−277.69	Nitrogen(II) oxide	NO(g)	+90.25
Sucrose	$C_{12}H_{22}O_{11}$(s)	−2222	Nitrogen(IV) dioxide	NO_2(g)	+33.18
Urea	NH_2CONH_2(s)	−333.51	Dinitrogen tetroxide	N_2O_4(g)	+9.16
Chlorine	Cl_2(g)	0	Ammonia	NH_3(g)	−46.11
Chlorine atom	Cl(g)	+121.68	Ammonium chloride	NH_4Cl(s)	−314.43
Hydrochloric acid	HCl(aq)	−167.16	Ammonium chlorate(VII)	NH_4ClO_4(s)	−295.8
Hydrogen chloride	HCl(g)	−92.31			
Copper	Cu(s)	0	Oxygen	O_2(g)	0

Substance	Formula and state	$\Delta_f H^\ominus$ /kJ mol^{-1}
Ozone	$O_3(g)$	+142.7
Hydroxide ion	$OH^-(aq)$	−229.99
Phosphorus (white)	P (white)	0
Phosphorus (red)	P (red)	−17.6
Phosphorus(III) chloride	$PCl_3(l)$	−319.7
Potassium	K(s)	0
Potassium hydroxide	KOH(s)	−424.76
Potassium chloride	KCl(s)	−436.75

Substance	Formula and state	$\Delta_f H^\ominus$ /kJ mol^{-1}
Sodium	Na(s)	0
Sodium chloride	NaCl(s)	−411.15
Sulfur (rhombic)	S(s) (rhombic)	0
Sulfur (monoclinic)	S(s) (monoclinic)	+0.33
Sulfuric acid	$H_2SO_4(l)$	−813.99
Sulfur dioxide	$SO_2(g)$	−296.83
Sulfur trioxide	$SO_3(g)$	−395.72
Hydrogen sulfide	$H_2S(g)$	−20.63

Think of water. According to the definition, the standard enthalpy of formation of water is the standard enthalpy change involved when pure hydrogen and pure oxygen react completely to give one mole of pure water – all at a pressure of 1 bar. At our chosen temperature of 25°C, both hydrogen and oxygen will be gases and water will be a liquid. The reaction will then be represented as

$$H_2(g) + {}^1/_2O_2(g) \rightarrow H_2O(l) \qquad \Delta_f H^\ominus = -285.83\ \text{kJ mol}^{-1}$$

The standard enthalpy change represented by the equation

$$2H_2(g) + O_2(g) \rightarrow 2H_2O(l) \qquad \Delta H^\ominus = -571.66\ \text{kJ mol}^{-1}$$

does not equal the standard enthalpy change of formation, since the equation indicates that two moles (not one mole) of water are made.

Although $\Delta_f H^\ominus(H_2O)$ may be measured experimentally, by studying the reaction of H_2 and O_2 in the laboratory, for many compounds direct measurement of $\Delta_f H^\ominus$ is impossible. For example, consider the thermochemical equation showing the formation of methanoic (formic) acid:

$$H_2(g) + O_2(g) + C(s) \rightarrow HCO_2H(l) \qquad \Delta_f H^\ominus(298\ \text{K}) = -\ 402.1\ \text{kJ mol}^{-1}$$

If we heat hydrogen, oxygen and graphite together, the reaction does not go to completion and so little methanoic acid is produced that it is impossible to directly measure its enthalpy change of formation. For this reason, the enthalpy change of formation is calculated using Hess's law.

Examples of thermochemical equations where the standard enthalpy changes are equal to the standard enthalpy of formation of compounds are

1 *ammonia*

$$^1/_2N_2(g) + {}^3/_2\ H_2(g) \rightarrow NH_3(g) \qquad \Delta_f H^\ominus = -\ 46.1\ \text{kJ mol}^{-1}$$

2 *benzoic acid*

$$7C(s) + 3H_2(g) + O_2(g) \rightarrow C_6H_5CO_2H(s) \qquad \Delta_f H^\ominus = -\ 385.1\ \text{kJ mol}^{-1}$$

3 *hydrogen iodide*

$$^1/_2H_2(g) + {}^1/_2I_2(s) \rightarrow HI(g) \qquad \Delta_f H^\ominus = +26.48\ \text{kJ mol}^{-1}$$

Standard enthalpy of formation of elements

The standard enthalpy of formation of diamond refers to the process

$$C_{(s)\ (graphite)} \rightarrow C_{(s)\ (diamond)} \qquad \Delta_f H^\ominus = +1.895\ \text{kJ mol}^{-1}$$

Not surprisingly, the standard enthalpy of formation of an element in its **reference state** is zero, since the 'reaction' involves no change:

$$C_{(s)\ (graphite)} \rightarrow C_{(s)\ (graphite)} \qquad \Delta_f H^\ominus = 0.000\ \text{kJ mol}^{-1}$$

This might appear trivial, but it will be important to keep this in mind when we use $\Delta_f H^\ominus$ values in calculations later.

EXERCISE **13E**

Enthalpy of formation

Complete the following thermochemical equations, which refer to the formation of one mole of the substances shown at 298K:

.... $\rightarrow H_2SO_4(l)$ $\Delta_f H^\ominus = -814\ kJ\ mol^{-1}$

.... $\rightarrow C_6H_{12}O_6(s)$ $\Delta_f H^\ominus = -1{,}268\ kJ\ mol^{-1}$

Calculation of $\Delta H^\ominus$ using standard enthalpies of formation

On page 222, we used Hess's law to calculate $\Delta H^\ominus$ for the reaction

$$HCl(g) + NH_3(g) \rightarrow NH_4Cl(s)$$

obtaining $\Delta H^\ominus = -176.0\ kJ\ mol^{-1}$. The same answer may be obtained by adding the standard enthalpies of formation of HCl(g) and $NH_3(g)$ together, and subtracting this sum from the standard enthalpy of formation of ammonium chloride:

$$\Delta H^\ominus = \Delta_f H^\ominus(NH_4Cl(s)) - [\Delta_f H^\ominus(HCl(g)) + \Delta_f H^\ominus(NH_3(g))]$$

The $\Delta_f H^\ominus$ values ($kJ\ mol^{-1}$) obtained from Table 13.2 are:

$\Delta_f H^\ominus(NH_4Cl(s)) = -314.4$; $\Delta_f H^\ominus(HCl(g)) = -92.3$; $\Delta_f H^\ominus(NH_3(g)) = -46.1$

so that

$$\Delta H^\ominus = (-314.4) - (-92.3 + (-)\ 46.1) = -176.0\ kJ\ mol^{-1}$$

This method is easier to use than the algebraic method (page 226) based upon Hess's law, and we will now generalize the calculation so that it applies to all reactions:

- To work out the standard enthalpy change ($\Delta H^\ominus$) of any reaction from $\Delta_f H^\ominus$ values, substitute into the following equation:

  $$\Delta H^\ominus = \Sigma\Delta_f H^\ominus(\text{products}) - \Sigma\Delta_f H^\ominus(\text{reactants}) \quad (13.15)$$

  where $\Sigma\Delta_f H^\ominus$(products) is the sum of the standard enthalpies of formation of the products in the reaction, and $\Sigma\Delta_f H^\ominus$(reactants) is the sum of the standard enthalpies of formation of the reactants.

- When summing the enthalpies of formation, the enthalpy of formation of each substance must be multiplied by its **coefficient** in the balanced equation for the reaction. This can be seen in Example 13.3.

EXAMPLE **13.3**

Calculate $\Delta H^\ominus$ at 298 K for the production of three moles of copper by the reaction:

$$3CuO(s) + 2NH_3(g) \rightarrow 1N_2(g) + 3H_2O(l) + 3Cu(s) \quad \textbf{(A)}$$

Answer

$$\Delta H^\ominus = \Sigma\Delta_f H^\ominus(\text{products}) - \Sigma\Delta_f H^\ominus(\text{reactants})$$

$$\Delta H^\ominus = (\underset{\uparrow}{1}\Delta_f H^\ominus(N_2(g)) + \underset{\uparrow}{3}\Delta_f H^\ominus(H_2O(l)) + \underset{\uparrow}{3}\Delta_f H^\ominus(Cu(s))] - [\underset{\uparrow}{3}\Delta_f H^\ominus(CuO(s)) + \underset{\uparrow}{2}\Delta_f H^\ominus(NH_3(g))]$$

The numbers indicated by arrows are the coefficients in the chemical equation, although we wouldn't normally show the 1 in front of N_2.

Using the standard enthalpy of formation data in Table 13.2,

$\Delta_f H^\ominus(H_2O(l)) = -285.83\ kJ\ mol^{-1}$

$\Delta_f H^\ominus(CuO(s)) = -157.3\ kJ\ mol^{-1}$

$\Delta_f H^\ominus(NH_3(g)) = -46.11\ kJ\ mol^{-1}$

And, remembering that the standard enthalpy of formation of the elements N_2 and Cu is zero,

$$\Delta H^\ominus = [0 + 3(-285.83) + 3(0)] - [3(-157.3) + 2(-46.11)]$$
$$= -293.4\ \text{kJ mol}^{-1}$$

Therefore the standard enthalpy change for reaction (A) is -293.4 kJ mol^{-1}.

EXERCISE **13F**

Standard enthalpy change

Use the $\Delta_f H^\ominus$ data in Table 13.2 to calculate the standard enthalpy change at 298 K of the following reactions:

(i) the burning of two moles of carbon monoxide gas:

$$2CO(g) + O_2(g) \rightarrow 2CO_2(g)$$

(ii) the reduction of one mole of iron(III) oxide by carbon monoxide:

$$Fe_2O_3(s) + 3CO(g) \rightarrow 2Fe(s) + 3CO_2(g)$$

(iii) the decomposition of one mole of ammonia gas:

$$NH_3(g) \rightarrow {}^1/_2N_2(g) + {}^3/_2H_2(g)$$

In the above examples, the calculated enthalpy changes have the units of 'kJ per mol' (kJ mol^{-1}) because the calculated enthalpy change applies to the reaction equation with the molar ratios of reactants (here 3CuO and $2NH_3$ so 3:2) as written, and we can think of it as applying 'per mol of reaction equation'.

You might be wondering why the calculation of $\Delta H^\ominus$ from $\Delta_f H^\ominus$ values by this method works. The answer lies in the definition of standard enthalpy of formation, which requires $\Delta_f H^\ominus$ values for elements in their reference states to be zero. Although the standard enthalpies (i.e. energies) of compounds and elements cannot be measured or calculated, the *difference* between the standard enthalpies of two substances is equal to the difference in the corresponding standard enthalpies of formation. This is the reason why equation (13.15) strongly resembles equation (13.2).

Table 13.3 Energy values

Fuel	Energy value/ kJ g^{-1}
Hydrogen ($H_2(g)$)	142
Methane ($CH_4(g)$)	56
Ethyne (acetylene, $C_2H_2(g)$)	50
Octane ($C_8H_{18}(l)$)	48
Coal (bituminous)	≈30
Pine wood	18
Methanol ($CH_3OH(l)$)	23
Apples	2
White bread	11
Butter	30
Cheese	4–18
Hard boiled eggs	7
Full cream milk	3
Baked potatoes	4
Chips	11
Nitroglycerine	7
TNT	5
Ammonium nitrate	2

Standard enthalpy of combustion

The standard enthalpy of combustion $\Delta_c H^\ominus$ is the

standard enthalpy schange when one mole of substance is burned in oxygen.

For example, methane (the main constituent of natural gas) burns as follows:

$$CH_4(g) + 2O_2(g) \rightarrow CO_2(g) + 2H_2O(l) \qquad \Delta_c H^\ominus\ (298\ \text{K}) = -890\ \text{kJ mol}^{-1}$$

(From now on we shall assume that all $\Delta_c H^\ominus$ values apply at 298 K.)

The heating power of fuels is often expressed as the **energy value**, defined as the heat produced when one gram of fuel burns completely. Its units are kilojoules per gram (kJ g^{-1}).

We work out the energy value of methane as follows. The molar mass of methane is 16.0 g mol^{-1}. When 16.0 g of methane burns, 890 kJ of heat is produced. When 1 g of methane is burned, 890/16.0 = 56 kJ of heat is produced, so that the energy value of methane is 56 kJ g^{-1}. The energy values of common fuels are shown in Table 13.3.

Determination of enthalpy changes of combustion in the laboratory

The determination of $\Delta_c H^\ominus$ is, in principle, similar to that for other enthalpy changes. In practice, the fuel (or food) is mixed with an oxidizing agent (such as solid sodium peroxide (Na_2O_2) or pure oxygen gas) and sealed in a **bomb calorimeter** (see Figure 13.4). The mixture is ignited by an electric current and the temperature rise recorded. The combustion reaction in a bomb calorimeter is not taking place at constant pressure, but this is easily allowed for in the final calculations.

EXERCISE **13G**

Energy from fuels

(i) Given the thermochemical equation

$C_{12}H_{22}O_{11}(s) + 12O_2(g) \rightarrow 12CO_2(g) + 11H_2O(l)$

$\Delta_c H^\ominus = -5{,}644\ \text{kJ mol}^{-1}$

calculate the energy value of sucrose in kJ g^{-1}.

(ii) From Table 13.3, work out the volume of hydrogen gas which, when burned completely, produces 100 kJ of heat (assume the molar volume of a gas at room temperature and pressure is 24 dm^3).

13.4 Nutrition

Energy value of food

Table 13.3 includes the energy value of common foodstuffs, since these can be thought of as 'human fuel'. If 1 g of apple were completely burned in a flame, it would give out about 2 kJ of heat energy. Energy is also produced when foods are digested in the body, but there are three important differences between the breakdown of foodstuffs (more properly called **food metabolism**) and the burning of a fuel in a flame:

1 In the human body, foodstuffs are consumed slowly in a complicated series of chemical reactions. About 40 per cent of the enthalpy change of combustion of a food is available to the body to be used in exercise.
2 Oxygen gas is not used as the oxidizing agent in the metabolism of food. The oxidizing agent used in the cell is called **nicotinamide adenine dinucleotide** (NAD).
3 Some of the energy released in the metabolism of foods is not given out as heat, but is used to make a substance called **adenosine triphosphate** (ATP). ATP may be converted to **adenosine diphosphate** (ADP) when required, and it is the ATP → ADP reaction that produces the energy upon which we depend for the synthesis of biologically important molecules, for the transport of vital nutrients within the body and for muscle action.

See **Extension 13** on the website at https://www.bloomsburyonlineresources.com/chemistry

Measuring energy use during bodily exercise

The metabolism of fats and glucose during prolonged exercise requires oxygen and produces carbon dioxide and water. The energy consumption of an individual during exercise can be estimated by measuring the ratio of the volume of CO_2 produced to the volume of O_2 used up. This ratio is called the **respiratory exchange ratio** (RER):

$$\text{RER} = \frac{\text{volume of CO}_2\text{ produced per minute}}{\text{volume of O}_2\text{ consumed per minute}}$$

Experimentally, the RER is obtained by analysis of the inhaled and exhaled air of the person exercising.

If a person's body is using only carbohydrates to provide energy, the RER is about 1.0. If only fats are being used as the source of energy, the ratio is about 0.7. In both cases, about 20 kJ of energy is being produced per dm^3 of oxygen consumed.

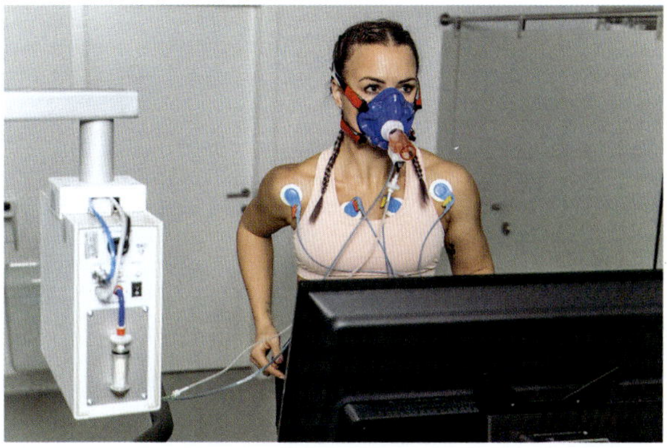

An athlete on a treadmill, showing the equipment used to measure the respiratory exchange of oxygen and carbon dioxide.

EXERCISE **13H**

Nutrition

Diet plans often tabulate the energy value of food in kilocalories (symbolized Cal) with 1 Cal = 4.18 kJ. A 20 g bar of chocolate possessed a calorific value of 120 Cal. What is the energy value of the chocolate, in kJ g^{-1}?

For an average person at rest (e.g. sitting in a chair) who lives on a mixed diet, the RER is about 0.8. Typically, an average male at rest consumes about 0.30 dm^3 O_2 per minute, or $0.30 \times 60 \times 24 \approx 430$ dm^3 per day. (The man would breathe in much more oxygen gas than this, but only a small fraction of the inhaled oxygen would be actually used up in the metabolism of foodstuffs.) Every day the typical man would produce $20 \times 430 \approx 8{,}600$ kJ (or 2,060 Cal) per day. This would also be the minimum daily energy requirement for such a man – if his diet was of a lower energy value his body would have to draw upon its food reserves. If the man engaged in walking, reading or writing, more than 8,600 kJ per day^{-1} would be required. In extreme cases, athletes can consume over 50,000 kJ per day! To lose weight, a person persistently ingests food with a calorific value which is less than their daily requirement – the practice is more difficult than the theory!

EXERCISE **13I**

Energy consumption

Estimate the minimum energy consumption per day of an average woman. Assume that her oxygen consumption is 270 $cm^3 min^{-1}$.

One commercially useful group of materials that produce heat energy extremely rapidly are explosives. These are discussed in Box 13.2.

BOX **13.2**

Explosives

An explosive is a substance which undergoes a very rapid chemical reaction in which gas is produced. The volume of gas produced is large compared with the bulk of the explosive, and the gas escapes so rapidly that a pressure wave is set up in the surrounding air – this causes the bang. For example, the explosive nitroglycerine (Figure 13.6) decomposes according to the overall equation

$$4C_3H_5(NO_3)_3(l) \rightarrow 12CO_2(g) + 10H_2O(g) + 6N_2(g) + O_2(g)$$

Although the earliest explosive was gunpowder (black powder), consisting of sulfur, potassium nitrate (as oxidizing agent) and charcoal, it was the discovery of nitroglycerine (in 1847), of trinitrotoluene (TNT, in 1863) and of dynamite (made in 1866 by Alfred Nobel by absorbing nitroglycerine into finely powdered earth) that revolutionized the explosives industry. The early investigations of explosives

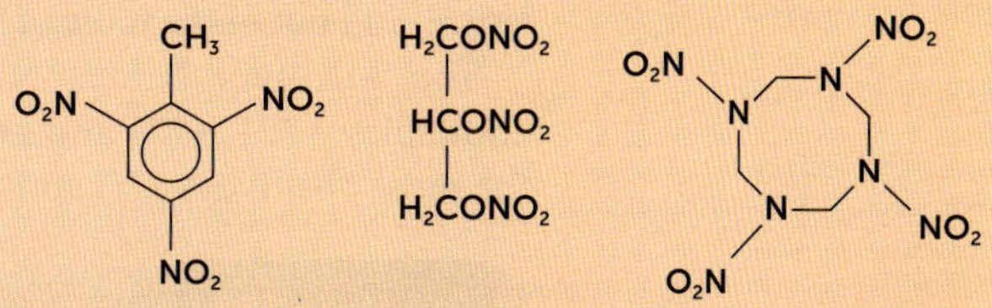

TNT Nitroglycerine HMX

Figure 13.5 Structures of the explosives TNT, nitroglycerine and HMX (tetramethylenetetranitramine). Nitroglycerine and HMX contain sufficient oxygen to burn completely, but TNT is blended with 80 per cent ammonium nitrate to ensure complete oxidation during the explosion.

BOX **13.2** *(continued)*

caused many fatalities, including Alfred Nobel's brother and father. Although we associate explosives with their military applications, most explosives are used in quarries, mines and for clearing land in road construction.

All common explosives contain nitrogen, and nitrogen gas is always produced as one of the gaseous products; the formation of the strong N≡N bond in N_2 makes combustion extremely exothermic. The explosion reaction is over in a very short time. For example, although the burning of nitroglycerine produces only 7 kJ g^{-1} of heat (Table 13.3), this heat energy is produced in less than 0.001 s. This is equivalent to a power of 7,000 kW, and the temperature of the escaping gas may reach 3,000°C and speeds of up to 7,000 m s^{-1}!

To illustrate the pressures involved, suppose that 5 g of nitroglycerine is exploded in a sealed container of volume 10 cm^3 (10^{-5} m^3). About 0.16 mol of gases are produced at 3,000°C. We can crudely estimate the pressure of the gas at the instant of the explosion using the ideal gas equation

$$P \approx \frac{nRT}{V} \approx \frac{0.16 \times 8.3145 \times (273 + 3000)}{10^{-5}}$$
$$\approx 4 \times 10^8 \text{ Pa (4000 atm)}$$

This is an enormous pressure and explains the ability of explosives to smash rocks and destroy buildings.

13.5 Lattice enthalpy

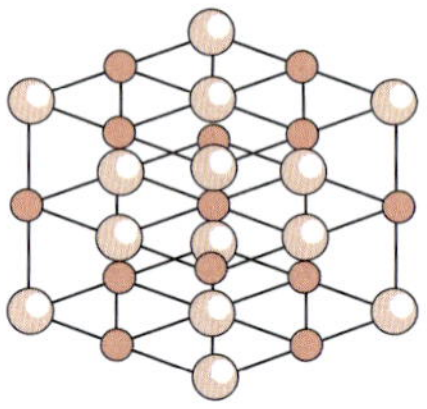

Na^+,Cl^- lattice

The standard enthalpy change in which 1 mol of a crystal lattice is broken up into isolated gaseous particles is called the **lattice enthalpy** ($\Delta_{latt}H^\ominus$) of the lattice. For example, 776 kJ of energy is needed to break up 1 mol of pure sodium chloride crystal lattice into gaseous sodium and chloride ions at 298 K and 1 bar pressure:

$$Na^+,Cl^-(s) \rightarrow Na^+(g) + Cl^-(g) \qquad \Delta_{latt}H^\ominus = +771 \text{ kJ mol}^{-1}$$

When 1 mol of gaseous sodium ions condenses with 1 mol of chloride ions, under the same conditions, 1 mol of sodium chloride crystals are made:

$$Na^+(g) + Cl^-(g) \rightarrow Na^+,Cl^-(s) \qquad \Delta H^\ominus = -\Delta_{latt}H^\ominus = -771 \text{ kJ mol}^{-1}$$

Generalizing:

Energy needs to be absorbed in order to break down a crystal lattice
Energy is given out when a crystal lattice is formed

Lattice enthalpy as a measure of the strength of ionic bonding

The lattice enthalpy of NaCl(s) is an approximate measure of the strength of the bonding between Na^+ and Cl^- ions in the sodium chloride crystal. Ions in a sodium chloride crystal lattice are not paired, and each sodium (or chloride) ion is attracted by six oppositely charged neighbouring ions and (less strongly) by more distant ions. This means that $\Delta_{latt}H^\ominus$ is a measure of the *total* attractive force between sodium and chloride ions within the lattice, and not of the force of attraction between isolated *pairs* of Na^+ and Cl^- ions.

The 'MX'-type lattices NaCl(s), NaBr(s) and NaI(s) possess the following lattice enthalpies: +771, +731 and +684 kJ mol^{-1}, respectively. Of these crystals, NaCl possesses the most positive $\Delta_{latt}H^\ominus$ and is said to be the 'most stable' of the series.

EXERCISE **13J**

Lattice enthalpies

Given that

$$Ca^{2+}(g) + 2Cl^-(g) \rightarrow Ca^{2+},2Cl^-(s) \qquad \Delta_{latt}H^\ominus = -2{,}237 \text{ kJ mol}^{-1}$$

how much energy is required to break up exactly 0.1 mol of calcium chloride crystal into the separate gaseous ions at 298 K?

Use of Hess's law in calculating the lattice enthalpy of an ionic crystal

The thermochemical equation showing the reaction in which sodium chloride is formed from its elements at 298 K,

$$Na(s) + {}^1/{}_2Cl_2(g) \rightarrow Na^+,Cl^-(s) \qquad \Delta_f H^\ominus = -411\ kJ\ mol^{-1} \qquad \textbf{(13.16)}$$

may be split up into several imaginary steps known as the **Born–Haber cycle** (Figure 13.7). The enthalpy change for each step has been defined in Table 13.1. As always, the endothermic steps cause the total enthalpy to increase (i.e. to move up the figure), whereas the exothermic steps cause the total enthalpy to decrease (i.e. to move down the figure). The stages, starting with Na(s) and $\frac{1}{2}Cl_2(g)$, are:

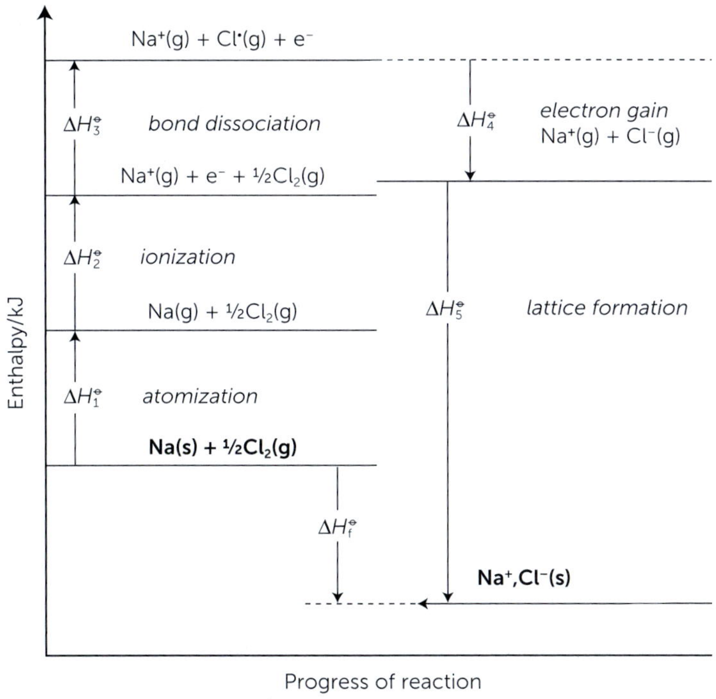

Figure 13.6 The Born–Haber cycle, showing the enthalpy changes involved in the formation of NaCl(s) from its elements. The enthalpy axis is not to scale. The cycle starts with 1 mol of metallic sodium and half a mole of chlorine gas. The cycle is named after Max Born and Fritz Haber.

1 The conversion of solid sodium to gaseous sodium (*atomization*):

$$Na(s) \rightarrow Na(g) \qquad \Delta H^\ominus_1 = +109\ kJ\ mol^{-1}$$

(i.e. the standard enthalpy of atomization of sodium).

2 The ionization of sodium atoms in the vapour:

$$Na(g) \rightarrow Na^+(g) + e^- \qquad \Delta H^\ominus_2 = +494\ kJ\ mol^{-1}$$

(i.e. the standard first ionization energy of sodium).

3 The breaking of the Cl–Cl bond to produce isolated chlorine atoms:

$$\tfrac{1}{2}Cl_2(g) \rightarrow Cl^\bullet(g) \qquad \Delta H^\ominus_3 = +121\ kJ\ mol^{-1}$$

(i.e. half the standard bond enthalpy of the Cl–Cl bond). The enthalpy change for this process is half the energy required to break one Cl–Cl bond because we only require one $Cl^\bullet$ atom to react with each Na. The standard bond enthalpy of Cl–Cl is +242 kJ mol^{-1}, so that $\Delta H^\ominus_3 = 242/2 = 121$ kJ mol^{-1}.

4 The gain of one electron by the chlorine atom:

$Cl^{\bullet}(g) + e^- \rightarrow Cl^-(g)$ $\qquad \Delta H^{\ominus}_4 = -364\ \text{kJ mol}^{-1}$

(i.e. the standard enthalpy of electron gain of the chlorine atom). The $Cl^{\bullet}$ atom is reaching a more stable configuration in gaining an electron.

5 The attraction of the sodium and chloride ions to make crystalline sodium chloride:

$Na^+(g) + Cl^-(g) \rightarrow Na^+,Cl^-(s)$ $\qquad \Delta H^{\ominus}_5 = -\Delta_{latt}H^{\ominus}$

Note that addition of the equations involving the $\Delta H^{\ominus}_1$, $\Delta H^{\ominus}_2$, $\Delta H^{\ominus}_3$, $\Delta H^{\ominus}_4$ and $\Delta H^{\ominus}_5$ terms gives equation (13.16). According to Hess's law,

$$\Delta_f H^{\ominus} = \Delta H^{\ominus}_1 + \Delta H^{\ominus}_2 + \Delta H^{\ominus}_3 + \Delta H^{\ominus}_4 + \Delta H^{\ominus}_5 \qquad \textbf{(13.17)}$$

The stages shown in Figure 13.7 are not meant to indicate the way that sodium and chlorine actually react, i.e. they do not represent the *mechanism* of the reaction.

Similar energy cycles may be set up for other ionic compounds, such as Ca^{2+},O^{2-}. The importance of equation (13.17) is that we can calculate one of the enthalpy terms knowing all the others. The unknown is usually the lattice enthalpy (which equals $-\Delta H^{\ominus}_5$), which cannot be measured directly by experiment.

Rearrangement of equation (13.17) gives

$$\Delta H^{\ominus}_5 = \Delta_f H^{\ominus} - \Delta H^{\ominus}_1 - \Delta H^{\ominus}_2 - \Delta H^{\ominus}_3 - \Delta H^{\ominus}_4$$

For example, for the sodium and chlorine cycle,

$$\Delta H^{\ominus}_5 = -411 - (+109) - (+494) - (+121) - (-364) = -771\ \text{kJ mol}^{-1}$$

Since

$$\Delta H^{\ominus}_5 = -\Delta_{latt}H^{\ominus}\,(NaCl)$$

$$\Delta_{latt}H^{\ominus}\,(NaCl) = -\Delta H^{\ominus}_5 = -(-771) = +771\ \text{kJ mol}^{-1}$$

EXERCISE **13K**

Enthalpy changes

(i) What is the standard enthalpy change for the process $Na(s) \rightarrow Na^+(g) + e^-$?

(ii) Estimate the amount of energy required for the following (hypothetical) process under standard conditions:

$Na^+,Cl^-(s) \rightarrow Na(g) + {}^1/_2Cl_2(g)$

EXERCISE **13L**

Lattice enthalpy of magnesium fluoride

Calculate the lattice enthalpy of magnesium fluoride given the following standard enthalpy data:

enthalpy of atomization of magnesium, $\Delta_{atom}H^{\ominus}$ (Mg(s)) = +148 kJ mol^{-1}
first ionization energy of magnesium, $\Delta_{ion}H^{\ominus}$ (Mg(g)) = +744 kJ mol^{-1}
second ionization energy of magnesium, $\Delta_{ion}H^{\ominus}$ (Mg^+(g)) = +1456 kJ mol^{-1}
bond enthalpy of F_2(g), $\Delta H^{\ominus}$ (F–F) = +158 kJ mol^{-1}
enthalpy of electron gain of F(g), $\Delta_{eg}H^{\ominus}$ = −339 kJ mol^{-1}
enthalpy of formation of MgF_2(s), $\Delta_f H^{\ominus}$ (MgF_2(s)) = −1121 kJ mol^{-1}.

BOX 13.3

Rocket propellants

Most spacecraft and rockets are propelled using the principle of a **reaction engine**, in which propulsion is achieved by expelling mass from the rear of the vehicle. In spacecraft, the mass being expelled is in the form of pressurized gas emerging through a supersonic nozzle. The waste gases are produced by burning a fuel in an oxidizing agent (such as liquid oxygen 'LOX') that is carried within the spacecraft (Figure 13.7). For hydrocarbon fuels, the waste contains superheated steam and CO_2.

(a)

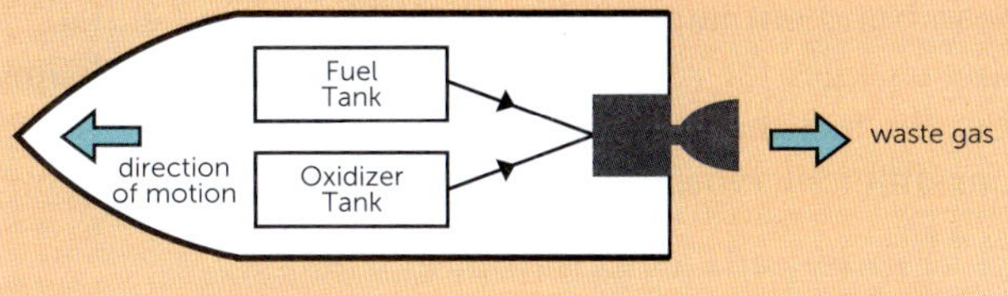

(b)

Figure 13.7 (a) A rocket engine, a type of reaction engine used in rocket propulsion. The blue arrows show the direction of the expelled gas (RHS) and of the motion of the rocket (LHS) as predicted by Newton's Third Law. (b) NASA's Atlas V rocket launching booster system uses about 3,000 tonnes of liquid oxygen and a fuel called refined petroleum (RP-1). It produces 4 million newtons of thrust for four minutes.

13.6 Bond energetics

If a covalent bond exists between two atoms, it will take energy to pull those atoms apart. For example, the dissociation of the A–B bond in the molecule AB may be represented by the equation

$$A{-}B(g) \rightarrow A(g) + B(g)$$

(Note that the reactants and products are all gases.) The amount of energy required to break a particular bond in a *gaseous molecule* under standard conditions is called the **(standard) bond enthalpy** of that bond, symbolized $\Delta H^{\ominus}$ (A–B). For example, the standard enthalpy change for the dissociation of chlorine molecules at 298 K,

$$Cl_2(g) \rightarrow Cl^{\bullet}(g) + Cl^{\bullet}(g)$$ *N.B. All species must be in the gas state*

is 242 kJ per mole of Cl–Cl bonds. Therefore, the bond enthalpy is 242 kJ mol^{-1}:

$$\Delta H^{\ominus}(Cl{-}Cl) = 242\ kJ\ mol^{-1}$$

Selected bond enthalpies are shown in Table 13.4. Covalent bonds where $\Delta H^{\ominus}$ (A–B) is greater than 400 kJ mol^{-1} are usually described as strong bonds.

When a reaction is reversed the energy changes are also reversed. We predict that the reaction

$$Cl(g) + Cl(g) \rightarrow Cl_2(g)$$

will involve a standard enthalpy change at 298 K of −242 kJ mol^{-1}, the released energy reflecting the fact that in this reaction the two chlorine atoms are achieving a stable electronic configuration through the formation of a covalent bond.

Bond enthalpies in polyatomic molecules

The standard enthalpy change for the reaction:

$$H{-}O{-}H(g) \rightarrow H(g) + OH(g) \qquad \Delta H^{\ominus} = +499\ kJ\ mol^{-1}$$

shows that the bond enthalpy of the HO–H bond, symbolized $\Delta H^{\ominus}$ (HO–H), is 499 kJ per mol of O–H bonds. However, if we now break the O–H bond in OH(g), only 428 kJ of energy per mol of O–H is required for complete dissociation:

$O{-}H(g) \rightarrow O(g) + H(g)$ $\Delta H^{\ominus} = \Delta H^{\ominus}(HO{-}H) = +428\ kJ\ mol^{-1}$

The **mean bond enthalpy** for the stepwise dissociation of both O–H bonds *in water* is $(499 + 428)/2 = 463.5\ kJ\ mol^{-1}$.

This does not mean that the O–H bonds in *water* are different. On the contrary, they are identical in every respect. The different values of bond enthalpies for the O–H bonds arise because they apply to very different species, H_2O and OH.

Mean bond enthalpies

The bond enthalpies of bonds such as C–H, C–Cl, C=O, N=N and O–H are approximately the same in different molecules. If the values of bond enthalpy for a bond between two atoms (A and B) in several different molecules are averaged, the resulting value is called the **mean bond enthalpy** (Table 13.4). Mean bond enthalpies are useful in estimating enthalpy changes for reactions for which standard enthalpies of formation are unavailable, but the fact that such values may have been obtained by averaging bond enthalpies from different types of molecule may lead to substantial errors in the calculated value of $\Delta H^{\ominus}$.

Table 13.4 Standard bond enthalpies for bonds in molecules in the gas phase

Bond	$\Delta H^{\ominus}$ (A–B) (298 K)/ kJ mol^{-1}
Bond enthalpies	
H–H	436
N≡N	945
O=O	497
F–F	158
Cl–Cl	242
H–F	565
H–Cl	431
H–Br	366
H–I	299
O**C=O**	531
H_3**C–Cl**	339
H_2**C=C**H_2	699
Mean bond enthalpies	
C–H	412
C–C	348
C=C	612
C=O	743
O–H	463
N–N	163
N–H	388
C–Cl	338

The atoms on either side of the bond are displayed in bold.

EXERCISE **13M**

Mean bond enthalpy of C=O in CO_2

The carbon dioxide molecule (O=C=O) contains two identical C=O bonds. If the bonds in CO_2 are broken in sequence it is found that $\Delta H^{\ominus}$(C=O) = 531 kJ mol^{-1} and $\Delta H^{\ominus}$(C=O, second bond being broken) = 1,075 kJ mol^{-1}. What is the mean bond enthalpy of the C=O bond in CO_2?

EXERCISE **13N**

Bond enthalpies 1 (use of data in Table 13.4)

(i) Table 13.4 shows that $\Delta H^{\ominus}$(C–H) = 412 kJ mol^{-1}. Estimate the enthalpy change for the following reaction under standard conditions:

$$CH_3(g) + H(g) \rightarrow CH_4(g)$$

(ii) The bond lengths (in pm) for the hydrogen halides are: H–F 92, H–Cl 128, H–Br 141 and H–I 160. Do these figures support the idea that long bonds are weak bonds?

(iii) Does the bond enthalpy data in Table 13.4 suggest that double and triple bonds are stronger than single bonds?

(iv) Suggest why so many explosives produce $N_2(g)$ as a product.

Use of bond enthalpies: fuels – where does the heat come from?

In introducing enthalpy changes (page 217) we stated that the size and sign of ΔH is decided by the relative bond strengths of the reactant and product molecules. When bonds are created, energy is *always* given out. Strong bonds release more energy than weak bonds. In order to break bonds, energy must *always* be absorbed (e.g. the reactant must be heated) and the stronger the bond, the greater the amount of energy that must be absorbed. We go into the 'bond strength arithmetic' in more detail below. Although we look at combustion, the principles apply to all reactions. If we have accurate values of bond energies we can easily calculate enthalpy changes and so predict whether or not a reaction will be exothermic or endothermic.

The burning of methane in air follows the equation:

$$CH_4(g) + 2O_2(g) \rightarrow CO_2(g) + 2H_2O(g) \quad \textbf{(13.18)}$$

We know that heat is given out during this reaction. But where exactly does the heat energy come from? We will now calculate the enthalpy change for this reaction using mean bond enthalpies. To do this we carry out the following operations:

1 We start by writing out the molecular structures for the reactants and products in full (Figure 13.8). We see that, going from left to right in the equation, the reaction involves the breaking of four C–H bonds and two O=O bonds and the making of two C=O bonds and four O–H bonds. The atoms on either side of the bond are displayed in bold.

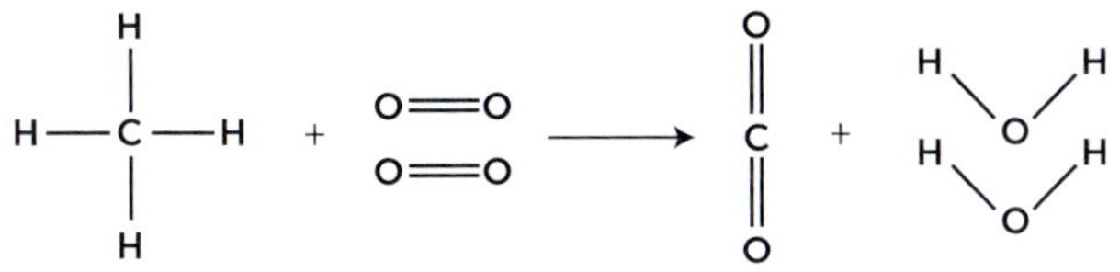

Figure 13.8 The burning of methane in air.

2 We imagine that the oxygen and methane molecules are *completely broken up into atoms* during the reaction, and that the resulting isolated atoms are then *completely assembled* into the product water and carbon dioxide molecules. (It does not matter whether the reaction really follows this mechanism because, according to Hess's law, the results of our 'energy count' are independent of how the molecules actually react or whether or not the products are made in several intermediate steps.)

3 The overall enthalpy change for reaction (13.18) is obtained by adding the enthalpy changes involved in the breaking up of the reactant molecules and in the subsequent formation of the product molecules. The breaking up of molecules involves positive enthalpy changes because energy needs to be absorbed in order to break apart atoms forming a covalent bond. The formation of bonds involves negative enthalpy changes because of the extra stability achieved by the atoms after combination.

Table 13.5 shows the arithmetic involved in using bond enthalpies to work out whether, overall, heat is taken in or given out in reaction (13.18). In this calculation, mean bond enthalpies have been used for O–H and C–H bonds. Overall, *we predict* that the reaction involves the net production of 696 kJ of heat per mol of methane burned. The reason that heat is given out is simply that, on balance, *the product molecules possess stronger bonds than the reactants*. The arithmetic shows that if the C–H bonds and O=O bonds were stronger, little or no heat would be given out!

Table 13.5 Estimating the enthalpy change for the burning of methane. it is assumed that the reaction goes to completion

Bonds	Energy change	Totals$^{\ominus}$/kJ mol^{-1}
Four C–H bonds broken	4 × 412 = 1648 kJ taken in	+1,648
Two O=O bonds broken	2 × 497 kJ taken in	+994
Two C=O bonds made	2 × 803 kJ given out	−1,606
Four O–H bonds made	4 × 463 kJ given out	−1,852
	Estimated enthalpy change $\Delta H^{\ominus}$ =	−816

* Positive value indicates heat taken in, negative value indicates heat given out.

We may generalize our results to all reactions:

- If the sum of the bond enthalpies of the product molecules is greater than the sum of the bond enthalpies for the reactants, the reaction will be exothermic.
- If the sum of the bond enthalpies of the product molecules is less than the sum of the bond enthalpies for the reactants, the reaction will be endothermic.

Whereas the calculated enthalpy change for reaction (13.18) is $-816\,\text{kJ mol}^{-1}$, the experimentally measured value (with methane and water as gaseous products) is $\Delta_c H^{\ominus} = -802\,\text{kJ mol}^{-1}$. The error would be much greater if we had used the mean bond energy of C=O in all common compounds ($743\,\text{kJ mol}^{-1}$) rather than the mean bond energy for CO_2 itself ($803\,\text{kJ mol}^{-1}$) – see Exercise 130. This example serves to remind us that the choice of bond enthalpies in such calculations should be carefully reviewed beforehand.

EXERCISE **130**

Bond enthalpies 2

Estimate the standard enthalpy change ($\Delta H^{\ominus}$) for the following reactions at 298 K:

(i) $CH_2{=}CH_2(g) + Cl_2(g) \rightarrow CH_2Cl{-}CH_2Cl(g)$

(ii) $2H_2O(l) \rightarrow 2H_2(g) + O_2(g)$ (Note: $H_2O(l) \rightarrow H_2O(g)\ \Delta H^{\ominus} = +44\,\text{kJ mol}^{-1}$).

13.7 Exothermic reactions

In many cases, the main 'driving force' for a chemical reaction starting from pure reactants – the dominant factor 'that makes it happen' – is the fact that the reaction is exothermic. In other words, the fact that the reactants lower their combined energy by being converted into products often results in that reaction being (thermodynamically) allowed, which (as we will see in Unit 16) also means that it has a large equilibrium constant, K. This allows us to make a prediction: if a reaction is exothermic, that is a valuable (but not universally reliable) indicator that a reaction *can* occur, and *strongly* exothermic reactions are *very likely* to occur. Many of the familiar chemical reactions we encounter in the laboratory (including the combustion of fuels and the oxidation of metals) are exothermic reactions.

Why should this be? As we find out in Unit 16, the criterion that determines whether a reaction can occur under constant external pressure is that the reaction in question has a *negative value* for the standard Gibbs energy change, $\Delta G^{\ominus}$. Theory shows that the more negative $\Delta H^{\ominus}$ is, the more negative $\Delta G^{\ominus}$ is. That is, reaction *exothermicity* (the release of energy into the surroundings) *reinforces* a negative $\Delta G^{\ominus}$ for the reaction. A strongly exothermic reaction is very likely to have a negative $\Delta G^{\ominus}$.

REVISION QUESTIONS

13.1 Which of the following is false?

(i) For exothermic reactions, $\Delta H^{\ominus}$ is always positive

(ii) $\Delta_c H^{\ominus}(H_2(g)) = \Delta_f H^{\ominus}(H_2(g))$

(iii) $\Delta_c H^{\ominus}$ (C(s) graphite) = $\Delta_f H^{\ominus}(CO_2(g))$

(iv) $\Delta H^{\ominus}$ for the reaction $2S(s) + 3O_2(g) \rightarrow 2SO_3(g)$ is the same as $\Delta_f H^{\ominus}(SO_3(g))$.

(v) $\Delta_f H^{\ominus}$ (C(s) graphite) = 0

(vi) $\Delta_c H^{\ominus}$ (C(s) graphite) = $\Delta_c H^{\ominus}$ (C(s) diamond)

(vii) The O–H bonds in the water molecule are equally strong.

13.2

(i) Define 'standard enthalpy of combustion'.

(ii) Calculate the standard enthalpy of combustion of methyl benzene (toluene), $C_7H_8(l)$, given that $\Delta_f H^{\ominus}$ (methyl benzene) = +12.0 kJ mol^{-1}. (For other data, use Table 13.2.)

(iii) Calculate the energy value of toluene fuel in units of kJ g^{-1}, and kJ cm^{-3} (the density of methyl benzene at room temperature is 0.86 g cm^{-3}).

13.3 Use the data in Table 13.2 to calculate the standard enthalpy changes for the following reactions at 298 K:

(i) P(s) (white) $\rightarrow$ P(s) (red)

(ii) the production of ethyl ethanoate from ethanoic acid and ethanol:

$$CH_3COOH(l) + C_2H_5OH(l) \rightarrow CH_3COOC_2H_5(l) + H_2O(l)$$

using $\Delta_f H^{\ominus}(CH_3COOH(l)) \approx -485$ kJ mol^{-1}, and $\Delta_c H^{\ominus}(CH_3COOC_2H_5(l)) \approx -481$ kJ mol^{-1}.

(iii) $6NH_4ClO_4(s) + 10Al(s) \rightarrow 3N_2(g) + 9H_2O(g) + 5Al_2O_3(s) + 6HCl(g)$ (This is the reaction used in the booster rockets of the space shuttle – the huge billowing clouds are made of $Al_2O_3(s)$.)

13.4 $\Delta_f H^{\ominus}$ for methanoic acid,

$$H_2(g) + O_2(g) + C(s) \rightarrow HCOOH(l) \quad \Delta_f H^{\ominus} = ?$$

cannot be determined directly. However, since methanoic acid readily burns in air, its enthalpy of combustion can be measured accurately. It is found that $\Delta_c H^{\ominus} = -254.6$ kJ mol^{-1}. Using this value of $\Delta_c H^{\ominus}$ (and the $\Delta_f H^{\ominus}$ data for CO_2 and H_2O contained in Table 13.2), calculate $\Delta_f H^{\ominus}$ (HCOOH(l)). (**Hint:** start by writing out the chemical equation for the combustion of methanoic acid in oxygen.)

13.5 Glucose ($C_6H_{12}O_6(s)$) is an important 'fuel' in the body.

(i) Write a balanced equation for the complete combustion of glucose in air to make liquid water and carbon dioxide gas.

(ii) Calculate the amount of energy generated in the body (assume that only 40 per cent of the enthalpy of combustion of the fuel is utilised) from the metabolism of 10 g of glucose ($\Delta_f H^{\ominus}(C_6H_{12}O_6(s)) = -2816$ kJ mol^{-1}).

13.6

(i) The third standard ionization energy of aluminium is 2751 kJ mol^{-1}. Write a thermochemical equation to which this enthalpy term applies.

(ii) The first, second and third standard ionization energies of aluminium are 584, 1823 and 2751 kJ mol^{-1}, respectively.
The standard enthalpy of atomization of aluminium is 326 kJ mol^{-1}. Calculate the enthalpy change for the process

$$Al(s) \rightarrow Al^{3+}(g) + 3e^-$$

13.7 The following standard enthalpy data (at 298 K) are provided:

(a) enthalpy change of atomization of calcium metal, $\Delta_{atom}H^{\ominus}$ (Ca(s)), = +193 kJ mol^{-1}

(b) first ionization energy of calcium, $\Delta_{ion}H^{\ominus}$ (Ca(g)) = +590 kJ mol^{-1}

(c) second ionization energy of calcium, $\Delta_{ion}H^{\ominus}(Ca^+(g))$ = +1150 kJ mol^{-1}

(d) bond enthalpy of $O_2(g)$, $\Delta H^{\ominus}(O{=}O)$ = +497 kJ mol^{-1}

(e) enthalpy of double electron gain of the oxygen atom (forming $O^{2-}(g)$), $\Delta_{eg}H^{\ominus}$, = +703 kJ mol^{-1}

(f) lattice enthalpy of CaO, $\Delta_{latt}H^{\ominus}$ = 3513 kJ mol^{-1}.

(i) Write down thermochemical equations that apply to the above enthalpy terms.

(ii) Use the data to calculate the standard enthalpy change of formation ($\Delta_f H^{\ominus}$) of calcium oxide (CaO).

13.8 This exercise involves the calculation of the bond enthalpy of the S–F bond. The standard enthalpy change for the formation of sulfur tetrafluoride (SF_4) from the gaseous elements at 298 K follows the thermochemical equation

$$2F_2(g) + S(g) \rightarrow SF_4(g) \qquad \Delta H^{\ominus} = -763 \text{ kJ mol}^{-1}$$

Given that $\Delta H^{\ominus}$(F–F) = 158 kJ mol^{-1}, estimate $\Delta H^{\ominus}$ (S–F) in SF_4.

14 Speed of Chemical Reactions

CONTENTS

OBJECTIVES

- This unit:
- Discusses the factors affecting reaction rate
- Defines the terms *rate constant* and *order of reaction*
- Introduces *activation energy* and *reaction mechanism*
- Looks at first-order reactions in detail
- Explains the role of catalysts

14.1 Reaction rate

Chemical reactions take place at different speeds. Rusting, the reaction of iron with oxygen and water, is a slow process, whereas the reaction of potassium metal with water is explosively fast. Reaction speeds are more properly referred to as 'reaction rates'. The study of the **rate of reactions** (and of the factors controlling reaction rates) is known as **chemical kinetics.**

A knowledge of the rate at which a reaction takes place is often crucial. In industry, reactions are economically profitable only if the yield of product is sufficient *and* if the products are made in a short enough time. In the chemistry of pollution, the rates at which a pollutant is formed and destroyed are important factors in assessing the hazard posed by the pollutant.

Definition of reaction rate

Table 14.1 Examples of 'rates'

Subject	Units
Oil production	barrels day^{-1}
Speed	$m\ s^{-1}$
Plant growth	cm $month^{-1}$
Population growth	people $year^{-1}$
Chemical reaction	concentration s^{-1}

The units are read as barrels per day, metres per second, etc.

The word 'rate' is part of everyday language (Table 14.1). The average rate of the chemical reaction:

$$M + N \rightarrow MN$$

(where M and N are reactants and MN is the product) is defined as the rate at which the *reactant* concentration *falls* between two times:

$$\text{rate of reaction} = \frac{\text{change in concentration of reactant}}{\text{change in time}}$$

$$= -\frac{\Delta[M]}{\Delta t}$$

where $\Delta[M]$ ('delta bracket M') is read as 'change in concentration of M' and Δt ('delta t') as 'change in time'. The minus sign shows that [M] is falling.

EXAMPLE **14.1**

The reaction of hydrogen and iodine to make hydrogen iodide at a particular temperature,

$H_2(g) + I_2(g) \rightarrow 2HI(g)$

was studied at various times by stopping the reaction (by cooling the mixture) and titrating any unreacted iodine with sodium thiosulfate. At 100.0 s after the start of the reaction, the iodine concentration had fallen from 0.010 mol dm^{-3} to 0.0080 mol dm^{-3}. What is the average rate of reaction during this period?

Answer

average rate = (0.010 – 0.0080)/100.0 = 2.0 × 10^{-5} mol dm^{-3} s^{-1}

Therefore, the average rate of reaction in the first 100s was 2.0 × 10^{-5} mol dm^{-3} s^{-1}.

The dimensions of reaction rate are

$$\frac{\text{units of concentration}}{\text{units of time}}$$

The units of reaction rate used in this book are mol dm^{-3} s^{-1}.

Alternatively, the average rate of a reaction may be defined as the rate at which the *product* concentration *rises* over a period of time:

$$\text{rate of reaction} = \frac{\text{change in concentration of product}}{\text{change in time}}$$

$$= +\frac{\Delta[\text{MN}]}{\Delta t}$$

Summarizing, the rate of reaction for reaction M + N → MN is given by the expressions

$$-\frac{\Delta[\text{M}]}{\Delta t} = -\frac{\Delta[\text{N}]}{\Delta t} = +\frac{\Delta[\text{MN}]}{\Delta t}$$

where the minus sign signifies that the concentration of M (and N) falls as time passes, and the plus sign signifies that the concentration of MN increases as time passes.

These equations need to be modified when we are dealing with chemical reactions in which the ratio of reactants to products is not 1:1. For example, consider the decomposition of dinitrogen pentoxide (N_2O_5) into nitrogen dioxide (NO_2) and oxygen:

$2N_2O_5(g) \rightarrow 4NO_2(g) + O_2(g)$

The equation shows that for every two molecules of N_2O_5 that decompose, four NO_2 molecules and one oxygen molecule are produced. This means that NO_2 is produced at *twice* the rate at which N_2O_5 decomposes, and also that O_2 is produced at *half* the rate at which N_2O_5 is destroyed.

Suppose that under a particular temperature and set of concentrations, the rate at which N_2O_5 decomposes at an instant was found to be 1 × 10^{-4} mol dm^{-3} s^{-1}. The rate of NO_2 and O_2 production at the same instant is 2 × 10^{-4} and 0.5 × 10^{-4} mol dm^{-3} s^{-1}, respectively. To avoid ambiguity in such cases, we should state to which substance a measured rate refers.

EXERCISE **14A**

Rate of reaction

The reaction between nitric oxide (NO) and hydrogen to make nitrous oxide (N_2O) and water obeys the equation:

$$H_2(g) + 2NO(g) \rightarrow N_2O(g) + H_2O(g)$$

If the rate (at a particular instant) at which hydrogen disappears was found to be 0.01 mol dm^{-3} s^{-1}, what is the simultaneous rate at which (i) nitric oxide disappears, (ii) nitrous oxide appears, (iii) water appears?

Experimental measurements of the reaction rate

In order to measure reaction rates experimentally, we monitor the concentration of reactants or products with time. Such concentrations are measured using standard analytical techniques such as titration, for those depending upon the degree of light absorption of a substance, or by changes in the electrical conductivity of the reaction mixture.

Occasionally we are able to follow the rate of a reaction because one of the products or reactants is coloured. An example is the reaction between bromine solution and methanoic acid:

$$\underset{\text{dark red}}{Br_2(aq)} + HCOOH(aq) \xrightarrow{H^+(aq)} 2Br^-(aq) + 2H^+(aq) + CO_2(g)$$

(H^+ is the catalyst.) It happens that bromine is the only coloured species involved in the reaction, and as the reaction proceeds its colour fades (Figure 14.1). The intensity of the colour is proportional to the concentration of bromine.

Change of rate of reaction with time

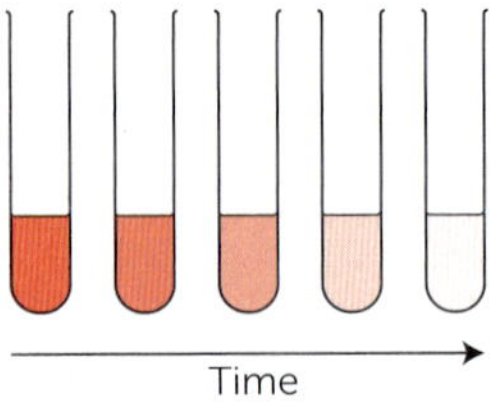

Figure 14.1 The reaction between bromine and methanoic acid. As the reaction continues, the red colour of bromine fades.

When two reactants are mixed together (for example, in a beaker) the rate of reaction is greatest at the start of the reaction. *As the reaction continues, the rate continues to fall.* If the concentration of reactant (such as bromine, Figure 14.1) or product is plotted against time, a graph like Figure 14.2(a) is produced. (If the reaction is an equilibrium reaction, the concentration of reactants in Figure 14.2(a) will level out when equilibrium has been achieved, but will not fall to zero.)

The rate of reaction, at an instant *t* seconds after the reactants are mixed, is calculated by working out the slope of the *tangent* to the concentration–time graph at that time. As time increases, the slope of the tangents at each instant of time decreases, confirming that the reaction rate is continuously falling.

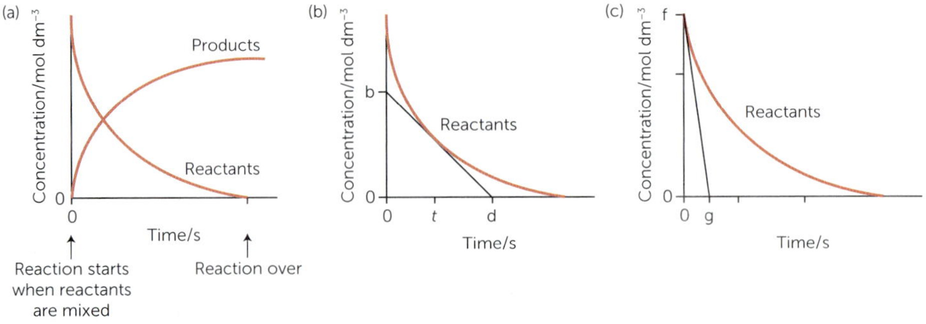

Figure 14.2 (a) How the concentrations of the reactants and products change with time. (b) Calculating the instantaneous rate of the reaction M + N → MN at *t* seconds after mixing. (c) Calculating the instantaneous rate of reaction M + N → MN at t = 0 (the 'initial rate').

Figure 14.2(b) shows the tangent to the reactant curve at time *t* seconds. The rate of reaction at this instant is

$$\frac{\textit{change in concentration}}{\textit{change in time}} = \frac{b-0}{d-0}\ \text{mol dm}^{-3}\text{s}^{-1}$$

To calculate the initial rate of reaction, we use the tangent to the curve at $t = 0$ (Figure 14.2(c)). The initial rate is then:

$$\frac{\textit{change in concentration}}{\textit{change in time}} = \frac{f-0}{g-0}\ \text{mol dm}^{-3}\text{s}^{-1}$$

(It may appear odd that we calculate the *initial rate* using $t = 0$ s, a time when it might be argued the reaction has not yet begun. The answer is that the initial rate being calculated is the rate a tiny fraction of second after mixing the reactants and when reaction has just begun but at which point the concentrations of M and N may be assumed to be equal to their starting concentrations).

EXERCISE **14B**

Calculating the rate of reaction from graphs (1)

The following graph shows how the concentration of M varies during the reaction M + N → products. Draw tangents at 0 and 25 s. Calculate the rate of reaction at these times.

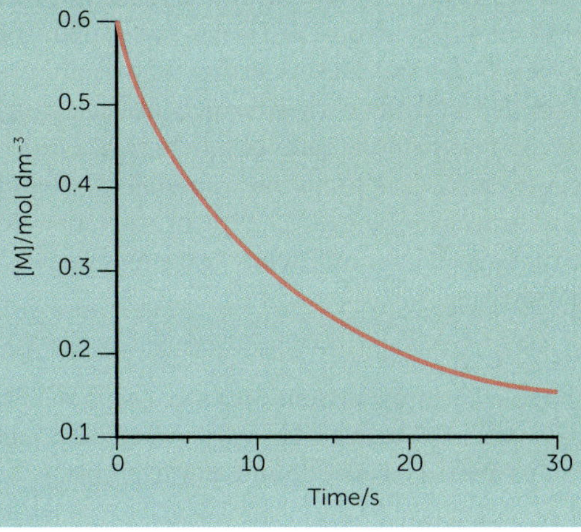

Why reaction rate falls with time

We have not explained why reaction rate should begin to fall as soon as the reactants are mixed. Our explanation depends upon two assumptions. The first is that for a reaction to take place, collisions must occur between reactant molecules (say, M and N to give the product molecule MN): 'no collisions, no reaction'. The second is that the probability of collisions between M and N falls as the concentrations of M and N fall. There are no product molecules (MN) at the start of the reaction, the concentrations of M and N are at their maximum and the reaction rate is at its highest. As soon as reaction starts, M and N molecules start to combine. The concentrations of M and N begin to fall, further reducing the number of collisions of M and N per second, reducing the reaction rate until it is virtually zero (Figure 14.3).

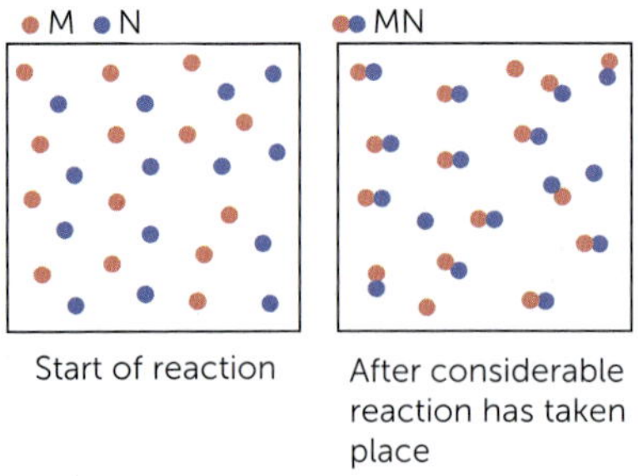

Figure 14.3 For the general reaction, $M + N \rightarrow MN$, reaction can only occur if molecules M and N collide. The chance of collisions between M and N molecules falls as more product (MN) molecules are made.

14.2 Factors affecting reaction rate

1 Nature of the reaction

Some reactions are naturally faster than others. Compare the reaction of hydronium and hydroxide ions,

$$H_3O^+(aq) + OH^-(aq) \rightarrow 2H_2O(l)$$

with the hydrolysis of bromoethane,

$$C_2H_5Br(l) + OH^-(aq) \rightarrow C_2H_5OH(aq) + Br^-(aq)$$

If in an experiment at 25°C we use the same initial reactant concentrations, the initial rate of the first reaction would be about 10^{15} times faster than the second reaction. In fact, the reaction between H_3O^+ and OH^- is so fast that it appears to be instantaneous. Special equipment is needed to study such extremely fast reactions.

Experimentally, we can compare the rates of different reactions by using reactants at the same initial concentrations. It is sometimes possible to predict whether a reaction will be naturally fast by simply looking at the type of reactants. Reactions in which the reactants are oppositely charged ions are usually extremely fast because the charges pull the ions towards each other e.g.

$$Ag^+(aq) + Cl^-(aq) \rightarrow AgCl(s) \qquad (14.1)$$

Reactions in which the colliding reactants are atoms or molecules with odd electrons (known as **free radicals**, such as H• and CH_3•) also occur extremely rapidly in an effort to use the odd electron to form a covalent bond. For example:

$$CH_3\bullet(g) + CH_3\bullet(g) \rightarrow C_2H_6(g) \qquad (14.2)$$

is fast, with a high efficiency per collision.

2 Concentration (or for gases, pressure)

Increasing the concentration of reactants increases the chances of collisions between reactant molecules. In other words, increasing concentration increases the number of collisions between molecules *per second*. This increases the reaction rate. The concentration of a gas increases with its partial pressure. So, we can increase the rate of gaseous reactions by increasing the partial pressure of the reactants.

Where one of the reactants is a solid, increasing the surface area of the solid also increases the chance of the particles of reactants colliding. For example, powdered carbon burns in oxygen faster than lumps of carbon:

$$C(s) + O_2(g) \rightarrow CO_2(g)$$

This is why, in power stations, coal is pulverized before allowing it to burn. Fine powders can sometimes react extremely rapidly. For example, stringent safety regulations need to be applied to grain and flour factories, since fine grain and flour can burn explosively fast in air (Exercise 14C).

EXERCISE **14C**

Surface area

A cube of marble 1 cm × 1 cm × 1 cm is cut into two smaller pieces along the dotted lines as shown.

(i) What is the increase in surface area after cutting?

(ii) How does this help explain why marble powder reacts more rapidly with hydrochloric acid than marble blocks?

3 Temperature

Increasing the temperature of a reaction mixture increases the rate of the chemical reaction. (This happens whether the reaction is exothermic or endothermic.) This increase is often dramatic (Figure 14.4).

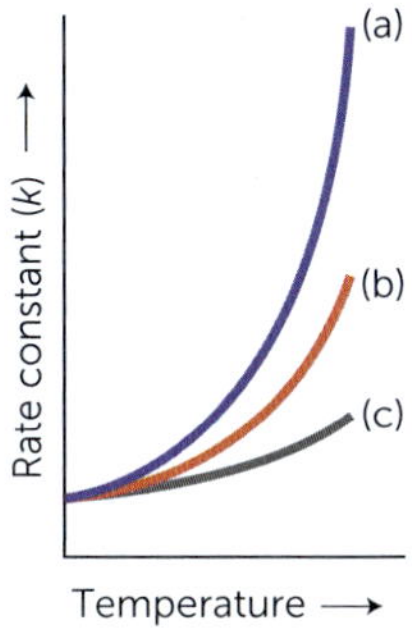

Figure 14.4 Variation of reaction rate with temperature for reactions which are (a) *very sensitive*, (b) *fairly sensitive*, (c) *relatively insensitive* to changes in temperature. (The **rate constant** is the reaction rate with the reactants at a concentration of 1.0 mol dm^{-3}.)

Although the number of collisions per second between reactant molecules rises with temperature, calculations show that this makes only a small contribution to the increase of reaction rate with temperature. The accepted explanation for the increase in reaction rate involves a key idea in chemical kinetics, that of **activation energy**.

Think of a chemical reaction in which molecules of M and N form the product molecule MN in a single step. In order for a collision between molecules to result in chemical change, old bonds need to be broken and new ones formed. Bond destruction and bond formation require a redistribution of the bonding electrons within the reactant molecules. However, as the molecules of M and N approach each of their outer electrons repel, so preventing the reactants from getting close enough for the bonding electrons to be rearranged. The reactant molecules then simply 'bounce off' each other, and the collision does not result in chemical change.

In order to overcome the repulsive forces between the molecules so that the molecules are forced together, the colliding molecules must between them possess a minimum value of kinetic energy, termed the **activation energy** (E_A). Unless the sum of the energies possessed by the reactant molecules involved in the collision equals (or exceeds) this minimum energy (the 'energy barrier'), no reaction can take place (Figure 14.5).

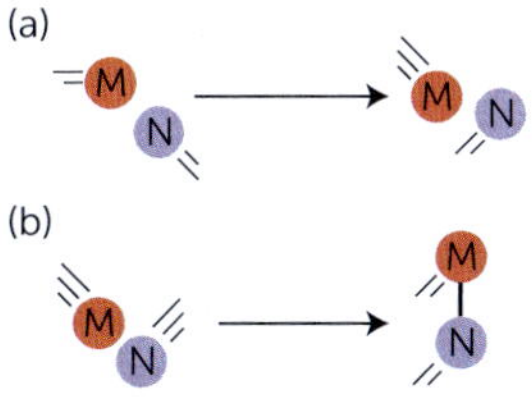

Figure 14.5 (a) Unsuccessful collision – the kinetic energy of M and N is less than E_A and so the reactants bounce apart. (b) Only energetic collisions can loosen old bonds so that new bonds can be formed. The formal requirement is that the kinetic energy of M and N must be greater than or equal to E_A.

The activation energy for a reaction *does not itself* vary with temperature, but increasing the temperature of a material does increase the average energy of its molecules. *It follows that the higher the temperature, the greater is the fraction of colliding molecules that possess at least the activation energy* (Figure 14.6). We then expect more collisions to result in chemical reaction at higher temperatures, and this explains the experimental observation of an increased reaction rate.

The rate of some reactions changes more rapidly with temperature than others (see Figure 14.4), and we are now in a position to explain this observation by assuming that *reactions whose rates are very sensitive to temperature possess relatively high activation energies*:

- The speed of a reaction which possesses a large activation energy changes rapidly with temperature because few of the reacting molecules will possess an energy equal to (or greater than) E_A at room temperature. Therefore, an increase in temperature will drastically increase the number of collisions that result in chemical change.

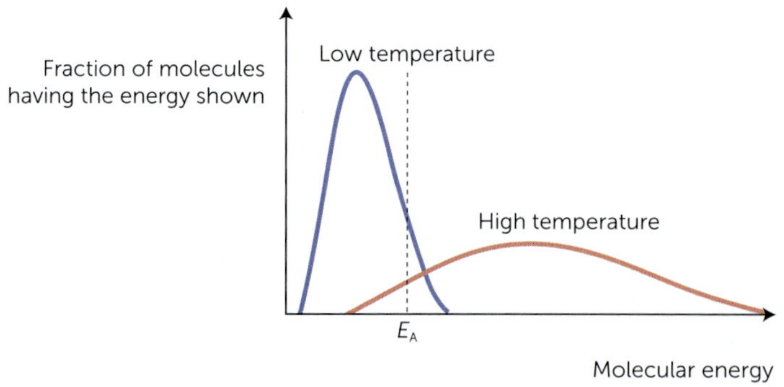

Figure 14.6 The graph shows the spread of molecular energy for a mixture of gases at low temperature and at high temperature. The higher the temperature, the greater the fraction of molecules possessing energy equal to (or greater than) the activation energy E_A.

The temperature of pressurized steam is above 100°C. A pressure cooker allows the safe production of pressurized steam, which then cooks food faster than the conventional method in which food is simply immersed in boiling water at 100°C in an open pot.

- A reaction that possesses a low E_A is relatively insensitive to temperature because most of the reacting molecules already possess the activation energy at room temperature.
- Reactions that are naturally fast (such as $CH_3\bullet(g) + CH_3\bullet(g) \rightarrow C_2H_6(g)$) possess activation energies that are virtually zero. As a consequence, the rates of such reactions show very little dependence upon temperature.

Provided that the reactant concentrations are not vastly different, a reaction with a low activation energy will always be faster than a reaction with a high activation energy, since more molecules possess the minimum energy needed for them to react.

EXERCISE **14D**

Increase in reaction rate at higher temperature

(i) The chemical reactions involved in the hardening ('boiling') of an egg roughly double in rate for every 10°C rise in temperature. How much faster will these reactions be at 100°C than at 20°C?

(ii) Very exothermic reactions are faster shortly after the reaction has started, even though the initial reactant concentrations have fallen – why ?

EXERCISE **14E**

Factors affecting reaction rate

Zinc reacts with hydrochloric acid to produce hydrogen gas:

$$Zn(s) + 2HCl(aq) \rightarrow ZnCl_2(aq) + H_2(g)$$

Experiments show that the rate of reaction decreases (i) if zinc powder is replaced by zinc nails, and (ii) if the following are added to the reaction mixture: (a) water, (b) ice, and (c) sodium carbonate. Carefully explain these effects.

BOX **14.1**

Transition states; activation energy

As the bonds in reacting molecules are rearranging during a collision, the energy of the molecules rises to a maximum above the starting energy of the reactants (Figure 14.7). This maximum is the **activation energy** E_A (in kJ mol^{-1}) of the (forward) reaction. The arrangement (or configuration) of atoms at this energy maximum is called the transition state.

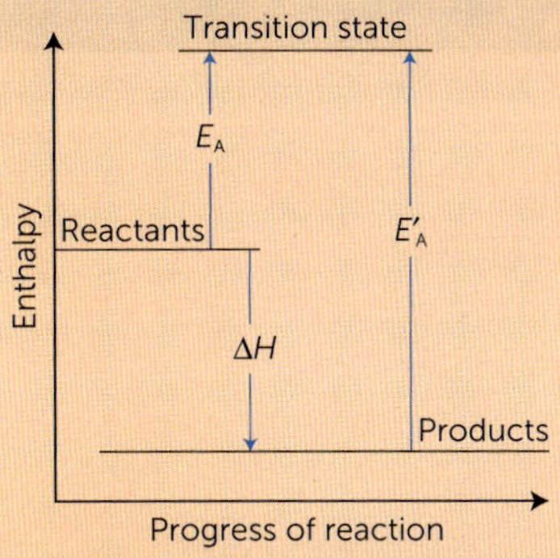

Figure 14.7 Connection between reaction activation energy E_A and the reaction enthalpy change ΔH for the exothermic single-step reaction M + N → MN.

The transition state is a 'half-way house' between reactants and products. It can never be isolated and exists for only a vanishingly short period of time. Once formed, the transition state either immediately 'falls back' to the reactants or else 'turns' into products. The significance of the activation energy is that if it is low, more transition states will be formed per second and so more transition states will break up into products per second. In other words, a low activation energy means a higher rate of reaction.

Figure 14.7 also shows the activation energy for the reverse reaction (MN → M + N), symbolized E'_A. When the forward reaction is exothermic, the activation energy of the forward reaction is always smaller than the activation energy for the reverse reaction, i.e. E_A. (The reverse is true for an endothermic reaction.) If a chemical reaction occurs in several steps, each step possesses its own transition state, with its own forward and back activation energies.

The reaction between 1-bromopropane and hydroxide ion:

$$RCH_2Br(l) + OH^-(aq) \rightarrow RCH_2OH(aq) + Br^-(aq)$$

(R = CH_3CH_2) occurs in one stage. The probable structure of the transition state is shown in Figure 14.8 with the negative charge of the OH^- being spread out over the structure.

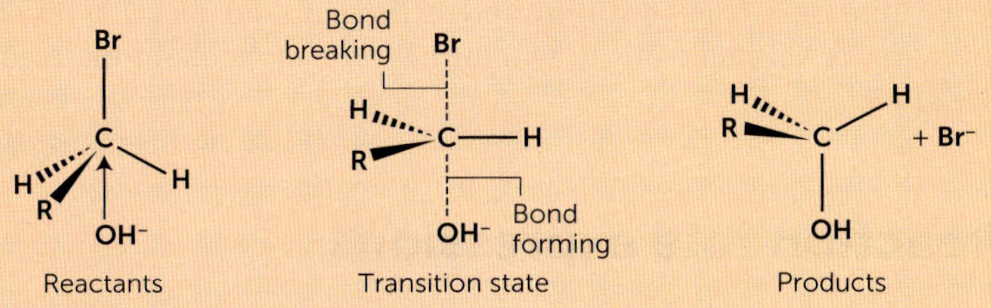

Figure 14.8 Reactants, transition state and products for the hydrolysis of 1-bromopropane.

4 Catalysts

Catalysts are used to speed up reactions. Their general characteristics are:

- Catalysts take part in chemical reactions but are regenerated during the reaction.
- Catalysts do not affect the enthalpy change or equilibrium constant of the reaction.
- Catalysts are often specific, i.e. they will only affect the rate of a particular reaction. This is particularly true of some biological catalysts (enzymes). Enzymes are proteins. They are often deactivated (or even destroyed) if the temperature rises much above 40°C (Figure 14.9). Most enzymes operate in a narrow band of pH, and many require the presence of a metal ion or another complex organic molecule as cofactor.

For more about enzymes, see **Extension 14** on the website https://www.bloomsburyonlineresources.com/chemistry

Rusting is the reaction of iron with oxygen in the presence of water. The rate of rusting is controlled by the surface area of the iron exposed to the oxygen, by the concentration of water vapour in the air and by temperature. Rusting proceeds more slowly at lower temperatures, but salt (often used on the roads in cold climates) acts as a catalyst and causes iron to rust faster.

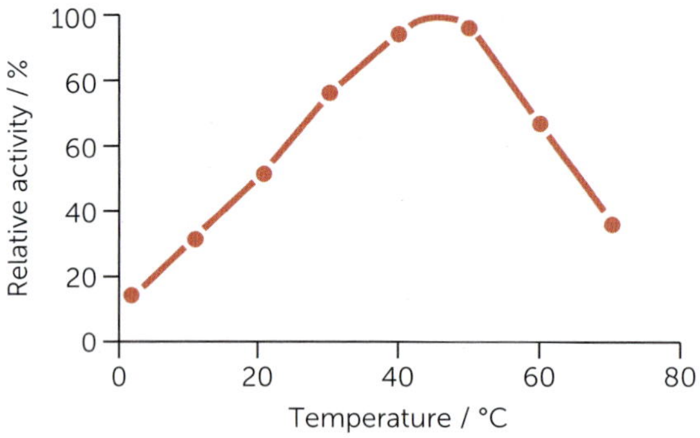

Figure 14.9 Effect of temperature on the catalytic activity of the enzyme *malt amylase* when hydrolysing starch to glucose.

Catalysts are important in industry. For example, iron is used to catalyse the production of ammonia from nitrogen and hydrogen, and nickel is used to speed up the conversion of vegetable oils to margarines. For more about the mechanism of catalysis, see s14.8.

14.3 Reaction rate expressions

More about the variation of rate with concentration

At a fixed temperature the rate of a chemical reaction

A + B + C . . . → products

(where A, B and C are the reactants) are usually found to follow the general **rate expression** in which reactant concentrations at an instant of time are multiplied together:

rate of reaction = $k[A]^x\,[B]^y\,[C]^z$ ...

where k is called the rate constant and the square brackets represent the concentrations (in mol dm^{-3}) of the reactants which are raised to the powers of x, y, z ... The concentration of reactants will begin to fall as soon as the reactants are mixed. Therefore, this equation defines the rate of the reaction at *one instant* – as [A], [B] and [C] continuously fall with time, so does the rate of reaction.

The powers x, y and z are said to be the *individual orders* of reaction for reactants A, B and C, respectively. The *overall order* of reaction is the sum of the individual orders x, y, and z ...

overall order = $x + y + z$...

Reactions with an overall order of one are said to be 'overall first order'; reactions with an overall order of two are said to be 'overall second order'; reactions with an overall order of zero are said to be 'overall zeroth order' and so on.

The values of x, y and z (etc.) are decided by experiment – they cannot be reliably predicted by simply looking at the chemical equation for the reaction.

The rate constant

- One way to look at k is to think of it as the rate of reaction when all of the reactant concentrations are exactly 1 mol dm^{-3}.
- k usually increases with the temperature of the reaction mixture. Therefore, the reaction rate also increases with temperature (Box 14.2).
- The size of k is partly decided by the activation energy for the reaction (a low E_A means a high k value and vice versa). Naturally fast reactions have high rate constants.
- The units of k depend upon the overall order of reaction (Table 14.2).

Table 14.2 Units of rate constants

Overall order of reaction	Units of rate constant
Zeroth	mol dm^{-3} s^{-1}
First	s^{-1}
Second	mol^{-1} dm^3 s^{-1}
Third	mol^{-2} dm^6 s^{-1}

EXAMPLE **14.2**

The reaction between A and B follows the rate expression

rate of reaction = k[A][B]

What is the overall order of reaction? What are the units of k?

Answer

We can rewrite this as

rate of reaction = $k[A]^x[B]^y$ where $x = y = 1$

The reaction is therefore first order with respect to the reactants A and B. Overall, the reaction order is $x + y = 1 + 1 = 2$, i.e. it is second order overall.

For the units of k, rearranging for k gives

$$k = \frac{\text{rate of reaction}}{[A][B]}$$

The units of k are therefore

$$\frac{mol dm^{-3} s^{-1}}{mol dm^{-3} mol dm^{-3}}$$

Cancelling

$$\frac{\cancel{mol}\ \cancel{dm^{-3}}\ s^{-1}}{\cancel{mol}\ \cancel{dm^{-3}}\ mol dm^{-3}} = mol^{-1} dm^3 s^{-1}$$

Note that a reaction which is *overall* second order would also be obtained if the rate expression was:

rate of reaction = $k[A]^2$ or rate of reaction = $k[B]^2$

EXERCISE **14F**

Working out the overall order from the rate expression

What is the overall order of reactions possessing the following rate expressions?

(i) rate of reaction = $k[A][B]^0$

(ii) rate of reaction = $k[D]^2$

(iii) rate of reaction = $k[E]^{0.5}[G]^{0.5}$

(iv) rate of reaction = $k[B][J][N]$.

For more about the calculation of Arrhenius activation energies, see **Extension 14** on the website at https://www.bloomsburyonlineresources.com/chemistry

BOX **14.2**

Dependence of the rate constant *k* upon temperature: the Arrhenius equation

Experiments show that the way that the *rate constant* for a chemical reaction varies with temperature can be described by the Arrhenius equation:

$$k = Ae^{-E/RT}$$

where k is the rate constant at temperature T (K), A is a constant for the reaction, R is the gas constant (8.3145 J mol^{-1} K^{-1}) and E (in J mol^{-1}) is the so-called **Arrhenius activation energy** for the reaction. The Arrhenius equation is useful because it is found to apply even if the reaction under study does not occur in one simple step.

Taking natural logs (ln) to both sides of the Arrhenius equation gives an equation of the form

$$y = mx + c$$

thus

$$\ln k = \ln A - \left(\frac{E}{R}\right)\frac{1}{T}$$

If we have measured the rate constants for a named reaction at various temperatures, we can plot ln k against $1/T$, giving a straight line graph of slope $-E/R$ from which E can be calculated.

The Arrhenius activation energy E is **not** the same as the reaction activation energy E_A (which is only meaningful for single-step reactions). It is better to look upon the Arrhenius activation energy simply as a quantity, found by experiment, which defines *how quickly* the rate constant of a reaction changes as the temperature is changed:

- If $E = 0$, the rate constant is independent of temperature.
- If E is small, the rate constant has only a slight dependence upon temperature (Figure 14.4(c)).
- If E is large, the rate constant changes dramatically with changes in temperature (Figure 14.4(a)).

If the rate constant of a reaction is k_1 at temperature T_1, and k_2 at temperature T_2, the Arrhenius equations for both temperatures may be combined to give the expression

$$\ln\left(\frac{k_1}{k_2}\right) = \frac{E}{R}\left(\frac{1}{T_2} - \frac{1}{T_1}\right)$$

Provided that we know the Arrhenius activation energy for the reaction, this equation may be used to calculate the rate constant at one temperature, knowing the rate constant at another temperature. For example, if for a particular chemical reaction,

$$k_1 = 1.0 \times 10^{-3}\ \text{mol}^{-1}\,\text{dm}^3\,\text{s}^{-1} \text{ at } 300\text{ K}$$

and

$$E = 50\text{ kJ mol}^{-1} \text{ (i.e. } 50\,000\text{ J mol}^{-1}\text{)},$$

then, at 320 K,

$$\ln\left(\frac{10^{-3}}{k_2}\right) = \frac{50\,000}{8.3145}\left(\frac{1}{320} - \frac{1}{300}\right) = -1.25$$

or

$$\left(\frac{10^{-3}}{k_2}\right) = e^{-1.25} = 0.287$$

so that $k_2 = 10^{-3}/0.287 \approx 3.5 \times 10^{-3}$ mol^{-1} dm^3 s^{-1}. In other words, the rate constant has increased by a factor of 3.5 for an increase in temperature of only 20 K.

14.4 Examples of rate expressions found by experiment

Decomposition of ethanal

Ethanal (CH_3CHO) molecules may be broken up by heating:

$$CH_3CHO(g) \rightarrow CH_4(g) + CO(g)$$

Experiments show the rate expression to be

$$\text{rate of reaction} = k[CH_3CHO]^{1.5}$$

That is, the order of reaction with respect to ethanal is 1.5. Since ethanal is the only reactant, the overall order is also 1.5. The order of reaction is something of a surprise, and we could certainly not predict it from the chemical equation for the reaction.

Hydrolysis of methyl ethanoate

The equation for hydrolysis of methyl ethanoate ester by sodium hydroxide is

$$CH_3COOCH_3(l) + OH^-(aq) \rightarrow CH_3COO^-(aq) + CH_3OH(aq)$$

The rate expression is found to be

$$\text{rate of reaction} = k[CH_3COOCH_3(l)]\,[OH^-(aq)]$$

The order of reaction is therefore one (i.e. first order) with respect to both methyl ethanoate and sodium hydroxide and the overall order of reaction is two (i.e. second order).

If the sodium hydroxide is in excess (at least ten times the concentration of ester), the *consumption* of sodium hydroxide during the reaction will be very small and may be neglected. This means that $[OH^-]$, the hydroxide ion concentration at an instant, may be replaced by the initial hydroxide concentration $[OH^-]_0$. k and $[OH^-]_0$ may now be lumped together as a new constant k'. Under these conditions the rate expression becomes

$$\text{rate of reaction} = k'[CH_3COOCH_3(l)]$$

where $k' = [OH^-]_0 \times k$. Experimentally, the reaction is now first order. We say that the reaction is **pseudo-first order** and that k' is the pseudo-first order rate constant. Here, the word 'pseudo' means 'falsely'. Deliberately making one of the reactant concentrations much higher than the other(s) is a useful tactic to force a reaction to show first-order kinetics.

14.5 Calculations using rate expressions

Calculating the rate constant from the initial rate of reaction

What is the rate of reaction at the start of a reaction? Immediately after mixing, the concentrations of reactants will not have changed very much from their starting values. We then define the *initial rate* of the reaction as

$$\text{initial rate} = k[A]^x{}_0[B]^y{}_0[C]^z{}_0 \ldots$$

where $[A]_0$, $[B]_0$ and $[C]_0$ are the initial concentrations of reactants A, B and C.

The rate constant k at a specific temperature may now be calculated by rearrangement:

$$k = \frac{\text{initial rate}}{[A]_0^x\,[B]_0^y\,[C]_0^z}$$

EXAMPLE **14.3**

The formation of nitrosyl chloride (NOCl) at 400 K

$$NO(g) + Cl_2(g) \rightarrow NOCl(g) + Cl(g)$$

follows the rate expression

$$\text{rate of reaction} = k[NO(g)][Cl_2(g)]$$

The initial rate of reaction (with both reactants at an initial concentration of 0.010 mol dm^{-3}) was found to be 3.2×10^{-6} mol dm^{-3} s^{-1}.

(a) What is the order of reaction?

(b) What is the value of k at 400 K?

EXAMPLE **14.3** *(continued)*

Answer

(a) The reaction is first order with respect to nitric oxide, first order with respect to chlorine, and second order overall.

(b) $$k = \frac{\text{initial rate of reaction}}{[NO(g)]_0[Cl_2(g)]_0}$$

Therefore,

$k = (3.2 \times 10^{-6}) / (0.010 \times 0.010) = 3.2 \times 10^{-2}\ mol^{-1}\ dm^3\ s^{-1}$.

EXERCISE **14G**

Calculating the rate of reaction

The reaction between nitric oxide and oxygen gas,

$$2NO(g) + O_2(g) \rightarrow 2NO_2(g)$$

is found to follow the rate expression

rate of reaction = $k[NO]^2[O_2]$

(i) What is the overall order of reaction?

(ii) At 300°C the rate constant k is 7,000 $mol^{-2}\ dm^6\ s^{-1}$. What is the rate of reaction at the instant when [NO] and $[O_2]$ are both 0.060 $mol\ dm^{-3}$?

EXERCISE **14H**

Calculation of rate constant

The hydrolysis of sucrose in the presence of acid catalyst is often represented as

$$C_{12}H_{22}O_{11} + H_2O \xrightarrow{[H^+(aq)]} 2C_6H_{12}O_6$$

The rate expression is found to be

rate of reaction = $k[C_{12}H_{22}O_{11}][H^+(aq)]$

(Notice that, in this example, the concentration of catalyst appears in the rate expression.) A reaction mixture at 298 K initially contained 0.20 $mol\ dm^{-3}$ sucrose and 0.10 $mol\ dm^{-3}$ H^+. The initial rate of reaction was found to be $4.0 \times 10^{-6}\ mol\ dm^{-3}\ s^{-1}$.

(i) What is the overall order of the reaction?

(ii) Calculate k at 298 K and state its units.

Finding the order of a reaction using the initial-rates method

We now look at the use of initial rates of reaction in helping us find the individual orders in a rate expression.

Suppose that we wish to find the overall order of the reaction

$$A + B \rightarrow \text{products}$$

This is another way of saying that we want to find out the values of x and y in the rate expression

rate of reaction = $k[A]^x[B]^y$

The initial rate of this reaction is given by the expression

$$\text{initial rate} = k[A]_0^x[B]_0^y$$

where $[A]_0$ and $[B]_0$ are the initial concentrations of A and B.

Now suppose that we carry out two experiments at the same temperature (so that k remains fixed) and using the same initial concentration of B but different initial concentrations (0.1 mol dm^{-3} and 0.2 mol dm^{-3}, respectively) of A.

For the first experiment,

$$\text{initial rate}_1 = k \times [0.1]^x \times [B]_0^y$$

For the second experiment

$$\text{initial rate}_2 = k \times [0.2]^x \times [B]_0^y$$

Both k and $[B]^y_0$ feature in both expressions. The initial rates in these experiments will be different only because the initial concentrations of A are different. We can see this more easily by writing the above equations in the form:

$$\text{initial rate}_1 = \text{constant} \times [0.1]^x$$
$$\text{initial rate}_2 = \text{constant} \times [0.2]^x$$

where

$$\text{constant} = k[B]_0^y$$

What if the initial rate doubles as [A] doubles?

If, in doubling [A], experiments show that the *initial rate* doubles, so that

$$\text{initial rate}_2 = 2 \times \text{initial rate}_1$$

i.e.

$$\text{constant} \times [0.2]^x = 2 \times \text{constant} \times [0.1]^x$$

the equation can only be true if the order x is one. This means that the reaction is first order with respect to A.

What if the initial rate increases by a factor of four as [A] doubles?

If, in doubling [A], the *initial rate* is found to have quadrupled, so that

$$\text{initial rate}_2 = 4 \times \text{initial rate}_1$$

i.e.

$$\text{constant} \times [0.2]^x = 4 \times \text{constant} \times [0.1]^x$$

then the order x *must* be two. This means that the reaction is second order with respect to A.

What if there is no change in initial rate as [A] doubles?

If experiments show that the initial rate of reaction remains unchanged, even though [A] in experiment 2 is double [A] in experiment 1, then

$$\text{initial rate}_2 = \text{initial rate}_1$$

i.e.

$$\text{constant} \times [0.2]^x = \text{constant} \times [0.1]^x$$

This can be true only if $x = 0$, since 'anything to the power of zero is one'. This means that the reaction is zeroth order with respect to reactant A.

Similar arguments can be applied to reactant B by carrying out experiments in which $[A]_0$ remains constant and in which the concentration of B is varied from experiment to experiment.

EXAMPLE **14.4**

The reaction of benzaldehyde with cyanide ions (as catalyst) at 50°C was studied by the initial-rates method:

$$2C_6H_5CHO(aq) \xrightarrow{CN^-(aq)} C_6H_5CH(OH)COC_6H_5(aq)$$

The results were:

	Initial concentration/mol dm^{-3}		Initial rate of reaction/mol dm^{-3} s^{-1}
Experiment number	$[C_6H_5CHO(aq)]_0$	$[CN^-(aq)]_0$	
1	0.500	0.500	1.00×10^{-4}
2	1.00	0.500	4.01×10^{-4}
3	1.00	1.00	7.98×10^{-4}

What is the overall order of reaction? What is the rate constant at this temperature?

Answer

The rate expression is:

$$\text{rate of reaction} = k[C_6H_5CHO(aq)]^x\,[CN^-(aq)]^y$$

Experiment 1 is used as a 'norm'. Doubling the concentration of benzaldehyde (experiment 2) quadruples the initial rate. This means that $x = 2$.

Doubling the concentration of cyanide ions (experiment 3) produces an initial rate which is double that of experiment 2. Therefore, $y = 1$ so that

$$\text{overall order} = x + y = 2 + 1 = 3$$

The reaction is third order overall and the rate expression becomes

$$\text{rate of reaction} = k[C_6H_5CHO(aq)]^2[CN^-(aq)]$$

Rearranging this equation and considering the initial rate:

$$k = \frac{\text{initial rate}}{[C_6H_5CHO(aq)]_0^2\,[CN^-(aq)]_0}$$

For experiment 1,

$$k = \frac{1.00 \times 10^{-4}}{[0.500]^2[0.500]} = 8.02 \times 10^{-4}\,\text{mol}^{-2}\text{dm}^{-6}\text{S}^{-1}$$

For experiment 2,

$$k = \frac{4.01 \times 10 - 4}{[1.00]^2[0.500]} = 8.02 \times 10^{-4}\,\text{mol}^{-2}\text{dm}^{-6}\text{s}^{-1}$$

For experiment 3,

$$k = \frac{7.98 \times 10^{-4}}{[1.00]^2[1.00]} = 7.98 \times 10^{-4}\,\text{mol}^{-2}\text{dm}^{-6}\text{s}^{-1}$$

The average rate constant at 50°C is

$$k = \frac{(8.00 + 8.02 + 7.98) \times 10^{-4}}{3} = 8.00 \times 10^{-4}\,\text{mol}^{-2}\text{dm}^{-6}\text{s}^{-1}$$

EXERCISE **14I**

Initial rates, 1

The reaction:

M + Y = products

was studied by the initial-rates method at 25°C. It was found that the initial rate of reaction halved when [M] was halved and that doubling [Y] had no effect upon the initial rate. Write down the rate expression for the reaction.

EXERCISE **14J**

Initial rates, 2

The kinetics of the reaction between bromide ($Br^-(aq)$), bromate ($BrO_3^-(aq)$) and $H^+(aq)$ ions:

$$5Br^-(aq) + BrO_3^-(aq) + 6H^+(aq) \rightarrow 3Br_2(l) + 3H_2O(l)$$

was studied by the initial-rates method and the following data obtained:

	Initial concentration/mol dm^{-3}			
Experiment	$[Br^-(aq)]_0$	$[BrO_3^-(aq)]_0$	$[H^+(aq)]_0$	Relative initial rate of reaction
1	0.01	0.01	0.01	1.00
2	0.02	0.01	0.01	2.00
3	0.01	0.04	0.01	4.00
4	0.01	0.01	0.02	4.00

Write down a rate expression for the reaction.

14.6 More about first-order reactions

If the simple reaction A $\rightarrow$ Product(s) is first order, then

rate of reaction = k[A]

For such a first-order reaction, the concentration of A at time t seconds after the start of the reaction, symbolized $[A]_t$, is related to the initial concentration of A, i.e. $[A]_0$, by the equation

$$[A]_t = [A]_0 \times e^{-kt}$$

where k is the first-order rate constant. Taking natural logs on both sides of this equation, followed by rearrangement, gives

$$\ln\left(\frac{[A]_0}{[A]_t}\right) = k \times t \qquad \textbf{(14.1)}$$

If the time at which exactly half of the concentration of A has disappeared is symbolized as $t_{1/2}$, then

$$[A]_{t_{1/2}} = [A]_0/2 \qquad \textbf{(14.2)}$$

Substituting equation (14.2) into equation (14.1) gives

$$\ln\left(\frac{[A]_0}{[A]_0/2}\right) = \ln 2 = k \times t_{1/2}$$

and, since ln 2 = 0.693,

$k \times t_{1/2} = 0.693$

Note that:

- These equations apply to any first-order reaction with only one reactant.
- The formal definition for the half-life (or reactant half-life) of a reaction, symbolized $t_{1/2}$, is that it is the time taken for the concentration of a reactant to fall by half.
- The half-life of a first-order reaction does not depend upon the initial concentration of the reactant. (This is not true for other orders of reaction.)

Example of half-life – the decomposition of azomethane

The half-life for the decomposition of azomethane ($CH_3N_2CH_3$):

$CH_3N_2CH_3(g) \rightarrow CH_3CH_3(g) + N_2(g)$

is first-order with respect to azomethane with $t_{1/2} = 2000$ s at 180°C.

rate of reaction = $k[CH_3N_2CH_3(g)]$

If the initial concentration of azomethane was 0.1 mol dm^{-3}, after 2,000 s the concentration would have fallen to 0.05 mol dm^{-3}, after another 2,000 s to 0.025 mol dm^{-3}, after a further 2,000 s to 0.0125 mol dm^{-3}, and so on.

Figure 14.10 shows a plot of $[CH_3N_2CH_3(g)]$ against t. The gradual fall in azomethane concentration follows equation (14.1) and is said to be *exponential*. The curvature of the plot is controlled by the value of k – the greater is k, the quicker the reactant concentration falls (Figure 14.11).

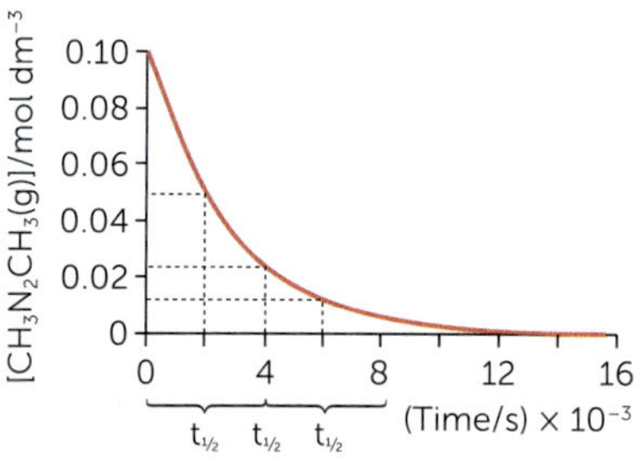

Figure 14.10 The fall in azomethane concentration at 180°C. Because the reaction is first order, its half-life does not depend upon the initial concentration of azomethane.

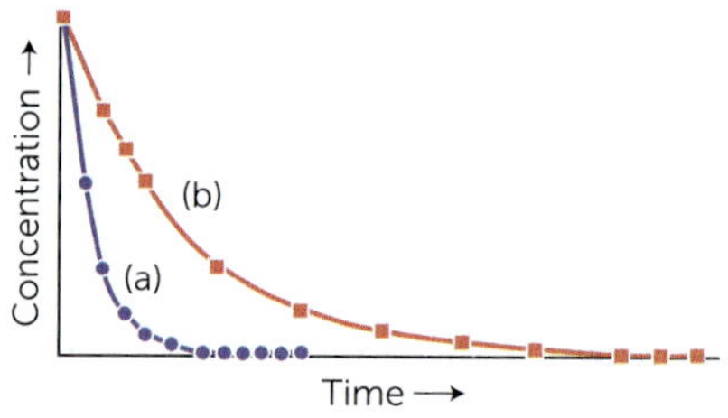

Figure 14.11 The exponential decay of a reactant in a first-order reaction where (a) k is large, (b) k is small. The larger the value of k, the faster is the decay.

The rate constant for the decomposition of $CH_3N_2CH_3(g)$ may be calculated directly from the half-life:

$$k = \frac{0.693}{t_{1/2}} = 0.693/2000 = 3.5 \times 10^{-4} s^{-1}$$

EXERCISE **14K**

Calculating half-life from *k*

Calculate the half-life of the first-order reaction

cis-CHCl=CHCl → *trans*-CHCl=CHCl

at a temperature of 825 K given that $k = 9.6 \times 10^{-3}$ s^{-1}.

EXERCISE **14L**

First-order reaction

The rearrangement of *N*-bromoacetanilide to 4-bromoacetanilide

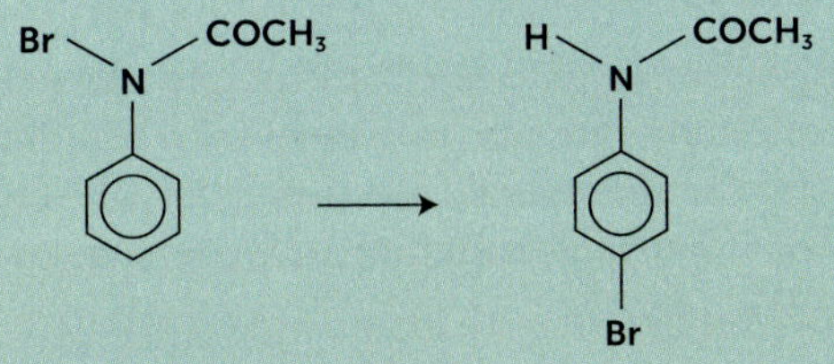

in chlorobenzene solvent is first order and $k = 6.5 \times 10^{-6}\ s^{-1}$ at 288 K. If the initial concentration of *N*-bromoacetanilide was 0.010 mol dm^{-3}, calculate the concentration of *N*-bromoacetanilide after 10 h.

EXERCISE **14M**

Half-lives of pesticides

The chemical degradation of a pesticide (P) in a field

$$P \rightarrow \text{byproducts}$$

was found to be first order:

$$\text{rate of reaction} = k[P]$$

with $k = 5.73 \times 10^{-7}\ s^{-1}$. After spraying, a pesticide was washed off plants into a pond. The initial concentration of pesticide in the pond was 0.10 mg dm^{-3}. What concentration of pesticide would be left in the pond after forty-two days? (Hint: Start by calculating the half-life of the pesticide.)

Examples of the use of half-lives

Radioactive decay

The decay of elements by radioactivity, although not a chemical reaction, follows first-order kinetics. For example, the half-life of $^{238}_{92}U$ is 4.5×10^9 years. This means that 4.5 billion years ago (when the Earth was formed), there was twice as much uranium-238 present than is the case now.

Degradation of organic compounds in the environment

Organic compounds (including solvents and pesticides) break up in the environment as a result of hydrolysis (reaction with water), photolysis (breakdown by light) or by the action of bacteria. These reactions are usually first order (or pseudo-first order).

Ideally, toxic organic compounds in waste should have a low half-life so that they break down quickly. The insecticide **parathion** slowly hydrolyses at room temperature:

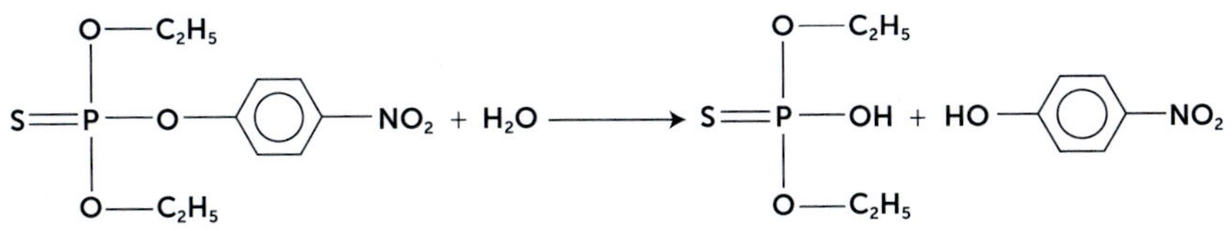

The $t_{1/2}$ of parathion in water due to hydrolysis is about 25 days – sufficiently long for it to pose a hazard.

14.7 Reaction mechanisms

Multi-step reactions

Many reactions do not proceed in one step (for example, via a single collision between reactants) but in several steps (known as elementary reactions) which collectively constitute the *mechanism* for the overall reaction. We will look at several examples below, but this raises an important question: why are sequential reactions so often preferred over one step reactions?

The explanation uses the idea of activation energy. A competing reaction with the lowest activation energy has the largest rate constant k, and is therefore a faster reaction. A sequence of reactions is preferred over a single step (reactants to products) provided that all of its reactions do *not* have activation energies that are larger than that of the reaction in which the reactants go to products in one stage. In a multi-step reaction, the slowest step (reaction) dictates the overall reaction rate and it is called the **rate determining step**; we might also call it the 'bottleneck' reaction.

Chlorination of methane

Chloromethane (CH_3Cl) is made by slowly reacting excess chlorine with methane using a low intensity UV lamp or direct sunlight. The overall equation for the reaction is

$$Cl_2(g) + CH_4(g) \rightarrow CH_3Cl(g) + HCl(g) \quad \textbf{(14.3)}$$

The reaction is known as a photochemical reaction because it is initiated with UV light. The overall equation, (14.3), suggests that chloromethane is produced on a molecular scale in one step by the collision of chlorine and methane molecules. This is not the case: the activation energy of reaction (14.3) is too high. Instead, the production of chloromethane occurs in several molecular steps known as elementary reactions. The activation energy of all these elementary reactions in the mechanism are less than that for reaction (14.3). The sequence of elementary reactions is called the reaction mechanism. For the chlorination of methane the elementary reactions include

$$Cl_2(g) + \text{heat/light} \rightarrow 2Cl\bullet(g) \quad \textbf{(14.4)}$$

$$Cl\bullet(g) + CH_4(g) \rightarrow CH_3\bullet(g) + HCl(g) \quad \textbf{(14.5)}$$

$$CH_3\bullet(g) + Cl_2(g) \rightarrow CH_3Cl(g) + Cl\bullet(g) \quad \textbf{(14.6)}$$

The UV light supplies the energy for reaction (14.4). Reactions (14.4) and (14.5) involve chlorine and methyl **free radicals** ($Cl\bullet$ and $CH_3\bullet$) as the reaction intermediates. Free radicals are chemically very reactive because they are eager to use their odd electrons (symbolized $^\bullet$) to covalently bond with other atoms.

Reactions (14.4), (14.5) and (14.6), are known as **propagation reactions** because they produce free radicals that undergo subsequent reactions (a chain reaction). Since $Cl\bullet$ is regenerated in (14.6), the sequence gets faster and faster.

A similar reaction to (14.5) may also occur but with $CH_3Cl(g)$ replacing $CH_4(g)$, producing the radical $\bullet CH_2Cl$ and hence (via a similar reaction to (14.6)) generating CH_2Cl_2. Subsequent and similar reactions also produce small quantities of $CHCl_3$ and CCl_4.

These reactions cannot go on indefinitely and eventually **termination reactions**, such as

$$CH_3\bullet(g) + CH_3\bullet(g) \rightarrow C_2H_6(g)$$

or

$Cl\bullet(g) + Cl\bullet(g) \rightarrow Cl_2(g)$

use up the free radicals and stop the overall reaction.

Chemical reactions that (like the chlorination of methane) occur in several steps are said to be **complex**. Reactions (14.4) and (14.5) are *endo*thermic – they use up energy, but this is compensated by reaction (14.6) which is *exo*thermic. In effect, once the reaction has been started by the spark or flash of light, reaction (14.6) 'drives' the reaction mechanism, providing the energy to sustain the chain reaction and allow the regeneration of free radicals.

Reaction intermediates are molecules, radicals or ions which are made (and then used up) during the reaction mechanism. Although they are often so reactive that they last for only a short time (and are present in very small concentrations at any moment during the reaction) they may often be detected spectroscopically, i.e. by the characteristic frequencies of light which they absorb.

DID YOU KNOW?

The mechanism of the reaction between hydrogen and chlorine also involves free radicals, but in this case the UV initiated reaction is explosively violent.

The use of reaction kinetics in helping to establish the mechanism of a reaction

The rate expression for a reaction often gives valuable information about the mechanism of a reaction. When we introduced rate expressions (page 246), we made the point that we could not reliably predict the rate expression from the equation for the reaction. The reason for this uncertainty is that we cannot usually say whether the chemical equation in question is elementary, i.e. whether it occurs in one step *or* in several steps.

It *is* possible to predict the rate expression for an *elementary reaction* from its chemical equation. For example, consider the elementary reaction (a bimolecular collision) represented by the equation

$A + B \rightarrow C$

Mathematical analysis shows that the rate at which randomly moving particles collide is directly proportional to their concentrations. We therefore expect the rate expression to take the form:

$\text{rate} = k[A][B]$

In practice, we usually look at the experimentally determined rate equation and see whether it is *consistent* with a proposed mechanism. As an example, consider the reaction of NO_2 and CO gases to produce NO and CO_2. Suppose we *propose* that the reaction occurs in one step:

$NO_2(g) + CO(g) \rightarrow NO(g) + CO_2(g)$

This is a bimolecular collision and the expected rate expression would be:

$\text{rate} = k[NO_2(g)][CO(g)]$

Experimentally it is found that the rate expression is

$\text{rate} = k[NO_2(g)]^2$

This provides evidence that the reaction of NO_2 and CO does not occur in one stage.

Now consider the reaction of 1-bromopropane with hydroxide ion. Let us propose that the reaction occurs in one step:

$CH_3CH_2CH_2Br(l) + OH^-(aq) \rightarrow CH_3CH_2CH_2OH(l) + Br^-(aq)$

Experiments show that the rate expression is:

$\text{rate} = k[CH_3CH_2CH_2Br(l)]\ [OH^-(aq)]$

Elementary reactions involving more than three atoms or molecules are unknown in chemistry. The probability of four or more molecules or atoms colliding is so small that it may be safely neglected.

Such a rate expression is consistent with (but does not absolutely prove) that the reaction occurs in one stage. (For more details of the bromopropane/hydroxide ion reaction, see page 374.)

Complicated rate expressions (such as those involving fractional orders of reaction) always indicate that the mechanism of the reaction is complex. This is illustrated by the decomposition of ethanal (page 248).

14.8 Catalysis

How catalysts work

The idea of an overall reaction as consisting of several stages helps to explain how catalysts work. Catalysts cause the reactants to make products through a *different* mechanism in which the *slowest* step in the new mechanism is faster (i.e. possesses a lower activation energy) than the slowest step (the rate determining step) in the mechanism that applies to the uncatalysed reaction. This has the effect of speeding up the rate of the overall reaction. See Figure 14.12.

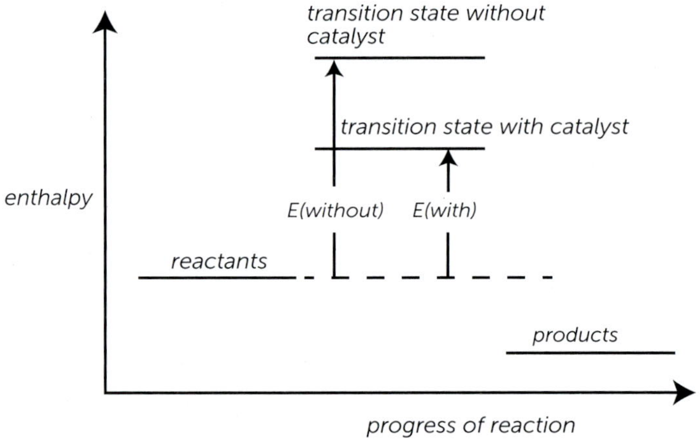

Figure 14.12 How catalysts work. Catalysts allow the *overall* reaction to proceed by a *different* mechanism in which the slowest step is faster (and has a lower activation energy, *E*) than the slowest step in the uncatalyzed reaction mechanism.

Heterogeneous catalysis

Catalysts that are in a different phase from the reactants (usually a solid catalyst speeding up reactants in a solution or gaseous mixture) are termed **heterogeneous catalysts**. Industrially important heterogeneous catalysts include solid nickel, platinum and rhodium. These catalysts work by allowing reactant molecules to stick (adsorb) to the surface of the metal catalyst, so causing the bonds between atoms in the reactant molecules to weaken, making molecular rearrangement easier.

Our example is the hydrogenation of *ethene* to produce *ethane*. In the direct reaction, an ethane molecule results from a single collision between $CH_2{=}CH_2$ and H_2 molecules:

$$\underset{\text{ethene}}{CH_2{=}CH_2(g)} + H_2(g) \rightarrow \underset{\text{ethane}}{CH_3CH_3(g)} \qquad \Delta H^\ominus = -137\ \text{kJ mol}^{-1} \qquad \textbf{(14.7)}$$

The reaction is extremely slow, even at 400°C. However, it proceeds rapidly in the presence of a heated nickel catalyst:

$$CH_2{=}CH_2(g) + H_2(g) \xrightarrow[\text{Ni}]{400\ ^\circ\text{C}} CH_3CH_3(g) \qquad \textbf{(14.8)}$$

(a) The stages involved in the catalysed route start with the starting materials, solid nickel and gaseous hydrogen and ethene (Figure 14.13(a)).

(b) The ethene and hydrogen molecules are attracted towards the Ni atoms (a process called **adsorption**), so weakening both the C=C and H–H bonds. Adsorption weakens the H–H bond so much that some of the H_2 molecules break up and the H atoms bond with the nickel atoms at the surface of the catalyst. The net result is that H *atoms* and ethene molecules are locked at the catalytic surface (Figure 14.13(b)).

(c) Some of the H atoms break away from the Ni atoms and react with the nearby ethene molecules. The reaction may be summarized as:

$$CH_2{=}CH_2\ \text{(on Ni surface)} + 2H\bullet \rightarrow CH_3CH_3\ \text{(on Ni surface)} \qquad \textbf{(14.9)}$$

with the product, ethane, already adsorbed onto the nickel catalyst (Figure 14.13(c)).

(d) The ethane **desorbs** (disconnects) from the metal surface, and moves into the gas phase (Figure 14.13(d)). The nickel is then able to adsorb fresh reactants.

We are now in a position to compare the uncatalysed and catalysed routes of ethane production. In the uncatalysed route, the molecular arrangement that leads to ethane being formed would occur after a collision of ethene and hydrogen molecules. The degree of molecular rearrangement needed to form ethane is high because both H–H bonds and C=C bonds have to be broken. As a consequence, producing ethane in a single step involves a high activation energy, which makes the reaction extremely slow.

Equation (14.8) is the overall reaction for the catalysed route of ethane production, but it conceals the fact that there are several stages in the reaction mechanism. In this mechanism, the molecular rearrangement in which ethane is formed (Equation (14.9)), involves the reaction of hydrogen *atoms* with ethene, with both reactants being held in place on the nickel surface. The activation energy of reaction (14.9) is relatively low because the strong H–H bonds were broken earlier (in stage (b)) leaving highly reactive hydrogen atoms as the weapon to attack the ethene molecule. This makes reaction (14.9) faster than the adsorption and desorption stages. Stage (b) is the slowest stage (the 'rate determining step') of the mechanism, but this stage is still faster than direct reaction of ethene and hydrogen according to equation (14.7).

Overall, the hydrogenation of ethene is very much faster with a catalyst than without.

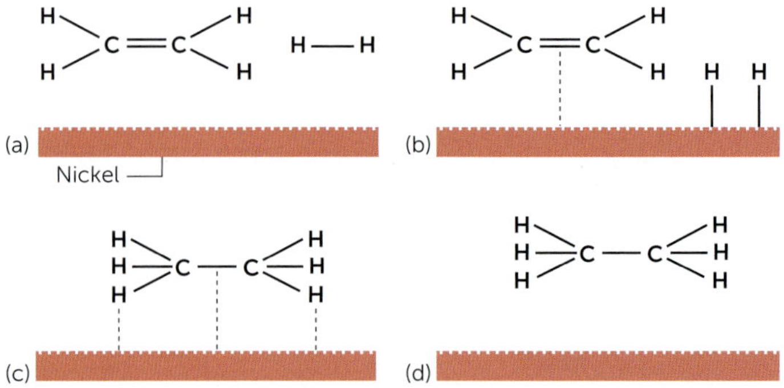

Figure 14.13 Catalytic hydrogenation of ethene (adsorption is shown by dotted lines).

14.9 Why we do not burst into flames

An important distinction is made between (a) whether a reaction 'can go' (i.e. whether it is thermodynamically 'allowed' to occur) and (b) 'how fast it goes'. Whether a reaction is 'allowed' to go is the subject of Unit 16. Just because a reaction *can go* doesn't mean that it occurs at a detectable rate. In this Unit we have been concerned with ways of speeding up reactions that are 'allowed to go' by heating, finely dividing a reactant, increasing concentrations or adding a suitable catalyst.

Chemical kinetics is important in industry as unless a reaction can be made reasonably fast the manufacturing process becomes uneconomical. But kinetics has a role in protecting us in a much more personal way. Organic compounds, including those in coal, wood and those from which our bodies are made, are thermodynamically allowed to combust in oxygen. The fact that we do not burst into flames is a matter of kinetics and not thermodynamics and we are only protected by nature because such oxidation reactions involve elementary steps with high activation energies and so they are exceedingly slow at room temperature. We should be grateful for this! Similarly, many geological formations are thermodynamically unstable but remain unchanged over millions of years because the reactions that might change them proceed at an extremely slow rate.

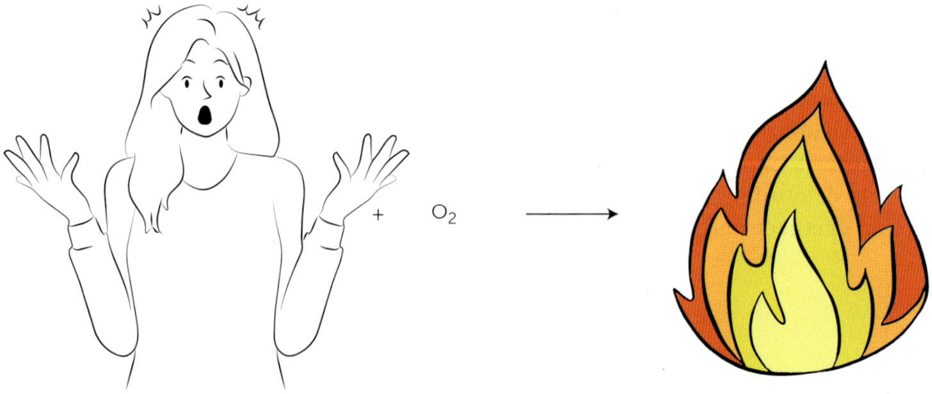

Nature (through thermodynamics) allows the combustion of the human body even at room temperature. Fortunately for the author and reader, such a chemical reaction is exceedingly slow!

REVISION QUESTIONS

14.1 Carefully explain the following key terms:
(i) rate of reaction
(ii) transition state
(iii) activation energy
(iv) order of reaction
(v) elementary reaction
(vi) reaction mechanism
(vii) reactant half-life.

14.2 Which of the following statements are true?
(i) Catalysts do not alter the enthalpy change (ΔH) of a reaction.
(ii) Reactions with high activation energies are usually very fast.
(iii) Activation energies vary with temperature.
(iv) Catalysts do not take part in the reactions they speed up.
(v) Reaction intermediates cannot be isolated.
(vi) Activation energies are always positive.

14.3 Which of the following diagrams shows the variation of reaction rate with temperature for (i) an enzyme-catalysed reaction, and (ii) an uncatalysed reaction?

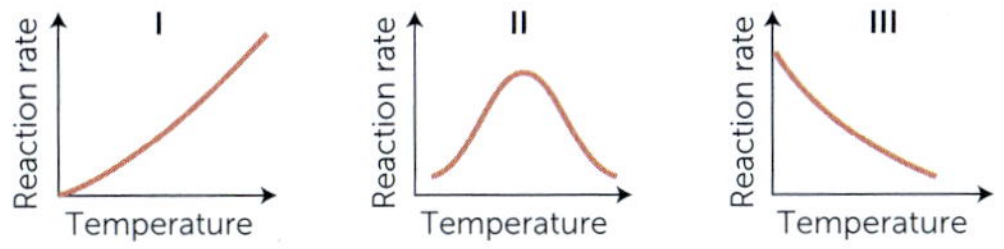

14.4 The decomposition of SO_2Cl_2,

$$SO_2Cl_2(g) \rightarrow SO_2(g) + Cl_2(g)$$

follows the rate expression

$$\text{rate of reaction} = k[SO_2Cl_2]$$

with $k = 2.00 \times 10^{-5}\ s^{-1}$ at 300°C. In an experiment, SO_2Cl_2 was heated to 300°C. The initial concentration of $SO_2Cl_2(g)$ was 0.100 mol dm^{-3}.
(i) What is the order of reaction?
(ii) What is the initial rate of reaction?
(iii) After 9 h the $SO_2Cl_2(g)$ was still decomposing, but its concentration had fallen to 0.0523 mol dm^{-3}. What is the rate of reaction at t = 9 h?
(iv) What is the half-life of $SO_2Cl_2(g)$ at 300°C?

14.5 The decomposition of dinitrogen pentoxide at 65°C:

$$2N_2O_5(g) \rightarrow 4NO_2(g) + O_2(g)$$

follows the rate expression

$$\text{rate of reaction} = k[N_2O_5(g)]$$

In an experiment carried out at 65°C, the concentration of N_2O_5 changed with time as follows:

$[N_2O_5]$/mol dm^{-3}	Time after start of reaction/h
10×10^{-2}	0 (i.e. at start)
8.6×10^{-2}	0.50
7.3×10^{-2}	1.0
6.3×10^{-2}	1.5
5.4×10^{-2}	2.0
4.6×10^{-2}	2.5
3.9×10^{-2}	3.0
3.4×10^{-2}	3.5
2.9×10^{-2}	4.0

(i) What is the overall order of reaction?
(ii) Plot $[N_2O_5(g)]$ against time. Join the points with a smooth curve.
(iii) From your graph, find the initial rate of reaction.
(iv) Calculate the rate constant for the reaction at 65°C and state its units.
(v) What is the half-life of N_2O_5 at this temperature?

14.6 The hydrolysis of bromomethane

$$CH_3Br(l) + OH^-(aq) \rightarrow CH_3OH(aq) + H_2O(l)$$

follows the rate expression:

$$\text{rate of reaction} = k[CH_3Br(l)][OH^-(aq)]$$

The rate constant k was found to be 3.0×10^{-4} mol^{-1} dm^3 s^{-1} at 300 K.
(i) With excess $OH^-(aq)$, the reaction was found to be pseudo first order. What does this mean? What would be the reaction order if the bromomethane (and not the $OH^-(aq)$) was in excess?
(ii) In an experiment $[OH^-(aq)]_0 = 1.0 \times 10^{-2}$ mol dm^{-3} and $[CH_3Br(l)]_0 = 1.0 \times 10^{-4}$ mol dm^{-3}. What is the value of (a) the initial rate of reaction, (b) the pseudo-first order rate constant, and (c) the pseudo-reaction half-life, at 300 K at this hydroxide concentration?

14.7 The reaction of chlorine oxide ($ClO^\bullet$) radicals with nitrogen dioxide,

$$ClO^\bullet(g) + NO_2(g) + N_2(g) \rightarrow ClONO_2(g) + N_2(g)$$

is an important 'sink' for chlorine oxide radicals in the atmosphere. (The nitrogen absorbs the excess energy of colliding molecules and acts as a type of catalyst.) The reaction was studied in the laboratory and showed the following dependence upon the concentrations of the reactants at 298 K:

	Initial concentration/mol dm^{-3}			
Experiment	$[ClO(g)]_0$	$[NO_2(g)]_0$	$[N_2(g)]_0$	Initial rate of reaction/ mol dm^{-3} s^{-1}
1	1.0×10^{-5}	2.0×10^{-5}	3.0×10^{-5}	3.5×10^{-4}
2	0.5×10^{-5}	2.0×10^{-5}	3.0×10^{-5}	1.8×10^{-4}
3	1.0×10^{-5}	4.0×10^{-5}	3.0×10^{-5}	7.1×10^{-4}
4	0.5×10^{-5}	2.0×10^{-5}	6.0×10^{-5}	3.6×10^{-4}

(i) Write down the rate expression for the reaction.
(ii) Calculate an average value of k (298 K).
(iii) Is the rate expression consistent with the formation of chlorine nitrate being a single-stage reaction? Explain.

14.8 Draw an energy profile, similar to Figure 14.7, for an endothermic reaction.

14.9 Show, starting from the equation $[A]_t = [A]_0e^{-kt}$, that the time (symbolized $t_{0.99}$) it takes for a first-order reaction to use up 99 per cent of the reactant is given by the expression

$$t_{0.99} = \frac{4.605}{k}$$

(**Hint:** After $t_{0.99}$ seconds, the ratio of initial to actual concentrations is 1/100.)

14.10 The reaction:

$$2CO(g) + O_2(g) \rightarrow 2CO_2(g)$$

is catalysed by powdered platinum in 'catalytic converters'. Draw sketches (similar to those found in Figure 14.13) which show a possible mechanism for the reaction. For simplicity, represent carbon monoxide as C=O.

14.11 Draw likely molecular arrangements for the two ethane-forming transition states in (a) the direct; and (b) the catalysed pathways for the $CH_2{=}CH_2$ and H_2 reaction (see Figure 14.13). For the catalysed pathway, assume that reaction (14.9) occurs in one step, as discussed on page 259.

DID YOU KNOW?

Chemists have reported making the most energetic molecule ever!

An allotrope of nitrogen, N_6, named **hexanitrogen**, (structure (a) below) has been synthesised by a team of chemists led by Peter Schreiner, along with Artur Mardyukov and Weiyu Qian at Justus Liebig University Giessen in Germany. The work was published in 2025. Hexanitrogen is electrically neutral overall, making it the second neutral allotrope of nitrogen, alongside the well-known N_2 molecule that is the most abundant gas in our atmosphere.

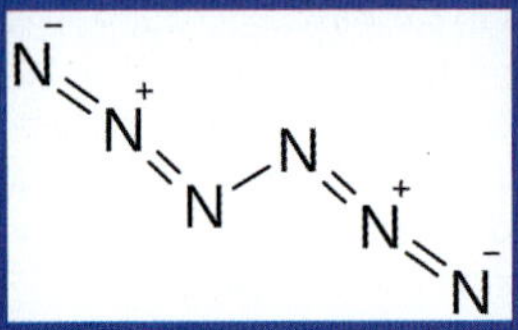

(a) The structure of hexanitrogen, N_6

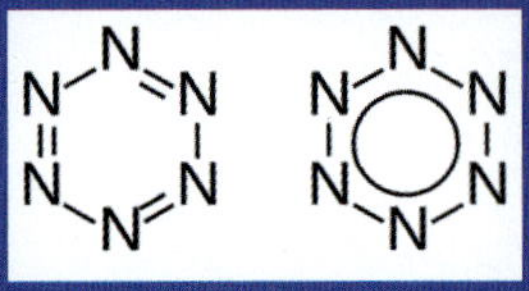

(b) Hexazine, also N_6

The decomposition of hexanitrogen is predicted to be very fast at room temperature:

$$N_6(g) \rightarrow 3N_2(g)$$

In their experiments, Schreiner and colleagues prepared Hexanitrogen in liquid nitrogen (-196 C). Although the activation energy for the decomposition is low, the very low temperature massively slows down the rate of decomposition. Since the N≡N bonds in N_2 are exceptionally strong, the decomposition is also predicted to be extremely exothermic. This had led chemists to speculate that compounds related to hexanitrogen, if they can be synthesised, would be exceptionally powerful explosives.

There is another possible allotrope of nitrogen, also with the formula structure for N_6, named **Hexazine**. So far, this has evaded synthesis, but if it was made, it could be represented by either of the structures shown in (b).

The mechanisms of organic chemical reactions are discussed further in Unit 20.

Chemical Equilibria

15

OBJECTIVES

This unit:

- Explains what is meant by chemical equilibrium
- Defines equilibrium constant
- Discusses factors that affect equilibrium concentrations
- Looks at calculations involving simple equilibria

CONTENTS

15.1 Introduction

Chemical reactions are examples of dynamic equilibria. Equilibrium reactions are sometimes called **reversible reactions**. As an example of an equilibrium reaction, consider the reaction of hydrogen and iodine to make hydrogen iodide in a sealed container at high temperature.

If we heat pure hydrogen and iodine the *forward* reaction takes place:

$$H_2(g) + I_2(g) \rightarrow 2HI(g) \quad \textbf{(15.1)}$$

Almost as soon as HI is formed, some HI molecules decompose in the *back reaction*:

$$2HI(g) \rightarrow H_2(g) + I_2(g) \quad \textbf{(15.2)}$$

At **equilibrium** the *rates* of both reactions are *equal*. This does not mean the *concentrations* of the reactants and products are equal, merely that the 'status quo' (the **equilibrium composition**) is being maintained by two opposite processes which are proceeding at the same rate (Figure 15.1). We summarize the forward and reverse reactions in one equation, using the double-headed arrows to mean 'equilibrium':

$$H_2(g) + I_2(g) \rightleftharpoons 2HI(g)$$

Figure 15.2 shows the general way that the concentration of reactants and products varies with time for an equilibrium reaction. Equilibrium is achieved *t* seconds after the start of the reaction.

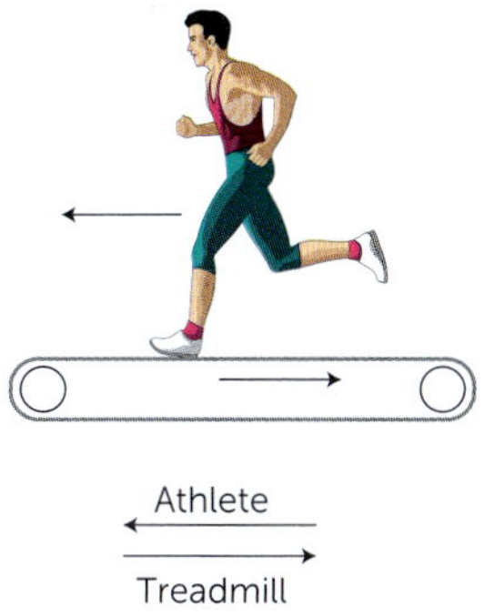

Figure 15.1 An illustration of a dynamic equilibrium – an athlete on a treadmill. The athlete does not fall off because he is moving at the *same* rate as the treadmill but in the *opposite* direction.

Picture of dynamic chemical equilibrium

The idea that, although the *rates* of the forward and back reactions are equal at equilibrium, the *concentrations* of the reactants and products are not necessarily equal is worthy of further explanation. To do this, we consider the simplest kind of chemical

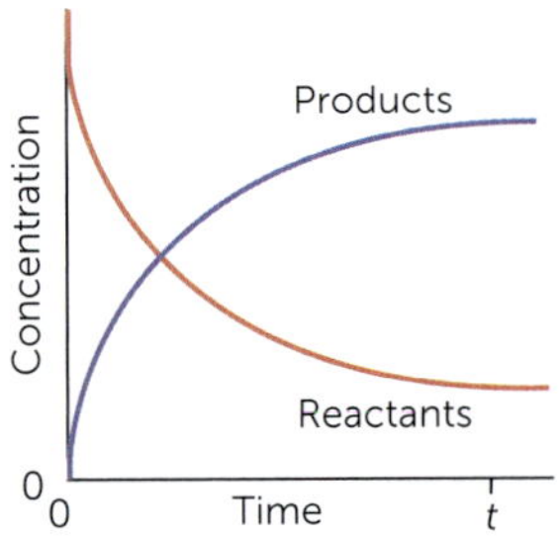

Figure 15.2 Concentration–time profile for an equilibrium reaction.

reaction in which there is only one reactant (A) and one product (B). The reaction is then represented as

$$A \rightleftharpoons B$$

Figure 15.3 represents the reaction as consisting of a store of A and B molecules on either side of a barrier. The volume of the container is fixed, and so the numbers of molecules represent the concentrations of those molecules. In our model, reaction has taken place when a molecule passes over the barrier, and we show this by changing the colour of the circles (● → ●) used to represent the molecules. We shall assume that the temperature is fixed.

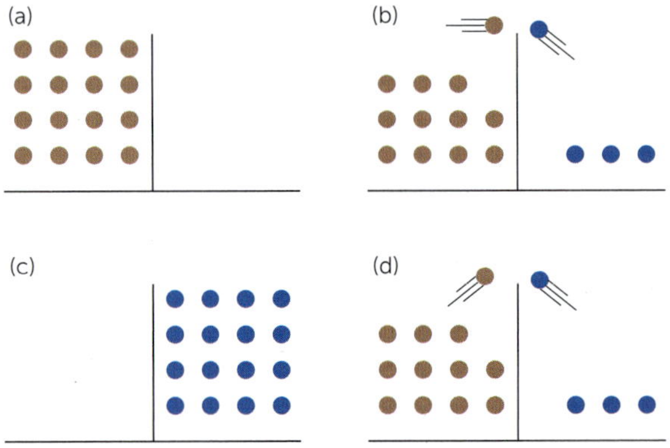

Figure 15.3 The chemical equilibrium $A \rightleftharpoons B$: ● represents A molecules; ● represents B molecules.

In Figure 15.3(a), we start with pure A. As the reaction proceeds, some of the A molecules change into B molecules, and soon after this some B molecules revert to A. At equilibrium (Figure 15.3(b)) the composition of the mixture is fixed, but A molecules are still changing into B and vice versa. The key point, though, is that at equilibrium the number of A molecules jumping over the barrier per second equals the number of B molecules jumping over the barrier per second. In our example, the equilibrium mixture contains 75 per cent of A (twelve out of sixteen molecules) and only 25 per cent of B. In chemical reactions, the composition of the reactant–product mixture at equilibrium depends upon the precise concentration of reactants and products (if any) at the start, and upon the temperature of the reaction mixture.

Our simple model may be used to illustrate another important feature of dynamic equilibria. Figure 15.3(c) shows the reaction starting with pure B with an initial concentration equal to that of A in Figure 15.3(a). As reaction starts, some B changes into A. Some A then reverts to B, and Figure 15.3(d) shows the composition of the equilibrium mixture. It is evident that it is the same as Figure 15.3(b), i.e. as was arrived at starting from pure A at the same temperature. We summarize this by saying that at the same temperature *the same equilibrium composition may be achieved from both forward and reverse directions.*

All reactions are, in principle, equilibrium reactions. However, for many reactions there is such a low concentration of limiting *reactants* at equilibrium that we may safely assume that the reactants have been completely converted to products, i.e. that the reaction 'has gone to completion'. We do not use the $\rightleftharpoons$ sign for reactions that go to completion, and the ordinary 'produces' sign ($\rightarrow$) is used instead.

15.2 Equilibrium law and equilibrium constant

Equilibrium law

For any chemical reaction at equilibrium we define an **equilibrium constant** K in terms of molar concentration as follows. For the generalized reaction at temperature T:

$$aA + bB + \ldots \rightleftharpoons cC + dD + \ldots$$

$$K = \frac{[C]^c[D]^d}{[A]^a[B]^b} \quad \textbf{(15.3)}$$

The fact that all equilibrium reactions can be fitted to this general equation is sometimes called the **equilibrium law**. The coefficients in the chemical equation are $a, b, c, d, \ldots$ and the square brackets represent the concentrations of the reactants and products *at equilibrium* (*not* the starting concentrations).

Applying the equilibrium law to the equation

$$H_2(g) + I_2(g) \rightleftharpoons 2HI(g)$$

gives

$$K = \frac{[HI(g)]^2}{[H_2(g)][I_2(g)]} \quad \textbf{(15.4)}$$

EXAMPLE **15.1**

Write down an expression for the equilibrium constant for the reaction

$$2SO_2(g) + O_2(g) \rightleftharpoons 2SO_3(g) \quad \textbf{(15.5)}$$

and state its units.

Answer

We might represent this equation more clearly as:

$$\mathbf{2}SO_2(g) + \mathbf{1}O_2(g) \rightleftharpoons \mathbf{2}SO_3(g)$$

in which the coefficients (2, 1 and 2, respectively) are displayed in bold. Comparison with equation (15.3) shows that, for this reaction,

$A = SO_2$ $B = O_2$ $C = SO_3$ $a = 2$ $b = 1$ $c = 2$

The expression for the equilibrium constant therefore becomes

$$K = \frac{[SO_3(g)^2]}{[SO_2(g)]^2[O_2(g)]}$$

The units of an equilibrium constant depend upon the number of concentration terms in the expression. Here the units are

$$K = \frac{(\text{mol dm}^{-3})^2}{(\text{mol dm}^{-3})^2(\text{mol dm}^{-3})} = \frac{\cancel{(\text{mol dm}^{-3})^2}}{\cancel{(\text{mol dm}^{-3})^2}(\text{mol dm}^{-3})}$$

i.e. K possesses the units of $\text{mol}^{-1}\,\text{dm}^3$.

K is practically important because it does not depend upon the concentrations of the reactants (or products) at the start of the reaction, and for a named reaction varies only with the temperature of the equilibrium mixture. Remember:

The value of an equilibrium constant for a particular reaction changes only with temperature.

The equilibrium constants of reactions involving gases may also be expressed in terms of the partial pressures (p, in bar) of the reactants and products. The equilibrium constant may then be symbolized K_p. For example, for reaction (15.5),

$$K_p = \frac{(p_{SO_3})^2}{(p_{SO_2})^2 \times p_{O_2}} = 3.0 \times 10^4 \text{bar}^{-1} \text{ at } 700 \text{ K} \quad \textbf{(15.6)}$$

Extension 15, on the website on the website shows how to convert K_p to K in molar concentrations.

EXERCISE **15A**

Reactions that go to completion

Study the equations

$CO(g) + H_2(g) \rightleftharpoons H_2O(g) + C(s)$

and

$NaOH(aq) + HCl(aq) \rightarrow H_2O(l) + NaCl(aq)$

Which of the following statements are correct?

(a) If we started with 0.1 mol of H_2 and 0.1 mol of CO, we would expect to produce 0.1 mol of H_2O and 0.1 mol of C.

(b) If we started with 0.1 mol of NaOH and 0.1 mol of HCl, we would expect to produce 0.1 mol of H_2O and 0.1 mol of NaCl.

EXERCISE **15B**

Equilibrium expressions

Write down expressions for the equilibrium constants of the following reactions:

(i) Oxidation of iron(II) ions by chlorine gas
$2Fe^{2+}(aq) + Cl_2(g) \rightleftharpoons 2Fe^{3+}(aq) + 2Cl^-(aq)$

(ii) Formation of chlorine nitrate gas
$ClO(g) + NO_2(g) \rightleftharpoons ClONO_2(g)$

(iii) Formation of nitrogen dioxide gas from dinitrogen tetroxide
$N_2O_4(g) \rightleftharpoons 2NO_2(g)$

What are the units of K in each case?

Equilibrium constants and reaction rate constants

First, a reminder – 'big K' symbolizes an equilibrium constant, whereas 'little k' symbolizes a rate constant.

Experiments show that the rate of reaction (15.1) follows the rate expression

rate of reaction = $k_f[H_2(g)]\,[I_2(g)]$

where k_f (subscript f for 'forward') is the rate constant (see Unit 14) and the brackets indicate the concentrations at that *instant*. Immediately after starting the reaction, the concentrations of hydrogen and iodine begin to fall and so the rate of the forward reaction falls.

The experimental rate expression for the *back reaction* (equation (15.2)) turns out to be equally straightforward:

rate of reaction = $k_b[HI(g)]^2$

As the concentration of HI builds up, the rate of the back reaction increases. At equilibrium, the rates of the forward and back reactions are equal:

$$k_b[HI(g)]^2 = k_f[H_2(g)]\,[I_2(g)]$$

Comparing this with equation (15.4) shows that

$$K = \frac{k_f}{k_b}$$

- This equation applies to all equilibrium reactions (and may be used whether or not the reaction occurs in a single stage).
- Remember that a rate constant is the rate of reaction when the substances in the rate expression are at unit concentration, i.e. 1 mol dm^{-3} (see page 247). The above equation shows that the size of the equilibrium constant depends simply upon how much faster the forward reaction is (at unit concentrations) than the back reaction (at unit concentrations) at that temperature. We need only to know the equilibrium constant and one of the rate constants in order to calculate the remaining rate constant. For example, for the reaction

$$H_2(g) + I_2(g) \rightleftharpoons 2HI(g)$$

k_f = 1.66 mol^{-1} dm^3 s^{-1} and K = 45.6 (unitless) at 764 K. Therefore,

$$k_b(764\text{K}) = \frac{1.66\,\text{mol}^{-1}\,\text{dm}^3\text{s}^{-1}}{45.6} = 0.0364\text{mol}^{-1}\text{dm}^3\text{s}^{-1}$$

- While it is true that if $k_f >> k_b$ then K will be large and the forward reaction product concentrations will dominate the equilibrium composition, if the absolute numerical value of k_f is very small, the time it takes to achieve equilibrium will be very large. This explains why many reactions with huge equilibrium constants proceed at a negligible rate unless conditions (increased temperature, catalyst) are changed.

The importance of stating the chemical equation to which the equilibrium constant applies

An equilibrium constant applies to a particular chemical equation. For example, whereas

$$K = \frac{[SO_3(g)]^2}{[SO_2(g)]^2[O_2(g)]} \quad \text{for } 2SO_2(g) + O_2(g) \rightleftharpoons 2SO_3(g)$$

we also write

$$K = \frac{[SO_3(g)]}{[SO_2(g)][O_2(g)]^{1/2}} \quad \text{for } SO_2(g) + \tfrac{1}{2}O_2(g) \rightleftharpoons SO_3(g)$$

The equilibrium constants for these expressions will have different numerical values. The moral is simple – never quote equilibrium constants unless you write down the chemical equation (or equilibrium expression) to which that value of K applies.

15.3 Meaning of equilibrium constants

Look back at Figure 15.3(b). At equilibrium, the reaction mixture consisted of twelve molecules of A and four of B. Since, for a fixed volume, the molar concentration is proportional to the number of molecules, we can write

$$K = \frac{[B]}{[A]} = \frac{4}{12} = 0.33$$

Suppose that for a different reaction, say C $\rightleftharpoons$ D, there were fifteen molecules of C and one of D at equilibrium at a particular temperature. Then,

$$K = \frac{[D]}{[C]} = \frac{1}{15} = 0.066$$

We see that the smaller the equilibrium constant, the smaller the relative number of molecules of product to reactant in the equilibrium mixture, and vice versa.

Where chemical equations contain more than one reactant or product molecule, the equilibrium constant for the reaction is not simply the ratio of a single product and single reactant concentration. (Equation (15.4), for example, contains a concentration that is squared.) Nevertheless, very *large equilibrium constants always mean that the product molecules dominate the equilibrium mixture*, whereas very small equilibrium constants show that reactant molecules dominate the equilibrium mixture.

As a guide, if K is greater than 10,000, the product dominates the equilibrium mixture and we would say that (for most purposes) the reaction goes to completion at that temperature. If K is less than 0.0001 we would say that (for most purposes) a negligible amount of product has been made and that 'no reaction has occurred' at that temperature. Summarizing,

if $K > 10{,}000$ **reaction goes to completion**
if $K < 0.0001$ **no reaction occurs**

If K is in the range 10000–0.0001, the equilibrium mixture may contain significant concentrations of reactants and products, but **the bigger is K the greater the dominance of the products.**

Reactions that go to completion include many of the reactions of inorganic ions we have already studied, for example

$Ag^+(aq) + Cl^-(aq) \rightarrow AgCl(s)$ $K_{(25\,°C)} >> 10{,}000$ $mol^{-1}\,dm^3$

The mole calculations discussed in previous units (including those involving titrations) have assumed that the reactions go to completion.

A list of K values for selected reactions is given in Table 15.1.

Table 15.1 Equilibrium constants

No	Reaction	K	Temperature/K	Units of K
1	$Cl_2(g) \rightleftharpoons Cl(g) + Cl(g)$	1.2×10^{-7}	1000	$mol\,dm^{-3}$
2(a)	$H_2(g) + Cl_2(g) \rightleftharpoons 2HCl(g)$	4.0×10^{31}	300	–
2(b)	$H_2(g) + Cl_2(g) \rightleftharpoons 2HCl(g)$	4.0×10^{18}	500	–
3	$2BrCl(g) \rightleftharpoons Br_2(g) + Cl_2(g)$	377	300	–
4	$CO_2(g) + H_2(g) \rightleftharpoons CO(g) + H_2O(g)$	1.56	1,073	–
5	$HF(g) + HCN(g) \rightleftharpoons HCN\cdots HF(g)$	1.04	298	$mol^{-1}\,dm^3$
6	$I^-(aq) + I_2(l) \rightleftharpoons I_3^-(aq)$	7.1×10^{-2}	298	$mol^{-1}\,dm^3$
7	$Cu^{2+}(aq) + 4NH_3(aq) \rightleftharpoons Cu[(NH_3)_4]^{2+}(aq)$	1.4×10^{13}	298	$mol^{-4}\,dm^{12}$
8(a)	$N_2(g) + 3H_2(g) \rightleftharpoons 2NH_3(g)$	3.5×10^8	298	$mol^{-2}\,dm^6$
8(b)	$N_2(g) + 3H_2(g) \rightleftharpoons 2NH_3(g)$	2.2	623	$mol^{-2}\,dm^6$
9	$2NH_3(g) \rightleftharpoons N_2(g) + 3H_2(g)$	0.46	623	$mol^2\,dm^{-6}$
10	$Cu^{2+}(aq) + Zn(s) \rightleftharpoons Zn^{2+}(aq) + Cu(s)$	1×10^{37}	298	–

Equilibrium constants as indicators of whether or not a reaction is allowed to occur

Chemical reactions starting from pure reactants with large K values are said to be 'thermodynamically allowed' (see page 290). However, equilibrium constants tell us nothing about whether the reaction under consideration will occur at a *fast enough speed* to be detectable. Many reactions possess large K values, but are extremely slow in reaching equilibrium at room temperature, for example:

- The formation of ammonia gas from hydrogen and nitrogen gases occurs at a negligible rate at room temperature.
- A mixture of hydrogen and oxygen gases at room temperature shows no signs of producing water, no matter how long it is kept.

Fortunately, the rates of reactions can (within certain limits) be controlled by altering the experimental conditions, and reactions may often be accelerated using a combination of high temperature, high reactant concentration and a catalyst.

EXERCISE **15C**

Equilibrium constants

(i) Which of the reactions shown in Table 15.1 go to completion at the temperatures listed?

(ii) The equilibrium constants for the reactions of sodium and magnesium ions with the complexing agent $EDTA^{4-}$ at room temperature are 50 $mol^{-1}\,dm^3$ and 5×10^8 $mol^{-1}\,dm^3$ respectively. Which ion cannot be satisfactorily analysed by titration with $EDTA^{4-}$, and why?

Values of equilibrium constants for reverse reactions

The equilibrium constant K for the formation of hydrogen iodide,

$$H_2(g) + I_2(g) \rightleftharpoons 2HI(g)$$

is

$$K_f = \frac{[HI(g)]^2}{[H_2(g)][I_2(g)]}$$

where 'f' stands for 'forward reaction'. The equilibrium constant for the reverse or 'back' reaction,

$$2HI(g) \rightleftharpoons H_2(g) + I_2(g)$$

is

$$K_b = \frac{[H_2(g)][I_2(g)]}{[HI(g)]^2}$$

from which we see that

$$K_b = \frac{1}{K_f}$$

We conclude that *if the equilibrium constant for the forward reaction is large, then the equilibrium constant for the reverse process is small.*

For example, K for the reaction

$$Cu^{2+}(aq) + Zn(s) \rightleftharpoons Zn^{2+}(aq) + Cu(s)$$

is 1×10^{37} at room temperature (Table 15.1). Thus K for the *reverse* reaction is 1×10^{-37}. The consequence of these values is clear cut: for all practical purposes the

EXERCISE **15D**

Equilibrium constants for reverse reactions

The equilibrium constant for the formation of nitric oxide (NO):

$$N_2(g) + O_2(g) \rightleftharpoons 2NO(g)$$

is 7×10^{-9} at 1,100 K. In an experiment, NO is heated to 1100 K. Will the NO decompose?

reaction of zinc metal and copper ions goes to completion, whereas the reaction of zinc ions with copper metal does not take place.

15.4 Effects of changing concentration, pressure and temperature upon equilibria

Detailed studies have been made of many different examples of equilibria. We look at two cases. In case 1, the hydrogen–iodine reaction, we confirm that *K* is independent of reactant concentrations. In case 2, an esterification reaction, we confirm that the equilibrium concentration of reactants and products (the 'equilibrium composition') depends upon their starting concentrations.

Case 1: Use of the reaction between hydrogen and iodine gases to show that *K* is independent of concentration

The reaction is

$$H_2(g) + I_2(g) \rightleftharpoons 2HI(g)$$

Table 15.2 shows the initial (i.e. at $t = 0$) and equilibrium compositions of three different mixtures at 490°C.

We start by checking whether or not the results obey the reaction equation. According to the equation, 1 mol of H_2 or I_2 produces 2 mol of HI. Therefore, the concentration of HI in Table 15.2 should be twice the *loss* in concentration of H_2 or I_2 that takes place by the time equilibrium has been reached. Within experimental error, this is seen to be the case. For example, in experiment 1, the equilibrium concentration of HI is 30.86 mol dm^{-3}. The reduction in I_2 concentration is (20.00 − 4.58) = 15.42 mol dm^{-3}, almost half of 30.86.

Table 15.2 The composition of mixtures containing hydrogen, iodine and iodide at 490°C (763 K)

Time	Concentrations/mol dm^{-3}		
	[H_2]	[I_2]	[HI]
Experiment 1			
$t = 0$	20.00	20.00	0
At equilibrium	4.58	4.58	30.86
Experiment 2			
$t = 0$	8.19	14.31	0
At equilibrium	0.72	6.84	14.94
Experiment 3			
$t = 0$	0	0	40.00
At equilibrium	4.58	4.58	30.86

Substitution of the equilibrium concentrations into equation (15.4) gives:

For experiments 1 and 3, $K = (30.86)^2/(4.58)^2 = 45.4$
For experiment 2, $K = (14.94)^2/(0.72 \times 6.84) = 45.3$

This confirms that although the starting concentrations of reactant are different, *K* remains the same (within experimental error) at that temperature.

Now look at experiment 3 more closely. Experiment 3 starts with 40.00 mol of pure HI. The HI decomposes into H_2 and I_2, and [HI] falls until equilibrium is reached;

40.00 mol of HI is the amount of HI that would be produced *if* we were able to convert all of the H_2 and I_2 in experiment 1 to HI. The equilibrium compositions of experiment 1 (in which we started with pure reactants) and experiment 3 (in which we started with pure product) are identical, confirming our earlier statement (page 264) that the same equilibrium composition may be achieved from both forward and back directions.

EXERCISE **15E**

Calculations involving equilibrium concentrations

(i) Pure iodine and hydrogen gases were mixed and allowed to reach equilibrium. The equilibrium concentrations of iodine and hydrogen gases at 490°C were each found to be 3.00 mol dm^{-3}. What is the equilibrium concentration of gaseous hydrogen iodide? What were the starting concentrations of reactants?

(ii) The equilibrium concentrations (in mol dm^{-3}) for a mixture of HI, I_2, and H_2 at 527°C were [HI(g)] = 19.1, [I_2(g)] = 3.50 and [H_2(g)] = 2.50. Does *K* for the reaction increase or decrease with increasing temperature?

Case 2: Esterification of ethanol used to illustrate the effect of concentration on equilibrium composition

The reaction of ethanol with ethanoic acid (in the presence of sulfuric acid catalyst) produces a sweet-smelling compound known as an ester. The production of an ester is called **esterification.** The name of the particular ester produced in this reaction is ethyl ethanoate:

$$\underset{\text{ethanoic acid}}{CH_3COOH(l)} + \underset{\text{ethanol}}{C_2H_5OH(l)} \rightleftharpoons \underset{\text{ethyl ethanoate}}{CH_3COOC_2H_5(l)} + H_2O(l)$$

From this we obtain

$$K = \frac{[CH_3COOC_2H_5(l)][H_2O(l)]}{[CH_3COOH(l)][C_2H_5OH(l)]} \tag{15.7}$$

Table 15.3 shows the composition of the equilibrium mixture for three initial mixtures of differing concentration. Substituting the equilibrium concentrations into equation (15.7) gives

for experiment 1 $K = 4.1$
for experiment 2 $K = 4.0$
for experiment 3 $K = 4.0$

We comment on these results on the next page.

Table 15.3 Equilibrium composition (at 25°C) of mixtures containing ethanoic acid, ethanol, water and ethyl ethanoate

Time	Concentrations/mol dm^{-3}			
	[CH_3COOH]	[C_2H_5OH]	[$CH_3COOC_2H_5$]	[H_2O]
Experiment 1				
$t = 0$	1.00	1.00	0	0
At equilibrium	0.33	0.33	0.67	0.67
Experiment 2				
$t = 0$	4.00	1.00	0	0
At equilibrium	3.07	0.070	0.93	0.93
Experiment 3				
$t = 0$	4.00	1.00	1.00	0
At equilibrium	3.13	0.13	1.87	0.87

1 Constancy of *K*

Within experimental error all three values of *K* are the same.

2 Effect of higher concentration of reactant upon product concentration at equilibrium

In experiment 2 the initial mixture contains a higher concentration of acid than was used in 1. This causes the equilibrium concentration of ester and water to be greater than that observed in experiment 1. This effect is observed in other equilibria:

Increasing the concentration of reactant(s) increases the equilibrium concentration of products.

3 Effect of the presence of product upon the gain in product concentration in the equilibrium mixture

Although the equilibrium concentration of ester in experiment 3 is greater than in experiment 2, we need to take into account that the initial reaction mixture in experiment 3 contained some ester. To do this, we define the **gain** in ester concentration (conc.):

gain in ester conc. = equilibrium ester conc. − initial ester conc.

The final concentration of ester in experiment 3 ($1.87\ \text{mol dm}^{-3}$) represents a gain of $1.87 - 1.00 = 0.87\ \text{mol dm}^{-3}$ ester. This is **less** than that obtained with experiment 2, where the gain was $0.93\ \text{mol dm}^{-3}$. The same effect is observed in other equilibria.

Generalizing,

The presence of some product at the start of a reaction reduces the gain of product in the equilibrium mixture.

In principle, the effects of increased reactant concentration, and of the presence of product in the initial mixture, are applicable to all equilibria. However, if the equilibrium constant is very large (so that the reaction goes to completion) or very small (so that reaction hardly occurs), changes in concentration are found to have an insignificant effect upon equilibrium concentrations.

We could come to the same conclusions as in points **2** and **3** by studying the expression for the equilibrium constant, equation (15.7). As an example, consider point **3**. Imagine that we add fresh ester to a mixture *already at equilibrium*. This causes $[CH_3COOC_2H_5(l)]$ to rise, and the ratio of concentrations,

$$\frac{[CH_3COOC_2H_5(l)][H_2O(l)]}{[CH_3COOH(l)][C_2H_5OH(l)]}$$

temporarily exceeds *K*. The ratio eventually reduces to *K* because some ester decomposes into ethanoic acid and ethanol:

$$CH_3COOC_2H_5(l) + H_2O(l) \rightarrow CH_3COOH(l) + C_2H_5OH(l)$$

The equilibrium mixture now contains fewer moles of ester than the sum of the number of moles of added ester and the number of moles of ester at equilibrium before the disturbance took place. Note that it may take several minutes for the reaction to re-establish equilibrium; see Figure 15.4.

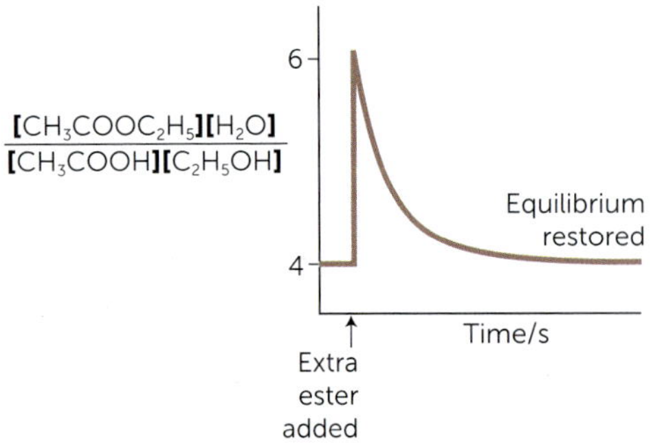

Figure 15.4 Esterification mixture at 25°C. If an equilibrium is disturbed (here by adding ester), the concentrations alter so as to re-establish equilibrium.

Chemistry under non-equilibrium conditions

In the laboratory, and in industry, we often deliberately disturb a chemical equilibrium. For example, the percentage of the number of moles of ester at equilibrium in experiment 2 in Table 15.3, is less than 20 per cent. However, if the ester is *physically removed* from the reaction mixture by distillation, the composition of the mixture will adjust to restore the yield of ester. The overall yield of ester (obtained by summing the ester from several distillations) is then very much greater.

EXERCISE **15F**

Working out whether or not a reaction has reached equilibrium

The decomposition of ammonia gas was studied at 700 K:

$$2NH_3(g) \rightleftharpoons N_2(g) + 3H_2(g)$$

It was found that $K = 0.30\ mol^2\ dm^{-6}$. In a separate experiment, some ammonia gas was heated to 700 K and the composition of the mixture analysed after a set time. The concentrations were: $[N_2(g)] = 0.04\ mol\ dm^{-3}$, $H_2(g) = 0.03\ mol\ dm^{-3}$, and $NH_3(g) = 0.02\ mol\ dm^{-3}$. Has the reaction reached equilibrium?

EXERCISE **15G**

Non-equilibrium conditions

K for the reaction of steam and heated solid iron,

$$2Fe(s) + 3H_2O(g) \rightleftharpoons Fe_2O_3(s) + 3H_2(g)$$

is very large at 600 K, yet it is quite easy to reverse this reaction by passing hydrogen gas over strongly heated iron oxide using the apparatus shown in Figure 15.5. How is this possible?

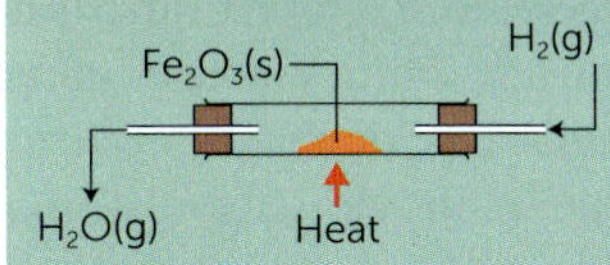

Figure 15.5 Heating Fe_2O_3 in a stream of H_2.

Effect of temperature on equilibrium constants

The effect of temperature upon *K* is summarized as follows:

1. **The equilibrium constant for an exothermic reaction falls with increasing temperature.**
2. **The equilibrium constant for an endothermic reaction rises with increasing temperature.**

For example, the formation of HI(g) from $H_2(g)$ and $I_2(g)$ is exothermic,[1] and *K* for this reaction falls with increasing temperature (Figure 15.6). This means that the percentage of HI(g) in the equilibrium mixture falls as the temperature is increased.

The variation of an equilibrium constant with temperature is often so dramatic that adjusting the operating temperature of a reaction is the most effective way of improving the yield of product. The following processes illustrate this.

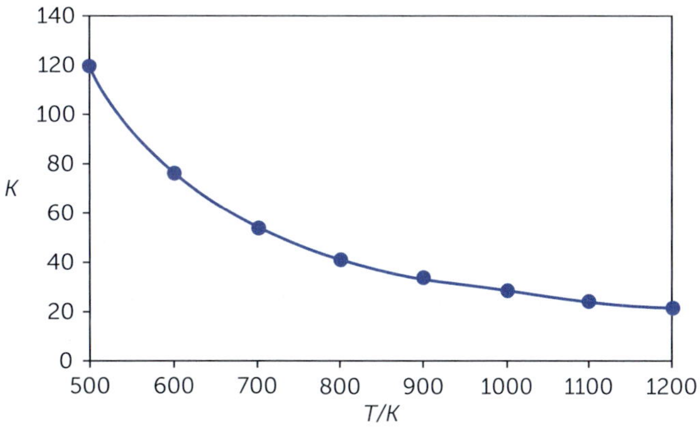

Figure 15.6 Variation with temperature of the equilibrium constant for the reaction $H_2(g) + I_2(g) \rightleftharpoons 2HI(g)$.

[1]The reaction is *endothermic* if iodine is a solid, but we are dealing with $T > 457$ K, when I_2 is gaseous.

EXERCISE **15H**

Effect of temperature upon equilibrium

Sketch a diagram, similar to Figure 15.4, showing a plot of

$$\frac{[HI(g)]^2}{[H_2(g)][I_2(g)]}$$

against time when a $HI/H_2/I_2$ mixture at equilibrium at 667 K (where $K = 61$) is suddenly warmed up to 764 K (when $K = 46$), and then allowed to slowly cool back to 667 K.

1 Formation of hydrogen-bonded compounds

The reactions in which *hydrogen-bonded compounds* are formed are exothermic and often possess very low equilibrium constants at room temperature. For example, K for the reaction between hydrogen fluoride and hydrogen cyanide gases,

$$HF(g) + HCN(g) \rightleftharpoons \underset{\text{hydrogen-bonded 'complex'}}{HCN\cdots HF(g)}$$

is only 1.04 $mol^{-1}\,dm^3$ at 298 K. For a starting mixture of 2.7×10^{-3} $mol\,dm^{-3}$ HCN and 2.7×10^{-3} $mol\,dm^{-3}$ HF, this corresponds to an equilibrium concentration of HCN···HF complex of only 7×10^{-6} $mol\,dm^{-3}$! For this reason it is beneficial to study such complexes at lower temperatures when K is larger, as the increased concentration of complex makes them easier to detect.

However, since reaction rates generally decrease with decreasing temperature, it is important to remember that a lower temperature *always* means that a reaction takes longer to *reach equilibrium*. This may not be a severe hindrance in the laboratory, but it is frequently a very important economic consideration in the chemical and metallurgical industries.

EXERCISE **15I**

Deducing whether a reaction is exothermic or endothermic

The association of nitrogen dioxide to form dinitrogen tetroxide involves a colour change:

$$\underset{\text{red brown}}{2NO_2(g)} \rightleftharpoons \underset{\text{colourless}}{N_2O_4(g)} \qquad \text{....(A)}$$

A sample of NO_2 and N_2O_4 is cooled to the temperatures shown in Figure 15.7.
Is reaction (A) exothermic or endothermic?

−196	0	23	35	50

T/°C →

Figure 15.7 Appearance of a mixture of NO_2 and N_2O_4 at different temperatures.

2 Metal extraction

The extraction of a metal (M) from its oxide (MO) using carbon as a reducing agent is generally endothermic. As expected, experiments show that the equilibrium constants of the general reaction,

$$MO(s) + C(g) \rightleftharpoons CO(g) + M(s)$$

increase with increasing temperature. This means that virtually *any* metal oxide may be reduced by carbon if the temperature is high enough. For example, in the reduction of zinc oxide,

$$ZnO(s) + C(g) \rightleftharpoons CO(g) + Zn(s)$$

the equilibrium yield of the free metal is only high at temperatures above 800°C. For aluminium, reduction takes place only above 2,000°C, and it is cheaper to produce this metal by electrolysis rather than attempt to maintain a furnace at such a high temperature.

BOX **15.1**

Explaining the variation of *K* with temperature

Why do the equilibrium constants of reactions vary with temperature? The answer lies in the enthalpy–progress of reaction diagram discussed in Box 14.1 (page 245).

For a reaction that is *exothermic* from left to right, the activation energy of the forward reaction is *less* than the activation energy of the back reaction. The smaller the activation energy, the less sensitive is the associated rate constant to changes in temperature. Thus, for an exothermic reaction, k_f changes more slowly with temperature changes than k_b. Put another way, for a particular *increase* in temperature (say 50°C), k_f increases by a smaller factor than k_b. Since $K = k_f/k_b$ (see page 269), this causes K to decrease with temperature. For reactions that are endothermic, $E_{A(f)} > E_{A(b)}$ and so k_f changes faster with temperature than k_b. This causes K to increase with increasing temperature.

Catalysts and equilibria

A catalyst increases the rate constant for the forward and back reaction by the same factor. For example, it might double both rate constants. It follows from the equation

$$K = k_f/k_b$$

that K is unaffected:

Catalysts do not alter the equilibrium constant of a reaction. They do not alter equilibrium concentrations or changes in enthalpy. They increase reaction rate so that equilibrium is reached in a shorter time.

Le Chatelier's principle

The effects of changing conditions upon chemical equilibria are summarized in Table 15.4. One way of remembering (but not explaining) such effects is to think of the chemical reaction as actively opposing (almost as if it were alive!) any changes in conditions that have been made. **Le Chatelier's principle** states:

The concentrations of reactants and products in an equilibrium mixture will alter so as to counteract any changes in pressure, temperature or concentration.

Predictions made using Le Chatelier's principle

1 If we add more reactant to an equilibrium mixture, the equilibrium composition shifts so that more reactants are used up, i.e. so that the equilibrium mixture

contains a higher concentration of product. (We arrived at the same conclusion by looking at experimental data for the ethanoic acid–ethanol reaction.)

2 The effect of temperature can be predicted by thinking of the heat evolved as a special kind of *product*. Suppose we increase the temperature of an equilibrium mixture in which the reaction is exothermic:

$$aA + bB \rightleftharpoons cC + dD + \text{heat}$$

Le Chatelier's principle predicts that the equilibrium composition alters so that less 'heat product' will be made, i.e. it alters in the direction in which the temperature will be lowered. This is, of course, in the direction of the endothermic (back) reaction. This means that the equilibrium concentration of reactants increases and that of the products falls. As we have seen (see Figure 15.6), this is indeed what happens.

Table 15.4 Summary of the effect of varying conditions on the equilibrium $aA+bB \rightleftharpoons cC+dD$

Change	Effect on percentage of product at equilibrium	Effect on equilibrium constant
Increasing the concentration* of one or both reactants	**increases**	no change
Decreasing the concentration* of one or both reactants	**decreases**	no change
For gas reactions, increasing the pressure of the *reactants* and *products* (e.g. if the mixture is in a piston, we apply extra force to the piston handle)	**1 decreases** if there are fewer molecules on the left-hand side of the reaction equation, i.e. if $(a + b) < (c + d)$	no change
	2 increases if there are more molecules on the left-hand side of the reaction equation, i.e. if $(a + b) > (c + d)$	no change
For gas reactions, allowing the mixture to expand, so reducing the pressure of the *reactants* and *products* (e.g. if the mixture is in a piston, we reduce the force on the piston handle)	**1 increases** if $(a + b) < (c + d)$	no change
	2 decreases if $(a + b) > (c + d)$	no change
For gas reactions, injecting an inert gas into the mixture while keeping the total gas volume constant	no change (this is true whatever the values of a, b, c and d)	no change
Increased temperature	**1 increases** for an endothermic reaction	**increases**
	2 decreases for an exothermic reaction	**decreases**
Catalyst	no change	no change

*For gases, the partial pressure.

3 Consider the gaseous reaction

$$H_2SO_4(g) \rightleftharpoons H_2O(g) + SO_3(g)$$

which we shall assume is occurring in a cylinder fitted with a piston maintained at a constant temperature. There is an increase in the number of molecules going from left (reactants) to right (products) in this reaction, and therefore the pressure of the mixture increases as reaction takes place, pushing up the piston and increasing the volume of the mixture until equilibrium is reached.

Imagine that the reaction mixture is allowed to reach equilibrium. Suppose we now increase the pressure of the equilibrium mixture by forcing the piston handle inwards. Le Chatelier's principle predicts that the reaction will attempt to resist such changes by shifting the equilibrium composition so that the pressure of the mixture decreases, i.e. so that fewer molecules are formed. Fewer molecules can only be made if the equilibrium concentration of product decreases. This is achieved by the combination of product molecules to form the reactant $H_2SO_4(g)$.

If the reaction involves a reduction in the number of molecules going from left to right, Le Chatelier's principle correctly predicts that increasing the pressure of the mixture will cause the equilibrium concentration of products to increase.

If the reaction involves an equal number of molecules on the reactant and product sides of the equation, altering the total pressure of the equilibrium mixture will not affect the equilibrium concentration of reactants or products.

Finally, imagine that an equilibrium mixture is maintained at constant volume (for example, the reaction is carried out in a sealed steel can) but that we subsequently increase the mixture pressure by injecting in an unreactive gas (such as argon). Because the *partial* pressures of the *reactants* and *products* are unchanged, Le Chatelier's principle successfully predicts that the equilibrium composition will be unaffected.

EXAMPLE **15.2**

Predict how the equilibrium yield of product(s) in the following reactions will be affected by (i) allowing the mixture to expand so that its pressure falls, and (ii) compressing the mixture so that its pressure increases:

$$H_2(g) + I_2(g) \rightleftharpoons 2HI(g) \quad \text{(A)}$$

$$2NH_3(g) \rightleftharpoons 3H_2(g) + N_2(g) \quad \text{(B)}$$

Answer

We take 'equilibrium yield' to mean the concentration of products in the equilibrium mixture. Applying the rules discussed above:

Reaction	Pressure decreases	Pressure increases
(A)	no change	no change
(B)	$[H_2(g)]$ and $[N_2(g)]$ increase	$[H_2(g)]$ and $[N_2(g)]$ decrease

Reaction (A) is unaffected because there is no change in the number of molecules going from left to right.

EXERCISE **15J**

Predicting the effects of changes on the equilibrium composition of a gaseous reaction

Ethene and hydrogen react in the presence of a catalyst, making ethane:

$$\underset{\text{ethene}}{CH_2{=}CH_2(g)} + H_2(g) \xrightleftharpoons{\text{(Ni catalyst)}} \underset{\text{ethane}}{CH_3CH_3(g)} \qquad \Delta H^\ominus = -137\,\text{kJ mol}^{-1}$$

An equilibrium mixture at 400 K was subjected to the following changes in separate experiments:

(i) The mixture was compressed into a smaller volume.

(ii) Hydrogen gas was pumped into the mixture while keeping the volume constant.

(iii) The mixture was maintained at constant volume, and nitrogen gas (which does not react with any of the reactants or products) was pumped into the mixture.

(iv) The mixture was cooled.

How will these changes affect (a) the equilibrium yield of ethane, and (b) the equilibrium constant?

15.5 Production of ammonia by the Haber–Bosch process

A piece of history

Up until 1913, the bulk of the world's artificial nitrogenous fertilizers was made from sodium nitrate that was mined in Chile. Calculations showed that this supply would soon be exhausted if demand continued to expand, and so Fritz Haber (1868–1934) attempted to make ammonia directly from nitrogen and hydrogen gases. Eventually, after eight years, Haber produced ammonia efficiently on a small scale. The brilliant chemical engineer Carl Bosch (1874–1940) enabled the production to be scaled up, and ammonia was first manufactured on a large scale in 1913. The world's production of ammonia now exceeds 180 million tonnes. About 80 per cent of all ammonia produced is used to make fertilizers with another 5 per cent being used to make nylon, and a further 5 per cent used to make explosives.

By the early 1900s, both the UK and Germany were seeking to manufacture nitrogenous fertilizers to increase food production without using naturally occurring nitrates. Sir William Crooke summed up the challenge: 'It is the chemist who must come to the rescue ... It is through the laboratory that starvation may ultimately be turned into plenty'. The Haber–Bosch process was described as achieving 'bread from air' because the nitrogen to make the ammonia was extracted from air. Fritz Haber (photo) received the Nobel Prize for Chemistry in 1918 but the award was controversial as he had led the development and use of poison gases by Germany in the First World War (1914–1918). In 1933 Haber was forced to leave Germany because of his Jewish descent and he died in Switzerland in 1934.

Haber–Bosch process

Ammonia gas is produced on a massive scale using the **Haber–Bosch process**. The reaction is

$$N_2(g) + 3H_2(g) \rightleftharpoons 2NH_3(g) \qquad \Delta H^{\ominus} = -92\,kJ\,mol^{-1}$$

Nitrogen gas is obtained from air, while hydrogen is produced by the reaction of methane with steam (see page 327).

To make the Haber–Bosch process viable we require (i) a high equilibrium concentration of ammonia, and (ii) that the equilibrium concentration of ammonia be produced in a short time (the 'kinetic' factor). In studying this reaction, Haber was aware of the following facts:

- Although K for this equilibrium is a massive $3.5 \times 10^8\ mol^{-2}\ dm^6$ at 298 K, the concentration of ammonia in a mixture of nitrogen and hydrogen gases at room temperature is negligible. This is because the rates of the forward and back reactions are extremely slow so that chemical equilibrium is never attained.
- Operating at higher temperatures increases reaction rate but because the formation of ammonia is exothermic the equilibrium constant falls extremely rapidly with rising temperature. For example, in increasing T from 25°C (298 K) to 527°C (800 K), K falls from 3.5×10^8 to $0.04\ mol^{-2}\ dm^6$.
- Using higher partial pressures of hydrogen and nitrogen increases the equilibrium concentration of ammonia. For example, using 75 atm of $H_2(g)$ and 25 atm of $N_2(g)$ at 500°C (773 K) produces an equilibrium mixture in which 10 per cent of all the molecules present are ammonia. The use of high partial pressures in this way also increases reaction rate, but (alone) is not enough to make the reaction fast enough.

Haber realized that a catalyst was needed to force the reaction to achieve equilibrium faster. In the commercial process, hydrogen and nitrogen in a 3:1 ratio and at 250 atm pressure are reacted in the presence of iron catalyst at a 'compromise temperature' of above 450°C (723 K). Below 450°C the reaction is too slow even with a catalyst. The use of pressures above 250 atm involves the use of very powerful compressor pumps, which are too expensive to run.

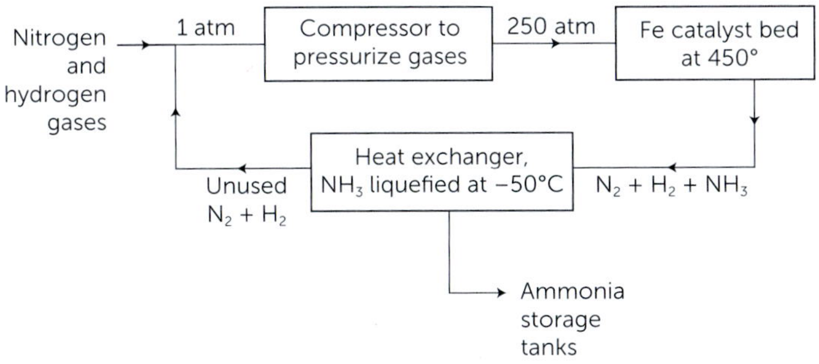

Figure 15.8 The Haber–Bosch process.

Sulfur trioxide, used in the production of sulfuric acid, is produced on a massive industrial scale by the equilibrium reaction of sulfur dioxide with oxygen. The effect of pressure on the yield of sulfur trioxide is calculated in Extension 16.

Figure 15.8 shows the main stages of the commercial process. The reaction of nitrogen and hydrogen takes place in the Fe catalyst bed. The equilibrium mixture (containing ammonia and unused hydrogen and nitrogen) is passed into a heat exchanger that cools down the mixture so that the ammonia (but not the reactants) become liquefied. The reactants are then recirculated to the catalyst bed. Taking the ammonia away by liquefying it allows the unused hydrogen and nitrogen to come back to equilibrium and produce yet more ammonia. The cycle is then repeated.

DID YOU KNOW?

- **The onset of the First World War caused the first Haber–Bosch factory to be diverted to produce explosives instead of fertilizers. By maintaining supplies of explosives the Haber–Bosch process extended the First World War.**
- **The impact of the Haber–Bosch process has been extraordinary, with fertilizers making possible an explosion in world population from about 1.6 billion in 1900 to about 8 billion today. It has been estimated that more than 50 per cent of the nitrogen found in human tissue as protein originates from the Haber–Bosch process.**

EXERCISE **15K**

State two major disadvantages of producing artificial fertilizers from the Haber–Bosch process.

15.6 Heterogeneous equilibria

Equilibria in which the reactants and products are all in the same phase (e.g. all gases) are known as **homogeneous equilibria**. Equilibria which involve substances in different phases are known as **heterogeneous equilibria**.

A simplification occurs with heterogeneous equilibria because the concentrations of pure solids and pure liquids are constant at a fixed temperature (see page 141). For example, the equilibrium expression for the production of carbon monoxide from coke and carbon dioxide,

$$C(s) + CO_2(g) \rightleftharpoons 2CO(g)$$

is

$$K = \frac{[CO(g)]^2}{[CO_2(g)][C(s)]}$$

Since [C(s)] is constant we define a new equilibrium constant:

$$K' = \frac{[CO(g)]^2}{[CO_2(g)]}$$

where $K' = K \times [C(s)]$.

Application of the equilibrium law to physical equilibria

The equilibrium law also applies to physical processes. For example, solid silver chloride is slightly soluble in water

$$AgCl(s) \rightleftharpoons Ag^+(aq) + Cl^-(aq)$$

and we can apply the equilibrium law expression for the dissolution of silver chloride

$$K_{(AgCl)} = \frac{[Ag^+(aq)][Cl^-(aq)]}{[AgCl(s)]} \quad \textbf{(15.8)}$$

However, since the concentration of a pure solid is constant, [AgCl(s)] = constant

$$K_{(AgCl)} \times [AgCl(s)] = [Ag^+(aq)][Cl^-(aq)] = \text{constant}$$

The constant is usually called the 'solubility product' and is symbolized as K_s This relationship may be expressed more simply as:

$$[Ag^+(aq)] \times [Cl^-(aq)] = \text{constant at that temperature}$$

The common ion effect (page 183) may now be explained using equation (15.8). For example, if chloride ions are added to a saturated solution of AgCl, this equation predicts that $[Ag^+(aq)]$ must be reduced in order to keep $[Ag^+(aq)] \times [Cl^-(aq)]$ constant. The reduction is achieved by precipitation of AgCl(s).

EXERCISE 15L

Application of the equilibrium law to vaporization

Show that application of the equilibrium law to the process

$$H_2O(l) \rightleftharpoons H_2O(g)$$

leads to the conclusion that the equilibrium vapour pressure of water is fixed at a particular temperature.

For an in-depth analysis of the $O_2/SO_3/SO_2$ equilibrium see Extension 15 at https://www.bloomsburyonlineresources.com/chemistry

REVISION QUESTIONS

15.1 What is missing from the statement '$K = 3.5 \times 10^4$' for the reaction of hydrogen and bromine gases at 2,040°C?

15.2 The table below shows the variation in the concentrations of $H_2(g)$, $I_2(g)$ and HI(g) with time when the following reaction was carried out at a fixed temperature:

$$H_2(g) + I_2(g) \rightleftharpoons 2HI(g)$$

(Concentrations/mol dm^{-3}) x 10^3				
Time/s	H_2(g)	I_2(g)	HI(g)	Q
0	1.000	1.000	0.000	
50	0.897	0.897	0.206	
100	0.813	0.813	0.374	
200	0.685	0.685	0.630	
300	0.592	0.592	0.817	
400	0.521	0.521	1.040	
500	0.465	0.465	1.069	
600	0.420	0.420	1.160	
700	0.383	0.383	1.234	
800	0.352	0.352	1.296	
850	0.333	0.333	1.334	
900	0.332	0.332	1.336	
950	0.331	0.331	1.338	

(Notice that the concentrations have been multiplied by 1,000. For example, the concentration of HI at t = 300 s is 0.817 x 10^{-3} = 8.17 x 10^{-4} mol dm^{-3}.)

(i) Calculate the value of the expression

$$Q = \frac{[\mathrm{HI(g)}]^2}{[\mathrm{H_2(g)}][\mathrm{I_2(g)}]}$$

at each time. If possible, do this using a suitable computer spreadsheet such as Microsoft Excel. Enter these values in the column labelled Q. (N.B. Q = K only when equilibrium has been reached)

(ii) Plot [H_2(g)] and [HI(g)] against time. Estimate the *initial rate* (a) at which H_2 is used up, and (b) at which HI is produced. Comment upon your answers.

(iii) Plot *Q* versus *t*. After what time is the equilibrium reached? What is the value of the equilibrium constant for the reaction at this temperature?

(iv) Use Figure 15.6 to estimate the temperature at which the experiments were carried out.

15.3

(i) Explain why adding water reduces the equilibrium yield of ester in the reaction

$$C_2H_5OH(l) + CH_3COOH(l) \rightleftharpoons CH_3COOC_2H_5(l) + H_2O(l)$$

(ii) In an experiment, ethanol and ethanoic acid were mixed and allowed to equilibrate at 25°C. The equilibrium concentrations of ethanol, ethanoic acid and ester were found to be 2.67, 1.93 and 4.48 mol dm^{-3}, respectively. Calculate (a) the equilibrium constant for this reaction, and (b) the initial concentrations of ethanol and ethanoic acid.

(iii) The rate constant for the formation of ethyl ethanoate from ethanol at 25°C is approximately 2.4 x 10^{-4} mol^{-1} dm^{-3} s^{-1}. Estimate the rate constant for the reverse reaction.

15.4 The formation of nitrosyl chloride from nitric oxide,

$$2NO(g) + O_2(g) \rightleftharpoons 2NOCl(g)$$

was investigated at a particular temperature. The pressures of gases at equilibrium were p_{NO} = 0.22 bar, p_{O2} = 0.11 bar, and p_{NOC1} = 0.32 bar. Calculate K_p and state its units.

15.5 *K* for the reaction of tin(II) ions with iron(III) ions,

$$Sn^{2+}(aq) + 2Fe^{3+}(aq) \rightleftharpoons Sn^{4+}(aq) + 2Fe^{2+}(aq)$$

is 1.0 x 10^{10} at 298 K. In an experiment at room temperature in which solutions of Sn^{2+}(aq) and Fe^{3+}(aq) were mixed, the equilibrium concentrations of tin ions were found to be

[Sn^{2+}(aq)] = 0.050 mol dm^{-3}
[Sn^{4+}(aq)] = 0.040 mol dm^{-3}

Calculate the concentrations of iron(III) and iron(II) ions at equilibrium.

15.6 The equilibrium constant for the decomposition of phosgene (a First World War poison gas) at 100°C is 2.2 x 10^{-10} mol dm^{-3}:

$$COCl_2(g) \rightleftharpoons CO(g) + Cl_2(g)$$

(Note how toxic the products are!) Phosgene is contained in a metal drum, whose outside surfaces are heated with steam. Would you expect all the phosgene to have decomposed once equilibrium had been achieved?

15.7 The exothermic reaction

$$CO(g) + 2H_2(g) \rightleftharpoons CH_3OH(g)$$

is allowed to reach equilibrium at 1,000 K. How would the yield of methanol and the value of the equilibrium constant be affected by the following individual changes?

(i) Increasing the partial pressure of methanol by injecting some methanol into the mixture.
(ii) Compressing the mixture so that its pressure increases.
(iii) Allowing the mixture to expand so that its pressure decreases.
(iv) Pumping in an inert gas (such as argon) which increases the total pressure without a change in volume.
(v) Lowering the temperature of the equilibrium mixture.

15.8 The formation of brown nitrogen dioxide gas (NO_2) from nitrogen and oxygen, involves two stages:

$$N_2(g) + O_2(g) \rightleftharpoons 2NO(g) \quad (1)$$
$$2NO(g) + O_2(g) \rightleftharpoons 2NO_2(g) \quad (2)$$

The net reaction is

$$N_2(g) + 2O_2(g) \rightleftharpoons 2NO_2(g) \quad (3)$$

15.9 Show that *K*(3) = *K*(1) x *K*(2).

(i) Henry's law (page 186), such as for the dissolution of oxygen in water, oxygen in air $\rightleftharpoons$ oxygen in water
(ii) The distribution ratio (page 184), such as for the solubility of I_2 between two solvents,

$$I_2 \text{ in } H_2O \rightleftharpoons I_2 \text{ in } CCl_4$$

both follow from an application of the equilibrium law.

15.10 Do very fast reactions always possess large equilibrium constants?

16 Gibbs Energy Changes

CONTENTS

OBJECTIVES

This unit:

- Introduces the idea of entropy and Gibbs energy
- Explains that a spontaneous process is a way of achieving a lower Gibbs energy
- Uses standard Gibbs energy changes of reaction to predict whether or not named chemical reactions are 'allowed to go' at that temperature
- Relates standard Gibbs energy changes to $E^{\ominus}$ and K

16.1 Spontaneous processes

In this unit we answer two questions: 'why are some reactions spontaneous?' and 'can we predict the equilibrium constant for a proposed reaction?' This unit builds upon the ideas of enthalpy change (Unit 13) and chemical equilibrium (Unit 15). The chemical reaction by which iron rusts in air, the dissolution of salt in water and the evaporation of ethanol are examples of processes that proceed *spontaneously* at room temperature. In this unit we are more interested in chemical rather than physical changes, although our conclusions apply to both. The following definition of thermodynamically spontaneous is used in this unit:

A spontaneous process is one which, once started, proceeds naturally in a particular direction without help.

If such a process *is happening* as we watch, or if we have evidence that it *has happened* (for example, there has been a change of colour), then the process is – or was – spontaneous. Spontaneous change is about movement in a particular direction and always towards equilibrium. A spontaneous process comes to a halt when equilibrium has been reached. Note that there is no mention of speed in the definition of a spontaneous process. In everyday language, a spontaneous process is one that is rapid and obvious. However, this is not part of the definition used in thermodynamics, and in some cases spontaneous processes proceed so slowly towards equilibrium that no change is visible even after millions of years!

If we state that a process is spontaneous at room temperature, this is equivalent to stating that the reaction is (thermodynamically) 'allowed to go' or that it is 'feasible'.

Examples of spontaneous reactions

If we are asking whether a reaction is *spontaneous* in a particular direction, we must always be clear which process we are talking about, and this is best achieved by writing down an equation. The direction is then from reactants to products.

Consider the reaction of ethanol and ethanoic acid studied in Unit 15:

$$CH_3CO_2H(l) + C_2H_5OH(l) \rightleftharpoons CH_3CO_2C_2H_5(l) + H_2O(l) \quad \textbf{(16.1)}$$

If pure ethanoic acid and ethanol are placed into a test tube, a drop of concentrated sulfuric acid is added as a catalyst and the mixture is warmed to 100°C, ethyl ethanoate ester is produced. We are then justified in stating that reaction (16.1) occurs spontaneously at 100°C. Experimentation shows it is spontaneous over a wide range of temperature.

We may be tempted to state that the back reaction (–16.1):

$$CH_3CO_2C_2H_5(l) + H_2O(l) \rightleftharpoons CH_3CO_2H(l) + C_2H_2OH(l) \quad \textbf{(–16.1)}$$

is non-spontaneous at 100°C, but this may not be true and the reservations about this statement apply to all equilibria. If the system has already reached equilibrium, then it is true that reaction (-16.1) does not occur because there is no net change in reactant or product concentrations, and so experimentally there is no evidence of continuing chemical change. However, if we start with *pure* ester and water only, the reactants will spontaneously react via reaction (–16.1), producing an equilibrium mixture which contains all four substances – ester, water, ethanoic acid and ethanol. (For a fuller discussion of how this equilibrium responds to changes, see Unit 15, page 271). In other words, the spontaneity of a named reaction depends on concentrations as well as temperature.

Our second example is the exothermic chemical reaction between hydrogen and oxygen gases:

$$2H_2(g) + O_2(g) \rightarrow 2H_2O(l) \quad \textbf{(16.2)}$$

This reaction is spontaneous at room temperature in the direction of water being formed, yet a mixture of hydrogen and oxygen gas will show no sign of reaction even if it is left untouched for years. One way to speed up the reaction is to expose the mixture to a flame, which provides the necessary activation energy. Once the reaction is started the flame is no longer needed and the reaction may be explosively fast, but it is important to remember that the reaction of hydrogen and oxygen is regarded as thermodynamically spontaneous, whatever its observed rate, because it proceeds naturally once started.

For more about spontaneous and non-spontaneous reactions see Extension 16 on the website.

16.2 Entropy

It is common experience that energy that is concentrated and localized tends to disperse. A hot poker shares its energy with the surrounding air as it cools; and once detonated, explosives spread out their energy in the form of the fast-moving molecules of gaseous products which then occupy a greater volume. *As a result of any spontaneous process the energy ends up being more dispersed.*

The degree to which energy[1] spreads is measured by a quantity named **entropy**, *S*. We can speak of the entropy of an object (such as a galaxy or a red hot filament) or of a known mass of a material (a pure substance or a mixture).

[1] At constant external pressure, the energy will be the enthalpy, *H*.

The units of entropy are joule per kelvin, abbreviated as J K^{-1}. The entropy of one mole of a substance when that substance is in its standard state is called its **standard molar entropy**, $S^{\ominus}$ and its units are joule per kelvin *per mol*, J K^{-1} mol^{-1}.

The entropy of a solid substance continues to increase as its temperature (T) is raised. Particularly big jumps in entropy occur when the substance melts and when it vaporizes (Figure 16.1). The increase of S with T has two causes. First, the average molecular energy of the substance increases as the temperature is raised. The molecules can now occupy higher (as well as lower) energy levels, and the net effect is that molecular energy is spread out over a wider range of energy levels. The second cause of the increase is the greater volume occupied by collections of molecules, whose motions are more energetic at the higher temperature. Once again, the molecular energy is more spread out, and we can think of the energy being more spread out in space.

Evidence for the spreading of molecular energy in gases upon heating is provided by the *greater spread* of molecular speed at higher temperatures (page 165). Even if a gas expands without being heated (think of air escaping from a balloon) the entropy of a fixed amount of gas still increases because it now occupies more volume.

Entropy is sometimes said to be an indicator of molecular 'disorder', meaning that the molecules are disordered but making no reference to the way that energy is dispersed. Connecting entropy with molecular disorder often allows correct predictions to be made of whether entropy is increasing or decreasing. However, whereas entropy is a precise thermodynamic quantity and is defined mathematically, the idea of disorder is not well defined and it is therefore of limited value.

Some writers go further, taking 'disorder' as being synonymous with entropy. This is misleading. For example, it is sometimes stated that if we take a pack of cards arranged in order (ace of spades, two of spades, three of spades ...) and shuffle them, the new (disordered) collection of cards possesses higher entropy. This is untrue. All of the usual changes associated with entropy change (energy flow, forming new molecules or changes in volume) are absent here. The entropy of a pack of cards is unaffected by simply scrambling them, and an untidy bedroom has the same entropy as a tidy one.

EXERCISE **16A**

(i) Arrange the following states of sulfur trioxide in order of increasing entropy: $SO_3(s)$, $SO_3(g)$, $SO_3(l)$.

(ii) 1 mol of oxygen gas is slowly pressurized in a cylinder. Will the pressurization change the entropy of the sample of oxygen molecules?

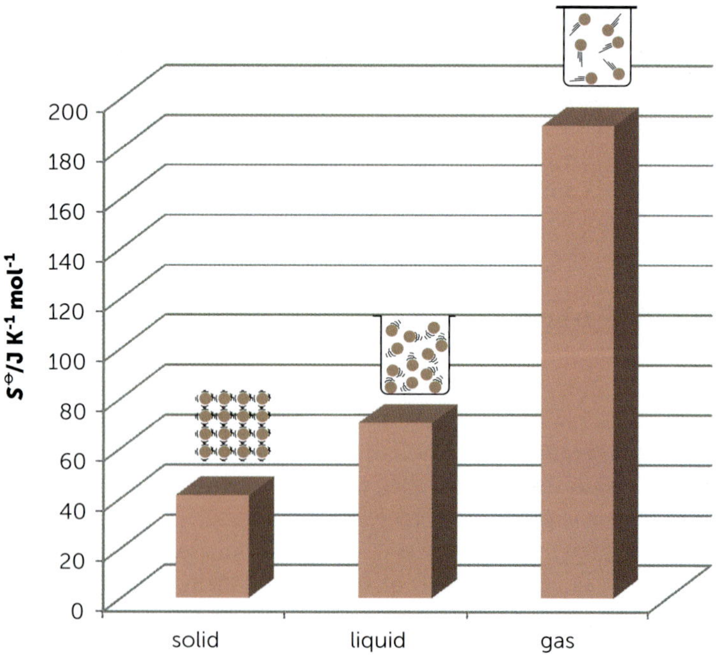

Figure 16.1 The standard molar entropy ($S^{\ominus}$) of water changes as it assumes the three physical states, ice, liquid water and water vapour.

DID YOU KNOW?

A consequence of the Third Law of Thermodynamics, also known as the unattainability principle, is that it is impossible to reach absolute zero (0 K or −273.15°C). In 2016, scientists at National Institute of Standards and Technology (NIST) in Boulder, Colorado, produced a temperature of 0.00036 K, 10,000 times colder than the near vacuum of interstellar space and tantalizingly close to absolute zero.

Entropy Change ΔS

The difference in total entropy accompanying a process is defined as the difference in the total entropy before and after the process:

$$\Delta S_T = S(\text{final})_T - S(\text{initial})_T \qquad (T \text{ for total}) \qquad \textbf{(16.3)}$$

Think of a chemical reaction taking place in a beaker. As a result of reaction, the entropy of the reacting system changes as reactants change into products, but the entropy of everything outside of the reaction system also changes, usually because the reaction either gives out or takes in energy ('heat') from its surroundings. ΔS_T includes the changes in the entropy of the reacting system *and* its surroundings.

Fortunately, we can make things simpler by considering only the entropy change of the reacting system and we will not be concerned with the entropy of the surroundings for the rest of this Unit. For the reaction A + B → C + D where reactants and products are in their standard states[2] the standard entropy change of the reacting system after the reaction goes to completion, $\Delta S^\ominus$, is defined as:

$$\Delta S^\ominus = (S^\ominus_C + S^\ominus_D) - (S^\ominus_A + S^\ominus_B) \qquad \textbf{(16.4)}$$

where $S^\ominus_C$, $S^\ominus_D$, $S^\ominus_A$ and $S^\ominus_B$ are the standard molar entropies of C, D, A and B at temperature T, obtained from NIST tables. $\Delta S^\ominus$ is simply referred to as the standard entropy change of reaction. (It might be helpful to make clear that we are concentrating on the reacting system by including a subscript r, $\Delta S^\ominus_r$, but this is rarely done.) Equation (16.4) follows the 'products minus reactants' format we used previously with enthalpy. If $\Delta S^\ominus$ is positive, the entropy of the reacting system has increased. Following the pattern of enthalpy changes (Unit 13), the entropy change for the reverse reaction at temperature T, C + D → A + B, is equal in size but opposite in sign.

The 'reacting system' or 'reaction' means the collection of molecules of reactants and products at each instant from the start to the end of the chemical or physical change. Everything outside of the reacting system (the surrounding air, the bench, solvent, air, the test tube ...) is termed 'the surroundings'. Although we take this way of theoretically dividing our scientific world for granted today, historically, this was a major intellectual step forward in the development of thermodynamics.

It is useful to be able to predict the *sign* of an entropy change, and to do so we use the idea of energy dispersal. For example, ammonium chloride dissociates upon heating:

$$NH_4Cl(s) \rightarrow NH_3(g) + HCl(g) \qquad \textbf{(16.5)}$$

Two molecules are produced from one unit of $NH_4Cl(s)$, so the energy of the products is dispersed over more particles (and therefore more widely) than was the case before reaction took place. The products are also gaseous (fast-moving molecules take up more volume), so that the dispersion of the energy of the products is even greater. On both counts we expect the entropy of the products to be greater than that of reactants and ΔS for the reacting system is (correctly) predicted to be positive.

[2] See page 218.

EXERCISE **16B**

Rationalizing the sign of $\Delta S^{\ominus}$

(i) The exothermic reaction:

$CuSO_4(s) + 5H_2O(l) \rightarrow CuSO_4.5H_2O(s)$ $\Delta H^{\ominus} = -2{,}279 \text{ kJ mol}^{-1}$ **(16.6)**

White anhydrous copper(II) sulfate; Blue hydrated copper(II) sulfate; $\Delta S^{\ominus} = -158.6 \text{ J K}^{-1} \text{ mol}^{-1}$

is spontaneous at room temperature: the addition of a few drops of water to anhydrous copper(II) sulfate causes it to turn blue and become hot.
Rationalize the negative sign of $\Delta S^{\ominus}$. What is $\Delta S^{\ominus}$ for the reverse reaction, (−16.6)?

(ii) Barium hydroxide reacts spontaneously with ammonium chloride at room temperature, producing ammonia. The reaction is endothermic and the reaction mixture gets very cold.

$Ba(OH)_2{\bullet}8H_2O(s) + 2NH_4Cl(s) \rightarrow 2NH_3(g) + 10H_2O(l) + BaCl_2(s)$ **(16.7)**

$\Delta S^{\ominus} = +591$ J K−1 mol−1 $\Delta H^{\ominus} = +164 \text{ kJ K}^{-1} \text{ mol}^{-1}$

Rationalize the sign of the standard entropy change for this reaction.

(iii) Suggest why the standard entropy change, $\Delta S^{\ominus}$, for the reaction:

$CH_4(g) + 2O_2(g) \rightarrow CO_2(g) + 2H_2O(l)$ $\Delta S^{\ominus} = -243 \text{ J K}^{-1} \text{ mol}^{-1}$ **(16.8)**

is strongly negative. Why is the entropy change for the following reaction so different?

$CH_4(g) + 2O_2(g) \rightarrow CO_2(g) + 2H_2O(g)$ $\Delta S^{\ominus} = -5.2 \text{ J K}^{-1} \text{ mol}^{-1}$ **(16.9)**

16.3 The Gibbs energy change, ΔG

Josiah Willard Gibbs (1839–1903) was a pioneer in the application of thermodynamics to chemistry.

The Gibbs energy,[3] G, of a substance at temperature T and constant external pressure is a combination of its enthalpy and entropy:

$$G = H - TS \tag{16.10}$$

G has the same units as H, i.e. energy per mole (J mol^{-1} or kJ mol^{-1}).

The changes in Gibbs energy (ΔG) of a chemical reaction or physical change can only be calculated if T and P remain constant. Constancy of T and P is commonplace when reactions take place in open vessels where bubbling gases and heat changes have a negligible effect upon the atmospheric pressure or temperature of the laboratory.

ΔG depends upon concentration. We are particularly interested in the change in the total Gibbs energy of the reaction system as the mole per cent of reactant and product (the 'reaction composition') change *en route* to equilibrium. The ΔG between two different chemical compositions 1 and 2 during the reaction is:

$$\Delta G = G_{comp\ 2} - G_{comp\ 1} \tag{16.11}$$

where $Gcomp_{\ 2}$ and $Gcomp_{\ 1}$ are the total Gibbs energy of the reacting system (mixture) at composition 2 and composition 1, respectively.

ΔG for a back reaction is equal but opposite in sign to ΔG for the forward reaction. For example, for the reaction A → B:

$$\Delta G(A \rightarrow B) = -\Delta G(B \rightarrow A) \tag{16.12}$$

[3] Gibbs energy is also known as 'the' Gibbs energy, 'free energy' or as the 'Gibbs free energy'.

16.4 ΔG and spontaneous change

We start with a very important statement about the Gibbs energy of a reacting system at constant temperature and pressure:

- **During a spontaneous reaction, the total Gibbs energy (*G*) continues to fall until it reaches equilibrium, after which G remains constant.**

A simple but universal rule applies:

- **Unless a chemical reaction involves a reduction in Gibbs energy, it will not be spontaneous**, and **it will not occur naturally because it is not thermodynamically allowed to do so** (Figure 16.2).

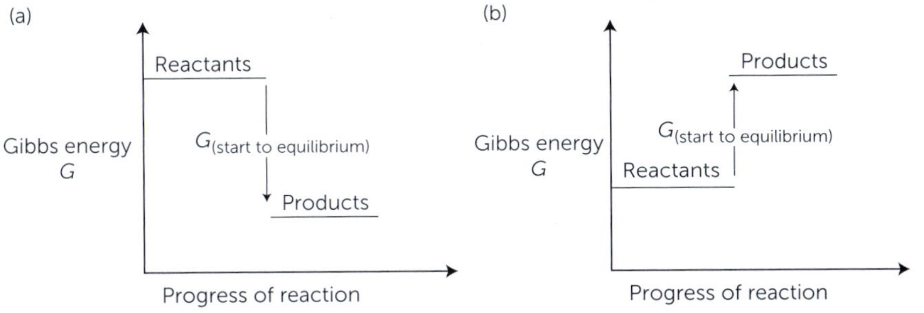

Figure 16.2 Changes in Gibbs energy at temperature *T*. (a) Spontaneous reaction A → Products: the reacting system moves from a higher Gibbs energy to a lower Gibbs energy. (b) In this case, if the reaction *were* to take place, *G* would increase. The reaction A → Products is therefore non-spontaneous and it will not occur of its own accord.

You might ask where these statements about *G* come from. They originate from nature's drive to spread out energy in the most statistically probable way. **An increase in total entropy is a consequence (and not a cause) of spontaneous change.** (see website, Extension 16).

How Gibbs energy, G, changes as a reaction moves towards equilibrium

We can think of a spontaneous process as a 'vehicle' (a mechanism) by which the reacting molecules move to lower Gibbs energy. This is illustrated in Figure 16.3 for the chemical reaction A → B, starting with 1 mol of A. The *y*-axis shows the total Gibbs energy, *G*, of the system (of reactants and products) as the reaction progresses. The *x*-axis shows the percentage mole composition of the reacting mixture.

As A is converted to product B, *G* falls continuously until the reacting mixture reaches point X, the lowest total Gibbs energy possible for any combination of reactant and product at that temperature. **At point X the reacting system has achieved chemical equilibrium.** The higher the percentage of product at equilibrium, the more the equilibrium point X lies to the right on the curve. The particular case shown in Figure 16.3 corresponds to an equilibrium concentration of about 70 per cent B. In this example, the most thermodynamically stable state of the system, the point of lowest *G*, is not achieved by the reaction going to completion, but by moving to a state where only 70 per cent of A is converted to B.

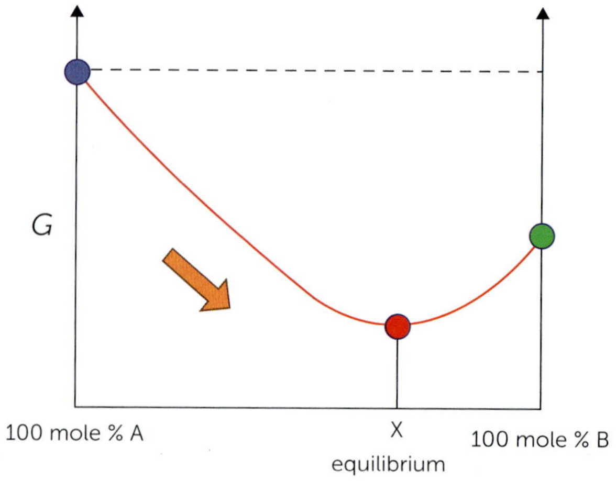

Figure 16.3 Fall in Gibbs energy at constant temperature and pressure for the spontaneous reaction $A \rightarrow B$.

The composition of the reacting mixture is represented by a circle. Initially, the mixture contains only reactant A (blue circle). The reaction takes place spontaneously and the change in composition is represented by sliding the blue circle down the curve in the direction shown by the arrow. *G* falls continuously until the reaction mixture reaches equilibrium, corresponding to the lowest total Gibbs energy possible (the red circle) for a mixture of A and B.

EXERCISE 16C

G and equilibria

Study Figure 16.3.

(i) Would the curve change: (a) if the temperature is changed; or (b) if we used a catalyst to speed up the reaction?

(ii) Suppose that Figure 16.3 applies to the generalized reaction $A \rightarrow B$ in solution and that initially the reaction vessel contains 1.0 mol pure A dissolved in 1 litre of water. Calculate the equilibrium constant, *K*, at that temperature?

(iii) It is found that when the reaction $A \rightarrow B$ is carried out at a low temperature, the equilibrium mixture contains virtually no product. Depict this on a *G* versus reaction composition diagram, using the same axes as in Figure 16.3.

An equilibrium may be reached from both forward and back directions

This is illustrated in Figure 16.4, where we once again look at the equilibrium $A \rightleftharpoons B$ at the same temperature as in Figure 16.3. This time we explore the reaction $B \rightarrow A$ and start with 1 mol of pure B. We start on the right side of the curve. The Gibbs energy of 1 mol of B is above that of X and so the system achieves minimum *G* by 'sliding down the curve' to the equilibrium point X, i.e. by making the reaction $B \rightarrow A$ spontaneous.

We end up at the same equilibrium composition (70 per cent B) as in Figure 16.3, confirming that at the same temperature, the same equilibrium composition may be achieved from both forward and back directions (see also page 271).

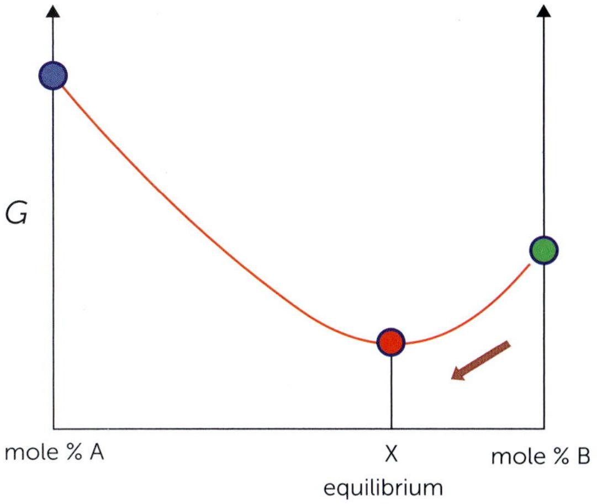

Figure 16.4 Reaching the equilibrium A $\rightleftharpoons$ B starting from pure B.

The reaction in question is the conversion of pure B to A, from right to left on the curve. The arrow shows that the direction of the spontaneous change is from pure B (green circle) to equilibrium at composition X (red circle), toward a lower total Gibbs energy.

EXERCISE **16D**

This question concerns the ethanoic acid/ethanol/ethyl ethanoate/water equilibrium. The addition of fresh ethyl ethanoate to an equilibrium mixture of the four substances causes ethyl ethanoate to spontaneously react with water. Use the idea of Gibbs energy to explain why this happens.

16.5 The standard Gibbs energy change of a reaction $\Delta G^{\ominus}$

The standard Gibbs energy change $\Delta G^{\ominus}$ for the reaction A $\rightarrow$ B is the Gibbs energy change when the reacting system goes to completion (i.e. from 100 per cent A to 100 per cent B), all substances being in their standard states at constant temperature and at a pressure of 1 bar:

$$\Delta G^{\ominus} = G^{\ominus}{}_{\mathrm{B}} - G^{\ominus}{}_{\mathrm{A}} \quad \textbf{(16.13)}$$

where $G^{\ominus}{}_{\mathrm{B}}$ and $G^{\ominus}{}_{\mathrm{A}}$ are the standard Gibbs energy of 1 mol of substances B and A (available from data tables). Since many reactions do not go to completion, $\Delta G^{\ominus}$ is often a purely theoretical quantity. Whereas we can calculate different values of ΔG by selecting different compositions along the curve in Figure 16.3 and subtracting the corresponding values of G, there is only *one* value of $\Delta G^{\ominus}$ for a reaction at that temperature.

For example, the standard Gibbs energy change for the reaction of 2 mol of hydrogen and 1 mol of oxygen gases to make liquid water:

$$\mathbf{2}H_2(g) + O_2(g) \rightarrow \mathbf{2}H_2O(l) \quad \textbf{(16.2)}$$

(in which the molar ratios are 2:1:2) is calculated using the expression:

$$\Delta G^{\ominus} = \mathbf{2}G^{\ominus}H_2O(l) - [\mathbf{2}G^{\ominus}H_{2}(g) + G^{\ominus}O_2(g)] \quad \textbf{(16.14)}$$

where $G^{\ominus}H_2O(l)$ is the standard Gibbs energy of 1 mol of liquid water etc.

16.6 $\Delta G^{\ominus}$ as an indicator of spontaneity

The standard Gibbs energy change, $\Delta G^{\ominus}$, allows a prediction to be made of whether that reaction is spontaneous (allowed to occur) at that temperature. Provided that we start with pure reactants only – so there are no products initially present – the rules are:

- **If $\Delta G^{\ominus}$ is negative, the reaction (starting from pure reactants) will take place spontaneously.**
- **If $\Delta G^{\ominus}$ is positive, the reaction (starting from pure reactants) will be non-spontaneous.**

For example, $\Delta G^{\ominus}$ (298 K) for reaction (16.2) is -513 kJ, showing that the formation of two moles of water from pure hydrogen and oxygen gases is spontaneous. $\Delta G^{\ominus}$ (298 K) for the reverse reaction, in which pure water decomposes into hydrogen and oxygen, is +513 kJ, showing that this is a non-spontaneous reaction at 298 K.

Equation (16.10) shows that Gibbs energy is a composite quantity, made up of enthalpy and entropy components. The same applies to changes in standard Gibbs energy for a reacting system (the reaction). In order to 'dig down' into the enthalpy and entropy changes of the reaction, we use the relationship:

$$\Delta G^{\ominus} = \Delta H^{\ominus} - T\Delta S^{\ominus} \quad \textbf{(16.15)}$$

where $\Delta G^{\ominus}$, $\Delta H^{\ominus}$ and $\Delta S^{\ominus}$ are the Gibbs energy, enthalpy and entropy changes of the reacting system only – not that of the surroundings. The θ symbol shows that all substances are in their standard states and $\Delta G^{\ominus}$, like $\Delta H^{\ominus}$, has the units of kJ mol^{-1}. Here, $\Delta S^{\ominus}$ has the units of kJ K^{-1} mol^{-1} (not J mol^{-1} K^{-1}) in order that we are consistent with the units of energy on both sides of the equation.

16.7 $\Delta G^{\ominus}$ and equilibrium constant K

We will now establish an important link between $\Delta G^{\ominus}$ for a reaction and its equilibrium constant.

The exact relationship between the *standard* Gibbs energy change, $\Delta G^{\ominus}$, of the forward reaction at temperature T:

$$A + B \rightarrow C + D$$

and its equilibrium constant at temperature T is:

$$\Delta G^{\ominus} = -RT \ln K \quad \textbf{(16.17)}$$

where $R = 8.3145$ J mol^{-1} K^{-1}, and ln is short for natural logarithm. Here, $\Delta G^{\ominus}$ *is in units of J mol*$^{-1}$. K is known as the **thermodynamic equilibrium constant** for the reaction of A and B and it is unitless, i.e. it is a number. K is numerically equal to the equilibrium constant in terms of partial pressure, K_p, with the pressure in the units of bar (1 bar = 0.987 atm). All the calculations based on equation (16.17) used in this book have equal numbers of molecules on the left and right of the reaction equation and in that case it can be shown that the thermodynamic equilibrium constant is K is numerically equal to K, the familiar equilibrium constant in terms of molar concentration.

To calculate K from $\Delta G^{\ominus}$ we rearrange (16.17) giving:

$$K = \exp\left(-\frac{\Delta G^{\theta}}{RT}\right) \quad \textbf{(16.18)}$$

where 'exp' means raised to the power of the mathematical constant e. To ensure consistency with R (= 8.3145 J mol^{-1} K^{-1}) in this equation $\Delta G^{\ominus}$ must be in units of J mol^{-1}, not kJ mol^{-1}.

Equation (16.18) shows that the more negative is $\Delta G^{\ominus}$ for the forward reaction **the greater is** K for the same forward reaction. This means that provided we start with pure reactants, a negative value of $\Delta G^{\ominus}$ results in an equilibrium in which there is a greater proportion of products to reactants at that temperature.

Table 16.1 displays the equilibrium constants (calculated using equation (16.18)) for different $\Delta G^{\ominus}$ values at 298 K. When looking at the table, it may be helpful to keep the generalized reaction A + B $\rightleftharpoons$ C + D in mind, although the K values apply to any reaction with the stated $\Delta G^{\ominus}$. If $\Delta G^{\ominus}$ is negative ($\Delta G^{\ominus} < 0$), then $K > 1$ and the equilibrium mixture contains more product than reactant. If we define a large equilibrium constant as one where $K > 10{,}000$, this requires $\Delta G^{\ominus}$ to be more negative than -22.8 kJ mol^{-1}. This corresponds to the reaction going (for most practical purposes) 'to completion', with only a very small concentration of products at equilibrium.

Table 16.1 The variation of equilibrium constants for different values of the standard free energy change of a reaction

$\Delta G^{\ominus}$ (298 K)/kJ mol^{-1}	K (298 K)	K Increasing	
−400	1.3×10^{70}		
−100	3.4×10^{17}		
−50	5.8×10^{8}		reaction allowed to go to completion
−10	57		
−1.0	1.5		
0	1		
+1.0	0.67		
+10	0.018		
+50	1.7×10^{-9}		reaction not allowed to occur to any measurable extent
	3.0×10^{-18}		
+400	7.7×10^{-71}		

EXAMPLE **16.1**

Calculating *K* from $\Delta G^{\ominus}$

Calculate the equilibrium constant in terms of concentration K for the reaction:

$$Zn(s) + Cu^{2+}(aq) \rightarrow Zn^{2+}(aq) + Cu(s) \qquad \Delta G^{\ominus}(298\ K) \approx -200\ kJ\ mol^{-1}$$

Answer

$\Delta G^{\ominus}$ (298 K) in J mol^{-1} = −200,000 J mol^{-1}.

$$(-\frac{\Delta G^{\theta}}{RT}) = -(-200{,}000/(8.3145 \times 298) = 80.72$$

$$K = \exp(-\frac{\Delta G^{\theta}}{RT}) = \exp(80.72) = K(298K) \approx 1.1 \times 10^{35}.$$

Comment

Although in principle all reactions produce an equilibrium of products and reactants, such a huge equilibrium constant confirms that for all practical purposes the reaction between zinc and copper ions can be considered to have gone 'to completion'. The value of Gibbs free energy change is shown to be approximate (hence the ≈ sign) and, because there is a logarithmic relationship between $\Delta G^{\ominus}$ and K, the calculated equilibrium constant is 'very sensitive' to $\Delta G^{\ominus}$, meaning that a small change in $\Delta G^{\ominus}$ makes a big difference to K. For example, making $\Delta G^{\ominus} = -210$ kJ mol^{-1}, a change of only 5 per cent, makes K about sixty times bigger.

EXERCISE **16E**

Adenosine triphosphate is one of the most important energy-providing molecules in the body. Its breaks down into adenosine diphosphate and phosphoric acid in the reaction for which $\Delta G^{\ominus} = -30.5$ kJ mol^{-1} at 298 K:

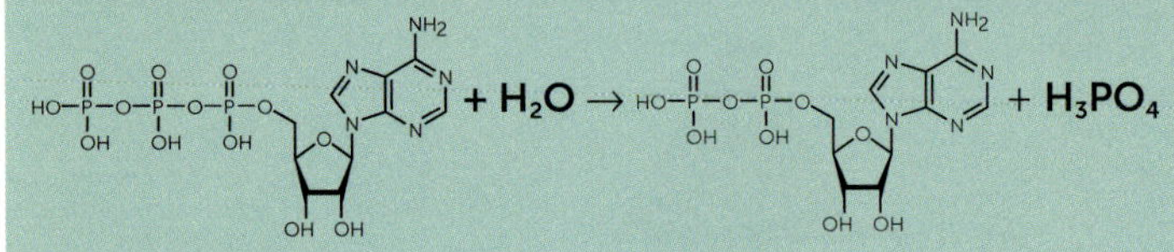

Calculate the equilibrium constant for the reaction at 298 K.

Calculations involving $\Delta G^{\ominus}$

The importance of $\Delta G^{\ominus}$ lies in its value in predicting, thermodynamically, which reactions are 'allowed' (spontaneous) at a certain temperature and which are not. Summarizing, if $\Delta G^{\ominus}$ for a proposed reaction starting with reactants only is negative (i.e. less than 0 kJ mol^{-1}), the equilibrium constant will be greater than one:

If $\Delta G^{\ominus} < 0$, $K > 1$ and the reaction is said to be 'allowed'
If $\Delta G^{\ominus} > 0$, $K < 1$ and the reaction is said to be 'not allowed'

If $\Delta G^{\ominus} >> 0$, the equilibrium constant (and the equilibrium product concentration) may be so small that the reaction is of no practical interest.

Remember that $\Delta G^{\ominus}$ does not allow us to predict how *quickly* a proposed reaction will occur.

To check whether a reaction is 'allowed to go' (is spontaneous) at a temperature T, we calculate the *standard* Gibbs energy change, $\Delta G^{\ominus}$, for the proposed reaction using the equation:

$$\Delta G^{\ominus} = \Delta H^{\ominus} - T\Delta S^{\ominus} \quad \textbf{(16.15)}$$

Inspection of equation (16.15) shows that **$\Delta G^{\ominus}$ will be large and negative if:**

- *$\Delta H^{\ominus}$ is large and negative; and*
- *$\Delta S^{\ominus}$ is large and positive and T is large (so that $-T\Delta S^{\ominus}$ is large and negative).*

$\Delta H^{\ominus}$ and $-T\Delta S^{\ominus}$ may therefore reinforce or oppose each other, depending on the sign of $\Delta H^{\ominus}$ and $\Delta S^{\ominus}$. **Spontaneity is a matter of arithmetic, a battle between the relative sizes and sign of ΔH and $T\Delta S$** and spontaneity is favoured by the reacting mixture losing energy (enthalpy) and also by increasing its entropy.

The size and sign of the $\Delta H^{\ominus}$ and $\Delta S^{\ominus}$ components that make up $\Delta G^{\ominus}$ can often be correlated with changes on a molecular scale. For example, the burning of fuels produces CO_2 molecules that contain very strong C=O bonds, making combustion strongly exothermic with a big and negative $\Delta H^{\ominus}$. This helps makes $\Delta G^{\ominus}$ for combustion reactions large and negative.

We have also seen that reactions that involve the production of smaller molecules (and preferably gases) make ΔS large and positive, and this also helps to make $\Delta G^{\ominus}$ negative unless counteracted by an unfavourably endothermic $\Delta H^{\ominus}$. The presence of the temperature, T, in the $-T\Delta S^{\ominus}$ component in equation (16.15) also has an important consequence. Unlike $\Delta H^{\ominus}$ and $\Delta S^{\ominus}$, which can often be assumed to be approximately independent of temperature if there are no changes in phase, $\Delta G^{\ominus}$ is strongly dependent upon T. In some cases, raising the temperature of reaction causes $\Delta G^{\ominus}$ to change sign and become positive, causing the reaction to move from 'being allowed/spontaneous' to 'not being allowed/non-spontaneous'.

EXAMPLE **16.2**

Calculation of free energy change $\Delta G^{\ominus}$ for a reaction

It is proposed that waste toxic carbon monoxide might be destroyed by converting it to CO_2:

$$2CO(g) + O_2(g) \rightarrow 2CO_2(g) \quad \textbf{(16.16)}$$

Will this reaction be 'allowed to go' at (a) 300 K (b) 1,000 K? You are given the following data:

$\Delta H^{\ominus} = -566$ kJ mol^{-1}; $\Delta S^{\ominus} = -173$ J K^{-1} mol^{-1}

Answer

We start by calculating the standard Gibbs energy change of reaction at (a) 300 K, (b) 1,000 K. We assume that $\Delta H^{\ominus}$ and $\Delta S^{\ominus}$ do not change with temperature. The units of $\Delta S^{\ominus}$ (J K^{-1} mol^{-1}) contain joules whereas the units of $\Delta H^{\ominus}$ are kJ mol^{-1}. To be consistent, we convert $\Delta S^{\ominus}$ to kJ K^{-1} mol^{-1} by dividing by 1,000.

$$\Delta S^{\ominus} = -173 \text{ J K}^{-1}\text{ mol}^{-1} = -0.173 \text{ kJ K}^{-1}\text{ mol}^{-1}.$$

We now use $\Delta G^{\ominus} = \Delta H^{\ominus} - T\Delta S^{\ominus}$

$$\Delta G^{\ominus}(300\text{ K}) = \Delta H^{\ominus} - T\Delta S^{\ominus} = (-566) - (300 \times -0.173) = -514.1 \text{ kJ mol}^{-1}$$

$$\Delta G^{\ominus}(1{,}000\text{ K}) = (-566) - (1{,}000 \times -0.173) = -393 \text{ kJ mol}^{-1}$$

Comment

1. Both $\Delta G^{\ominus}$ values are negative, so that reaction (16.19) is thermodynamically 'allowed to go' at both temperatures.
2. Note that because the reaction is moving in the direction of fewer gas molecules, $\Delta S^{\ominus}$ is negative and so $-T\Delta S^{\ominus}$ is positive. Since $\Delta H^{\ominus}$ is strongly negative, the entropy and enthalpy terms are working against each other, making $\Delta G^{\ominus}$ less negative than it would otherwise be.

EXERCISE **16F**

Calculate $\Delta G^{\ominus}$ at (a) 0°C, (b) 5°C, (c) 100°C for the strongly endothermic reaction:

$$Ba(OH)_2{\bullet}8H_2O(s) + 2NH_4Cl(s) \rightarrow 2NH_3(g) + 10H_2O(l) + BaCl_2(s) \quad \textbf{(16.7)}$$

for which $\Delta H^{\ominus} = +164$ kJ mol^{-1} and $\Delta S^{\ominus} = +591$ J K^{-1} mol^{-1}
Is reaction (16.7) 'allowed to go' at the three temperatures?

Strongly exothermic reactions are often spontaneous over a wide range of temperatures

Strongly exothermic reactions are often thermodynamically allowed over a wide range of temperatures because it is the enthalpy term in equation (16.15) that often has the biggest impact on the size and sign of ΔG°. Even if ΔS° is unfavourably negative, it is often only at high temperatures that the $-T\Delta S^{\circ}$ term offsets a negative ΔH° to produce an unfavourably positive ΔG°. For example, in the case of reaction (16.16), it is only above 3271 K that ΔG° would become positive, causing the *reverse* reaction ($2CO_2(g) \rightarrow 2CO(g) + O_2(g)$) to have an equilibrium constant greater than one. In this example, such a high temperature is not only unattainable in most laboratories, but it would also allow alternative reactions to compete, including those in which molecules directly dissociate into their atoms.

BOX **16.1**

Protein folding

Proteins consist of long chains of amino acids that reside in the water-based cells of living organisms. Their 3-D shape (a 'protein bundle') is essential for them to do their job. But why do proteins fold up in the first place? When the protein folds, there is little or no space for the water molecules that were previously attracted to the polar groups of the unfolded protein. The water molecules are displaced and dispersed into the greater body of the surrounding water. Accordingly, we can think of the folding as a process that is similar to a chemical reaction, with a reactant (unfolded protein molecules that are attached to a number of water molecules – call the number *n*) and two products (an optimally folded protein and excluded water molecules dispersed in solution):

(A) Unfolded protein....$nH_2O \rightleftharpoons$ Folded protein + nH_2O

In general, it is believed that protein folding is only marginally favourable, i.e. ΔG° for reaction (A) is only slightly negative. Nevertheless, folding *is* spontaneous and folding continues until an optimum degree of folding has been achieved. The critical importance of the (albeit small) preference of proteins to assume the folded state may be simply stated. Without the highly organized 3-D structure of proteins, life as we know it would be impossible and you would not be reading this box! Several diseases, including cystic fibrosis, Alzheimer's, Huntington's and Parkinson's arise from defects in protein folding.

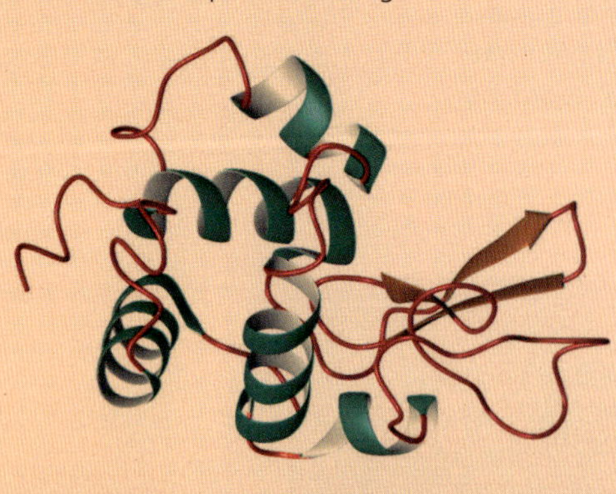

A diagram of ribbon-like coiled structure of the protein enzyme lysozyme, which is part of our immune system.

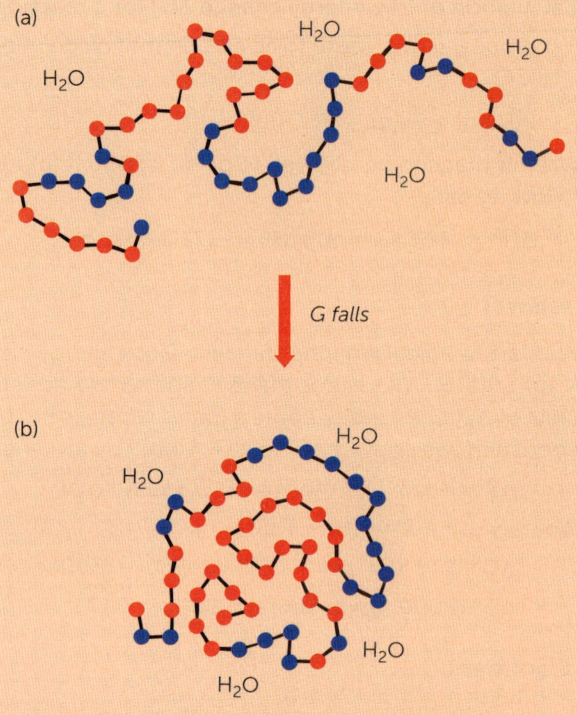

Figure 16.5 The folding of a protein in an aqueous environment. Since the number of possible conformations of a twisted protein is vast, getting the right folding to allow the protein to do its job within a cell cannot be a random selection. Instead, the geometry and chemical nature of the amino acids making up the protein chain forces the folding to occur in a definite and unique way. The Gibbs energy (*G*) of the combined protein falls marginally when the protein is folded. The non-polar amino acids (red circles) congregate in the centre of the fold.

Extension material is available on the website at https://www.bloomsburyonlineresources.com/chemistry.

16.8 Gibbs energy and stability

We can define relative (thermodynamic) stability using $\Delta G^\ominus$. If we state that in a reaction the products are collectively 'thermodynamically more stable' or 'thermodynamically preferred' than the reactants, this simply means that the products have a lower Gibbs energy than the reactants. $\Delta G^\ominus$ provides a thermodynamic reason why the atoms in molecules rearrange themselves in the way that they do. For example, we have explained the fact that the water molecule is bent in shape (and not linear) because of the repulsion of electron pairs (see page 68) but *thermodynamically* the justification is that a bent water molecule has a lower Gibbs energy than a linear one, i.e. it has a more negative $\Delta G^\ominus$ of formation. Similarly, the delocalization within the carbonate ion (see page 60) or benzene (page 338) can be rationalized by stating that the resulting delocalized species has a lower Gibbs energy than the non-delocalized structures.

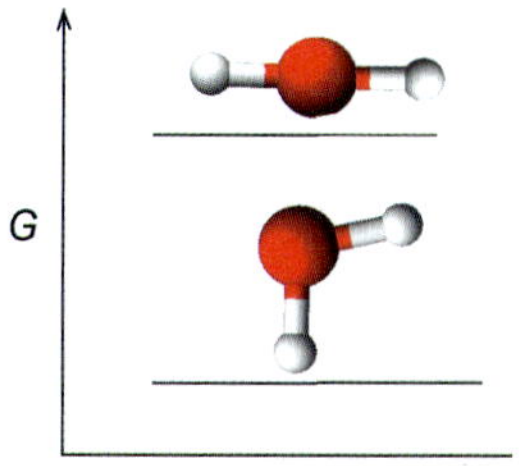

Unconfined water molecules are always bent in shape, because a mole of water molecules that are bent has a lower standard Gibbs energy at 298 K than a mole of water molecules that are linear.

16.9 Standard electrode potentials, $\Delta G^\ominus$ and K

In the unit on reduction and oxidation (page 115), we showed how *differences* in standard electrode potentials, $E^\ominus$, may be used to predict whether an ionic reaction (starting from pure reactants) is allowed to occur – although we cannot predict how fast the reaction would proceed. By 'occur' we mean a large equilibrium constant. We should not be surprised that the difference in $E^\ominus$ for the proposed pair of half-reactions and $\Delta G^\ominus$ for the net redox reaction are related. Indeed, this is necessary or else they might make different predictions! It turns out that the bigger the difference in $E^\ominus$, the more negative is $\Delta G^\ominus$ for the reaction under consideration. This is discussed further in Extension 16 on the website.

16.10 Thermodynamically and kinetically controlled reactions

Whether we are in the laboratory or scaling up a chemical reaction on an industrial scale, a good percentage yield of product is always desirable. A good yield of product depends on favourable thermodynamics and kinetics. By thermodynamics we mean a large equilibrium constant. By kinetics we mean the speed at which equilibrium is achieved. If K is small then the rate of reaction is generally irrelevant because the yield will be small anyway[4] at the temperature being used, but even when K *is* large, a slow reaction may not be commercially viable.

In Unit 2, s.2.5, we introduced the idea of competing reactions – giving an example where two possible products are produced from the same reactants. It is necessary that both competing pathways are thermodynamically spontaneous ($\Delta G^\ominus$ negative and so $K > 1$) so that both products are theoretically possible under set conditions. The equilibrium constants for the two possible pathways will not be equal and so we might assume that the product produced by the pathway with the biggest K will dominate: that product is said to be 'thermodynamically preferred'. The extra dimension introduced in this section is that the product that dominates the product mixture is decided by both

[4] There is an exception to this rule. It is possible to produce good yields even if K is low, by physically removing the product from the reaction mixture, as in the Haber–Bosch process (page 279) when ammonia is liquefied, or in some organic chemistry preparations when the product is volatile and is distilled away. In these cases, since the product is being removed from the reaction mixture, the flask re-establishes equilibria and so produces more product. The cycle is repeated and the product samples collected.

thermodynamic and kinetic factors and it is possible to use a lower temperature to produce more of a less thermodynamically stable product.

Our example is a type of organic reaction known as an addition reaction (see page 368) in which the H and Br atoms of hydrogen bromide (HBr) may become bonded to 1,3-butadiene in two ways:

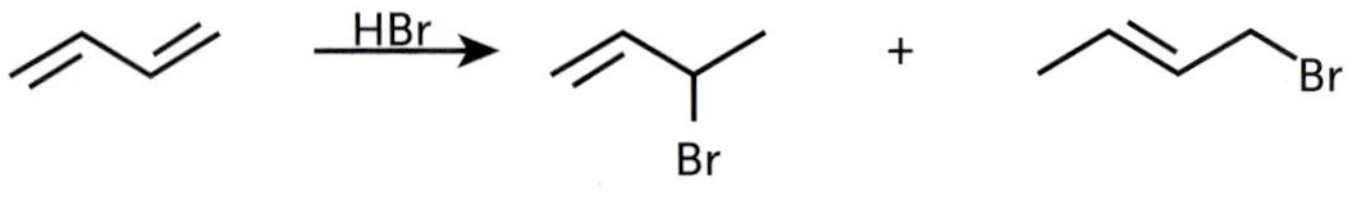

A B C

Specifically, a mixture (labelled **A**) of 1,3-butadiene and hydrogen bromide may produce one of two isomers: either **B** (3-bromo-1-butene) or **C** (1-bromo-2-butene). To make the molecular structures clearer, we may depict the reaction using 'ball-and-stick' diagrams in which the bromine atoms are coloured red:

A B C

The formation of **C** involves a more negative $\Delta G^{\ominus}$ over a wide range of temperature (this makes **C** more thermodynamically stable than **B**) and provided that equilibrium is reached, **C** is expected to dominate the product mixture. This is the case at room temperature. The thermodynamics of the reaction are depicted in Figure 16.6.

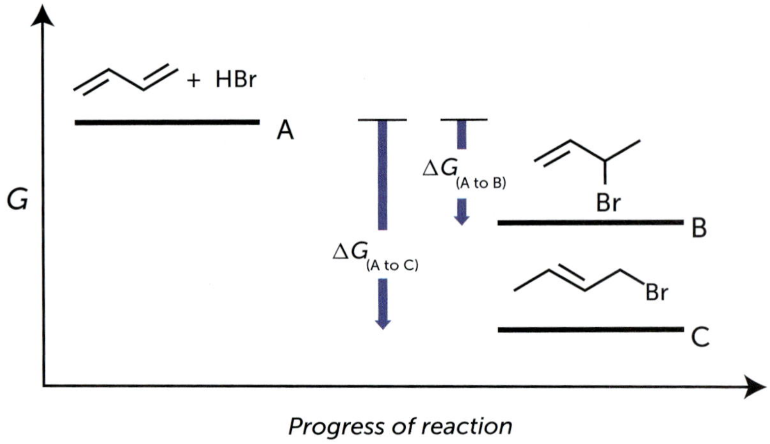

Figure 16.6 The changes in Gibbs energy for two possible products (**B** and **C**) from the reaction mixture (labelled **A**) of 1,3-butadiene and HBr.

At lower temperature things are different. Reaction **A** → **B** has a lower activation energy than that of reaction **A** → **C** and the effect of this difference on the relative reaction rates becomes more marked at low temperatures when there is less energy available to 'push' the reactant molecules over the energy barrier caused by the activation energy requirement. At −10°C the **A** → **C** reaction is much slower than **A** → **B** and equilibrium is not achieved over the timescales of the experiment, so even though **C** is thermodynamically preferred, the product mixture contains mostly **B** ('high yield'). If the mixture at -10°C is allowed to warm up to room temperature, the product mixture changes: equilibrium is slowly reached and the equilibrium mixture is dominated by **C**.

We summarize this as follows. At the low temperature, the reaction of 1,3-butadiene with HBr is **kinetically controlled**, but at higher temperature it is **thermodynamically controlled**.

EXERCISE **16G**

When cyclopentadiene and furan react, the resulting mixture contains two isomers, A and B:

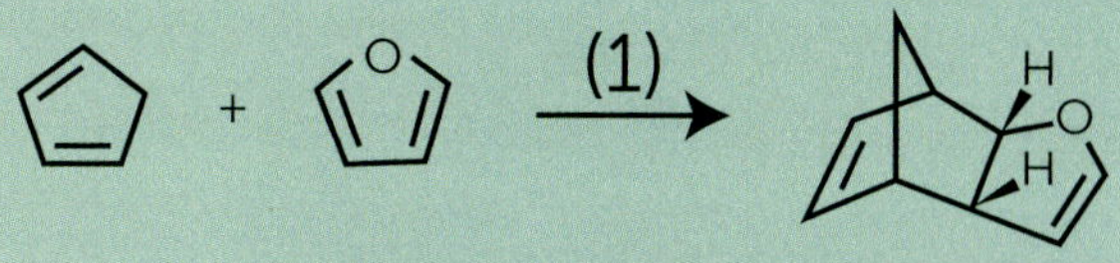

A

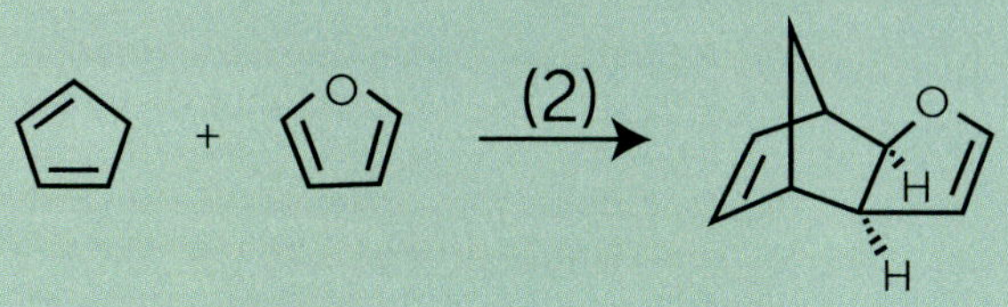

B

The isomers differ in the position of the hydrogen atoms, either pointing upwards (as in **A**) or downwards (**B**). $\Delta G^{\ominus}$ for reaction (2) is more negative than for reaction (1) and this is true over a wide range of temperature. Use this information to suggest an explanation for the following:

(a) The product mixture contains more A than B at 25°C, but at 80°C, B dominates.

(b) The mixture at 25°C very slowly converts to a mixture dominated by B.

REVISION QUESTIONS

16.1 Which of the following is true?

(a) A spontaneous process is one that is thermodynamically allowed and which, once started, proceeds naturally at a detectable rate.

(b) The units of standard molar entropy are J mol^{-1}.

(c) *R*, the universal gas constant, has the same units as molar entropy.

(d) A catalyst does not affect the ΔG of a reaction.

(e) For a reaction to be spontaneous *and* to proceed at an observable rate, ΔG(reaction) must be negative and the reactants must have enough energy to overcome the activation energy for the reaction.

16.2 The molar entropy of straight-chained alcohols is listed below. Plot the number of carbon atoms against $S^{\ominus}$ and comment on the graph.

Substance	State	Formula	Standard molar entropy $S^{\ominus}$/J K^{-1} mol^{-1}
Methanol	Liquid	$CH_3OH(l)$	126.8
Ethanol	Liquid	$CH_3CH_2OH(l)$	160.7
1-Propanol	Liquid	$CH_3(CH_2)_2OH(l)$	193.6
2-Propanol	Liquid	$CH_3(CH_2OH)CH_3(l)$	181.1
1-Butanol	Liquid	$CH_3(CH_2)_3OH(l)$	225.8
1-Hexanol	Liquid	$CH_3(CH_2)_5OH(l)$	287.4

16.3 Calculate the standard entropy change, $\Delta S^{\ominus}$, of the reaction:

$$\bullet CH_3(g) + \bullet H(g) \rightarrow CH_4(g)$$

given the following $S^{\ominus}$ data in J K^{-1} mol^{-1}: $S^{\ominus}(CH_4(g)) = 186.13$; $S^{\ominus}(\bullet CH_3(g)) = 194.2$; $S^{\ominus}(H\bullet(g)) = 114.7$. Comment on the sign of $\Delta S^{\ominus}$. What is $\Delta S^{\ominus}$ for the reverse reaction?

16.4

(i) Look back to the 'cold pack' of Box 13.1 (p. 218), which is based upon the endothermic dissolution of ammonium nitrate. Predict the sign of $\Delta S^{\ominus}$. Why is the dissolution process 'allowed to go'?

(ii) Explain why the decomposition of silver chlorite:

$$AgClO_2(s) \rightarrow Ag(s) + \tfrac{1}{2}Cl_2(g) + O_2(g)$$

'goes' at room temperature, even though ΔH is approximately zero.

16.5 The thermodynamic data for the reaction in which two nitric oxide molecules decompose:

$$2NO(g) \rightarrow N_2(g) + O_2(g)$$

is as follows: $\Delta H^{\ominus} = -180.58$ kJ mol^{-1}; $\Delta S^{\ominus} = +0.020$ kJ mol^{-1}

Calculate *K* at (i) 298 K (ii) 600 K. Is the dissociation of NO thermodynamically allowed at both temperatures?

16.6 (i) Calculate $\Delta G^{\ominus}$ for the following reaction at (i) 300 K and (ii) 600 K:

$2H_2(g) + O_2(g) \rightarrow 2H_2O(l)$ (16.3)

given that $\Delta H^\circ(16.3) = -483.64$ kJ mol^{-1} and $\Delta S^\circ(6.3) = -96$ J K^{-1} mol^{-1}. Is the reaction 'allowed to go' at these temperatures? What is ΔG° for the back reaction, (–16.3), at 300 K and 600 K? Calculate the equilibrium constant, K, at both temperatures.

(ii) What effect does increasing T have on K? What is predicted by Le Chatelier's principle?

16.7 Use the following data to calculate the standard free energy change at 298 K for the reaction of two hydrogen atoms to make $H_2(g)$:

$\bullet H(g) + \bullet H(g) \rightarrow H_2(g)$

$\Delta H^\circ = -435.94$ kJ mol^{-1} and $\Delta S^\circ = -0.0987$ kJ mol^{-1}

Comment on your result.

16.8 It is proposed that ethanol could be manufactured from the reaction of ethene with steam:

$CH_2{=}CH_2(g) + H_2O(g) \rightarrow CH_3CH_2OH(g)$

(i) Given the following data, calculate the temperature at which $\Delta G^\circ = -1.0$ kJ mol^{-1}. Does K increase or decrease with increasing temperature? You may assume that ΔH° and ΔS° do not change with temperature.

(ii) Increasing T is a trade-off because, although the reaction rate increases, K becomes smaller. How could you produce a greater yield without increasing the temperature?

Substance	$CH_2{=}CH_2(g)$	$H_2O(g)$	$CH_3CH_2OH(g)$
$\Delta_f H^\circ$ (298 K)/kJ mol^{-1}	52.3	–242.8	–234.0
S° (298 K)/J mol^{-1}	219.5	188.8	282.6

16.9 Two experiments are carried out on two different spontaneous reactions at room temperature, starting with 1 mol of pure reactant (J and M respectively):

J → K   M → N

The following diagrams show the variation in Gibbs energy with reaction mixture composition as both reactions proceed toward equilibrium.

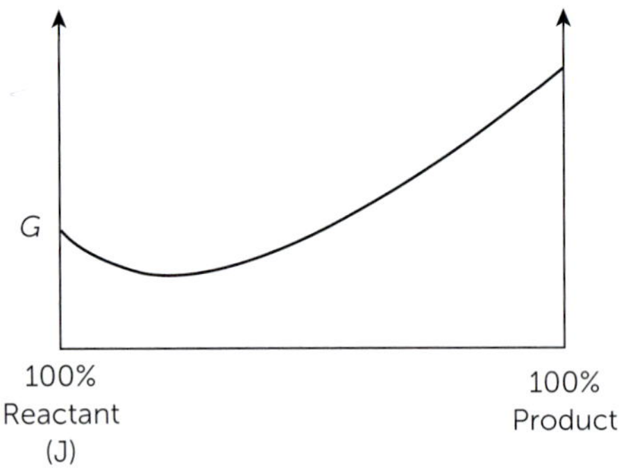

J → K

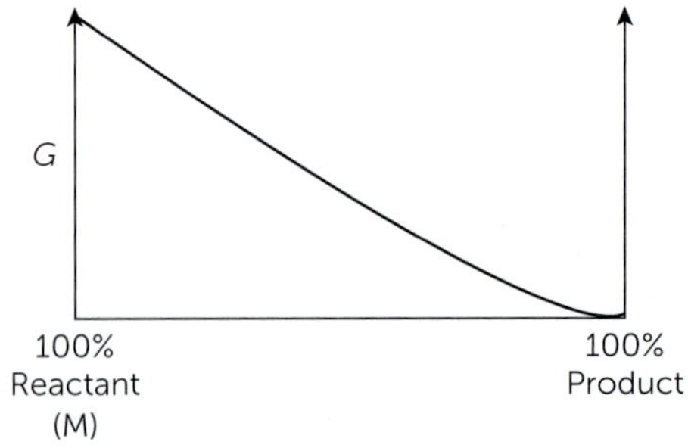

M → N

For both cases:

(i) Mark the composition of the equilibrium mixture. What can you state about the Gibbs energy, G, at equilibrium?

(ii) From the graphs, estimate the composition of the equilibrium mixture for both reactions.

(iii) Reaction M→ N is exothermic. How would the equilibrium constant K for M → N change with increasing temperature?

16.10 Lysozyme is an antimicrobial enzyme that is part of the body's immune system. For the folding of lysozyme:

unfolded lysozyme → folded lysozyme $\Delta G^\circ(298\ K) = -62$ kJ mol^{-1}

Calculate the equilibrium constant at 298 K. Assume that lysozyme contains 100 amino acid 'residues'. What is $\Delta G^\circ(298\ K)$ per residue? What is $\Delta G^\circ(298\ K)$ for the unfolding of lysozyme?

16.11 In Unit 15 (page 266) we showed that the equilibrium constants for forward (f) and back reactions (b) at the same temperature are related as follows $K_b = 1/K_f$.

(i) Show that this equation may be derived from equation (16.18).

(ii) Data for the reaction in which hydrogen iodide (HI) is formed is shown below:

$H_2(g) + I_2(g) \rightleftharpoons 2HI(g)$ $K(764\ K) = 45.6$

$\Delta G^\circ(764\ K) = -24.26$ kJ mol^{-1}

Calculate $K(764\ K)$ and $\Delta G^\circ(764\ K)$ for the reverse reaction:

$2HI(g) \rightleftharpoons H_2(g) + I_2(g)$

16.12 Consider the reaction of hydrogen and hydroxide ions in water:

$H^+(aq) + OH^-(aq) \rightarrow H_2O\ (l)$

ΔG° (298 K) = –80.1 kJ per mole of water formed

Is it likely that water molecules will decompose into $H^+(aq)$ and $OH^-(aq)$ ions at room temperature?

16.13 Are there more exothermic reactions in nature than endothermic reactions?

Acid–Base Equilibria

17

OBJECTIVES

This unit:

- Demonstrates the use of the ionic product constant of water in calculations
- Defines pH, pOH, pK_a and pK_b
- Discusses salt hydrolysis
- Looks at calculations involving weak acids and bases
- Explains the use of buffers and acid–base indicators

CONTENTS

17.1 Ionic equilibria in water

Ionization of water

Pure water is a very poor conductor of electricity. This shows that there are virtually no ions present. However, there are some, and these play a very important role in the equilibria of aqueous solutions.

At room temperature about one water molecule in every 10,000,000 is ionized:

$$H_2O(l) + H_2O(l) \rightleftharpoons H_3O^+(aq) + OH^-(aq) \qquad \textbf{(17.1)}$$

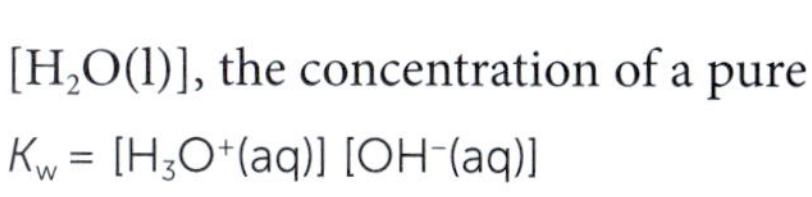

The self-ionization of water

In this reaction, a proton is being transferred from one water molecule to another. This is called **self-ionization** and the equilibrium expression in terms of concentration is:

$$K = \frac{[H_3O^+(aq)]\,[OH^-(aq)]}{[H_2O(l)]^2} \qquad \textit{at equilibrium}$$

But $[H_2O(l)]$, the concentration of a pure liquid, is constant (page 141). This gives

$$K_w = [H_3O^+(aq)]\,[OH^-(aq)] \qquad \textbf{(17.2)}$$

where K_w, which equals $K \times [H_2O(l)]^2$, is known as the **ionic product constant** (or **autoionization constant**) of water.

We can express equation (17.2) in words:

When the concentrations of hydroxide and hydronium ions present in water are multiplied together, the product is fixed at that temperature.

EXERCISE **17A**

Variation of K_w with temperature

The data in Table 17.1 suggest that the dissociation of water is endothermic – why?

Experiments show that equation (17.2) applies to water and also to *all* aqueous solutions, such as sodium chloride solution, hydrochloric acid and sodium hydroxide solution.

Note that reaction (17.1) is sometimes expressed in a simplified form as:

$$H_2O(l) \rightleftharpoons H^+(aq) + OH^-(aq)$$

$H^+(aq)$ and $H_3O^+(aq)$ are often used interchangeably.

Sophisticated kinetic experimental measurements have shown that reactions in which protons are lost or gained by molecules in solution are extremely fast. This means that we may safely assume that all the reactions discussed in this unit have already reached equilibrium.

Table 17.1 shows the values of K_w at various temperatures.

Table 17.1 The Ionic product constant of water at various temperatures

Temperature/°C	$10^{14} \times K_w$/mol² dm⁻⁶
0	0.114
10	0.293
20	0.681
25	1.008
30	1.471
40	2.916
50	5.476
100	51.3

From Table 17.1, at 25°C

$$K_w \approx 1.0 \times 10^{-14}\ \text{mol}^2\ \text{dm}^{-6}$$

(this is easy to remember) and we can now say that:

When the concentrations of hydroxide and hydronium ions in water or in aqueous solutions are multiplied together, the product is always 1.0×10^{-14} mol² dm⁻⁶ at 25°C.

Equation (17.1) shows that two water molecules produce one hydronium and one hydroxide ion; therefore, the concentration of these ions in water is equal. A solution containing equal numbers of each ion is said to be **neutral**. At 25°C,

$$1.0 \times 10^{-14} = [H_3O^+(aq)]\,[OH^-(aq)]$$

Therefore,

$$[H_3O^+(aq)] = [OH^-(aq)] = \mathit{1.0 \times 10^{-7}\ mol\ dm^{-3}}$$

Calculations of pH were introduced in Unit 9.

$$pH = -\log\,[H^+_{(aq)}],\ \text{or}\ pH = -\log\,[H_3O^+_{(aq)}]$$

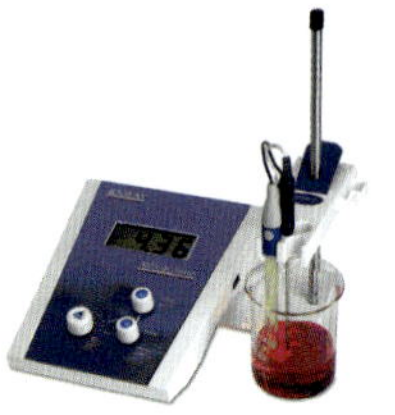

pH measurements require a meter, a measuring probe called a glass electrode, and a reference electrode. These are often packed together in a compact case. pH meters are used to provide rapid and accurate measurements of the pH of soils, of waste waters and of swimming pools.

We can now calculate the pH of pure water (water which contains no dissolved solids, liquids or gases):

$$\text{pH of pure water} = -\log\,(1.0 \times 10^{-7}) = 7.00$$

(Pure water is rarely encountered. Even distilled or deionized water contains dissolved carbon dioxide which gives it a pH of about 6. The pH of river water or lake water may be above 7 if it contains dissolved limestone. Acid rain has a pH as low as 4.)

Definitions of neutral, acidic and basic solutions

1 Neutral solution

A **neutral solution** is one where the concentrations of hydroxide and hydronium ions are equal. *At 25°C, the pH of a neutral solution is 7.*

2 Acidic solution

An **acidic solution** is one where the concentration of hydronium ions is greater than the concentration of hydroxide ions. *At 25°C, the pH of an acidic solution is less than 7.*

3 Basic solution

A basic solution (or **alkaline solution**) is one where the concentration of hydroxide ions is greater than the concentration of hydronium ions. *At 25°C, the pH of a basic solution is greater than 7.*

Calculations using the ionic product constant of water

The ionic product constant of water (K_w) may be used to calculate the hydroxide ion concentration in solutions of acids. It may also be used to calculate the hydronium ion concentration in solutions of bases.

EXAMPLE **17.1**

A 0.456 mol dm^{-3} solution of hydrochloric acid (HCl(aq)) contains 0.456 mol dm^{-3} of hydronium ions. Calculate the pH of the solution of acid. What is the concentration of hydroxide ion in the acid? (T = 25°C)

Answer

First, calculate the pH:

$$\text{pH} = -\log(0.456) = -(0.341) = 0.341$$

Then, calculate the hydroxide ion concentration from K_w:

$$[H_3O^+(aq)] \times [OH^-(aq)] = 1.0 \times 10^{-14}\ \text{mol}^2\ \text{dm}^{-6}$$

or

$$[OH^-(aq)] = \frac{1.0\times10^{-14}}{[H_3O^+(aq)]} = \frac{1.0\times10^{-14}}{0.456} = 2.3 \times 10^{-14}$$

EXAMPLE **17.2**

A 0.020 mol dm^{-3} solution of sodium hydroxide at 25°C contains 0.020 mol dm^{-3} of hydroxide ion. Calculate its pH.

Answer

First, calculate the hydronium ion concentration:

$[OH^-(aq)] = 0.020\ \text{mol dm}^{-3}$. We know that
$[H_3O^+(aq)] \times [OH^-(aq)] = 1.0 \times 10^{-14}\ \text{mol}^2\ \text{dm}^{-3}$

Rearranging,

$$[H_3O^+(aq)] = \frac{1.0\times10^{-14}}{[OH^-(aq)]} = \frac{1.0\times10^{-14}}{0.020} = 5.0\times10^{-13}\ \text{mol}^2\ \text{dm}^{-3}$$

To calculate the pH:

$$\text{pH} = -\log[H_3O^+(aq)] = -\log(5.0 \times 10^{-13}) = 12.30$$

Comment

The pH of **strongly basic solutions** is above 12, so this is a reasonable answer. Now try Exercise 17C.

EXERCISE **17B**

Calculations involving solutions of acids

(i) What is (a) the hydronium ion concentration and (b) the hydroxide ion concentration in a 0.10 mol dm^{-3} solution of nitric acid (HNO_3)? What is the pH of this solution? (Nitric acid is a strong acid and therefore completely dissociated into ions in solution.)

(ii) A solution of hydrochloric acid (a strong acid) has a pH of 2.20. What is the concentration of hydronium ions in the solution?

EXERCISE **17C**

Calculations involving solutions of alkalis

(i) 0.10 g of calcium hydroxide ($Ca(OH)_2$) is dissolved in water and made up to 200 cm^3. Calculate the pH of the solution at 25°C. (Calcium hydroxide is a 'strong base', i.e. it is completely dissociated into hydroxide and calcium ions in solution.)

(ii) A solution of sodium hydroxide (also a strong base) has a pH of 11.10. What is the concentration of (a) hydroxide ion and (b) hydronium ion in the solution? (Take the temperature to be 25°C.)

pH, pOH and pK_w

Taking logs on both sides of equation (17.2) gives

$$\log K_w = \log [H_3O^+(aq)] + \log [OH^-(aq)]$$

Multiplying throughout by –1 gives

$$-\log K_w = -\log [H_3O^+(aq)] -\log [OH^-(aq)]$$

We know that pH = –log [H_3O^+(aq)]. We similarly define pOH and pK_w as

$$pOH = -\log[OH^-(aq)] \text{ and } pK_w = -\log K_w$$

from which we see that

$$pK_w = pH + pOH$$

This expression is just another form of equation (17.2), and also applies to all aqueous solutions. At 25°C, pK_w = –log (1.0 × 10^{-14}) = 14, so that

$$pH + pOH = 14$$

Now try Exercises 17D and 17E.

EXERCISE **17D**

Relationship between pH and pOH

The pH of a solution is 5.80 at 25°C. Calculate its pOH and [OH^-(aq)].

EXERCISE **17E**

pH and pOH scale

The following table shows the hydronium and hydroxide ion concentrations (in mol dm^{-3}) for 14 aqueous solutions. Complete the table: the first entry is complete.

[H_3O^+(aq)]	[OH^-(aq)]	pH	pOH
10^{-14}	1	14	0
10^{-13}			
10^{-12}			
10^{-11}			

$[H_3O^+(aq)]$	$[OH^-(aq)]$	pH	pOH
10^{-10}			
10^{-9}			
10^{-8}			
10^{-7}			
10^{-6}			
10^{-5}			
10^{-4}			
10^{-3}			
10^{-2}			
10^{-1}			

17.2 Acids and bases in aqueous solution

Strong acids

Hydrochloric acid is made by dissolving hydrogen chloride gas in water. The dissolved HCl (symbolized HCl(aq)) then reacts with water as follows:

$$HCl(aq) + H_2O(l) \rightarrow H_3O^+(aq) + Cl^-(aq)$$

Hydrochloric acid consists almost entirely of hydronium and chloride ions and the concentration of covalent acid molecules is negligibly small. For this reason hydrochloric acid is said to be a **strong acid**:

a strong acid is completely ionized in solution

This means that if 0.1 mol of HCl(g) is dissolved in water and made up to 1 dm^3, the concentration of both H_3O^+ and Cl^- ions is also 0.1 mol dm^{-3}.

There are only a few strong acids. Apart from HCl(aq), the list includes:

- perchloric acid ($HClO_4$(aq)), and cheerfully referred to as the 'strongest known acid' in TV quizzes;
- nitric acid (HNO_3(aq)), whose ancient name, *aqua fortis*, means 'strong water';
- hydrobromic acid (HBr(aq)) and hydriodic acid (HI(aq)).

The case of sulfuric acid is discussed on page 307.

The term *strong acid* has nothing to do with the corrosive properties of an acid. It is also important to distinguish between acid strength (strong acid, weak acid) and the acidity of a solution. The *strength of an acid* in water is a property of the acid *molecule*, whereas the *acidity of a solution* is a qualitative description of its pH. A **strongly acidic solution** is usually regarded as one with a pH of below 3. A weakly acidic solution has a pH of 3–7. A low pH might result from a concentrated solution of a weak acid or from a very dilute solution of a strong acid. But hydrochloric acid is a strong acid whether its concentration is 0.5 mol dm^{-3} or 0.0005 mol dm^{-3}.

Weak acids

Ethanoic acid is only partially ionized in aqueous solution:

$$CH_3COOH(aq) + H_2O(l) \rightleftharpoons \underset{\text{ethanoate (or acetate) ion}}{CH_3COO^-(aq)} + H_3O^+(aq) \quad (17.3)$$

and is therefore termed a **weak acid**:

a weak acid is incompletely ionized in solution

If we apply the equilibrium law to equation (17.3) the equilibrium expression becomes

$$K_a = \frac{[CH_3COO^-(aq)][H_3O^+(aq)]}{[CH_3COOH(aq)]}$$

where K_a is the equilibrium constant ('a' is for acid) at temperature *T*. As usual, the square brackets refer to the equilibrium (*not* starting) concentrations. Once again, the concentration of water (a constant) has been incorporated into K_a – this will be taken for granted in similar expressions below.

The term K_a is commonly known as the **acidity constant** (or the **dissociation constant**) of the acid. For ethanoic acid, $K_a(25°C) = 1.8 \times 10^{-5}$ mol dm^{-3}, the low value confirming that very few of the acid molecules have ionized. Hydrogen fluoride:

$$\underset{\text{hydrogen fluoride}}{HF(aq)} + H_2O(l) \rightleftharpoons \underbrace{F^-(aq) + H_3O^+(aq)}_{\text{hydrofluoric acid}}$$

is also a weak acid with $K_a(25°C) = 3.5 \times 10^{-4}$ mol dm^{-3}.

Table 17.2 shows the acidity constants for selected acids at 25°C. Symbolizing the acids as AH, the acidity constants are the equilibrium constants for the general reaction

$$AH(aq) + H_2O(l) \rightleftharpoons A^-(aq) + H_3O^+(aq) \quad (17.4)$$

(If AH is a solid, liquid or gas at room temperature (e.g. $C_6H_5COOH(s)$, $CH_3COOH(l)$, or $HCl(g)$) we assume that AH first dissolves in water making AH(aq), which then reacts according to this equation.)

The higher the K value, the stronger the acid.

The acidity constants of strong acids are difficult to determine experimentally because the concentration of un-ionized molecules is minute, but there is no doubt that the acidity constants of strong acids are much higher than those of the weaker acids listed in Table 17.2. This reflects the fact that the acid molecules are almost entirely dissociated in solution.

We may also define **pK_a**, where

$$pK_a = -\log K_a$$

For example, the pK_a of ethanoic acid at 25°C is $-\log(1.8 \times 10^{-5}) = 4.75$ and the pK_a of lactic acid is $-\log(8.4 \times 10^{-4}) = 3.08$.

The stronger the acid, the lower is its pK_a.

In equation (17.4), the acid is donating a hydrogen ion (a proton) to a water molecule to produce a $H_3O^+(aq)$ ion. This means that we can look at a K_a value as an indicator of the **proton-donating power** of that acid molecule toward the water molecule:

The greater the K_a value, the greater the proton donor ability of the acid towards water.

The species $A^-(aq)$ in equation (17.4) is referred to as the **conjugate base** of the acid AH(aq). This is because the $A^-(aq)$ ion acts like a base in that it accepts a proton from $H_3O^+(aq)$ in the reverse of reaction (17.4). For example, the conjugate base of ethanoic acid is the ethanoate ion $CH_3COO^-(aq)$.

Table 17.2 K_a and K_b values for selected weak acids and weak bases in aqueous solution at 25°C. These values change only slightly within the range 15–30°C. The acidic hydrogen or basic nitrogen atom is highlighted in bold

Acids	K_a/mol dm^{-3}
Trichloroethanoic acid $CCl_3CO_2\mathbf{H}$	3.0×10^{-1}
Chloroethanoic acid $CH_2ClCO_2\mathbf{H}$	1.4×10^{-3}
Sulfur dioxide SO_2 (sulfurous acid $\mathbf{H}_2SO_3$*)	1.6×10^{-2}
Lactic acid $CH_3CH(OH)CO_2\mathbf{H}$	8.4×10^{-4}
Hydrogen fluoride **H**F (hydrofluoric acid)	3.5×10^{-4}
Benzoic acid $C_6H_5COO\mathbf{H}$	6.5×10^{-5}
Ethanoic acid $CH_3COO\mathbf{H}$	1.8×10^{-5}
Carbon dioxide CO_2 (carbonic acid $\mathbf{H}_2CO_3$*)	4.3×10^{-7}
Hydrogen cyanide **H**CN (hydrocyanic acid)	4.9×10^{-10}
Phenol (carbolic acid) $C_6H_5O\mathbf{H}$	1.3×10^{-10}
Bases	K_b**/mol dm^{-3}**
Urea $\mathbf{N}H_2CONH_2$*	1.3×10^{-14}
Phenylamine (aniline) $C_6H_5\mathbf{N}H_2$	4.3×10^{-10}
Morphine $C_{17}H_{19}O_3\mathbf{N}$	1.6×10^{-6}
Ammonia $\mathbf{N}H_3$	1.8×10^{-5}
Methylamine $CH_3\mathbf{N}H_2$	3.6×10^{-4}
Dimethylamine $(CH_3)_2\mathbf{N}H$	5.4×10^{-4}
Triethylamine $(C_2H_5)_3\mathbf{N}$	1.0×10^{-3}

*K_a or K_b given for first ionization.

Calculating the pH of solutions of weak acids

We shall use ethanoic acid as an example of a weak acid that contains one acidic hydrogen atom. We will symbolize the initial concentration of ethanoic acid in a solution as C_A mol dm^{-3}. (For example, suppose we make a solution of ethanoic acid by dissolving 0.10 mol of pure ethanoic acid in water and making the solution up to 1 dm^3. Then $C_A = 0.10$ mol dm^{-3}.) The difficulty in calculating the pH of the resulting solution is that, unlike strong acids, the equilibrium concentration of hydronium ion in the solution is not the same as the concentration of the acid originally added, i.e. $[H_3O^+(aq)] \neq C_A$.

The $H_3O^+(aq)$ concentration of a very weak acid with one acidic hydrogen can be estimated by rearranging the following equation:

$$K_a \approx \frac{[H_3O^+(aq)]^2}{C_A} \qquad \textbf{(17.5)}$$

This equation is derived in Box 17.1. The equation is a good approximation provided the percentage of acid molecules that are ionized,

$$\text{percentage of acid molecules ionized} = \frac{[H_3O^+(aq)]}{C_A} \times 100 \text{ does not exceed 5\%.}$$

EXAMPLE **17.3**

Some white vinegar contains 0.50 per cent by mass of ethanoic acid, i.e. 100 g of vinegar contains 0.50 g of acid. Estimate its pH at 25°C. What percentage of the ethanoic acid molecules are ionized? (1.0 cm^3 of vinegar has a mass of 1.0 g.)

Answer

Consider exactly 1 dm^3 of vinegar. 1 dm^3 of vinegar contains 5.0 g CH_3COOH. The molar mass of ethanoic acid, $M(CH_3COOH)$, is 60 g mol^{-1}.

$$\text{number of moles of acid} = \frac{5.0\,\text{g}}{60\,\text{g}\,\text{mol}^{-1}} = 0.083 = 0.083$$

$$\text{concentration of acid} = \frac{0.083\,\text{mol}}{1\,\text{dm}^3} = 0.083\,\text{mol}\,\text{dm}^{-3}$$

Rearranging equation (17.5) gives

$$[H_3O^+(aq)] \approx \sqrt{K_a \times C_A}$$

$C_A = 0.083\,\text{mol}\,\text{dm}^{-3}$ and $K_a = 1.8 \times 10^{-5}\,\text{mol}\,\text{dm}^{-3}$ (Table 17.2), so that

$$[H_3O^+(aq)] \approx \sqrt{(1.8 \times 10^{-5} \times 0.083)} = 1.2 \times 10^{-3}\,\text{mol}\,\text{dm}^{-3}$$

$$\text{pH} = -\log\,[H_3O^+(aq)] = -\log\,(1.2 \times 10^{-3}) = 2.92$$

The percentage of ethanoic acid molecules that are ionized in vinegar is calculated as

$$\frac{[H_3O^+(aq)]}{C_A} \times 100 = \frac{1.2 \times 10^{-3} \times 100}{0.083} = 1.4\%$$

Comment

The percentage of ionized acid molecules is less than 5 per cent, and this confirms that we are able to use equation (17.5) to calculate $[H_3O^+(aq)]$ for this acid.

EXERCISE **17F**

Calculations involving weak acids

(i) Calculate the percentage of HCN molecules ionized in a HCN solution of concentration 0.020 mol dm^{-3}. What is the pH of the solution? (Use the data in Table 17.2.)

(ii) Benzoic acid is a natural bacteriostatic preservative, i.e. it helps prevent the growth of bacteria. Some fruit contains 0.050 per cent by mass of benzoic acid. Estimate the pH of 1.0 dm^3 of the solution produced from 4.0 kg of the fruit. (**Hint:** Start by calculating the concentration of benzoic acid.)

EXERCISE **17G**

Weak acid equilibria

(i) 25.0 cm^3 of 0.10 mol dm^{-3} HCl requires 25.0 cm^3 of 0.10 mol dm^{-3} NaOH for neutralization. Even though only 1.3 per cent of ethanoic acid molecules are ionized at room temperature, 25.0 cm^3 of 0.10 mol dm^{-3} CH_3COOH also requires 25 cm^3 of 0.10 mol dm^{-3} NaOH for neutralization. Explain this.

(ii) The enthalpy change of neutralization ($\Delta_{neut}H^{\ominus}$) is the standard enthalpy change when one mol of water is made in the neutralization of an acid by a base. Values are:

Acid–base pair	$\Delta_{neut}H^{\ominus}$/kJ mol^{-1}
HNO_3–KOH	−57.3
HCl–NaOH	−57.1
CH_3COOH–NaOH	−55.2

Why are the values for the first two pairs approximately the same? Why is $\Delta_{neut}H^{\ominus}$ for the neutralization of ethanoic acid less negative than the others? (Hint: the ionization of an acid is endothermic.)

BOX **17.1**

An expression to calculate the pH of a weak acid

We need the two following relationships to derive equation (17.5):

1 The equilibrium concentration of un-ionized ethanoic acid will be the difference between C_A and the equilibrium hydronium ion concentration:

$$[CH_3COOH(aq)] = C_A - [H_3O^+(aq)] \quad \textbf{(17.6)}$$

2 For every molecule of ethanoic acid that ionizes, one hydronium and one ethanoate ion is produced. Thus,

$$[CH_3COO^-(aq)] = [H_3O^+(aq)] \quad \textbf{(17.7)}$$

Using the equilibrium expression

$$K_a = \frac{[CH_3COO^-(aq)][H_3O+(aq)]}{[CH_3COOH(aq)]}$$

and substituting the right-hand sides of equations (17.6) and (17.7) gives

$$K_a = \frac{[H_3O^+(aq)]^2}{C_A - [H_3O^+(aq)]}$$

For acids that are only slightly ionized (< 5 per cent) we are justified in assuming that C_A – $[H_3O^+(aq)]$ is approximately equal to C_A. We then obtain equation (17.5):

$$K_a \approx \frac{[H_3O^+(aq)]^2}{C_A}$$

Sulfuric acid – a diprotic acid

Sulfuric acid possesses two acidic hydrogens and is said to be **diprotic**. There are two ionization steps, with sulfuric acid itself being the acid in the first step, and the hydrogensulfate ion being the acid in the second step:

First ionization

$$H_2SO_4(aq) + H_2O(l) \rightleftharpoons H_3O^+(aq) + \underset{\text{hydrogensulfate ion}}{HSO_4^-(aq)} \qquad K_a(298\ K) = \text{large}$$

Second ionization

$$HSO_4^-(aq) + H_2O(l) \rightleftharpoons H_3O^+(aq) + SO_4^{2-}(aq) \qquad K_a(298\ K) = 0.012\ mol\ dm^{-3}$$

The net reaction is the sum of both steps:

$$H_2SO_4(l) + 2H_2O(l) \rightleftharpoons 2H_3O^+(aq) + SO_4^{2-}(aq)$$

The K_a values show that sulfuric acid is a strong acid and that the hydrogensulfate ion is a relatively weak acid (although much stronger than most of the acids listed in Table 17.2).

Dilute sulfuric acid of concentration 0.0100 mol dm^{-3} contains an $H_3O^+(aq)$ concentration of 0.0145 mol dm^{-3}, giving a pH of 1.84. Of the total $[H_3O^+(aq)]$, 0.0100 mol dm^{-3} is contributed by the first ionization, and only 0.0045 mol dm^{-3} of the total [H_3O^+(aq)] is provided by the second ionization. If both K_a values were large, each reaction would contribute a hydronium ion concentration of 0.0100 mol dm^{-3}, and the pH of the dilute acid would then be –log [0.0200] = 1.70.

The HSO_4^- ion is more than strong enough to be used as an acid in its own right, and sodium hydrogensulfate crystals (Na^+, HSO_4^-) are used as a powerful disinfectant. They are also provided in chemistry sets as a safe substitute for sulfuric acid.

Strong and weak bases

The commonest **strong bases** are the water-soluble hydroxides of sodium, potassium, calcium, barium and lithium. They are all ionic solids. Since they are strong bases they are completely ionized in water. For example, NaOH(s) breaks up completely in water producing Na^+(aq) and OH^-(aq):

$$\underset{\text{sodium hydroxide}}{Na^+,OH^-(s)} \xrightarrow{H_2O} Na^+(aq) + \underset{\text{hydroxide ion}}{OH^-(aq)}$$

For example, a solution containing 0.5 mol of dissolved NaOH per dm^3 of solution contains 0.5 mol dm^{-3} of OH^-(aq).

A second category of bases consists of those which produce hydroxide ions in solution by *reaction* with water. An example is ammonia:

$$NH_3(aq) + H_2O(l) \rightleftharpoons NH_4^+(aq) + OH^-(aq)$$

Ammonia is a **weak base** because it is incompletely ionized in solution, and

$$K_b = \frac{[NH_4^+(aq)][OH^-(aq)]}{[NH_3(aq)]}$$

where K_b is the **basicity constant** (or **dissociation constant**) of ammonia at temperature *T*. Generalizing, the basicity constant for a base B is the equilibrium constant for the reaction:

$$B(aq) + H_2O(l) \rightleftharpoons BH^+(aq) + OH^-(aq) \qquad \textbf{(17.8)}$$

Any solution which contains the hydroxide ion (such as NaOH(aq) or NH_3(aq)) will neutralize acids in the exothermic reaction:

$$H_3O^+(aq) + OH^-(aq) \rightarrow 2H_2O(l)$$

Selected K_b values are included in Table 17.2, with triethylamine being the strongest base listed because it has the *biggest* K_b *value*. By coincidence, K_b for ammonia (one of the commonest weak bases) is numerically equal to K_a for ethanoic acid at 25°C.

Books often tabulate **pK_b** values where

$$pK_b = -\log K_b$$

For example, the pK_b of ammonia is – log (1.8×10^{-5}) = 4.74. The stronger the base, the *lower* is its pK_b.

In reaction (17.8), the base is accepting a proton from a water molecule. This enables us to use K_b values as indicators of the *proton-accepting ability* of bases in water, in the same way that we used K_a values as indicators of the proton-donating ability of acids in water. A large K_b value shows that the base is a strong proton acceptor.

In equation (17.8), the species BH^+(aq) is referred to as the **conjugate acid** of base B(aq). This is because BH^+(aq) donates a proton to the OH^-(aq) in the *reverse* of reaction (17.8).

The self-ionization of water,

$$H_2O(l) + H_2O(l) \rightleftharpoons H_3O^+(aq) + OH^-(aq)$$

involves one molecule of water donating a proton to another. This makes water both a proton donor and a proton acceptor.

When calculating the pH of a weak base, we follow the same pattern as for an acid,

$$K_a \approx \frac{[OH^-(aq)]^2}{C_B}$$

where C_B is the initial molar concentration of base in the solution. If C_B and K_b are known, we can then calculate $[OH^-(aq)]$. $[H_3O^+(aq)]$ is then worked out using K_w. Now try Exercise 17H.

EXERCISE **17H**

Calculations involving weak bases

(i) pK_b for nicotine, $C_{10}H_{14}N_2$, is 5.98. What is K_b?

(ii) Estimate the pH of a solution of ammonia of concentration 0.200 mol dm^{-3} at 25°C.

(iii) Write down an equation showing the reaction of phenylamine with water to produce hydroxide ions. What is the conjugate acid of phenylamine? Some phenylamine was dissolved in pure water. The pH of the solution was found to be 7.80 at 25°C. What was the concentration of phenylamine in the solution?

17.3 Hydrolysis of salts

pH of solutions of salts

A salt is produced when a base and acid neutralize each other. On this basis, four classes of salts are possible:

1 a salt of a strong acid and a strong base (SA–SB);
2 a salt of a weak acid and a strong base (WA–SB);
3 a salt of a strong acid and a weak base (SA–WB);
4 a salt of a weak acid and a weak base (WA–WB).

Sodium chloride solution (made with pure water) is neutral, with a pH of 7 at 25°C. This may lead us to suppose that solutions of all ionic salts are neutral. In fact, *only solutions of salts made from strong acids and strong bases are always neutral.* Solutions of weak acids or weak bases are usually either acidic or basic because:

1 These salts react with the water *producing a weak acid or a weak base.*
2 Since the weak acid or weak base is only partially ionized in water, formation of these molecules ties up hydroxide or hydronium ions.
3 This produces unequal concentrations of hydronium and hydroxide ions, and the resulting solution of salt is then acidic or basic.

Reactions with water are called **hydrolysis**, and the reactions of a salt with water are referred to as **salt hydrolysis**. We now look at one example of this effect.

Hydrolysis of a salt of a weak acid and a strong base

Using sodium ethanoate (CH_3COO^-,Na^+) as an example, the ions present in an aqueous solution of this salt are

$$CH_3COO^-,Na^+(s) \xrightarrow{H_2O} CH_3COO^-(aq) + Na^+(aq) \quad \text{(from the } salt\text{)}$$

$$2H_2O(l) \rightleftharpoons H_3O^+(aq) + OH^-(aq) \quad \text{(from the } water\text{)}$$

The potential products of any reaction between the ions in solution are NaOH (from the reaction of $Na^+(aq)$ and $OH^-(aq)$) and CH_3COOH (from the reaction of $CH_3COO^-(aq)$ and $H_3O^+(aq)$). The NaOH is fully ionized in solution, but the ethanoic acid is only partially ionized. This means that $[OH^-(aq)] > [H_3O^+(aq)]$, and the solution is basic.

Generalizing,

Solutions of salts of weak acids and strong bases are basic.

Consideration of the possible reactions in solutions of the other classes of salts leads to the conclusions given in Table 17.3.

Table 17.3 The pH of solutions of the four classes of salt

Class of salt	Example	Formula	Acidic	Basic	Neutral
SA–SB	sodium chloride	NaCl			✓
WA–SB	sodium ethanoate	CH_3COONa		✓	
SA–WB	ammonium chloride	NH_4Cl	✓		
WA–WB	ammonium cyanide	NH_4CN		✓*	

* If K_a(WA) > K_b(WB) the solution is acidic. If K_a(WA) < K_b(WB) the solution is basic. If K_a(WA) = K_b(WB) – as in the case of ammonium ethanoate – the solution is neutral.

EXERCISE **17I**

Salt hydrolysis

Use Table 17.3 to complete this table:

Salt	Formula	Parent acid and base	Acidic, basic or neutral?
Iron(III) nitrate			
Calcium chloride			
Sodium sulfate			
Ammonium benzoate			

17.4 Buffer solutions

Adding small amounts of acids or alkalis may lead to drastic changes in solution pH

The addition of even one drop of dilute hydrochloric acid to water drastically changes its pH (Box 17.2). Such changes in pH can be troublesome in the laboratory, and catastrophic in living cells. Changes in pH due to trace contamination with acids or bases can be prevented using a **buffer solution** (usually simply called a 'buffer' (Figure 17.1)).

A buffer solution resists changes in pH when it is diluted or when acid or base is added.

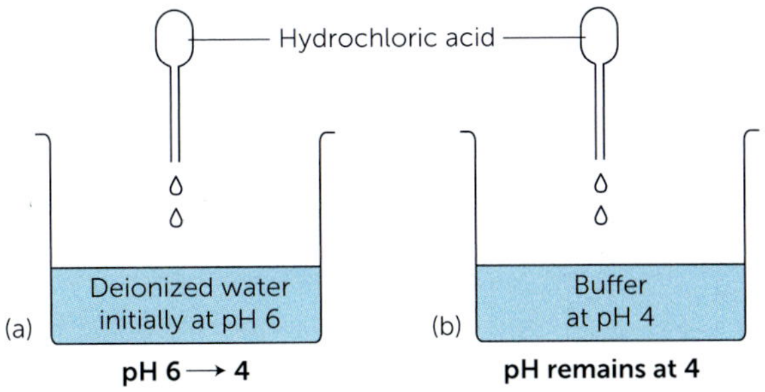

Figure 17.1 A buffer in action.
(a) When a drop of 2.0 mol dm^{-3} hydrochloric acid is added to water, the pH falls drastically.
(b) There is no change in pH when an acid is added to the buffer solution.

BOX **17.2**

pH of a jug of water contaminated with a drop of dilute hydrochloric acid

Suppose we add 1 drop of 'bench hydrochloric acid' (concentration = 2.0 mol dm^{-3}) to 1 dm^3 of deionized water. What is the pH of the resulting solution?
The volume of the drop (≈ 0.050 cm^3) is negligible compared with the water, so we may take the solution volume as 1 dm^3. The number of moles of HCl in the 0.05 cm^3 drop is

$$\frac{0.050}{1000} \times 2.0 = 1.0 \times 10^{-4}\ \text{mol}$$

and the concentration of HCl in the solution is therefore

$$\frac{1.0 \times 10^{-4}\ mol}{1 dm^3} = 1.0 \times 10^{-4}\ \text{mol dm}^{-3}$$

The pH of the solution is calculated as follows:

$$\text{pH} = -\log\,[H_3O^+(aq)] = -\log\,[1 \times 10^{-4}] = 4.0$$

The pH of deionized water in the laboratory is about 6 because it contains dissolved carbon dioxide. Therefore, *the addition of only one drop of dilute acid has reduced its pH from 6 to 4.*

Figure 17.2 shows the pH values in separate experiments in which (a) 0.10 mol dm^{-3} HCl, (b) deionized water and (c) 0.10 mol dm^{-3} NaOH were added to a sodium ethanoate–ethanoic acid buffer. In this case, adding 3 cm^3 or less of (a) or (c) produces only a slight change in pH. If we add more acid or alkali, the buffer becomes exhausted and it can no longer resist changes in pH. However, dilution of the buffer by adding water causes virtually no change in buffer pH.

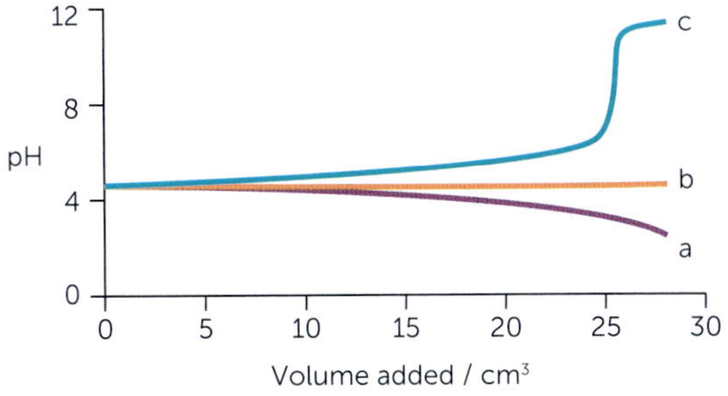

Figure 17.2 Changes in pH when (a) 0.10 mol dm^{-3} HCl, (b) water and (c) 0.10 mol dm^{-3} NaOH are added to a buffer consisting of 25 cm^3 of 0.10 mol dm^{-3} ethanoic acid and 25 cm^3 of 0.10 mol dm^{-3} sodium ethanoate (temperature 25°C; data kindly supplied by Keith Morgan).

How buffers work

Buffers consist of either a weak base and one of its salts, or a weak acid and one of its salts. As an example, we look at a mixture containing ethanoic acid and sodium ethanoate. The ions present in such a mixture are shown by the following equations:

$$CH_3COO^-,Na^+(s) \xrightarrow{H_2O} CH_3COO^-(aq) + Na^+(aq)$$

$$CH_3COOH(aq) + H_2O(l) \rightleftharpoons CH_3COO^-(aq) + H_3O^+(aq) \quad \textbf{(17.9)}$$

The salt, sodium ethanoate, is fully ionized (since it is an ionic compound), and generates a relatively high concentration of ethanoate ions.

Now consider the ionization of ethanoic acid in the presence of sodium ethanoate. The expression for the acidity constant of ethanoic acid is:

$$K_a = \frac{[CH_3COO^-(aq)]\,[H_3O^+(aq)]}{[CH_3COOH(aq)]} \quad \textbf{(17.10)}$$

If ethanoate ions are added to an aqueous solution of ethanoic acid, the equilibrium composition will shift in order to keep the right-hand side of this expression equal to K_a. As a result of the shift, $[H_3O^+(aq)]$ becomes very low, meaning that very little ethanoic acid is now ionized. Calculations show that in the presence of sodium ethanoate, the ionization of ethanoic acid is so small that $[CH_3COO^-(aq)]$ may be taken to be equal to the initial concentration of sodium ethanoate salt solution, C_s.

How does the buffer mixture resist changes in pH? The explanation is as follows.

- Suppose we add some hydrochloric acid to the mixture. The equilibrium composition shifts, with ethanoate and hydronium ions combining together in the *reverse* of equation (17.9). This mops up the added $H_3O^+(aq)$ ions, and there is no change in pH. (The same conclusion is arrived at using Le Chatelier's principle.)
- If we add hydroxide ions (e.g. by adding sodium hydroxide solution), the tiny amount of $[H_3O^+(aq)]$ at equilibrium in the buffer mixture reacts with $OH^-(aq)$ in the neutralization reaction

$$H_3O^+(aq) + OH^-(aq) \rightarrow 2H_2O(l).$$

More ethanoic acid ionizes to restore the equilibrium concentration of $H_3O^+(aq)$, and ionization continues until just enough extra $H_3O^+(aq)$ has been made to convert all the $OH^-(aq)$ to water. In this way, the addition of a base does not lead to a change in the pH of the buffer.

Calculating the pH of a buffer solution

We can easily obtain an expression with which to estimate the pH of a buffer, starting with equation (17.10). Since virtually all the ethanoate ions in the buffer come from the sodium ethanoate, we are justified in substituting C_s for $[CH_3COO^-(aq)]$. Since so little ethanoic acid is ionized in the buffer, we are also justified in taking the equilibrium concentration of ethanoic acid, $[CH_3COOH(aq)]$, to be equal to the initial concentration of acid, C_A. Substitution of this information into equation (17.10) gives

$$K_a \approx \frac{C_s \times [H_3O^+(aq)]}{C_A}$$

Rearranging,

$$[H_3O^+(aq)] \approx \frac{C_A \times K_a}{C_s} \quad \textbf{(17.11)}$$

This equation shows that dilution of the buffer mixture does not affect its pH because C_A and C_s are affected equally. For example, if we add 50 cm^3 of water to 50 cm^3 of buffer, *both* the acid (C_A) and salt (C_s) concentrations in the buffer solution are halved, and the pH predicted by equation (17.11) is unaffected by the dilution.

EXERCISE **17J**

Calculating acidity constants

The acidity constant of an acid may be calculated from the way that the pH of a buffer changes as the concentration of acid and salt are changed. This involves using a *graphical* method. Can you see from equation (17.11) how this might be done?

Buffer solutions may also be prepared by mixing a weak base and one of its salts (such as ammonia solution and ammonium chloride). The $[OH^-(aq)]$ of such a mixture may be calculated using the expression

$$[OH^-(aq)] \approx \frac{C_B \times K_b}{C_s}$$

where C_B is the concentration of base, C_s is the concentration of salt and K_b is the basicity constant.

Data for the sodium ethanoate–ethanoic acid buffer

Let us use equation (17.11) to calculate the pH of an ethanoate–ethanoic acid buffer at 25°C made by mixing 25 cm^3 of 0.100 mol dm^{-3} sodium ethanoate solution with 25 cm^3 of 0.100 mol dm^{-3} ethanoic acid solution. Since the volume doubles, the concentrations are halved and $C_A = C_s = 0.050$ mol dm^{-3}. K_a for ethanoic acid at 25°C = 1.8×10^{-5} mol dm^{-3}. Substituting into equation (17.11) gives

$$[H_3O^+(aq)] = \frac{0.050 \times 1.8 \times 10^{-5}}{0.050} = 1.8 \times 10^{-5}\,mol\,dm^{-3}$$

The pH of the buffer mixture is

$$pH = -\log\,[1.8 \times 10^{-5}] = 4.74$$

(This is approximately the initial pH of the buffer in Figure 17.2.) By using different acids (different K_a values) and their salts, buffers operating at different pH values may be obtained. To adjust the pH of a buffer more exactly, the ratio of the initial concentrations of acid and salt in the mixture is carefully controlled.

Buffer capacity

The amount of acid or alkali that needs to be added before the pH of a buffer changes is called the **buffer capacity** of the buffer. The buffer capacity of a buffer containing a relatively high number of moles of acid and salt is greater than the capacity of a buffer with a lower number of moles of acid and salt. As we have already noted, the buffer capacity of the buffer in Figure 17.2 is equivalent to roughly 3 cm^3 of 0.1 mol dm^{-3} HCl or 3 cm^3 of 0.1 mol dm^{-3} NaOH.

EXERCISE **17K**

Calculating the pH of buffer solutions

Investigate the effect of using different volumes of 0.100 mol dm^{-3} ethanoic acid and 0.100 mol dm^{-3} sodium ethanoate salt, by repeating the pH calculation starting with **(i)** 25 cm^3 of acid and 50 cm^3 of salt solution, and **(ii)** 50 cm^3 of acid and 25 cm^3 of salt solution.

Alternative form of equation (17.11) – the Henderson–Hasselbalch equation

Equation (17.11) is sometimes used in a different form. Taking logs on both sides gives

$$log[H_3O^+(aq)] = log K_a + log\frac{c_A}{c_S}$$

Multiplying by – 1,

$$-log[H_3O^+(aq)] = -log K_a - log\frac{c_A}{c_S}$$

Remembering that

$$-\log\frac{c_A}{c_S} = \log\frac{c_S}{c_A}$$

and using the definitions of pH and pK_a, we arrive at the relationship

$$pH = pK_a + \log\frac{c_S}{c_A}$$

which is known as the **Henderson–Hasselbalch** equation (Exercise 17L).

17.5 Acid–base indicators

Acid–base indicators, such as methyl orange, phenolphthalein and litmus, show two extreme colours, one at lower pH and the other at higher pH.

For example, in the case of methyl orange, the two extreme colours are red (lower pH) and yellow (higher pH). As with most indicators, the change in colour does not occur over a very small change of pH. The range of pH over which an indicator changes colour is called its **pH range**. The pH range of methyl orange is 3.2–4.4. Below pH 3.2 methyl orange is red, but as the pH is increased *above* 3.2, the observed colour contains an increasing amount of yellow until, beyond pH 4.4, it is entirely yellow.

Indicators are sometimes absorbed into paper strips (e.g. litmus paper). The indicator known as 'universal indicator' is a mixture of selected indicators that displays different colours at different pH values. (To demonstrate this, place a large crystal of tartaric acid in a small flask. Add dilute universal indicator solution and one drop of NaOH solution. Slow and gentle swirling of the flask causes the acid to slowly dissolve, and the indicator passes through an impressive sequence of colours.)

EXERCISE 17L

Using the Henderson–Hasselbalch equation

A buffer solution contains 0.10 mol of one of the acids listed in Table 17.2 mixed with 0.10 mol of its potassium salt. The observed pH of the buffer was 4.18. What is the name of the acid?

Methyl orange indicator is red at lower pH and yellow at higher pH.

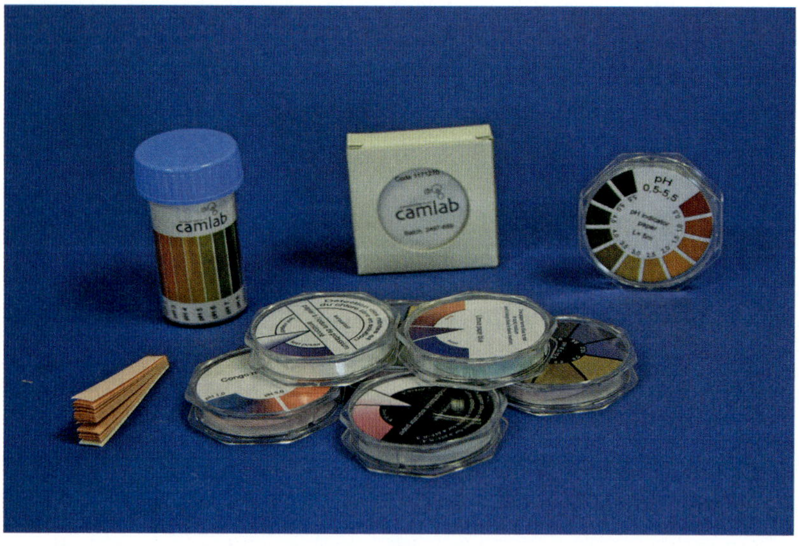

Acid–base indicators are available in easy to use strips.

EXERCISE **17M**

Colours of acid–base indicators

Table 17.4 shows the pH range of selected acid–base indicators. Use this table to help you decide the colour of phenolphthalein and bromocresol green at a pH of:

(i) 2, **(ii)** 3.7, **(iii)** 5,
(iv) 6, **(v)** 8, **(vi)** 11.

Table 17.4 Acid–base indicators

Indicator	Extreme colours		pH range
	Lower pH	Higher pH	
Alizarin yellow	colourless	yellow	10.1–12.0
Bromocresol green	yellow	blue	4.0–5.6
Litmus	red	blue	4.7–8.2
Methyl orange	red	yellow	3.2–4.4
Methyl red	yellow	red	4.8–6.0
Phenolphthalein	colourless	pink	8.2–10.0

How acid–base indicators work

Acid–base indicators consist of molecules that have different molecular structures at lower pH and at higher pH. These higher-pH and lower-pH forms have different colours.

Consider methyl orange. Above pH 4.4, methyl orange exists entirely as the ion

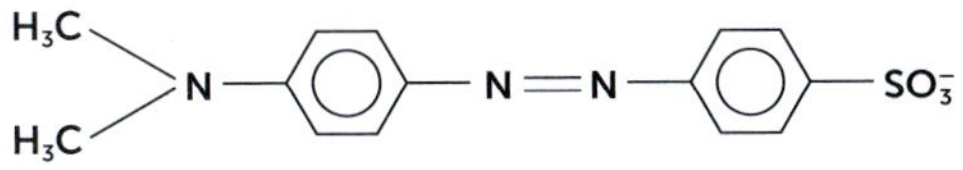

This ion is the higher-pH form. It is coloured yellow. It is represented as $R–SO_3^-$. Below pH 3.2, methyl orange exists entirely as

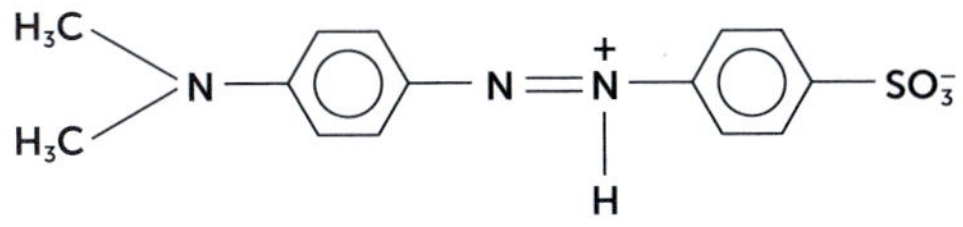

in which one of the nitrogen atoms has been bonded to a proton (i.e. protonated) as a result of reaction with $H_3O^+(aq)$. This ion is the lower-pH form. It is coloured red and is represented as $HR^+–SO_3^-$. These changes are summarized by the equilibrium equation

$$\underset{\text{yellow}}{R–SO_3^-(aq)} + H_3O^+(aq) \rightleftharpoons \underset{\text{red}}{HR^+–SO_3^-(aq)} + H_2O$$

All acid–base indicators undergo similar molecular changes with changing pH. The general equation representing these changes is

$$\text{higher-pH form} + H_3O^+(aq) \rightleftharpoons \underset{\text{('acid form')}}{\text{lower-pH form}} + H_2O$$

17.6 Variation of pH during an acid–base titration

In an acid–base titration, base solution is added to the acid solution until the acid has been exactly neutralized. This is known as the **equivalence point** (or **stoichiometric point**) of the reaction.

If we are using indicators in an acid–base titration, we are making the assumption that the **endpoint** of the indicator (the point at which it appears to change colour) is the same as the equivalence point. In order to look at this assumption further, we look at the way that the pH of a reaction mixture changes during an acid–base titration.

Table 17.5 pH changes during the titration of 0.100 mol dm^{-3} NaOH with 25.00 cm^3 0.100 mol dm^{-3} HCl

Vol of added NaOH (cm^3)	pH
0.00	1.00
5.00	1.18
10.00	1.37
15.00	1.60
20.00	1.95
24.00	2.69
24.90	3.70
24.95	4.00
25.00	7.00
25.05	10.00
25.10	10.30
25.50	11.00
26.00	11.29
30.00	11.96
35.00	12.22
40.00	12.36
45.00	12.46
50.00	12.52

Titration of a strong acid and a strong base (SA–SB)

Suppose that we slowly add 0.100 mol dm^{-3} NaOH to 25.00 cm^3 of 0.100 mol dm^{-3} HCl, and use a pH meter to record the pH of the mixture after each addition of base. The reaction is

$$HCl(aq) + NaOH(aq) \rightarrow NaCl(aq) + H_2O(l)$$

Table 17.5 shows the changes in pH during the titration. The initial pH of the acid is 1.00, and the pH of the mixture at the equivalence point (after addition of exactly 25.00 cm^3 of NaOH, when only pure water and sodium chloride are present) is 7.00. When these data are plotted, Figure 17.3(a) is obtained – virtually identical curves are obtained using any pair of strong acids and strong bases. Upon further addition of NaOH, the pH rises slowly and, suddenly, shoots up near the equivalence point. After this, the pH slowly rises as the amount of excess NaOH increases.

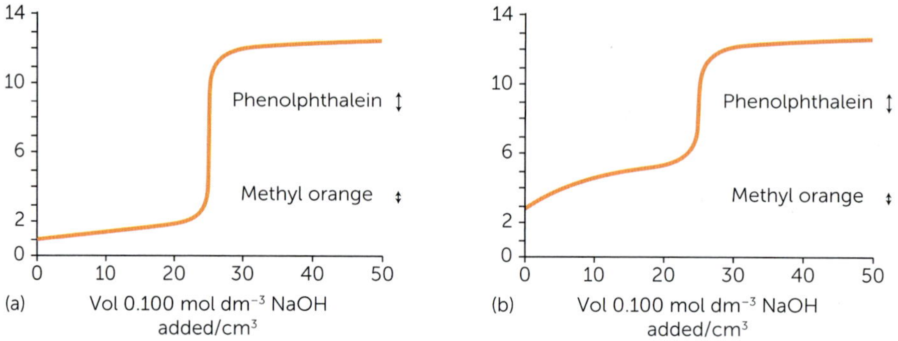

Figure 17.3 pH curves for titrations of (a) a strong acid and a strong base (HCl and NaOH), and (b) a weak acid and a strong base (CH_3COOH and NaOH). All initial concentrations are 0.100 mol dm^{-3}. The initial volume of acid was 25.0 cm^3 in both cases.

Points to note:

- The very rapid change in pH near the equivalence point has a simple explanation. We have already noted (see Box 17.2) that the addition of even one drop of acid to a neutral solution causes a massive change in pH, and the same applies to the addition of base to a neutral solution. This is exactly what happens during the titration *just before* and *just after* the equivalence point. This is seen from the highlighted data in Table 17.5, where the addition of only 0.10 cm^3 of base (about two drops) causes the pH to change from 4 to 10.
- The approximate pH ranges of phenolphthalein (8–10) and methyl orange (3–4) are also shown on Figure 17.3(a). Methyl orange begins to change colour fractionally before, and the phenolphthalein very fractionally after, the equivalence point. In practice, because the pH changes so rapidly near the equivalence point, the colour changes for phenolphthalein will start and finish over the addition of less than one drop (< 0.05 cm^3) of added base. The error involved in assuming that the indicated endpoint of phenolphthalein is the same as the equivalence point is negligible. The use of methyl orange will produce a slight titration error with an endpoint at about 24.95 cm^3.

Titration curve for a weak acid and a strong base (WA–SB)

Figure 17.3(b) shows the pH curve for the titration of a strong base (NaOH) against a weak acid (ethanoic acid):

$$CH_3COOH(aq) + NaOH(aq) \rightarrow CH_3COONa(aq) + H_2O(l)$$

Similar pH curves are obtained with all titrations involving weak acids and strong bases. Beyond the equivalence point (25.00 cm^3 of NaOH), the CH_3COOH–NaOH pH curve is identical to the HCl–NaOH curve, but before the equivalence point the curve differs from Figure 17.3(a) in two important respects:

- The pH at the equivalence point is 8.7 (not 7) because the salt, sodium ethanoate, hydrolyses in solution.
- The pH of the acid at the beginning of the titration is higher than for the HCl–NaOH case because $CH_3COOH(aq)$ is a weak acid and is only partially ionized. A mixture of ethanoic acid and sodium ethanoate is also a buffer solution. This has the effect of shortening the vertical part of the graph in the region of the equivalence point. This, in turn, means that *methyl orange is unsuitable for use in this titration, since it changes colour well before the equivalence point*. However, phenolphthalein may still be used.

EXERCISE **17N**

Strong acid–strong base titration

(i) Which of the indicators listed in Table 17.4 are unsuitable for use in a strong acid–strong base titration?

(ii) Which of the following pairs will give a pH curve virtually identical to Figure 17.3(a):

(a) HNO_3–KOH **(b)** HBr–NaOH **(c)** HCl–NH_3?

EXERCISE **17O**

Weak acid–strong base titration

(i) Citric acid, a weak acid, is a constituent of fizzy drinks. Would you use bromocresol green indicator in titrations involving citric acid and potassium hydroxide?

(ii) Refer to the NaOH–CH_3COOH titration shown in Figure 17.3(b). What simplified form does the Henderson–Hasselbalch equation take when 12.50 cm^3 of NaOH has been added? Use this relationship to estimate $K_a(CH_3COOH)$ from Figure 17.3(b).

17.7 Buffering action of carbon dioxide in water

Carbon dioxide in water

Carbon dioxide is fairly soluble in water, and a small amount of the dissolved CO_2 reacts with water as follows:

$$CO_2(aq) + 2H_2O(l) \rightleftharpoons \underset{\text{hydrogencarbonate ion}}{HCO_3^-}(aq) + H_3O^+(aq) \qquad \textbf{(17.12)}$$

where $K = 4.3 \times 10^{-7}$ mol dm^{-3} at 25°C. The production of hydronium ions causes water in contact with air (for example rain water, tap water and laboratory deionized water) to be *slightly acidic*, and experiments show that such water possesses (at 25°C) a pH of about 5.6.

However, many natural waters (lakes, ponds and reservoirs) are basic due to the presence of dissolved solids. For example, water saturated with CO_2 and limestone (calcium carbonate) at 25°C contains a higher concentration of hydrogencarbonate ions than water saturated with CO_2 alone, and is *slightly basic* with a pH of 8.3.

If acid is added to a pond, the added H_3O^+ ions cause the equilibrium concentrations of water and carbon dioxide to increase. The CO_2 then bubbles off into the air. Without such buffering, natural waters would be easily acidified by acid rain. However, if too much acid is present, the buffer is overwhelmed and the pH of the water does change.

Buffers in the body

The CO_2–water equilibrium is also the main buffer present in blood. Excess $H_3O^+(aq)$ in blood, due to the presence of lactic acid produced by the metabolism of food, is removed by the reaction of HCO_3^- and H_3O^+ (i.e. by the reverse of equation (17.12)). In this way, the pH of blood is maintained at 7.4. Variations in arterial blood pH of more than about 0.4 can be fatal.

The hydrogencarbonate buffer in blood would soon become exhausted if there was not a way for the body to get rid of waste. Excess carbon dioxide and excess acid are removed by the lungs and kidneys, respectively. In this way, the hydrogencarbonate ions in the blood are released to act as a buffer once again.

EXERCISE **17P**

Calculation of the pH of unpolluted rain water

This question is concerned with the concentration of dissolved CO_2 in water, $[CO_2(aq)]$. Measurements show that provided the water is in equilibrium with the CO_2 in the air, $[CO_2(aq)] = 1.2 \times 10^{-5}$ mol dm^{-3} at 25°C.

(i) $CO_2(aq)$ is a weak acid. Using the expression

$$K_a \approx \frac{[H_3O^+(aq)]^2}{[CO_2(aq)]}$$

show (by calculation) that the pH of water in equilibrium with CO_2 in air is about 5.6.

(ii) Interestingly, concentrated salty water possesses a pH closer to 7 than that of deionized water in the laboratory – why?

For more on ionic equilibria and buffers, including the haemoglobin system in the blood, see **Extension 17** at https://www.bloomsburyonlineresources.com/chemistry.

REVISION QUESTIONS

17.1 Label the following as true (T) or false (F):

(i) The pH range of an indicator is the band of pH in which the indicator is coloured.

(ii) If methyl orange shows a yellow colour, this indicates that the solution is basic.

(iii) The pH of a solution of hydrochloric acid depends upon its concentration.

(iv) The acidity constant of an acid is the equilibrium constant for its ionization in solution.

(v) 25 cm^3 of 0.02 mol dm^{-3} $H_2SO_4(aq)$ requires 50 cm^3 of 0.02 mol dm^{-3} NaOH for neutralization.

17.2 Calculate the pH at 25°C of

(i) 0.080 mol dm^{-3} HNO_3,

(ii) 0.080 mol dm^{-3} KOH.

17.3

(i) Calculate the pH of water at 100°C.

(ii) 11.0 cm^3 of 0.040 mol dm^{-3} sodium hydroxide and 9.0 cm^3 of 0.040 mol dm^{-3} HCl are mixed in a beaker. Calculate the pH of the solution at 25°C.

(iii) Write down an equation showing the ionization of the (weak base) morphine ($C_{17}H_{19}O_3N$) in water. Calculate the pOH and pH of an aqueous solution containing 1.0 mg of morphine per dm^3 of water at 25°C (K_b(morphine) = 1.6×10^{-6} mol dm^{-3}; ignore dissolved CO_2).

(iv) What volume of 0.0010 mol dm^{-3} HCl would be required to exactly neutralize 100 cm^3 of the morphine solution? Write an equation for the neutralization reaction involved.

17.4 The ionization of phenol (a weak acid) in water is represented by the equation:

$$C_6H_5OH(aq) + H_2O(l) \rightleftharpoons C_6H_5O^-(aq) + H_3O^+(aq)$$

Estimate the pH at 25°C of a solution of phenol of concentration 5.00×10^{-3} mol dm^{-3} (ignore dissolved CO_2).

17.5 The reaction of ammonia with water is represented by the equation

$$NH_3(aq) + H_2O(l) \rightleftharpoons NH_4^+(aq) + OH^-(aq)$$

Show, by substitution, that the **acidity constant** of the conjugate acid NH_4^+(aq), symbolized $K_a(NH_4^+(aq))$, is related to $K_b(NH_3(aq))$ by the expression

$$K_a(NH_4^+(aq)) \times K_b(NH_3(aq)) = K_w$$

Calculate $K_a(NH_4^+(aq))$ at 25°C ($K_b(NH_3(aq)) = 1.8 \times 10^{-5}$ mol dm^{-3}).

17.6

(i) Solutions of carbon dioxide are weakly acidic:

$$CO_2(aq) + 2H_2O(l) \rightleftharpoons HCO_3^-(aq) + H_3O^+(aq)$$

Derive the following expression for a solution containing dissolved CO_2(g):

$$K_a \approx \frac{[H_3O^+(aq)]^2}{[CO_2(aq)]}$$

(ii) Carbonated water (fizzy water) consists of water *saturated* with CO_2. Water *saturated* with CO_2 at 20°C contains 4.0×10^{-2} mol of dissolved gas per dm^3 of solution. Calculate the pH of carbonated water at this temperature.

17.7 Predict whether solutions of the following salts will be basic, acidic or neutral:

(i) phenylamine hydrochloride ($C_6H_5NH_3^+,Cl^-$),
(ii) ammonium ethanoate,
(iii) ammonium nitrate,
(iv) potassium bromide,
(v) sodium carbonate.

17.8 0.100 g of benzoic acid, C_6H_5COOH, was added to 0.080 g of sodium benzoate salt (C_6H_5COONa), the mixture was dissolved in water and made up to 100 cm^3. Estimate the pH of the resulting buffer, and use equations to explain how the buffer responds upon

(i) the addition of acid,
(ii) the addition of alkali.

17.9 The following figure shows the way the pH changes when 0.100 mol dm^{-3} NH_3(aq), a weak base, is titrated with 25.00 cm^3 of 0.100 mol dm^{-3} HCl(aq) at room temperature.

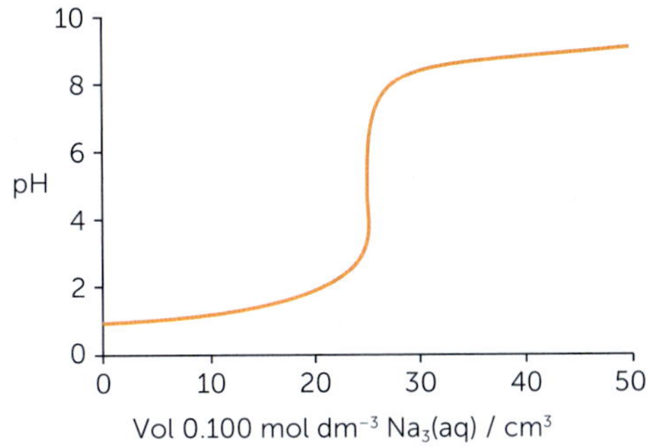

Briefly discuss the suitability of the indicators listed in Table 17.4 for this titration.

17.10 Scientists have carried out experiments to see whether athletes (for example, sprinters) might be able to improve their performance by ingesting sodium hydrogencarbonate before exercise. Explain the reasoning behind these experiments.

17.11 Is the water molecule an acid or a base?

DID YOU KNOW?

It has been predicted that if CO_2 emissions continue at their current rate the average pH of the oceans would fall from 8.2 to 7.7. This corresponds to a three-fold increase in [H^+]. Check this calculation for yourself.

18 Organic Chemistry: Hydrocarbons

CONTENTS

OBJECTIVES

This unit:

- Explains what is meant by 'organic chemistry'
- Introduces you to some families of hydrocarbons
- Describes the sources of hydrocarbons and how hydrocarbons form the starting materials for many synthetic organic compounds
- Distinguishes between 'aliphatic' and 'aromatic' compounds

One element in the Periodic Table, carbon, forms so many compounds that it has an entire branch of chemistry devoted to it. Most of the chemistry of this element comes under the heading **organic** chemistry. The reason that carbon forms such a huge number of compounds is that it has the ability to form chains of carbon atoms (or the ability to **catenate**) – the carbon atoms are covalently bonded to one another and to other elements, such as hydrogen, oxygen and nitrogen.

The name 'organic' originates from the days when chemical compounds were divided into two classes, depending on their origin: 'inorganic' and 'organic'. Inorganic compounds were obtained from mineral sources, whereas organic chemicals were derived from living things. It was believed that organic compounds contained a 'vital force' and could not be made from inorganic compounds, until F. Wöhler (1800–1882) made the organic compound urea, $CO(NH_2)_2$, from the inorganic salt ammonium cyanate, NH_4OCN. Urea is formed as a waste product when proteins are metabolized. Today, organic chemistry refers to most of the chemistry of carbon compounds. Some exceptions include the chemistry of metal carbonates, carbon dioxide and carbon monoxide, which are included within inorganic chemistry.

Organic compounds can be organized into families of compounds with similar structural formulae and similar properties. This makes the study of organic chemistry easier, since there are billions of individual organic compounds!

18.1 Alkanes

The first family we shall look at is called the **alkanes**. They were originally called the paraffins (which comes from the Latin 'little affinity') because apart from combustion reactions they are not particularly chemically reactive. They are found in crude oil and natural gas and are used extensively as fuels, or as starting materials for a wide variety of organic compounds. The first four members of the alkane family, which are all gases at 25°C, are shown in Table 18.1. The alkanes belong to a larger classification of compounds, the **hydrocarbons**. Hydrocarbons are organic compounds that contain the elements carbon and hydrogen *only*. The alkanes have the general formula C_nH_{2n+2} where *n*, the number of carbon atoms, is 1, 2, 3, 4, 5 and so on. A family of compounds whose formulae differ by $-CH_2$ is called a **homologous series**.

Table 18.1 The first four alkanes

Name	Molecular formula	Structural formula
Methane (natural gas)	CH_4	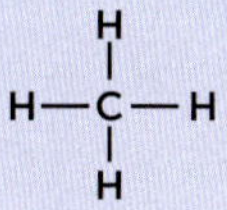
Ethane	C_2H_6	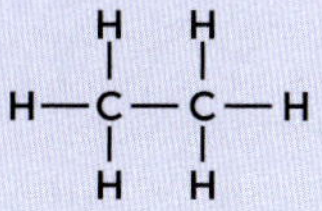
Propane (camping gas)	C_3H_8	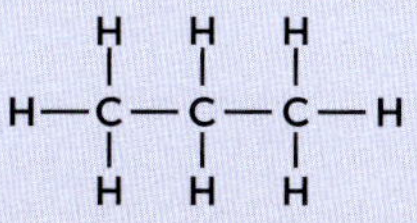
Butane (lighter fuel)	C_4H_{10}	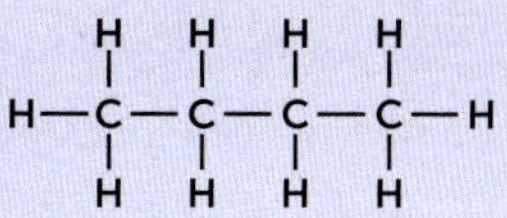

Two other ways of drawing butane

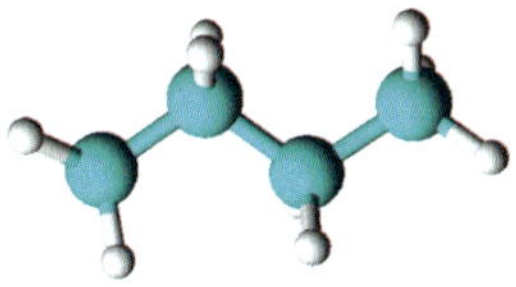

The ball-and-stick representation

The skeletal structure representation

After butane, the names of higher alkanes are taken from the Latin or Greek prefix for the number of carbon atoms in one molecule of the alkane:

Name	Number of carbon atoms
Pentane	5
Hexane	6
Heptane	7
Octane	8
Nonane	9
Decane	10
Undecane	11
Dodecane	12

Writing the structural formulae of the higher alkanes can take up a great deal of space, so we often write **condensed structural formulae**. For example, the structural formula of pentane can be written in four different ways. These are shown below:

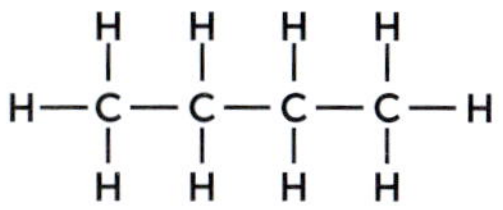 or $CH_3CH_2CH_2CH_2CH_3$ or $CH_3(CH_2)_3CH_3$ or 

The last is a skeletal formula, where each end represents a carbon joined to three hydrogens and each point a carbon joined to two hydrogens. The second and third representations are condensed structural formulae; they show all the atoms but not all the bonds.

EXERCISE **18A**

Alkanes

(i) The alkanes are covalent compounds. Write Lewis structures for ethane and propane.

(ii) Write the molecular and structural formulae for octane.

(iii) What is the molecular formula for dodecane, which has twelve carbon atoms?

More about skeletal structures

These structures show just the carbon skeleton of an organic compound. Note that the hydrogens in the carbon skeleton are not shown, but you must assume they are there. The structures also give a better representation of the angles in the carbon skeleton.

Hydrocarbon examples

Name	Structural formula	Skeletal formula
Ethane	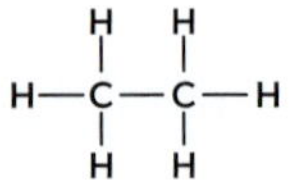	
Butane	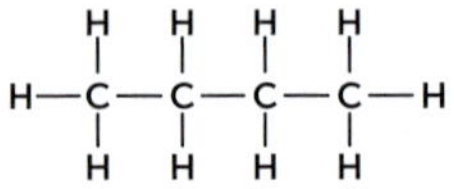	
Propene	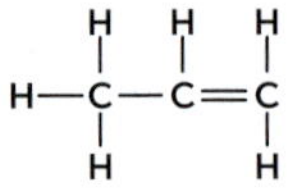	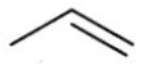
Methylpropane	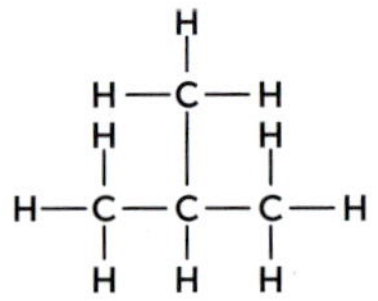	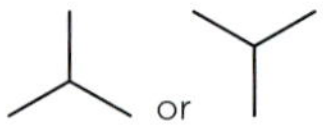

Some examples involving cyclic hydrocarbons are shown in Box 18.1.

Where other atoms or groups are substituted for hydrogen or carbon, they are added to the skeletal structure. For example:

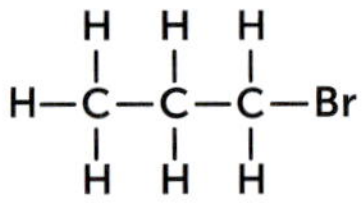 **becomes** 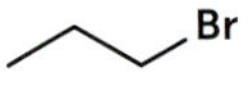

and

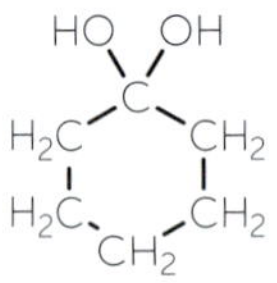 **becomes** 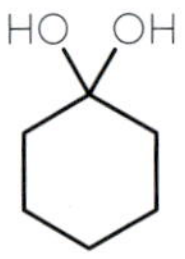

A tip: Don't combine aspects of a condensed structural formula with a skeletal structure in the same molecule. This can be very confusing – stick to one or the other!

Shapes of alkane molecules

Although we often draw the methane molecule (CH_4) as being flat, the shape of the molecule is *tetrahedral*, as drawn in Figure 18.1.

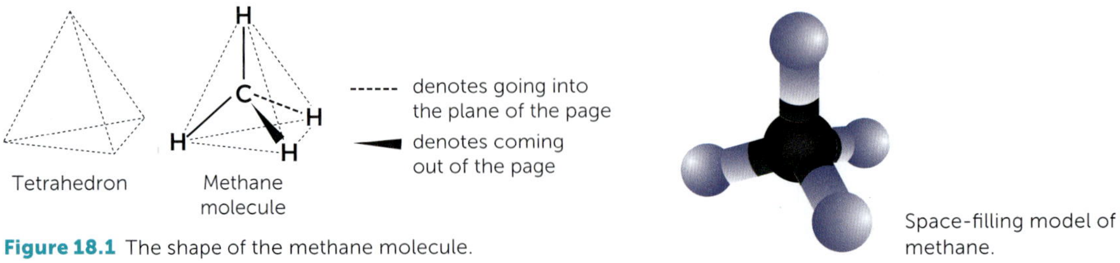

Figure 18.1 The shape of the methane molecule.

Isomers

There are two possible structures for C_4H_{10}:

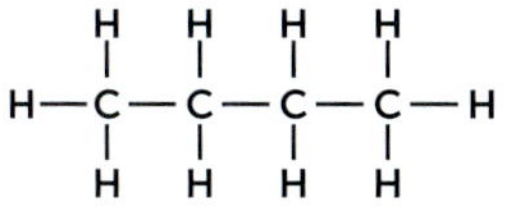

butane

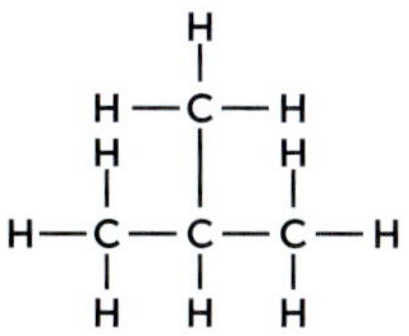

methylpropane

They have a different structure, although they both have the same molecular formula. One is called butane, the other is called methylpropane – the latter name describes the structure of the compound, as you will see later. These compounds are isomers of one another. They are *not the same compounds* and have different melting points, boiling points and solubilities. **Isomers are compounds that have the same molecular formula, but different molecular structures.** After butane, the longer the carbon chain of an alkane, the more **structural isomers** are possible for a particular molecular formula. For example, there are seventy-five **isomers of decane** ($C_{10}H_{22}$) and more than 300,000 isomers of eicosane ($C_{20}H_{42}$)!

EXERCISE **18B**

Shapes of the alkanes

Draw the three-dimensional structure of ethane (C_2H_6). Note that the methyl groups can freely rotate about the C—C bond; this is easy to see if you build a model.

EXERCISE **18C**

Isomers of pentane

Draw the structures of the three isomers of pentane. It may help to make models of the isomers. If you do not have a molecular modelling kit, use straws to represent bonds and Plasticine® balls, of different colours, to represent atoms of hydrogen and carbon.

Naming alkanes

The rules for naming alkanes are as follows:

1 Identify the longest carbon chain.

2 Identify the 'branches' on the longest carbon chain and name them according to the number of carbon atoms they contain, together with the ending 'yl': methyl (1), ethyl (2), propyl (3), butyl (4), and so on. Note that these groups are called alkyl groups and have the general symbol R–. For example, R-H could mean CH_3–H, CH_3CH_2–H or $CH_3CH_2CH_2$–H, etc.

3 Number the carbon atoms on the longest carbon chain to describe the positions of the branches – *use the lowest numbers possible.*

4 Write the branches in alphabetical order.

5 If there are more than one branch with the same name, use the prefixes di-, tri-, tetra- etc.

EXAMPLE **18.1**

Name the compound

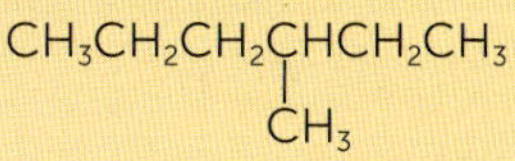

EXAMPLE **18.1** *(continued)*

Answer

1 Identify the longest carbon chain:

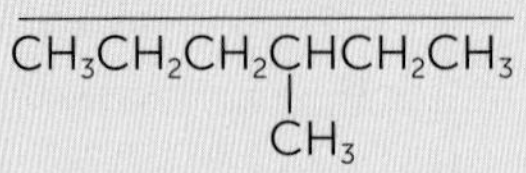

hexane

So, the compound is derived from hexane, the alkane with six carbon atoms.

2 Identify the branch:

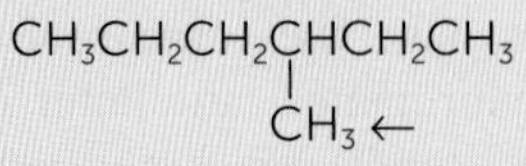

methylhexane

The chain has a branch with one carbon atom – a methyl group.

3 Number the position of the branch:

This can be done as

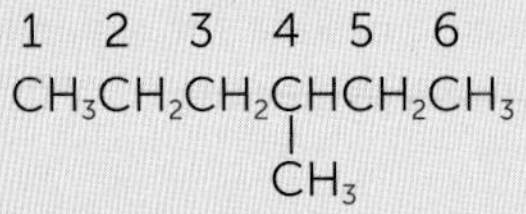

or as

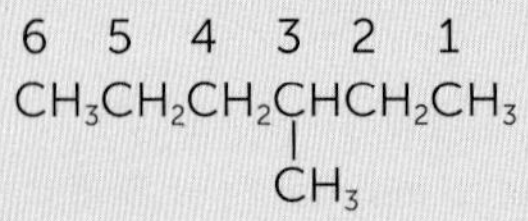

3-methylhexane

In the last structure, the methyl group is on the carbon atom with the lowest number. The name of the compound is therefore 3-methylhexane *and not* 4-methylhexane.

EXAMPLE **18.2**

Name the compound

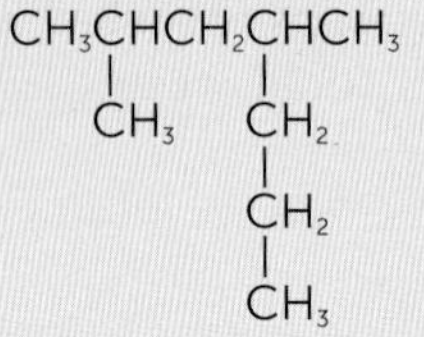

Answer

1 Identify the longest chain – a heptane:

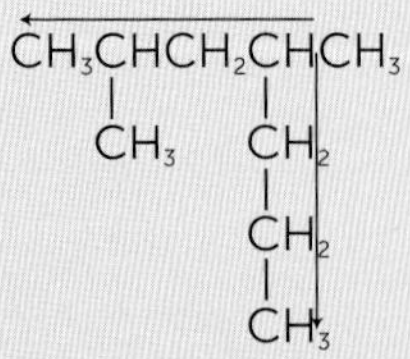

2 Identify the branches – there are two methyl branches (dimethyl):

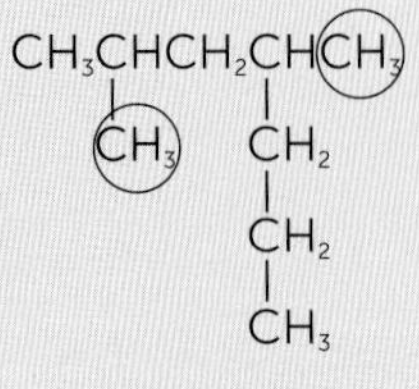

3 Number the carbon atoms: a methyl on C_2 and one on C_4:

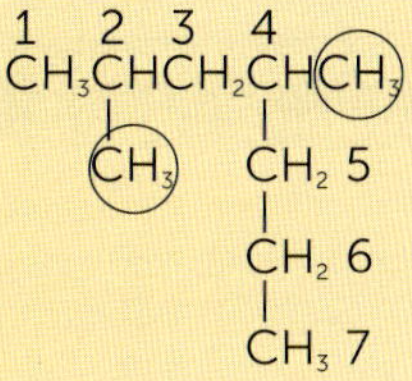

Therefore, the name is 2,4-dimethylheptane.

EXERCISE **18D**

Naming alkanes

Name the following:

(i) 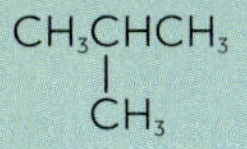

(ii) 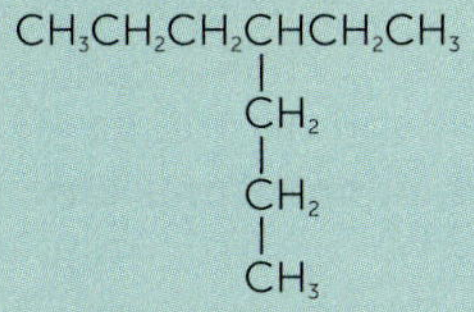

(iii) 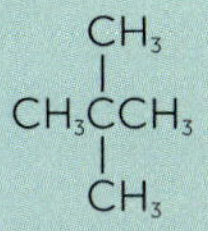

(iv) 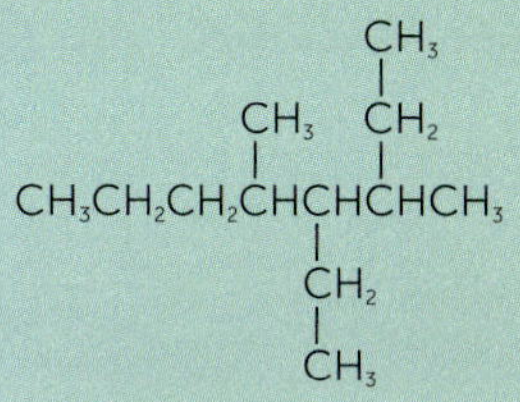

EXERCISE **18E**

Writing structural formulae from names of alkanes

Write the structural formulae of:

(i) 3-methylpentane

(ii) 2,2,5-trimethyloctane

(iii) 4-ethyl-2-methylheptane

(iv) 2,3,4-trimethylheptane

Physical properties of the alkanes

1 The alkane molecules contain carbon and hydrogen atoms covalently bonded together. There is very little difference between the electronegativities of these atoms and so the C–H bond is only weakly polar. Furthermore, in a symmetrical molecule such as methane, the weak polarities of the C–H bonds cancel out and the molecule is non-polar overall. Therefore, alkane molecules are either non-polar or weakly polar and the intermolecular forces between them are weak van der Waals forces, so that the lower alkanes are volatile. As the molecular masses of the alkanes increase (and thus the number of electrons in the molecules), the intermolecular forces between the molecules of the heavier alkanes become increasingly stronger. As the molecules of the alkanes get larger, therefore, their melting and boiling points increase in a reasonably regular fashion.

2 Because they are non-polar and covalent compounds, the alkanes are soluble in non-polar solvents such as benzene or ethoxyethane (diethyl ether) and insoluble in water.

EXERCISE **18F**

Physical properties of alkanes

Using data in the following table, plot a graph of the number of carbon atoms contained in the straight-chain alkane (*x* axis) against boiling point of the alkane (*y* axis). Draw a curve to join up the points.

(i) Which alkane with the most number of carbon atoms is a gas at room temperature?

(ii) Estimate the boiling point of octane.

Alkane	Number of carbon atoms	Boiling point/K
Methane	1	109
Ethane	2	185
Propane	3	231
Butane	4	273
Pentane	5	309
Hexane	6	342
Heptane	7	372
Octane	8	***
Nonane	9	424
Decane	10	447

Sources of alkanes

Alkanes are obtained from crude oil and natural gas. These relatively simple organic chemicals can be used to make other, more complicated, organic compounds.

Oil

An oil refinery is combination of chemistry and engineering on a vast scale

Oil and gas are trapped in rock and are the decayed remains of tiny marine plants and animals. Natural gas contains 60–90 per cent methane. Crude oil is a mixture of many carbon compounds with different chain lengths and is not very useful if left untreated. The oil is therefore separated into groups of compounds with roughly the same chain length in an **oil refinery**. The technique that is used to refine the oil is called **fractional distillation**. The oil is heated and passed into a tall fractionating **column** (Figure 18.2). Compounds with large molecules and high boiling points are collected at the bottom of the tower, whereas the more volatile components, with smaller molecules, are collected near the top. Each group of compounds collected is called a **fraction**. Each fraction is still a complicated mixture, since it contains alkanes of a range of carbon numbers and isomers of each carbon number. In addition to alkanes, crude oil contains some **cycloalkanes** and **aromatic compounds** – the proportion of each present depends upon the source of the oil. Aromatic compounds will be discussed later. Most synthetic chemicals produced by industry originate from the alkanes derived from crude oil: plastics, pharmaceuticals, adhesives and synthetic fibres are just a few examples.

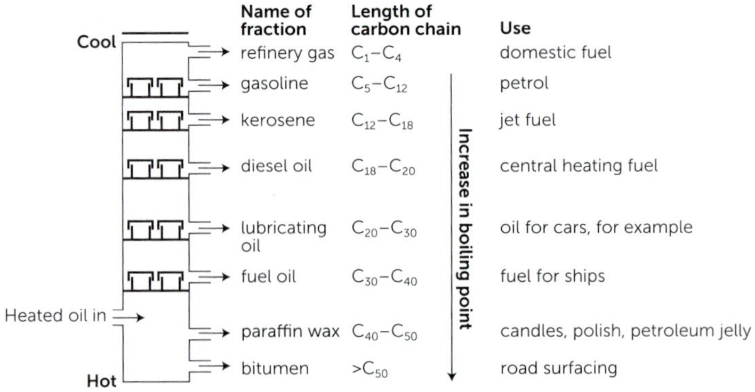

Figure 18.2 A fractionating column.

Natural gas is used to produce hydrogen and carbon monoxide by reacting it with steam at 700°C at high gas pressure and in the presence of a nickel catalyst. There are two stages. In the first, that of steam reforming (SR), a mixture of CO and H_2 (named 'synthesis gas') is produced. In the second, more H_2 is generated but CO_2 is also produced:

$$CH_4(g) + H_2O(g) \rightleftharpoons \underbrace{CO(g) + 3H_2(g)}_{\text{synthesis gas}} \qquad (A)$$

$$CO(g) + H_2O(g) \rightleftharpoons CO_2(g) + H_2(g) \qquad (B)$$

Reaction (B) is called the water–gas shift reaction (WGSR). The hydrogen produced is used in oil refining and in the manufacture ammonia in the Haber–Bosch process (see page 278 and 490).

Coal

Coal is formed by the decomposition of plants under pressure. Some coals contain 2–6 per cent sulfur. Burning these coals has led to pollution of the atmosphere and acid rain.

When coal is heated in the absence of air, a process called destructive distillation, it yields three products: coal gas (mainly CH_4 and H_2 with some toxic carbon monoxide); coal tar (a range of organic materials which can be separated by distillation); and coke (a grey solid fuel used in the making of steel).

At present, most organic chemicals are derived from petroleum and produced in oil refineries. It is important that sulfur is removed from petroleum products and this is achieved using hydrogen, which converts sulfur to hydrogen sulfide, H_2S, a process called hydrotreating.

Cracking

The very long-chain alkanes are less useful than the smaller alkanes. For example, there is much more of a demand for petrol than for lubricating oil. Long-chain alkanes can be broken into smaller molecules, using a process called **cracking**. In **thermal cracking**, alkanes are heated to a high temperature. Smaller alkanes, hydrogen and small **alkenes**, such as **ethene**, are produced. Alkenes are members of another family of organic compounds, which are discussed later on in this unit; they are useful starting materials for the synthesis of many organic chemicals. **Catalytic cracking** involves heating higher boiling point petroleum fractions in the presence of a silica–alumina catalyst. Shorter alkanes that are suitable for use in petrol are made.

BOX **18.2**

Quality of petrol

Long-chain hydrocarbons tend to burn unevenly in car engines and so cause a rattling noise (knocking). Branched alkanes, produced from straight-chain alkanes by catalytic cracking, are added to the fuel to prevent knocking because they burn more evenly.

The alkane 2,2,4-trimethylpentane has good antiknock properties, whereas the antiknock properties of heptane are poor.

The octane number of petrol is a measure of its quality – an octane number of 0, for a particular petrol, means that it burns as well as pure heptane. If the octane number is 100, then the fuel is equivalent to pure 2,2,4-trimethylpentane. Many petrols have an octane number of 70; they burn as would a mixture of 70 per cent 2,2,4-trimethylpentane and 30 per cent heptane.

The burning of fuel in a car engine produces a variety of pollutants, such as carbon monoxide, unburned hydrocarbons and nitrogen oxides. Catalytic converters change these compounds into more environmentally friendly compounds.

BOX **18.1**

Cycloalkanes

Molecules such as $CH_3CH_2CH_2CH_3$ have their carbon atoms connected in a chain. Carbon atoms can also be joined together in rings, in which case a cyclic molecule is formed. **Cyclic** molecules are often represented by polygons – the corners of the polygon represent a carbon atom together with the hydrogens joined to it. Some examples are:

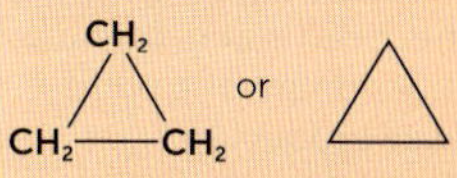

Cyclopropane

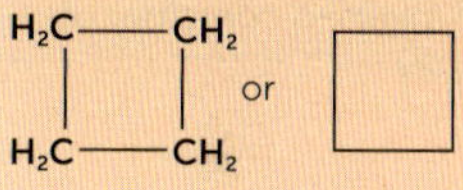

Cyclobutane

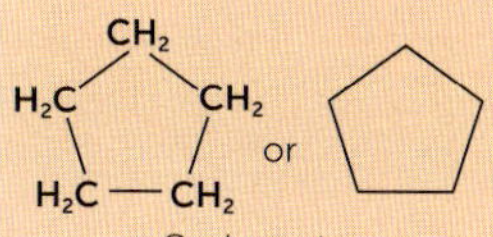

Cyclopentane

EXERCISE **18G**

Cycloalkanes

(i) Can you write a general molecular formula for the cycloalkanes?

(ii) Cyclopentane is a much more stable and less reactive compound than cyclopropane. Can you give a reason why? (Hint: Make models of both molecules.)

(iii) What are the molecular and structural formulae of hexane and cyclohexane?

BOX **18.3**

Putting a smell in dangerous gases

Crude propane is also known as LP (liquid petroleum) gas. It is generally used in rural areas that are not served by gas pipelines. LP contains about 80 per cent propane and the rest butane. It can be stored as a liquid under normal temperatures, provided it is stored under pressure. The liquid propane occupies a much smaller volume than the gaseous propane. When pressurized tanks containing LP are opened, the gaseous fuel is released.

Propane is extremely flammable and like methane (natural gas) has no taste, colour or smell. To warn users of an accidental leak, both gases are mixed with a strong-smelling substance such as ethyl mercaptan, C_2H_5SH (found in the aroma of skunks and rotting meat).

Anaerobic production of methane

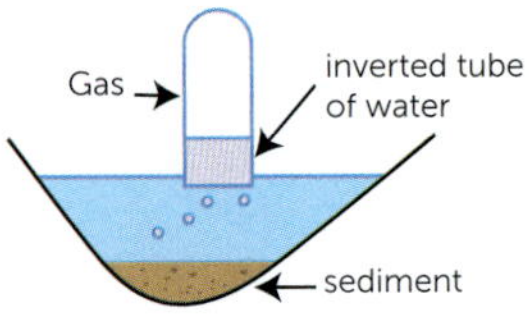

Anaerobic digestion. Over time, dead plants forming the sediment of a pond decay producing a gas mixture (mainly methane) which may be collected in an inverted container of water.

If a stagnant pond is stirred with a stick, bubbles of gas move to the surface. If the gas is collected by displacing water in a tube and the tube exposed to a flame back in the lab, a flame is observed. The flammable gas is commonly called marsh gas and it is mainly methane with some CO_2 and trace amounts of hydrogen sulfide, which has the smell of rotten eggs. The process by which methane is produced in the oxygen-free depths of a pond or bog is natural and is called **anaerobic digestion** (AD) and it involves the breakdown of organic material (such as dead plants) by bacteria in the absence of oxygen. For example, if the organic material is glucose:

$$C_6H_{12}O_6 \rightarrow 3CO_2 + 3CH_4$$

AD is maximized by design in landfill sites and in commercial vats ('digesters') of food and agricultural waste. In both cases the methane (known as 'biogas') may be harvested for use as an energy source.

Chemical properties of alkanes

Alkanes are termed **saturated hydrocarbons** because their molecules contain the maximum number of hydrogen atoms, and other atoms or groups cannot 'add on' to alkane molecules.

1 Combustion

Alkanes are not particularly reactive; they do not readily react with dilute acids, alkalis or oxidizing agents such as potassium permanganate and sodium dichromate. They do, however, burn in a plentiful supply of oxygen to form carbon dioxide and water:

$$CH_4(g) + 2O_2(g) \rightarrow CO_2(g) + 2H_2O(l)$$

The reaction is very exothermic, which is why alkanes are used as fuels. The heavier alkanes, however, tend to burn in air with a sooty flame because there is insufficient oxygen in the air for all the carbon in the alkane to be converted to carbon dioxide (**complete combustion**), so some carbon and carbon monoxide are formed (**incomplete combustion**) (see page 479).

2 Halogenation

Chlorine or bromine react with alkanes to form chloroalkanes or bromoalkanes (the general term is **halogenoalkanes**). The reaction with halogens is called **halogenation** and, if it takes place in the presence of light, it is termed a **photochemical** reaction.

When methane, for example, reacts with excess chlorine gas in low intensity UV (e.g. sunlight) chlorine atoms substitute for hydrogen atoms and **chloromethane (methyl chloride)** is produced. Depending on the amounts of methane and chlorine present, further substitution can occur to form CH_2Cl_2, $CHCl_3$ and eventually CCl_4. If propane is chlorinated, then the first substitution may be on an end carbon atom or a middle carbon atom so that the first product is a mixture of two possible **isomers**, namely

$CH_3CH_2CH_2Cl$ or $CH_3CHClCH_3$
1-chloropropane 2-chloropropane

Because a mixture of products may be produced by the halogenation of alkanes, comprising different isomers and/or molecules with different degrees of substitution, the reactions are not suitable for the laboratory preparation of halogenoalkanes. The process is more useful on an industrial scale where separation of individual compounds can be economically worthwhile.

For the mechanism of halogenation of alkanes see page 368.

DID YOU KNOW?

The combustion of methane in oxygen is well known, but it is less well known that methane also burns in chlorine, producing carbon (soot) and hydrogen chloride (HCl):
$CH_4(g) + Cl_2(g) \rightarrow C(s) + 2HCl(g)$

BOX **18.4**

Naming halogenoalkanes

The rules for naming halogenoalkanes are very similar to those for naming alkanes; the name of the halogenoalkane is derived from the parent alkane. Remember, however, *to choose the longest carbon chain that also contains the halogen(s)* as a starting point.

The halogenoalkane is prefixed fluoro-, chloro-, bromo-, iodo-, dichloro-, tribromo- etc. Substituents are again arranged in alphabetical order:

$CHCl_3$
trichloromethane

$CH_3CH_2CHCH_2CHCH_3$ (with CH_3 on C-4 and Br on C-2)
2-bromo-4-methylhexane

EXERCISE **18H**

The mild chemical reducing powers of methane

The following diagram shows the equipment used to demonstrate the mild reducing property of methane.

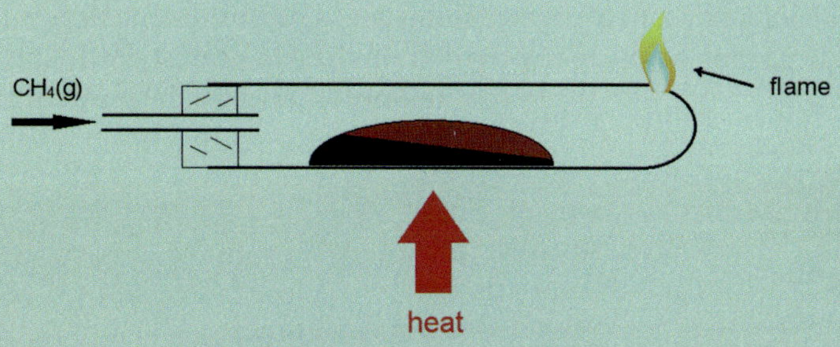

Methane is passed over hot (black) copper oxide, which slowly turns red/brown. Unused methane is burnt through a hole in the test tube, through which product gases also pass.

(i) Identify the red/brown solid and the two product gases.
(ii) Write the overall equation for the reduction reaction.
(iii) The experiment is repeated with hydrogen gas instead of methane. The reaction is very much faster. Write an equation for the reaction.

EXERCISE 18I

Ethene molecule

(i) Draw a Lewis structure to show the arrangement of the outer electrons of the atoms in an ethene molecule.

(ii) Make a model of the ethene molecule. Is there free rotation about the carbon–carbon double bond? How does this compare with ethane?

18.2 Alkenes

Another family of hydrocarbons is the **alkenes**. Alkene molecules contain a **double (C=C) bond** that cannot be rotated. They have the general formula C_nH_{2n} and their names end in 'ene' (Table 18.2). The first three are gases at 25°C.

Shape of the ethene molecule

In ethene, each carbon atom lies at the centre of a triangle. The corners of each triangle are the two attached hydrogen atoms and the other carbon atom. The bond angles are approximately 120°:

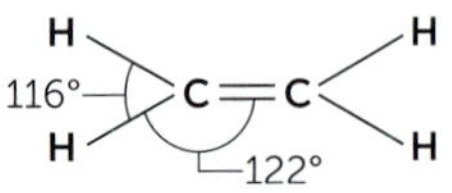

Table 18.2 The first three alkenes

Name	Molecular formula	Structural formula
Ethene (Old name is ethylene)	C_2H_4	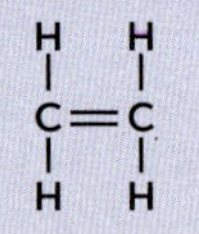
Propene	C_3H_6	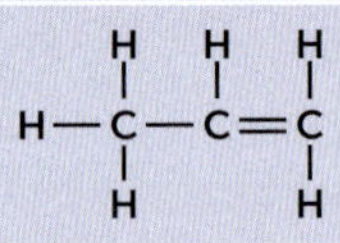
But-1-ene	C_4H_8	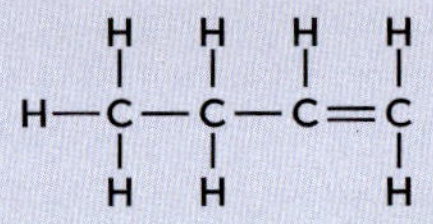

Naming alkenes

The following additional rules apply when naming alkenes:

1 Choose the longest chain of carbon atoms *that contains the double bond.*
2 Number the carbon atoms in the chain from the end nearest the double bond.
3 Pick the carbon atom with the lowest number to describe the position of the double bond.

EXAMPLE 18.3

4 3 2 1
$CH_3CH_2CH{=}CH_2$ is called but-1-ene *not* but-2-ene

EXAMPLE 18.4

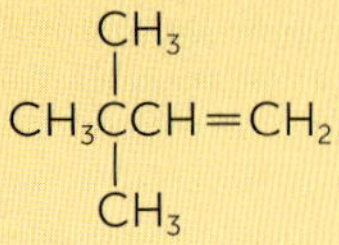

is called 3,3-dimethylbut-1-ene

EXERCISE 18J

Naming alkenes 1

Name the following:

(i) $CH_3CH_2CH_2CH_2CH{=}CH_2$

(ii) $CH_3CH{=}CHCH_3$

(iii)

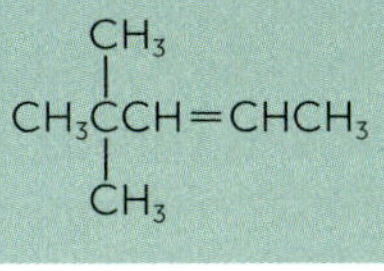

(iv) $CH_2{=}CHCH_2Br$

EXERCISE 18K

Naming alkenes 2

Write structural formulae for:

(i) hex-3-ene

(ii) 2-methylpropene

(iii) 2,3-dimethylhex-2-ene

(iv) 3-chloro-2-methylpropene

(v) 3-bromohex-1-ene.

Isomerism in alkenes

There are a number of possible structural arrangements for a compound of formula C_4H_8:

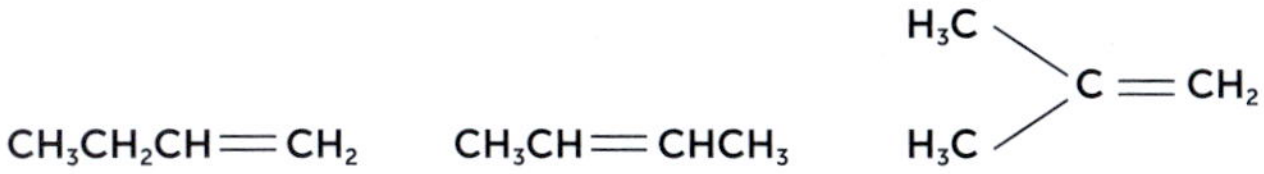

but-1-ene but-2-ene methylpropene

Experiments show that there are *four* isomers of C_4H_8. How can we account for the fourth isomer? In fact, there are two types of but-2-ene – they have different boiling points and melting points. In each butene, the atoms are arranged as follows:

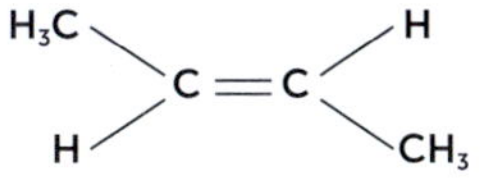

trans-but-2-ene

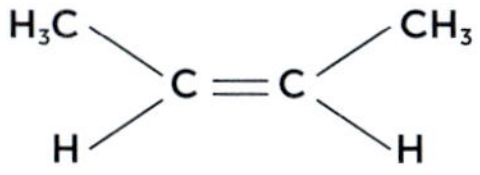

cis-but-2-ene

Both molecules are flat (planar). In *trans*-but-2-ene the methyl groups are on opposite sides of the double bond (*trans* means 'across'), whereas in *cis*-but-2-ene, the methyl groups are on the same side of the double bond (*cis* means 'on the side'). The two molecules are not the same because *no free rotation can occur around the carbon–carbon double bond*. These two butenes differ from one another only in the way that the atoms are orientated in space. They are **geometric** isomers.

The prefixes *cis* and *trans*, however, are only useful when two atoms, or groups, are the same on adjacent carbons forming the double bond.

What happens when you have to distinguish between two isomers like this? Can you say which one is *cis* or which one is *trans*?

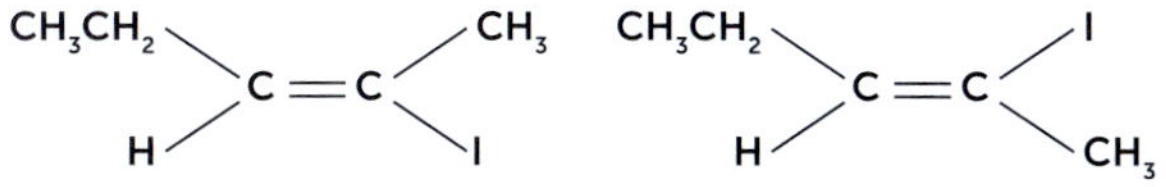

Not really. We have to find a system that works for these geometric isomers. This is called the ***E*–*Z*** system.

You look at what is attached to each end of the double bond, in turn, and assign each a 'priority' according to a set of rules called the Cahn–Ingold–Prelog rules:

- Look at the atoms directly attached to each carbon of the double bond. Rank them according to decreasing atomic number.

priority of common atoms: I>Br>Cl>S>F>O>N>C>H

(1) (9)

high priority - - - - - - -➤ low priority

If both high priority atoms are on the same side of the double bond, it is designated *Z* (from the German *zusammen* = together). If the high priority atoms are on opposite sides of the double bond, it is *E* (*entgegen* = across).

What happens in groups of atoms where the connected atom is the same, but the groups are different?

If there are two atoms with equal rank, go along the two substituent chains, working away from the double bond, until there is a difference. Find which of the chains has the first connection to an atom with the highest priority (the highest atomic number). That chain has the higher priority.

So $-CH_2CH_3$ is higher priority than $-CH_3$ and $-CH_2Br$ is higher than $-CH_2Cl$.

So, in the example above:

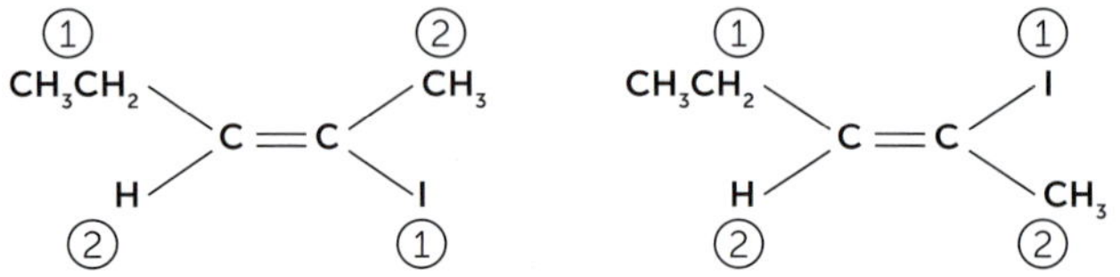

It gets a little more complicated if you have double (or triple) bonds **in the chain**. Treat the atoms as if they were bonded to two (or three) single bonded atoms.

So C=O has a higher priority than C–O (think of C=O as two C–Os)

Similarly, C≡N has a higher priority than C–N

These rules are also useful in labelling the three-dimensional structure (or configuration) of enantiomers. See page 358.

BOX **18.5**

More about isomerism

There are two major types of isomerism: *structural isomerism* and *stereoisomerism*.

In **structural isomerism**, first met in the discussion of alkanes and which exists between but-1-ene, but-2-ene and methylpropene, the compounds have their atoms bonded to different groups of atoms. For example, consider the underlined carbon atoms in the structural isomers of butene (C_4H_8):

$CH_3CH_2\underline{C}H{=}CH_2$

$\underline{C}$ attached to: — H
— CH_2CH_3
= CH_2

$CH_3CH{=}\underline{C}HCH_3$

$\underline{C}$ attached to: — H
— CH_3
= $CHCH_3$

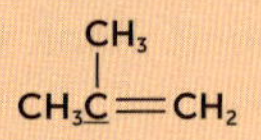

$\underline{C}$ attached to: — CH_3
= CH_2

In **stereoisomerism** all atoms in the compounds are attached to the same partners, but the partners are arranged differently in space. Geometric isomerism – the isomerism shown by some alkenes because there is no free rotation about the double bond – is one type of isomerism in this class. The geometric isomers of 1,2-dichloroethene are shown below:

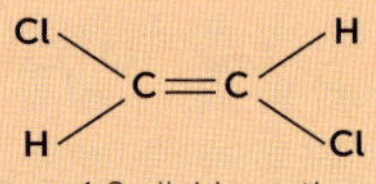

trans-1,2-dichloroethene

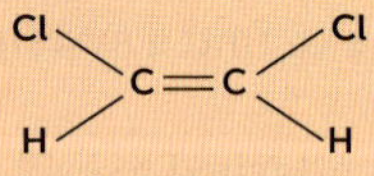

cis-1,2-dichloroethene

Here, equivalent atoms in the two molecules are attached to the same atoms, but these atoms are arranged differently in space.

Another type of stereoisomerism is optical isomerism, which we will meet later.

EXERCISE **18L**

Structural and geometric isomerism

Below are two isomers with the formula $C_2H_2Br_2$:

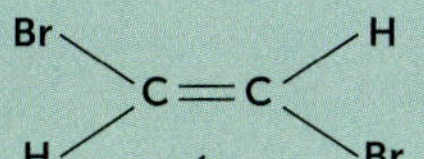

1

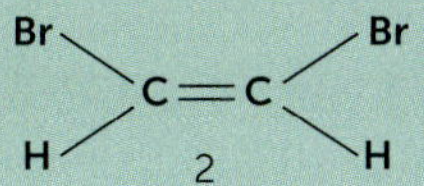

2

(i) A third isomer exists. Draw its structural formula and label it '3'.

(ii) What type of isomerism exists between the isomers 1 and 2?

(iii) What type of isomerism exists between isomers 2 and 3?

(iv) Name compounds 1, 2 and 3.

Chemical properties of alkenes

The physical properties of the alkenes are very similar to those of the alkanes. Chemically, however, they are much more reactive than the alkanes because they contain double bonds. Alkenes are said to be **unsaturated hydrocarbons**. Because of the double bond, their molecules can add on more atoms of hydrogen. The unsaturation of alkenes gives rise to reactions with substances other than hydrogen. The chemical properties of alkenes are as follows:

1 Combustion

Like the alkanes, the alkenes burn in a plentiful supply of oxygen to form carbon dioxide and water:

$$C_2H_4(g) + 3O_2(g) \rightarrow 2CO_2(g) + 2H_2O(l)$$

2 Addition

Because alkenes are unsaturated, they will undergo reactions in which the carbon–carbon double bond becomes a single bond and other atoms bond to the carbons of the original double bond. A reaction in which two molecules combine to yield a single molecule of product is called an **addition reaction**. For ethene the general reaction may be written as

$$CH_2{=}CH_2 + X{-}Y \rightarrow \underset{X}{CH_2}{-}\underset{Y}{CH_2}$$

Some examples of the reagents 'X–Y' are given below:

Bromine (Br_2) Br–Br

Alkenes react rapidly with bromine at room temperature to form colourless dibromoalkanes. For example, if ethene gas is bubbled into **bromine water** (an aqueous solution of bromine) the colour of the solution turns from reddish-brown to colourless. This reaction is often used to test for unsaturation in a hydrocarbon:

$$CH_2{=}CH_2 + Br{-}Br \rightarrow \underset{Br}{CH_2}{-}\underset{Br}{CH_2}$$

Chlorine (Cl_2) Cl–Cl

With chlorine, a similar reaction occurs to that with bromine but without the distinctive colour change:

$$CH_2{=}CH_2 + Cl{-}Cl \rightarrow \overset{\displaystyle Cl}{\overset{|}{C}}H_2{-}\overset{\displaystyle Cl}{\overset{|}{C}}H_2$$

Hydrogen (H_2) H–H

The addition of hydrogen to an alkene is called hydrogenation and the product is an alkane. Hydrogenation is carried out in the presence of a catalyst (nickel, palladium or other platinum metals) and at a high pressure of hydrogen:

$$RCH{=}CH_2 + H{-}H \xrightarrow[500°C]{Ni} R\overset{\displaystyle H}{\overset{|}{C}}H{-}\overset{\displaystyle H}{\overset{|}{C}}H_2$$

Hydrogen halides – (HCl) H–Cl, (HBr) H–Br and (HI) H–I

The gaseous hydrogen halides react directly with the alkene to produce halogenoalkanes:

$$RCH{=}CH_2 + H{-}Cl \longrightarrow R\overset{\displaystyle Cl}{\overset{|}{C}}H{-}\overset{\displaystyle H}{\overset{|}{C}}H_2$$

Sulfuric acid H–OSO_3H (H_2SO_4)

Alkenes react with cold, concentrated sulfuric acid to form alkyl hydrogensulfates:

$$RCH{=}CH_2 + H{-}OSO_3H \longrightarrow R\overset{\displaystyle OSO_3}{\overset{|}{C}}H{-}\overset{\displaystyle H}{\overset{|}{C}}H_2$$

3 Oxidation

Alkenes can be oxidized by cold, slightly alkaline potassium permanganate (manganate(VII)) to make substances called 1,2-diols, containing two OH groups on adjacent carbon atoms. The potassium permanganate solution turns from purple to brown. This is another test (the **Baeyer test**) that is often used to detect unsaturation (i.e. the presence of double or triple carbon bonds) in a hydrocarbon:

$$CH_2{=}CH_2 \xrightarrow{KMnO_4} \underset{\text{ethane-1,2-diol}}{\overset{\displaystyle OH}{\overset{|}{C}}H_2{-}\overset{\displaystyle OH}{\overset{|}{C}}H_2}$$

4 Polymerization

Under the right conditions, alkenes will undergo addition reactions with each other. The reaction is called **polymerization**. For example, many ethene molecules will join together to form polythene – the double bonds break and the molecules link up:

$$\underset{\substack{\text{where } n \text{ = a large number} \\ \text{(typically 20 000)}}}{n(CH_2{=}CH_2)} \xrightarrow[\text{or catalyst}]{O_2\text{, heat, pressure}} \cdots CH_2{-}CH_2{-}CH_2{-}CH_2{-}CH_2{-}CH_2\cdots$$

The product can also be written as

$$\cdots(CH_2{-}CH_2{-})_n\cdots$$

Long-chain molecules called **polymers** are formed from the small ethene molecules, the **monomers**. The long, thin molecules of a polymer may also be called **linear macromolecules**. Polythene is an unreactive solid; it can be easily moulded and is used for plastic bags, bottles, washing-up bowls and plastic piping. The word 'plastic' means 'easily moulded'. Substituted ethene molecules can undergo the same addition reaction to give other polymers:

Styrene

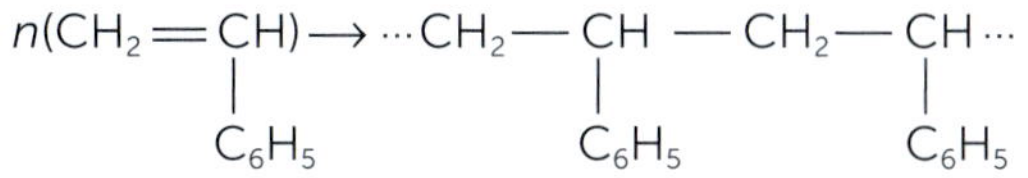

styrene   polystyrene

Vinyl chloride

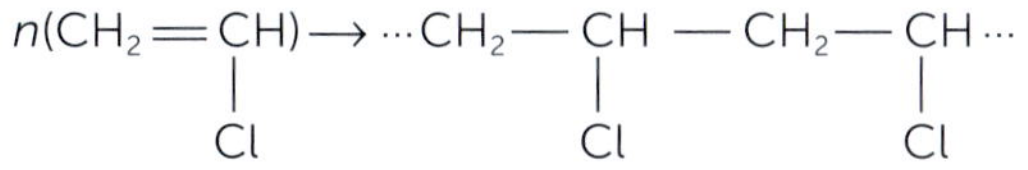

vinyl chloride   poly(vinyl chloride) (PVC)

Note that the starting molecules (monomers) are named by their common, older names so that you can see how the names of the polymers arose.

BOX **18.6**

Better model for bonding in alkenes

Although Lewis structures can be used to describe bonding in alkenes, they do not really explain why alkenes are so much more chemically reactive than alkanes. In order to explain the reactivity of alkenes, a more sophisticated model of bonding must be used. In the Lewis structure model for ethene, two electron pairs are shared by the carbon atoms:

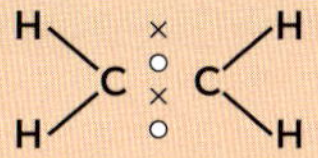

The Lewis structure suggests that both bonds in this double bond are of the same type. However, in our 'improved' model for the bonding in ethene, you will see that they are not the same.

First, we need to consider that the bonding electrons in the atoms concerned are in orbitals. Covalent bonds are formed by overlap of these orbitals. Each carbon in ethene forms three 'normal' covalent bonds by overlap with the orbitals containing electrons on the other carbon and two hydrogen atoms. The electron cloud

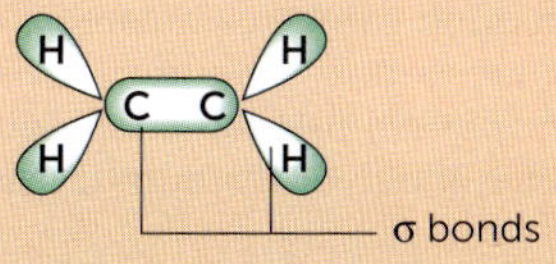

Figure 18.3 σ bonds in ethene.

between the nuclei of the atoms in each bond 'glues' the atoms together. Normal covalent bonds of this type are called σ (sigma) bonds (Figure 18.3).

One electron is left over on each carbon atom. The 'left over' electrons are in *p* orbitals, as shown in Figure 18.4.

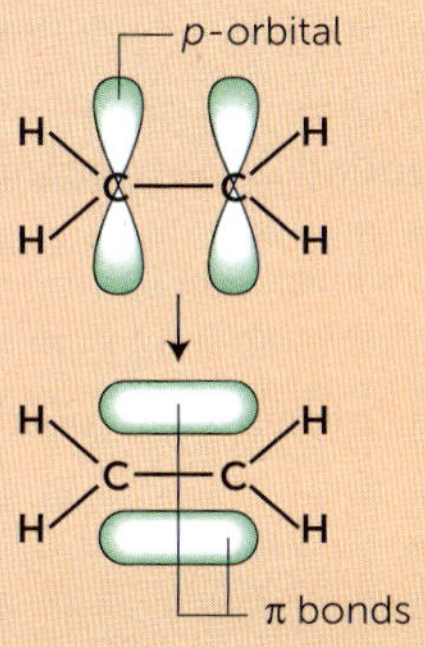

Figure 18.4 Formation of a π bond in ethene.

A π (pi) bond is formed by sideways overlap of these *p* orbitals on the carbon atoms. In this type of bond, the electron cloud is concentrated above and below the horizontal plane joining the two carbon atoms. The double bond between the carbon atoms in ethene therefore consists of a σ bond and a π bond. The exposed negatively charged electron cloud in the π bond is open to attack by positively charged species and this is why alkenes are so reactive.

EXERCISE **18M**

Reactions of alkenes

(i) Write an equation for the reaction of ethene with hydrogen bromide. Name the product.

(ii) Write an equation for the reaction of ethene with cold, concentrated sulfuric acid. Name the product.

(iii) Write an equation for the reaction of propene with bromine. Name the product.

(iv) Write the structural formula of the product of the reaction of propene with cold, alkaline potassium permanganate.

(v) Write an equation for the formation of the polymer Orlon, formed by polymerization of the monomer $CH_2{=}CHCN$ (acrylonitrile).

18.3 Alkynes

The **alkynes** are a family of hydrocarbons that contain a carbon–carbon triple bond, with the general formula C_nH_{2n-2}. The molecule cannot rotate about its C≡C bond. The first member of the family is ethyne (old name acetylene). Ethyne is a linear molecule:

$$H-C\equiv C-H$$

When it burns in oxygen, it burns with a sooty flame unless there is extra oxygen provided, in which case a great deal of heat is generated – oxy-acetylene torches are used for cutting and welding metals:

$$2C_2H_2(g) + 5O_2(g) \rightarrow 4CO_2(g) + 2H_2O(l)$$

The next member of the family, propyne, is obtained by replacing one of the hydrogen atoms in ethyne by a $-CH_3$ group:

$$CH_3-C\equiv C-H$$

Alkynes are unsaturated and undergo addition reactions in a similar fashion to alkenes. This time it is possible to obtain two addition products, depending on the number of moles of reagent reacting with the alkyne. If one mole of hydrogen, for example, is added to one mole of ethyne, the product is ethene:

$$H-C\equiv C-H(g) + H_2(g) \xrightarrow[150°C]{\text{Ni catalyst}} H_2C{=}CH_2(g)$$

Two moles of hydrogen produce ethane:

$$H-C\equiv C-H(g) + 2H_2(g) \xrightarrow[150°C]{\text{Ni catalyst}} H_3C-CH_3(g)$$

EXERCISE **18N**

Alkynes

(i) Give the structural formula of but-1-yne, a third member of the family of alkynes.

(ii) What are the names and formulae of the two possible products formed when propyne is reacted with hydrogen in the presence of a nickel catalyst?

(iii) Write an equation for the reaction that occurs when ethyne is reacted with *excess* bromine. Name the product.

BOX **18.7**

Analysis of organic compounds

Often, one of the first steps in the identification of an unknown organic compound is to submit the compound for quantitative elemental analysis. This type of analysis will determine the percentage by mass of the elements present in the compound. From the results of the analysis, the empirical formula of the compound can be calculated. To find out the relative amounts of carbon and hydrogen in a hydrocarbon, a weighed sample of the hydrocarbon is passed through a tube packed with copper(II) oxide at a temperature of about 700°C. The copper(II) oxide oxidizes the carbon in the hydrocarbon to carbon dioxide and the hydrogen to steam:

'CH_2' from organic compound + 3CuO → 3Cu + CO_2 + H_2O

The gases pass out of the combustion tube and through two weighed tubes containing first a drying agent, to absorb the water, and then a strong base (such as sodium hydroxide) to absorb the carbon dioxide formed. The increase in the mass of each tube gives the mass of each product from the combustion of the hydrocarbon.

For example, a sample of methane with a mass of 7.25 mg produced 19.90 mg of carbon dioxide and 16.17 mg of water.

$$\text{Mass of carbon present} = \text{mass of carbon dioxide} \times \frac{M(\text{C})}{M(\text{CO}_2)}$$

$$= 19.90 \times \frac{12}{44} = 5.4\ \text{mg}$$

$$\text{Mass of hydrogen present} = \text{mass of water} \times \frac{M(2\text{H})}{M(\text{H}_2\text{O})}$$

$$= 16.17 \times \frac{2}{18} = 1.8\ \text{mg}$$

The percentage composition of the methane is therefore

$$\text{Carbon: } \frac{5.4}{7.25} \times 100 = 75 \text{ per cent;} \quad \text{Hydrogen: } \frac{1.8}{7.25} \times 100 = 25 \text{ per cent}$$

The empirical formula of an unknown compound can be worked out from its percentage composition using the calculations in Unit 8.

Note that it is possible, using various methods, to analyse for other elements in an organic compound, such as nitrogen, sulfur and halogens.

EXERCISE 18O

Quantitative analysis of compounds

(i) A 7.02 mg sample of a hydrocarbon X gave 21.99 mg of carbon dioxide and 8.95 mg of water on combustion. Calculate the percentage composition of the elements in the hydrocarbon and determine its empirical formula.

(ii) X has molecular mass 42. X reacts with excess hydrogen in the presence of a nickel catalyst to form Y. X decolorizes bromine water to form a compound Z. Write structural formulae for X, Y and Z.

18.4 Aromatic hydrocarbons

The families of hydrocarbons studied so far, those with linear chains of carbon as the 'backbone' of the compounds, are classed as **aliphatic** hydrocarbons. There is another class of hydrocarbons – **aromatic** hydrocarbons. These are compounds that contain a benzene ring, or behave chemically like benzene. Benzene has the molecular formula C_6H_6, and its structural formula, with three double bonds, was first suggested by Kekulé in 1865:

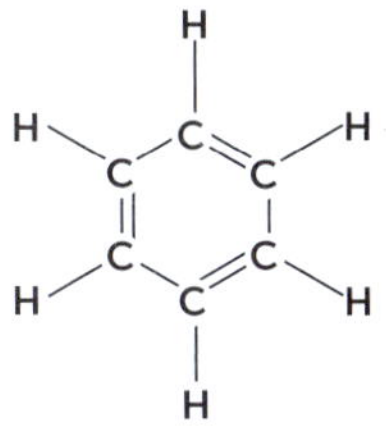

This was usually abbreviated to:

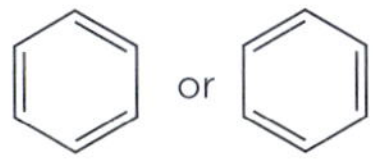

In fact, none of these structures correctly describes benzene. All the carbon–carbon bond lengths in benzene have been found to be the same, but the structures above do not predict this since double bonds are shorter than single bonds between the same atoms. Also, and unlike ethene, benzene does not readily undergo addition reactions, which we would not expect in a compound that contains double bonds. These days the benzene ring is often written as below, although the Kekulé structure is favoured when describing more complex molecules and reaction mechanisms (see Unit 20):

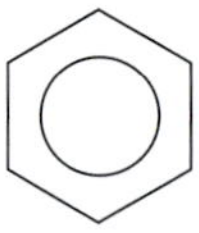

This structure – a hexagon with a 'hole'! – represents the fact that electrons from the carbon–carbon double bonds are 'spread out' or **delocalized** over the whole molecule, which makes the molecule thermodynamically more stable. A Lewis structure is not sophisticated enough to describe the bonding in benzene. In order to obtain a more satisfactory description of the bonding, the electrons must be treated as 'clouds' (rather than 'crosses' or 'dots'). (See Box 18.8.)

BOX **18.8**

The bonding in benzene in greater detail

The observed properties of benzene can be explained using a more sophisticated model than that proposed by Kekulé for the bonding in benzene. Each carbon atom in the ring is bonded to a hydrogen and to one carbon on either side of it by σ bonds. Every carbon has one unused *p* orbital, containing a single electron, perpendicular to the plane. This gives a *planar* 'skeleton' as shown in Figure 18.5.

The *p* orbitals overlap sideways with their neighbours to form π clouds or 'doughnuts' of electron density, above and below the molecule, as shown in Figure 18.6. The electrons no longer 'belong' to the carbon atom from which they originate; they are free to 'wander' around the π clouds and are delocalized.
Delocalization of the 'left over' *p* electrons is believed to make benzene more stable than would be expected if the Kekulé structures were correct, and explains why benzene does not react as if it had three double bonds. The model predicts that all the carbon–carbon bond lengths in benzene would be of the same length, which they are. The length of each bond is between those normally observed for C=C and C–C.

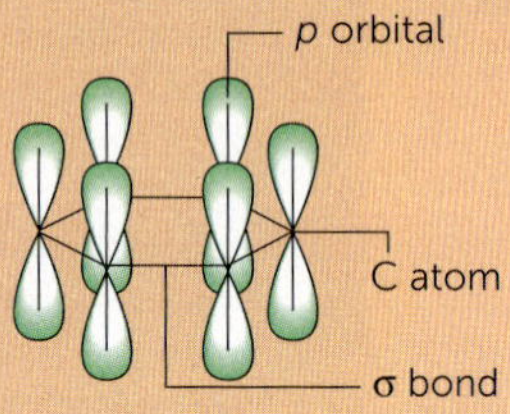

Figure 18.5 Carbon skeleton in benzene (hydrogen atoms have been omitted for clarity).

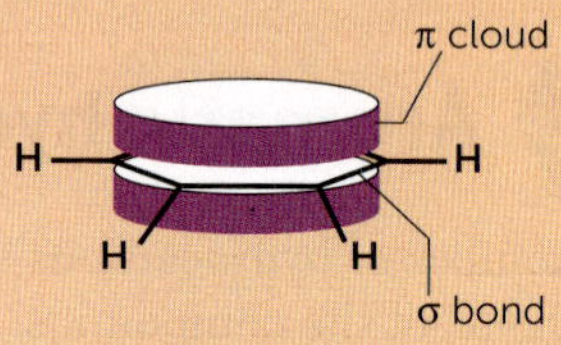

Figure 18.6 The bonds in benzene.

Note that 'Ar–' stands for **aryl** and is used to represent an aromatic group, including a benzene ring. For example, Ar–Cl (chlorobenzene) can be represented on paper in two ways (Figure 18.7).

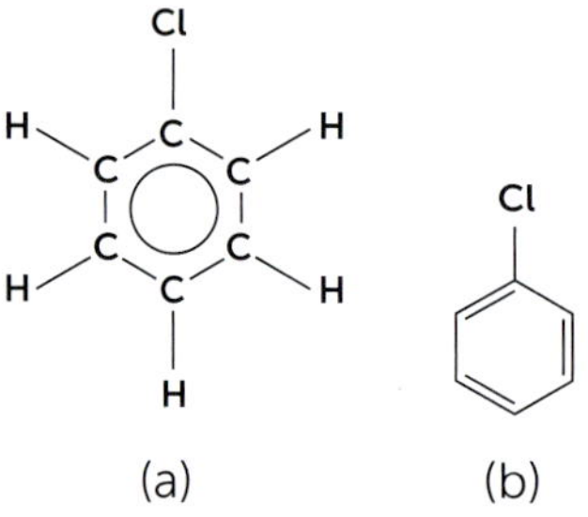

Figure 18.7 Two ways of representing chlorobenzene.

The symbol 'Ph–' stands specifically for the **phenyl** group (C_6H_5–), so chlorobenzene might also be written C_6H_5–Cl or Ph–Cl.

Physical properties of benzene

Benzene is a colourless liquid that is insoluble in water. It was widely used as a solvent, but now this use is limited since benzene has been found to be toxic – inhalation of its vapour can cause leukaemia. Toluene (see next section) is used as a safer alternative.

Chemical properties of benzene

1 Combustion

Benzene burns in air with a smoky, luminous flame. In addition to carbon dioxide and water, particles of carbon are formed because of incomplete combustion.

2 Substitution

Benzene undergoes **substitution reactions** rather than addition reactions (it does not behave chemically like the alkenes). In these reactions, hydrogens attached to the ring carbons are substituted by other atoms or groups. Some examples are shown below:

- Benzene can be **nitrated** by reaction with a mixture of concentrated sulfuric and nitric acids at 60°C. Nitrobenzene, a pale yellow liquid, is formed:

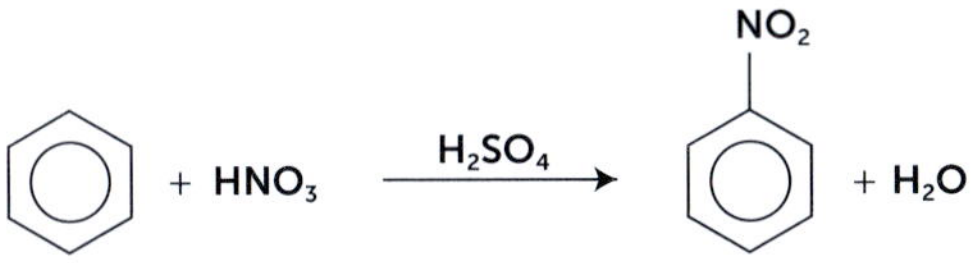

- Benzene is **halogenated** at room temperature by reaction with chlorine or bromine, in the presence of a catalyst. For example, chlorine reacts with benzene in the presence of iron(III) chloride. A chlorine atom is substituted for hydrogen in the benzene ring and chlorobenzene is formed:

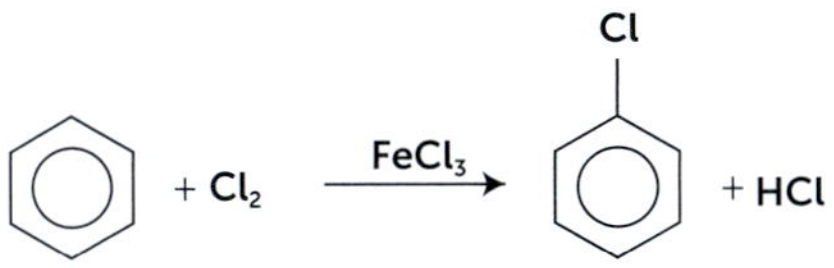

- Benzene is **alkylated** by reaction with a halogenoalkane, in the presence of a catalyst. For example, chloroethane reacts with benzene in the presence of the catalyst aluminium chloride to form ethylbenzene:

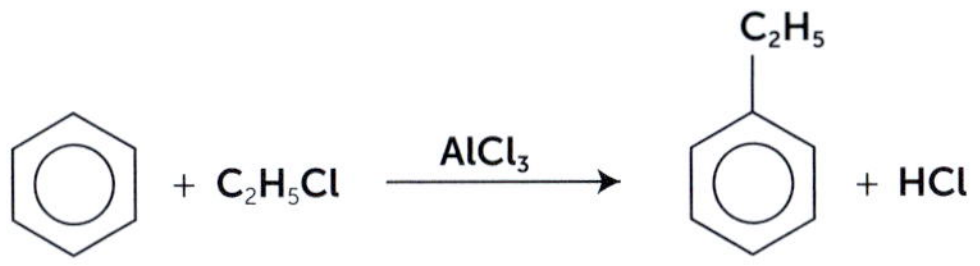

  This type of reaction is often referred to as a **Friedel–Crafts alkylation**.

- Benzene is acylated by reaction with an acid halide in the presence of a catalyst. An example of an acyl group is

  where R is an alkyl group such as methyl, ethyl etc. The reaction of ethanoyl chloride with benzene in the presence of aluminium chloride gives phenylethanone:

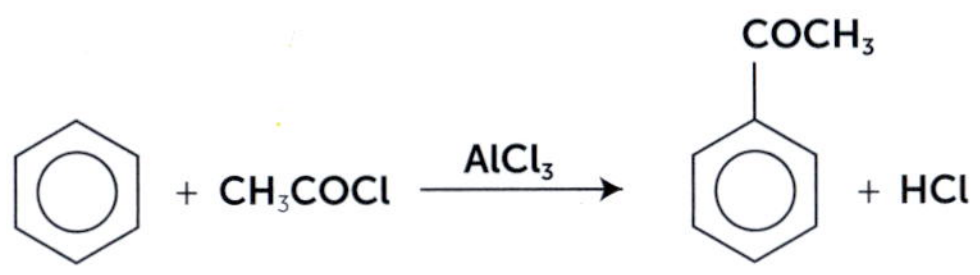

The mechanism of the substitution reactions of benzene is described on page 370.

This type of reaction is called a **Friedel–Crafts acylation**.

3 Addition

Benzene does undergo some addition reactions:

- It can be hydrogenated to cyclohexane by reaction with hydrogen in the presence of a nickel catalyst, but at a higher temperature than that needed to hydrogenate simple alkenes.

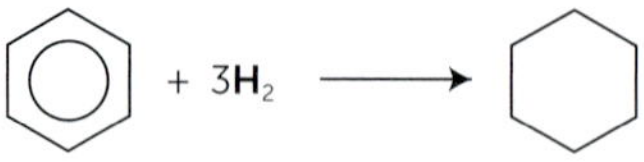

- In ultraviolet light, chlorine adds to benzene and eventually 1,2,3,4,5,6-hexachlorocyclohexane is formed. The numbers show that one chlorine atom is attached to each carbon atom of the ring.

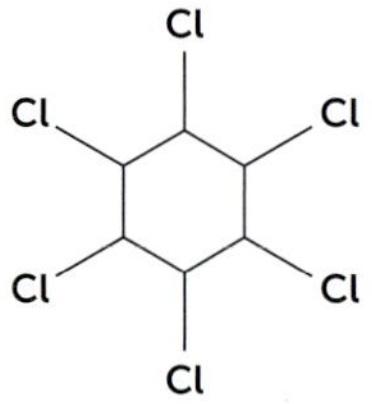

Sources of benzene

The main source of benzene is petroleum, although it can be obtained from coal tar.

EXERCISE **18P**

Enthalpy of hydrogenation of benzene

The catalytic hydrogenation of gaseous cyclohexene, in the presence of a nickel catalyst, proceeds according to the equation

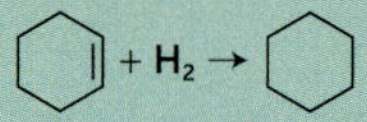

$\Delta H = -119$ kJ mol^{-1}

(i) If benzene had the Kekulé structure, write an equation that would represent the complete hydrogenation of benzene.

(ii) According to your equation, what would be the expected value of its standard enthalpy change of hydrogenation?

(iii) The experimental value of the enthalpy of hydrogenation of benzene to cyclohexane is -208 kJ mol^{-1}. Why is this value different from the one you have calculated?

Arenes

Arenes are aromatic hydrocarbons (most commonly based on benzene rings). The structures of some arenes are

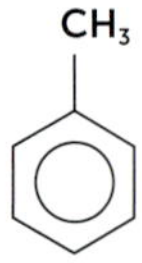

methylbenzene (common name toluene)

ethylbenzene

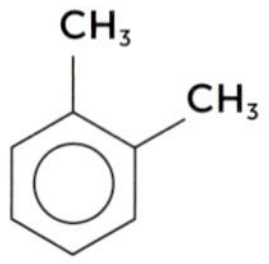

1,2-dimethylbenzene (old name *o*-xylene)

Notice that there can be different structural isomers where more than one carbon atom in the ring is attached to separate groups. There are three possible xylenes, for example, the other two being:

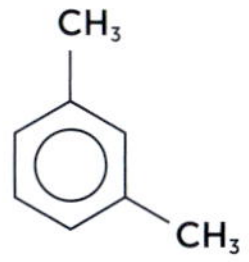

1,3-dimethylbenzene
(old name *m*-xylene)

1,4-dimethylbenzene
(old name *p*-xylene)

Notice that they are named by numbering the different carbon atoms in the ring. The first carbon atom to be attached to an atom or group is given the number 1:

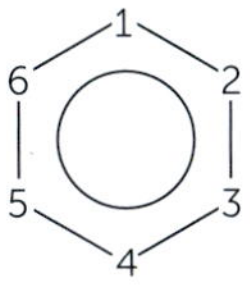

Fused ring aromatic compounds

Aromatic rings that share a pair of carbon atoms are said to be fused. Some examples are

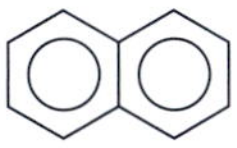

naphthalene

anthracene

phenanthrene

Naphthalene has been used in mothballs. All these compounds are obtained from coal tar. They are also called **polyaromatic hydrocarbons** (PAHs). Polyaromatic hydrocarbons are found abundantly in the environment. They originate from many sources including cigarette smoke and car exhaust gases. There is cause for concern because some PAHs are considered to be toxic.

EXERCISE **18Q**

Arenes

Write down the structural formulae of all arenes with the molecular formula C_9H_{12} (there are eight).

BOX **18.9**

Friedrich Kekulé

German chemist Friedrich Kekulé (1829–1896) was Professor of Chemistry at the universities of Ghent and Bonn. He proposed the concept of the 'carbon chain', or the *catenation* of carbon, as the basis of many organic compounds. In the 1860s, he turned his attention to aromatic compounds. Benzene had been identified by Faraday in 1825. It was known that the empirical formula of benzene was C_6H_6, but its structure remained a mystery until Kekulé proposed his cyclic structure in 1865.

Some possible structures that have the same molecular formula are shown alongside. None of these was acceptable. One reason is because of the reluctance of benzene to undergo addition reactions. See if you can think of any more possibilities for the structure of C_6H_6.

Kekulé claimed that the answer to the puzzle of the structure of benzene came to him in a dream. In his dream, groups of six carbon atoms turned into snakes that swirled around his laboratory. Suddenly, the snakes formed circles. In this way the 'benzene rings' were revealed to the scientist.

A postage stamp showing Kekulé and his proposed structure of benzene. He claimed that the structure came to him in a dream.

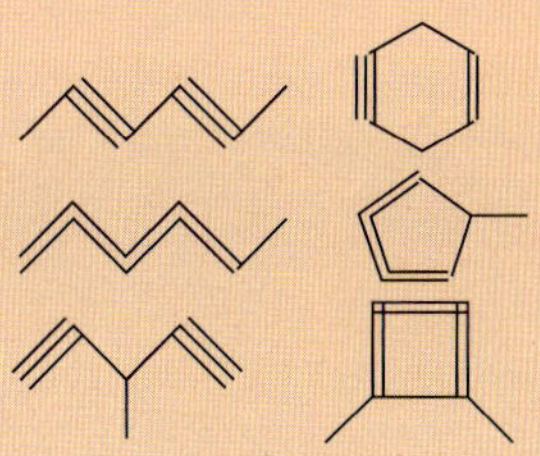

REVISION QUESTIONS

18.1 Write the formula of hexane in three different ways.

18.2 Write the structural formulae of the five isomers of hexane and name them.

18.3 Name the following compounds:

(i) $CH_3CH{=}CHCH_2CH_3$

(ii)

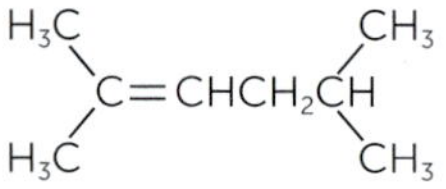

18.4 Write structural formulae for:

(i) pent-1-ene

(ii) 5-chlorohex-2-ene

(iii) 4,4-dimethylpent-2-ene

(iv) buta-1,3-diene

(v) methylcyclopentane

(vi) 5-methyloct-3-yne.

18.5 Which of the following can exist as geometric isomers (*cis* and *trans* forms)?

(i) $(CH_3)_2C{=}CH_2$

(ii) $CH_3CH{=}CHCH_3$

(iii) $CH_3CH{=}CHCH_2CH_3$.

18.6 Write the structural formulae of, and name, the following compounds:

(i) The product of the reaction of pent-1-ene with bromine water.

(ii) The product of the reaction of but-1-ene with alkaline potassium permanganate solution.

(iii) The product of the reaction of *trans*-but-2-ene with hydrogen.

18.7 Give the structural formulae of:

(i) *trans*-1,2-dibromoethene

(ii) *trans*-1-chloroprop-1-ene

(iii) *cis*-hex-3-ene

(iv) *E*-2-bromopent-2-ene

(v) *Z*-1-bromo-2-chloro-2-fluoro-1-iodoethene

18.8 Write the structure of tetrafluoroethene. This alkene polymerizes to form polytetrafluoroethene (PTFE). Write the equation for the reaction. See if you can find what PTFE is used for in the home.

18.9 A compound is composed of 88.9 per cent C and 11.1 per cent H by mass.

(i) What is the empirical formula of the compound?

(ii) If the molecular mass of the compound is 54 u, what is the molecular formula of the compound?

(iii) Draw the structure of, and name, all possible isomers of the compound.

18.10 Benzopyrene, $C_{20}H_{12}$, is a fused ring aromatic compound. Write its structural formula.

18.11 Use the average bond enthalpies listed in Table 13.4 (page 234) to calculate the standard enthalpy of formation of benzene – assume that the atoms are arranged according to the Kekulé structure

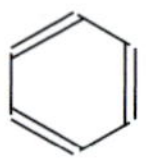

You are given the additional information (at 298 K)

$C_6H_6(l) \rightarrow C_6H_6(g)$ $\Delta H^{\ominus} = 31\ \text{kJ mol}^{-1}$

$C(s) \rightarrow C(g)$ $\Delta H^{\ominus} = 717\ \text{kJ mol}^{-1}$

and the enthalpy diagram

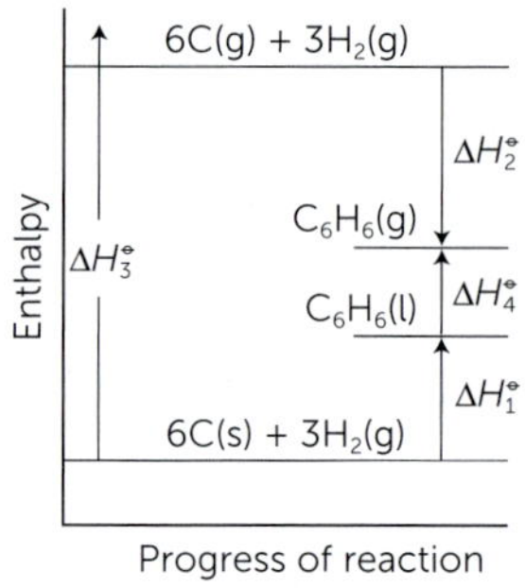

Compare your calculated value for the standard enthalpy of formation of benzene with the experimental value of 50 kJ mol^{-1}. Why is your value different?

18.12

(a) Write the following formulae in skeletal form:

(i) but-1-yne

(ii) ethyl benzene

(iii) 3-chloro-2-methylpropene

(b) Give structural formulae for the following:

DID YOU KNOW?

Ethyne (acetylene, C_2H_2) was used as a lamp by miners and in cars and bicycles and even for lighting in the home. Ethyne was produced by slowly dripping water on solid calcium carbide (CaC_2):.

$$CaC_2 + 2H_2O(l) \rightarrow Ca(OH)_2 + C_2H_2$$

Calcium carbide is ionic and is better represented as Ca^{2+}, $^{-}C{\equiv}C^{-}$. The engineering of acetylene lamps was ingenious. See https://www.youtube.com/watch?v=7nvXSbtwxQU

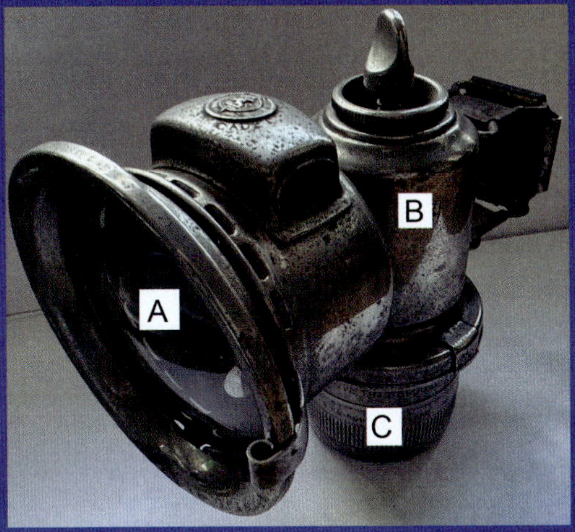

An acetylene lamp for a bicycle. These lamps were in use from about 1895 to 1935, replacing earlier oil lamps. The solid calcium carbide solid is placed inside C. Water is stored in B, and the screw above controls the rate at which water drips onto the carbide. The lens (A) is then opened and the acetylene travelling through the outlet valve is lit with a match. (Image: Hetty Simpson).

For an introduction to strange and unusual alkane molecules, see Extension 18 at https://www.bloomsburyonlineresources.com/chemistry

19 Common Classes of Organic Compounds

CONTENTS

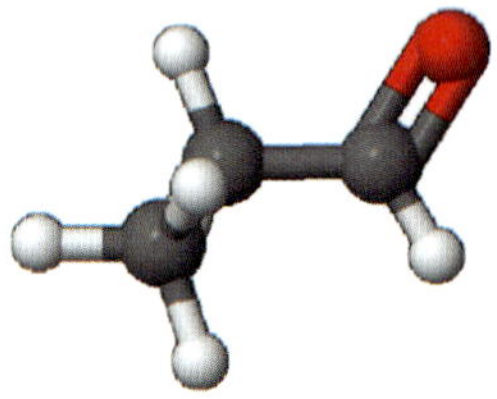

The compound ***propanal*** (CH_3CH_2CHO) belongs to the aldehyde group of compounds because it contains the functional group–CHO. Here are two ways in which the molecule is represented, by skeletal structure (top) and by ball and stick (below).

Beware! The names of organic compounds can easily be confused. Propanol, propanal and propenal are different molecules!

OBJECTIVES

This unit:

- Explains what is meant by functional group
- Shows you how organic compounds can be grouped into families, depending upon the functional groups they contain
- Describes the main characteristics of common organic families
- Introduces optical isomerism

A great many organic chemicals can be derived from hydrocarbons, by replacing one or more hydrogens by other atoms or groups (called **functional groups**). To try and bring some order into this vast number of compounds, we can again divide the organic compounds into 'families'. Members of each family have the same functional group(s) and the reactions that occur because of the presence of that group are characteristic of the family. It is a good idea to *learn* these functional groups before you go any further. See Table 19.1.

Table 19.1 Common functional groups

General formula of compounds	Name of functional group	Example
R–**X** (X = a halogen)	halogenoalkane	CH_3Cl chloromethane
R–**OH**	alcohol	CH_3CH_2OH ethanol
R–$\mathbf{NH_2}$	amine	CH_3NH_2 methylamine (or methanamine)
R–**COOH**	carboxylic acid	CH_3COOH ethanoic acid
R–**COCl**	acid chloride	CH_3COCl ethanoyl chloride
R–$\mathbf{CONH_2}$	amide	CH_3CONH_2 ethanamide
R–**COOR′**	ester	CH_3COOCH_3 methyl ethanoate
R–**CHO**	aldehyde	CH_3CHO ethanal
R**CO**R′	ketone	CH_3COCH_3 propanone
R**O**R′	ether	$CH_3CH_2OCH_2CH_3$ diethyl ether
R–**CN**	nitrile	CH_3CN ethanenitrile

Representation of functional groups using skeletal formulae

Remember that only hydrogen atoms and bonds to hydrogen atoms attached to carbon can be left out. Hydrogen atoms attached to other elements must be shown. Some common functional group skeletal structures are shown below:

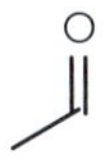

ethanal

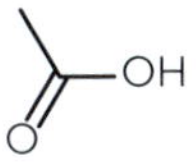

ethanoic acid

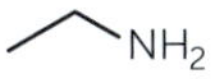

ethanamine

dimethylamine

19.1 Halogenoalkanes (or alkyl halides)

We have already met these compounds in Unit 18. They are formed by replacing the hydrogen atom of a hydrocarbon by a halogen (F, Cl, Br or I) and have the general formula R–X where X = a halogen. Thus:

CH_3F fluoromethane

CH_3Cl chloromethane

CH_3Br bromomethane

CH_3I iodomethane

Halogenoalkanes are denser than water and are insoluble in water. If added water, they form the bottom layer.

Important halogenoalkanes

Trichloromethane (or chloroform, $CHCl_3$) is non-flammable, and was used as an anaesthetic before it was found to cause liver damage. Tetrachloromethane (or carbon tetrachloride, CCl_4) also has anaesthetic properties, but it is even more toxic. Bromochlorotrifluoroethane (common name halothane) is now widely used. It also has anaesthetic properties and is much safer.

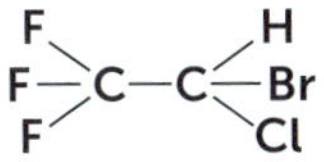

halothane

Substitution of the halogen in halogenoalkanes

A carbon–halogen bond is polar ($C^{\delta+}–X^{\delta-}$) and the carbon can be attacked by groups that carry an unshared pair of electrons. The result is a substitution reaction in which one atom, ion or group is substituted for another. If some halogenoalkanes are heated with aqueous sodium hydroxide, the halogen is substituted by an –OH group, producing an alcohol:

$CH_3CH_2CH_2Br$	$+ OH^-$	$\rightarrow CH_3CH_2CH_2OH$	$+ Br^-$
1-bromopropane	+ hydroxide ion	→ propan-1-ol	bromide ion

For more information on the mechanism of this reaction, see page 374.

19.2 Alcohols

The formulae of these compounds can be represented by R–OH. Their names end in 'ol'. The ending is sometimes prefixed by a number to indicate the carbon atom to which the –OH group is attached. Examples of **alcohols** are shown in Table 19.2.

EXERCISE 19A

Halogenoalkanes

(i) Write the structural formulae of the following alkyl halides:

(a) chloroethane
(b) 2-bromopropane
(c) 1,2-dichloro-3-fluorobutane
(d) 1,3,5-tribromobenzene.

(ii) Name the following:

(a) 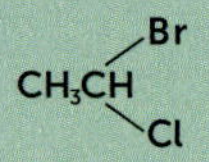

(b) $CH_3CH_2CHICH_2CH_3$

(c) CCl_3F

(d) 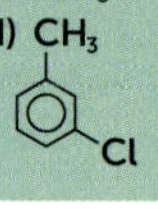

Alcohols with more than one –OH group are named using diol, triol etc., depending on how many such groups are present. A 1,2-diol is often also referred to as a **glycol**.

Table 19.2 Some members of the alcohol family

Name	Formula	Boiling point/°C	State at room temperature
Methanol	CH_3OH	65	l
Ethanol	CH_3CH_2OH	79	l
Higher alcohol members can exist in isomeric forms – notice how they are named			
Propan-1-ol	$CH_3CH_2CH_2OH$	98	l
Propan-2-ol	$CH_3CHOHCH_3$	83	l
Butan-1-ol	$CH_3CH_2CH_2CH_2OH$	117	l
2-Methylpropan-2-ol	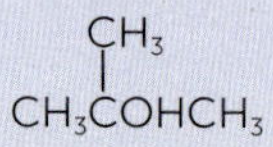	82	s/l

EXAMPLE 19.1

Name the compound:

OH OH
| |
CH_2CH_2

Answer

ethane-1,2-diol (old name ethylene glycol)

EXERCISE 19B

Nomenclature of alcohols

(i) Name and write the structural formulae for the different isomers of pentanol, $C_5H_{11}OH$. There are eight.

(ii) Glycerol, or propane-1,2,3-triol, is used in many cosmetic preparations to make them feel soft. Write its structural formula.

Physical properties of alcohols

1 Ability to hydrogen bond

The –OH group in an alcohol is polar:

$R—O^{\delta-}—H^{\delta+}$

Alcohols are therefore able to **hydrogen bond** with themselves or with water:

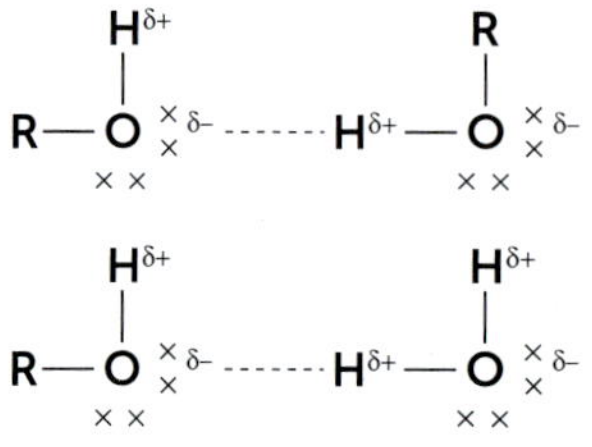

Because alcohols form hydrogen bonds with other alcohol molecules, the boiling points of the alcohols tend to be higher than those of other organic compounds of similar molecular mass. Their ability to hydrogen bond with water allows the lower molecular mass alcohols to dissolve in, or be miscible with, water. Ethane-1,2-diol, or ethylene glycol, is used as a component in antifreeze because it has a high solubility in water, it contains two hydroxy groups that can hydrogen bond with the water, and it has a high boiling point and low freezing point.

2 Alcohols are good solvents

The presence of an –OH group means that alcohols of low molecular mass have some of the solvent properties of water and can dissolve ionic compounds. The R– group allows them to behave as solvents to organic, covalent substances.

EXERCISE **19C**

Properties of alcohols

Chloroethane, butane and propan-1-ol have boiling points of 286 K, 272 K and 371 K, respectively.

(i) Calculate the molecular mass of each compound.

(ii) Assuming all their melting points are lower than room temperature, what is the state of each substance under conditions of 20°C and 1 atm pressure?

(iii) Explain why the boiling point of propan-1-ol is much higher than those of the other two compounds.

Ethanol

Although there are many alcohols, when the non-chemist refers to 'alcohol' they generally mean ethanol. Ethanol can be made by the decomposition of **carbohydrates**, a process known as **fermentation.**

Fruit juices, such as grape juice, contain the sugar glucose (molecular formula $C_6H_{12}O_6$). Glucose can be fermented into alcohol by the addition of yeast. Yeast is a fungus that contains the enzyme **zymase,** which catalyses the decomposition of glucose into ethanol. **Enzymes** are biological catalysts. The fermentation reaction is carried out at about 20°C – high temperatures kill the yeast and at lower temperatures the process is too slow. The equation for the reaction may be written as

$$\underset{\text{glucose}}{C_6H_{12}O_6(aq)} \rightarrow \underset{\text{ethanol}}{2C_2H_5OH(aq)} + 2CO_2(g)$$

Fermentation ceases when the ethanol content of the mixture reaches about 14 per cent (v/v); higher concentrations of ethanol kill the yeast. If a higher concentration of alcohol is required, the mixture is **fractionally distilled.** 'Spirits' such as whisky or brandy may contain about 40 per cent ethanol. Although fermentation is used to produce alcoholic beverages, it is rarely used to synthesize the chemical ethanol on a large scale. Alcohols, such as ethanol, can be produced commercially by the addition of water across an alkene double bond (a **hydration reaction**). The reaction is catalysed by phosphoric acid:

$$\underset{\text{ethene}}{CH_2=CH_2} + \underset{\text{steam}}{H_2O} \xrightarrow[\text{300°C, 70 atm}]{H_3PO_4} \underset{\text{ethanol}}{CH_3CH_2OH}$$

Dilute aqueous solutions of ethanol, when drunk as beverages, cause a pleasant relaxed feeling; however, too much 'alcohol' leads to headaches and vomiting. Long-term abuse of ethanol can lead to liver damage.

Alcohol (ethanol) containing drinks have been known for thousands of years. The best wines, like this one, have been actively developed and their taste is dependent on many factors, including soil composition, climate and winemaking technique. (Image by courtesy of the Brecon Estate California.)

Methanol

Originally, methanol was produced when wood was distilled and, for this reason, was known as wood alcohol. Today, methanol can be produced from the hydrogenation of carbon monoxide using a mixture of copper, zinc and aluminium oxides as catalyst:

$$CO(g) + 2H_2(g) \rightarrow CH_3OH(l)$$

Methanol is used as a solvent and as an antifreeze for car radiators. It is much more toxic than ethanol – drinking methanol can lead to blindness and then to death. Methanol can be used as fuel for cars.

Methylated spirits

Alcoholic beverages are heavily taxed in many countries, but besides being used to make drinks, ethanol has many useful industrial applications. It is used to make other organic chemicals and is an extremely useful industrial solvent. In order to avoid heavy taxation of the ethanol used for industrial purposes it is rendered unfit to drink (**denatured**). Small amounts of methanol are added to ethanol to produce 'industrial methylated spirit'. A purple dye is often added to methylated spirit sold for domestic use. The dye serves as a warning that the mixture is undrinkable.

Chemical properties of alcohols

1 They burn in air, giving out heat, and can be used as fuels:

$$C_2H_5OH(l) + 3O_2(g) \rightarrow 2CO_2(g) + 3H_2O(l)$$

2 Alcohols can be oxidized to carboxylic acids. For example, ethanol is slowly oxidized to ethanoic acid (*acetic acid* found in *vinegar*) by bacteria present in the air. This is why an open bottle of wine tastes sour when left in the air for a few days.

$$C_2H_5OH(l) + O_2(g) \xrightarrow{\text{bacteria}} CH_3COOH(l) + H_2O(l)$$

Strong oxidizing agents, such as acid potassium permanganate (manganate(VII)) solution or chromic acid (acidified sodium dichromate(VI) solution) will also oxidize alcohols.

3 Alcohols react with carboxylic acids to make esters (see some reactions of carboxylic acids, page 353).

> EXERCISE **19D**
>
> **Primary, secondary and tertiary alcohols**
>
> Classify the isomers of pentanol in Exercise 19B(i) as primary, secondary or tertiary.

Primary, secondary and tertiary alcohols

Apart from methanol, alcohols may be classified as primary, secondary or tertiary:

primary alcohols contain a $-CH_2OH$ group

secondary alcohols contain a >CHOH group

tertiary alcohols contain a >COH group

19.3 Carbonyl compounds

Carbonyl compounds contain the carbonyl group, >C=O. The carbonyl group is part of the structural formulae of a number of organic families. The name of the family is determined by the other atoms bonded to the carbonyl carbon. If one of the atoms bonded is hydrogen, the compound belongs to the class of organic compounds known as **aldehydes**. However, if two carbon atoms are bonded to the carbonyl carbon, the compound is a **ketone**. To summarize,

aldehydes contain the R—C=O group (sometimes written R—CHO)
(with H attached to C)

ketones contain the R—C=O group (where R and R′ are alkyl groups)
(with R′ attached to C)

The name of an aldehyde is derived from the number of carbon atoms present and ends in 'al'. For example:

HCHO	methanal (old name formaldehyde)
CH_3CHO	ethanal (old name acetaldehyde)
CH_3CH_2CHO	propanal

Ketones are named by changing the end of the name of the parent alkane (the one with the same number of carbon atoms) to 'one'. Sometimes it is necessary to number the position of the carbonyl group to avoid confusion:

CH_3COCH_3	propanone (commonly called acetone)
$CH_3COCH_2CH_3$	butanone
$CH_3COCH_2CH_2CH_3$	pentan-2-one

Aldehydes and ketones can be isomeric with one another. For example, CH_3CH_2CHO and CH_3COCH_3 both have the molecular formula C_3H_6O. These **structural isomers** have some different chemical properties as well as different physical properties.

EXERCISE **19E**

Aldehydes and ketones 1

Classify each of the following compounds as an aldehyde or a ketone:

(i) C_6H_5CHO

(ii) $CH_3CH_2COCH_2CH_3$

(iii) $CH_3CH_2CH_2CHO$

(iv) $CH_3CH_2C(=O)CH_3$

(v) 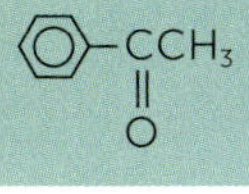

Physical properties of aldehydes and ketones

The carbonyl group is polar and has unshared electron pairs on the oxygen atom:

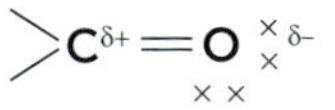

The group can therefore take part in hydrogen bonding with water molecules:

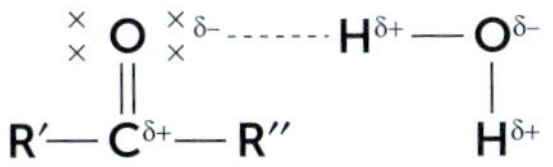

Aldehydes and ketones with a low molecular mass are therefore soluble in water. Propanone (acetone) is widely used as a solvent. It is soluble in water and dissolves many polar and non-polar molecules. Examples include paint, nail varnish and plastics.

A 37 per cent solution of methanal (common name formaldehyde) dissolved in water, with some methanol added, is called **formalin** and is used for preserving biological specimens.

Ethanal is formed when ethanol is oxidized in the liver. A buildup of ethanal in the bloodstream causes a hangover.

Ketones can have distinctive sweet smells. Spearmint-flavoured chewing gum gets its flavour from the ketone carvone:

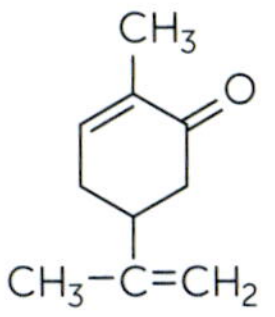

Carvone

Acetone (propanone) is a colourless, smelly solvent and a member of the ketone family of compounds. It is used domestically to remove nail varnish.

A dramatic example of acetone's solvent power. This expanded polystyrene sphere was soon destroyed by adding ten drops of acetone. The acetone solvent softens the polystyrene, releasing trapped air, so that the original structure collapses.

EXERCISE **19F**

Aldehydes and ketones 2

(i) Why is there no such compound as ethanone?

(ii) Write down the structural formula of the aldehydes that are isomeric with $CH_3CH_2CH_2COCH_3$. What are their names?

(iii) Write down the names and structures of two aldehydes and one ketone with the molecular formula C_4H_8O.

Preparation of aldehydes and ketones by oxidation of alcohols

Primary or secondary alcohols can be oxidized to aldehydes or ketones. Aldehydes are easily further oxidized to the corresponding carboxylic acid, and oxidation of a **primary alcohol** usually continues through the aldehyde to the acid. The oxidation of a **secondary alcohol** usually yields a ketone. Ketones are more resistant to further oxidation because C–C bonds would have to be broken and can only be further oxidized using very strong oxidizing conditions. **Tertiary alcohols** are resistant to any oxidation, again because C–C bonds would have to be broken. These changes can be summarized as follows:

$$\underset{\text{ethanol}}{\overset{\text{Primary alcohol}}{CH_3CH_2OH}} \xrightarrow{\text{Oxidation}} \underset{\text{ethanal}}{\overset{\text{Aldehyde}}{CH_3CHO}} \xrightarrow{\text{Oxidation}} \underset{\text{ethanoic acid}}{\overset{\text{Carboxylic acid}}{CH_3COOH}}$$

$$\underset{\text{propan-2-ol}}{\overset{\text{Secondary alcohol}}{CH_3CHOHCH_3}} \xrightarrow{\text{Oxidation}} \underset{\text{propanone}}{\overset{\text{Ketone}}{CH_3COCH_3}}$$

EXERCISE **19G**

Oxidation of alcohols

Write the names and formulae of the products of the oxidation of the following compounds:

(i) butan-2-ol **(ii)** propan-1-ol **(iii)** octan-2-ol.

Acidified sodium dichromate is an example of an oxidizing agent that will bring about the oxidation of primary alcohols through to carboxylic acids and secondary alcohols to ketones. If an aldehyde is required as the product of oxidation of a primary alcohol, the apparatus must be designed so that the aldehyde will be distilled off as it is formed.

Carbonyl compounds can be reduced back to alcohols by using the reducing agents lithium aluminium hydride ($LiAlH_4$) or sodium borohydride ($NaBH_4$).

Tests to distinguish between aldehydes and ketones

The fact that aldehydes are easily oxidized to carboxylic acids, whereas ketones are not, can be used to distinguish between them:

1 Fehling's solution may be regarded as containing Cu^{2+} ions in basic solution. When this solution is boiled with an aldehyde, a brick-red precipitate of copper(I) oxide is formed – the aldehyde is oxidized to a carboxylic acid by the copper(II) ions. Ketones are *not* oxidized by this reagent.

$$RCHO + 6H_2O + 2Cu^{2+} \rightarrow RCOOH + 4H_3O^+ + Cu_2O$$

2 Tollens' reagent contains Ag^+ ions dissolved in aqueous ammonia. If the reagent is warmed with an aldehyde, the silver ions are reduced to silver metal and a distinctive 'silver mirror' is deposited on the reaction container. Again, ketones do not react with this reagent.

$$RCHO + 3H_2O + 2Ag^+ \xrightarrow{\text{aqueous ammonia}} RCOOH + 2H_3O^+ + 2Ag$$

Identification of aldehydes and ketones

The carbonyl group in aldehydes and ketones will react with **hydrazine** derivatives. Hydrazine is related to ammonia:

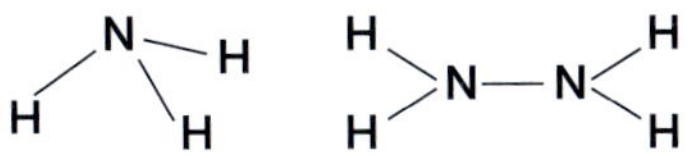

ammonia hydrazine

The products of the reaction of hydrazine with an aldehyde or ketone are known as **hydrazones:**

a hydrazone

Because water is 'eliminated' in the reaction, the reaction is known as a **condensation** reaction. A slight variation of this reaction can be used to identify aldehydes or ketones using 2,4-dintrophenylhydrazine because the hydrazones formed are orange solids that can be recrystallized easily.

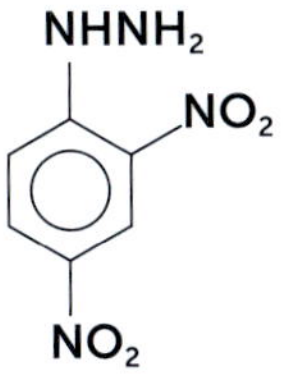

2,4-dintrophenylhydrazine

An aldehyde or ketone is converted to the corresponding dinitrophenylhydrazone (DNPH). The DNPH is purified by recrystallization and its melting point compared with those reported in data books. In this way the original carbonyl compound can be identified.

EXERCISE **19H**

Reactions of aldehydes and ketones

Draw the structural formulae of the organic products of the following reactions:

(i) The reaction of butanal on heating with Fehling's solution.

(ii) The reaction of propanone with 2,4-dinitrophenylhydrazine.

(iii) The reaction of propanal with hydrazine.

(iv) The reaction of propanal on warming with Tollens' reagent.

BOX **19.1**

Carbohydrates

Carbohydrates are naturally occurring compounds of carbon, hydrogen and oxygen, with hydrogen and oxygen atoms in the ratio 2:1 as in water. They are **polyhydroxy** (they contain many –OH groups) aldehydes and ketones, or compounds that can be hydrolysed to make polyhydroxy aldehydes or ketones. The chemical names of many carbohydrates end in 'ose'. Starch, table sugar (sucrose) and cotton and paper (cellulose) are all composed of carbohydrates.

The simplest carbohydrates are the **monosaccharides** (or simple sugars). They cannot be split up into simpler carbohydrate units. Glucose ($C_6H_{12}O_6$) and fructose ($C_6H_{12}O_6$) are monosaccharides:

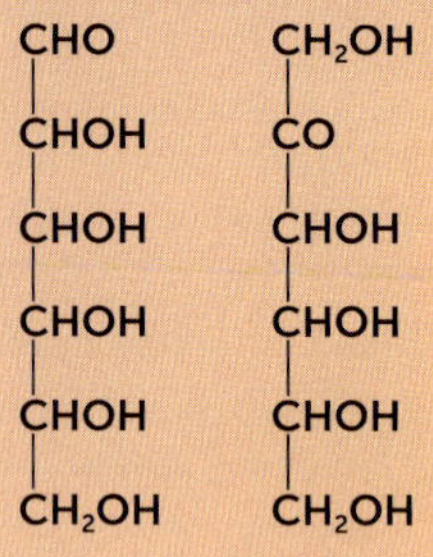

glucose fructose

Glucose is a sugar found in grapes, whereas fructose is the very sweet-tasting sugar found in honey and fruit. Because simple carbohydrates have a large number of –OH groups they are able to hydrogen bond and are therefore very soluble in water. Sugars also undergo some reactions that are typical of carbonyl compounds; for example, glucose reacts with DNPH to give a crystalline derivative.

In addition to the 'open chain' form described above, glucose can exist in a ring form:

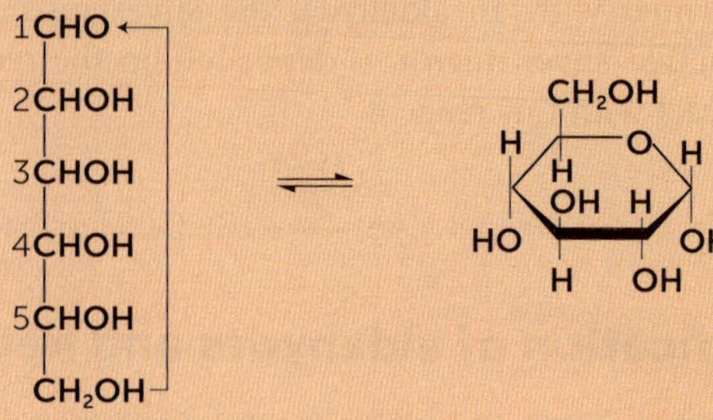

open chain ring

The –OH group on carbon atom 5, reacts with the carbonyl group 1, and a ring is formed. In aqueous solution an equilibrium exists between the two forms.

Monosaccharide units can join together by eliminating water: in twos (**disaccharides**), between two and eight units (**oligosaccharides**) and more than eight units (**polysaccharides**). Sucrose is a disaccharide that can be broken up into glucose and fructose units, whereas starch (from flour) and cellulose (from plants) are both polysaccharides. Our bodies convert sucrose and starch to glucose, which we either use for energy or store as **glycogen** (another polysaccharide). Glycogen is reconverted to glucose when energy is needed.

EXERCISE **19I**

Naming carboxylic acids

Name the following acids:

(i) $CH_3CH_2CH_2COOH$

(ii) $CH_3(CH_2)_5COOH$

(iii)

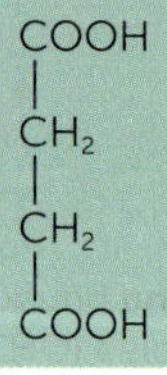

19.4 Carboxylic acids

Carboxylic acids contain a **carboxyl** group. The carboxyl group is written as –COOH or $-CO_2H$ and has the structure

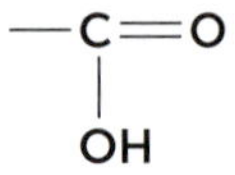

The systematic names of the acids end in 'oic acid'. Many of the acids, however, are still referred to by their old names. Some simple carboxylic acids are shown in Table 19.3.

The prefixes 'di', 'tri' etc. are used to show more than one carboxylic acid group. Ethanedioic acid (common name **oxalic acid**) is found in rhubarb leaves:

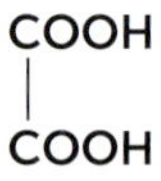

oxalic acid

Table 19.3 Some carboxylic acids and their common names

Formula	Systematic name	Common name	Where found
HCOOH	methanoic acid	formic acid	ant or nettle stings
CH_3COOH	ethanoic acid	acetic acid	vinegar
CH_3CH_2COOH	propanoic acid	propionic acid	milk, butter and cheese
C_6H_5COOH	benzoic acid	benzoic acid	preservatives

Physical properties of carboxylic acids

Pure ethanoic (acetic) acid is a liquid at room temperature, and is referred to as **glacial ethanoic acid**. The melting and boiling points of carboxylic acids are relatively high because of hydrogen bonding. Pairs of carboxylic acid molecules link up to produce hydrogen bonded dimers:

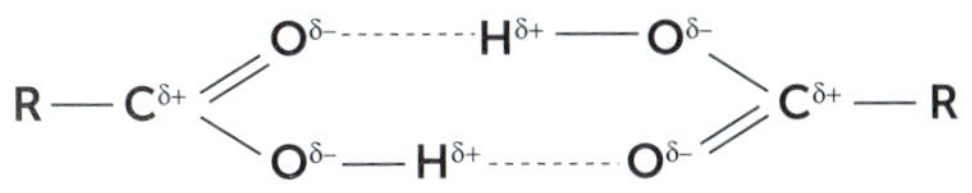

Because of their ability to hydrogen bond, the lower molecular mass carboxylic acids are soluble with water.

Methanoic, ethanoic and propanoic acids all have sharp, 'vinegary' smells. The smell of butanoic acid is detected in rancid butter and human sweat. Each person's sweat glands produce a characteristic blend of carboxylic acids, and the ability of dogs to track humans stems from the fact that the animals can detect and distinguish these different combinations of carboxylic acids.

A weak solution of ethanoic acid, along with a few crystals of sodium chloride, are delicious with fish and chips ....

Acid strengths of carboxylic acids

Carboxylic acids are weak acids compared with dilute sulfuric, nitric or hydrochloric acids:

$$CH_3COOH(l) + H_2O(l) \rightleftharpoons CH_3COO^-(aq) + H_3O^+(aq)$$

If an electronegative species, such as chlorine, is present in the acid molecule that is close to the carboxyl group, the chlorine pulls electrons towards it. This stabilizes the anion produced when the acid ionizes. For example, consider the reaction of chloroethanoic acid, $CH_2ClCOOH$, with water:

$$ClCH_2COOH + H_2O \rightleftharpoons ClCH_2COO^- + H_3O^+$$

The $ClCH_2COO^-$ ion is more stable than CH_3COO^-, so K_a for chloroethanoic acid is much greater than that for ethanoic acid. Additional electron-withdrawing groups increase this effect, but the influence of the electron-withdrawing group on acid strength falls off with distance.

Some reactions of carboxylic acids

1 Carboxylate salts are formed when a carboxylic acid reacts with a base:

$$CH_3COOH(aq) + NaOH(aq) \rightarrow \underset{\text{sodium ethanoate}}{CH_3COO^-Na^+(aq)} + H_2O(l)$$

2 Carboxylic acids react with alcohols in an acid-catalysed, reversible reaction. The compounds formed are called **esters**. An ester contains the $-COOR$ (or $-CO_2R$) group:

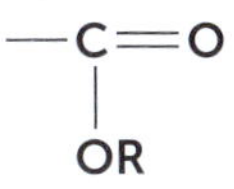

In a typical example of an esterification reaction, glacial ethanoic acid is heated with ethanol in the presence of concentrated sulfuric acid, which acts as the catalyst. The sweet-smelling ester ethyl ethanoate is formed:

$$CH_3COOH(l) + CH_3CH_2OH(l) \rightleftharpoons \underset{\text{ethyl ethanoate}}{CH_3COOCH_2CH_3(l)} + H_2O(l)$$

The general reaction is

$$RCOOH + R'OH \rightleftharpoons RCOOR' + H_2O$$

The salts of ethanoic acid are called ethanoates or (more commonly) acetates. The image shows a brooch produced by the acclaimed French artist Lea Stein (b. 1936). Stein worked with her husband, chemist Fernand Steinberger, to layer and 'bake' very thin sheets of cellulose acetate along with different fabrics and colours.

EXERCISE **19J**

Strength of carboxylic acids

Arrange the following in order of *increasing* acid strength:

$CHCl_2COOH$, CF_3COOH, CH_3COOH, CCl_3COOH, $CH_2BrCOOH$, $CH_2ClCOOH$

Note that these reactions are reversible. In order to maximize the yield of ester, excess of one of the reactants is used (usually the cheaper one!). Concentrated sulfuric acid, besides acting as the catalyst, removes the water produced, so shifting the equilibrium composition so that the concentration of ester increases. Because the reaction is reversible an ester can be **hydrolysed**, or reacted with water, in the presence of an acid catalyst to make the parent alcohol and acid.

3 Carboxylic acids are reduced to primary alcohols using lithium aluminium hydride:

$$CH_3COOH \xrightarrow{LiAlH_4} CH_3CH_2OH$$

EXERCISE **19K**

Esterification

(i) Name and write the structural formulae of the esters produced from the reactions of the following acids and alcohols:
(a) ethanoic acid and methanol
(b) octanoic acid and ethanol
(c) methanoic acid and methanol
(d) oxalic acid and ethanol

(ii) Which alcohols and acids would be obtained on hydrolysis of the following esters?
(a) methyl propanoate
(b) ethyl methanoate
(c) $CH_3COOCH_2CH_2CH_3$
(d) $CH_2ClCOOCH_3$

BOX **19.2**

Fatty acids

Carboxylic acids with long hydrocarbon chains can be obtained from hydrolysis of fats or oils. They are called fatty acids. Stearic acid is a fatty acid and its sodium salt (sodium stearate) is used as soap. The hydrocarbon chain in a fatty acid may be saturated, as in stearic acid, or it may be unsaturated (contain a double bond). Oleic acid, obtained from corn oil, contains one double bond, whereas fatty acids with more than one double bond can be prepared from vegetable oils.

Common name	Formula	Systematic name
Stearic acid	$CH_3(CH_2)_{16}COOH$	octadecanoic acid
Oleic acid	$CH_3(CH_2)_7CH{=}CH(CH_2)_7COOH$	octadec-9-enoic acid

Derivatives of carboxylic acids

Esters are members of one family of compounds that can be derived from carboxylic acids. These and some other derivatives are shown below:

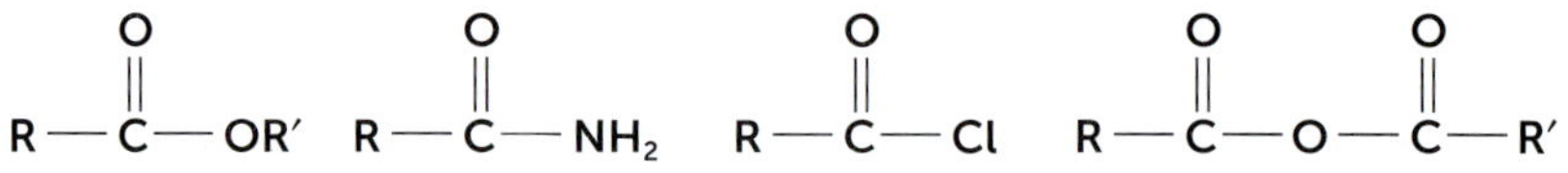

an ester   an amide   an acid chloride   an anhydride

All these derivatives contain the **acyl** group or RCO–

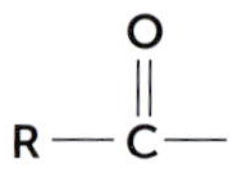

the acyl group

BOX 19.3

Some examples of carboxylic acid derivatives

Nylon

The man-made fibre nylon 6,6 is an example of a polymer that consists of long-chain molecules, containing repeating –CONH– groups. Its formula is

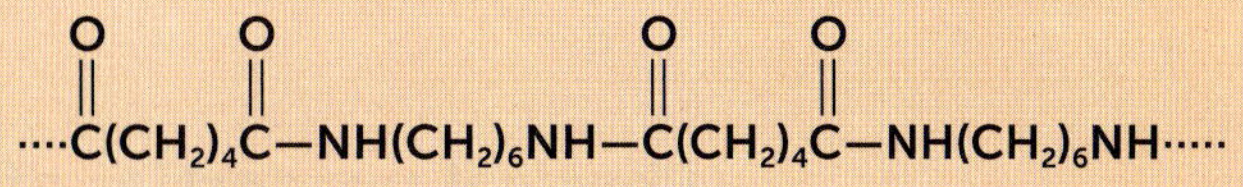

The –CONH– group is also present in amides:

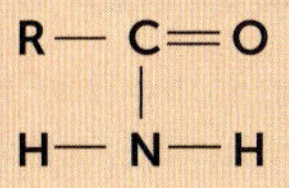

Nylon is therefore a **polyamide**. Proteins are also examples of polyamides.

Nylon 6,6 chains are formed by the reaction of many hexanedioic acid molecules with hexan-1,6-diamine molecules (a member of the amine family). Molecules of water are produced so that the chains can form. The formation of long-chain molecules from two different monomers, with elimination of water, is called a **condensation polymerization** reaction:

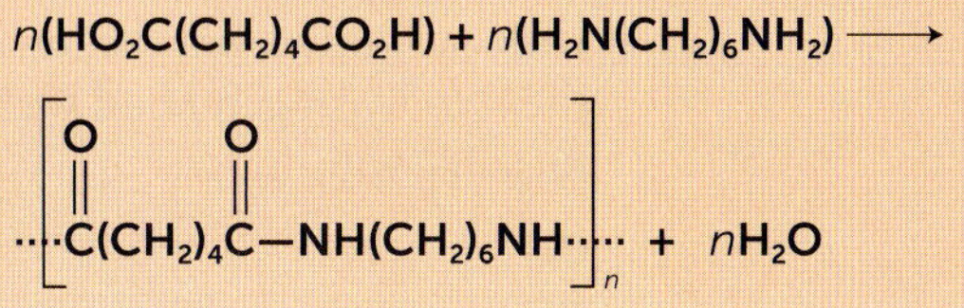

where n = a large number.

Fats and oils

These are naturally occurring esters of the triol glycerol (propane-1,2,3-triol). A fat is a substance that is solid at room temperature, whereas an oil is liquid. In vegetable oils, the hydrocarbon chains have many double bonds (they are **polyunsaturated**). Fats, however, tend to have very few or no double bonds. Removal of most of the double bonds in a vegetable oil, by reaction with hydrogen (**hardening**), will convert it to a solid fat. Hardened corn oil, for example, is used to make margarine.

an oil

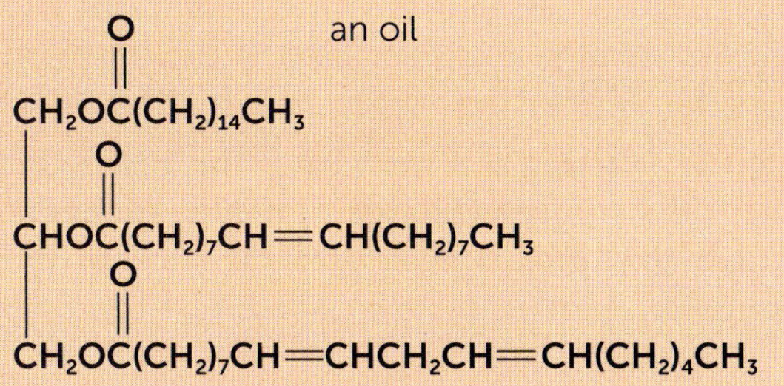

a fat

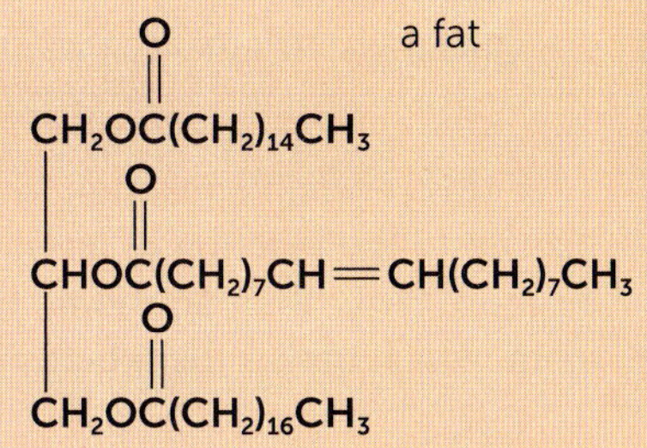

Hydrolysis of these fats or oils gives glycerol and fatty acids as products.

19.5 Amines and nitriles

Amines

The formulae for primary **amines** are of the form R–NH_2, where –NH_2 is the amino group. Examples of primary amines are shown below. These are still generally known by their older names; newer systematic names are shown in brackets.

CH_3NH_2	methylamine (methanamine)
$CH_3CH_2NH_2$	ethylamine (ethanamine)
$CH_3CH_2CH_2NH_2$	propylamine (propan-1-amine)
$H_2NCH_2CH_2CH_2NH_2$	propan-1,3-diamine

Secondary amines have the general formula

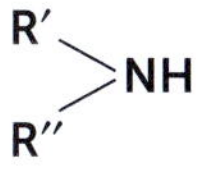 e.g. dimethylamine 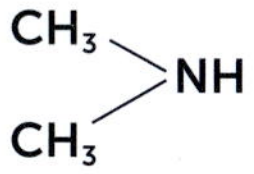

Tertiary amines have the structure

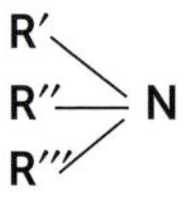
e.g. trimethylamine
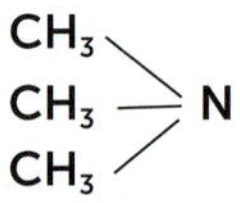

Physical properties of amines

1 Amines have very offensive fishy smells (trimethylamine smells like rotting salmon!). Amines occur in decomposing protein material. The amines $NH_2(CH_2)_4NH_2$ (putrescine) and $NH_2(CH_2)_5NH_2$ (cadaverine) are found in decaying animal flesh.

2 Primary and secondary amines have N–H bonds and therefore undergo hydrogen bonding with each other and with water. Tertiary amines cannot hydrogen bond with each other, but the lone pair of electrons present on the nitrogen atom enables them to hydrogen bond with water. Amines with low molecular mass are therefore miscible with water:

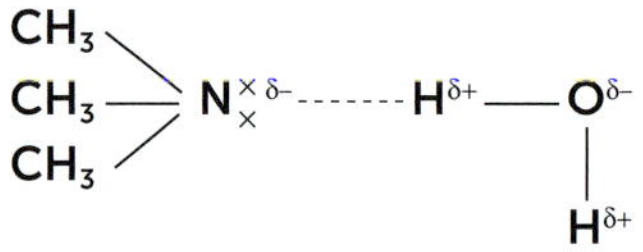

EXERCISE **19L**

The basicity of amines

Arrange the following compounds in order of *increasing* basicity:

CH_3NH_2, CH_3NHCH_3, NH_3.

Basic properties of amines

An unshared pair of electrons on the nitrogen of an amine can accept a proton from an acid. The product is an ionic amine salt:

$$CH_3NH_2 + HCl \rightleftharpoons \underset{\text{methyl ammonium chloride}}{CH_3NH_3^+, Cl^-}$$

In this reversible reaction, the amine is acting as a weak base. This reaction can be likened to the reaction of ammonia with hydrogen chloride. However, amines tend to be more basic than ammonia because alkyl groups push electrons onto the nitrogen atom, which makes them better proton acceptors.

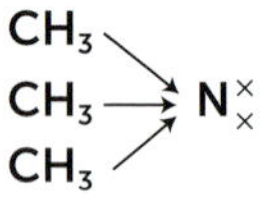

The 'electron push' of alkyl groups is known as an **inductive effect**.

Nitriles

Nitriles are organic compounds containing the cyanide (also known as cyano) group -C≡N. The general formula of nitriles is R-CN. Examples are:

CH_3CN	ethanenitrile (also called methyl cyanide or acetonitrile)
CH_3CH_2CN	propanenitrile
$CH_3CH_2CH_2CN$	butanenitrile

All three are colourless liquids at room temperature. Simple nitriles are made by reacting a haloalkane with potassium cyanide KCN (K^+,CN^-):

$$R\text{-}X + KCN \rightarrow R\text{-}CN + KX$$

For example:

$$CH_3CH_2Cl + KCN \rightarrow CH_3CH_2CN + KCl$$

This reaction is important because it allows an extra carbon atom to be added to the haloalkane: this is referred to as 'ascending the homologous series'. The nitrile can then be converted to an amide and then to a carboxylic acid by hydrolysis using acid. For example:

$$CH_3CH_2CN \xrightarrow{H^+} CH_3CH_2CONH_2 \xrightarrow{H^+} CH_3CH_2COOH$$

The connection between nitriles and amines is that nitriles can be reduced to amines using lithium aluminium hydride ($LiAlH_4$) in the laboratory, or using hydrogen gas and hot palladium catalyst in industry. For example:

$$CH_3CH_2CN + 2H \text{ (from } LiAlH_4 \text{ or } H_2) \rightarrow CH_3CH_2NH_2$$

Notice that the number of carbon atoms is reduced in this reaction.

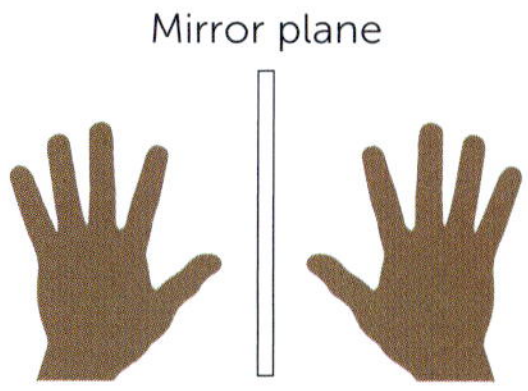

Figure 19.1 A glove and its mirror image.

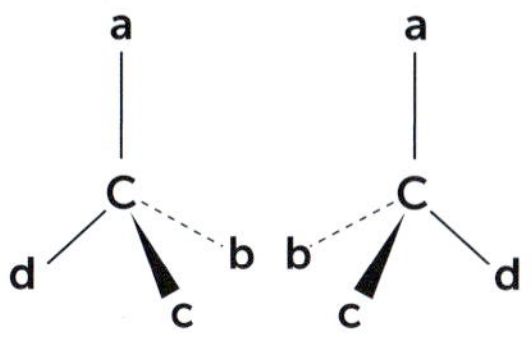

Figure 19.2 Two non-superimposable isomers (a, b, c and d are different atoms or groups).

19.6 Optical isomerism

Some organic molecules can exist in two forms that are mirror images of one another. Think of a pair of gloves – the left and right hand gloves cannot 'fit over' each other or be **superimposed**. The right glove has the shape of the image of the left glove as seen in a mirror (Figure 19.1).

Similarly, where a carbon atom is bonded to four *different* atoms or groups, then two such non-superimposable isomers are possible, as shown in Figure 19.2 (make two models and try to superimpose one upon the other). The carbon atom is known as a **chiral centre**.

The mirror image isomers are called **enantiomers**. They generally have very similar chemical and physical properties, but differ in the way that they rotate **plane polarized light**. The isomers are **optically active** – they will rotate the light by equal amounts, but in opposite directions.

BOX **19.4**

Plane polarized light

Ordinary light travels in waves and the wave vibrations are at right angles to the direction of travel. Plane polarized light is light that vibrates in only one plane and is obtained by passing ordinary light through a 'Polaroid' lens or polarizer. If the plane polarized light is passed through a solution of one enantiomer of a compound, the light will be rotated, say to the right, by a certain angle. This is shown in Figure 19.3. The light is rotated by an equal angle, but to the left, by a solution of the other enantiomer of the compound with the same concentration.

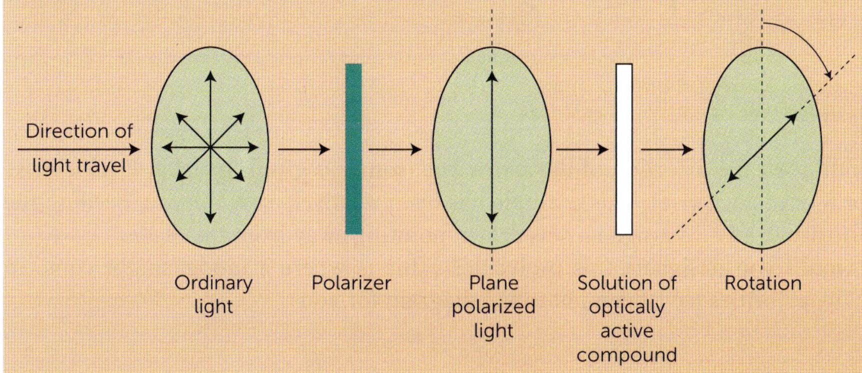

Figure 19.3 Rotation of plane polarized light by an optically active compound.

EXERCISE **19M**

Optical isomers

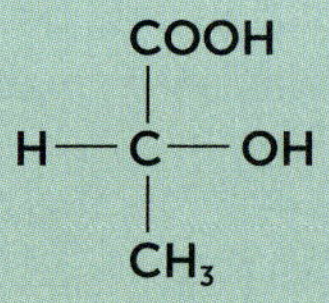

(i) Draw, or make models of, the two isomeric forms of lactic acid, (shown opposite), the substance that gives milk its sour taste. Mark the chiral centre with an asterisk.

(ii) Which of the following compounds can exist in two enantiomeric forms?

2-methylbutan-1-ol, 1-chloropentane, 2-bromo-1-chlorobutane.

Naming optical isomers

When naming these isomers, how can we distinguish between them?

Again, you can use the Cahn–Ingold–Prelog rules, described on page 331. We have to consider the directions of the bonds (and making your own models here can help you a great deal). Look at these two isomers of chlorobromoiodomethane and number the atoms according to their priorities (highest atomic number first):

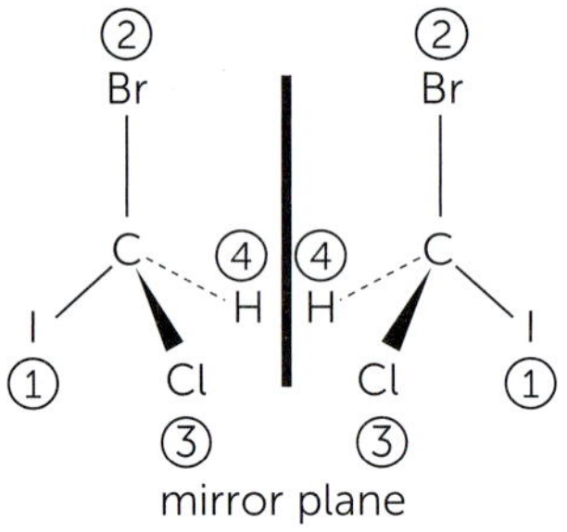

Now, making sure the very lowest priority atom or group points away from you, determine whether the order of priority – lowest number to highest – is clockwise or anticlockwise using a curved arrow (you now ignore the one that is pointing away):

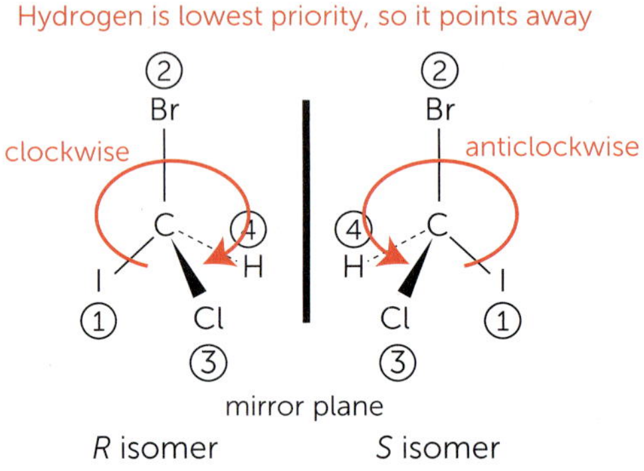

Depending on the direction of the arrow, the isomer is given the label *R* (clockwise arrow) or *S* (anticlockwise). This example has been relatively easy for you, because the lowest priority atom – hydrogen – was drawn pointing away from the reader.

How would you designate this molecule? (This is where a model might come in handy!) The priorities have been put in the diagram for you.

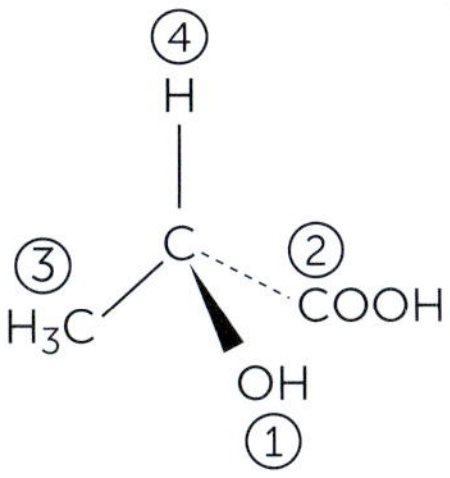

The lowest priority atom is, again, hydrogen; however this time it is not pointing away from you. You have to 'turn' the molecule so it does:

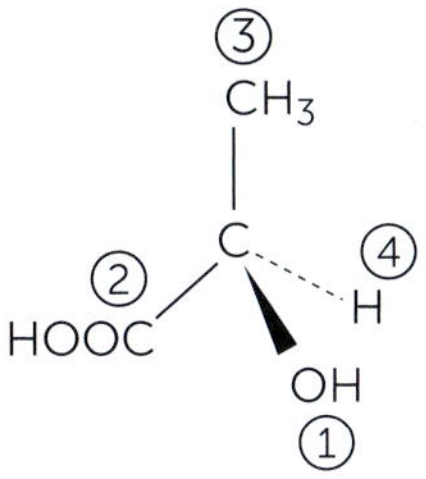

Then draw the arrow:

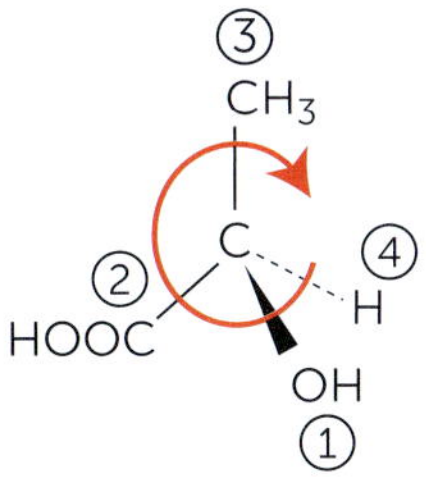

It is the *R* isomer

The full name of the molecule is *R*-2-hydroxypropanoic acid.

19.7 Amino acids and proteins

Amino acids

Amino acids contain an amino group ($-NH_2$) and a carboxylic acid group ($-COOH$). Since their formulae can be quite complicated they usually are referred to by their common names rather than their systematic names. For example, H_2NCH_2COOH is called **glycine** rather than aminoethanoic acid. Some common amino acids are shown in Table 19.4.

Amino acids are non-volatile crystalline solids that are very soluble in water. Many of them are optically active. They also have large dipole moments. Because the molecules contain both an acidic and a basic group, the following equilibrium exists in solution:

$$H_2NCHRCOOH \rightleftharpoons {}^{+}H_3NCHRCOO^{-}$$

The 'acidic end' of the molecule provides a proton to protonate the 'basic' end. The ion that is formed is known as a **dipolar ion** or **zwitterion** and, because of the ionic charges at both ends, the amino acid has many properties of a salt.

EXERCISE 19N

Structure of amino acids

(i) Most naturally occurring amino acids can exist in two optically active forms. Draw the two possible structures of the amino acid alanine, CH_3CHNH_2COOH.

(ii) What is the formula of the dipolar ion formed by alanine?

Table 19.4 Amino acids

Name	Formula	Abbreviation
Alanine	$CH_3CHCOOH$ NH_2	Ala
Phenylalanine	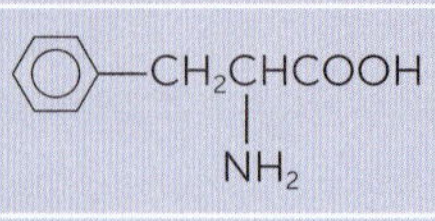	Phe
Cysteine	$HSCH_2CHCOOH$ NH_2	Cys
Glutamine	O $H_2NCCH_2CH_2CHCOOH$ NH_2	Gln
Lysine	$H_2NCH_2CH_2CH_2CH_2CHCOOH$ NH_2	Lys

EXERCISE 19O

Peptides

(i) Write the structural formula of the dipeptide glycylalanine (Gly-Ala).

(ii) How many different tripeptides can be made from different combinations of the three amino acids Gly, Ala and Cys?

Peptides

Peptides are formed when the amino group of one amino acid interacts with the carboxyl group of another. Water is eliminated in the reaction:

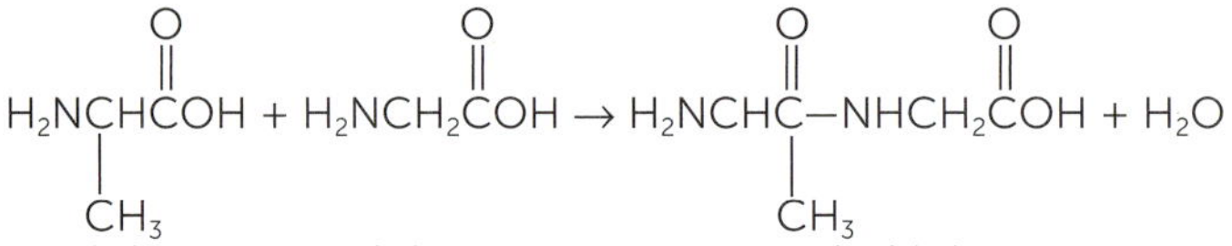

alanine glycine alanylglycine

The group –CONH– that is formed in such compounds is a **peptide linkage** or **peptide bond**. **Dipeptides** and **tripeptides** are derived from the combinations of two and three amino acids respectively. **Polypeptides** is a term that describes peptides of molecular mass up to 10,000 u. Molecules larger than this are known as **proteins**. Polypeptides and proteins can also be thought of as polyamides made from amino acids. The name of a peptide is constructed from the names of the amino acid units as they appear left-to-right, starting with the amino acid fragment that has a free $–NH_2$ group:

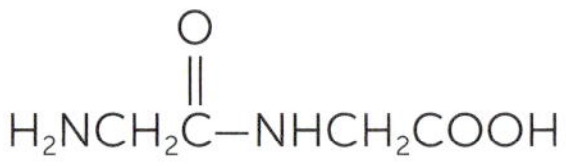

glycylglycine
(Gly-Gly, a dipeptide)

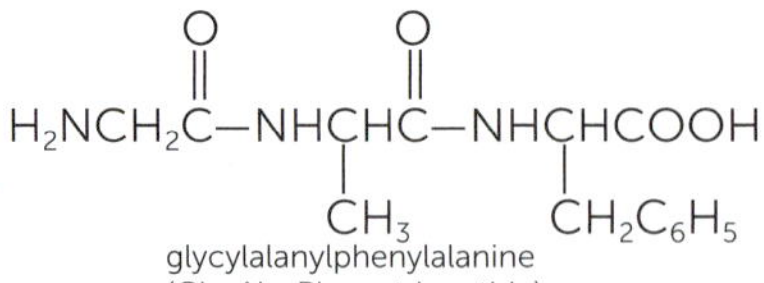

glycylalanylphenylalanine
(Gly-Ala-Phe, a tripeptide)

Proteins are found in all living cells. They make up a large part of an animal's body and are found, for example, in skin, blood, enzymes and hormones. A single protein molecule contains many, many amino acid units joined by peptide linkages. Proteins are divided into two groups:

1 Fibrous proteins, in which the molecules are long, tough and thread-like. They make up keratin in nails, wool, hair and feathers. These proteins are insoluble in water.
2 Globular proteins consist of molecules that are folded into compact shapes. They are found, for example, in enzymes and hormones, and in haemoglobin, which transports oxygen in the body.

Structure of proteins

The amino acids found in proteins are α-aminocarboxylic acids. The **prefix α** refers to the fact that the amino group is attached to the same carbon atom bonded to the carboxylic acid group:

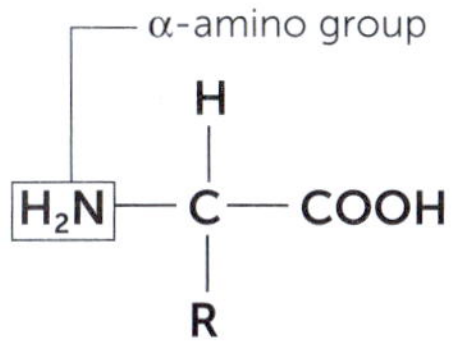

DID YOU KNOW?

If hair clippings or hair is heated with sodium hydroxide solution, the amide groups are converted to pungent smelling ammonia gas.

The structures of proteins can be studied on a number of levels:

1 The **primary structure** of a protein is the order in which the amino acid residues are linked together to form long chains.
2 The **secondary structure** of a protein is the way in which the chains are arranged in space to form coils, sheets or almost spherical shapes. The chains are held together by hydrogen bonds (see page 80). X-ray diffraction has proved to be a valuable technique in determining the often complicated secondary structure of proteins.

   An example of a protein that has a well studied secondary structure is keratin, found in hair. Each protein molecule in keratin is arranged in the shape of a spiral, called a **helix**. Hydrogen bonding holds the helix together, by linking to different sections of the same chain. The arrangement of the spiral in keratin is called an **alpha helix**. All alpha helixes have spirals with a *right-handed* turn, as in Figure 19.4.
3 The **tertiary structure** of a protein is a description of further interactions that occur, such as further folding of the secondary structure.

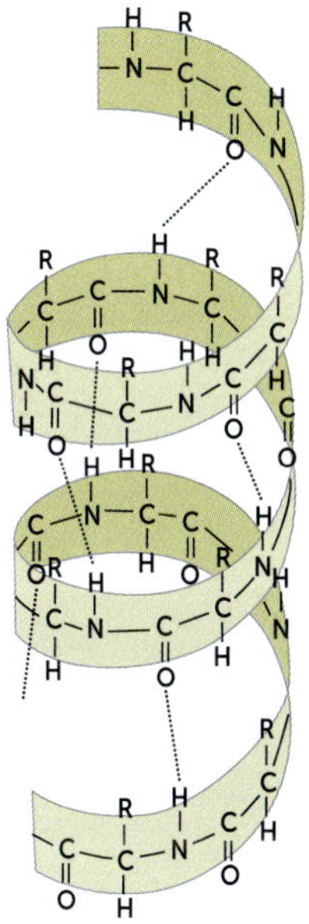

Figure 19.4 The arrangements of the protein chains in keratin.

19.8 Substituted benzene derivatives

So far, we have discussed families of aliphatic compounds containing different functional groups. Benzene derivatives containing the same functional groups often show very different properties when the functional group is attached directly to the benzene ring. They are also sometimes named differently from their aliphatic counterparts.

Phenols

Phenols are compounds with hydroxy groups bonded to an aromatic ring (Ar–OH). The simplest member of this family is phenol, C_6H_5OH.

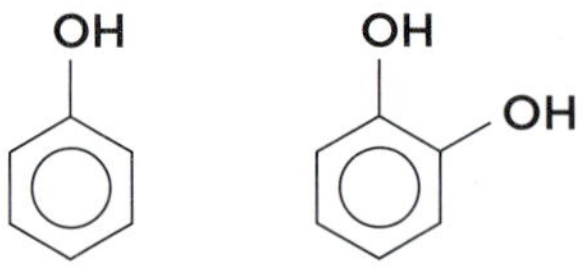

phenol

catechol
(modern name
2-hydroxyphenol)

Some phenols are obtained from coal tar, but nowadays most phenols are made by oxidation of cumene, C_9H_{12}. Cumene is produced from benzene and propene.

Phenols have acidic properties. For example, they react with aqueous sodium hydroxide to produce sodium phenoxides:

$$C_6H_5OH(s) + NaOH(aq) \rightarrow C_6H_5O^-,Na^+(aq) + H_2O(l)$$

Phenol is therefore much more soluble in sodium hydroxide solution than it is in water. Phenol is only weakly acidic, however, and it *does not react with carbonates to form carbon dioxide.*

BOX **19.5**

Phenol

Phenol was originally called carbolic acid. In the nineteenth century, Joseph Lister (1827–1912) used it to prevent wounds going septic and recommended that the compound should be used as an antiseptic in hospitals. Before this, amputees often had their fresh wounds dipped in coal tar (which contains phenol) and about half of patients operated on died from infections gained during the operations. The compound is a severe irritant to the skin, however, so it had to be replaced by alternatives. Many modern antiseptic molecules, however, still contain phenolic groups:

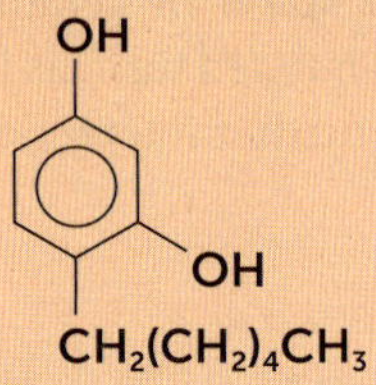

n-hexylresorcinol
used in antiseptics

Joseph Lister proposed the use of phenol as an antiseptic. He was a Scottish surgeon who rightly suspected that germs might be responsible for infection in wounds.

Benzoic acid

In benzoic acid, a carboxyl group is attached to the benzene ring:

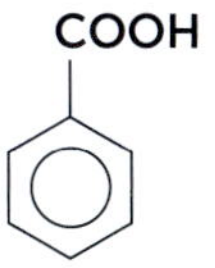

benzoic acid

The solid acid is slightly soluble in water and undergoes the typical reactions of carboxylic acids described previously. It is used as a preservative in food and drinks including iced lollies.

Aniline

Aniline is an aromatic amine with the structure

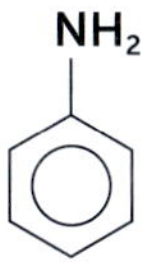

aniline
(modern name phenylamine)

The presence of the aromatic ring makes aniline a much weaker base than ammonia. Aniline is an *extremely toxic* liquid that readily enters the body through the skin or by inhalation and should be handled with extreme care.

Aniline undergoes an interesting reaction with sodium nitrite and ice-cold aqueous acid. The compound formed is called a **diazonium salt**:

$$C_6H_5NH_2(l) + NaNO_2(aq) + 2HCl(aq) \rightarrow \underset{\text{diazonium salt}}{C_6H_5{-}N{\equiv}N^+, Cl^-(aq)} + NaCl(aq) + 2H_2O(aq)$$

Diazonium salts can be reacted to form many classes of compounds but decompose, even at cold temperatures, so they must be used immediately. They also react with certain aromatic compounds to form **azo compounds**. The reaction is called a coupling reaction, for which the general equation is

$$Ar{-}N{\equiv}N^+ + Ar'H \rightarrow \underset{\text{azo compound}}{Ar{-}N{=}N{-}Ar'} + H^+$$

The double bond between the nitrogen atoms makes azo compounds strongly coloured. Because of their intense colours they are widely used as dyes. Many acid–base indicators are azo dyes.

Some examples of azo dyes are

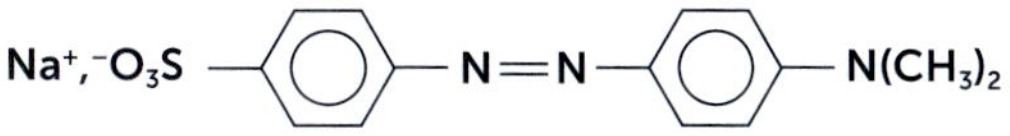

methyl orange
(sodium salt)
(red in acid, yellow in base)

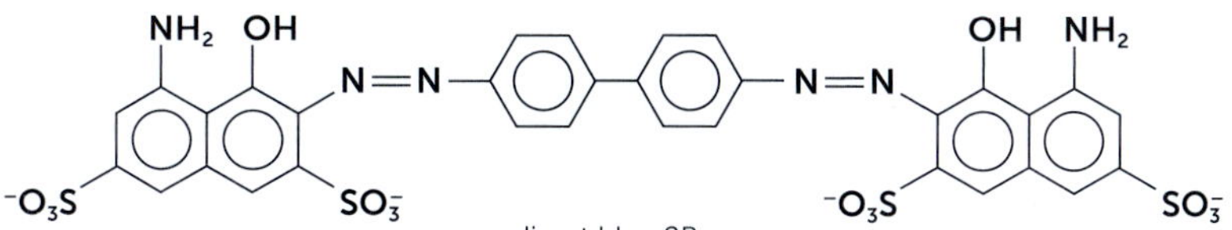

direct blue 2B

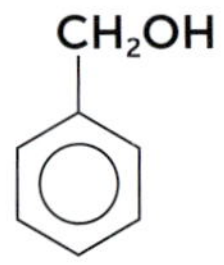

benzyl alcohol
(modern name phenylmethanol)

Benzyl compounds

The benzyl group, $C_6H_5CH_2$–, is an **aralkyl group**. It contains an aromatic group (the benzene ring) and an alkyl group. If functional groups are attached to the alkyl part of the group and *not* to the ring, the behaviour of the new compound tends to resemble similar alkyl compounds. For example, benzyl alcohol is classed as an **alcohol** and does not behave chemically as a phenol.

BOX **19.6**

Examples of reactions of phenol, ethanol and benzyl alcohol

Phenol	Ethanol	Benzyl alcohol
No ester formed when reacted with an organic acid	Ester formed	Ester formed
No simple product formed on oxidation	Aldehyde/acid formed on oxidation	Aldehyde/acid formed on oxidation
Reacts with neutral $FeCl_3$ to give a violet complex	No colour produced with the same reagent	No colour produced with the same reagent

REVISION QUESTIONS

19.1 Write the structural formulae of the following:

(i) hexan-2-ol
(ii) 2-methylpropan-2-ol
(iii) butane-1,3-diol
(iv) 3-chlorophenol.

19.2 Arrange in order of *increasing* acidity: ethanoic acid, ethanol, phenol, water.

19.3 Write the structural formulae for

(i) 3-methylhexanal
(ii) pentane-2,4-dione
(iii) phenylethanone
(iv) 2-methylpropanoic acid
(v) butane-1,4-dioic acid.

19.4 What would be the likely product of the oxidation of ethane-1,2-diol with excess acidified dichromate?

19.5 Write the structures of the organic products of the following reactions:

(i) The reaction when propanoic acid is heated with ethanol in the presence of concentrated sulfuric acid.
(ii) The reduction of propanoic acid with lithium aluminium hydride.
(iii) The oxidation of pentanal with sodium dichromate.
(iv) The oxidation of ethanal with Fehling's solution.

19.6 Write the structural formula of the following derivatives of methanoic acid:

(i) the propyl ester
(ii) the amide
(iii) the acid chloride.

19.7 The compound 2-amino-3-hydroxypropanoic acid is more commonly known as serine and is an amino acid:

(i) Write the structural formula of serine.
(ii) Does serine exist as two different optical isomers?
(iii) What would be the structure of the product formed when serine is treated with acidified sodium dichromate solution?
(iv) Write the structural formula of the dipeptide Ser-Ser.

19.8 Write an equation for the acid-catalysed reaction of methanol with benzoic acid.

19.9 Write the structural formula of a fat derived from the esterification reaction of one molecule of glycerol and three molecules of octadecanoic acid ('octadec' = 18).

19.10 Identify the functional groups in the following well-known compounds:

(i) the synthetic fibre Dacron

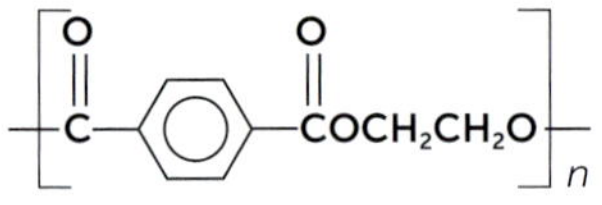

(ii) amphetamine

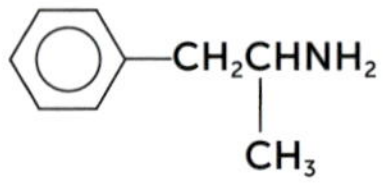

(iii) aspirin

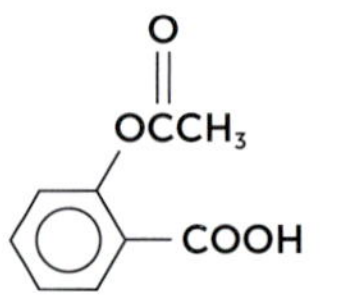

(iv) urea

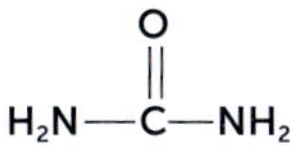

(v) adrenaline

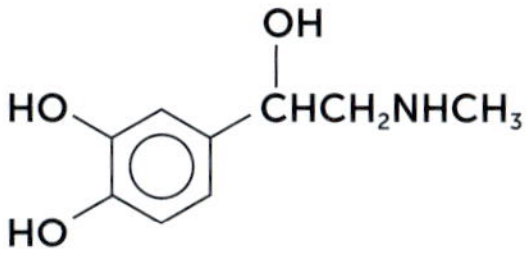

(vi) the polymer Saran (used for kitchen wrap).

$$\text{---}[CH_2CHCl—CH_2CCl_2]_n\text{---}$$

19.11 A carbonyl compound X has been found by analysis to have the molecular formula C_3H_6O:

(i) Describe a chemical test to prove that X contains a carbonyl group.
(ii) X is an aldehyde. Describe how you would experimentally confirm this fact.
(iii) Write down the structural formulae of the compound(s) that X could be. How could you confirm the identity of X?

19.12 Asparagine is an amino acid that is found in many plant proteins. It was first isolated in asparagus juice. The structural formula of asparagine is shown below:

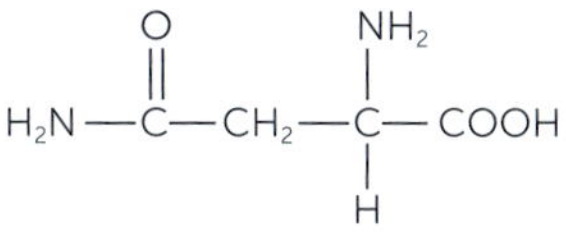

Represent this formula as a skeletal structure.

19.13 Identify the functional group in the following molecule.

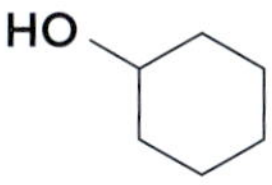

(a) amine **(b)** phenol **(c)** alcohol **(d)** aldehyde
(e) ketone

19.14 A molecule has the structure:

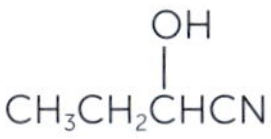

It exists as two enantiomers. Draw a three-dimensional representation of each and label them *R* or *S*.

DID YOU KNOW?

For thousands of years, humans have used natural substances – mainly derived from plants – as dyes.

In 1856 18-year-old William Henry Perkin accidentally discovered mauveine ('Perkin's mauvre'), the first synthetic dye. He was attempting to synthesize quinine (which was used in the treatment of malaria) from aniline and, instead, produced a purple dye. The following year, Perkin opened a factory in Middlesex to produce the synthetic dye in large quantities. Mauve was used to dye clothes and curtains and was even used to colour British 6d postage stamps from 1867–1880.

Original Perkin dye bottle

Extension 19 includes information on the class of organic molecules called ethers, see https://www.bloomsburyonlineresources.com/chemistry.

20 Organic Mechanisms

CONTENTS

OBJECTIVES

This unit:

- Explains what mechanisms are and why we need to know about them
- Describes types of organic reactions and important steps in mechanisms
- Discusses named reactions involving: free radicals, electrophilic substitution and nucleophilic substitution
- Introduces the idea of aromaticity
- Reveals the strange molecules found in outer space

20.1 Mechanisms and types of organic reactions

A reaction mechanism is a step-by-step description of the order in which bonds break and atoms rearrange during a reaction. A mechanism proposed by a chemist *must* take into account all the known facts about a reaction – if a mechanism does not fit in with any new facts discovered it is discarded or refined. Many mechanisms, however, have been sufficiently 'tested' to become part of the theory of organic chemistry.

Why do we study mechanisms in organic chemistry?

There are very many reactions in organic chemistry and studying mechanisms helps us to:

- make sense of (i.e. see patterns in) the reactions;
- change the experimental conditions in order to increase the yield of products; and
- predict what might happen in a reaction.

There are four types of reaction:

1 Substitution reactions (where one atom, ion or group is substituted for another). An example is $X + AB \rightarrow AX + B$.
2 Addition reactions (where one reagent is added to another without the loss of any other atoms). Alkenes undergo addition reactions, where a reagent adds across the double bond:

3 Elimination reactions (the reverse of addition; atoms or groups are removed from a molecule creating a multiple bond):

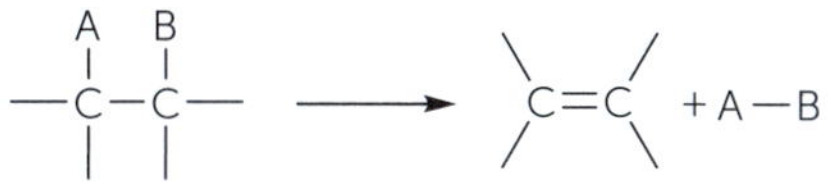

4 Rearrangements (where an atom or group goes from one position in a molecule to another):

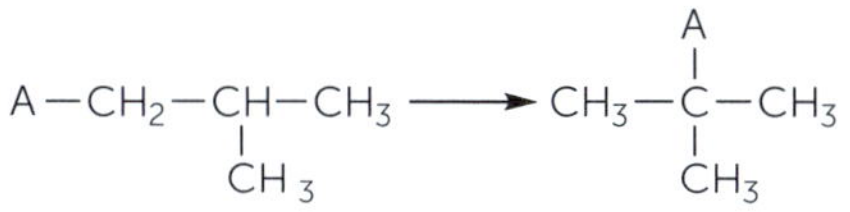

20.2 Key factors in reaction mechanisms

Bond-breaking

A covalent bond between two atoms can be broken in three ways:

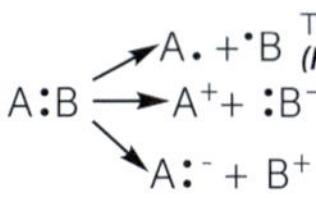

Free radicals are atoms, or groups of atoms, having an odd unpaired electron.

Attack! (Or love at first sight?)

A reagent that is attracted towards a region of high electron density (an electron pair) is called an **electrophile** ('electron lover'). Electrophiles are often positively charged, e.g. H_3O^+, NO_2^+ (the nitronium ion). Sulfur trioxide, SO_3 is also an electrophile. A reagent that possesses a pair of electrons that it is anxious to share with a nucleus that is 'short' of electrons is called a **nucleophile** ('nucleus lover'), e.g. OH^-, H_2O, Br^-, NH_3. Because these two reagents find what they are looking for in each other, they have a natural attraction:

X+	+	:Y–	→	XY
electrophile	+	nucleophile	→	product

Curly arrows are used in equations to show the movement of electrons, e.g. in heterolytic fission:

The tail of an arrow shows where an electron pair moves from, the head of the arrow shows where it moves to. If a covalent bond breaks, the tail of the arrow starts at the middle of the bond. A 'half arrow' or 'fish hook' is used to show the movement of one electron:

20.3 Mechanism of the chlorination of methane

Methane and chlorine do not react together at room temperature when they are in darkness, but in sunlight they undergo a series of substitution reactions. The overall equation is:

$CH_4 + Cl_2 \rightarrow CH_3Cl + HCl$

The reaction proceeds by the following steps. In the presence of sunlight (ultraviolet light), the chlorine molecule undergoes homolytic fission and two free radicals are formed:

$Cl_2 \rightarrow 2Cl^{\bullet}$ (the **initiation** step).

This step could also be written

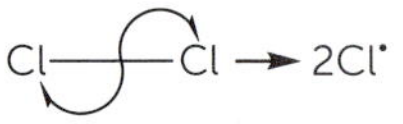

The chlorine radical then takes a hydrogen atom from methane, so hydrogen chloride and a methyl radical are formed:

$CH_4 + Cl^{\bullet} \rightarrow HCl + CH_3^{\bullet}$

The methyl radical then takes a chlorine atom from a chlorine molecule, and chloromethane and another chlorine radical are produced:

$CH_3^{\bullet} + Cl_2 \rightarrow CH_3Cl + Cl^{\bullet}$

The reaction mechanism is also discussed on page 256. Because a mixture of products (C_2H_6, CH_3Cl, CH_2Cl_2, $CHCl_3$ and CCl_4) can result from these reactions, this reaction is not really a suitable method to prepare chloromethanes in the laboratory.

20.4 Addition reactions of alkenes

With bromine

Ethene and bromine react to form 1,2-dibromoethane:

1,2-dibromoethane

A bromine molecule is non-polar, but as it approaches the ethene molecule it becomes polarized; the electrons in the double bond of ethene repel the electrons in the bromine molecule:

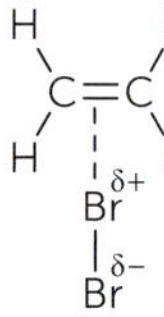

The positive end of the bromine molecule attacks the double bond and a bridged bromonium ion is formed. This addition reaction therefore starts by the electrophilic attack of the positive end of the bromine molecule on ethene.

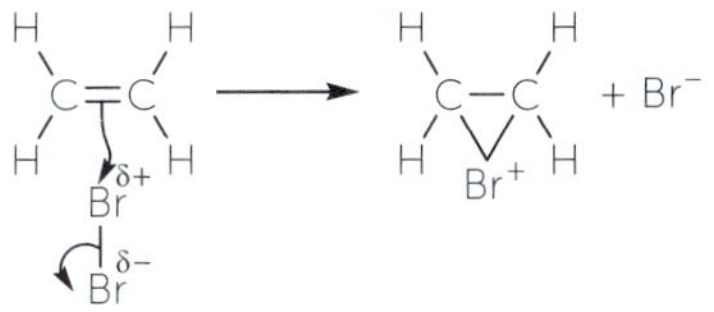

The positive bromonium ion is rapidly attacked by the negatively charged bromide ion and 1,2-dibromoethene is formed.

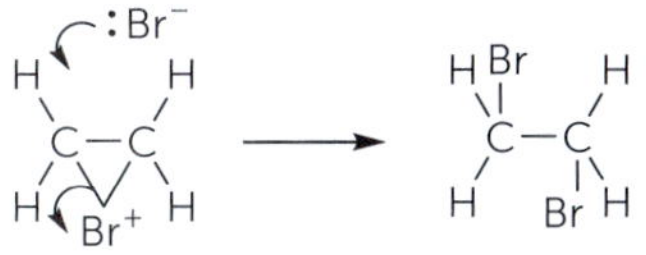

EXERCISE 20A

(i) Why do you think the bromide ion attacks from the side it does?

(ii) Predict (and name) the product from the addition of one mole of bromine to one mole of ethyne

With hydrogen bromide

When hydrogen bromide reacts with ethene, bromoethane is formed:

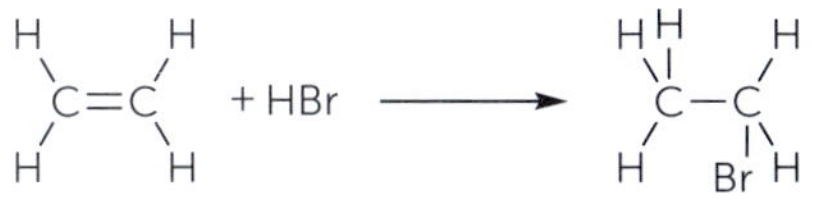

Hydrogen bromide is a polar molecule, $H^{\delta+}–Br^{\delta-}$. The first step in the reaction is electrophilic attack of a proton on the π-bond of ethene and the H–Br bond breaks. A positively charged carbocation is formed, which reacts with the bromide ion to form bromoethane.

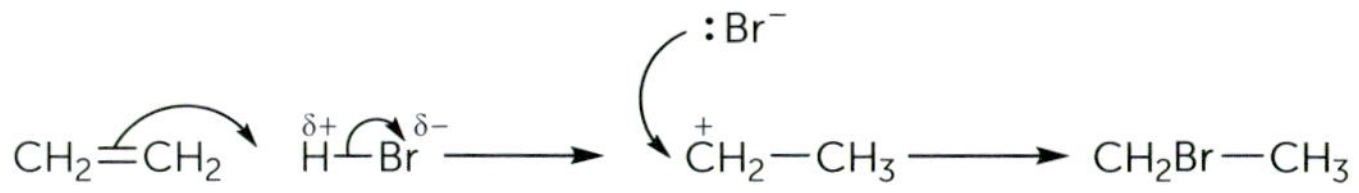

carbocation

20.5 The stability of carbocations

Carbocations are organic cations with the positive charge carried on the carbon. If propene reacted with HBr, there would be two possible carbocations:

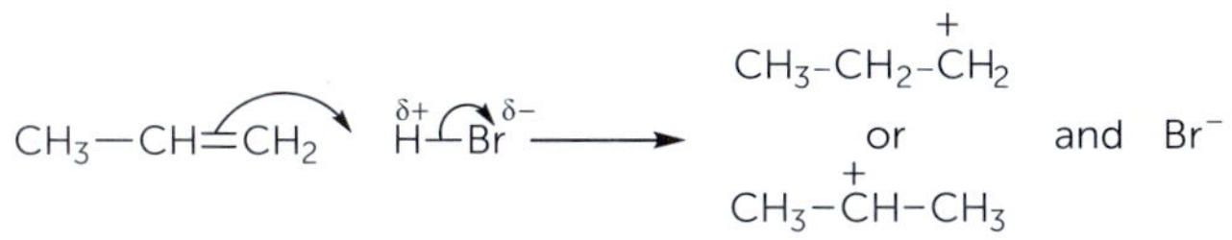

giving two possible products:

$CH_3–CH_2–CH_2Br$	or	$CH_3–CHBr–CH_3$
1-bromopropane		2-bromopropane

In fact, much more 2-bromopropane is formed, because carbocation $CH_3–C^+H–CH_3$ is the more stable of the two possibilities. It contains two alkyl groups ($–CH_3$) that tend to 'push' electron density on to a carbon atom joined to them. The more electron density

EXERCISE **20B**

When 2-methylpropene reacts with HBr, what will be the major product?
(You should work out the structure of the two possible intermediate carbocations that could be formed, then consider which would be more stable.)

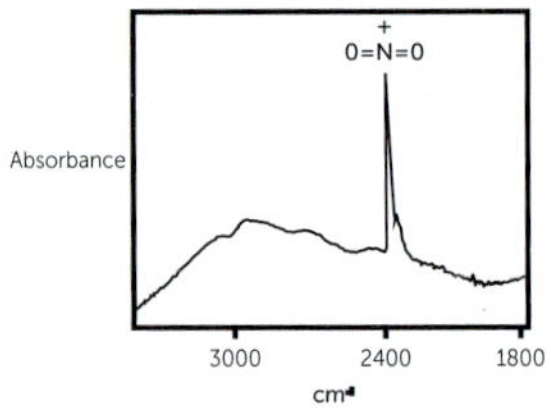

Infrared spectroscopy provides experimental evidence for the nitronium ion. A mixture of concentrated nitric and sulfuric acids generates a peak in its infrared spectrum at about 2360 cm^{-1}. This peak is absent in the spectrum of either acid alone. The peak belongs to the nitronium ion NO_2^+. (Image by author).

that is pushed on to the carbon with the positive charge, the more stable is the cation.

two alkyl groups stabilizing the carbocation

one alkyl group stabilizing the carbocation V

The addition of a hydrogen halide to a double bond is described by Markovnikov's rule:

Markovnikov's rule predicts the major product when HX (X = Cl or Br) reacts with an alkene: the major product is the one in which the hydrogen atom attaches itself to the carbon atom carrying the larger number of hydrogen atoms.

20.6 Substitution reactions of benzene

The nitration of benzene

A mixture of concentrated sulfuric and nitric acids added to benzene results in the formation of nitrobenzene. This is a substitution reaction – the nitro group is substituted for a hydrogen atom in the benzene ring:

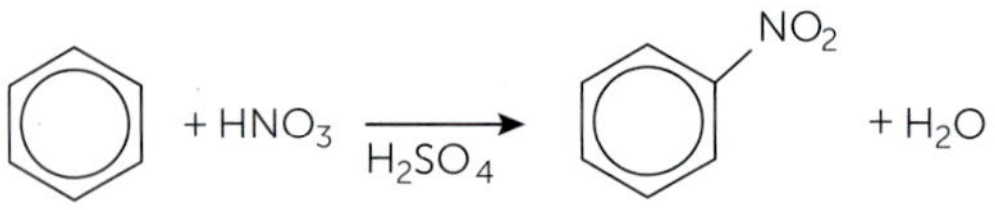

The reaction between concentrated sulfuric and nitric acids is shown by the following equation:

$$HNO_3 + 2H_2SO_4 \rightleftharpoons NO_2^+ + H_3O^+ + 2HSO_4^-$$

The powerful electrophile NO_2^+ is produced, which attacks the clouds of electron density above and below the benzene ring then bonds to one of the carbon atoms in the ring. The 'doughnuts' of delocalized electrons are disrupted and the intermediate cation formed immediately breaks down into nitrobenzene so that the more stable delocalized system can be re-formed. In the process it gives up a proton to HSO_4^-, producing H_2SO_4 again.

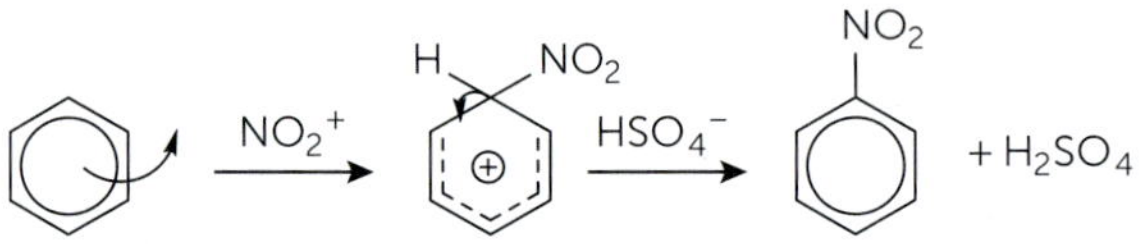

cationic intermediate

The reaction is classified as an electrophilic substitution reaction. Other electrophilic substitution reactions include halogenation, alkylation and acylation.

Halogenation

Here a halogen (e.g. chlorine) is substituted for a hydrogen atom in a benzene ring. The reaction takes place in the presence of a catalyst, such as iron(III) chloride, which 'draws' electron density from the halogen molecule towards it.

$$Cl{-}Cl + FeCl_3 \longrightarrow \overset{\delta+}{Cl}{-}\overset{\delta-}{Cl}......FeCl_3$$

a 'complex'

The halogen molecule becomes polarized, its positive end acts as an electrophile and attacks the benzene ring:

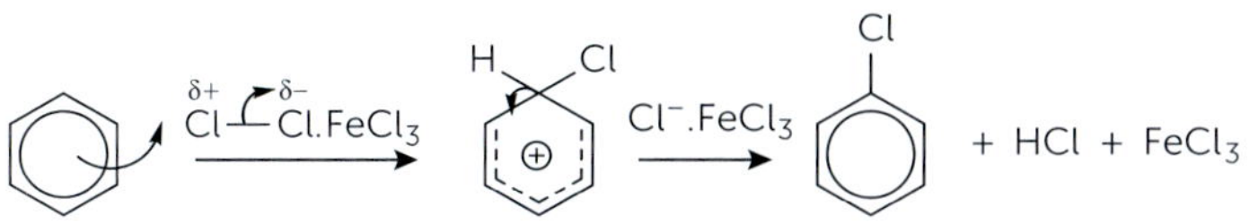

Alkylation (a Friedel–Crafts reaction)

A halogenoalkane reacts with benzene in the presence of a catalyst, and one of the hydrogen atoms on the benzene ring is replaced by an alkyl group. For example, chloromethane reacts with benzene in the presence of catalyst aluminium chloride and toluene is formed:

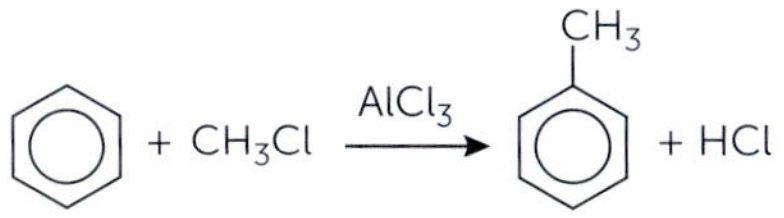

The mechanism is similar to that of the electrophilic substitution reactions already discussed and involves attack by the positive end of the complex $CH_3^{\delta+}{-}Cl^{\delta-}.AlCl_3$. Aluminium chloride draws electron density from chloromethane and further polarizes the molecule.

Acylation (another Friedel–Crafts reaction)

An acid chloride reacts with benzene, in the presence of aluminium chloride, and a ketone is formed (a hydrogen atom on benzene is replaced with a –COR group). For example:

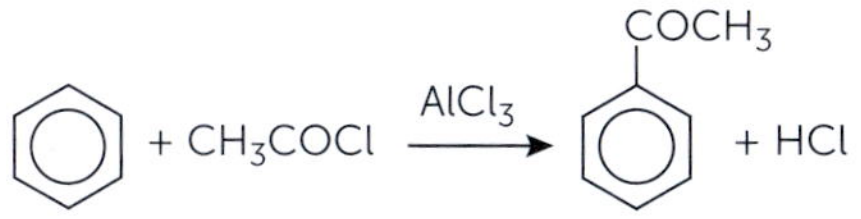

The attacking electrophile is the polarized complex:

$$H_3C{-}CO^{\delta+}{-}Cl^{\delta-}.\ AlCl_3$$

EXERCISE 20C

Give equations for the steps involved in the mechanism of the reaction of chloroethane with benzene in the presence of aluminium chloride.

Reactivity of substituted benzenes

What happens when substituted benzene undergoes further substitution?

Suppose we take a substituted benzene (G is a symbol for an atom or group) and try to substitute another hydrogen in the ring. The next substituent can go into one of the following positions:

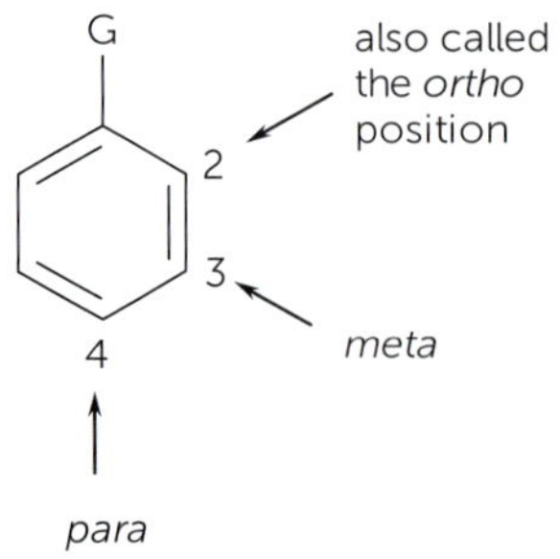

Where will the next substituent go? It depends on the nature of G.

Some atoms, or groups, are activating – they donate electron density into the benzene ring and increase its reactivity with respect to electrophilic substitution. Other atoms, or groups, are deactivating – they withdraw electron density from the ring and make it less reactive.

In general, activating groups direct substituents to either the 2 or 4 (*ortho* or *para*) position and deactivating groups direct to the 3 (*meta*) position. **An important exception to this rule is when G is Cl, Br or I**. Although the halogens are deactivating – remember they are electronegative and pull electron density in a covalent bond towards them – they are *ortho* and *para* directing. The reason for this is that a lone pair on the halogen delocalizes over the ring, which makes the *ortho* and *para* positions more attractive to incoming electrophiles.

Table 20.1 summarizes the influence of common groups that could be G.

Table 20.1 Effects of common substituents in a benzene ring

Substituent	Activating/ deactivating	Directional influence towards further substituents	Reactivity towards electrophilic substitution
$-NH_2$	activating	2,4 or *o,p*	↑ increasingly more reactive to electrophiles
−OH, −OR	activating	2,4 or *o,p*	
−R	activating	2,4 or *o,p*	
−Ph	activating	2,4 or *o,p*	
−H			benzene
−Cl, Br or I	deactivating	2,4 or *o,p*	↓ increasingly less reactive to electrophiles
−CHO	deactivating	3 or *m*	
$-CO_2H$, $-CO_2R$	deactivating	3 or *m*	
$-NO_2$	deactivating	3 or *m*	

EXAMPLE **20.1**

What will be the main product(s) of

(i) the chlorination of nitrobenzene

(ii) the nitration of chlorobenzene?

The nitro group in nitrobenzene is *meta* directing:

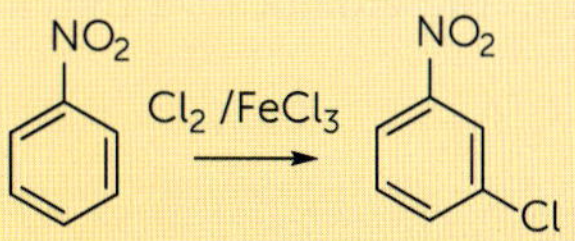

3-chloronitrobenzene
or *m*-chloronitrobenzene

The chloro group in chlorobenzene is *ortho* or *para* directing:

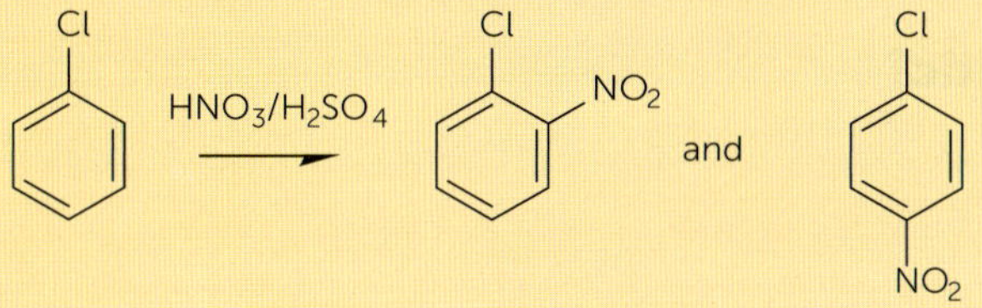

2-chloronitrobenzene	4-chloronitrobenzene
or *o*-chloronitrobenzene	or *p*-chloronitrobenzene
35%	64%

So the order in which you introduce the substituents into benzene can result in different isomers, and there are two products from the second reaction. In practice, there are two *ortho* positions for one *para* position in the ring, so you might expect more of the *ortho* isomer than the *para* isomer in reaction (ii). This can change if one of the substituents is bulky; then the *para* isomer is preferred (note that the percentages in reaction (ii) do not add up to 100 per cent; a very small amount of the meta isomer is formed).

20.7 Aromatics

Aromaticity is a molecular phenomenon in which a planar ring of alternating single and double bonds (according to its Kekulé structure) has more stability than would be expected. Benzene is the 'parent compound' of this class, but there are others and lone pairs from atoms other than carbon can be included. The stability comes from all the atoms in the ring having an equal share of the electrons that are delocalized in an electron cloud.

Hückel's Rule

Hückel's Rule states that:

> *a planar molecule, in which all atoms contribute electrons, that has $4n + 2$ delocalized electrons (where n is a whole number) will have aromatic properties.*

For example benzene has a planar framework (see page 338) in which each carbon uses three of its valence electrons to bond to two other carbons in the ring and a hydrogen atom:

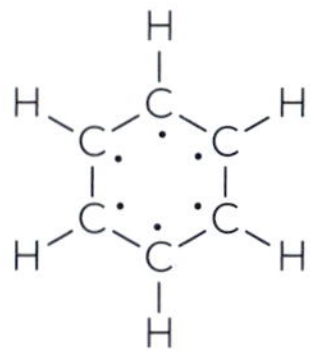

There is one p electron 'left over' on each carbon atom and these delocalize into an electron cloud above and below the ring:

This is in accordance with Hückel's Rule, where there are six delocalized electrons, i.e. $4n+2$ where $n = 1$.

Is naphthalene aromatic?

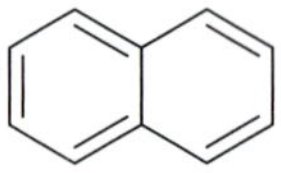

The Kekulé structure of naphthalene is shown above. Here, we have another system of alternating single and double bonds with ten electrons (one on each carbon) available for delocalization. If $n = 2$, then $4n + 2 = 10$ so naphthalene is aromatic.

When a lone pair is involved, e.g. furan

Carbon can form rings in which an atom of another element is involved. These are called heterocyclic molecules. Furan is an example:

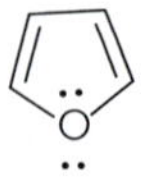

It is aromatic and obeys Hückel's Rule; here oxygen has two lone pairs and one of those pairs delocalizes with the single electrons available on each of the four carbons, i.e. $4 + 2 = 6$ or $4n+2$ when $n = 1$.

Reminder

The forms of alkyl halides are:
Primary CH_3X, $-CH_2X$
Secondary >CHX
Tertiary

20.8 Nucleophilic substitution reactions

A halogenoalkane is converted into an alcohol after reaction with aqueous sodium hydroxide (the halogen is substituted by an –OH group). Because of the electronegativity difference between carbon and the halogen, a halogenoalkane is polar, with a partial positive charge on one of its carbon atoms:

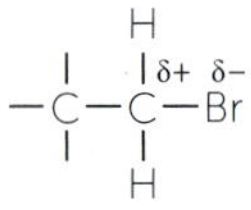

The partially positive carbon atom is open to attack by a nucleophile, such as the OH^- from sodium hydroxide. This can happen in one of two ways:

1 The halide ion leaves (a carbocation is formed) before the attack by OH^-:

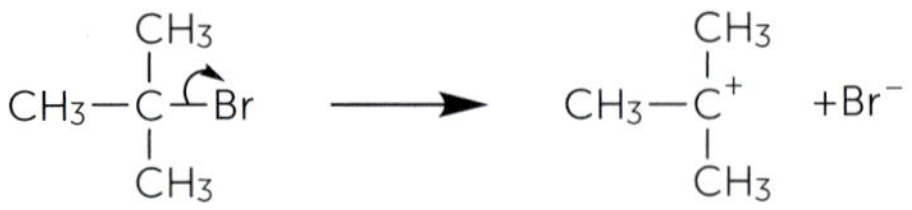

then:

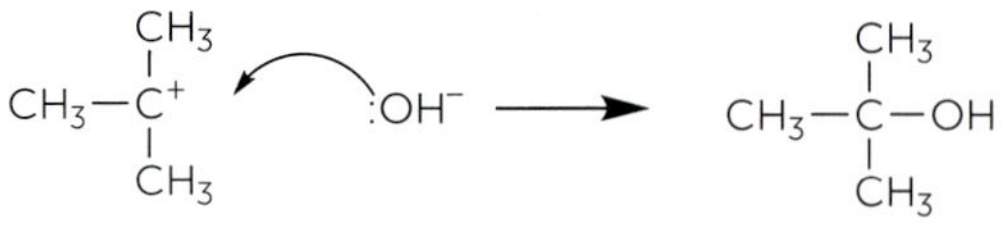

This is the case for the reactant $(CH_3)_3CBr$. In general, the reaction takes place by this mechanism if the positive carbon of the carbocation formed can be stabilized by electron-releasing alkyl groups (here, three methyl groups). The mechanism for this reaction is called S_N1 which stands for substitution nucleophilic unimolecular. The rate of the reaction depends on the slowest step (rate determining step), which is the first step (bond-breaking), and this involves only one molecule:

$$\text{Rate} = k\,[(CH_3)_3CBr]$$

2 Option 2 is that the attacking OH^- becomes attached to the partially positive carbon atom before Br^- leaves it; a transition state is formed.

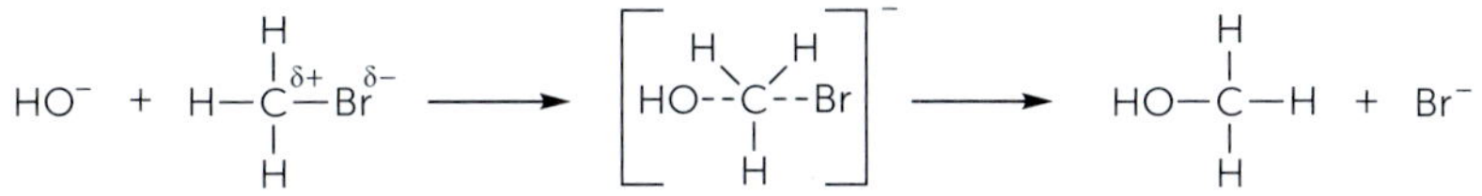

transition state

This is the case with the molecule CH_3Br. In the transition state, part of the energy needed to break the C–Br bond is produced from the formation of the HO–C bond. The dotted lines show these 'partial bonds'. This type of substitution is termed S_N2 (substitution nucleophilic bimolecular), because two molecules are involved in the step that determines the rate of reaction and:

$$\text{Rate} = k\,[CH_3Br]\,[OH^-]$$

As a general rule (although there are other factors that influence the mechanism): primary alkyl halides tend to react via S_N2, tertiary alkyl halides by S_N1 and secondary alkyl halides often proceed by a mixture of both pathways.

EXERCISE 20D

Would you expect the reaction of benzyl bromide with sodium hydroxide to proceed via an S_N1 or S_N2 mechanism? Write the structural formula of the product.

benzyl bromide

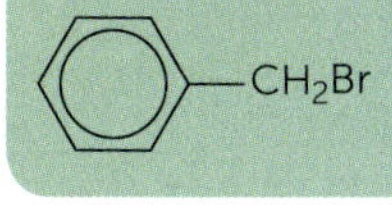

20.9 Strange molecules in outer space

The vast space between the stars is called the interstellar medium (ISM). The ISM contains dust clouds of very fine particles of silicate, carbon or ice dispersed in gas at such an extremely low gas pressure that we would describe it as a near perfect vacuum. At such low gas pressures, molecular collisions are extremely rare. Temperatures range from 10 K (only 10 degrees above absolute zero) to 100 K (still extremely cold).

The microwave spectroscopy of the ISM has been studied using the huge dishes of radio telescopes and the absorption spectra of about 300 molecules have been detected at extremely low concentrations. The molecules include H_2, CO, NH_3 and HCN (hydrogen cyanide), CH^+, CH_3CN (methyl cyanide) and C_2H_5OH (ethanol). Also observed are some strange and unexpected species with multiple bonds including HC_8CN (cyanotetraacetylene), more clearly written as:

$$H{-}C{\equiv}C{-}C{\equiv}C{-}C{\equiv}C{-}C{\equiv}C{-}C{\equiv}N$$

This is a remarkable linear molecule with four triple carbon–carbon bonds and a cyanide group (also with a triple bond, C≡N) at its end. The molecule does not occur on Earth and was unknown before its identification in space in 1971. If it were to exist on Earth it would be very unstable and it would readily undergo addition reactions like its parent ethyne (acetylene):

$$H{-}C{\equiv}C{-}H$$

Equally unexpected is the detection of the giant molecule C60 (see page 74). The biggest molecule found in the ISM so far is C70, which looks more like a rugby ball than a football.

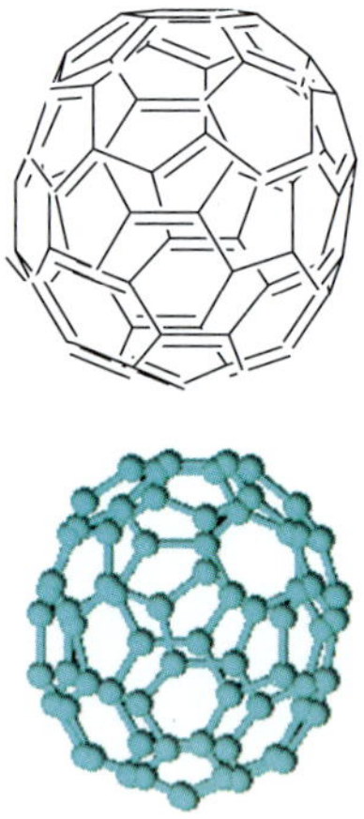

Skeletal structure and ball-and-stick representation of the molecule C70, also called *rugbyballene*

The existence of these strange molecules is puzzling. If molecules were exposed to strong UV light, they would dissociate into smaller fragments, suggesting that the strange molecules are protected from UV light once they are formed. One reaction mechanism being considered involves the frozen interstellar dust grains behaving as 'chemical factories', with the molecules within the grain reacting slowly. The material of the grain shields the product molecules from UV light so that the intermediate and final products do not decompose; this makes long reaction times immaterial. Different molecules may be produced by different mechanisms. One thing is clear. The chemistry of the ISM is very different from that of Earth, and the mechanisms by which molecules are formed at such low temperatures are also very different from the familiar ways we make molecules in our laboratories and factories!

DID YOU KNOW?

Dimethyl sulfide, abbreviated to DMS, is a component in the smell of boiled cabbage or beetroot. Since the only known natural source of DMS on Earth is from marine algae such as phytoplankton, DMS is often regarded as an indicator (or 'marker') for life. In 2025 DMS was detected in a planet (K2-18b) beyond our solar system. This is consistent with similar life existing on K2-18b, but it is not proof. One of the most challenging and complex reaction mechanisms involves explaining how life was formed from simpler molecules. The difficulties are outlined in Extension 20 at https://www.bloomsburyonlineresources.com/chemistry.

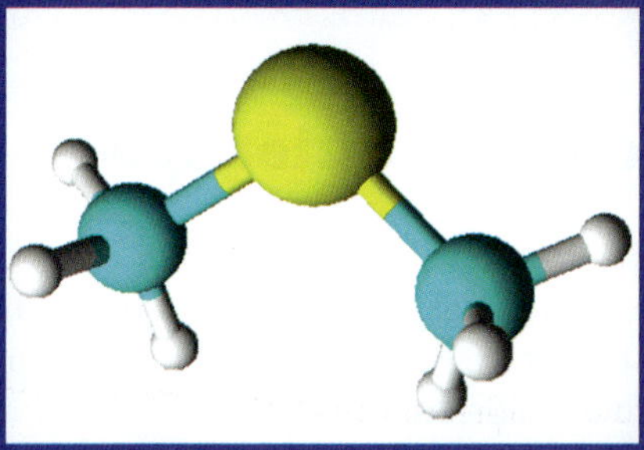

Dimethyl sulfide CH_3-S-CH_3

REVISION QUESTIONS

20.1 A compound found in the gasoline fraction of crude oil is decane, $C_{10}H_{22}$.

(i) Write a balanced equation for the complete combustion of decane in oxygen.

(ii) What products would you expect to find as a result of the incomplete combustion of decane?

20.2 But-1-ene was reacted with a solution of bromine, the bromine was in excess.

(i) Assuming that dibromoalkane is the sole product of this reaction, write a balanced equation for the reaction. Give the systematic name of the product.

(ii) If 15 g of but-1-ene produced 35 g of halogenated product, calculate the percentage yield of the reaction.

20.3

(i) Write the structural formula of propene.

(ii) Write the structural formula and name the product formed by polymerization of this alkene.

20.4 Explain why ethene reacts with bromine very rapidly, but the reaction of benzene with bromine is slow unless $AlBr_3$ is present.

20.5 Which of the following molecules are aromatic?

(i) 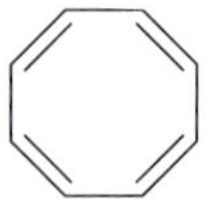

(ii) 

(iii) 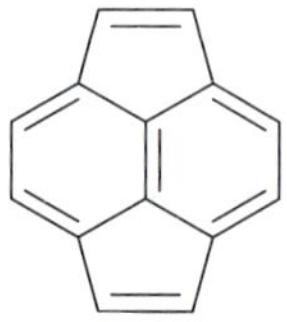

(iv) 

20.6 Which of the following heterocyclic molecules are aromatic?

(i) 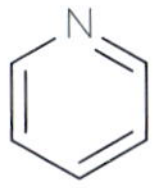 (ii)  (iii) 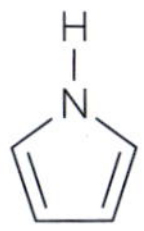 (iv) 

20.7 What would be the products of the following Friedel–Crafts reaction?

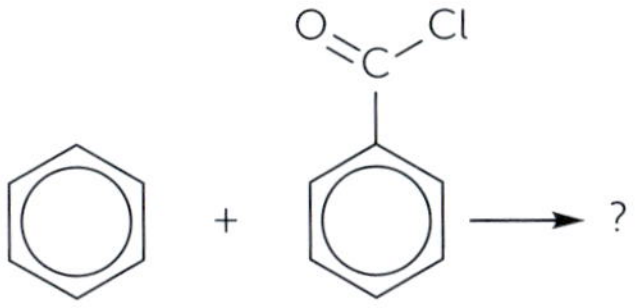

20.8 What is the expected order of reactivity of the following compounds in electrophilic nitration? (more reactive > less reactive)

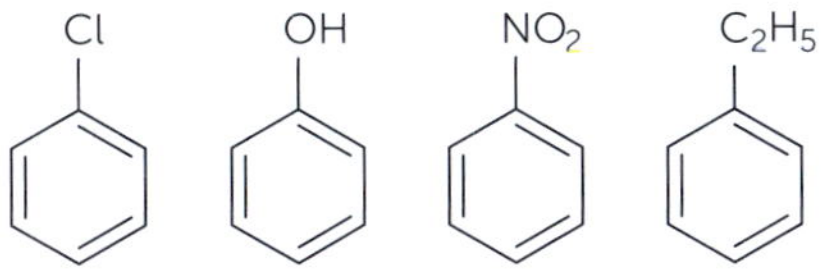

A B C D

20.9 Give the structure of the products formed from the bromination of cumene. Which product do you think is formed in the highest percentage and why?

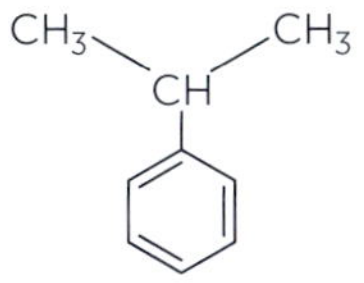

Cumene (2-phenylpropane)

20.10 You are required to synthesize the compound below from benzene. Which reactants would you use and in what order would you carry out the reactions?

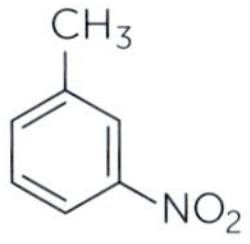

20.11 The cyanide ion CN^- is a powerful nucleophile. Draw the mechanism for the reaction of NaCN with $CH_3CH_2CH_2Cl$.

20.12 The reaction:

$$CH_3CH(Br)CH_3 + OH^- \longrightarrow CH_3CH(OH)CH_3 + Br^-$$

can proceed by both S_N2 and S_N1 mechanisms.
Using curly arrows, draw the two steps of the S_N1 mechanism.

DID YOU KNOW?

It might be said that Henry Edward Armstrong (1848–1937) was the first to investigate organic mechanisms, because he proposed something very close to what we now know as the electrophilic substitution mechanism in benzene in 1890. He also spent some time investigating the directing effect of substituents on further substitution in benzene and offered an explanation from what he called six 'affinities', which existed within a circle in benzene. This work was all the more remarkable because the electron had not been discovered at that time!

Extension material is available on the website at https://www.bloomsburyonlineresources.com/chemistry.

21 Separating Mixtures

CONTENTS

OBJECTIVES

This unit:

- Describes the techniques used by chemists to separate mixtures
- Demonstrates how to separate solids from solutions or suspensions, liquids from mixtures of liquids, and solids from solid mixtures
- Explains what is meant by the terms 'ion exchange' and 'solvent extraction'
- Discusses the various types of chromatography

Often in practical chemistry, only one substance is needed from a mixture of substances. The substance has to be **separated** from the mixture.

21.1 Separating a solid from a liquid

A solid that is mixed with a liquid may be **insoluble** in the liquid, or it may have **dissolved**. When an insoluble solid floats in a liquid, the mixture is called a **suspension**.

Insoluble solids

Insoluble solids can be separated by **filtration**. The solid is trapped by the filter paper, while the liquid passes through it. A simple example of this technique is the separation of a mixture of fine sand in water. A residue of sand collects in the filter paper and the liquid that drips through (in this case water) is called the **filtrate**. This type of filtration is called **gravity filtration** (Figure 21.1).

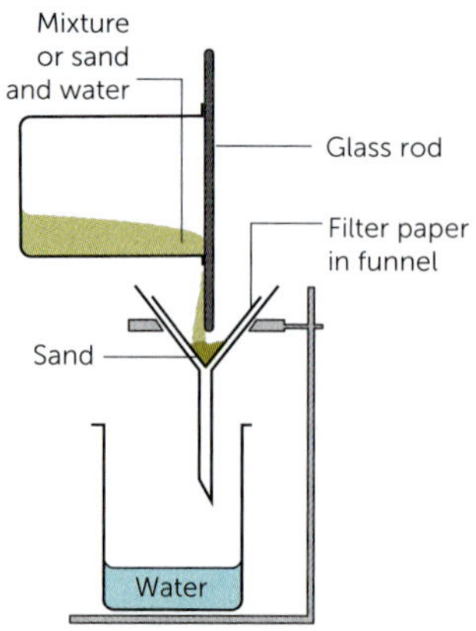

Figure 21.1 Gravity filtration.

Small amounts of a fine suspension can be separated in a **centrifuge**. Tubes of the suspension are spun around very quickly and the solid collects at the bottom of the tube, leaving clear liquid at the top. The clear liquid can then be removed with a pipette, leaving the solid in the tube (Figure 21.4).

BOX **21.1**

Folding a filter paper

The diagrams in Figure 21.2 show how a filter paper is folded for gravity filtration.

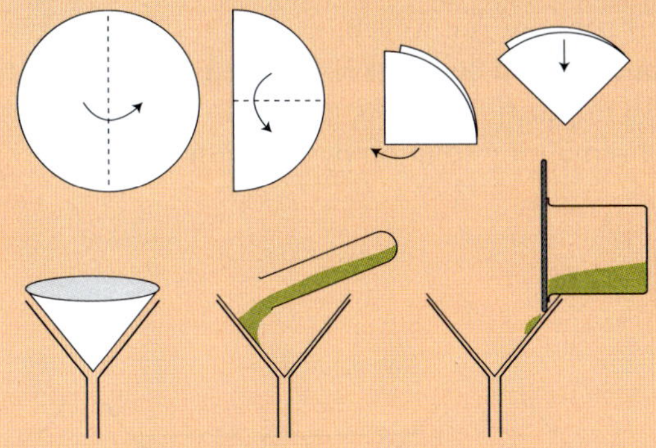

Figure 21.2 Folding a filter paper.

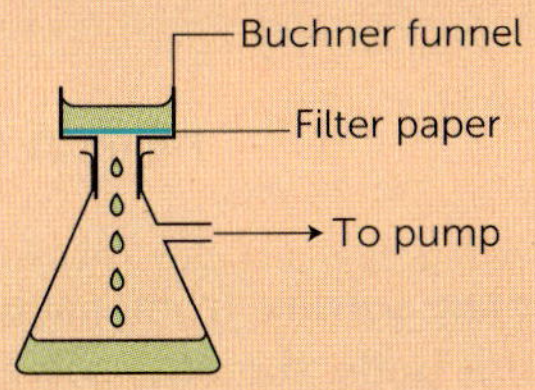

Figure 21.3 Vacuum filtration.

Gravity filtration has the disadvantage of being rather slow. The process of filtration is speeded up using **vacuum filtration**, but more sophisticated equipment is required. A vacuum is used to 'pull' the liquid through the filter paper (Figure 21.3).

A special funnel (called a **Buchner funnel**) is used that is connected, via a rubber seal, to a thick-walled flask with a side arm. A pump is attached to the side arm of the flask, and when the vacuum is applied the solution in the funnel is filtered rapidly. The filter paper lies flat inside the bottom of the funnel. The liquid is sucked through the filter paper, leaving insoluble solid behind. If the liquid is volatile, for example ethoxyethane (diethyl ether), sucking air through the filter paper for a few minutes after all the liquid has been pulled through is sufficient to dry the solid.

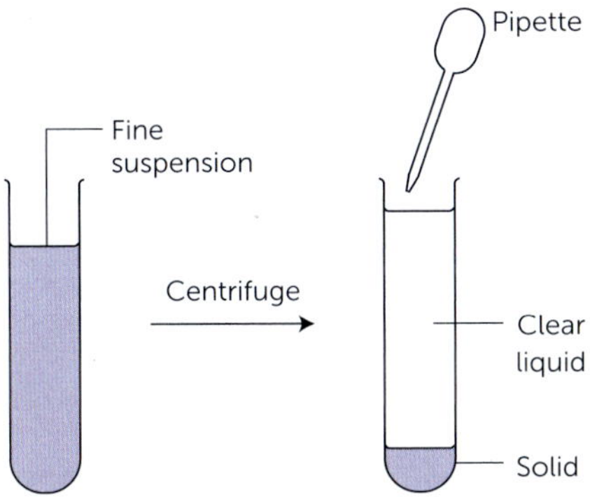

Figure 21.4 Removing the liquid from a centrifuge tube.

Solid dissolved in a liquid (in solution)

The solid cannot be separated from the solvent by filtering or centrifuging, because now the solid particles are too small.

The solvent can be removed from the solid by **evaporation**. The solution is heated until the solvent boils off, leaving the solid behind (Figure 21.5). This is often used for separating salt from salt solution. However, there are disadvantages with this method. If impurities are in the solvent they will be left behind and contaminate the solid. Also, some solids may decompose during heating to dryness.

A better method of recovering a solid from a solution is **crystallization**. The solution is heated and some of the solvent evaporates, so the solution becomes more concentrated. The concentrated solution is then left to cool and the solid crystallizes out (Figure 21.6), with impurities tending to remain in solution. The crystals can then be separated from the remaining solution by filtration.

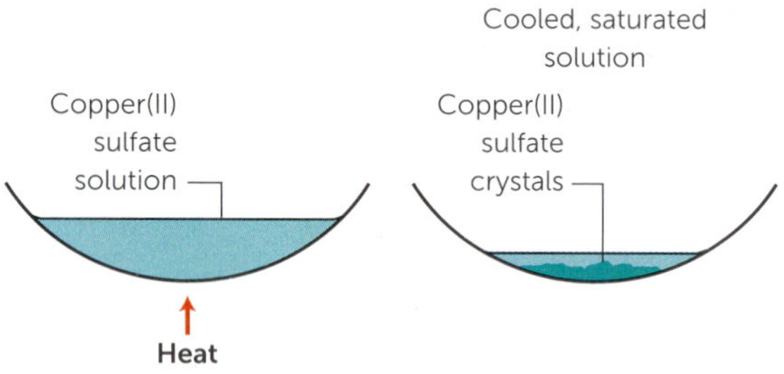

Figure 21.5 Solvent removal by evaporation.

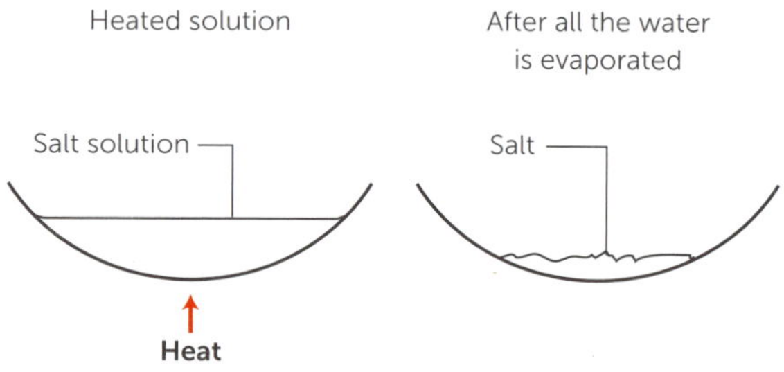

Figure 21.6 Crystallization.

In order to recover the solvent from a solution, **distillation** is used. The solution is heated in a flask, until it boils, then the solvent vapour is passed through a **condenser** where it is cooled and condensed. The apparatus is shown in Figure 21.7. The condenser consists of a glass tube surrounded by a glass jacket through which cold water is circulated. In this example, steam from the boiling salt solution passes through the condenser and condenses into water, which drips into the beaker. Salt is left behind in the flask.

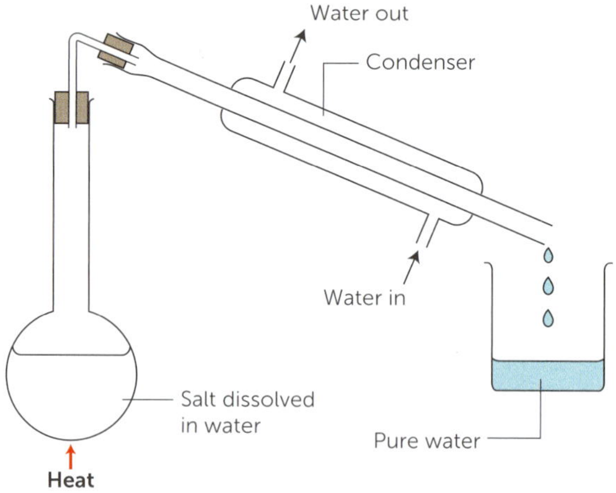

Figure 21.7 Simple distillation. Water enters the condenser from the bottom so as to avoid 'hotspots' consisting of air pockets.

EXERCISE **21A**

Separation of salt from seawater

Seawater contains a great deal of salt. Describe how you would separate a sample of salt from seawater. Do you think the sample would be pure NaCl?

BOX 21.2

Recrystallization

Recrystallization is used to purify a product. Many chemical preparations give 'dirty' products in which the product is contaminated with, for example, excess reactants or products from side reactions. A solvent is found that dissolves the desired product when the solvent is hot, but the product is not soluble when the solvent is cold. This choice of solvent is the difficult part and involves some 'trial and error' experiments. The impure product is dissolved in the **minimum** of the hot solvent and the hot solution is filtered very quickly to remove insoluble impurities before the mixture cools down – Buchner filtration (*using warm apparatus*) is useful here. The filtered solution is allowed to cool and the product crystallizes out. It is not necessary that the impurities are more soluble than the desired product as soluble impurities (since they are present at low concentration) tend to remain in solution and the desired product (which dominates the mixture) crystallizes out in a more pure form. If the product is required to be very pure, several recrystallizations may be necessary.

Marie and Pierre Curie, Nobel Prize-winning discoverers of the elements radium and polonium, carried out laborious recrystallization experiments in order to eventually isolate very small amounts of a radium compound from the uranium ore pitchblende, obtained from Bohemian mines. Eventually, in 1902, about 0.1 g of radium chloride ($RaCl_2$) was obtained from several tonnes of ore. The radioactive compound ionized the surrounding air and glowed in the dark!

Marie Sklodowska Curie. She eventually died of radiation poisoning.

21.2 Separating two liquids

Two liquids may be **immiscible** (they do not mix and two layers are formed) or **miscible** (they mix completely). For more about miscibility, see Unit 11.

Immiscible liquids

Immiscible liquids can be separated using a separating funnel. For example, consider the separation of oil and water. This mixture will separate out with the less dense oil floating on top of the water. If the mixture is poured into a separating funnel (Figure 21.8) and allowed to settle, two layers are formed. The tap is opened to let the water run out and closed before the oil layer runs through the tap. In this way, the two liquids are separated.

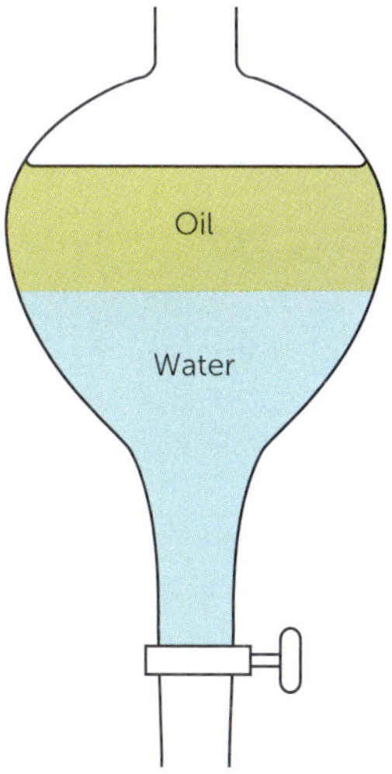

Figure 21.8 A separating funnel. The less dense liquid floats on top.

Miscible liquids

Two miscible liquids can be separated by simple distillation (Figure 21.7) if the boiling point of one liquid is at least 25°C higher than the other. If the difference between their boiling points is less than 25°C, then miscible liquids can be separated by **fractional distillation**. The apparatus is similar to that for ordinary distillation, except that a **fractionating column** is inserted between the flask and the condenser (Figure 21.9).

A fractionating column is a glass column packed with glass beads, or some other unreactive material that has a high surface area, upon which vapour can condense.

The less volatile a liquid is, the higher is its boiling point. When the mixture of liquids is heated, the vapour of both liquids passes up the column but the vapour of the liquid with the *higher boiling point* (water in this example) will tend to condense in the column and run back down into the flask. The vapour of the *lower boiling point liquid* tends to enter the condenser, condense and pass into the receiving beaker. If the boiling points of the two liquids are not too close, good separations can be achieved. In the example chosen (ethanol and water), a liquid much richer in ethanol (the **distillate**) is collected in the receiving flask, because ethanol (boiling point 78°C) is the more volatile component.

In distillation, the separation depends on equilibrium being established between descending liquid and rising vapour, and is best achieved by heating the mixture as slowly as possible. Fractional distillation can be used to separate more than two components. It is used in industry to separate crude oil into several fractions. Each fraction contains components with boiling points that are close together.

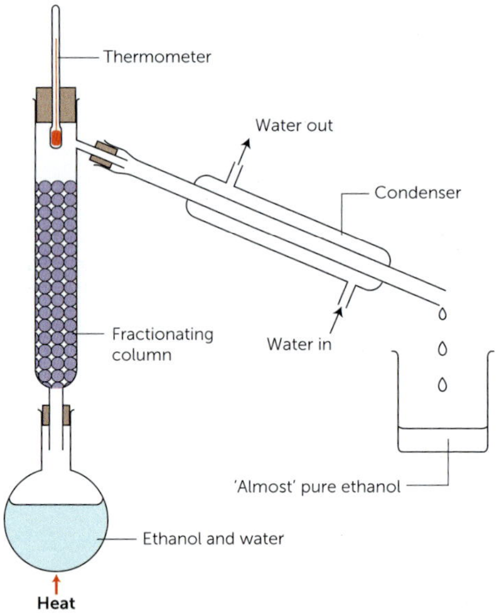

Figure 21.9 Fractional distillation being used to separate a mixture of ethanol (bp 78°C) and water (bp 100°C)

EXERCISE **21B**

Separations

(i) Water and ethoxyethane (diethyl ether) are immiscible liquids. What apparatus would you use to separate them?

(ii) A mixture of sugar and sand needs to be separated. What techniques would you use?

(iii) The solubilities of potassium chloride and potassium chlorate in water are as follows:

Substance	Solubility in water/grams per 100 cm^3 water	
	at 20°C	at 100°C
Potassium chloride	**35**	**57**
Potassium chlorate	**8**	**54**

Given a mixture of 8 g potassium chloride and 25 g of potassium chlorate, how would you obtain a pure sample of potassium chlorate?

(iv) How would you separate two miscible liquids, with boiling points of 80°C and 60°C, respectively?

(v) What technique would you use to obtain pure water from seawater?

21.3 Separating solids

If a solid **sublimes** (passes from the solid state directly to the vapour state), this fact can be used to separate the solid from a mixture in which there are other solids. The mixture is heated and the sublimed material condensed on a cool surface.

A simple form of the apparatus, used to separate a mixture of salt and iodine, is shown in Figure 21.10.

Often the process is carried out under reduced pressure, so that the substance can be sublimed at a lower temperature (Figure 21.11).

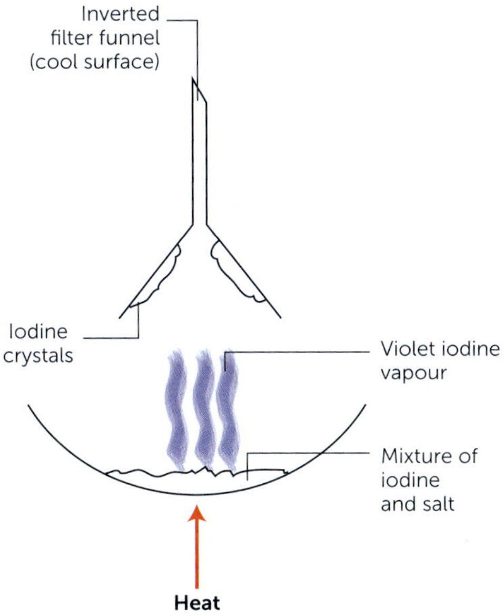

Figure 21.10 Separation by sublimation.

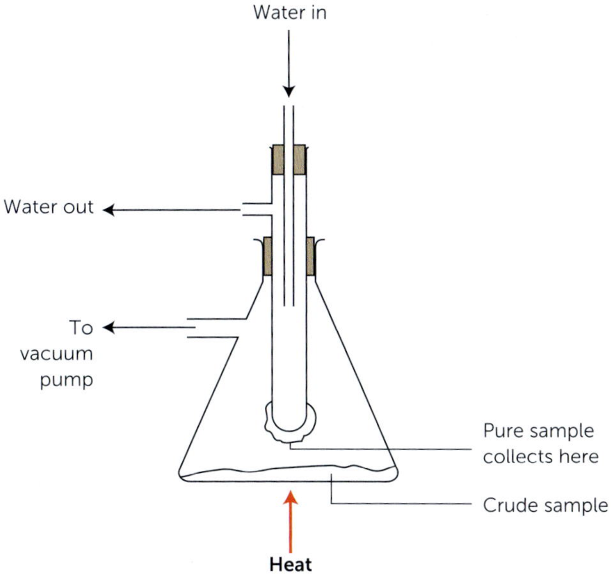

Figure 21.11 Sublimation under reduced pressure.

Ammonium chloride is an example of a solid that is *said* to sublime, but this is untrue; it actually decomposes when heated and reforms on a cold surface:

$$NH_4Cl(s) \rightleftharpoons NH_3(g) + HCl(g)$$

21.4 Steam distillation

Steam distillation is used to separate volatile organic substances that are immiscible with water from a mixture of involatile substances. It makes use of the fact that the total vapour pressure of two **immiscible** liquids is the sum of the vapour pressures of the pure liquid components, so that the boiling point of the mixture is lower than that of either of the components. This is useful for separating organic compounds that decompose at temperatures near their boiling point and therefore cannot be separated from a mixture by ordinary distillation. Steam from a steam generator is passed into the impure mixture. Vapours of the desired organic component and water condense in the condenser and are collected in the receiving flask (Figure 21.12). The organic component can be separated from the water layer using a separating funnel.

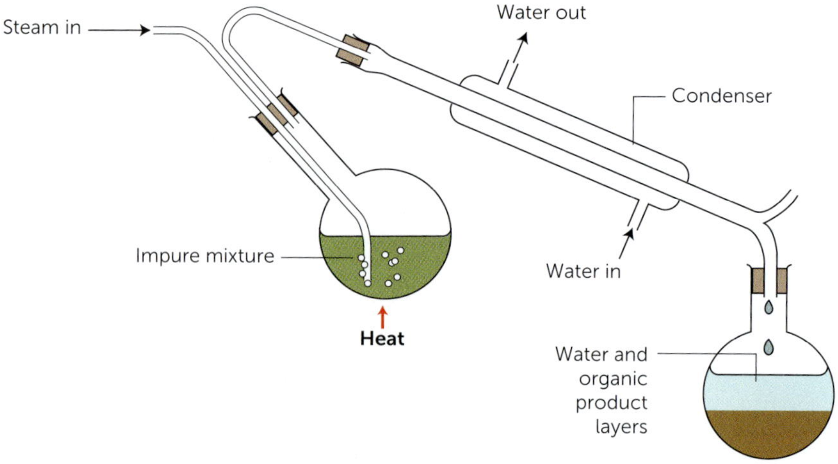

Figure 21.12 Steam distillation.

21.5 Ion exchange

In the process of **ion exchange**, ions are removed from a solution and replaced with other ions from a solid phase. An ion exchanger is a column filled with small insoluble beads made of ion exchange resin (Figure 21.13). The beads have ions loosely bound to them which are exchangeable with ions in a solution when it is passed through the column. The ions in the solution are replaced with the ions that were originally bound to the resin. After a time, the resin becomes exhausted because all its exchangeable ions have been replaced by ions from the solution. At this point the resin has to be **regenerated** and this is usually done by passing a concentrated solution of the ions that were originally bound to the resin through the column. Ions in the concentrated solution replace the ions now bound to the exhausted resin and the column is ready for use again.

For example, **hard water** is caused by the presence of calcium and magnesium ions in solution. Sodium aluminosilicate is a water softener that replaces the calcium and magnesium ions in hard water with sodium ions:

$$2NaAlSi_2O_6(s) + Ca^{2+}(aq) \rightleftharpoons Ca(AlSi_2O_6)_2(s) + 2Na^+(aq)$$

To regenerate the aluminosilicate a concentrated solution of sodium chloride is poured through the resin. Sodium ions replace the calcium and magnesium ions. The reaction is now the reverse of the one shown in the above equation.

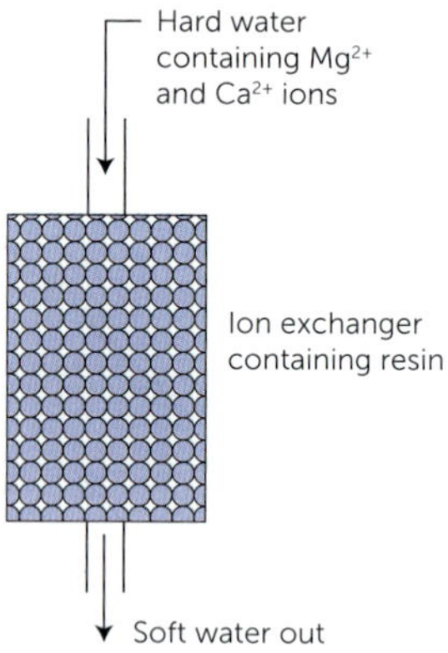

Figure 21.13 An ion exchanger.

21.6 Solvent extraction

Frequently in organic preparations a mixture of compounds is produced. Inorganic compounds may be present and, to dissolve these, it is necessary to wash the reaction mixture with water. Now the required organic product may be in solution or a suspension in water. If this mixture is shaken with an organic solvent, which is immiscible with water but in which the organic product is soluble, the organic product can be extracted from the water. Ethoxyethane (diethyl ether) is a good organic solvent for extracting organic products from aqueous solutions for the following reasons:

1 It is immiscible with water.
2 It is a good solvent.
3 It has a low boiling point and can therefore be removed easily by evaporation.

Ethoxyethane is shaken with the aqueous mixture in a separating funnel. The desired organic product moves from the aqueous layer into the ethoxyethane layer. The ethoxyethane layer can then be separated off (Figure 21.14(a)). Note that after one separation, the water layer often still contains some of the desired organic product and, for this reason, the water layer is usually extracted with ethoxyethane a further one or two times with fresh solvent in order to collect as much organic product as possible (Figure 21.14(b)). The separated layers are combined, then dried by shaking with, for example, a little anhydrous magnesium sulfate and then filtered (Figure 21.14(c)). The pure, dry organic product is recovered by evaporating off the ethoxyethane (boiling point 35°C).

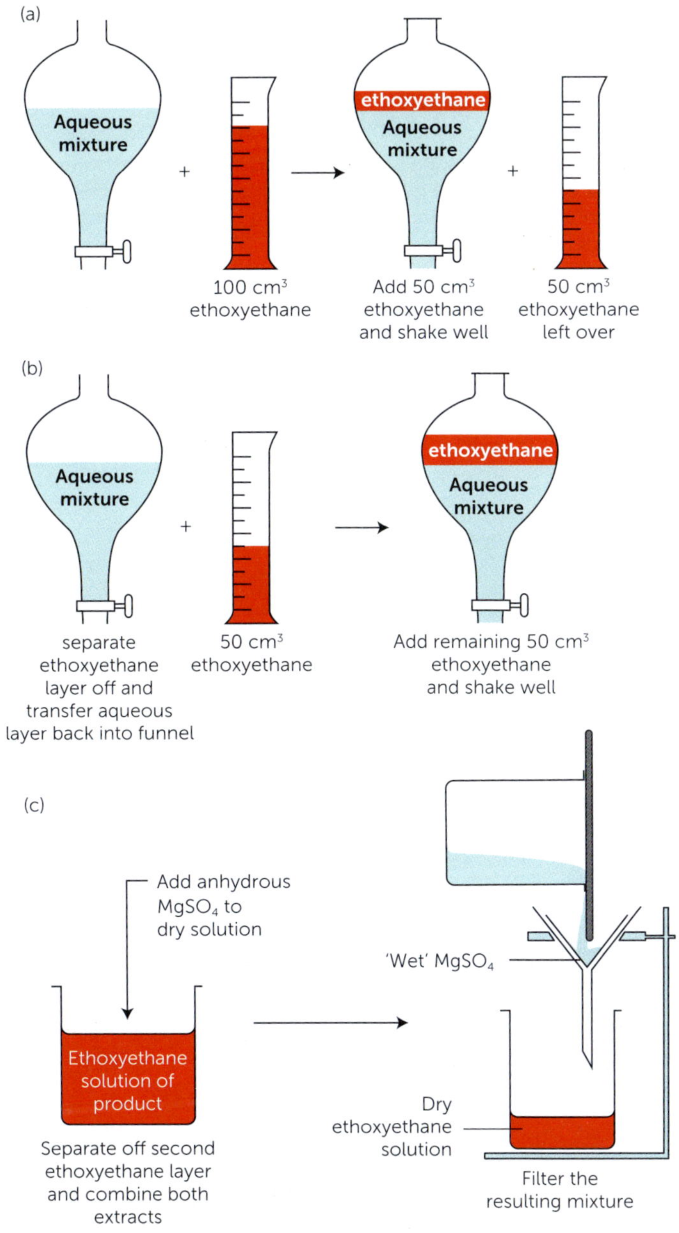

Figure 21.14 Steps (a)-(c) in the extraction of an organic product from a reaction mixture. The process ends up producing an ethoxyethane solution that contains the product. Ethoxyethane is readily evaporated, producing the pure product. For extra clarity, the ethoxyethane and its solution have been coloured red and the aqueous (watery) mixture has been coloured blue, but water and ethoxyethane are actually colourless.

BOX **21.3**

Zeolites

Some aluminosilicates are known as **zeolites**. Zeolites are part of a class of compounds known as **molecular sieves** and consist of an aluminosilicate frame with group 1 or 2 cations trapped inside tunnels or cages. In addition to their ability to act as ion exchangers, they can selectively remove small molecules, such as water, from mixtures – hence the term 'molecular sieves' (Figure 21.15).

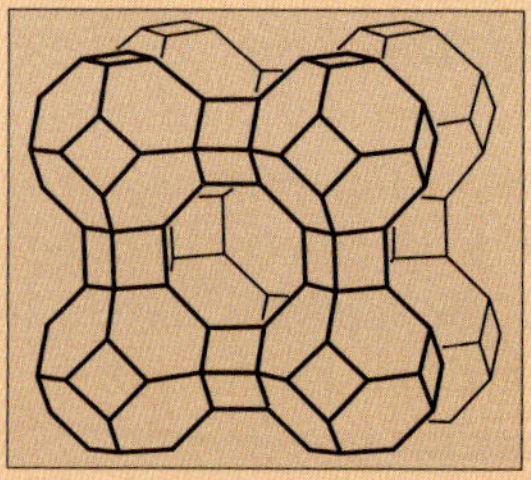

Figure 21.15 Cage structure of a typical zeolite.

EXERCISE **21C**

Multiple extractions

(i) The distribution ratio of an organic compound X between toluene (methylbenzene) and water is 7.0. An aqueous mixture (100 cm^3) contains 1.0 g of X. The mixture is extracted once with 200 cm^3 of toluene. What mass of X is extracted into the organic layer?

(ii) Repeat the calculation but for a two-stage extraction, using 100 cm^3 of toluene each time. Do you extract more X using 200 cm^3 of toluene in one extraction, or 2 × 100 cm^3 of toluene in two successive extractions?

For more information on calculations of multiple solvent extractions, see Extension 21 at https://www.bloomsburyonlineresources.com/chemistry.

BOX **21.4**

Reasoning behind multiple extractions

If a compound is shaken in a separating funnel with two immiscible solvents, such as ethoxyethane and water, some of the compound will dissolve in the ethoxyethane and some in the water. The amount of compound that dissolves in each depends upon how soluble it is in the solvents. K_d is defined as the distribution ratio

$$K_d = \frac{\text{concentration of compound in ethoxyethane}}{\text{concentration of compound in water}}$$

Note that K_d is a constant at a particular temperature. For the ethoxyethane–water system, K_d = 4 at 20°C.

Supposing 10.0 g of organic product are in 100 cm^3 of water at 20°C and 100 cm^3 of ethoxyethane are available for extraction of the compound. Is it better to use the 100 cm^3 of ethoxyethane in one go, or to perform two extractions using 50 cm^3 of ethoxyethane each time?

1 All 100 cm^3 of ethoxyethane used in one extraction

Let *x* g of the organic product be extracted into the ethoxyethane layer. The amount of organic product remaining in the aqueous layer is now (10 – *x*) g.

Using the equation

$$K_d = \frac{\text{concentration of compound in ethoxyethane}}{\text{concentration of compound in water}}$$

$$4 = \frac{x/100}{(10 - x)/100}$$

Solving this equation gives *x* = 8 g. So, 8 g of product can be recovered from the ethoxyethane layer.

2 Two extractions are performed, each using 50 cm^3 ethoxyethane

For the first extraction let *y* g of the organic product be extracted into the ethoxyethane layer. The amount of organic product left in the water layer will be (10 – *y*).

As before,

$$4 = \frac{y/50}{(10 - y)/100)}$$

Therefore, y = 6.7g.

Before the second extraction, 10 – 6.7 = 3.3 g of product remains in the aqueous layer. Let *z* g of this be extracted into the ether layer. Then, the amount of organic product left in the water layer will be (3.3 – *z*).

As before,

$$4 = \frac{z/50}{(3.3 - z)/100)}$$

Therefore, *z* = 2.2 g.

The total amount of organic product obtained from the two extractions is

(*y* + *z*) = 6.7 + 2.2 = 8.9g.

This is nearly one more gram of product than that obtained from one extraction. More product would be obtained with more extractions, i.e. with smaller amounts of ethoxyethane, but the increase in amount of product obtained becomes smaller with each successive extraction.

Soxhlet extraction

A compound can be isolated from a crude, solid product by repeated extraction using a hot solvent. The apparatus usually used for this purpose is a Soxhlet extraction apparatus (Figure 21.16).

The crude material is put into a porous thimble at A, and the extracting solvent is placed in a flask at the bottom of the apparatus. The solvent in the flask is gently boiled and its vapour passes into the condenser via the side arm of the apparatus. Solvent vapour condenses and drips into the thimble, where the hot solvent dissolves some of the desired product present. When the level of solvent reaches the level of the top of the fine glass tube, the whole volume of solvent, containing dissolved product, siphons into the solvent flask at the bottom of the apparatus. The process continues automatically until all the desired product has been extracted and is present, in solution, in the solvent flask. If the substances to be extracted are coloured, the process is stopped when the liquid being siphoned into the flask becomes pale or colourless.

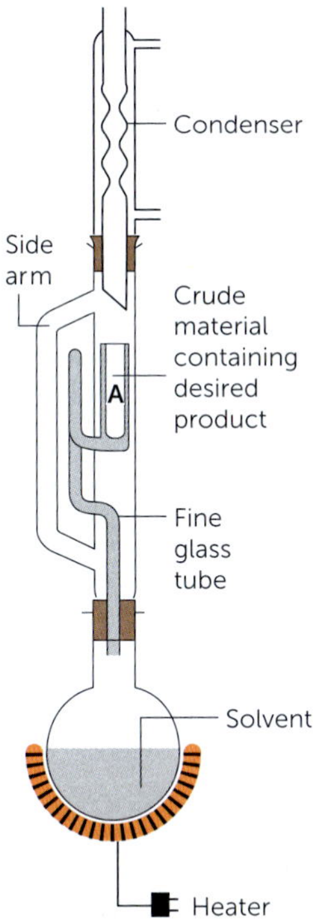

Figure 21.16 Soxhlet extraction apparatus.

BOX **21.5**

Extraction of caffeine using supercritical fluid extraction

Caffeine has the molecular formula $C_8H_{10}N_4O_2$ and is present in tea, coffee and cola. It can be slowly extracted from tea or coffee by Soxhlet extraction, using trichloromethane (chloroform) as a solvent but **supercritical fluid extraction** (SFE) is a faster extraction technique. The solvent used is supercritical CO_2 (see Unit 11) which possesses excellent solvating (dissolving) powers. Supercritical carbon dioxide is the preferred solvent in producing decaffeinated coffee. It will not react with the substances that it is used to extract, and (unlike hexane, dichloromethane or chloroform) is non-toxic.

21.7 Chromatography

Chromatography has become one of the most important methods of separating substances from a solution. In chromatography, the solution is passed over (or through) a carefully chosen solid, such as paper or aluminium oxide. Chromatography depends upon the ability of the solid to slow down the movements of the substances in the solution. Different substances are separated because they are 'held back' by differing amounts. The 'holding back' is caused by the differing degrees of adsorption between the surface of the solid and the different substances in the mixture, or by the differing solubilities of the substances in a liquid around the surface of the solid, or a mixture of both effects. Chromatography is like a race, with the runners being the substances in the mixture. At the beginning of the race, or the start of the separation, the competitors are together, whereas by the end of the race the field is split and the individual runners are well separated.

The word chromatography literally means 'colour writing'. Chemists and biologists both use chromatography. For example, chromatographic separations are used extensively in forensic science to detect and identify trace amounts of substances. There are many types of chromatography, of which the simplest is probably **paper chromatography**.

Paper chromatography

Black ink is a mixture of several coloured substances. It can be separated into its components by **paper chromatography.** A small drop of black ink is spotted onto a rectangular sheet of filter paper. When the ink has dried the paper is folded into a cylinder shape and secured with a clip. The paper is then placed into a beaker containing a solvent, in this case propanone (acetone), making sure that the level of propanone is below the ink spot (Figure 21.17). The solvent gradually rises up the filter paper by capillary action and separates the ink into a number of coloured spots. This separation happens because the different dyes present in the ink have different solubilities in the water surrounding the cellulose particles of the paper (the **stationary** phase) and the solvent (the **moving** phase). The dyes most soluble in the moving phase are carried furthest up the filter paper in a given period of time, whereas those most soluble in the water on the cellulose tend to lag behind. The different dyes distribute (**partition**) themselves between the water on the cellulose and the solvent moving over the cellulose. When the solvent reaches the top of the filter paper the different dyes in the ink will have separated out into different coloured spots. The paper can be taken out and dried – the resulting pattern is called a **chromatogram** (see Figure 21.18).

This technique is generally used to identify the components in a mixture. For example, the dyes that it is thought an ink *might* contain can be spotted on the filter paper along with the ink.

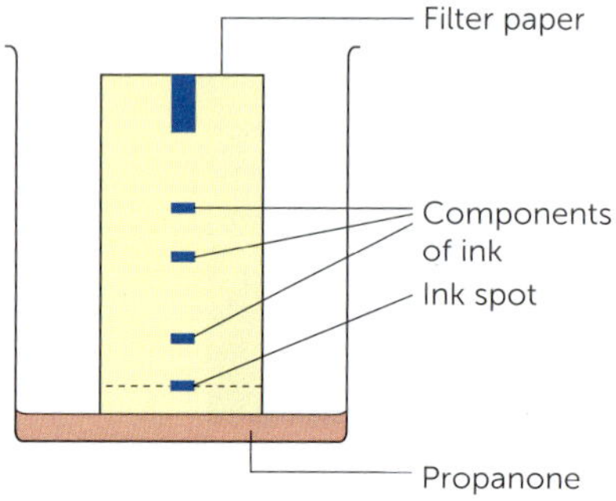

Figure 21.17 Paper chromatography.

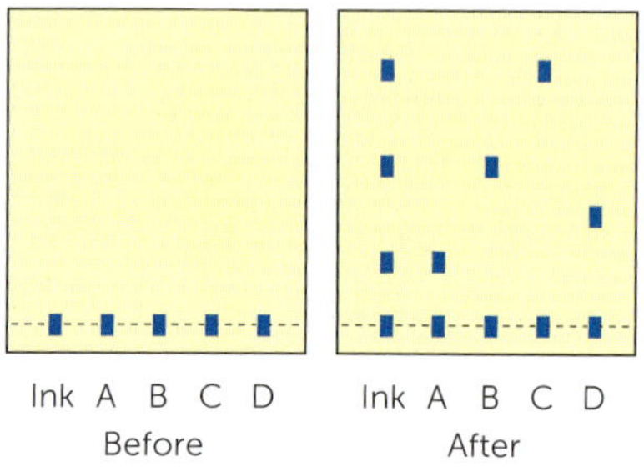

Figure 21.18 Running a chromatogram.

The chromatogram obtained in Figure 21.18 shows that the ink contains components A, B and C but not D because the ink has separated into three spots, which are at the same height as the spots obtained for A, B and C.

Column chromatography

This is used to separate mixtures of compounds, so that each compound can be collected. A glass column is packed with an **adsorbent**, through which a solution of the mixture is passed. Alumina (**aluminium oxide**) or silica gel (prepared by adding concentrated hydrochloric acid with sodium silicate) are commonly used as adsorbents. The column must be uniformly packed so that solvent flows freely through it. The mixture is added at the top of the column and solvent is passed through the column to wash (or **elute**) the mixture down through the column. As the mixture moves through the column it separates into coloured bands of its components because the components move through the column at different rates. For example, a compound that is not strongly adsorbed but is very soluble in the eluting solvent, will move through the column rapidly. The bands are finally washed out at the bottom of the column and each one is collected in a different flask (Figure 21.19).

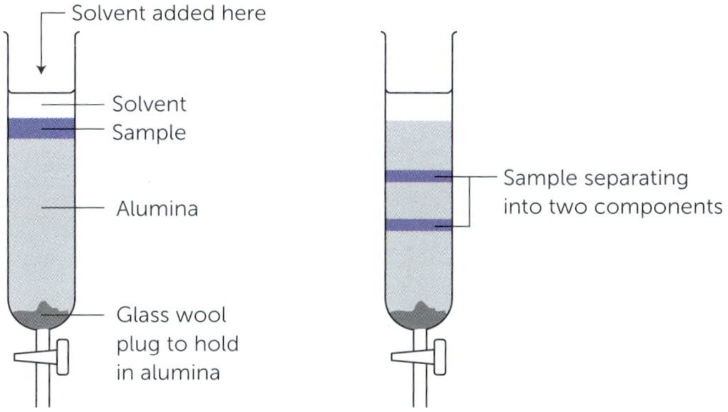

Figure 21.19 Column chromatography.

Column chromatography becomes a little more difficult if the separation involves colourless compounds because the different bands cannot be detected by the eye. A series of fractions of the same volume can be collected (say 25 cm^3) and the purity of the fractions monitored by other chromatographic methods, such as thin-layer or gas chromatography.

Thin-layer chromatography

Thin-layer chromatography (TLC) is a variation of column chromatography. A strip of glass is coated on one side with a thin layer of adsorbent (e.g. alumina). The substance to be tested is spotted near one end of the plate and the plate is put into a **developing jar** with a small amount of the solvent (to a level below the spot) and a cap put on the jar (Figure 21.20). The solvent rises up the plate, separating the sample into its components. At the end of the experiment, a spot for each component is seen on the plate. When the solvent is near the top of the plate, *the position the solvent has reached is marked with a pencil* and the plate removed and dried. If the components are colourless, a number of techniques can be used to show them up:

1 Some compounds show up under ultraviolet light.
2 The dry plate can be placed in a stoppered jar with a few crystals of iodine – the crystals sublime and most organic compounds show up as coloured spots when exposed to iodine vapour.
3 The plate can be sprayed with a reagent that will react with a compound on the plate to form a coloured product.

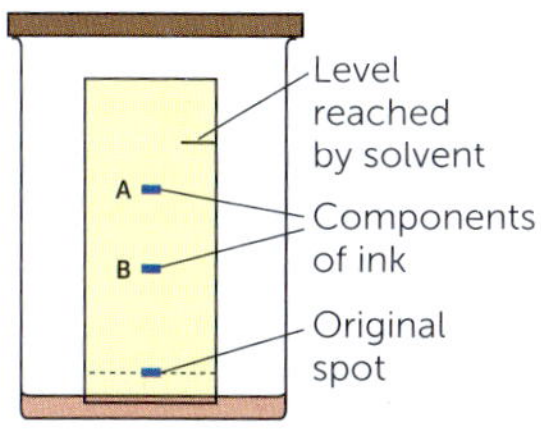

Figure 21.20 Running a TLC plate.

TLC needs only small amounts of sample and is generally used to check the purity of the product of a chemical reaction. TLC can also be used to confirm the identity of an unknown sample. The **retention factor** (R_f value) of a compound is

$$R_1 = \frac{\text{the distance travelled by the compound}}{\text{the distance travelled by the solvent}}$$

The R_f value is constant for a compound, on the same plate, if the temperature and solvent are kept the same. If a spot from an unknown substance is developed on a TLC plate together with a spot from a substance that is suspected to be the unknown one, and the two substances are found to have the same R_f value, they are probably the same substance. A substance that just produces one spot when subjected to TLC is pure.

EXERCISE **21D**

R_f values

Using the diagram of a TLC plate below, write expressions, involving x, y and z, for:

(i) the R_f value for component A

(ii) the R_f value for component B

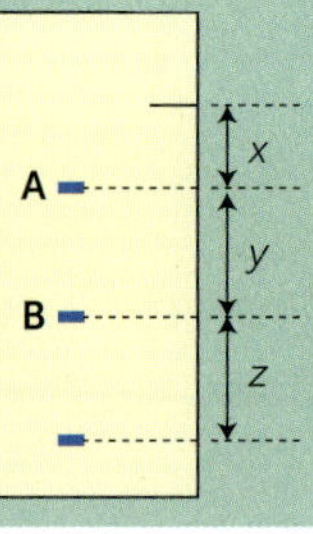

Gas chromatography

Gas chromatography (GC) is used for checking the purity or identity of volatile liquid samples. In most experiments small quantities of sample are used and mixtures are separated, and the components identified, rather than collected. The sample is injected into the instrument, vaporized and carried in a stream of carrier gas (helium or nitrogen) onto a chromatography column. The components of the sample move through the column at different rates and reach a detector at different times. When a component hits the detector an electric signal is sent to a recorder, or computer, which produces a graph, or chromatogram, of the different components passing through the detector. The time taken for a compound to pass through the apparatus and reach the detector is called the compound's retention time. The retention times of the components of a mixture can be compared with the **retention times** of known compounds obtained under the same conditions. In this way, the components of the mixture may be identified. The trace produced for each component also gives a rough idea of the relative quantity of the component. The *area* under each peak (*not* the height of the peak) is proportional to the percentage of that component present in the sample.

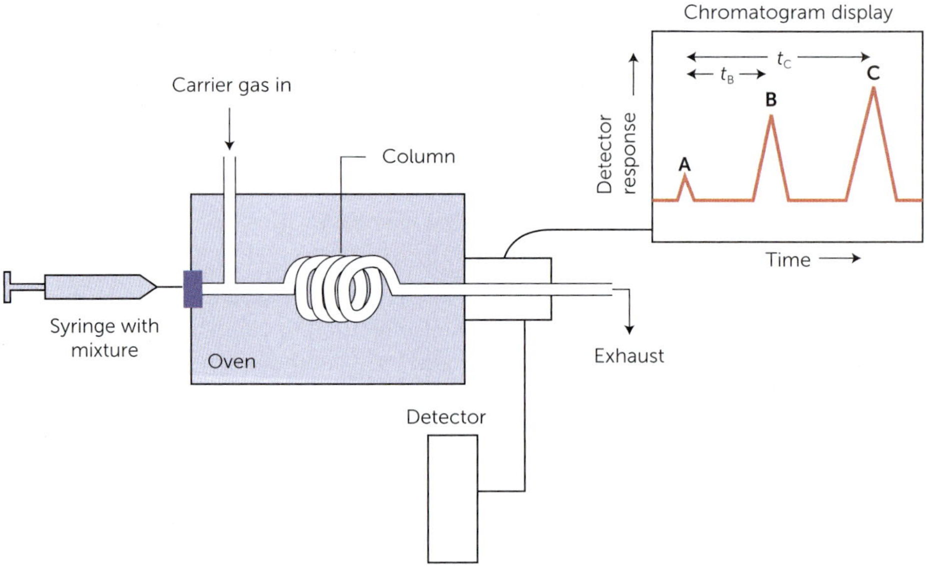

Figure 21.21 Gas chromatogram.

BOX **21.6**

Gas chromatography–mass spectrometry (GC–MS)

It is possible to 'add on' another technique to analyse the components once they are separated out from a mixture using GC. GC–MS combines gas chromatography with mass spectrometry (MS). The previously unknown organic components of a mixture are separated by GC, then passed into a MS unit where they are ionized and produce a characteristic fragmentation pattern. The fragmentation pattern of a substance can be compared with a computer library of patterns of different substances, leading to the identification of the components of the original mixture. For example, this technique has been successfully employed in identifying trace amounts of drugs in the blood of athletes. For further details and some practice in MS interpretation, see Extension 3 on the website https://www.bloomsburyonlineresources.com/chemistry.

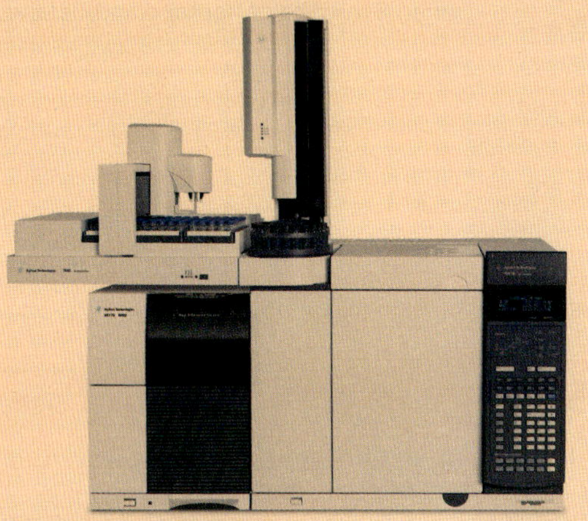

Combined GC–MS instrument. The sample injector is at the top left, the GC column sits in an oven (behind the black panel, bottom left) and the MS sits behind the control panel (bottom right).

The chromatograph in the diagram (Figure 21.21) shows three peaks. One of these peaks – A – is a reference material, injected along with the sample (in this case air, from an air bubble in the syringe). Two components, B and C, have separated out from the mixture. The retention time for B, for example, is the time taken for the recorder to mark from the tip of the reference compound (A) to the tip of peak B (t_B).

A summary of separation techniques is shown in Table 21.1.

Table 21.1 Separation techniques

Mixture	Separation technique
Insoluble solid in solvent	Filtration (or, if a fine precipitate, centrifugation)
Solution of solid in liquid	Evaporation or (better) crystallization to recover solid; distillation to recover liquid
Miscible mixture of liquids	Fractional distillation
Immiscible mixture of liquids	Separating funnel
Mixture of two solids, one of which sublimes	Sublimation
An organic substance that is immiscible with water from a mixture of involatile substances	Steam distillation
Mixture of substances in solution	Paper chromatography or TLC for identification; column chromatography to collect the substances
Removal of ions (e.g. calcium or magnesium) from aqueous solution	Ion exchange
Separation and identification of a mixture of volatile liquids	Gas chromatography and GC–MS
Separation of organic components from an aqueous mixture	Solvent extraction

REVISION QUESTIONS

21.1 What technique would you use to separate the following?

(i) Solid potassium chloride from an aqueous solution of potassium chloride.
(ii) Pure water from lemonade.
(iii) Hexane from crude oil.
(iv) Oxygen from liquid air.
(v) The colouring substance in a coloured drink.
(vi) Petrol from petrol and water.
(vii) Salt from a mixture of salt and chalk.

21.2 A green substance in grass dissolves in propanone (acetone). Describe how you would show that this green pigment contains more than one substance.

21.3 How would you separate a fine mixture of ammonium chloride and potassium chloride solids?

21.4 Screened methyl orange is a mixture of dyes that are soluble in ethanol.

(i) What techniques could you use to show that the substance is a mixture?
(ii) What technique could you use to separate and collect the dyes?

21.5 Amino acids, which are colourless and water soluble,

amino acid + ninhydrin → blue/lilac product react with a substance called ninhydrin according to the reaction:

amino acid + ninhydrin → blue/lilac product
(colourless) (colourless)

What technique could you employ to show that urine contains a mixture of amino acids?

21.6 Whisky contains a number of chemicals that belong to the alcohol family of organic compounds. If you were given a bottle of a particular whisky, what technique(s) could you use to separate, and possibly identify, the constituent alcohols?

21.7 You are a forensic scientist and have been asked to find out whether the ink from a suspect's pen could have been used to write the signature on a forged cheque. How would you proceed?

21.8 At room temperature, the distribution ratio of butanoic acid between ethoxyethene (diethyl ether) and water is 3.5. In an experiment, 150 cm^3 of water containing 0.50 g of butanoic acid was shaken with 50 cm^3 of ethoxyethane. What mass of acid did the ethoxyethane extract?

21.9 The diagram shows the gas chromatogram obtained when a mixture of heptane and octane was analysed.

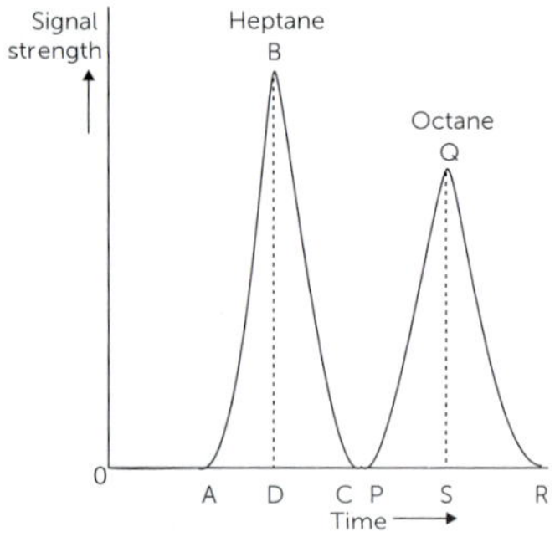

(i) Assume the peaks are approximately triangular in shape; calculate the areas ABC and PQR. *The area of a triangle* = ½ × *base of triangle* × *height of triangle*.
(ii) What can you say about the relative amounts of heptane and octane in the original mixture?
(iii) Can you think of a method of comparing the area under the peaks that does not involve calculation and could be used even if they were not shaped like triangles?

21.10 How would you show that the colours of Smarties (the coated chocolate sweets) are one single dye, or a mixture of dyes?

DID YOU KNOW?

Electrophoresis involves the separation of charged molecules in an electric field. The molecules move over paper or through a gel. Negative ions move towards an anode and positive ones towards a cathode. The rate of movement of different ions depends upon their masses and charges.

Protein molecules can be hydrolysed into amino acid fragments and those fragments ionized by using buffer solutions of a suitable pH. Using electrophoresis on fragments of DNA from body fluids produces a pattern that is unique to an individual, but certain bands are common in families. DNA fingerprinting is therefore a valuable tool, both for the forensic scientist and for establishing paternity cases.

Spectroscopy

22

OBJECTIVES

This unit:

- Looks at the absorption and emission of light
- Explains how spectrometers work
- Discusses the emission spectrum of the hydrogen atom
- Outlines the uses of IR and NMR spectroscopy in analysis
- Introduces the Beer–Lambert law

CONTENTS

The interaction of light with matter may result in physical changes (such as reflection, absorption, diffraction or emission) or in chemical changes (such as dissociation or isomerization). Absorption and emission form the basis of the **spectroscopic analytical techniques** and the subject specialism is called **spectroscopy**. The chemical changes are no less important – the formation of smog, the depletion and formation of ozone, and photosynthesis itself, are all initiated by light.

22.1 Electromagnetic spectrum

We use the word 'light' to mean any radiation that is part of the **electromagnetic spectrum** (Figure 22.1), although only a small wavelength region of the spectrum (roughly between 400 and 700 nm) is visible to the human eye.

All **light** travels at 3.00×10^8 metres per second (186,000 miles per second) in a vacuum. This velocity is given the symbol c.

Frequency and wavelength

Light may be thought of as being made up of waves or, alternatively, as being composed of particles of pure energy called **photons**.

The **wavelength** of a wave is literally the 'length of one wave' – see Figure 22.2. It is symbolized λ and has units of metres.

The **frequency** of a wave is the number of waves passing a point every second. Frequency is symbolized ν (nu) and has units of hertz (Hz). (1 Hz is one wavelength

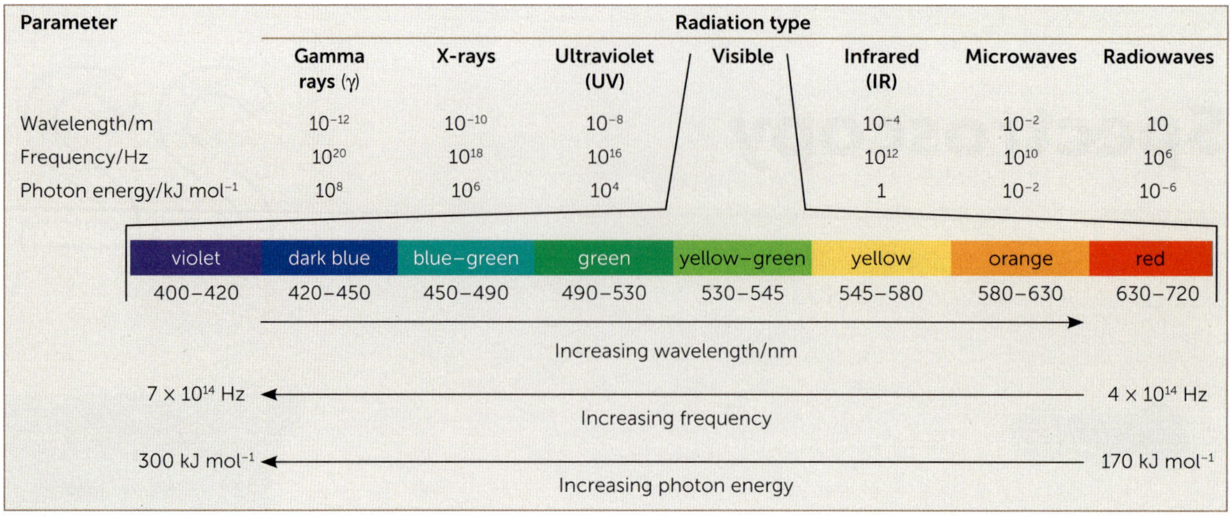

Figure 22.1 The main parts of the electromagnetic spectrum and the orders of magnitude of their wavelengths, frequencies and photon energies. The visible region is shown expanded with the wavelengths expressed in nanometres (1 nm = 10^{-9} m).

per second.) The relationship between the wavelength, frequency and velocity of light in a vacuum is

$$c = \nu\lambda$$

or

$$\nu = \frac{c}{\lambda}$$

This equation shows that

a wave with a high frequency possesses a short wavelength (and vice versa).

The **quantum theory** regards light as consisting of particles of energy, called **photons**. The energy E of one photon is given by the equation

$$E = h\nu$$

where h is a universal constant called the **Planck constant**, which has the value

$$h = 6.626 \times 10^{-37} \text{ kJ s}$$

It is sometimes more convenient to consider the energy of one mole of photons, and to use the Planck constant per mol of photons:

$$h \text{ per mol} = h \times N_A = 6.626 \times 10^{-37} \times 6.022 \times 10^{23} \approx 3.99 \times 10^{-13} \text{ kJ s mol}^{-1}$$

Substituting

$$\nu = \frac{c}{\lambda}$$

in the equation $E = h\nu$ gives

$$E = \frac{hc}{\lambda}$$

It follows from these equations that

High-energy photons correspond to high frequency (short wavelength) light.
Low-energy photons correspond to low frequency (long wavelength) light.

EXAMPLE **22.1**

Violet light has a wavelength of about 400 nm. What is the energy of one mole of photons of violet light?

Answer

If we substitute hN_A for h in the equation $E = h\nu$, then E will have the units of kJ per mole of photons.

$$E = \frac{hN_A c}{\lambda} = \frac{3.99 \times 10^{-13}\,\text{kJ s mol}^{-1} \times 3.00 \times 10^{8}\,\text{m s}^{-1}}{400 \times 10^{-9}\,\text{m}} = 299\,\text{kJ mol}^{-1}$$

Therefore, the energy of the photons is about 300 kJ mol^{-1}.

Comment

The energy of chemical bonds is typically 200–400 kJ mol^{-1}. So we should not be surprised if taking in violet (and the even more energetic UV) light by molecules sometimes breaks chemical bonds and so brings about *chemical* change. Now try Exercise 22B.

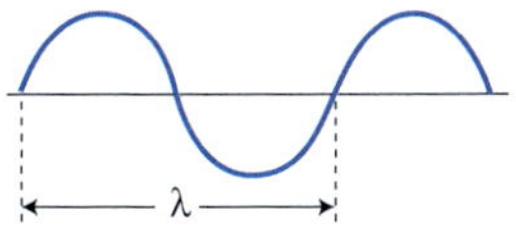

Figure 22.2 Definition of wavelength.

EXERCISE **22A**

Frequency and wavelength

To receive BBC Radio 1 you require a radio tuned to a frequency of about 98 megahertz (MHz).

What wavelength is this?

EXERCISE **22B**

Photon energies

Calculate the energy of a mole of photons of **(i)** red light of wavelength 700 nm and **(ii)** yellow light of wavelength 560 nm.

Energy transitions

The quantum theory restricts the energy of a molecule or atom to certain definite energy levels (Figure 22.3). Movements between levels are called **transitions**.

Movement up the 'ladder' of energy levels (an upward transition) is only possible if the molecule or atom **absorbs** a photon whose energy *exactly* equals the energy gap (ΔE) involved in the jump. (This requirement is called the **Bohr condition**.) This process (termed **absorption**) is shown in Figure 22.3. During absorption, the photon vanishes. The relationship between the energy jump and the frequency of the absorbed light is

$$\Delta E = h\nu$$

Emission, also shown in Figure 22.3, is the loss of energy by the emission of photons. The equation $\Delta E = h\nu$ also applies to emission, with ΔE being the energy lost in the emission and ν being the frequency of the emitted light.

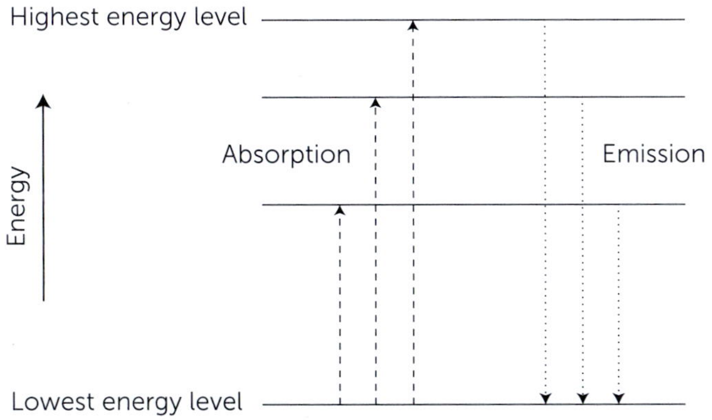

Figure 22.3 The energy levels of an isolated atom or molecule. Three examples of transitions involving photon absorption (dashed lines) and photon emission (dotted lines) are shown.

EXERCISE **22C**

Bohr condition

Two energy levels in an atom are 100 kJ mol^{-1} apart. What frequency of light needs to be absorbed in order for one mol of atoms to move from the lower to the upper level?

22.2 Energy levels of atoms and molecules

We now look at the types of energy possessed by molecules in more detail (Figure 22.4). Atoms and molecules possess **electronic energy** – energy due to the electrons in the atomic or molecular orbitals. In addition to electronic energy, molecules (but not isolated atoms) also possess **vibrational energy** – energy which causes the atoms in molecules to vibrate.

Most of the energy of an atom or molecule is in the form of electronic energy. Making an analogy with money, electronic energy is 'big money' (pounds), whereas vibrational energy is 'small money' (pence).

The quantum theory allows molecules to possess only certain values of electronic and vibrational energy. The energy of a molecule may be looked upon as a 'ladder' of allowed electronic energy levels or 'states' (labelled E_0, E_1 …) with sublevels according to the vibrational energy (labelled v_0, v_1, v_2 …) possessed by the molecule. For example, a molecule in its second electronic energy level and its third vibrational level would be labelled E_1, v_2.

The lowest energy level of a molecule or atom is called its **ground state**, whereas higher levels are referred to as **excited states**. The making of excited states (by photon absorption or by raising the temperature of the sample) is called **excitation**. At room temperature nearly all atoms and molecules lie in their ground electronic and (for molecules) in their ground vibrational energy levels, i.e. they lie at E_0 and v_0. This means that, except at very high temperatures, molecular or atomic transitions involving absorption always start from the lowest energy level.

Transitions involving changes in electronic energy result in electrons moving from one orbital to another, the electrons being promoted to a higher energy orbital in absorption and demoted to a lower energy orbital in emission. The patterns observed in a **spectrometer**, which are caused by these transitions, are referred to as **electronic spectra**. The transitions involve so much energy (typically > 100 kJ per mole of absorbing atoms or molecules) that photons of ultraviolet or visible light are involved, and so the spectra are also called UV–visible spectra. In Figure 22.4, transition 'a' involves the absorption of **UV** or **visible light** – this is what happens when a molecule of dye absorbs visible light. Transition 'c' involves the emission of UV or visible light.

Transitions in which the vibrational energy (only) of molecules is changed involve photons of infrared light, and the resulting spectroscopic patterns are referred to as **infrared spectra**. The energy jumps (ΔE) involved in infrared (IR) spectra are smaller than those involved in UV–visible spectra, and are typically 10 kJ per mole of absorbing molecules. As a consequence of infrared absorption, the atoms in the molecule move farther in each vibration (the amplitude of the vibration increases). Emission of infrared light causes the atoms in a molecule to vibrate with reduced amplitude. In Figure

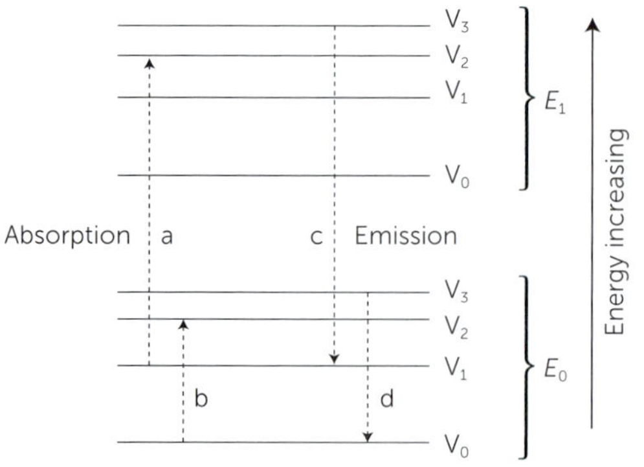

Figure 22.4 Electronic and vibrational energy levels of an isolated molecule. This diagram is a more detailed version of Figure 22.3. The transitions labelled a and c involve the absorption and emission (respectively) of ultraviolet or visible light. The transitions labelled b and d involve the absorption and emission (respectively) of infrared light.

22.4, transition 'b' involves the absorption of infrared light, and transition 'd' involves the emission of infrared light. All hot objects emit infrared light. The warming of a beaker of water placed close to a heated electric filament occurs partly by absorption of infrared light.

22.3 Spectrometers

Instruments which measure and record the wavelengths at which samples emit or absorb light are called **spectrometers**. The basic components of a UV–visible or infrared spectrometer are shown in Figure 22.5. The components include

- A **source** that continuously emits a broad range of wavelengths of light.
- A **diffraction grating** or prism that separates the wavelengths of light by spreading them at different angles and which (ideally) allows only one wavelength of light to strike the cell.
- A **cell** that is transparent to the incident light (quartz for UV, glass or perspex for visible light and solid sodium chloride for infrared). The cell contains the sample whose spectrum is required.
- An electronic **detector** (or, historically, a photographic emulsion) which allows measurement of the intensity of the light beam emerging from the cell. Commercial spectrometers are 'computer driven', and spectra are displayed on a monitor.

The range and intensity of wavelengths emitted by the light source (its **emission spectrum**) are found using an uninterrupted beam (i.e. an empty cell, Figure 22.5(a)) and may be displayed either as a graph of light intensity versus light wavelength, or in the form of a spectrum directly produced by a prism or diffraction grating.

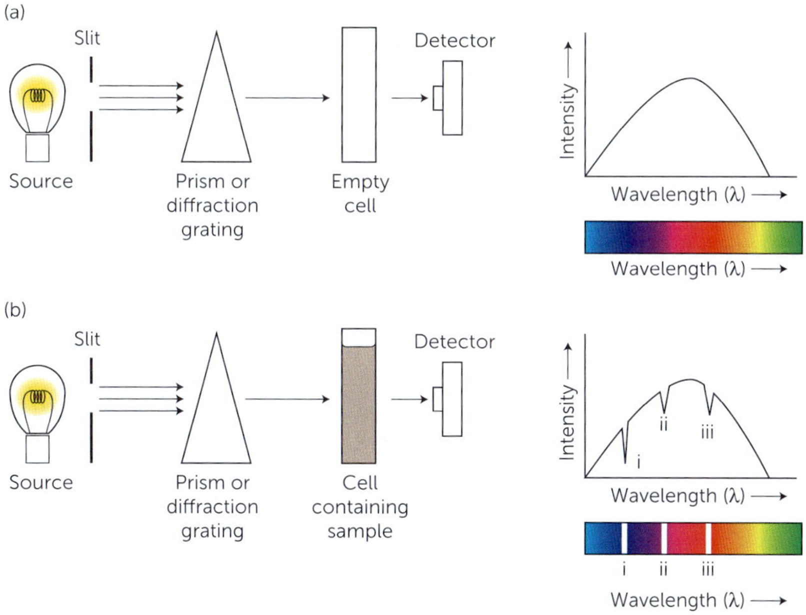

Figure 22.5 (a) Obtaining the emission spectrum of the source. (b) If an absorbing sample is placed in the light beam, intensity measurements over the wavelength range of the source show that only certain wavelengths are absorbed. These appear as troughs in the graphical profile or, if observed directly without an electronic detector (by eye, using a hand spectroscope, or using photographic film), as lines or bands. In this example the absorbed wavelengths occur in the visible region but the same principles apply in the UV and IR regions.

If a sample is exposed to the light source, and some of the photons emitted by the light source have energies equal to the gaps between the energy levels found in the atoms or molecules in the sample, absorption may take place. The spectrum in Figure 22.5(b) shows reduced light intensity reaching the detector at three wavelengths – (i), (ii) and (iii) – indicating that three different energy transitions have taken place within the atoms or molecules of the sample. If the absorption spectrum is observed directly by eye or recorded on photographic film, the absorptions appear as lines or bands of reduced intensity.

22.4 Absorbance and transmittance of a sample

The light source in the spectrometer continuously bombards the sample cell with photons. If the sample cell is empty of absorbing material, the maximum intensity of light (at that wavelength) reaches the detector. We symbolize the maximum intensity as $I_{0(\lambda)}$ where λ is the wavelength of light selected by the grating or prism. If an absorbing sample is introduced into the beam and the sample absorbs light at that wavelength, excited states are formed. The excited states lose their energy by collisions or by light emission, and the intensity of light reaching the detector along the cell axis is *reduced*. We symbolize the new value of intensity as $I_{(\lambda)}$. The **per cent transmittance** of the sample, symbolized $\%T$, at wavelength λ is defined as

$$\%T = \frac{100 \times I_{(\lambda)}}{I_{0(\lambda)}} \qquad (22.1)$$

Alternatively, the reduction in beam intensity at that wavelength may be expressed as an **absorbance** A_λ. Absorbance is a unitless quantity defined as

$$A_\lambda = \log\left(\frac{I_{0(\lambda)}}{I_{(\lambda)}}\right) \qquad (22.2)$$

In commercial spectrometers, the instrument adjusts the diffraction grating so that successively lower (or higher) wavelengths are isolated, and then allowed to irradiate the sample. This is called a **scan** over the desired wavelength range and gives values of I at each λ. Spectrometers also obtain I_0 at each λ, either (as in **double-beam instruments**) by periodically diverting the incident beam through an empty cell during the scan, or (as in **single-beam instruments**) by computer storage of the variation of I from a previous scan using an empty cell. Either way, the spectrometer has now sufficient information to produce a plot of either $\%T$ or A against wavelength. Such a plot is called an **absorption spectrum**. The absorption spectrum of the empty sample cell will be a straight line since, by definition, I equals I_0 at each λ (Figure 22.6(a)). Scanning an absorbing sample will result in the spectrum showing peaks (Figure 22.6(b)) which indicate the wavelengths of light absorbed.

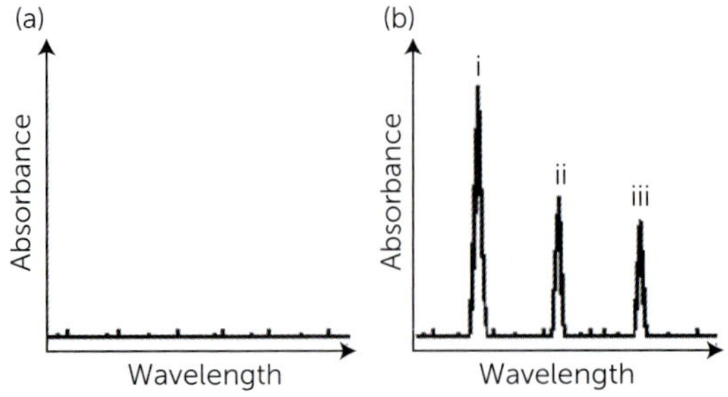

Figure 22.6 Absorbance spectra for the case shown in Figure 22.5. (a) The absorption spectrum of the spectrometer with an empty sample chamber (or with a non-absorbing sample). The small peaks are due to electronic 'noise' from the detector. (b) The absorption spectrum of a sample that absorbs at three wavelengths.

EXAMPLE **22.2**

40 per cent of the light emitted by a spectrometer source at a wavelength of 460 nm is absorbed by a sample. What is the absorbance of the sample at 460 nm?

Answer

If 40 per cent of the light intensity is absorbed, 60 per cent is transmitted. Since

$$\%T = \frac{100 \times I_{(\lambda)}}{I_{0(\lambda)}}$$

then

$$60\% = \frac{100 \times I_{(\lambda)}}{I_{0(\lambda)}}$$

or,

$$\frac{I_{0(\lambda)}}{I_{(\lambda)}} = \frac{100}{60} = 1.7$$

Therefore, the absorbance is log (1.7) = 0.23.

Comment

Absorbances greater than 2.0 involve so little transmitted light that measurements become inaccurate. If the sample is a solution, the remedy is simple: dilute the sample so that the absorbance is reduced!

EXERCISE **22D**

Absorbance

(i) Calculate the absorbance of a sample where 10 per cent of the light is absorbed at the monitoring wavelength.

(ii) The absorbance of a solution is 3.00. What percentage of the incident light is transmitted by the sample at this wavelength?

Use of spectra

Spectra are useful for two reasons:

1 They may be used as a 'fingerprint' of a compound or element. This is useful in chemical identification (**qualitative analysis**). Infrared and nuclear magnetic resonance absorption spectra are widely used in the laboratory for this purpose.

2 The absorbance of a substance at a particular wavelength is proportional to the concentration of that substance – a relationship called the Beer–Lambert law (see page 412). Similarly, the intensity of light emission by a substance increases with the concentration of emitter. These relationships are useful in finding out the concentration of compounds or elements in mixtures (**quantitative analysis**).

For example, steel is analysed by subjecting it to a powerful electric spark. This causes the atoms in the steel to emit certain wavelengths of light ('spectral lines'), which are used by the operators to deduce the elements (e.g. Mn, Cr), in addition to iron, which are present in the steel. The intensity of the lines depends upon the percentage of the element present in the sample of steel.

The remainder of this unit discusses the different types of spectra (and their applications) in greater detail.

In an atomic absorption spectrometer, a sample is turned into atoms (atomized) using a hot flame. UV or visible radiation at a suitable wavelength is passed through the atomized sample. The concentration of absorbing atoms in the sample is proportional to the observed absorbance at that wavelength. The technique allows accurate measurements to be made of the concentrations of trace metals (e.g. Cu, Pb, Cr or Mn) even where the concentrations are below 1 mg per dm^3.

22.5 More about ultraviolet and visible spectra

Lyman emission series of the hydrogen atom

The simplest type of electronic spectra results from transitions within the simplest atom, hydrogen. The emission spectrum of the hydrogen atom at ultraviolet wavelengths consists of a series of emission peaks called the **Lyman series**.

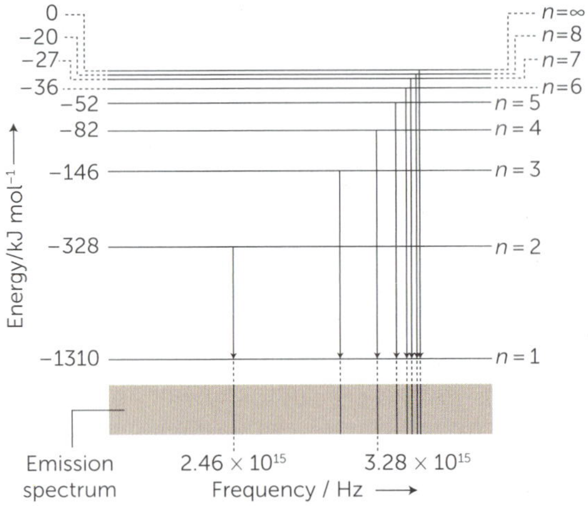

Figure 22.7 Schematic diagram of the transitions and the observed emission lines in the Lyman series of the hydrogen atom.

Experimentally, the emission spectrum of the hydrogen atom is obtained by passing an electric spark through a quartz tube containing hydrogen gas. The energy from the spark dissociates some of the hydrogen molecules into atoms. Some of the dissociated hydrogen atoms have so much energy that they are in an excited electronic state.

Although many of these excited atoms lose their excess energy by collisions with other atoms or molecules, some lose the radiation as light. The emission spectrum is obtained by passing the emitted light through a spectrometer. Measurements of the Lyman series must be made in equipment from which air has been evacuated. This is because the emission lines appear in the range 90–120 nm and both nitrogen and oxygen gases absorb UV radiation at these wavelengths.

Transitions that cause the Lyman series

The hydrogen atom contains only one electron. The electronic energy of a hydrogen atom depends upon the orbital in which the electron is found. If the electron lies in the lowest energy orbital ($n = 1$), the atom is in its ground electronic energy level (its ground state). The energy of one mol of hydrogen atoms in this level is −1310 kJ. (A negative energy seems strange, but this is just a convention. It means that a stationary electron just outside the attraction of the hydrogen nucleus will have zero energy.) If the electron lies in higher orbitals ($n > 1$), the atom is in an excited electronic state and its energy becomes bigger (i.e. less negative). At very high values of n the energy of one mol of hydrogen atoms approaches zero (see Figure 22.7).

The transitions involved in the Lyman series involve hydrogen atoms in higher electronic energy levels losing energy by emitting light so that the hydrogen atom ends up in its ground state. We can represent these transitions as follows:

$$H_{(n>1)} \rightarrow H_{(n=1)} + h\nu$$

$H_{(n>1)}$	$\rightarrow H_{(n=1)}$	$+ h\nu$
electronically excited state	ground state	emitted light

where n in the excited state takes the values 2, 3, 4, 5 etc.

These transitions are shown by downward arrows in Figure 22.7.

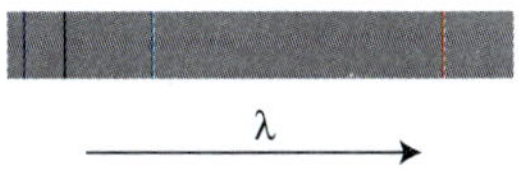

Emission lines of H atom in the visible region (Balmer series), showing the red H_α line on the far right.

Key facts about the Lyman series

- As with all spectra, the greater the energy gap involved in a transition, the higher is the frequency (and the lower is the wavelength) of the emitted light.
- Successive energy levels in the H atom get closer together. This is shown in Figure 22.7. In consequence, the frequencies at which the lines appear in the spectrum get closer together. Eventually, the emission lines converge at 3.283×10^{15} Hz. This is the frequency of the light that would be emitted from the transition

$$H_{(n=\infty)} \rightarrow H_{(n=1)} + h\nu$$

The energy gap for this transition is equal in size to the (**first**) **standard ionization energy** of the H atom and is easily calculated:

$$\Delta E = h\,N_A\nu$$
$$= 3.99 \times 10^{13} \times 3.283 \times 10^{15} = 1310\ \text{kJ mol}^{-1}$$

If a mole of hydrogen atoms in their ground states are supplied with exactly 1310 kJ of energy, the electron in each atom will (just) be pushed out of range of the attraction of the nucleus, and all the atoms will be ionized:

$$H(g) \rightarrow H^+(g) + e^- \qquad \Delta H^\ominus = +1310\ \text{kJ mol}^{-1}$$

If more than 1310 kJ of energy is supplied, the extra energy is channelled into the kinetic energy of the escaping electrons.

- There are other emission series for the H atom, one of which (the **Balmer series**) involves sufficiently small energy jumps that it produces a spectrum in the visible region. In the Balmer series, all the transitions result in the hydrogen atoms ending up in the $n = 2$ level:

$$H_{(n>2)} \rightarrow H_{(n=2)} + h\nu$$

The transition from $n = 3$ to $n = 2$ gives rise to an emission at 656.3 nm called the hydrogen atom 'alpha line' (Hα). The Hα line is easily observed in the **emission** from a hydrogen discharge lamp as a red line, but it is *not* easily observed in the laboratory in **absorption** ($n = 2$ to $n = 3$) because so few hydrogen atoms populate the $n = 2$ level at room temperature that the line is too weak to detect (Exercise 22E).

EXERCISE **22E**

Absorbance

The Balmer series (including the Hα line) is readily observed in absorption in the spectra of stars (including our sun) – why?

EXERCISE **22F**

Lyman series

Use the energies given in Figure 22.7 to calculate the frequency of the light emitted in the Lyman transition $H_{(n=4)} \rightarrow H_{(n=1)} + h\nu$.

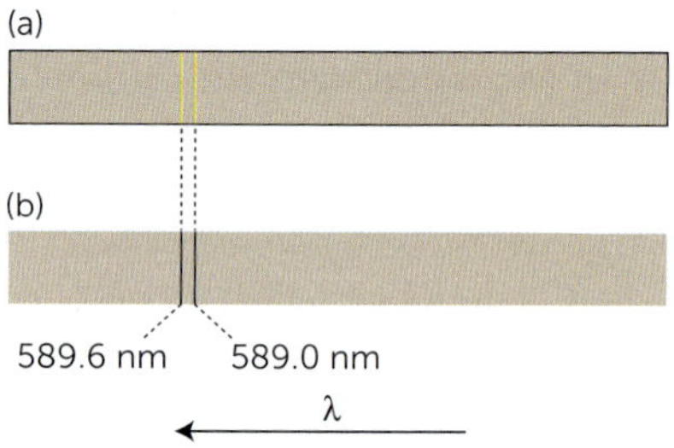

Figure 22.8 The two strongest lines in the visible region in (a) the emission and (b) the absorption spectra, of sodium atoms. The black lines in spectrum (b) represent the absence of light at those wavelengths.

Electronic spectra of other atoms

The emission spectrum of sodium atoms is produced by passing an electric discharge through sodium vapour (easily produced by heating sodium metal). Figure 22.8(a) shows how the two main emission lines in the sodium atom spectrum would appear when examined with a hand-held prism. When the human eye is used as a detector, as in a hand-held spectroscope, both lines are seen to be an intense yellow colour. The lines are responsible for the yellow colour of sodium streetlights and the yellow colour observed when sodium compounds (such as salt) are sprinkled into flames. Although the lines are separated only by 0.6 nm (0.6 nanometres), they can be distinguished even with hand-held spectroscopes.

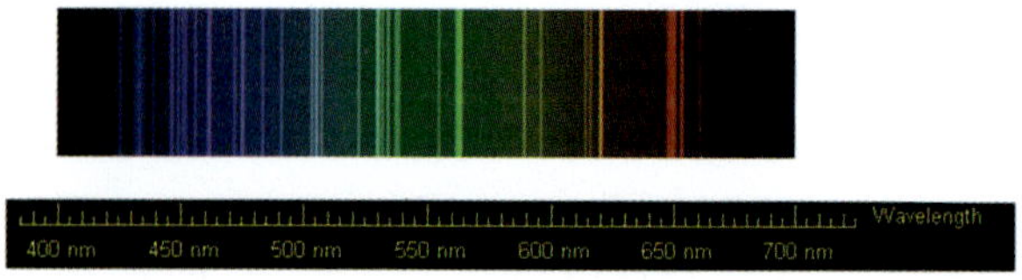

Hand-held spectroscope. The light to be examined enters the right hand side and is diffracted by a small prism. The human eye, pressed to the left-hand side of the tube, acts as detector. On the right-hand side is the emission spectrum of calcium (Image: adapted from NASA).

The absorption spectrum (Figure 22.8(b)) of sodium atoms is produced by measuring the light wavelengths absorbed when white light is passed through sodium vapour. The black lines represent a *reduction* in the intensity of light at those wavelengths.

The absorption and emission spectra of metals such as copper, lithium, caesium, calcium and potassium involve lines at different wavelengths from those observed in the sodium spectrum. As with sodium, a few lines in each spectrum dominate the colours of compounds of these metals in the gas flame. This is the basis of the flame tests that are used to indicate the presence of compounds of these metals (see page 196).

Electronic spectra of molecules

Changes of electronic energy in *simple molecules*, such as HCl or H_2, are colourless because they do not absorb or emit visible light. H_2O is also almost entirely colourless (but see page 426). Many larger organic molecules (including dyes) and some inorganic ions are coloured because they display part of their electronic spectra in the visible region. Figure 22.9 shows the electronic spectrum of potassium manganate(VII) (potassium permanganate, $KMnO_4$), in aqueous solution between 400 and 600 nm. Water and potassium ions do not absorb in this wavelength range – the absorption (and the purple colour) is solely due to the MnO_4^- ion.

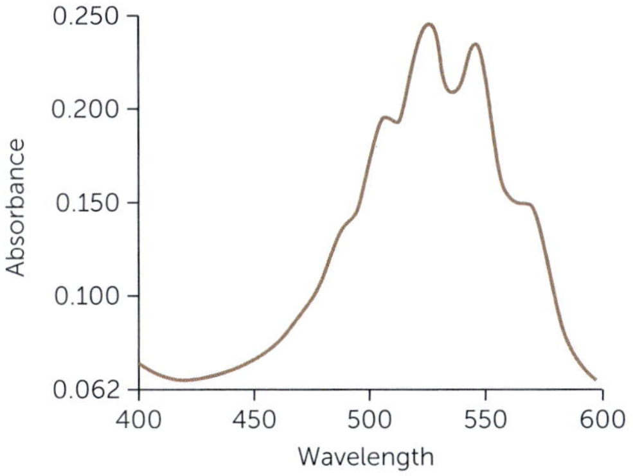

(a)

(b)

Figure 22.9 (a) Electronic spectrum of the manganate(VII) ion between 400 and 600 nm (perspex 1.0 cm cell). (b) Colour of aqueous potassium manganate(VII) solution.

Fluorescence

Some molecules absorb ultraviolet light and immediately emit light at a slightly longer wavelength. This is called **fluorescence**. For example, quinine (which gives tonic water its bitter taste) is colourless, but upon irradiation with UV light it emits a **blue** glow. The emission stops as soon as the light is turned off. Other common fluorescent molecules include caffeine, benzene and 'fluorescein'. Fluorescent materials are used to label valuables ('security tagging') – the written information can only be seen in UV light.

Tonic water contains quinine which fluoresces under UV light, giving the liquid a pale blue cloudy appearance. The tonic water is colourless in the absence of UV light.

Chemiluminescence

Chemical reactions that produce light are said to be **chemiluminescent**, and the light is emitted by electronically excited molecules or atoms produced during the reaction mechanism. Examples of chemiluminescence in nature are to be found in the firefly, glow-worms, gas flames and in some types of fish found in the (otherwise dark) ocean depths. The familiar glow of air when subjected to high voltage discharges (including lightning) is also due to chemiluminescence.

Chemiluminescent 'fun sticks' contain reactants in separate tubes. On bending the stick, the tubes are broken and the reactants mix, producing the chemiluminescent emission. The observed colour of the emission depends upon the type of dye in the reaction mixture. The same principle is used in 'light sticks' designed to be used in emergencies by cavers and walkers, some of which continue to emit light for several hours.

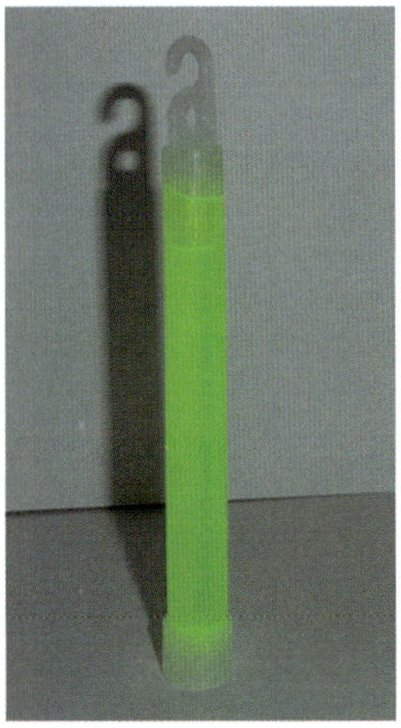

Chemiluminescent emergency 'cave stick'

EXERCISE **22G**

Chemiluminescence

The rate of chemiluminescent emission is controlled by the speed at which the excited electronic states are produced by the chemical reaction. With this in mind, what would you expect to happen to the intensity of emission from a fun stick or cave stick if the stick was placed in **(i)** iced water, **(ii)** warm water?

22.6 Absorption spectra and colour

Colour of compounds

The colour of a solution is controlled by the light that is *transmitted* through the solution.

Think of the manganate(VII) ion. Its solution, viewed under white light, is violet. From its spectrum (Figure 22.9) we see that the most intense absorptions in the visible wavelengths lie between 520 and 550 nm. This is yellow-green light. The remaining colours of white light pass through the solution, and it happens that the human eye interprets this combination of transmitted wavelengths (and their relative intensities) as violet in colour (Figure 22.10).

Similarly, a coloured solid absorbs only some wavelengths of light. The observed colour of the solid is due to the wavelengths of light that are unabsorbed, and which are *reflected* into the eyes of the observer.

It follows that the colour of a compound can be *predicted* from its absorption spectrum in the visible region.

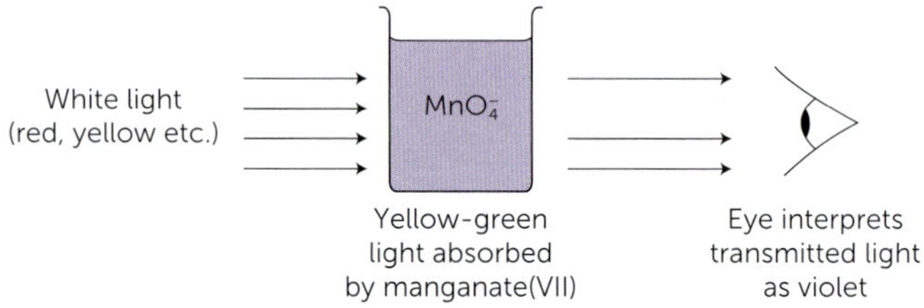

Figure 22.10 Potassium manganate(VII) solution is violet because it absorbs yellow-green light.

Colour cheese

The 'colour cheese' summarizes the relationship between the wavelength of light most strongly absorbed by a substance or solution when illuminated by white light, and its *observed* colour. To use the 'colour cheese' (Figure 22.11):

1 Locate the approximate wavelength of light that is absorbed. The colour of the compound is shown in the opposite segment. Applying this to manganate(VII) ions, the absorbed light is yellow-green and so the solution appears violet. Reversing the procedure, the coloured component of carrots (β-carotene) must absorb blue-green light in order to appear orange.

2 Remember that absorptions outside the visible range (i.e. not listed in the cheese) are irrelevant to the colour of the compound. It follows that colourless substances (such as water) do not absorb very much in the visible wavelength range (400–720 nm).

3 Note that the colour cheese only works if one colour is absorbed more strongly than the others. If very different colours are absorbed equally strongly, the observed colour depends upon the sensitivity of our eyes to the colours involved.

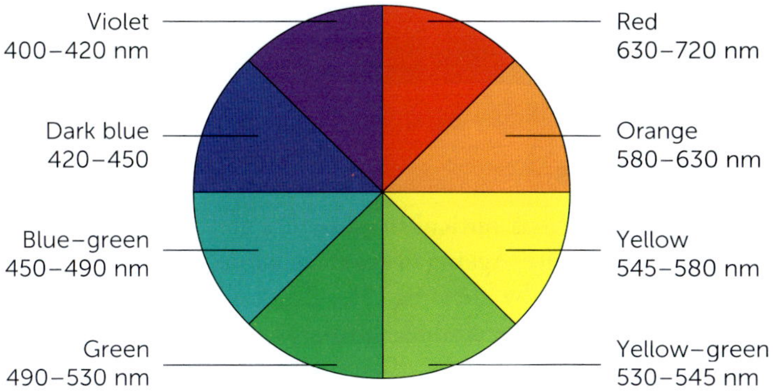

Figure 22.11 The colour cheese.

EXERCISE **22H**

Predicting colours using the colour cheese

(i) Chlorophyll compounds are green coloured. What colour of light do they strongly absorb?

(ii) Copper(II) sulfate solution absorbs orange light strongly. What colour would you expect its solution to be?

(iii) Benzene shows a strong absorption between 200 and 250 nm, but does not absorb between 400 and 720 nm. Predict its colour.

Although we are often concerned with the case of absorption for solutions or gases where the wavelengths of *transmitted* light are of interest, the colour cheese also applies to *reflection* by solids. For example, a rock that is red (in sunlight) to our eyes appears so because the reflected light is red, and the colour cheese shows that the material in the rock's surface has mainly absorbed green light. If the solid did not absorb any visible light, it would appear white (this is the case with zinc oxide). If it absorbed all of the incident visible sunlight, it would appear black (as with copper oxide). In summary, both transmitted and reflected light provides us with 'information' about the wavelengths of visible light that are absorbed.

Sunscreens are creams containing compounds that absorb the harmful ultraviolet radiation emitted by the sun. Skin protection is also provided by UV reflective materials (such as zinc oxide paste). Both absorption and reflection do the same job: they stop harmful UV light from reaching the skin.

EXERCISE 22I

Predicting colour from spectra

Predict the colour of the dichromate(VI) ion from its spectrum (Figure 22.12).

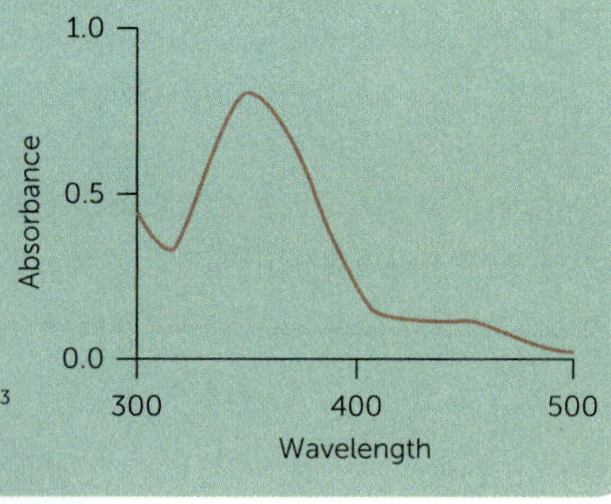

Figure 22.12 Ultraviolet–visible spectrum of $2.5 \times 10^{-4}\,dm^{-3}$ potassium dichromate(VI) solution (1.0 cm quartz cell).

Optical brighteners

Fluorescent molecules known as **optical brighteners** are used in washing powders to make clothes 'whiter than white'. Ageing fibres from white garments acquire a yellowish tinge that cannot be removed by cleaning. The fluorescent molecules in the washing powder attach themselves to the clothes during the washing cycle. The fluorescent molecules absorb UV light from the sun and emit blue light, and such emissions (when added to the reflected light of the fabric) cause the fabric to appear whiter than it would otherwise be. This is sometimes called 'blue whiteness'!

22.7 Infrared spectroscopy

Absorption of infrared radiation by compounds

Infrared light is invisible to the human eye. To illustrate infrared absorption, we require a simple IR detector connected to a voltmeter (a suitable detector is the 'infrared sensor', sold by the Philip Harris Company). There is usually sufficient IR radiation reflected by benches and walls to provide a substantial background laboratory source of infrared light (Figure 22.13(a)). If a watch glass of volatile solvent (e.g. ethoxyethane or methanol) is placed below the path of the infrared light, the intensity of detected infrared light falls (Figure 22.13(b)). This reduction is caused by invisible solvent molecules in the vapour absorbing some of the infrared light. It is customary to use **wavenumber** instead of wavelength in infrared spectroscopy (Box 22.1).

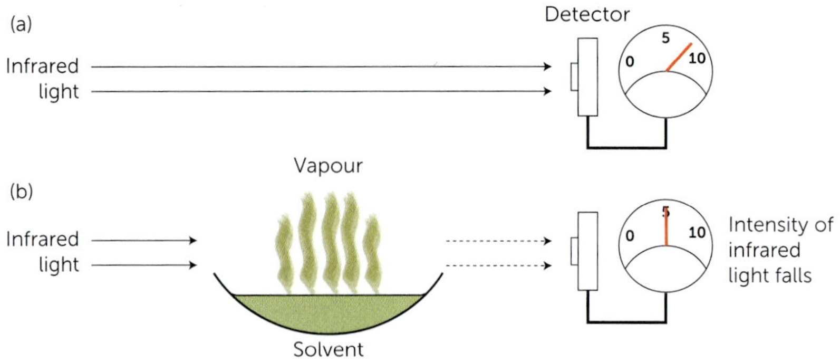

Figure 22.13 Demonstration of infrared absorption. Vapour from the solvent absorbs some of the infrared light so that the intensity of light recorded by the detector falls.

BOX **22.1**

Wavenumbers

In the infrared region, chemists depart from the practice of plotting absorbance against frequency or wavelength. Instead, spectra usually consist of absorbance values plotted against wavenumber. This gives more conveniently sized values. The wavenumber, symbolized $\bar{\nu}$, is the number of complete wavelengths per centimetre. This means that it is connected to wavelength by the relationship

$$\bar{\nu} = \frac{1}{\lambda}$$

when λ is expressed in centimetres, $\bar{\nu}$ is in cm^{-1} – its usual units.

Wavenumbers are directly proportional to the frequency $\bar{\nu}$ (in Hz):

$$\bar{\nu}\ (cm^{-1}) = 3.333 \times 10^{-11} \times \nu$$

1 cm^{-1} is equivalent to light of frequency 3.000×10^{10} Hz. A high frequency means a high wavenumber. In fact, some chemists in conversation speak of frequency when they mean wavenumber! Although incorrect, this serves to remind us that only a simple conversion factor separates the two quantities.

Spring model of molecules

A diatomic molecule, such as HCl, can be thought of as two tiny masses (the H and Cl atoms) connected by a microscopic spring (the covalent bond) (Figure 22.14).

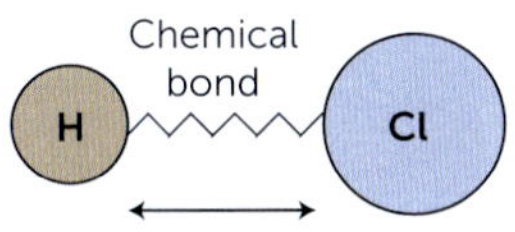

Figure 22.14 The vibration of a hydrogen chloride molecule is caused by the stretching of its chemical bond. The molecule vibrates about 10^{14} times a second. Absorption of a photon increases the amplitude of the vibration, but does not alter its frequency.

The vibration of a diatomic molecule involves the compression and extension of its covalent bond. The number of 'to and fro' movements per second is called the **vibrational frequency** of the molecule. The greater the vibrational energy of a molecule, the higher the amplitude of its vibration, i.e. the greater the distance travelled by the atoms during the vibration.

All molecules consist of atoms in continuous vibration – even at 0 K. Infrared absorption spectra are produced when molecules increase their vibrational energy by absorbing infrared light. The *electronic* energy of molecules does not change in these transitions (see Figure 22.4). When a hydrogen chloride molecule absorbs infrared light, it increases its vibrational energy from the ground vibrational energy level to the second vibrational energy level (v_1). The transition involved is

$$HCl(E_0, v_0) + \text{infrared photon} \rightarrow HCl(E_0, v_1)$$

The $v_0 \rightarrow v_1$ transition is the only important transition in the infrared spectra of most diatomic molecules at room temperature. This and the fact that diatomic molecules have only one way of vibrating – a simple stretching of the bond between the two atoms – explains why diatomics have only one major absorption peak in their infrared spectrum (Figure 22.15).

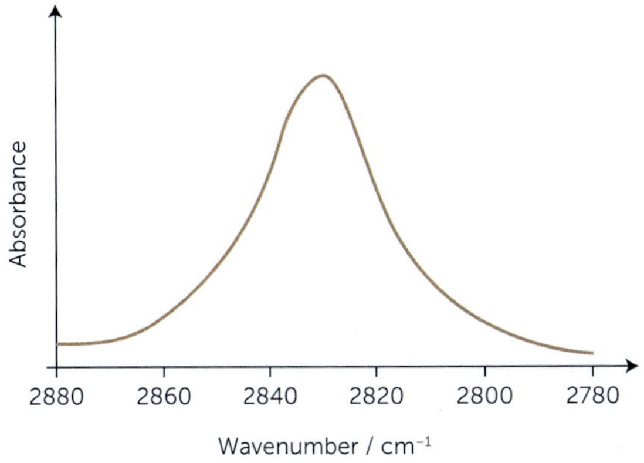

Figure 22.15 The infrared absorption peak of hydrogen chloride (HCl) at low resolution. (The gas is dissolved in CCl_4, but the solvent does not absorb in this wavenumber range.)

Polyatomic molecules

Molecules containing more than two atoms (polyatomic molecules) may also be thought of as consisting of atoms connected by springs. For example, the propanone molecule

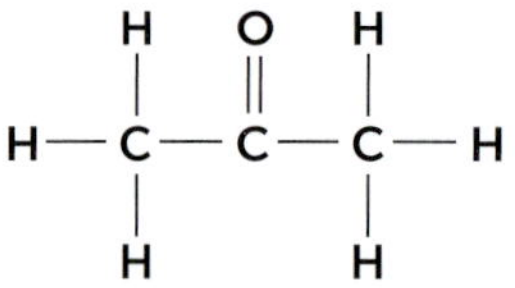

may be represented as

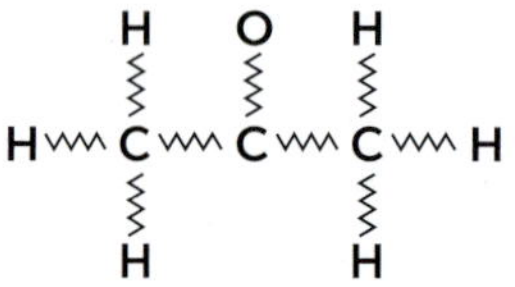

where ∿∿ are the 'springs'. The atoms on either side of bonds in polyatomic molecules may undergo **stretching vibrations** (Figure 22.16(a)), but they may also bend against the skeleton of the molecule. These are **bending vibrations** (Figure 22.16(b)).

If a polyatomic molecule absorbs an infrared photon, the whole molecule (like some microscopic jelly) vibrates with a higher amplitude. However, it is often found that only a few atoms in the molecule vibrate appreciably (i.e. suffer *large* compressions, extensions or bends). (Returning to the jelly analogy, although the whole jelly is vibrating at a particular frequency, only parts of it move substantial distances during a single vibration.)

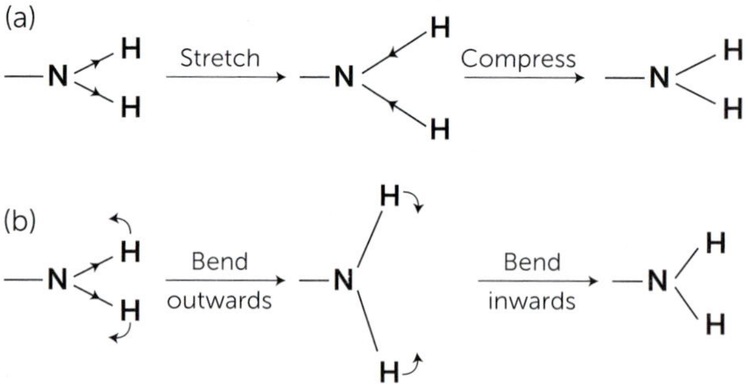

Figure 22.16 (a) Stretching vibration and (b) bending vibration of the NH_2 group.

For example, propanone molecules absorb infrared light at precisely 1715 cm^{-1}. As a result of this absorption, the C and O atoms of the C=O group undergo a stretching motion at a higher amplitude. The motion of the other atoms in the molecule is largely unaffected by photon absorption, i.e. the vibration of the molecule is more or less localized around the carbonyl group. However, if propanone molecules absorb infrared light at exactly 2950 cm^{-1}, the C–H stretching motion becomes more pronounced, and the rest of the molecule – including the C=O group – moves very little.

An infrared spectrometer exposes the molecule under investigation to infrared light with a range of different wavenumbers during a 'scan'. Peaks in the infrared spectrum occur at the frequencies at which the different groups of atoms within the molecule become 'activated' at their higher amplitudes. This is evident from the spectra in Figures 22.17–22.25, where the vibrations associated with key peaks are labelled.

A simple pattern is often observed in infrared spectra in that *molecules containing the same groups of atoms give rise to peaks that are at roughly the same wavenumber irrespective of the molecule in which they are found.* Typical of such groups are C=O, C–H, $-NH_2$ and –OH. For example, the higher amplitude vibration of the carbonyl group (C=O), due to the transition v_0 to v_1, is activated by infrared radiation between 1700 and 1800 cm^{-1}, the exact wavenumber varying with the molecule in question. This transition gives rise to a 'carbonyl peak' at 1715 cm^{-1} in propanone (CH_3COCH_3, Figure 22.17), 1725 cm^{-1} in ethanal (CH_3CHO, Figure 22.18), 1742 cm^{-1} in the ester ethyl ethanoate ($CH_3COOC_2H_5$, Figure 22.19) and 1720 cm^{-1} in ethanoic acid (CH_3COOH, Figure 22.20). The high intensity (peak height) and relatively narrow wavenumber range over which the C=O peak is observed in compounds makes it easy to recognize in the infrared spectrum of an unknown compound. This gives us a very useful way of identifying compounds that contain carbonyl groups. For example, photographic film consists of the ester cellulose triacetate and a spectrum of the film shows an intense absorption at about 1745 cm^{-1}.

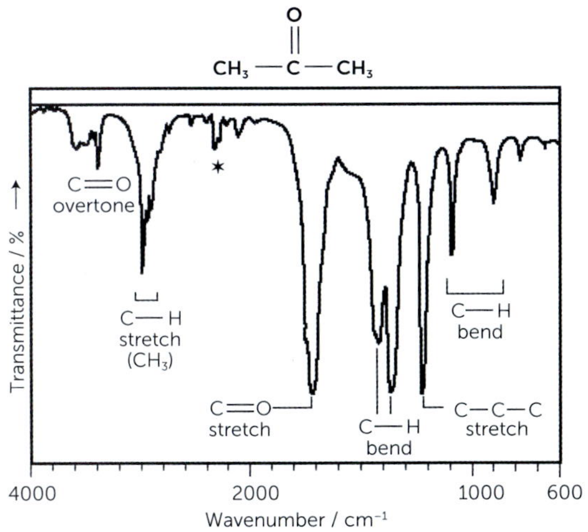

Figure 22.17 Infrared spectrum of propanone CH_3COCH_3. (The peak marked* is due to CO_2 in the spectrometer.)

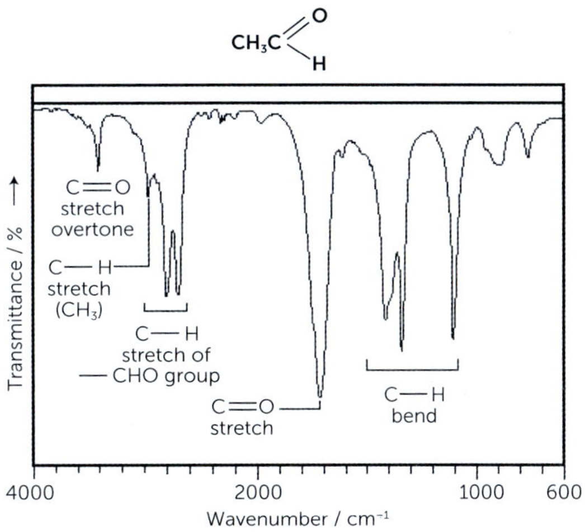

Figure 22.18 Infrared spectrum of ethanal CH_3CHO.

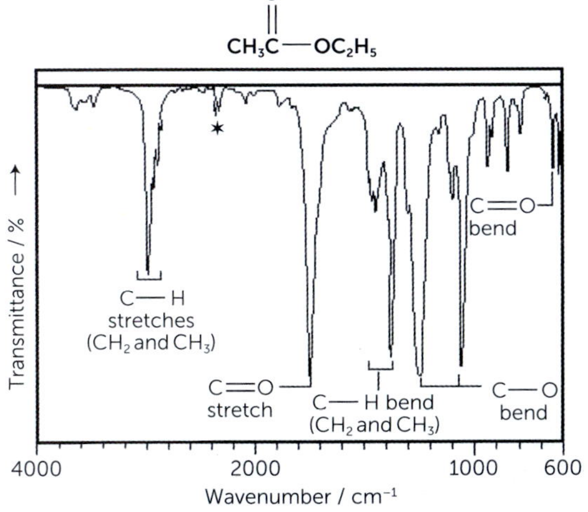

Figure 22.19 Infrared spectrum of ethyl ethanoate $CH_3CO_2C_2H_5$.

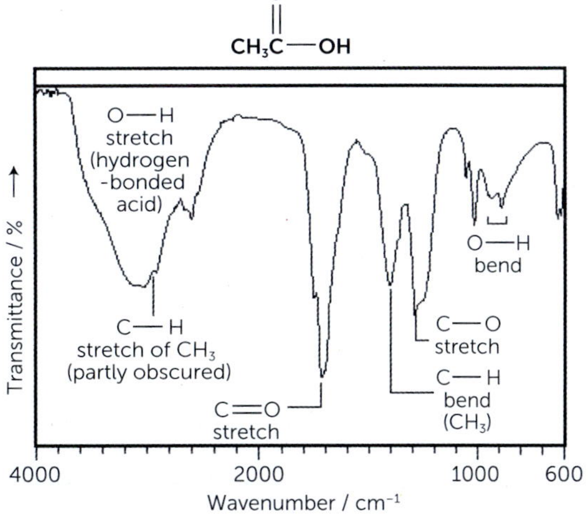

Figure 22.20 Infrared spectrum of ethanoic acid CH_3COOH.

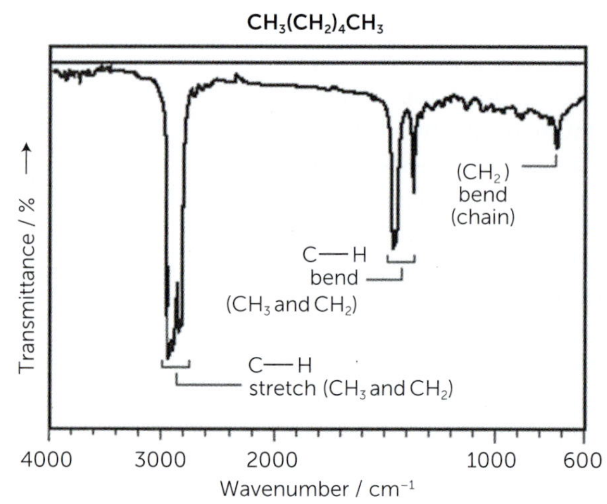

Figure 22.21 Infrared spectrum of hexane $CH_3(CH_2)_4CH_3$.

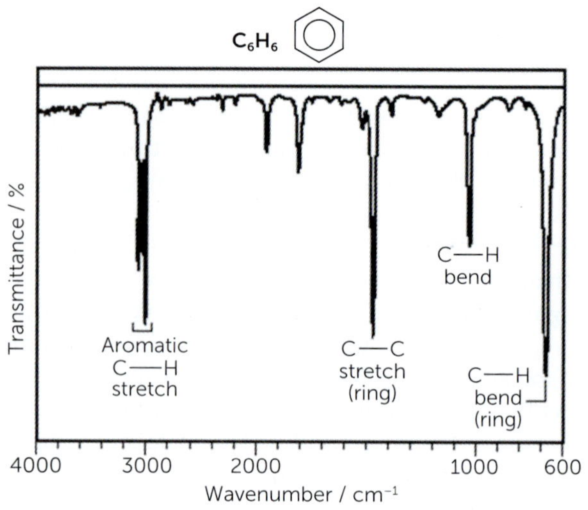

Figure 22.22 Infrared spectrum of benzene C_6H_6.

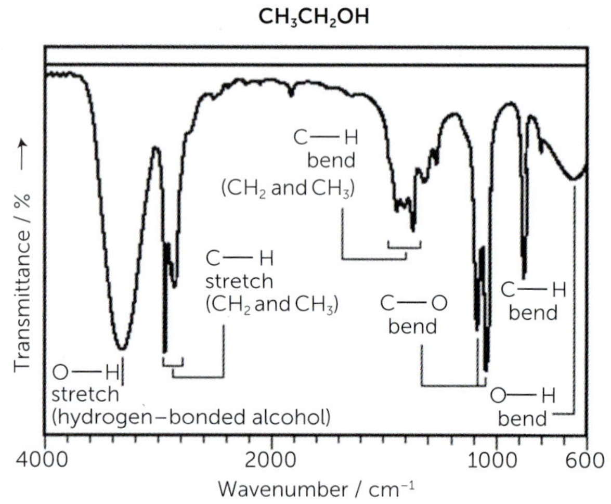

Figure 22.23 Infrared spectrum of ethanol CH_3CH_2OH.

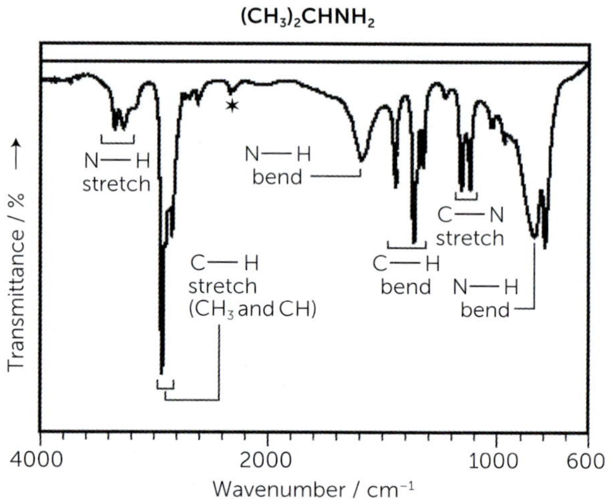

Figure 22.24 Infrared spectrum of propan-2-amine $(CH_3)_2CHNH_2$.

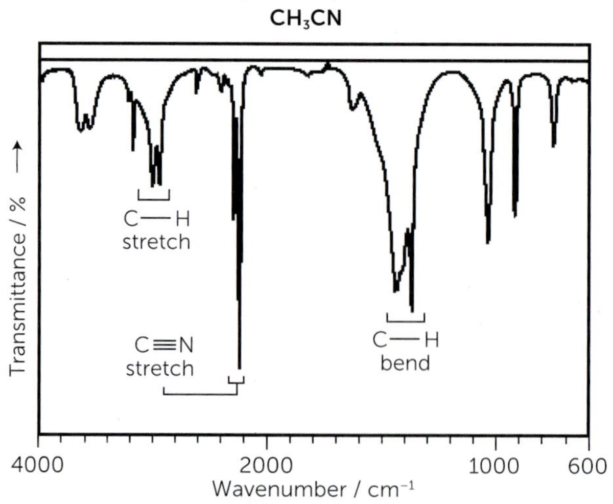

Figure 22.25 Infrared spectrum of ethanenitrile CH_3CN.

Other groups of atoms also give identifiable absorptions:

- C–H vibrations in molecules where the carbon atom of the C–H group is linked to other carbon atoms using single C–C bonds, as in

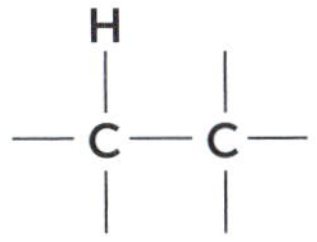

  are observed as sharp peaks in a relatively narrow wavenumber band 3000–2850 cm^{-1} (see hexane, Figure 22.21). The exceptions are C–H vibrations within the CHO group of aldehydes, which usually give two peaks between 2850 and 2750 cm^{-1} (Figure 22.18). These are useful to distinguish aldehydes from ketones.
- C–H vibrations involving carbon atoms bonded to other carbon atoms by double or triple bonds, such as

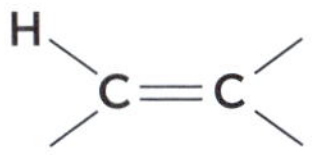 and 

  give rise to peaks above 3000 cm^{-1}. C–H peaks above 3000 cm^{-1} are also observed where the vibrating C–H group is part of a benzene ring (see benzene, Figure 22.22).
- The hydrogen-bonded –OH group of alcohols and phenols gives rise to a characteristic and very broad absorption in the range 3500–3200 cm^{-1} (see ethanol, Figure 22.23). The –OH bond in carboxylic acids, also very broad, is found between 3500 and 2400 cm^{-1} (see ethanoic acid, Figure 22.20). The spectra of carboxylic acids are easily distinguished from those of alcohols and phenols, since acid spectra also contain the characteristic C=O peak.
- The NH_2 group gives rise to two weak absorptions (due to the N–H stretch) in the range 3500–3100 cm^{-1} (see propan-2-amine, Figure 22.24).
- Nitriles (also known as cyanides), show a CN stretch near 2250 cm^{-1} (see ethanenitrile, Figure 22.25). However, this peak is extremely weak in HCN(g) itself.

The spectra in Figures 22.17 to 22.25 show the potential of infrared spectroscopy for distinguishing between common types of organic molecule, particularly if combined with information from other techniques such as mass spectrometry, nuclear magnetic resonance (NMR), and with the results of chemical tests.

EXAMPLE **22.3**

A compound P contains the following elements (percentage by mass):

C (30.6 per cent), H (3.8 per cent), O (20.3 per cent), Cl (45.2 per cent)

Its mass spectrum contains a parent ion at m/z = 78. Its infrared spectrum contains a very strong peak at 1800 cm^{-1}. P does not reduce Fehling's solution. Suggest a molecular formula for P.

Answer

The simplest formula of the compound is first calculated from the percentage of elements by mass:

Element	Carbon	Hydrogen	Oxygen	Chlorine
Percentage by mass	30.6%	3.8%	20.3%	45.2%
Molar ratio	$\frac{30.6}{12.0} = 2.55$	$\frac{3.8}{1.0} = 3.80$	$\frac{20.3}{16.0} = 1.27$	$\frac{45.2}{35.5} = 1.27$
Simplest molar ratio	$\frac{2.55}{1.27} = 2.00$	$\frac{3.80}{1.27} = 3.00$	$\frac{1.27}{1.27} = 1.00$	$\frac{1.27}{1.27} = 1.00$
Simplest whole number ratio	2	3	1	1

EXERCISE **22J**

Using IR spectra in organic analysis 1

Which of the following compounds would be expected to show **(i)** a peak due to C=O, **(ii)** a broad band due to –OH, **(iii)** a peak due to C–H?

(a) butanol

(b) propanoic acid

(c) ethanal

EXAMPLE **22.3** *(continued)*

Therefore, the simplest formula of the substance is C_2H_3OCl. The mass spectrum gives the mass of the parent ion as 78 u (^{35}Cl isotope), and we conclude that the molecule has the formula C_2H_3OCl.

The peak at 1800 cm^{-1} suggests the presence of a carbonyl group. Taking 'C=O' out of the formula C_2H_3OCl gives CH_3Cl. P does not reduce Fehling's solution, which excludes the aldehyde CH_2ClCHO. The structure for P is then most likely to be CH_3COCl:

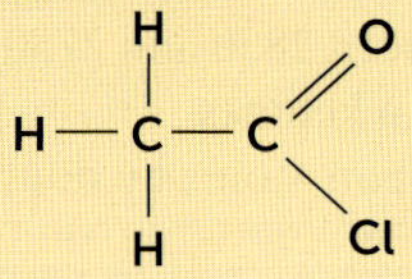

EXERCISE **22K**

Using IR spectra in organic analysis 2

A compound Y contains (by mass) 19.2 per cent nitrogen, 15.1 per cent hydrogen and 65.8 per cent carbon. Its mass spectrum contained an intense peak, due to the parent ion, at *m/e* = 73. Its infrared spectrum contained two peaks of medium intensity at 3300 cm^{-1} and 3370 cm^{-1}. Suggest a formula for Y.

22.8 Beer–Lambert law

Beer–Lambert equation

The relationship between the **absorbance** of a sample at a wavelength λ, symbolized A_λ, and the concentration of absorber c, is known as the **Beer–Lambert law** and is summarized by the equation

$$A_\lambda = \varepsilon_\lambda \times c \times b$$

where ε_λ (read as 'epsilon at wavelength lambda') is a constant for the absorbing substance known as the **molar absorption coefficient** of the substance at wavelength λ, and b is the thickness of the sample (the cell pathlength).

The Beer–Lambert equation shows that if the concentration of *absorber* doubles, the *absorbance* of the solution also doubles. The fact that the absorbance is also proportional to cell pathlength can be useful. If the absorbance of a solution or gas is low because the concentration of absorber is low, this may be compensated for by an increase in the cell pathlength. For example, the analysis of trace gases in car exhausts by infrared spectroscopy is often accomplished using a cell with a pathlength of 10 m, giving 100 times the absorbance of a standard 10 cm gas cell (see Figure 24.4, page 440).

The **Beer–Lambert law** applies to all wavelengths, although different types of light sources, cells and detectors are used to monitor the absorption of UV, visible and IR light. In practice, the absorption of radio waves by material (as in nuclear magnetic resonance, NMR) is not sufficiently sensitive to concentration to make it useful as a *quantitative* technique, although it remains a powerful *qualitative* technique. In other words, NMR is excellent at telling us 'what compound is present' but not at revealing 'how much of it is present'.

More about ε

- The easiest way to look upon ε_λ is to think of it as simply the absorbance of the sample when the concentration of the absorber is 1 mol m^{-3} and when the pathlength is 1 m. The bigger the value of ε_λ, the more *intense* is the absorption of the compound at that wavelength.
- If an absorbing compound chemically reacts in any way (e.g. it dissociates or associates by forming hydrogen bonds), ε_λ changes too. Even changes of solvent may alter ε_λ.

- If the units of c are mol m^{-3} and the units of b are metres then the units of ε_λ are mol^{-1} m^2.
- An alternative unit of ε_λ in common use is mol^{-1} dm^{-3} cm^{-1}. This unit is convenient because the concentration term in the Beer–Lambert equation is now expressed in mol dm^{-3} and the cell pathlength is in centimetres. The conversion is

  1 mol^{-1} m^2 = 10 mol^{-1} dm^3 cm^{-1}

For example, 90 mol^{-1} m^2 = 900 mol^{-1} dm^3 cm^{-1}.

Beer–Lambert plots

The Beer–Lambert equation is of the form $y = mx$, and predicts that a plot of the absorbance (A_λ) of a compound at a particular wavelength against the concentration (c) of a compound should be a straight line, provided we use the same cell (i.e. same b) throughout the experiment. The slope of the graph is equal to $\varepsilon_\lambda \times b$.

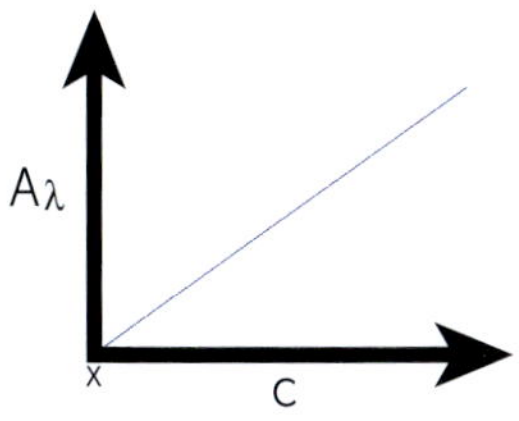

Beer–Lambert equation plot. Point x represents zero absorbance and zero concentration.

EXAMPLE **22.4**

Use the spectrum in Fig. 22.12 (Exercise 22I) to estimate the molar absorption coefficient of potassium dichromate(VI) at 350 nm.

Answer

Figure 22.12 shows that the absorbance of the sample at 350 nm is approximately 0.78:

$c = 2.5 \times 10^{-4}$ mol dm^{-3} = 0.25 mol m^{-3}; b = 1.0 cm = 0.010 m

Rearranging the Beer–Lambert equation,

$\varepsilon_{350} = A_{350}/cb = 0.78/(0.25 \times 0.010) = 310$ mol^{-1} m^2 (to two significant figures)

Therefore, ε_{350} for potassium dichromate(VI) solution is 310 mol^{-1} m^2.

EXERCISE **22L**

Beer–Lambert law calculations

(i) Chlorophyll 'a' shows an absorption maximum at 680 nm with $\varepsilon \approx 1.1 \times 10^4$ mol^{-1} m^2. Estimate the concentration of chlorophyll 'a' in a leaf assuming (a) that chlorophyll 'a' is evenly distributed throughout a leaf of thickness 0.010 cm, (b) that the leaf possesses an absorbance of 1.1, and (c) that all the absorption at this wavelength is entirely due to chlorophyll 'a'.

(ii) The molar absorption coefficient of DNA (deoxyribonucleic acid) is about 1000 mol^{-1} m^2 (expressed per mole of base) at 260 nm. What concentration of the compound would be required to produce a solution of absorbance 0.010 in (a) a 1.0 cm cell and (b) a 4.0 cm cell?

(iii) A solution of a drug (concentration 0.0020 mol dm^{-3}) was placed in a 1.0 cm cell. The sample absorbance at 250 nm was 0.40. Calculate the molar absorption coefficient of the drug at 250 nm.

EXERCISE **22M**

Beer–Lambert law with two components

The molar absorption coefficient of 2-nitrophenol and 4-nitrophenol at 312 nm is 1.2×10^2 and 9.6×10^2 mol^{-1} m^2 respectively. A solution contains 2-nitrophenol and 4-nitrophenol, both at a concentration of 1.0×10^{-3} mol m^{-3}. What is the absorbance of the mixture at 312 nm in a 1.0 cm cell? Assume that the two phenols do not react with each other. (Hint: The mixture absorbance is the sum of the individual absorbances.)

EXAMPLE **22.5**

An example of a Beer–Lambert plot is shown in Figure 22.26. This shows the absorbance of 4-nitrophenol ($C_6H_4OHNO_2$) in ethanol solution (concentration between 0 and 12 x 10^{-3} mol m^{-3}) at a wavelength of 312 nm. The cell pathlength was 1.0 cm. Calculate the molar absorption coefficient of 4-nitrophenol at 312 nm.

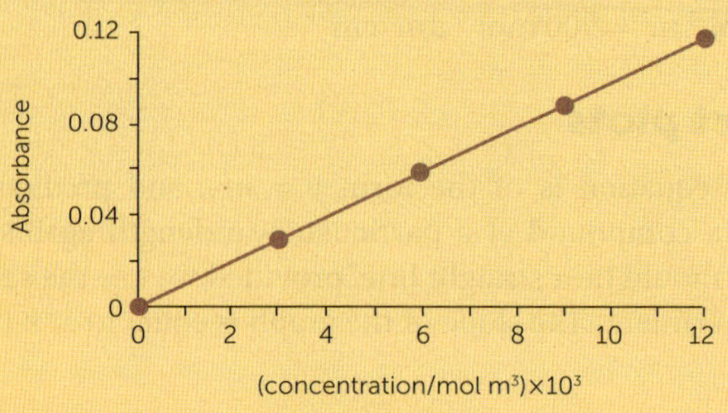

Figure 22.26 Beer–Lambert plot for 4-nitrophenol in ethanol solution at 312 nm.

Answer

The slope of the graph is

$$\frac{0.115-0.000}{12.0\times10^{-3}-0.00}=9.60\ \text{mol}^{-1}\text{m}^{-3}=\varepsilon\times b$$

Since b = 1.0 cm = 0.010 m,

$$\varepsilon=\frac{9.60\ \text{mol}^{-1}\text{m}^{-3}}{0.010\text{m}}=9.6\times10^{2}\ \text{mol}^{-1}\text{m}^{2}$$

Therefore, ε_{312nm} for 4-nitrophenol is 9.6 × 10^2 mol^{-1} m^2.

EXERCISE **22N**

Beer–Lambert plot

The absorbance of different concentrations of manganate(VII) ions at 526 nm in a 1.0 cm glass cell was found by experiment:

[MnO_4^- (aq)]/mol dm^{-3}	Absorbance at 526 nm
0.0	0.00
8.7 x 10^{-5}	0.25
12.2 x 10^{-5}	0.35
14.0 x 10^{-5}	0.40
17.5 x 10^{-5}	0.50

(i) Convert the concentrations to mol m^{-3}.

(ii) Plot absorbance against concentration and hence calculate the molar absorption coefficient of the MnO_4^-(aq) ion at 526 nm.

22.9 Photosynthesis

What is photosynthesis?

Photosynthesis is a complicated sequence of chemical reactions, driven by absorbed sunlight, in which carbohydrates are made. Photosynthesis is one of the most important of all reactions that involve light, because plants (either directly or indirectly) are the main source of our food.

The overall reaction involved in photosynthesis may be represented as

$$6CO_2(g) + 6H_2O(l) + h\nu \rightarrow \underset{\text{glucose}}{C_6H_{12}O_6(s)} + 6O_2(g)$$

Key points about photosynthesis

Photosynthesis occurs within the leaves of plants. For thousands of years the growth of plants was regarded as mysterious and caused by magical or divine forces. Painstaking research has shown that photosynthesis involves an extremely complicated set of chemical reactions.

- This reaction is highly endothermic with $\Delta H^{*} \approx +2800\,kJ\,mol^{-1}$. The reaction does not occur directly but in many steps, and in effect the energy for the conversion is obtained from sunlight ($h\nu$). This energy is initially trapped (i.e. absorbed) by chlorophyll, forming excited chlorophyll molecules. The excited chlorophyll molecules then pass on this energy to enable the sequence of reactions to occur.
- Cellulose (which provides plants with their bulk and mass) is manufactured within the plant by bonding together a large number (typically 300–2500) of glucose molecules in a straight chain, with the elimination of one water molecule from each glucose unit. (Taking H_2O from the formula for glucose, $C_6H_{12}O_6$, gives $C_6H_{10}O_5$, so that cellulose can be represented by the formula $(C_6H_{10}O_5)_n$.)
- Starch (which is used by plants as a 'fuel' to supply energy for respiration and growth) is also made from glucose molecules linked together although, unlike cellulose, both straight and branched chains are present.
- In photosynthesis, the carbon (cellulose-based) skeleton of plant tissue and its stored carbohydrates are 'chemically assembled' from carbon dioxide gas. The production of carbon-based solids in plants from a carbon-containing gas was something of a mystery to early scientists, who referred to photosynthesis as 'carbon fixation' – a phrase that is still in use today. Because of the widespread occurrence of plants using photosynthesis on land and in the sea, the scale of carbon fixation is enormous, with about 10^{11} tonnes of carbon being taken out of the atmosphere per year.
- Photosynthesis consumes CO_2 and also returns O_2 to the atmosphere. Photosynthesis is the most important way in which CO_2 levels (increased by the burning of fuels and by respiration) are reduced, and in which oxygen levels in the air are replenished.

22.10 Nuclear magnetic resonance: 1H-NMR spectroscopy

In **nuclear magnetic resonance** the absorption of radio waves by atomic nuclei surrounded by a powerful magnet produces spectra that are used to help decide the structural formulae of compounds.

Nuclei that have an odd number of protons or an odd number of neutrons (or both of these) behave like tiny bar magnets; they have a **magnetic moment** (they are also said to have **spin**). Just as when a small bar magnet is placed in the field of a second magnet it will be attracted or repelled, in an applied magnetic field such nuclei will have their magnetic moments aligned with the field or opposed to it. These arrangements will be of different energies, as shown in Figure 22.27, and the difference in energy between the two will depend upon the strength of the external field.

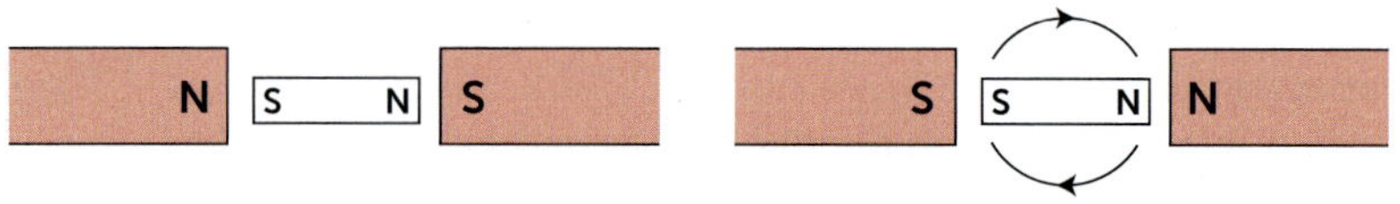

Figure 22.27 Different alignments of bar magnets lead to different energy states.

NMR depends upon the fact that energy in the radio-frequency region of the electromagnetic spectrum can be absorbed by the sample and cause the energy of the nuclei in the lower energy state to jump to a higher energy. The frequencies at which the nuclei absorb radiation are referred to as **resonance frequencies** or (simply) as **resonances** or **signals**. The earliest (and still the most important) nuclei to be studied by NMR were hydrogen nuclei (i.e. protons), and it is proton NMR (symbolized as ^{1}H-NMR) that we will first deal with, followed by carbon-13 NMR (page xxx).

Chemical shifts in proton NMR (1H-NMR)

Protons are sensitive to their chemical environment – electrons moving near them produce their own magnetic field, which changes the external field experienced by the proton. Protons in different chemical environments therefore experience slightly different magnetic fields and absorb at different frequencies.

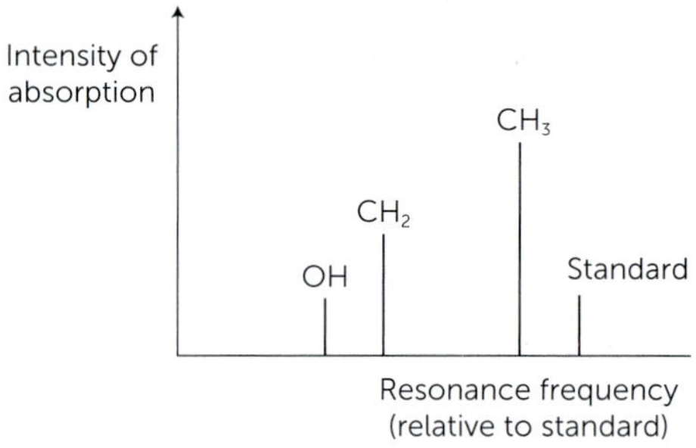

Figure 22.28 Low-resolution NMR spectrum of ethanol, CH_3CH_2OH.

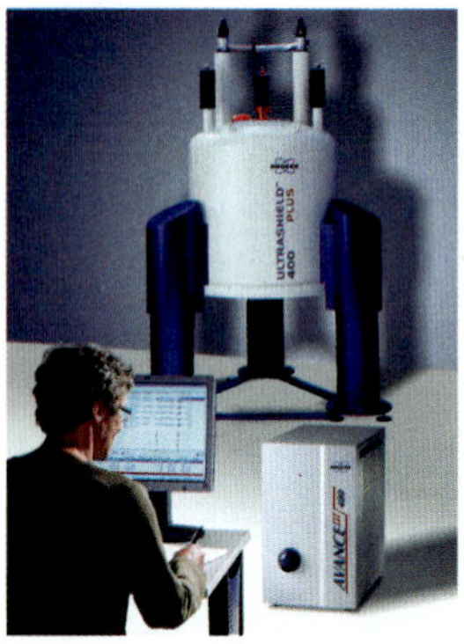

NMR spectrometer. The magnetic field is produced by a helium-cooled superconducting magnet stored in a giant container shown in the background. The sample is lowered into the magnet before scanning takes place.

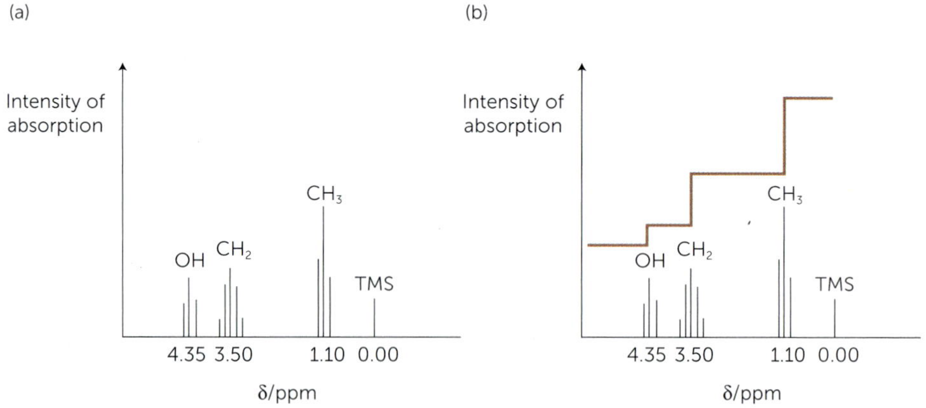

Figure 22.29 (a) High-resolution NMR spectrum of pure ethanol, CH_3CH_2OH. (b) The same spectrum with integration.

Figure 22.28 shows the low-resolution NMR spectrum of ethanol, CH_3CH_2OH. There are three signals because there are protons in three different environments: those present in CH_3, in CH_2 and the single **H** attached to the oxygen.

The resonance frequencies of the different protons are expressed as **chemical shifts** (δ) relative to a standard. **Tetramethylsilane** (TMS), or $(CH_3)_4Si$, is widely used as a standard because it is inert and has a spectrum with a single absorption (there is only one type of hydrogen in the molecule). Different NMR spectrometers have magnets that generate different fields so, in order that scientists can compare data from different instruments, the chemical shift of a proton (δ) is calculated as

$$\delta = \frac{\text{shift from TMS for a particular proton in Hz}}{\text{spectrometer frequency in MHz}}$$

On this scale the resonance of TMS is exactly 0.00. As a result of using this system, values of frequencies for a given proton will always be the same, no matter what instrument is used, and data are easily compared.

If the spectrum of ethanol is run under conditions of high resolution, the original peaks are seen to be split into multiplets as shown in Figure 22.29(a). This is called **spin–spin splitting.** *Each type of proton experiences the protons on neighbouring atoms* and its resonance peak is split into $n + 1$ components, where n is the number of protons on the nearest neighbouring atom(s).

Look again at Figure 22.29(a). The resonance at δ 0.00 is due to the protons from TMS and the resonances centred at δ 4.35, δ 3.50 and δ 1.10 are due to the protons of ethanol. The ethanol resonances are split. At δ 1.10 the resonance is split into three (**a triplet**) because the CH_3 protons experience two protons on the next carbon atom (the CH_2 group and $n = 2$) and so their resonance splits into $n + 1$, or $2 + 1 = 3$, peaks. At δ 3.50 the resonance is split into five (**a quintet**): the CH_2 protons experience four neighbouring protons (three from CH_3 and one from OH) so the resonance is split into $n + 1$, or five, peaks.

You should now be able to work out why the resonance for the OH proton is a triplet. Notice that the intensities of the peaks of the split resonances are not all the same – these approximately follow a set of numbers known as Pascal's triangle in which each number is the sum of the left and right numbers above it (Table 22.1 gives the first five lines of the triangle).

EXERCISE **220**

NMR signals

How many different signals will there be in the low-resolution ^{1}H-NMR spectrum of:

(i) CH_3CH_2CHO

(ii) 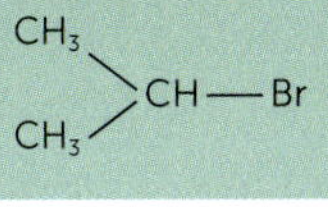

Table 22.1 Pascal's triangle

Number of peaks	Intensities as shown by Pascal's triangle
1 (singlet)	1
2 (doublet)	1 1
3 (triplet)	1 2 1
4 (quartet)	1 3 3 1
5 (quintet)	1 4 6 4 1

The area under each resonance (whether split or not) is proportional to the number of protons responsible for that resonance. The NMR instrument will **integrate** each resonance and produce a vertical line over it, the height of which is proportional to its area. We are now in a position to consider the complete NMR spectrum of ethanol, with integration, in Figure 22.29(b). Notice that the heights of the vertical lines over the OH, CH_2 and CH_3 resonances are in a ratio of 1:2:3, i.e. in accordance with the number of protons on each group.

Table 22.2 shows typical values of chemical shifts for protons in organic compounds. Notice that protons attached to a carbon atom also bonded to a halogen tend to have relatively high chemical shift (δ) values; the electronegative atoms reduce the electron density around the proton so that it experiences more of the external magnetic field (the proton is **deshielded**) and absorbs radiation at a higher frequency. Similarly, if a group or atom pushes electron density on to a proton (**shields** it), the proton absorbs radiation at a lower frequency (its chemical shift is lower).

Figures 22.30 to 22.34 show some typical NMR spectra.

EXERCISE 22P

Integration and splitting

(i) Draw the integration traces that would be obtained on the spectra in Figures 22.30 and 22.32.

(ii) In Figure 22.31, the resonance at δ 2.90 is split into a septet. What would the ratio of intensities be in a septet?

Table 22.2 Typical values of chemical shifts for protons in organic compounds

Group	Type of compound	Approximate chemical shift δ/ppm
CH_3–C	Alkane	0.9
C–CH_2–C	Alkane	1.3
CH_2=C	Alkene	2.6
C_6H_5–	Arene	7.3
CH_3–N	Amine	2.3
CH_3COO–	Ester, acid	2.0
CH_3CO–	Ketone	2.1
CH_3Br	Bromoalkane	2.6
CH_3Cl	Chloroalkane	3.1
CH_3F	Fluoroalkane	4.3

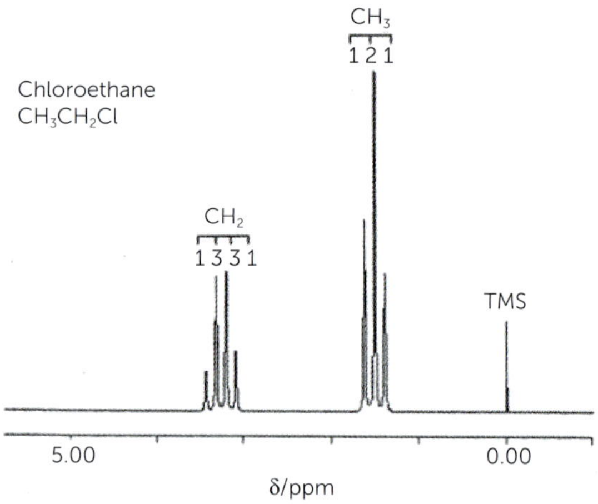

Figure 22.30 ¹H-NMR spectrum of chloroethane at 60 MHz.

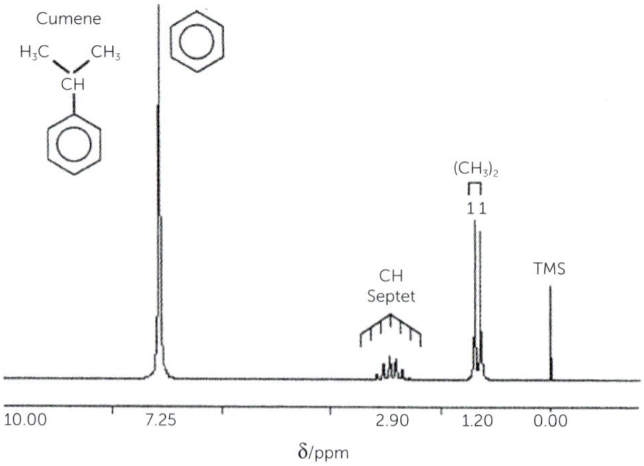

Figure 22.31 ¹H-NMR spectrum of cumene at 60 MHz.

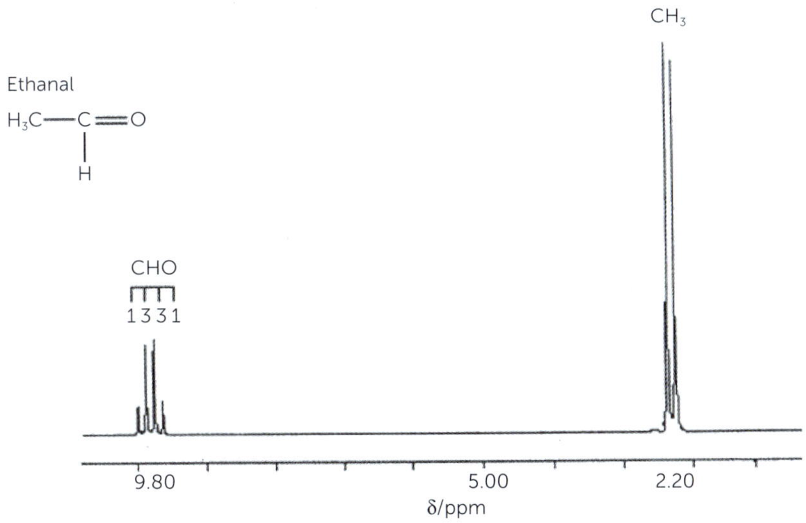

Figure 22.32 ^{1}H-NMR spectrum of ethanal at 60 MHz.

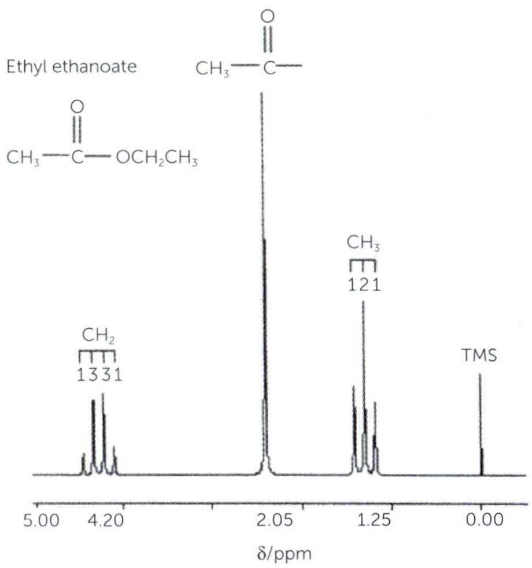

Figure 22.33 ^{1}H-NMR spectrum of ethyl ethanoate at 60 MHz. Notice that the resonance at δ 2.05 is a singlet – the methyl protons in the acyl group are too far from the other protons for their resonance to be split.

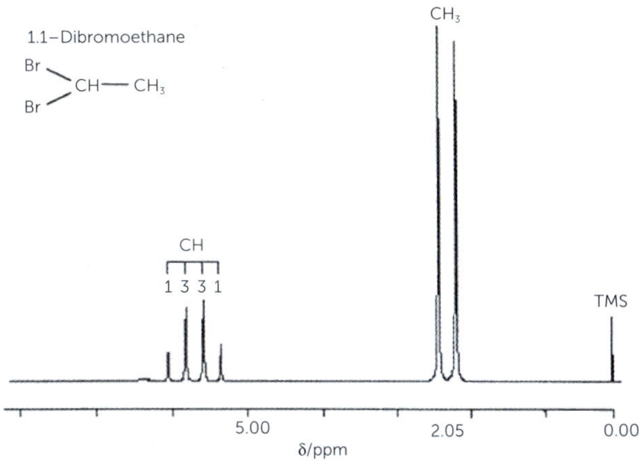

Figure 22.34 ^{1}H-NMR spectrum of 1,1-dibromoethane at 60 MHz.

Human beings are composed of organic material, and therefore contain a large amount of combined hydrogen in many different compounds. This makes the human body a possible 'sample' for an NMR instrument. In hospitals, the phrase 'nuclear magnetic resonance' has been dropped in favour of **magnetic resonance imaging (MRI)** (because patients might wrongly believe the technique involves ionizing radiation) and the instruments used are commonly simply called **scanners**, but the principle of the instruments is the same as their laboratory-based cousins.

EXERCISE **22Q**

NMR spectra of compounds

See if you can match the following formulae with their spectra:

(i) 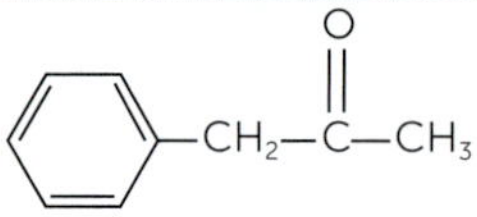

(ii) CH_3CH_2Br

(iii) Cl_2CHCH_2Cl

(a)

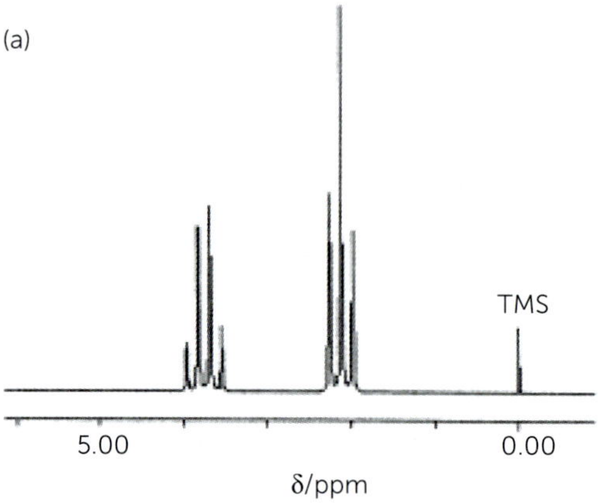

(b)

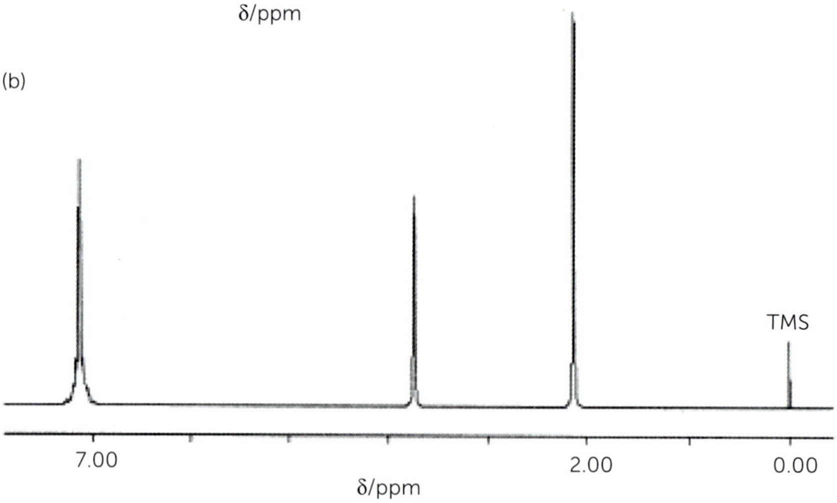

(c)

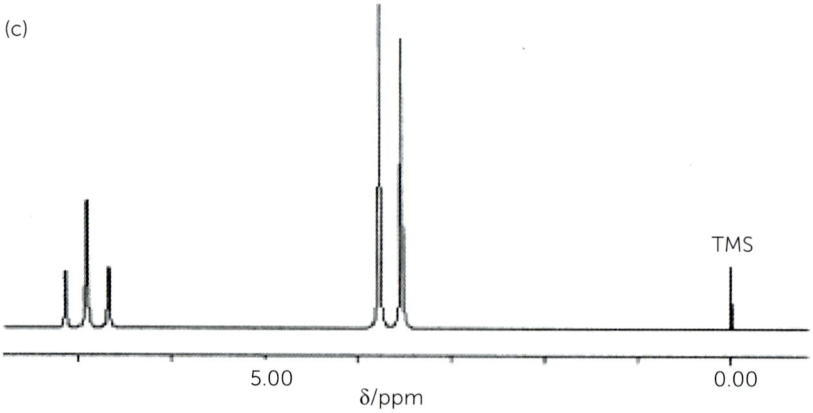

In 'whole body MRI systems', the patient lies inside a tube with the main magnetic field lying along the longitudinal axis of the body, and the radio field acting upwards. Smaller magnetic fields are used to select a 'slice' of body for examination. Such instruments are able to produce 'pictures' of the chemical compositions of slices of human tissue (including organs), with different chemical compositions being highlighted in different shades or colours. The patient is not harmed by the scan. Changes in the composition of tissue are often valuable clinical indicators of a medical condition, and scanners are therefore used in the diagnosis of a range of illnesses.

22.11 ^{13}C-NMR spectroscopy

So far, we have discussed ^{1}H-NMR spectra. NMR spectroscopy can also be employed to investigate the surroundings of nuclei other than hydrogen. ^{13}C-NMR is used to identify the carbon nuclei in different environments in a molecule. Like ^{1}H, the ^{13}C isotope also has spin. Both ^{1}H-NMR and ^{13}C-NMR are often used on the same molecule to provide a more comprehensive picture of the structure.

Key differences between the two techniques are:

- In naturally occurring carbon, only 1.1 per cent is ^{13}C – the other isotopes (^{12}C, ^{14}C) are not magnetically active and do not provide NMR signals. Compare this with hydrogen, where over 99 per cent is ^{1}H. This means that ^{1}H-NMR is a much more sensitive technique than ^{13}C-NMR.
- ^{1}H and ^{13}C nuclei absorb radio-frequency radiation of different energies, and the range values of their chemical shifts are very different. The spectra have to be run separately – they cannot be done at the same time.
- Because ^{1}H and ^{13}C both have spin, they can couple to each other and an adjacent hydrogen can split a ^{13}C signal, much as splitting occurs in ^{1}H-NMR spectra. However, ^{13}C spectra are often recorded **as proton decoupled**. In simple terms, this involves irradiating the sample so that the spins of the protons are randomized – now they do not couple to ^{13}C. This makes the spectra simpler and much easier to understand (we do not tend to see coupling between ^{13}C nuclei, because of the low abundance of this isotope).

Look at the ^{13}C-NMR spectrum of ethyl ethanoate in Figure 22.35. There are four carbons in different chemical environments and each one is represented by a single line. Notice that two lines represent the carbons in methyl groups – these carbons are not in *identical* environments; one is also attached to a carbonyl group, whilst the other is attached to $-CH_2$. So which is which?

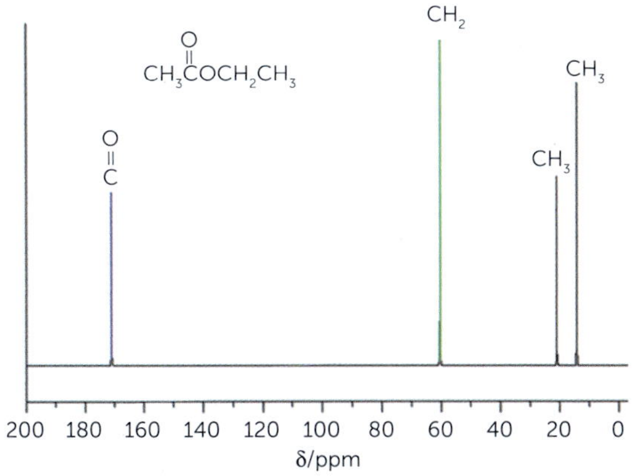

Figure 22.35 ^{13}C-NMR spectrum of ethyl ethanoate.

Just as in ^{1}H-NMR, the external magnetic field experienced by the carbon nucleus depends upon the electronegativity of the atoms near to it. The carbon attached to the carbonyl group experiences a 'pull' of electron density towards the carbonyl group and is deshielded as compared with the carbon attached to $-CH_2$, so it appears at a higher value of δ.

Table 22.3 Typical values for chemical shifts of some carbon nuclei (bold) in ^{13}C-NMR

Carbon environment	Approximate chemical shift δ/ppm
R**C**H_3	10–15
CH_3CO-	20–30
R_3**C**H	25–35
R**C**H_2NH_2	37–45
R**C**H_2OH	50–65
-**C≡C**-	67–85
>**C=C**<	115–140
aromatic ring **C**	125–150
>**C**=O (acids)	170–185
>**C**=O (esters)	170–185
>**C**=O (aldehydes)	190–200
>**C**=O (ketones)	205–220

DID YOU KNOW?

Proton NMR studies of a type of penicillin were reported in 1961. Since then, NMR has become essential in helping to understand the structure of large organic molecules.

EXERCISE **22R**

Using ^{13}C-NMR spectra

Match the following compounds with their ^{13}C-NMR spectra.

A

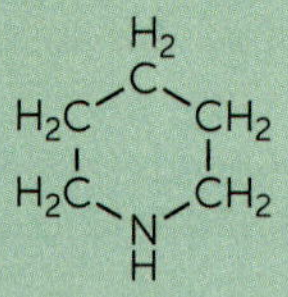

B

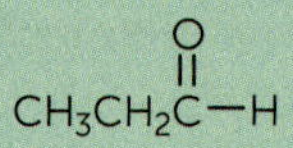

C

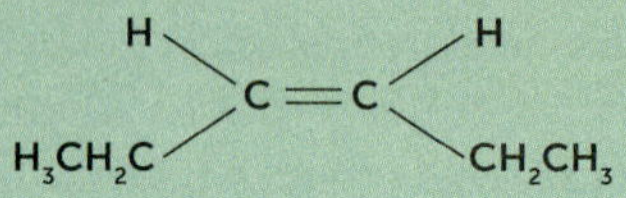

D

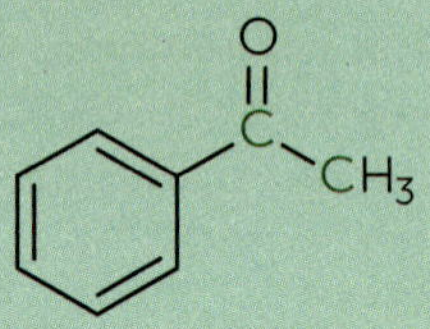

I

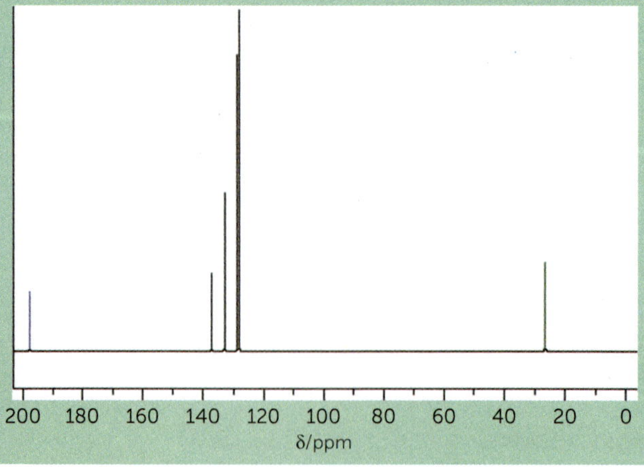

II

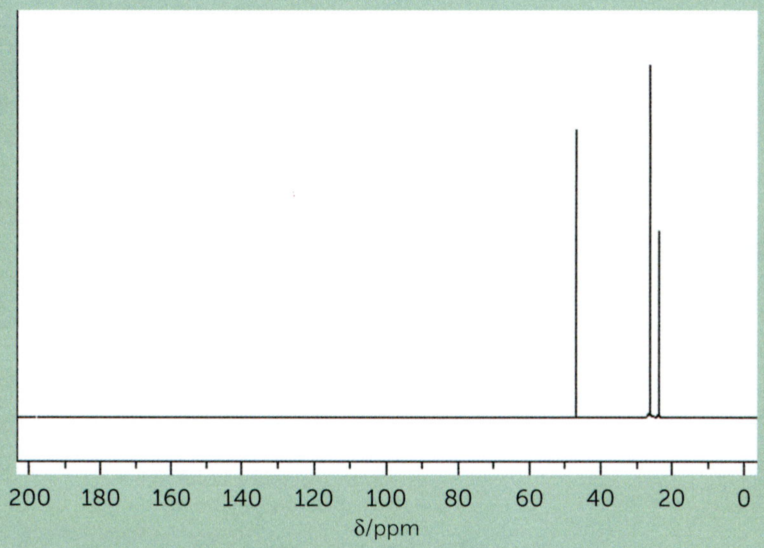
200
180
160
140
120
100
80
60
40
20
0
δ/ppm

III

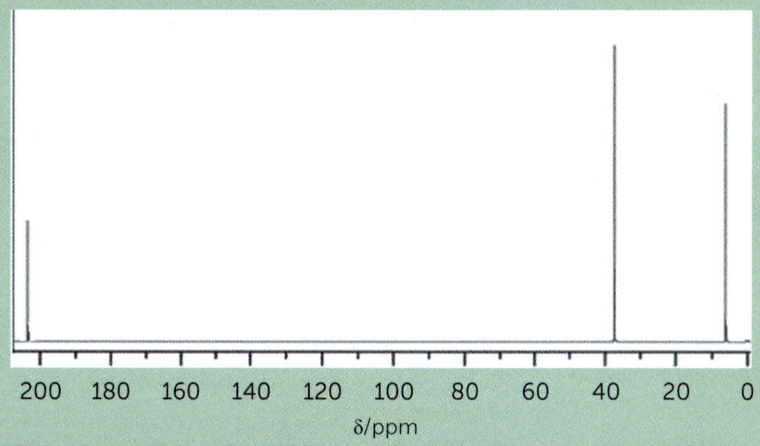
200
180
160
140
120
100
80
60
40
20
0
δ/ppm

IV

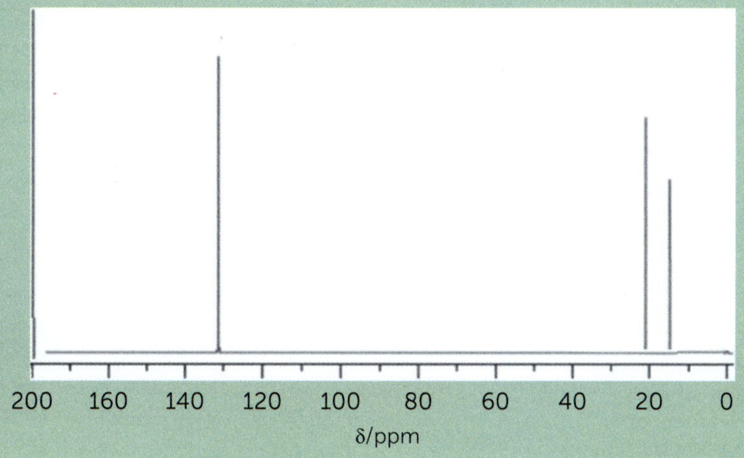
200
160
140
120
100
80
60
40
20
0
δ/ppm

REVISION QUESTIONS

22.1 Convert the following to frequencies (Hz):
(i) 2×10^{-7} m,
(ii) 500 nm,
(iii) 1000 cm^{-1}.

22.2 Complete the table below (the first entry has been completed).

Definition	Word
Gain of energy of an atom or molecule by collision with a photon	absorption
Loss of energy by an atom or molecule by the giving out of light	
The logarithm of the ratio of the incident to transmitted intensities	
The production of light by chemical reaction	
The absorbance of a sample at a wavelength, at unit pathlength and unit concentration	

22.3 Calculate the energy of 1 mol of photons of light with an equivalent wavelength of 1,000 nm.

22.4 The energy gap between two energy levels in a molecule is 30 kJ mol^{-1}. What is the frequency of the light required to be absorbed or emitted in a transition?

22.5
(i) In a spectrometer, 84 per cent of the incident light at 240 nm is absorbed by a solution. What is the absorbance of the solution?
(ii) The molar absorption coefficient of a dye at 450 nm is 1000 mol^{-1} m^2. Calculate the percentage of light transmitted through exactly 1 cm of dye solution of concentration 1.00×10^{-5} mol dm^{-3} at the same wavelength. By how much will the dye solution need to be diluted to give a solution with an absorbance of 0.010?

22.6 The (first) standard ionization energy of Na(g) is 494 kJ mol^{-1}. What wavelength of light would need to be absorbed in order to ionize sodium atoms?

22.7 The energy E (in kJ mol^{-1}) of the electronic levels in the hydrogen atom are easily calculated using the equation

$$E = -\frac{1310}{n^2}$$

where n is the principal quantum number. (Test this formula by applying it to $n = 2$. As Figure 22.7 shows, $E = -328$ kJ mol^{-1}.)
(i) Calculate the energy of the $n = 50$ level.
(ii) What frequency of light will be emitted in the transition $n = 50 \rightarrow n = 3$?

22.8
(i) The sun consists of an incredibly hot centre (above 10^6 K) surrounded by a relatively cool layer at about 5,800 K. If the sun's light is examined by a hand spectroscope, the solar spectrum is seen to contain some dark lines, Fraunhofer lines, superimposed upon a continuous blend of 'rainbow' colours. What causes the dark lines?
(ii) The following diagram shows the simplified electronic energy levels of an atom or molecule:

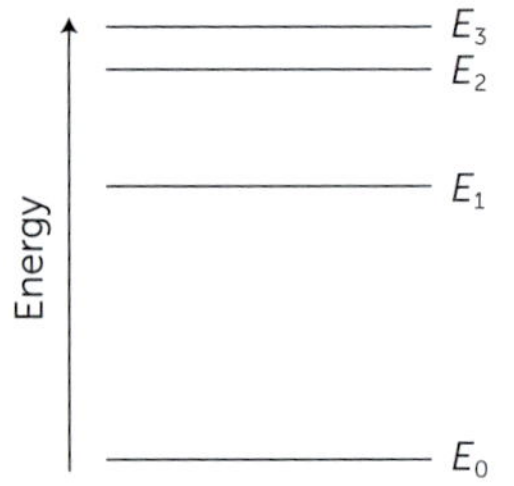

Draw in all the possible transitions corresponding to absorption and emission at room temperature. Use this diagram to explain why the emission spectra of atoms and molecules may contain more lines than their absorption spectra.

22.9
(i) Sketch the expected absorption spectrum for a red filter.
(ii) Study the following UV–visible spectrum of a compound and predict the colour of its solution.

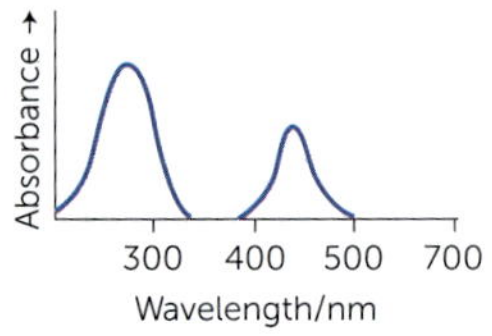

(iii) The colour of paints may also be explained using the colour cheese to predict which frequencies of light are absorbed and which are reflected. Explain why a black colour results when lots of different paints are mixed.

22.10 The following table lists the molar absorption coefficient ε, of ozone, O_3(g), at 298 K in part of the ultraviolet region.

Wavelength/nm	Molar absorption coefficient/mol^{-1} m^2
210	11
220	35
230	101
240	204
250	298
254	369
260	298
270	204
280	101
290	35
300	11

Plot ε against wavelength. Calculate the absorbance at 254 nm of a 10.0 cm gas cell containing pure ozone at a pressure of 100.0 Pa at 298 K. (Hint: Use the ideal gas equation to convert gas pressure to gas concentration.)

22.11 The simplified mass spectrum (a) and infrared spectrum (b) of an unknown compound are shown below.

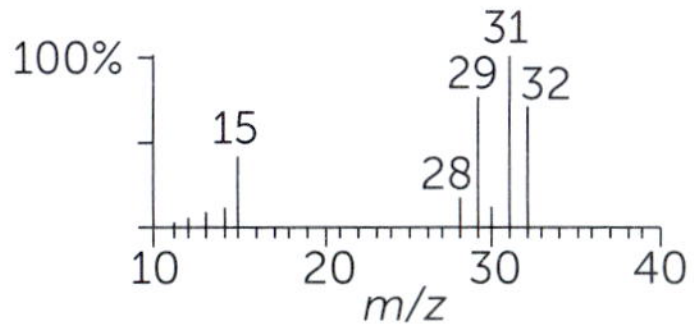

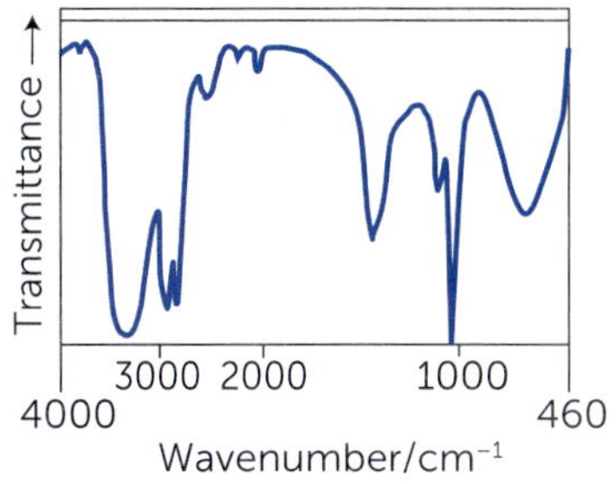

Additional information:

(a) The low-resolution ^{1}H-NMR spectrum (not shown) contains two signals.

(b) The compound contains 37.5 per cent C and 12.5 per cent H; testing for other elements was not carried out.

Suggest a structure for the unknown compound.

22.12 Hydrocarbon concentrations in water can be measured by extracting the hydrocarbons from the water with a suitable solvent (such as trifluorotrichloroethane (CF_3CCl_3)), and subsequently measuring the absorbance of the hydrocarbon solution in the C–H stretch region at 2860 cm^{-1}.

(i) Why should chloroform $CHCl_3$ not be used as the solvent in such measurements?

(ii) In an experiment, octane from a water sample was completely extracted using an equal volume of CF_3CCl_3 solvent. The absorbance of octane in solution in a 20 mm cell was found to be 0.050 at 2860 cm^{-1}. Given that $\varepsilon(C_8H_{18})$ in $CF_3\,CCl_3$ at 2860 cm^{-1} is 5.0 mol^{-1} m^2, calculate the concentration of octane in the original water sample in (a) mol m^{-3}, and (b) mg dm^{-3}.

22.13 Chromate(VI) ions (CrO_4^{2-}(aq)) may be completely reduced to chromium(III) ions (Cr^{3+}(aq)) in the presence of thiourea catalyst and a sodium ethanoate–ethanoic acid buffer. The reaction may be simply expressed as

$$CrO_4^{2-}(aq) + 8H^+(aq) + 3e^- \rightarrow Cr^3(aq) + 4H_2O(l)$$

(i) By working out the oxidation numbers of chromium in the two species, show that reduction has taken place.

(ii) At 440 nm, ε (CrO_4^{2-}(aq)) = 200.0 mol^{-1} m^2 and ε(Cr^{3+}(aq)) = 1.0 mol^{-1} m^2. In an experiment, a solution initially containing 5.0×10^{-4} mol dm^{-3} chromate(VI) ion was completely reduced to chromium(III) ions. Assuming a cell pathlength of 1.0 cm, what will be the initial and final absorbance of the reaction mixture at 440 nm? (Assume that the thiourea and buffer do not absorb at 440 nm.)

22.14 Flame atomic absorption spectroscopy (AAS) is a specialized spectroscopic technique widely used in analytical laboratories to find the levels of trace metals in solution. By consulting a textbook on analytical chemistry, find out **(i)** what light sources are used in AAS, and **(ii)** the purpose of the flame.

22.15 Identify the compound, of molecular formula $C_2H_4Br_2$, whose ^{1}H-NMR spectrum is shown below:

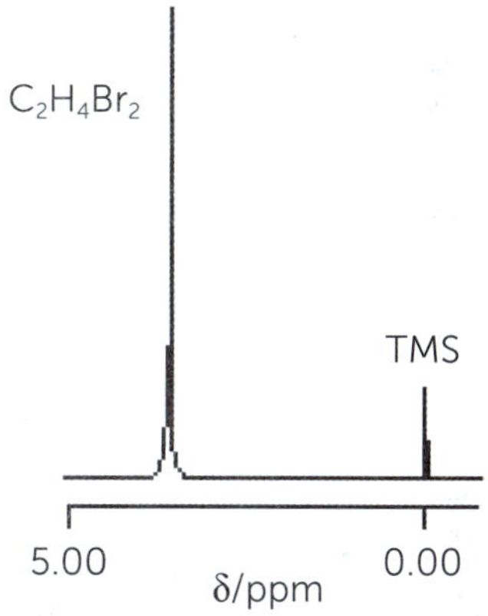

22.16 A compound contains C, H and O. Can you work out its identity from the following spectra?

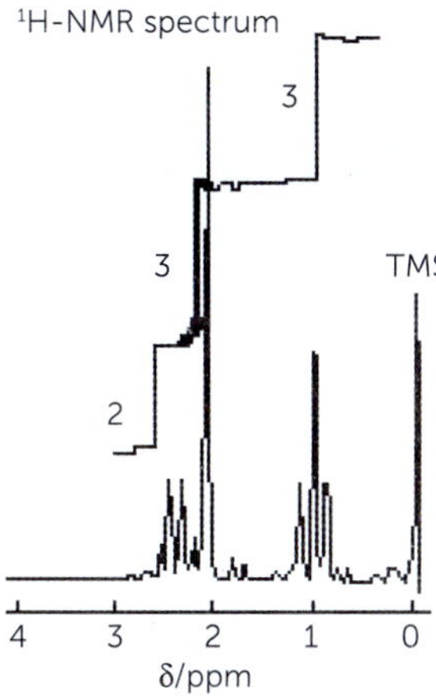

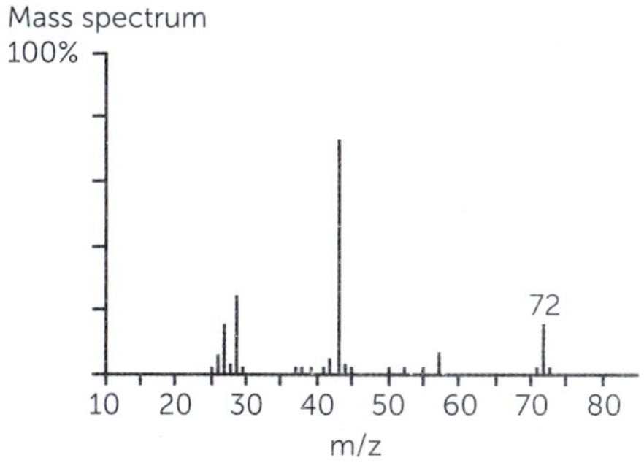

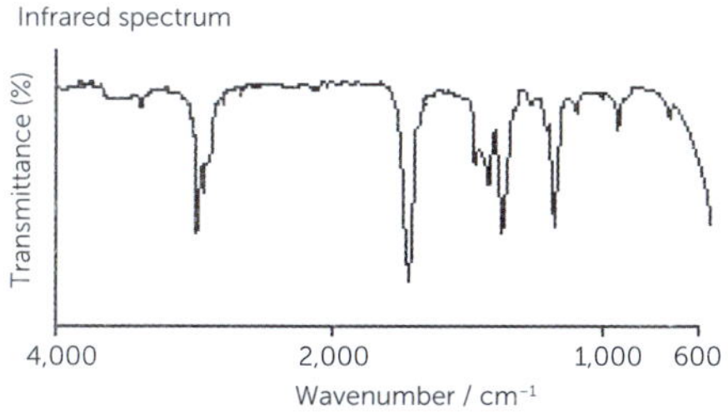

22.17

Four isomeric alcohols have the formula $C_4H_{10}O$.

(i) Write structural formulae for each of them.

(ii) Which one would be represented in the following ^{13}C-NMR spectrum?

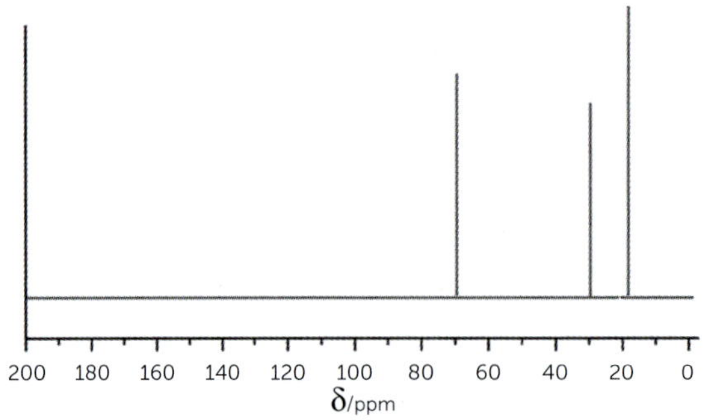

22.18 Should mass spectrometry be regarded as a type of spectroscopy?

22.19 Explain why the emission intensity of a firefly increases between 28 and 45°C but decreases at higher temperature.

DID YOU KNOW?

Although water appears transparent much of the time, it absorbs very weakly in the blue region and as predicted by the Beer–Lambert Law, absorbance is only appreciable (and its true colour can only be seen by the eye), if the pathlength is greatly increased. This is accomplished by shining a white light (positioned at one end of the tube) through a 3 metre glass tube containing water and sealed at both ends with quartz windows (Figure 22.36(a)).

When an observer looks through the tube, the water appears slightly blue (Figure 22.36(b)). If the experiment is repeated with 'heavy water', water in which the hydrogen atoms are replaced with deuterium, the water is colourless even with the longer pathlength because the heavier $^{2}_{1}H$ atoms alter the wavelength of the allowed transitions so that they only involve light outside the visible light range.

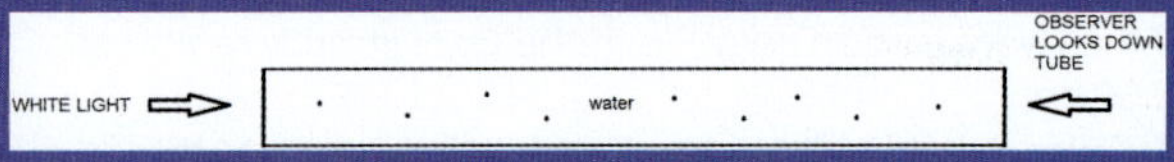

(a)

(b)

Extension 22 on the website at https://www.bloomsburyonlineresources.com/chemistry contains an article on the use of spectroscopy in astronomy..

Nuclear and Radiochemistry

23

OBJECTIVES

This unit:

- Defines radioactivity
- Looks at the properties of nuclear radiation
- Gives examples of calculations involving radionuclides and isotopic half-lives

CONTENTS

The **radioactivity** of naturally occurring and artificially produced isotopes, **nuclear fission** (the reactions involved in commercial nuclear reactors) and **nuclear fusion** (the source of the sun's power) are all examples of nuclear reactions (Figure 23.1).

Unlike chemical reactions, in which atoms are simply *rearranged*, all the nuclear reactions discussed here involve the production of nuclei with different atomic numbers, and therefore of *new elements*. Nuclear reactions also involve much greater energy changes per atom than is the case for chemical reactions and the energy changes may be calculated using the famous Einstein equation $E = mc^2$.

The first type of nuclear reaction to be discovered was natural radioactivity, which was initially regarded as a mere curiosity. From these humble beginnings came nuclear generated electricity power and the awesome power of nuclear warheads.

Nuclear reactions
- Nuclear fusions
- Nuclear fissions
- Radioactive decay

Figure 23.1 The main types of nuclear reaction.

Marie Curie (1867–1934), who carried out pioneering research into radioactivity. She earned two Nobel Prizes, in 1903 and 1911. Her daughter, Irene Joliot-Curie, also won the Nobel Prize, in 1935.

23.1 Radioactivity

Radioactivity was discovered accidentally in 1896, when Antoine Henri Becquerel (1852–1908) noticed that a uranium ore caused a photographic plate to blacken. The uranium gave off invisible 'rays' which were similar to the X-rays discovered the previous year. Forty-nine years later, doctors in Hiroshima reported that previously unexposed X-ray plates in hospital vaults had been fogged by the radiation emitted by the first atomic bomb.

Pioneering work in radioactivity was carried out by the husband and wife team of Pierre and Marie Curie, who (in 1898) extracted the elements polonium and radium from an ore called pitchblende. The harmful effects of nuclear radiation were then unknown, and even today the Curies' laboratory notebooks remain dangerously radioactive.

This is the internationally accepted symbol warning for ionizing radiation. (Image by Cary Bass).

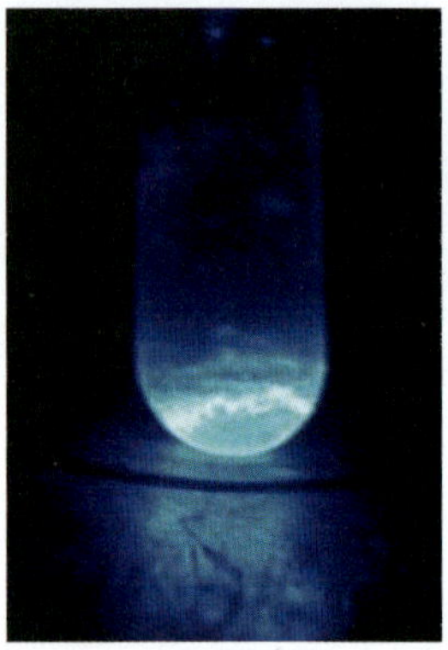

The isotope einsteinium 253 (atomic number 99, half-life 20.5 days) does not occur in nature and is produced in dedicated nuclear reactors. It is such an intense emitter of alpha radiation that it glows in air. This sample contains only 0.3 mg of ^{253}Es. (Image by R. G. Haire, US Department of Energy).

Definition of radioactivity

Atomic nuclei are said to be **unstable** when they *spontaneously disintegrate* and simultaneously **emit nuclear radiation**. This phenomenon is called **radioactivity** or radioactive decay. The key ideas in this definition are:

1 Spontaneous

Radioactivity is thermodynamically spontaneous. It does not require any help to start or to continue. Because radioactivity results from the spontaneous disintegration of a nucleus, no collisions are involved and so – in contrast to chemical reactions – the rate of radioactive decay is unaffected by temperature, pressure or by the presence of other atoms and it cannot be catalysed. For example, uranium-235 decays at the same speed (and into the same products) whether it is pure uranium, combined as uranium oxide ($^{235}UO_2$) or as uranium fluoride ($^{235}UF_6$). Radioactive decay is a statistical event and although we cannot say which atom of (say) U-235 in a sample will decay in the next second, the overall rate of decay of the trillions of U-235 atoms in the sample will occur at a remarkably constant rate – the sample has its own 'internal clock'.

2 Disintegration

In radioactive decay, one **nuclide** changes into another. We can represent this as

parent nuclide → daughter nuclide + nuclear radiation

where the term 'nuclide' means an **atom** (and not a nucleus) of a particular isotope. As a result of this change, the number of protons or neutrons (or both) in the parent nuclide changes. Where the number of protons changes, a new element is formed.

3 Nuclear radiation

'Nuclear radiation' originates from within the nucleus that is undergoing change. It includes particles, such as alpha particles and high-energy electrons and also high-frequency light (gamma rays). See Table 23.1 and Figure 23.2.

Table 23.1 The main types of nuclear radiation

Name	Symbol	Nature	Penetration
Alpha particle only	$^{4}_{2}He^{2+}$ or α	helium nuclei (i.e. helium ions)	relatively massive but travel only a few cm in air and are stopped by thin card
Beta particle	$^{0}_{-1}e^{-}$ or β	highly energetic (and fast moving) electrons	travel up to 5 m in air; stopped by thin metal plate
Positron	$^{0}_{-1}e^{+}$ or β^{+}	highly energetic positively charged electrons	as beta particle
Gamma rays	γ	very short wavelength electromagnetic radiation; travels at the speed of light	very penetrating; stopped by thick lead plate and/or thick concrete shielding
Neutrons*	$^{1}_{0}n$	relatively massive particles	as gamma rays
Fission fragments*	–	highly ionized nuclei of medium and large atoms, e.g. U^{20+}	relatively massive and possess enormous energies; stopped only by thick lead and thick concrete shielding

* Produced during nuclear fission.

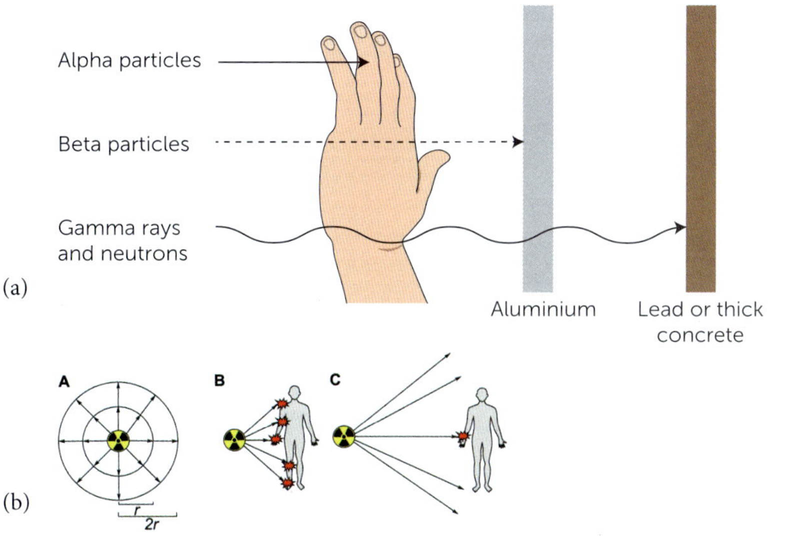

Figure 23.2 (a) Penetrating powers of nuclear radiations; (b) Inverse square law. The measured intensity of all types of nuclear radiation falls with the square of the distance, *r*, from the radioactive source (**A**). The dangers of exposure to a source (**B**) are reduced with increased distance (as in (**C**)). (Courtesy of American Chemical Society, *J. Chem. Educ.* 2021, 98, 12, 3831–40.)

23.2 Radionuclides and radioisotopes

The term **radionuclide** means an atom that undergoes radioactive decay; such an atom is said to be (radioactively) **unstable**. Isotopes which are radioactive are called **radioisotopes**.

There are about 350 different nuclides (including all the isotopes of the elements) in nature. Of these, fifty-five are *radio*nuclides. In addition, there are hundreds of radionuclides that have been artificially manufactured, mainly by smashing nuclei together at very high speed in giant devices called **particle accelerators**. For example, in 1937, technetium (Tc) was made by firing deuterium atoms at molybdenum (Mo) atoms:

$$^{96}_{42}\text{Mo} + ^{2}_{1}\text{H} \rightarrow ^{97}_{43}\text{Tc} + ^{1}_{0}\text{n}$$

Elements with atomic numbers greater than that of uranium (92) are known as **transuranium elements**. They do not occur naturally and have all been made in particle accelerators or in nuclear reactors.

Table 23.2 shows the half-lives and the mode of decay of selected radionuclides.

The Aboriginal people of Australia were familiar with uranium ore, which they called 'sickness rocks' because they realized that if mishandled it could cause illness. Naturally occurring uranium is chemically poisonous because it interferes with the chemistry of the body when ingested or inhaled. It is also poisonous because it is mildly radioactive: it emits nuclear radiation and prolonged exposure damages the cells of the body (radiation poisoning).

Table 23.2 Data on selected radionuclides

Nuclide	Main type of radiation emitted	Half-life
$^{137}_{55}\text{Cs}$	β, γ	30.17 years
$^{14}_{6}\text{C}$	β	5730 years
$^{131}_{53}\text{I}$	β, γ	8.04 days
$^{226}_{88}\text{Ra}$	α, γ	1600 years
$^{60}_{27}\text{Co}$	β, γ	5.27 years
$^{241}_{95}\text{Am}$	α, γ	432 years
$^{90}_{38}\text{Sr}$	β	28 years
$^{32}_{15}\text{P}$	β	14.3 days

Nuclide	Main type of radiation emitted	Half-life
$^{40}_{19}K$	β, γ	1.25×10^9 years
$^{238}_{92}U$	α	4.46×10^9 years
$^{235}_{92}U$	α	7.04×10^8 years
$^{3}_{1}H$ (tritium, T)	β	12.26 years
$^{222}_{86}Rn$	α, γ	3.82 days
$^{35}_{16}S$	β	87.2 days
$^{239}_{94}Pu$	α, γ	2.41×10^4 years

23.3 More about nuclear radiation

Alpha particles

Alpha particles are doubly ionized helium atoms, i.e. helium nuclei. Since a helium nucleus consists of two protons and two neutrons, the loss of an alpha particle from a nucleus reduces the number of protons by two and the number of neutrons by two. Therefore the mass number *A* falls by four and the atomic number *Z* falls by two. The emission of alpha particles is represented by the general equation

$$^{A}_{Z}X \rightarrow {}^{A-4}_{Z-2}Y^{2-} + {}^{4}_{2}He^{2+}$$

where X is the parent nuclide and Y is the daughter nuclide. Both Y^{2-} and $^{4}_{2}He^{2+}$ become electrically neutral by losing or gaining electrons from the container or the surrounding air.

An example of alpha emission is provided by the decay of radium-226:

$^{226}_{88}Ra$	$\longrightarrow$	$^{222}_{86}Rn^{2-}$	+	$^{4}_{2}He^{2+}$
88 protons		86 protons		2 protons
138 neutrons		136 neutrons		2 neutrons
88 electrons		88 electrons		0 electrons

Notice that the total charge, the sum of the mass numbers and the sum of the atomic numbers, are equal on both sides. This is true of all nuclear equations.

Alpha particles are so bulky that they do not usually travel more than 10 cm in air. They are easily stopped by card, brick or thin metal sheet. However, their mass and speed (typically about 10^7 m s^{-1}) mean that they possess considerable kinetic energy and so cause substantial ionization, and if brought into close contact with human tissue (for example, by ingestion) they usually cause much more damage than beta or gamma radiation.

EXERCISE **23A**

Alpha decay

Complete the equation

$$^{238}_{92}U \rightarrow ?? + {}^{4}_{2}He^{2+}$$

Beta particles

Beta particles are high-energy electrons that are ejected from the nucleus. Since there are not normally any electrons in the nucleus, the beta particles must have been produced in the nucleus during radioactive decay. It is now known that during this process, a neutron changes into a proton and an electron. From this it follows that the atomic number (the number of protons) of the nuclide increases by one, while the mass number (the number of protons and neutrons) remains the same. The electron escapes from the nucleus and is now called a beta particle. The general equation is

$$^{A}_{Z}X \rightarrow {}_{Z+1}^{\ \ A}Y^{+} + {}^{\ 0}_{-1}e^{-}$$

where the atomic number of the electron is (for the purposes of a balanced nuclear equation) regarded as −1. An example of beta emission is the radioactive decay of strontium-90:

$^{90}_{38}Sr$	$\longrightarrow$	$^{90}_{39}Y^{+}$	+	$^{0}_{-1}e^{-}$
38 protons		39 protons		0 protons
52 neutrons		51 neutrons		0 neutrons
38 electrons		38 electrons		1 electron

The positively charged yttrium ion will gain an extra electron from the surroundings.

Beta particles vary greatly in their energies and speeds, and (therefore) in their penetration of matter. Some beta particles travel at speeds of 99 per cent of the speed of light, and as a consequence penetrate several metres of air.

EXERCISE **23B**

Beta decay

Complete the equation

$^{215}_{83}Bi \rightarrow ?? + ^{0}_{-1}e^{-}$

DID YOU KNOW?

The murder of Alexander Litvinenko

Former spy Alexander Litvinenko was murdered by radioactive poisoning in London in 2006. His killers used a minute quantity (about 10 millionth of a gram = 10 μg) of an intense alpha particle emitter, polonium-210 (≈ 140 days):

$$^{210}_{84}\mathbf{Po} \rightarrow ^{206}_{82}\mathbf{Pb}^{2-} + ^{4}_{2}He^{2+}$$

Polonium is not chemically toxic and if ingested its danger arises from the intense alpha radiation it produces. The Po-210 was administered in Litvinenko's tea. Doctors failed to diagnose radioactive poisoning until a few hours before Litvinenko's death, previously believing that the toxin was the metal thallium. Once suspected, intensely radioactive materials are easily located even if their concentrations are tiny, and detectives used alpha particle monitors to detect Po-210 in several London locations linked to the murder. The murder was the subject of a public enquiry chaired by Sir Robert Owen, the results of which were published in January 2016.

EXERCISE **23C**

Alpha and beta decay

An element P radioactively decays to an element Q with the emission of an alpha particle. Q decays to element R by beta emission and R decays to element S by beta emission. What is the relationship between P and S?

Gamma radiation

Gamma radiation is very short wavelength electromagnetic radiation, very similar to X-rays. Their wavelengths are usually 10^{-10}–10^{-13} m, corresponding to about 10^{6}–10^{9} kJ per mole of photons. Gamma rays are usually emitted by radionuclides following the emission of alpha or beta particles.

EXERCISE **23D**

Energy of gamma rays

Confirm that a gamma ray wavelength of 10^{-10} m corresponds to an energy of 10^{6} kJ per mole of photons.

23.4 Mathematics of radioactive decay

The number of atoms of a radionuclide N_t remaining after an interval of t seconds may be calculated from the formula

$$N_t = N_0 \times e^{-kt}$$

where N_0 is the number of atoms at $t = 0$ and k is the first-order rate constant (often called the **decay constant**) with the units of s^{-1}.

Since the number of atoms of a radionuclide is proportional to its mass, N_t and N_0 may be replaced by the mass of radionuclide at time t (symbolized m_t) and $t = 0$ (symbolized m_0) respectively:

$$m_t = m_0 \times e^{-kt}$$

These equations show that the fall in the number of atoms of a radionuclide with time is exponential (Figure 23.3(a)). In fact, any quantity that is proportional to the number of atoms (such as the number of counts per minute registered on a Geiger counter) may also be substituted in these equations.

The **half-life** ($t_{1/2}$) of a radionuclide is the time that elapses before half the nuclei of the radionuclide have disintegrated. The factor k is related to the half-life of the decay by the equation

$$t_{1/2} = \frac{0.693}{k}$$

(see page 254). If k is required in the units of s^{-1}, $t_{1/2}$ must be in the units of seconds.

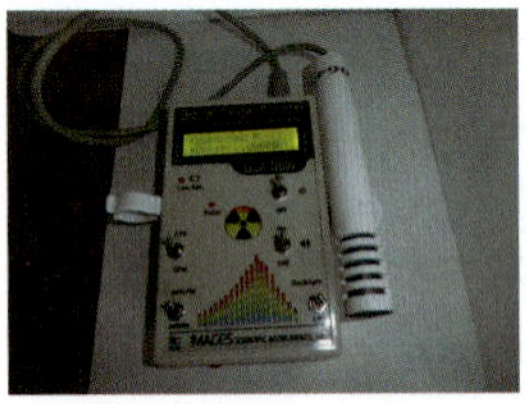

A Geiger counter is used to detect nuclear radiation. The detector tube is on the right.

> **EXERCISE 23E**
>
> **Half-life 1**
>
> Sketch a graph showing the decay of 1 g of uranium-238 on the same time scale as shown in Figure 23.3.

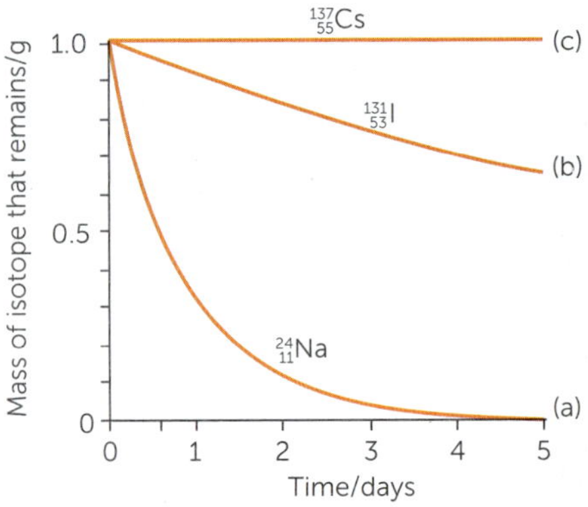

Figure 23.3 The decay of 1 g of each of three radionuclides: curve (a) $^{24}_{11}$Na ($t_{1/2}$ = 15.0 h); curve (b) $^{131}_{53}$I ($t_{1/2}$ = 8.04 days); curve (c) $^{137}_{55}$Cs ($t_{1/2}$ = 30.17 yr).

Key points about half-life

- A long half-life indicates that it will be a long time before the radioactivity of a sample containing atoms of that nuclide fades to a near-zero level. Such a radionuclide is said to be **long-lived.** Similarly, a radionuclide with a short half-life is said to be **short-lived**. Most of the world is slowly becoming less radioactive, as the radioactive elements brought together when the Earth was formed (about 4.5 billion years ago) decay away.

- Short-lived radionuclides emit nuclear radiation more rapidly than long-lived radionuclides, and a sample of a short-lived radionuclide will (statistically) undergo more disintegrations per second than the same number of nuclei of a stable radionuclide. It follows that short-lived radionuclides give out *more intense* nuclear radiation than long-lived radionuclides, but for a shorter time.
- The disintegration of radionuclides is *random*. Think of 100 radium-226 atoms decaying with a half-life of 1,600 years. Although it is statistically likely that fifty Ra atoms will have disintegrated after 1,600 years, we cannot predict *which* individual Ra atoms *will* disintegrate during this time.

Plot of radionuclide mass against time

Figure 23.3 shows the fall in mass of samples of three radionuclides (with an initial mass of 1 g each) over five days, calculated using the equation $m_t = m_0 \times e^{-kt}$.

- All three curves are drawn on the same scale. The longest lived radionuclide ($^{137}_{55}Cs$) has undergone a negligible amount of decay over five days.
- In general, the early part of an exponential plot is straight (linear).
- 99.9 per cent of any radionuclide has decayed after ten half-lives has elapsed.

EXERCISE **23F**

Half-life 2

(i) In nature, uranium atoms consist of approximately 0.7 per cent uranium-235 and the rest is uranium-238. From Table 23.2, explain why the percentage of uranium-235 was much higher than this millions of years ago.

(ii) $^{201}_{87}Fr$ decays by alpha emission, with a half-life of 0.0048 s. $^{231}_{90}Th$ decays by beta emission, with a half-life of 25 h.

(a) Write equations showing the decay of Fr and Th. What elements are produced in the decay of francium and thorium, respectively?

(b) Suppose that a mixture contains 0.0100 g of $^{201}_{87}Fr$ and 0.0100 g of $^{231}_{90}Th$. How much of each radionuclide will be left after 100 h?

Nuclear fission and nuclear fusion

Nuclear fission (sometimes called *induced* nuclear fission) is the thermodynamically spontaneous reaction of a relatively 'slow' neutron with a suitable and *heavy* nucleus, such as that of U-235:

$$^{235}_{92}U + ^{1}_{0}n(\text{slow}) \rightarrow ^{139}_{54}Xe + ^{95}_{38}Sr + 2^{1}_{0}n \qquad \textbf{(23.1)}$$

The target ^{235}U nuclide is broken roughly in half, producing two new nuclei (new elements, specifically intensely radioactive isotopes of xenon and strontium), more neutrons and vast amounts of thermal energy (exothermic process). The fact that more neutrons are produced than consumed allows nuclear fission to undergo a chain reaction but since fission requires a collision between a neutron and U-235 nucleus, there is scope to control the rate of fission by making it harder for the neutron to successfully strike the target ^{235}U. Such control is essential for the safe running of a nuclear (fission) power station; in nuclear fission bombs ('atomic bombs'), the chain reaction is intentionally *un*controlled. Once a sample of U-235 has undergone fission, the sample becomes contaminated by-product nuclides with short half-lives, so that once fission is started in a nuclear power station, the fuel mixture is intensely radioactive and so requires thick concrete shielding to protect workers.

In nuclear *fission*, reaction (23.1), the lighter product nuclei are more thermodynamically stable than the target (U-235) nucleus. In cases where nuclear *fusion* is thermodynamically spontaneous (reaction (23.2)), stability is achieved through the coalescing of two nuclei. Experiments show that nuclear fusion is only spontaneous when it involves *light* nuclei, such as deuterium and tritium (both isotopes of hydrogen) to make a heavier nucleus:

$$^{2}_{1}\mathrm{H} + ^{3}_{1}\mathrm{H} \rightarrow ^{4}_{2}\mathrm{He} + ^{1}_{0}\mathrm{n} \quad \textbf{(23.2)}$$

Nuclear fusion is massively exothermic and it is responsible for the energy produced by the sun and in the most powerful nuclear weapons (the 'hydrogen bombs'). Perhaps, in the future, nuclear fusion will be used to generate electricity on Earth.

The applications of radionuclides, including a detailed discussion of nuclear power, are to be found in **Extension 23** on the website at https://www.bloomsburyonlineresources.com/chemistry

Applications of radioisotopes

Apart from their use in nuclear power, radionuclides are used as means of dating artefacts (e.g. C-14 dating), in the treatment of cancer, in measuring the thickness of metal and as radioactive tracers (e.g. in following the pathway of drugs throughout the body).

Nuclear fission provides about 20 per cent of the electricity generated within the US, 65 per cent in France and 15 per cent within the UK. Economic worries about the decommissioning of old nuclear stations, and major accidents, such as Windscale, UK (1957), Three Mile Island, USA (1979), Chernobyl, Ukraine (1986) and Fukushima, Japan (2011), have caused many people to question whether or not more nuclear plants should be built but in modern times nuclear power has been proposed as an important way to help decarbonize electricity production. Nuclear power is regarded as green energy but it is not renewable energy because the uranium fuel is consumed. The photograph above shows the fourth reactor at the Chernobyl nuclear power plant where an explosion resulted in the world's worst nuclear accident.

Albert Einstein realized that through nuclear reactions even small masses could, in principle, be converted into huge amounts of energy ($E = mc^2$). But if every gram of material contains this tremendous energy, why did this fact go unnoticed for so long? Einstein's answer was simple enough: 'So long as none of the energy is given off externally, it cannot be observed. It is as though a man who is fabulously rich should never spend or give away a cent; no-one could tell how rich he was.'

DID YOU KNOW?

The Radium Girls

Radium dial painters working in a factory.

If radium is mixed with zinc sulfide, the nuclear radiation produced by the radium causes the zinc sulfide to glow in the dark. This effect is known as radioluminescence and the mixture was used in the paint on the dials of watches. In 1917, the first large factory to produce glowing watches was built in New Jersey, USA. The radium was expensive and so the work required young workers with good eyes. All the painters were female and the painters re-shaped their delicate paint brushes by wetting them on their tongue ('lip, dip and paint'), so exposing themselves to dangerous alpha-emitting radium-226. Since radium and calcium are in the same group in the Periodic Table and have similar chemical reactions, the human body incorporates radium in bone, so that exposure to radiation becomes permanent. Unaware of the risk, some workers used the radium paint as a cosmetic, painting their lips, teeth, eyelids and fingernails. Painters began to suffer from 'radium jaw', a collection of symptoms caused by anaemia, bone fractures and a breakdown of the jaw. Initially, the link with radium was denied. The painters started court litigation and the cases were well publicized by the media in the 1930s. Some painters had died by the time the workers won their case.

REVISION QUESTIONS

23.1 In 1919, Lord Rutherford made the first artificial radioisotope by bombarding nitrogen atoms with alpha particles:

$$^{14}_{7}N + ^{4}_{2}He^{2+} \rightarrow ?? + ^{1}_{1}H$$

Complete the equation and identify the radioisotope he made.

23.2 Tritium decays by emitting beta particles with $t_{1/2}$ = 12.26 yr.

(i) Write an equation for the radioactive decay of tritium.

(ii) Tritium is prepared commercially by reacting lithium-6 with neutrons from a nuclear reactor. Write an equation for the reaction.

23.3 The time for 99 per cent of a sample of a radioisotope to decay is given by the equation

$$t_{0.99} = \frac{4.605}{k} \text{seconds}$$

Similarly, the time for 99.9 per cent of a sample of a radioisotope to decay is given by

$$t_{0.999} = \frac{6.908}{k} \text{seconds}$$

A sample of pure polonium-210 ($t_{1/2}$ = 140 days) is used in a laboratory experiment. Calculate the time it takes for (a) 99 per cent, (b) 99.9 per cent of the mass of the polonium to decay.

23.4 Radon (specifically, Rn-222) is a colourless, odourless radioactive gas formed by the radioactive decay of the tiny amounts of uranium found naturally in all rocks and soils. Rn-222 decays via alpha emission into Po-218; it has a half-life of 3.8 days.

(a) Use the Periodic Table to find the atomic number (proton number) of Rn and Po. Write an equation for the alpha decay of Rn-222.

(b) 4.0 mg radon is trapped in a sealed container for 7.6 days. How much radon is in the box after this interval of time?

23.5

(i) From government publications, obtain a list showing the various contributions to the background radiation to which the population is exposed. What is the biggest source of background radiation?

(ii) List four factors that control the degree of harm caused by nuclear radiation to an individual.

23.6 Contrary to the Law of Conservation of Mass, significant losses in mass are observed in all nuclear reactions, including radioactive decay, nuclear fission and nuclear fusion. Einstein's famous equation, $E = mc^2$, allows a calculation of the energy (E) produced when a mass (m) is converted into energy. c is the speed of light. (E is measured in joules (J), m in kg and $c = 3 \times 10^8$ m s^{-1}). Use this equation to calculate the energy released when a nuclear fuel rod loses 1 g of mass. Compare this with the amount of heat produced when 1 g of coal burns in air.

For more on nuclear changes and the uses of radionuclides see **Extension 23** on the website https://www.bloomsburyonlineresources.com/chemistry.

The Chemistry of Pollution

24

CONTENTS

OBJECTIVES

This unit:

- Defines the meaning of the word 'pollutant'
- Discusses examples of atmospheric and water pollution
- Explains the difference between 'good' and 'bad' ozone

24.1 Pollutants

A substance becomes a **pollutant** when it is present in a concentration that is high enough for it to have a harmful effect on living things or the natural environment.

Often, only highly toxic substances are thought of as pollutants, but even substances that are normally considered harmless may pollute if they are present in high enough concentrations and in the *wrong place* at the *wrong time*. For example, nitrate fertilizer is added to soil in order to increase plant growth, but an excessive concentration of nitrate present in drinking water can be toxic, especially to young children.

Pollution originates from a **source**. The pollutant is then **transported** by air, or by water or dumped on land. Some of the pollutant may be absorbed (**assimilated**) or **chemically changed** by the environment; the rest builds up to a concentration that enables it to damage organisms or buildings, or to upset the balance of environmental processes (Figure 24.1).

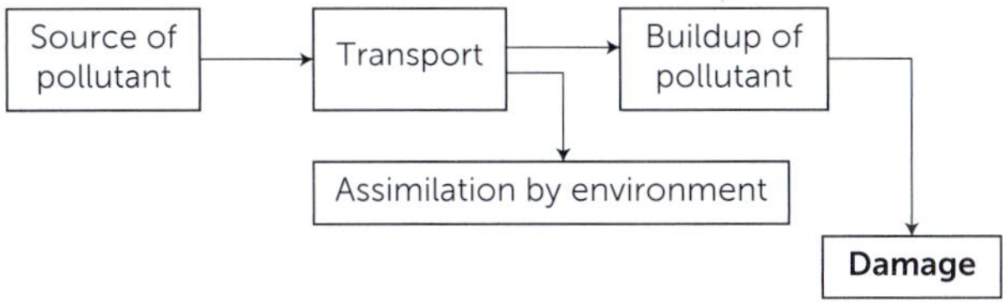

Figure 24.1 An environmental pollution model.

24.2 The depletion of ozone in the stratosphere

We will discuss atmospheric pollution in two parts. First, pollution of the upper atmosphere (specifically the depletion of the ozone layer) by chlorine-containing compounds. Second, pollution in the lower atmosphere, often simply called 'air pollution'.

Ozone depletion in the stratosphere: what's the problem?

Ozone (O_3) exists in the part of the atmosphere called the **stratosphere**, 20–50 km above sea-level. At this altitude, air pressure is one-thousandth of the pressure at sea-level, but a tiny concentration of the molecule ozone (O_3) protects the world by absorbing UV radiation that would otherwise harm living things. In other words, *ozone acts as a global sunscreen*. The effects of ozone loss could include: increased human cataracts and skin cancer, reduction of plankton in ocean waters and the destruction of plants, including crops. Ozone layer destruction in the Antarctic was reported in 1985, a major cause of which was the release of chlorofluorocarbon compounds (CFCs or 'freons') such as CCl_2F_2. These compounds are almost chemically unreactive, non-toxic and odourless, properties that at one time caused them to be used as solvents, aerosol propellants, refrigerant fluids and blowing agents for expanded plastic foams. They are so chemically stable, however, that they persist in the atmosphere for years and eventually move up to the stratosphere, where they are broken down by the powerful UV radiation emitted by the sun. Their decomposition products can then destroy ozone as follows:

1 The decomposition of the chlorofluorocarbon produces a highly reactive chlorine atom (**free radical**). This reacts with ozone:

$$Cl\bullet + O_3 \rightarrow ClO\bullet + O_2 \qquad \textbf{(24.1)}$$

2 Oxygen molecules split up into atomic oxygen:

$$O_2 + h\nu\ (<240\ \text{nm}) \rightarrow 2O\bullet \qquad \textbf{(24.2)}$$

3 Reaction with atomic oxygen produces another chlorine free radical:

$$ClO\bullet + O\bullet \rightarrow Cl\bullet + O_2 \qquad \textbf{(24.3)}$$

The sum of reactions (24.1) and (24.3) is:

$$O_3 + O\bullet \rightarrow 2O_2 \qquad \textbf{(24.4)}$$

Notice that Cl• radicals are regenerated in reaction (24.3), and are free to react with more ozone in (24.1). It has been shown that more than 100,000 ozone molecules can be destroyed by one Cl•.

What has been done about ozone pollution?

Although initially predicted purely theoretically, measurements by the British Antarctic Survey in 1985 confirmed that the protection of the ozone layer against UV radiation had been seriously weakened. Discovery of a gigantic 'hole' in the ozone layer in the Antarctic showed that $[O_3]$ had fallen by two-thirds since the 1970s. This stimulated action. The 1987 Montreal Protocol led to the phasing out of CFCs and their replacement by hydrogen-containing hydrochlorofluorocarbons (HCFCs) such as $CHCl_2CF_3$. These compounds are more likely to be destroyed by atmospheric reactions before they reach the ozone layer. Also, since hydrofluorocarbons do not contain chlorine, they cannot produce chlorine free radicals, and for this reason CH_2FCF_3 is now used as a domestic refrigerant. The recovery of the ozone in the stratosphere is slow because the movement of molecules within the layers of the atmosphere is so slow, but the good news is that it is predicted that the ozone layer will be mostly recovered by 2040.

DID YOU KNOW?

CFCs (chlorofluorocarbons) and HCFs (hydrochlorofluorocarbons) are also powerful greenhouse gases.

BOX **24.1**

The filtering of UV light by ozone: a spectroscopic interlude

Ultraviolet radiation (roughly any radiation in the wavelength range 50–400 nm) can be classified into one of three regions, depending upon its wavelength:

Wavelength/nm	Sub-classification
<290	UV-C (very damaging to life)
290–320	UV-B (damaging to life)
320–400	UV-A (less damaging to life)

The ultraviolet light wavelengths absorbed by ozone molecules are shown in Figure 24.2. Ozone, O_3, has a different molecular structure, and therefore a different set of energy levels and spectrum, from diatomic oxygen, O_2.

No component of the atmosphere absorbs UV-A radiation to any great extent, so most of this radiation reaches the Earth's surface. However, UV-A radiation is the least harmful of light within the ultraviolet range. Oxygen and ozone molecules filter out all of the sun's radiation in the dangerous UV-C range. Oxygen absorbs only weakly in the UV-B range. Ozone absorbs more strongly in the UV-B range and most of the absorption of UV-B light below 300 nm is due to ozone (Figure 24.2). The unabsorbed UV-B radiation reaches the surface of the Earth. The key point is that a reduction in ozone in the atmosphere due to CFCs would lead to more UV-B radiation reaching the surface of the Earth and increased sunburn of human skin and damage to crops. This can lead to skin cancer because UV-B is absorbed by DNA, and increased exposure to UV-B may

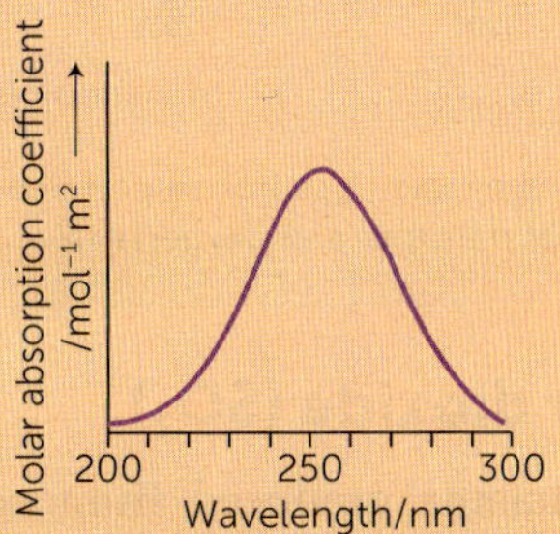

Figure 24.2 The absorption spectrum of ozone.

also lead to more eye cataracts and reduce the effectiveness of our immune systems.

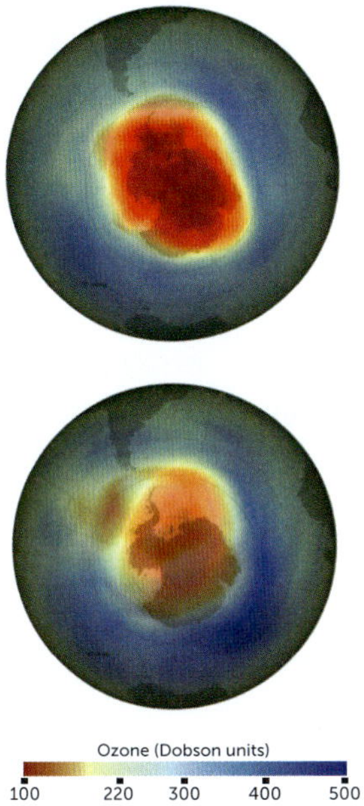

A success story: the recovery of ozone over the Antarctic between September 1994 and September 2019. The lowest ozone concentrations (representing the greatest damage) are in the top image and coloured in red and yellow. (NASA Earth Observatory: https://earthobservatory.nasa.gov/world-of-change/Ozone).

24.3 Air pollution

We now move from the stratosphere to the part of the atmosphere from the ground to 20 km, termed the troposphere, the place of clouds, water and land; and of plants and animals. The contribution of greenhouse gases (e.g. carbon dioxide CO_2, methane CH_4 and ground-level ozone O_3) to climate change will be discussed separately in Unit 28. In this section we concentrate on pollutants that are directly or indirectly harmful to living things. As we shall see, many air pollutants are emitted from the burning of coal or oil, petrol (gasoline) or diesel-powered transport.

Infrared spectroscopy can be used to study the composition of car exhaust gases. See Figure 24.3.

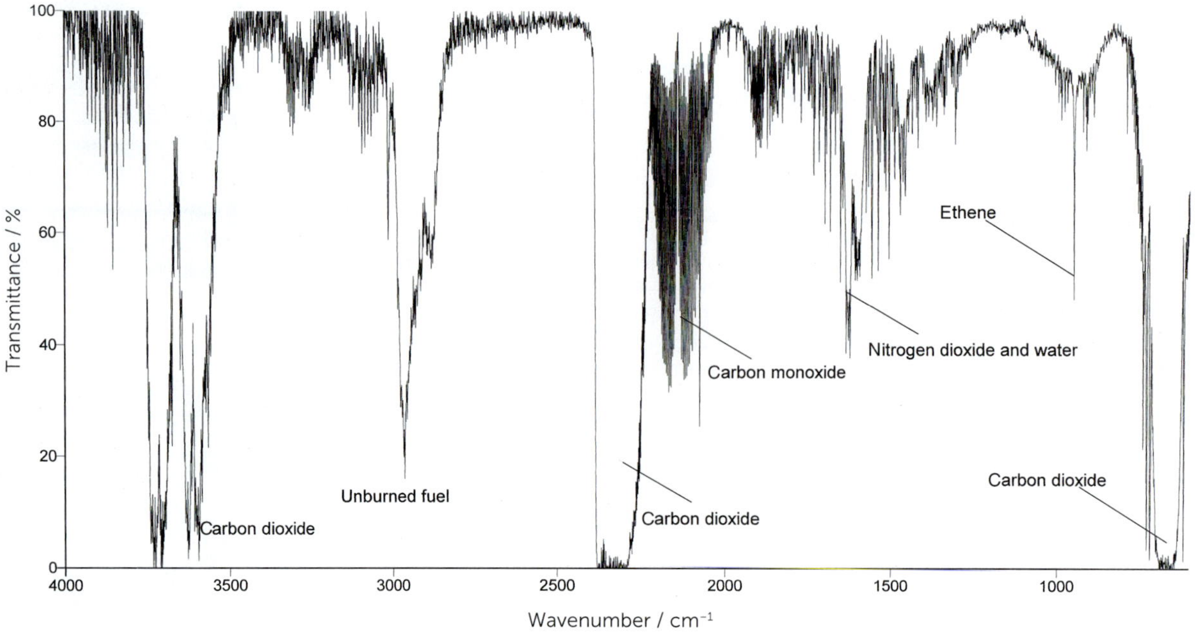

Figure 24.3 An infrared spectrum of the exhaust gas from a petrol-powered car. Cell pathlength = 10 m. Important peaks are labelled (image by the author). Note that nitrogen and oxygen molecules (in the air) do not absorb IR radiation and so do not exhibit an IR spectrum.

Sulfur dioxide (SO_2)

The pollutants produced from coal or oil combustion include sulfur dioxide gas (SO_2), which is transported long distances by the wind. SO_2 dissolves in water making sulfuric acid (see page 96) and is deposited as 'acid rain', acidifying lakes, corroding limestone buildings and causing reduced growth in trees ('die back'). The use of low sulfur containing fuels have helped European countries reduce SO_2 emissions, but such emissions are still substantial in countries that use coal powered power stations, including China, India, the Czech Republic, Poland and Serbia.

Nitrogen oxides (NO_x)

Nitrogen oxides (abbreviated as NO_x, meaning nitric oxide (NO) and nitrogen dioxide gas NO_2) are emitted from car exhausts with some contribution from coal and oil combustion. Most sources emit NO (a colourless gas) in the first instance. Even though the concentration of NO is low, NO is rapidly converted to brown NO_2 in air by reaction with another pollutant of the troposphere, ozone:

$$NO(g) + O_3(g) \rightarrow NO_2(g) + O_2(g) \quad \textbf{(24.5)}$$

(The ozone is produced via (24.7), discussed later). The reaction of NO with ozone is much faster than the three molecule reaction of NO_2 with oxygen.

$$2NO(g) + O_2(g) \rightarrow 2NO_2(g) \quad \textbf{(24.6)}$$

NO_2 is very harmful to living things – much more than NO. NO_2 also dissolves in water, making nitric acid (see page 97) and when it originates from power stations it also contributes to acid rain. However, NO_X is quickly broken down and so is a localized pollutant.

Volatile organic compounds

Volatile organic compounds (VOC)s are organic compounds that are gases or that readily vaporize at room temperature. They are found in petrol and diesel and leak out of engines, storage tanks or are emitted from vehicle exhausts. They are also emitted from manufacturing and from the use of solvents and paints. They include aromatic compounds (benzene, xylene and toluene), most aldehydes (e.g. ethanal) and ethanol, hydrocarbons from vehicle exhausts and isoprene, C_5H_8, emitted naturally from trees. VOCs are essential precursors for the formation of tropospheric ozone. Methane is the most abundant hydrocarbon in the troposphere. Because of its low reactivity (which is responsible for its 'long chemical half-life' of ≈10 years), methane has plenty of time to mix in the troposphere globally. It therefore contributes more to a global background of tropospheric ozone ('bad ozone') rather than to local spikes in ozone caused by other VOCs and NO_x. Although we are most concerned here with the role of VOCs in the production of tropospheric ozone, many are hazardous in themselves, particularly indoors when they are produced, along with particulates, from the combustion of gas or solid-fuel fires.

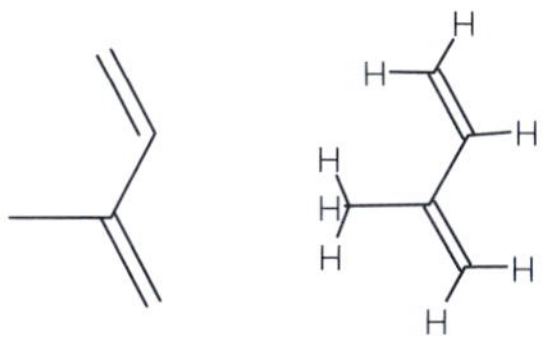

Two ways of representing the molecule isoprene C_5H_8, a VOC from trees

Bad ozone (O_3) and photochemical SMOG

Ozone in the stratosphere is often called 'good ozone' because it protects us from harmful UV radiation, but tropospheric ozone (also called ground-level ozone) is very harmful ('bad ozone') because it causes lung and cardiovascular disease. Bad ozone can travel hundreds of miles from its source. Ground-level ozone is not produced rapidly, and it (and its by-products, such as PAN (see below)) are slowly generated during the day.

There are no direct emissions of ground-level O_3 from vehicles or industry. The ozone is produced solely by the reaction of a wide range of volatile organic compounds (VOCs) with NO_x in the presence of sunlight. The mechanism of reaction is complex, and involves the radicals •OH, •HOCO and •O, but it may be summarized as:

$$VOC + NO_2 + \text{sunlight} \rightarrow O_3 \quad \textbf{(24.7)}$$

(i.e. VOC is being oxidized to ozone)

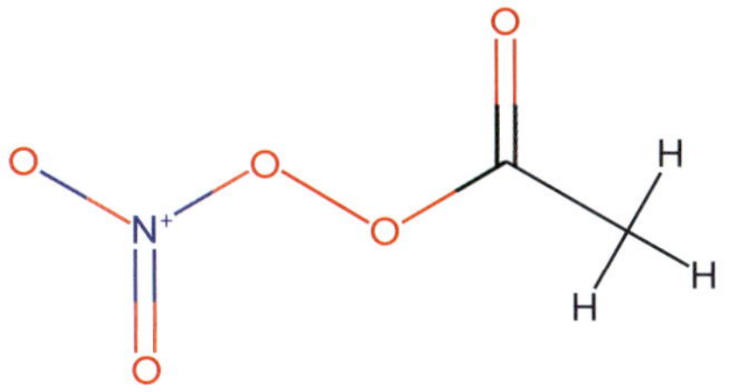

Peroxyacetyl nitrate (PAN) $C_2H_3NO_5$ is a harmful compound found in photochemical smog. It contains a O—O bond. Like tropospheric ozone, PAN is considered a secondary pollutant because it is produced from the primary pollutants NOx and VOCs.

Photochemical SMOG (the word SMOG comes from smoke and fog), involves the production of a compound called peroxyacetyl nitrate (PAN). Again, the reaction is complex, but it requires NO_x, sunlight, oxygen and specific VOCs, such as ethanal and ketones that can generate peroxyacetyl radicals (RCOO•). It is infamously connected with the irritating haze in major cities such as New Delhi and Los Angeles. PAN is a very strong irritant – even worse than ozone – and makes existing respiratory conditions much worse. Note that photochemical SMOG is distinguished from general SMOG, (such as the infamous London Smog of the 1950s) often associated with the burning of coal, and which also includes SO_2 and particulates.

Photochemical SMOG in a city.

Particulate matter

Particulate matter (PMs) are minute specks of solid or liquid material suspended in an aerosol than cause serious long-term harm to humans. They also act as nuclei for water, creating fog. Particles are classified by their diameter in microns (e.g. PM10), smaller particles being more damaging to health. Particles emissions from exhausts are particularly linked with older diesel vehicles but with newer vehicles the greatest amount of particulates arises from the wear of tyres on vehicles, particularly from heavy vehicles. Heavy lorries create the most tyre wear, but although battery electric vehicles (BEVs) are getting lighter, they are generally heavier than fossil fuel cars and so produce more particulates from tyre wear.

The good news, but with more to do

In many developed countries, health concerns about pollution have led governments to legislate and set targets for reduced emissions of pollutants:

- Coal use has dropped dramatically, so reducing SO_2.
- Low-emission zones in cities, improved vehicle catalysts (which convert NO_x to nitrogen gas) have reduced NO_x (which, in turn, reduces ground-level O_3). Particulate filters for diesel vehicles have reduced roadside particulates. Despite this, roadside emissions sometimes still exceed the permitted level.
- Electrification (e.g. of steel furnaces) and the adoption of more electric vehicles will reduce emissions in the future.
- Methane is both a major cause of tropospheric ozone and a greenhouse gas (see Unit 28) but in many countries methane emissions have been greatly reduced. For example, in the UK, methane emissions have dropped by over 62 per cent since 1990.

Figure 24.4 shows how the emissions of three other pollutants in the UK have been reduced in recent years. Reductions have also been achieved in many other countries including the USA, Germany and France.

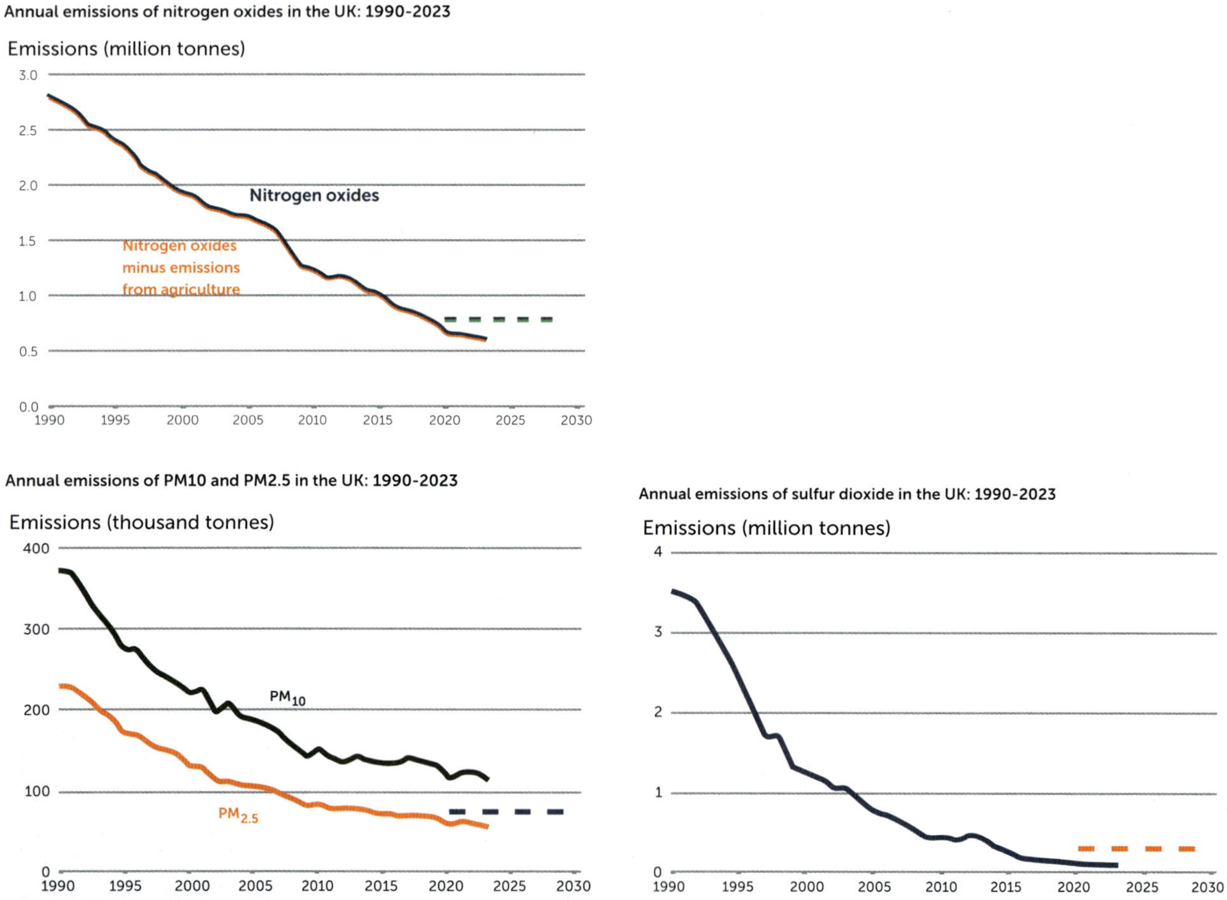

Figure 24.4 Average levels of three pollutants in the UK since 1990. Graphs adapted from UK Government Statistics (Department of Food, Environment and Rural Affairs). The horizontal dashed lines are the current targets for emissions.

EXERCISE **24A**

Removing polluting gases from car exhausts

Catalytic converters in car exhaust systems, which contain mixtures of transition metals and their oxides, convert toxic CO (from incomplete combustion of hydrocarbons) and NO, present in the exhaust gases, into CO_2 and N_2 respectively.

Write an equation for these two reactions and classify each reaction as catalytic oxidation, or catalytic reduction.

24.4 Water pollution

The quality of drinking water is very important to human welfare. The pollution of water by sewage has been linked to the spread of diseases such as cholera and typhoid fever. Elimination of these diseases has been a direct result of purification of such water, including by disinfection using chlorine.

However, waste from industry can contaminate water supplies with toxic chemicals:

1 Heavy metals

Metals such as Cd, Pb and Hg may be present in industrial or mining waste. These metals can prove poisonous to humans – cadmium and mercury can cause kidney damage, and lead poisoning can cause damage to the kidneys, liver, brain and central nervous system. All of these metals are **cumulative poisons** – the body does not excrete them and their concentration builds up. For this reason, the dumping of toxic metal waste is prohibited by law.

2 Detergents, fertilizers

Detergents and fertilizers may contain **phosphates** as additives. The addition of phosphorus to water, in the form of the phosphate anion PO_4^{3-}, encourages the formation of algae, which reduces the dissolved oxygen concentration of the water. This process is known as **eutrophication** and threatens the development of higher life forms, such as fish. The 'run off' of additives into river water is carefully monitored by government agencies.

In 1969–1970 the famous Norwegian explorer Thor Heyerdahl led an expedition across the Atlantic on a papyrus raft named Ra II. During the journey, the crew recorded clumps of oil and discarded plastics in the ocean, bringing a knowledge of water and oceanic pollution to a much wider audience. (Image by Ralf Roletschek.)

BOX **24.2**

Heavy metals and aluminium concentrations in river water

The term 'heavy metal', to a chemist, means one of a group of the metals in the lower right hand corner of the Periodic Table (metals in Groups 13–16 that are in periods 4 and greater). Examples include Fe, Cu, Pb, Cd and Hg. One reason that these metals are considered toxic is that they have a great affinity for sulfur and attack sulfur bonds in enzymes, making them unable to function properly.

Aluminium is considered to be a light metal, but it can still be toxic at high concentrations. Concentrations of the aqueous aluminium ion have been found to be higher in 'acid' (pH 4–5) river and lake waters, and this has been found to adversely affect the trout population in those waters, particularly if the calcium concentration is low.

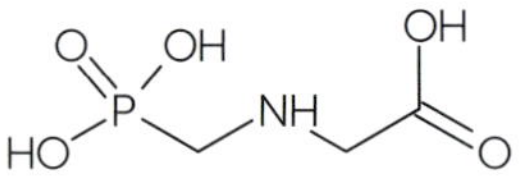

Glyphosate ($C_3H_8NO_5P$), used as a pesticide (specifically, a herbicide, to kill weeds).

3 Pesticides

Two million tons of pesticides are used world-wide each year to kill weeds and to increase the yield of crops by killing insects. All countries use pesticides, but the biggest users include China, the USA, Argentina, Brazil, France, Italy and Thailand. The danger is that pesticides will be washed into rivers and ground water, causing poisoning outside of the targeted habitat. Pesticide concentrations are often monitored regularly and some are removed in sewage treatments. One commonly used pesticide, glyphosate, may cause cancer in humans, and its use in agriculture and in maintaining urban paths free from weeds is under review.

BOX **24.3**

Rachel Carson (1907–1964) was a marine biologist and conservationist whose book *Silent Spring* (1962) is often credited with initiating the international environmental movement. Her work highlighted the harmful effects of pesticides, including the weakening of the eggshells of large birds that reduced their populations. The use of DDT in the USA was in effect banned in 1972. (Image by US Fish and Wildlife Service.)

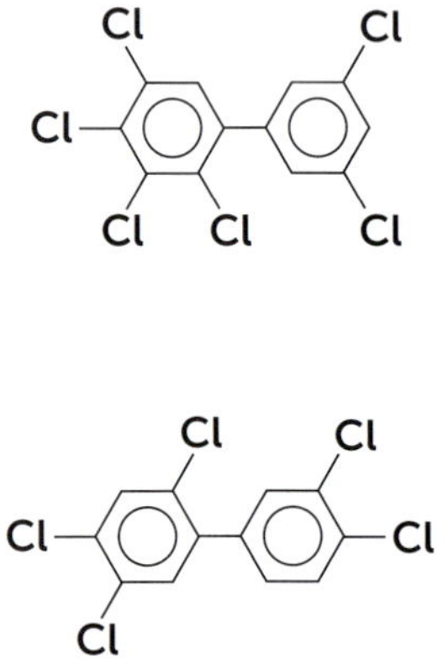

Examples of polychlorinated biphenyl molecules.

4 Acid-polluted water (pH < 3)

This is deadly to most forms of aquatic life. Water downstream from a mine may be contaminated by **acid mine drainage**, the result of microbial oxidation of discarded waste material at the mine site. Acid mine water principally contains sulfuric acid produced by the oxidation of iron pyrites (FeS_2). Industrial wastes and acid rain may also contribute to the acidity of natural waters.

5 PCBs (polychlorinated biphenyls)

Peak production of these chemicals occurred in the twentieth century. They have high stabilities and this led to their being used in many applications, for example as fluids in transformers and capacitors. Although their manufacture has been stopped, PCBs are resistant to oxidation when released into the environment and so persist for a long time and can cause skin disorders in humans. They may be carcinogenic to humans.

Importance of dissolved oxygen in water

The dissolved oxygen concentration in water is vital for the support of fish. The lower the level, the more polluted is the sample. The dissolved gas is used by micro-organisms to oxidize organic matter in sewage. Unless the water can restore its dissolved oxygen concentration – for example by the turbulent flow of shallow waters – the water will no longer support many organisms. Typically, fish growth is inhibited below a dissolved oxygen concentration of 6 mg dm^{-3}.

It is not generally realized how little oxygen is present in a sample of 'pure' water. A litre of water at 5°C, in free contact with the atmosphere, contains only about 9 cm^3 of oxygen, weighing 13 mg. As the temperature rises, the oxygen concentration falls so that, at 20°C, it is only about two-thirds of the level it was at 5°C (see page 186).

Oxygen reaches the water in two main ways. First, it dissolves at the surface from the atmosphere. Still water takes up oxygen slowly, whereas turbulent water takes it up more rapidly. The second source of oxygen in water is from photosynthesis. Where there are many green plants present, the water often becomes **supersaturated** with oxygen during the hours of daylight. However, after dark, photosynthesis stops but the plants continue to respire and actually reduce the amount of dissolved oxygen.

Drinking water

Fluoride

Soluble fluoride is often added to drinking water to bring it up to a concentration of 1 ppm or 1 mg dm^{-3}. This concentration is within agreed safety limits and has been shown to protect teeth against decay.

High concentrations of fluoride are poisonous, and are harmful to bones and teeth at levels over 10 ppm (mg dm^{-3}).

Lead

The limit for the concentration of lead ions in drinking water is 10 ppb (μ g dm^{-3}). The 'first draw' of water from the taps of homes that use lead pipes might exceed this, especially in areas where the water is relatively acidic. It is a good idea not to drink the water that has been standing overnight in an older plumbing system, until you have run out the water for at least a minute.

pH

The pH of drinking water should be between 5.5 and 9.5. A decrease in the pH of the water increases the solubility of metal ions.

Other metals

The maximum recommended levels of impurities for drinking water, set by the EU, are as follows

Impurity	Concentration/ppm or mg dm^{-3}
Fe	0.2
Mn	0.05
Cu	2
Cd	0.005
Al	0.2

Sulfate

Sulfate is harmless at moderate levels, but excessive sulfate (>500 ppm) is thought to have a laxative effect.

Nitrate

Excess nitrate in drinking water can lead to methemoglobinaemia ('blue-baby' syndrome). It also may be linked to stomach cancer, although this link has not been proved. The EU has set a maximum limit of 50 ppm in drinking water.

EXERCISE **24B**

Food waste

Food wastes, although they are considered 'natural', can have a very damaging effect on the environment if they are simply discharged into rivers and lakes. This is because they have a high demand for oxygen as they decompose ('high biochemical oxygen demand'). Suppose that the effluent from a food manufacturing industry has a glucose ($C_6H_{12}O_6$) content of 600 mg dm^{-3}.

(i) Write the chemical equation for the complete oxidation of glucose.

(ii) What is the oxygen requirement (in mg dm^{-3}) for the complete oxidation of 1.0×10^5 dm^3 of this waste?

BOX **24.4**

The Bhopal disaster

The worst industrial accident ever recorded occurred at Bhopal, India, on 3 December 1984, when forty tons of methyl isocyanate, a colourless and highly poisonous liquid used in the manufacture of carbamate pesticides, was accidently released from a storage tank, killing about four thousand residents within hours of the leak and injuring many more. It has been estimated that there were 15,000–20,000 premature deaths over the twenty years following the accident.

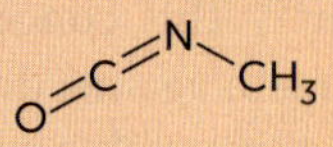

Methyl isocyanate C_2H_3NO

REVISION QUESTIONS

24.1

Given the following molecular structures:

(i) 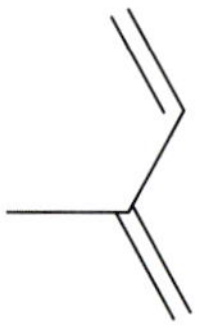

(ii) 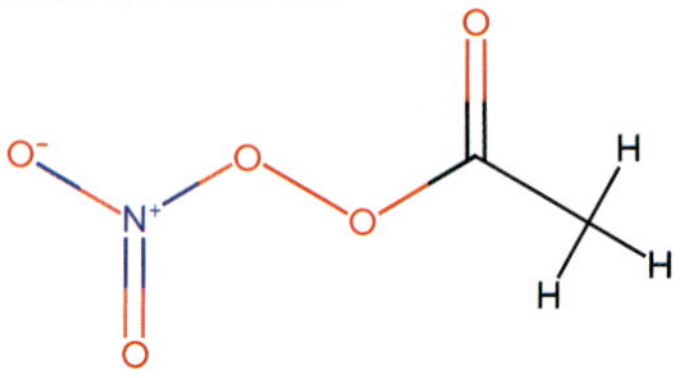

(iii) 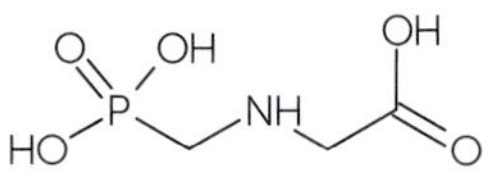

(iv) 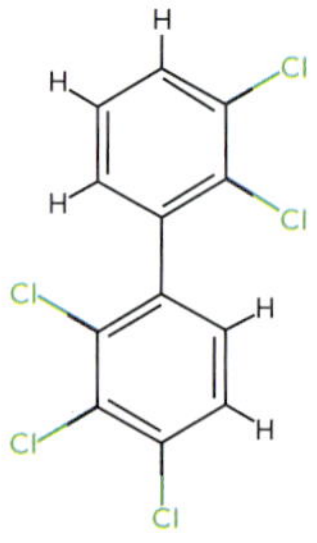

(v) 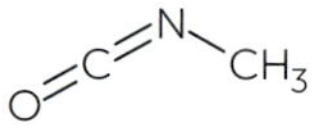

Give the letter of the molecule that is:

(a) A member of the PCB family.

(b) The killer of thousands of people in Bhopal.

(c) Used as a pesticide.

(d) Abbreviated PAN.

(e) Is emitted by trees.

24.2

(a) Explain why the presence of even minute concentrations of chlorine-containing compounds in the stratosphere can cause substantial reduction in ozone concentration.

(b) Chlorine atoms (radicals) can react with nitrogen dioxide in the atmosphere to form chlorine nitrate:

$$ClO + NO_2 \rightarrow ClONO_2$$

What would be the implications for stratospheric ozone if this reaction was very fast?

24.3

(a) Distinguish between 'good' and 'bad' ozone.

(b) A mixture of inert gas (N_2), with traces of NO_2, NO and propene (C_3H_6) gases, is placed in a quartz glass vessel. The mixture is irradiated with artificial sunlight and the resulting mixture analysed using a mass spectrometer. The mass spectrum shows peaks at the following *m*/z values: 30, 42, 46, 48, 44. What ions might be responsible for these peaks?

24.4

The maximum safe limit of nitrate in drinking water is a concentration of 45 ppm (mg dm^{-3}). A sample of water (volume 250 cm^3) was analysed and found to contain 0.012 g of NO_3^-.

(i) What is the concentration of NO_3^- in the sample in mol dm^{-3}?

(ii) Would the water be considered safe to drink?

(iii) Ammonium nitrate (NH_4NO_3) is a common, solid fertilizer. It supplies nitrogen to the soil, which is an essential plant nutrient. What percentage of nitrogen by mass does it contain?

24.5 A factory discharges waste containing lead ions [Pb^{2+}(aq)] into a river at the rate of 10 ppm (mg dm^{-3}) per minute. Convert this rate to mol dm^{-3} min^{-1}.

24.6

(i) Catalytic converters in car exhausts allow complete oxidation of octane in oxygen to occur at a low temperature. Write an equation for this reaction.

(ii) A converter operates at 95 per cent efficiency. How much CO_2 is formed from the combustion of 200 g of octane?

24.7 The release of aluminium ions as a result of acid rain is less significant in areas containing limestone rock. Why?

24.8 Carbon monoxide is a very poisonous gas. A flask of volume 1 dm^3 contains pure CO at a pressure of 1.5 atm at 298 K. The flask is dropped and broken by a careless laboratory worker and the gas is released into a small sealed room of dimensions 5 m × 5 m × 4 m which contains air at 1 atm pressure and 298 K.

(i) Estimate the concentration of CO in the room in mg m^{-3}.

(ii) Assuming that the concentration of CO is hazardous above 55 mg m^{-3}, will the laboratory worker be exposed to a dangerous level of the gas?

24.9 The vapour of tetrahydrofuran (C_4H_8O), a solvent for PVC, is detectable by odour alone at concentrations of 10 mg m^{-3} or above. What mass of the solvent would need to be allowed to evaporate in a room of volume 100 m^3 in order to achieve this concentration? (Assume no loss of vapour.)

24.10 A bottled water had the following analysis on its label:

Ca^{2+}	80 ppm
Mg^{2+}	15 ppm
K^+	4 ppm
Na^+	6 ppm
SO_4^{2-}	85 ppm

(i) If the volume of the bottle was 1.25 L, how many grams of Mg^{2+} did it contain?

(ii) What is the concentration of sulfate in the bottle, in units of mol dm^{-3}? (Remember that 1 L = 1 dm^3.)

24.11 Acid pollution causes the pH of a lake to decrease from 8.40 to 8.10. By how much has the [H^+(aq)] increased?

DID YOU KNOW?

Tiny particles of industry-made plastics are widely dispersed across the world and inside us. In 2025 scientists reported the first evidence of trace micro and (even smaller) nano plastic buildup in the lungs of birds. It is believed that the unique respiratory systems of birds make them especially vulnerable to accumulating plastics. In one study, fifty-one species of birds were studied – many with different feeding and nesting habits. Plastics were found in the lungs of every bird. Plastics identified include polyethylene, polyurethane, polyvinyl chloride (PVC) and butadiene rubber. The rubber may have originated from aircraft and road vehicle tyres.

Extension material to support this unit is available in Extension 24 on our website at https://www.bloomsburyonlineresources.com/chemistry.

25 Forensic Chemistry

CONTENTS

OBJECTIVES

This unit:

- Introduces calculations on drugs and alcohol
- Defines the 'drink driving' legal limit
- Highlights the importance of analytical measurements in forensic science
- Explains the advantages of analysing hair for drugs

25.1 Prerequisites

The ideas which form the background to this unit are listed in the following table:

Topic	See page(s)
Half-life	253
Organic chemistry (especially of alcohols)	Unit 19, 347
IR spectrum of ethanol	Figure 22.23, page 410
Chromatography	387

25.2 The scope of forensic chemistry

Forensic chemistry is the application of chemistry to the legal system. It includes the analysis of unknown substances from the scene of a crime or from samples of body tissue. Forensic chemistry uses the full range of analytical techniques available to the chemist, as shown in Table 25.1.

Forensic chemistry has strong links with forensic toxicology, the 'science of poisons'. Examples of molecules of interest to the forensic chemist are given in Figure 25.1.

Table 25.1 Examples of analytical techniques used in forensic analysis

Technique	Examples of use
Gas chromatography (GC)	(a) Analysis of alcohol in blood (b) Identification of fuels at the scene of an arson attack
Infrared spectroscopy (IR)	(a) Analysis of alcohol in breath (b) Identification of unknown solids from a suspected illicit drug factory
Liquid chromatography (HPLC)	(a) Analysis of suspected doping agents in race horses (b) Identification of paracetamol in blood
Atomic absorption spectroscopy (AAS)	(a) Analysis of arsenic in human tissue (b) Identification of alloys used in making bullets
Gas chromatography with mass spectrometry (GC-MS)	(a) Detection of the presence of cocaine metabolites (breakdown products) in hair (b) Identification of strychnine in blood
Ultraviolet spectroscopy (UV)	(a) Analysis of pen inks in forgery cases (b) Identification of fibre dyes at the scene of a crime

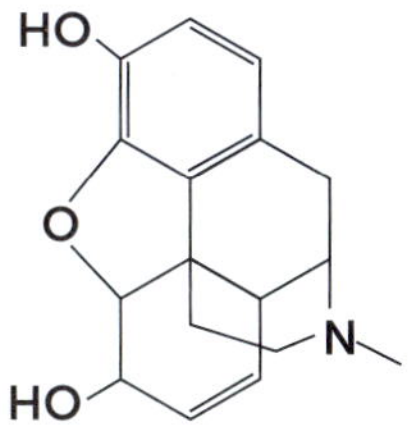

Morphine (a drug used in pain relief)

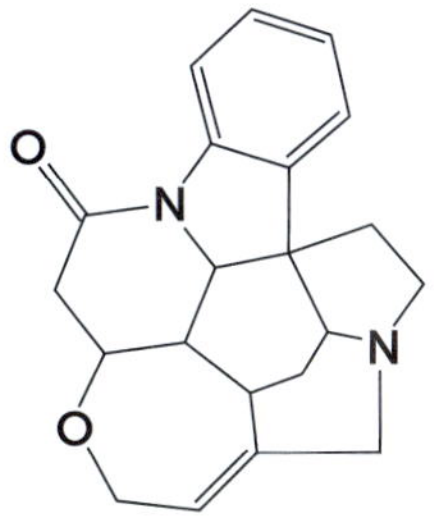

Strychnine (a poison)

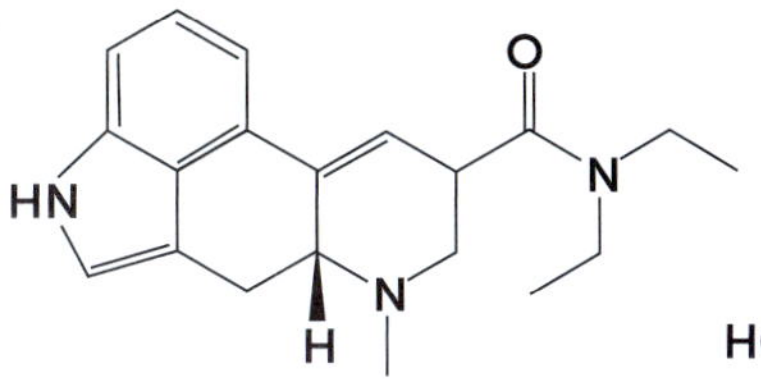

Lysergic acid diethylamide (LSD)

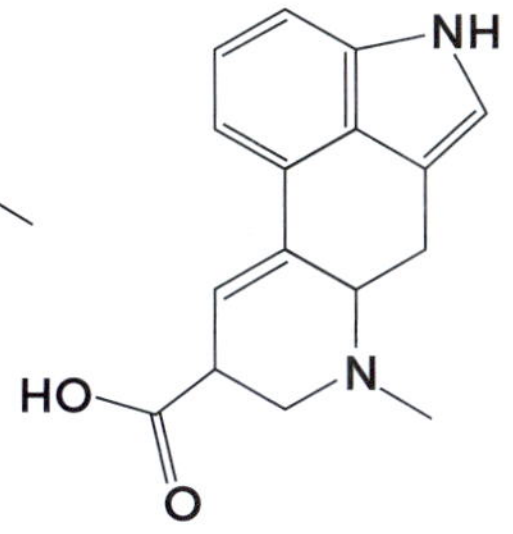

Lysergic acid (the parent acid of LSD)

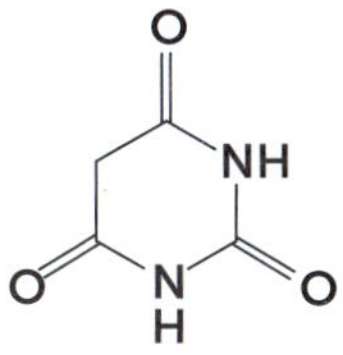

Barbituric acid (parent acid of barbiturate drugs, which are nerve depressants)

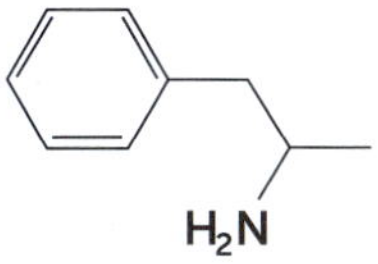

Amphetamine (stimulant)

Figure 25.1 Examples of molecules of interest to forensic chemists.

25.3 Variation of the concentration of a poison with time

Many poisons that are absorbed into the bloodstream are broken down by enzymes within the body's organs. The concentration of the poison in the blood then falls from its peak value. Figure 25.2 illustrates two cases where the fall in concentration is *first-order* (as with paracetamol) and *zeroth-order* (as with ethanol).

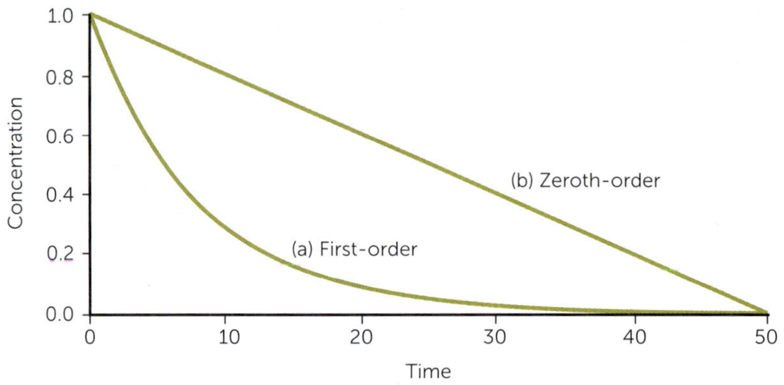

Figure 25.2 Variation of concentration of poison in blood where the poison decays by (a) first-order kinetics, (b) zeroth-order kinetics.

Not all poisons follow simple laws of kinetics. For example, the poisonous gas *carbon monoxide* permanently binds itself to haemoglobin in the blood and is not eliminated from the body.

25.4 Calculations using first-order kinetics

The **first-order decay** of poisons follows the equation first introduced on page 253:

$$[A]_t = [A]_0 \times e^{-kt} \qquad \textbf{(25.1)}$$

where $[A]_0$ is the peak concentration of poison A (the time of the peak concentration is assigned as $t = 0$), k is the first-order rate constant (in units of s^{-1} or h^{-1}, depending on whether or not time is in seconds or hours) and $[A]_t$ is the concentration of poison A at time t. This equation was used to plot Figure 25.2(a). $t_{1/2}$ is the half-life of the poison.

Exercise 25A

The drug paracetamol decays by first-order kinetics in the body with a half-life of 4.0 h. A person's blood is found to contain 50 mg dm^{-3} of paracetamol 8.0 h after ingestion of the drug.

Estimate the peak blood concentration of drug.

EXAMPLE **25.1**

A person is suspected to have taken an overdose of morphine. The half-life of morphine in blood is about 3.0 h. If the blood plasma concentration 5.0 h after ingestion was 6.0 mg dm^{-3} of blood, estimate the peak concentration of morphine at the time of ingestion.

Answer

We assume that the peak concentration is rapidly achieved so that it equals the initial concentration, $[A]_0$, of the first-order decay.

Rearranging equation (25.1) gives

$$[A]_0 = [A]_t / e^{-kt} \quad \textbf{(1)}$$

The first-order rate constant k is calculated using the equation:

$$k = 0.693/t_{1/2}$$

As $t_{1/2} = 3.0$ h, $k = 0.693/3.0 = 0.23\ h^{-1}$ (the units are 'hours to the minus one'). Substituting k and t (= 5.0 h) into equation (1) yields

$$[A]_0 = 6.0 / e^{-(0.23 \times 5.0)}$$

We start by working out $0.23 \times 5.0 = 1.15$. The quantity $e^{-1.15}$ is evaluated using the e^x function on a calculator, giving $e^{-1.15} = 0.317$. Finally

$$[A]_0 = 6.0 / 0.317 = 18.9\ mg\ dm^{-3}$$

In conclusion, we have estimated the peak concentration of morphine in blood as about 19 mg dm^{-3}.

25.5 The zeroth-order kinetics of ethanol decay

Ethanol breaks down into ethanal (CH_3CHO) in the body. Ethanol (common alcohol) decay does not follow first-order kinetics. Instead, the variation of **blood alcohol concentration** (symbolized 'BAC') with time follows *zeroth-order kinetics* in which the reaction rate is independent of the concentration of A:

$$[A]_t = [A]_0 - kt \quad \textbf{(25.2)}$$

giving the concentration/time curve shown in Figure 25.2(b).

Calculations using equation (25.2) are straightforward. BAC is commonly expressed in units of milligrams of ethanol per 100 cm^3 of blood (mg/100 cm^3). At room temperature k is approximately 15 milligrams of ethanol per 100 cm^3 of blood *per hour*, and we will apply this value to all individuals irrespective of gender and health. (For more accurate work, k values are adjusted for gender and body mass index.) Remember, though, that if we use k in these units, t must be in units of *hours*.

EXAMPLE **25.2**

The BAC of a person on finishing their drinks was 90 mg of ethanol per 100 cm^3 of blood. Calculate the BAC of the person four hours later.

Answer

We symbolize the blood alcohol concentration at $t = 0$ as $[A]_0$.

$$[A]_t = [A]_0 - kt \qquad \text{where } k = 15.$$

$$[A]_t = 90 - (15 \times 4) = 30 \text{ mg of ethanol per 100 cm}^3 \text{ of blood.}$$

Therefore, BAC after four hours = 30 mg of ethanol per 100 cm^3 of blood.

Exercise 25B

A person is found to have a BAC of 100 mg of ethanol per 100 cm^3 of blood 6 hours after their last drink.

Estimate their BAC on leaving the pub.

Table 25.2 The physiological effects of alcohol on humans

Blood alcohol concentration (mg/100 ml)	Effects
<50	No obvious effect, except perhaps a tendency to a feeling of well-being and to become more talkative.
50–100	First obvious effects start to show. Some loss of coordination and sensory perception, possibly some slurred speech.
100–150	More marked loss of coordination, poor sensory perception, possibly nausea and desire to lie down.
150–200	Drunkenness and probably nausea.
200–300	General nausea, inability to stand, vomiting, probably coma.
>300	Approaching danger limit, coma and anaesthesia, impaired circulation and respiration, possible death.
>450	Probable death due to respiratory paralysis.

25.6 Drinking and driving

Excess alcohol is a major cause of traffic accidents and all countries have taken steps to legislate against driving whilst over a set BAC limit, although the precise limit varies. The limit is eighty milligrams of alcohol per 100 ml of blood in England and Wales but this is under review. In many European countries the limit is fifty.

Typically, the law allows for a police officer to conduct a screening test (the 'roadside breathalyser') on a driver who is suspected of driving whilst 'over the limit'. The screening test was originally based on the reduction of dichromate ions but this has been replaced by a 'fuel cell' that oxidizes the alcohol to ethanoic acid. If the screening test proves positive, the suspect is arrested and taken to a police station where an instrument based on an infrared spectrometer that measures ethanol absorbance, evidence that provides may be used in court.

The background to the use of the IR spectrometer is as follows. Alcohol is volatile (page 170) and some of the alcohol in the blood soon makes its way into human breath. The infrared instrument used in police stations analyses the *breath* of a suspect. Because of the way that gases inhaled through the lungs are exchanged within fine capillaries, the alcohol in blood and breath are assumed to be in equilibrium (Figure 25.3). The concentration of alcohol in the blood of a person is about 2300 times greater than the concentration of alcohol in that person's breath:

$$\frac{[\text{ethanol in blood}]}{[\text{ethanol in breath}]} = 2300$$

Breath concentrations are usually expressed in units of micrograms (μg) of alcohol per 100 cm^3 *of breath*. The breath equivalent of a blood alcohol concentration of 80 mg of alcohol per 100 ml of blood is 35 μg of ethanol per 100 cm^3 of breath.

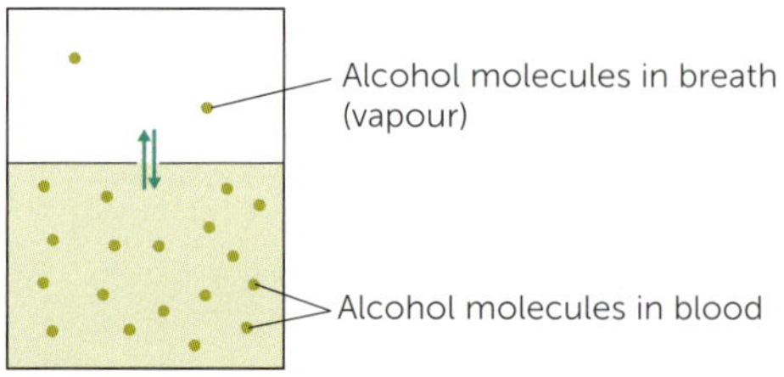

Figure 25.3 Alcohol in breath and blood are in dynamic equilibrium.

Alcohol may also be detected in urine (Table 25.3).

Table 25.3 Summary of equivalent prescribed alcohol limits in england and wales

Substrate	Concentration limit
Blood (BAC)	80 mg of ethanol/100 ml of blood
Urine (UAC)	107 mg of ethanol/100 ml of urine
Breath (BrAC)	35 μg of ethanol/100 ml of breath

The Laboratory of the Government Chemist has estimated the uncertainty in the measurement of ethanol in blood via breath analysis as

$$\pm 2\,\text{mg}/100\,\text{ml}$$

A measured value of 81 mg/100 ml, determined using an evidential analyser, may therefore represent a true value in the range 79–83 mg/100 ml. In practice, a more generous allowance of 6 mg/100 ml is subtracted by the analyst in his report. For our measured value of 81 mg/100 ml, the reported concentration would be 81 − 6 = 75 and the formal statement from the forensic scientist would be that 'the sample contained not less than 75 mg of ethanol per 100 ml of blood'.

Drivers who are affected by other drugs (such as amphetamines or cocaine) also cause many traffic accidents but because these drugs are not volatile their presence can only be confirmed by analysis of blood or urine. Such measurements cannot usually be conducted by the roadside and drivers suspected of driving under the influence of drugs are first arrested and taken to the police station where samples of blood or urine are obtained for subsequent chromatographic analysis.

Figure 25.4 Measuring alcohol concentration in breath by infrared absorption. The instrument is a Lion Intoxilyzer 6000, manufactured by Lion Laboratories, Barry, South Wales.

25.7 Analysis of drugs in hair

Many job applicants are required to provide blood or urine samples as part of an interview. If habitual drug users refrain from using drugs in the week before the test, they will avoid detection because the drug has long passed through their bodies.

Analysis of hair avoids these difficulties because hair retains within it tiny traces of drugs that a person may have consumed for many months (Figure 25.5). Further, since hair grows about 1 cm per month, the analysis of segments of hair allows an estimate to be made of drug consumption over the previous months.

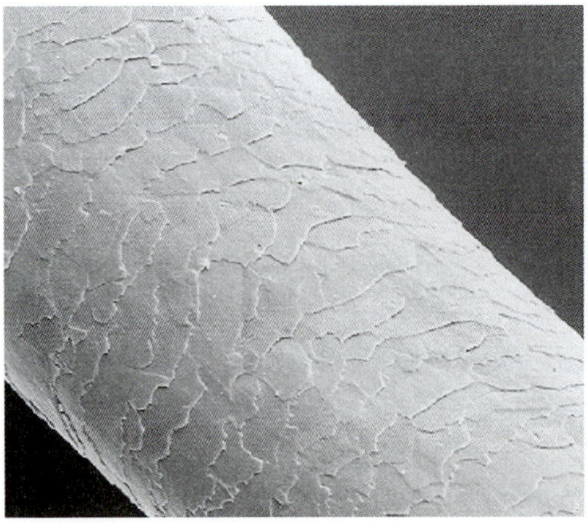

Figure 25.5 A single hair. It has long been known that poisons, including heavy metals, are deposited in hair. The observation that minute quantities of drugs such as cocaine and ecstasy are also stored in hair offers the possibility of identifying individuals with a history of drug addiction. (Photograph reproduced courtesy of Dr John Gray.)

BOX **25.1**

Dame Agatha Christie – the world's most famous crime author

Agatha Christie (1890–1976) wrote crime novels that have been translated into more than 100 languages. She had experience as a nurse in a hospital in the First World War (1914–1918), and this gave her a useful background when devising the many different murder weapons for her stories. In her novels, Christie's murderers employ a range of poisons including: arsenic, white phosphorus, morphine, belladonna, monkshood, strychnine, cyanide and thallium.

Agatha Christie in c.1926.

DID YOU KNOW?

Hydrogen cyanide gas (HCN, also known as *Prussic acid*) is the parent compound of the group of compounds known as cyanides. Both HCN and cyanides are highly toxic and they kill by combining with the iron atom in a cellular enzyme so that the individual is starved of oxygen. Curiously, low concentrations of hydrogen cyanide solution ('hydrocyanic acid') were declared to be medicinally beneficial and prescribed by some doctors between about 1820 and 1900. At these concentrations it may have a sedative effect but it was subsequently superseded by better (opioid) pain killers.

Medically prescribed hydrogen cyanide solution

The most infamous use of HCN came in the Second World War (1939–1945). HCN was used as a pesticide and sold under the trade name Zyklon-B. The war saw Zyklon-B used to horrifically kill more than a million captives in German extermination camps. Stacks of empty metal cans of Zyklon-B can be seen by visitors to these camps today.

For the production of HCN in fires see page 480.

REVISION QUESTIONS

25.1 The following table shows the concentration of strychnine in the blood of a patient at various times. (Measurements started as soon as possible after hospital admission of the patient.)

Time from ingestion/h	Strychnine concentration/mg dm^{-3}
16.0	1.02
18.0	0.93
20.0	0.85
24.0	0.71
26.0	0.65
30.0	0.54
32.0	0.49
34.0	0.45
36.0	0.41
38.0	0.37
40.0	0.34
51.0	0.20

Assume first-order kinetics. Estimate **(i)** the half-life of strychnine in the blood, **(ii)** the concentration of strychnine in the patient at t = 0.

25.2 For the equilibrium

$$\text{ethanol in breath} \rightleftharpoons \text{ethanol in blood}$$

it is found that a concentration of 35 µg of ethanol per 100 cm^3 of *breath* corresponds to 80 mg per 100 cm^3 of *blood* at 25°C. Use these data to calculate the molar concentration ratio:

$$\frac{[\text{ethanol in blood}]}{[\text{ethanol in breath}]}$$

25.3

(i) A person is suspected of being involved in a drink-related road accident. The blood alcohol concentration in the subject two hours after the accident was 79 mg per 100 ml of blood. Estimate the blood alcohol concentration at the time of the accident.

(ii) The concentration of alcohol in urine is about 1.34 times that of blood. The concentration of alcohol in saliva is about 9 per cent higher than that of blood. Calculate the urine and saliva alcohol concentrations in a person with a blood alcohol concentration of 80 mg per 100 ml.

25.4 The structure of cocaine is:

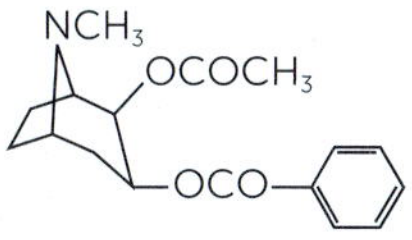

(i) How many carbonyl groups (C=O) are present in cocaine?

(ii) What is the molecular formula of cocaine?

(iii) What is the molecular mass of cocaine?

(iv) The IR spectrum of cocaine contains an intense peak at about 1,700 cm^{-1}. Explain this.

(v) The mass spectrum of cocaine shows a peak at m/z = 303. What ion causes this peak?

25.5

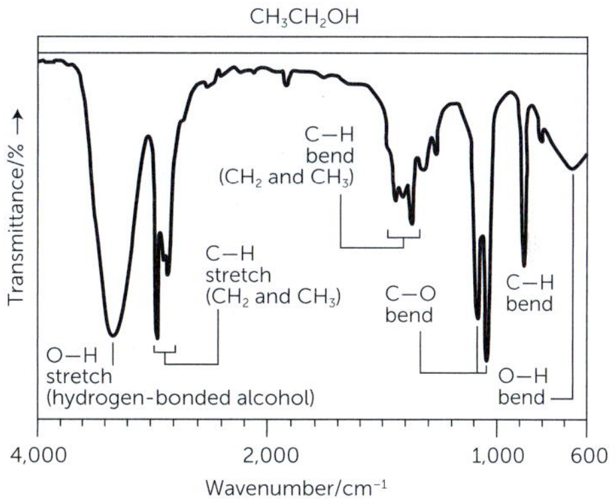

The IR spectrum of liquid ethanol is shown above. The spectrum of ethanol vapour is similar. Study the spectrum and identify two peaks that you think may be used to measure the ethanol concentration in breath. Will **(a)** the absorbance, **(b)** the % transmittance of the peaks be proportional to C_2H_5OH concentration?

25.6 A person is arrested by police and taken to the police station where the alcohol concentration in their breath is measured as equivalent to 100 mg ethanol per 100 ml of blood. **(a)** What official statement of alcohol concentration would be made by a forensic scientist? **(b)** Estimate the BAC three hours after the measurement.

25.7 Find out the origin of the saying 'As mad as a hatter'. (Hint: there is a poison involved!)

More information about cyanide can be found in extension 25 on our website at https://www.bloomsburyonlineresources.com/chemistry

DID YOU KNOW?

In the 1800s, sugar was expensive, and it was commonplace to use plaster of Paris (dehydrated calcium sulphate, a white powder) to dilute sugar in sweets. In 1858, in Bradford, England, (white) arsenic trioxide was accidentally sold instead of plaster of Paris, leading to the poisoning of customers who believed they were eating harmless sweets. Twenty people died, including children, and the Bradford Sweet Poisonings helped shape future legislation to make such incidents less likely.

Cartoon by John Leech in the *Punch Magazine* 1858.

BOX **25.2**

The famous Marsh Test for arsenic

Accidental poisoning by arsenic is one thing, but premeditated murder is another! The most infamous poison used by murderers is arsenic, often administered in form of arsenic trioxide (As_2O_3), a white, deadly powder. In France, arsenic came to be known as *poudre de succession* ("inheritance powder") because it was frequently used where the motive for murder was financial gain. The use of arsenic as a poison spans many centuries. One of many murderers who used arsenic was the solicitor Herbert A Armstrong, who lived in Cusop near Hay on Wye, UK. Armstrong had bought arsenic to kill dandelions in his garden, but he used some of it to kill his wife. Armstrong was hanged in 1922.

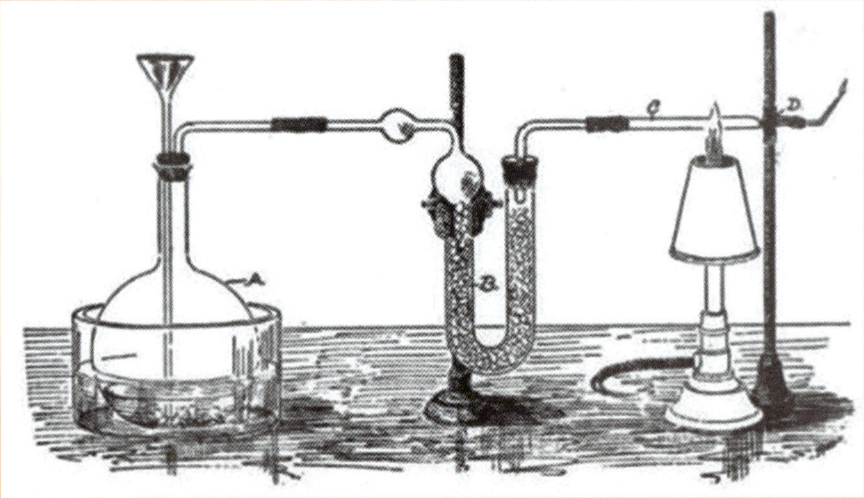

The equipment used in the Marsh Test. The steps in the test referred to below occur at positions A, B, C and D. From 'An Introduction to Chemical Pharmacology' by Hugh McMuigan (1921).

Nowadays, a suspected sample of arsenic would be tested using modern spectroscopic techniques, but in Victorian times such technology was a distant dream. One chemist, James Marsh, who worked at the Royal Arsenal in Woolwich, near London, developed known chemistry to produce, in 1836, a reliable test for arsenic. The test became known as the 'Marsh Test'.

The stages used in the Marsh Test use the equipment shown above. The suspected arsenic (a suspect powder, food or human tissue thought to contain arsenic) is reacted with zinc and sulfuric acid in flask A, producing the poisonous arsine gas (arsenic trihydride, $\mathbf{AsH_3}$):

$$As_2O_3(s) + 6Zn(s) + 6H_2SO_4(aq) \rightarrow 2\mathbf{AsH_3}(g) + 6ZnSO_4(aq) + 3H_2O(l) \quad (1)$$

The mixture of excess zinc and sulfuric acid also produces hydrogen gas:

$$Zn + H_2SO_4 \rightarrow ZnSO_4(aq) + H_2(g) \quad (2)$$

The hydrogen gas from reaction (2) is ignited as a flame at the extreme right of the apparatus, at the very tip of the tube. The arsine is dried by passing it through drying agent (B) and the arsine heated just beyond position C when the arsine decomposes into the element arsenic and more hydrogen.

$$2AsH_3(g) \rightarrow 2As(s) + 3H_2(g) \quad (3)$$

The arsenic produced by the decomposition of arsine deposits on the glass tube at D as a silvery-black 'stain'. The intensity of the 'arsenic stain' was then compared to stains using samples containing known concentrations of arsenic. This allowed a calculation to be made of the mass of arsenic in the sample. In the hands of a skilled chemist, this measurement proved surprisingly accurate.

The Marsh test was first used in court in France in 1840, providing crucial evidence that convicted Marie Lafarge of poisoning her husband. It is also believed that the Marsh test acted as a deterrent. Now that the presence of arsenic could be proved after death, murderers might well think twice before using arsenic as their choice of murder weapon.

26 Medicinal Chemistry

CONTENTS

OBJECTIVES

This unit:

- Illustrates the contribution that chemistry has made to medicine
- Explains the stages involved in developing a medicine
- Describes how selected drugs were discovered and used

26.1 Medicinal chemistry

Medicinal (or pharmaceutical) chemistry is concerned with designing, synthesizing and developing pharmaceutical drugs. The pharmaceutical industry imposes strict controls over the raw materials and methods (the 'recipe') to make those drugs. The analysis of drug samples assures users of the quality of the manufactured product.

A drug may be described as a substance or mixture that has properties that modify bodily functions. A medicine uses these properties for treating or preventing disease. Four examples are shown in Figure 26.1.

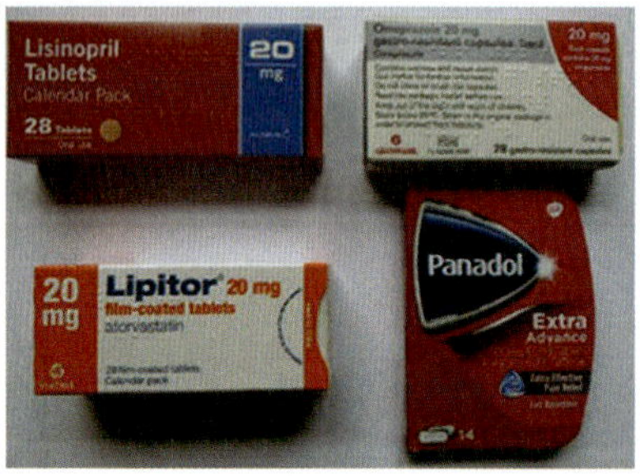

Figure 26.1 Examples of medicinal drugs. Lisinopril is an angiotensin-converting enzyme inhibitor that is used to reduce blood pressure. Omeprazole is a medication used in the treatment of gastroesophageal reflux disease. Atorvastatin (here sold under the brand name Lipitor), is a statin medication used to treat high cholesterol. Paracetamol with some caffeine (here sold under the brand name Panadol), is used to treat light or moderate pain. (Photo by Rufus Townsend.)

Compounds used as medicines are usually complicated organic compounds with long systematic names, so in the industry they are known by shorter generic names. Often, they will also have a brand name. Warfarin (**generic name**), for example, is used as an anti-coagulant. You will be prescribed it if you have a blood clot. It is sold under different **brand names** (one is Coumadin) and its **chemical name** is 4-hydroxy-3-(3-oxo-1-phenylbutyl)coumarin. Now carry out an internet search to complete Exercise 26A.

EXERCISE **26A**

Names of pharmaceutical compounds

A popular pain reliever is sold under the brand name ADVIL. What is:

(i) the generic name,

(ii) the chemical name of the active ingredient?

Drug development

In this unit we look at selected examples of drugs that have saved the lives or treated disease suffered by many millions of people. Medicinal drugs are big business, and their development can take many years and cost millions or even billions of dollars. Fewer drug discoveries are made by serendipity (that is, by chance) nowadays, and highly sophisticated computer modelling is used to identify molecules that might be effective in treating a disease. Ultimately, potentially useful drugs are tested in clinical trials.

A drug might be very effective in treating a condition, but if it severely poisons the patient at the concentrations used then it cannot usually be prescribed for use. The safety of drugs is an important part of drug research. Two important measurements of the acute toxicity for any potential poison are LD50 and LDLo, both expressed as the mass of the substance per unit of body mass (mg/kg). LD50 stands for 'Lethal dose 50' and it is the dose of a substance that, when administered in a single shot, is lethal to 50 per cent of a group (often a group of small animals) within a specified period. LDLo stands for 'Lethal Dose Low' and represents the lowest dose of a substance that is known to have caused death in a particular animal species under controlled conditions; in other words, it is the lowest dose that results in mortality. Note that LD50 > LDLo.

The delivery of drugs into the human body

The effectiveness of a drug once it gets into the body is one thing, but of equal importance is how that drug is administered, i.e. how it is absorbed by the human body. This might be by inhalation (as in a nasal spray), or by injection (for example, insulin injections) but the commonest delivery mechanism is by ingestion via the mouth. Ingenious methods are used to ensure that the contents of a drug-containing capsule are slowly released into the bloodstream.

EXERCISE **26B**

The solubility of drugs and their slow-release

Most drugs are big organic molecules.

(a) Why are many big drug molecules insoluble in water?

(b) Many drugs are only effective if they reach the bloodstream after ingestion. This requires them to be soluble in water (and therefore in blood). Suggest one way in which drugs can be made to be soluble in water.

(c) How can a pharmaceutical drug capsule be designed to slowly release its contents?

(d) Briefly outline how you would design a series of experiments to determine how quickly an ingested encapsulated drug (drug X) is released in the body. (Start with a large supply of identical capsule drugs.)

EXERCISE **26C**

Lethal dose of paracetamol

The oral LDLo of paracetamol in humans has been documented as 143 mg kg^{-1}. A tablet of paracetamol typically contains 0.5 g. How many paracetamol tablets would deliver this dose to a 43 kg person?

It is important to note that harm might be caused at doses well below the LDLo.

BOX **26.1**

Thalidomide

Thalidomide is a drug that was prescribed to treat morning sickness in pregnancy, from the late 1950s to 1961. During this time, a large number of children were born with serious deformities and it was eventually realized that prescription of thalidomide was a common factor. The drug was withdrawn from the market after being found to be a cause of birth deformities in humans. Initial tests had been carried out on rats and the drug hadn't had the same effect – rats metabolize it in a different way.

The molecule has a chiral centre (Figure 26.2) and exists as enantiomers. It was originally administered as an equal mixture of both (a racemic mixture), but now we know that whilst one enantiomer is effective as a sedative, the other is teratogenic – meaning that it interferes with the normal development of a foetus.

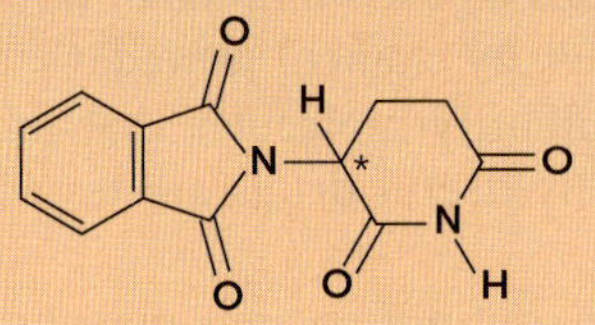

Figure 26.2 Structure of thalidomide, the chiral centre is marked with an asterisk.

The seemingly obvious solution to the problem would be to separate the two enantiomers and just use the 'safe' isomer, but the two isomers exist in equilibrium in solution and in a very short space of time there would again be a 50/50 mixture of each in the bloodstream.

26.1 Analgesics (painkillers)

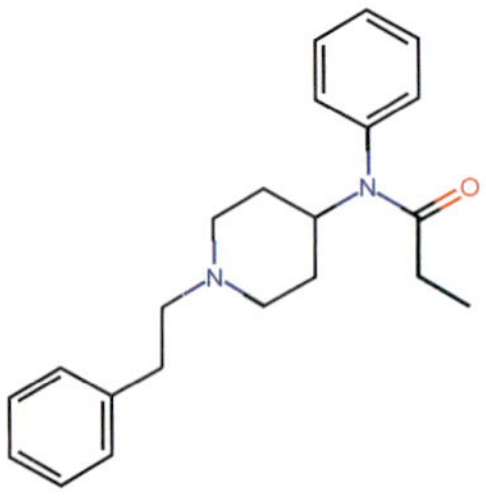

Figure 26.3 Fentanyl $C_{22}H_{28}N_2O$. Fentanyl is a powerful analgesic and a member of the opiate family of drugs. Highly addictive, it does not occur naturally and it is synthesized in laboratories.

Analgesics[1] range from aspirin, paracetamol, ibuprofen, co-codamol (a mixture of paracetamol and a moderately strong opioid called codeine) to the strong and addictive opioid painkillers like tramadol, morphine, diamorphine (heroin, made from morphine) and fentanyl. The mechanism by which these drugs act varies, but they all affect the nervous system by reducing the intensity of 'pain signals' the body sends to the brain.

Hippocrates, who lived around 400 BCE, left records of pain relief treatments, including the use of powdered leaves of the willow tree to relieve and to help heal pains and fevers. In the early nineteenth century it was discovered that the compound called salicin in willow plants was the ingredient that effected the pain relief. It turns out that salicin doesn't itself relieve pain, but the human body converts salicin into salicylic acid (Figure 26.4(a)) which is the painkiller. As demand for salicylic acid increased, the pharmaceutical company Bayer began to manufacture it synthetically, but it soon became apparent that it could have unpleasant side effects – irritation of the mouth and stomach. The German chemist Felix Hoffman prepared a number of compounds, each with a slight structural modification to Figure 26.4(a) and eventually found that the molecule shown in Figure 26.4(b), aspirin, had the pain relieving properties of salicylic acid but with reduced irritating side effects.

[1] **Analgesics** are different from **anaesthetics**, which eliminate sensation and often cause unconsciousness.

EXERCISE **26D**

Old bottles of aspirin

Very often bottles of aspirin that have been kept in a humid bathroom for a long time smell 'vinegary'. Can you suggest why this might be and write an equation for the reaction?

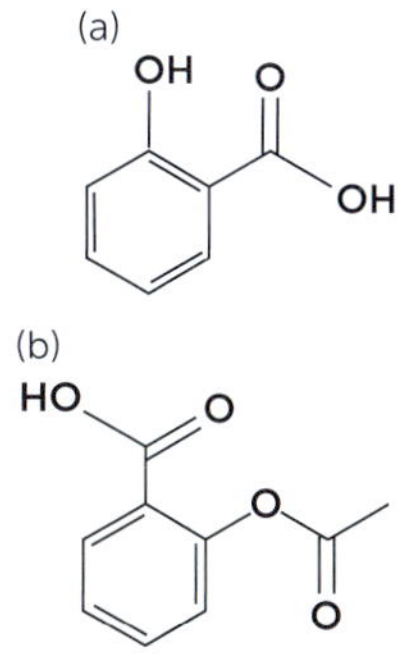

Figure 26.4 Structures of (a) salicylic acid or 2-hydroxybenzoic acid, and (b) aspirin, common chemical name acetylsalicylic acid, systematic name 2-acetoxybenzoic acid.

26.3 Sulfa drugs and antibiotics

Antibacterials are substances that stop bacteria from growing and causing disease (like sulfa drugs) or kill bacteria (like antibiotics).

We start with sulfa drugs. In 1932 it was discovered that mice injected with a red azo dye, named Prontosil, resisted bacterial infection. The structure of Prontosil is shown in Figure 26.5(a). The azo group is the –N=N– group in the middle. When Prontosil is broken down (**metabolized**) in the body, it forms a much simpler molecule named sulfanilamide and it was realized that this was the active molecule in inhibiting bacterial growth. Unfortunately, sulfanilamide was initially sold in the USA in ethylene glycol solvent (page 346), which proved lethally toxic to patients: seventy-six patients died before it was realized that it was the ethylene glycol that was poisoning them.

Sulfanilamide-type drugs ('sulfa' drugs) changed medicine. More sulfanilamide-type drugs were created and one of the most effective was found to be sulfapyridine, which was originally labelled as '**May and Baker 693**'. Sulfapyridine was often celebrated as a wonder drug and was even the subject of sermons in churches. Although there were toxic side effects, sulfapyridine was used to alleviate infections in the Second World War (1939–1945), including the treatment of dysentery in the Far East campaigns. It was also used to save the life of Winston Churchill, who contracted pneumonia in North Africa in 1943.

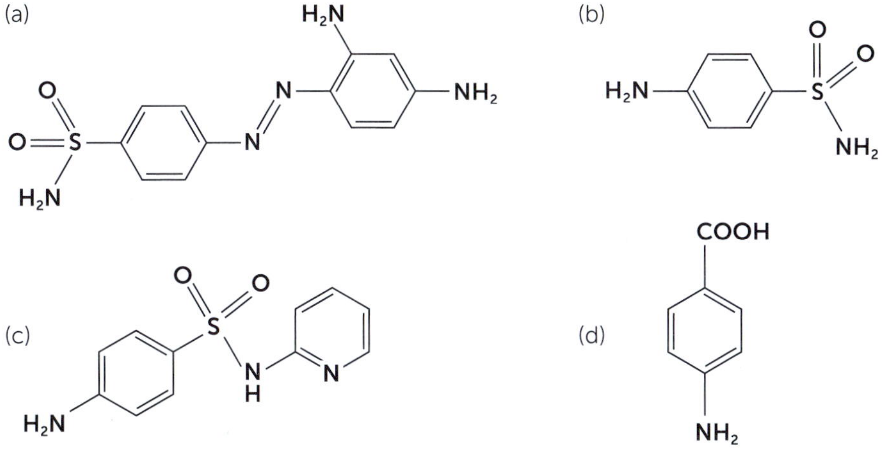

Figure 26.5 Molecular structures for (a) Prontosil, (b) sulfanilamide, (c) sulfapyridine ('May and Baker 693'), (d) 4-aminobenzoic acid (also known as para-aminobenzoic acid, PABA), a related molecule, is shown for comparison.

EXERCISE **26E**

Sulfanilamide

Explain why, despite their mechanism of action, the consumption of sulfa drugs does not produce a shortage of folic acid in humans.

Sulfa drugs are bacteriostatic; they slow down the growth of bacteria but do not kill them outright. They do not affect viruses. The mechanism of action is related to the fact that, as can be seen in Figure 26.5, sulfa drugs are chemically similar to 4-aminobenzoic acid (PABA). Sulfa drugs take the place of 4-aminobenzoic acid (PABA) on an enzyme within the bacterial cell and so prevent the bacterium from making folic acid (vitamin B), required for cell replication. Nowadays, antibiotics (which kill bacteria) have largely replaced sulfa drugs in treating infections, but sulfa drugs are still used in the treatment of other conditions, including diabetes.

BOX **26.2**

Tu Youyou and the fight against malaria

Tu Youyou (born 1930) was born in China and studied Chinese herbal medicine. Spurred on by ancient books on herbal medicine, she extracted an antimalarial drug called artemisinin from a plant called sweet wormwood. She volunteered to be the first test-subject for the drug and its use has saved the lives of millions of people. Tu Youyou was awarded the Nobel Prize for Medicine and Physiology in 2015.

Antibiotics: one of the most important scientific discoveries ever made

The term antibiotic is used when one substance or organism attacks and kills another. Antibiotics are **bacteriocidal** – they kill bacteria (but not viruses). Antibiotics can be used to treat a wider range of bacterial infections than sulfa drugs and for many years bacteria were found to be less likely to become resistant to their action.

The first antibiotic was penicillin. Alexander Fleming, born in Scotland, was a doctor during the Great Influenza Pandemic ('Spanish Flu') of 1918–1919 in which 20 million people died, principally from post-influenza infection by bacteria. In 1928, in what was subsequently seen as a chance discovery and the product of a particular set of laboratory and weather conditions, a mould on a petri dish of nutrient was observed to have destroyed staphylococcus bacteria. Fleming named the mould 'penicillin'. The preparation of larger amounts of penicillin was achieved only in 1940. Thereafter, American companies, aware of the potential of penicillin to save the lives of many servicemen, set up large-scale production plants to produce the 'miracle drug'. The 1949 film *The Third Man*, starring Orson Welles, is centred on the black market for penicillin in Vienna immediately after the Second World War.

As Figure 26.6(a) shows, a penicillin contains a ring consisting of three carbon atoms and one nitrogen atom. This is called a β-lactam ring, a ring that is also present in some molecules involved in the building of bacterial cell walls. Penicillin works by fooling a bacterial enzyme into attacking penicillin, so preventing the construction of the bacterium's cell wall – without which cell division is impossible.

By 1946 bacteria were becoming strongly resistant to penicillin. The response of scientists was to produce semi-synthetic penicillins, so-called because they were obtained by a mixture of chemical synthesis and fermentation. One example is amoxycillin. An unrelated group of antibiotics, called the cephalosporins, which are also obtained from moulds, has also proved remarkably successful. Nevertheless, resistant bacteria, particularly methicillin-resistant *Staphylococcus aureus* (MRSA), continue to cause concern.

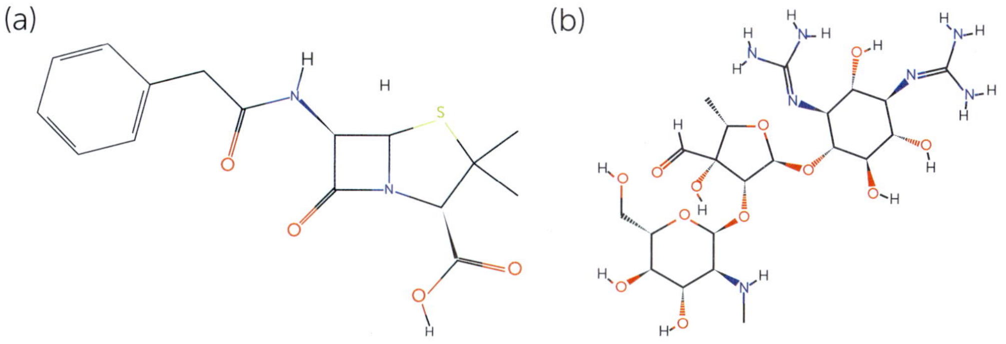

Figure 26.6 (a) An example of a penicillin called Penicillin-G (Benzylpenicillin). (b) Streptomycin (National Institute of Health, USA).

The antibiotic streptomycin occurs in natural bacteria and was discovered by Selman Waksman, Albert Schatz and Elizabeth Bugie in 1943. Streptomycin is effective against a large number of bacteria that do not respond to penicillin and its influence in virtually eradicating tuberculosis ('TB', 'Consumption', or 'the White Plague') in the post-Second World War period is one of the finest examples of the benefits of modern science.

Antibiotics or sulfa drugs are not effective against infections caused by viruses. This is because viruses are much simpler than bacteria, have no cell walls and do not make their own DNA.

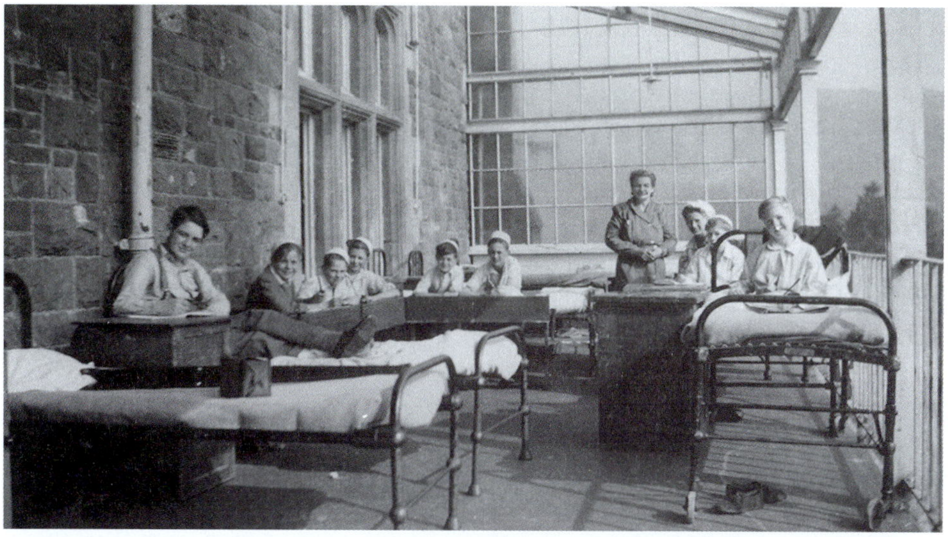

It has been estimated that TB has killed over a billion people in the last two centuries. Here, children are seen recovering in the Craig-Y-Nos tuberculosis sanatorium in 1951 (courtesy of Ann Shaw and Carole Reeves).

26.4 Drugs used to treat mental health disorders

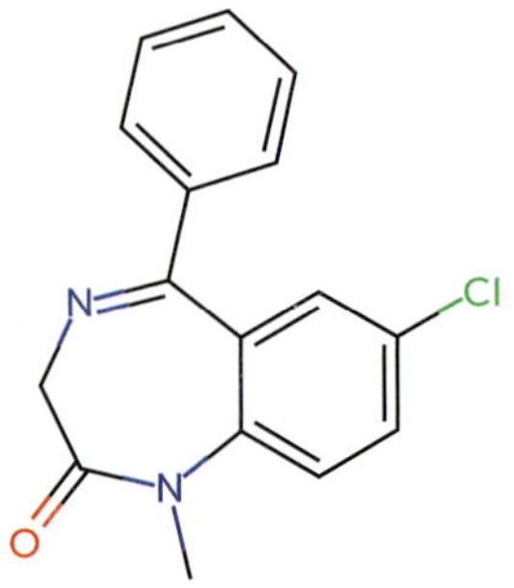

Figure 26.7 Diazepam, a chlorine containing drug.

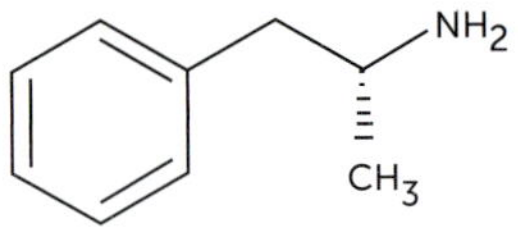

Figure 26.8 Structure of amphetamine.

There are many types of mental health disorders and many drugs used to treat them. We will pick out two to illustrate their value.

Diazepam (Figure 26.7), a member of the benzodiazepine group, made synthetically and first marketed as **Valium**. It is commonly used for treating anxiety, panic attacks, insomnia and seizures and in treating alcohol withdrawal symptoms.

How does diazepam work? In the human body, gamma-aminobutyric acid (GABA), blocks the stimulation of nerve cells in the brain. Diazepam amplifies, by chemical reaction, GABA's effect, leading to a lowering of activity of the central nervous system.

Amphetamine (1-phenylpropan-2-amine, Figure 26.8) is a stimulant that is known to produce increased concentration and decrease appetite. Amphetamine works by increasing the levels of dopamine and norepinephrine in the brain, inducing euphoria. Brand names of the drugs that contain, or metabolize into, amphetamine include Adderall, Dexedrine, as well as **Desoxyn**. The drug is also used illegally for recreation. A common street name for amphetamine is '**speed**'.

EXERCISE **26F**

Enantiomers of amphetamine

Amphetamine actually exists as two enantiomers. The one is given in Figure 26.8. Can you identify the second? Mark the chiral carbon atom in the molecule with a star.

26.5 Antihypertensives and cholesterol reducing statins

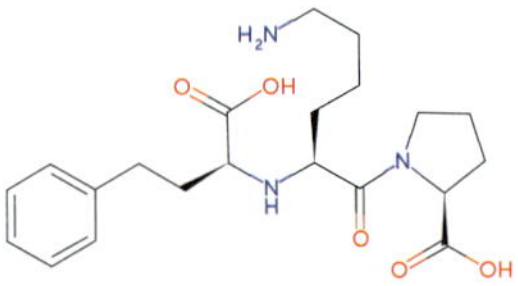

Figure 26.9 Lisinopril, used to control blood pressure.

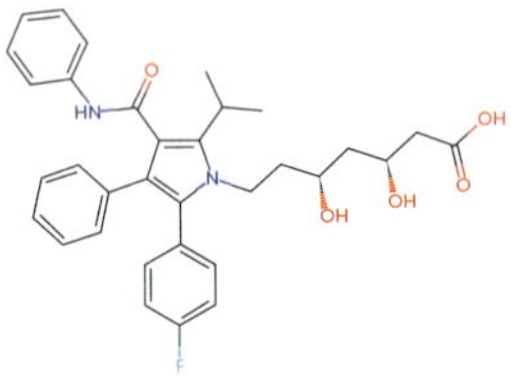

Figure 26.10 Atorvastatin, one of the biggest selling pharmaceutical drugs in history.

High blood pressure and high blood cholesterol levels are risk factors for heart disease. Drugs called antihypertensives are used to lower blood pressure. An example is Lisinopril (Figure 26.9). A hormone in the body called angiotensin II is known to increase blood pressure, and Lisinopril works by blocking the production of angiotensin II.

Atorvastatin (Figure 26.10) is used to reduce LDL ('bad') cholesterol in blood by blocking an enzyme that is essential for cholesterol production in the liver.

DID YOU KNOW?

85 per cent of all drugs approved by the USA Food and Drugs Administration contain nitrogen, and of the top 200 selling drugs, over 75 per cent have molecular structures with nitrogen atoms forming part of a ring (termed nitrogen heterocyclic compounds) that strongly resemble parts of biologically important molecules that occur widely in nature. The ability of N compounds to form hydrogen bonds also allows them to bind with enzymes and receptors in biological targets.

For an impressive inventory of heterocyclics, most of which contain N, see https://en.wikipedia.org/wiki/Heterocyclic_compound.

26.6 Reducing acidity in the stomach

An **antacid** is generally a base which neutralizes stomach acidity. Gastric acid has a pH of 1–2 and consists mostly of hydrochloric acid. Heartburn is a burning feeling in the middle of the chest caused by acid refluxing upwards from the stomach into the oesophagus, which doesn't have a protective lining as in the stomach, so irritation results. Antacids include compounds such as aluminium hydroxide, sodium hydrogencarbonate and magnesium hydroxide. Magnesium hydroxide occurs as a bulky, white powder that is practically insoluble in water and in alcohol. Milk of Magnesia is a suspension containing 80 mg of magnesium hydroxide per cm^3 of water.

Proton Pump Inhibitors (PPIs) are drugs that go further than simply neutralizing acids in the stomach; they reduce gastric acid production in the stomach before it becomes a problem. Their name comes from the fact that they slow down the pumping of hydrogen ions (protons) into the body. A commonly used PPI is Omeprazole (Figure 26.11).

> EXERCISE **26G**
>
> **Milk of Magnesia**
>
> Write a balanced equation for the reaction of gastric acid with magnesium hydroxide.

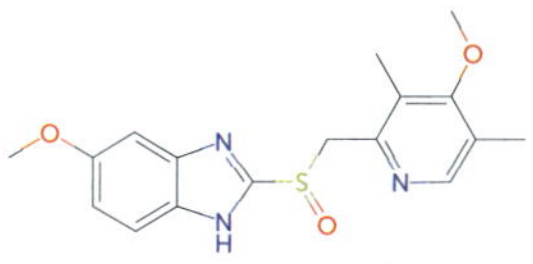

Figure 26.11 Omeprazole.

26.7 Chemotherapeutic drugs (anti-cancer drugs)

Until comparatively recently in human history, the average human lifespan was perhaps as low as thirty or forty years. As people lived longer, new diseases, including cancerous tumours and cancerous diseases of the blood, became more commonplace. It is now known that the group of diseases known as cancer is characterized by a rapid and uncontrolled cell division resulting in cell mutations. In addition, evidence began to accumulate of a connection between some chemical substances and the incidence of cancer, suggesting that cancer can be initiated by some materials, such as tobacco tar and some aromatic dyes.

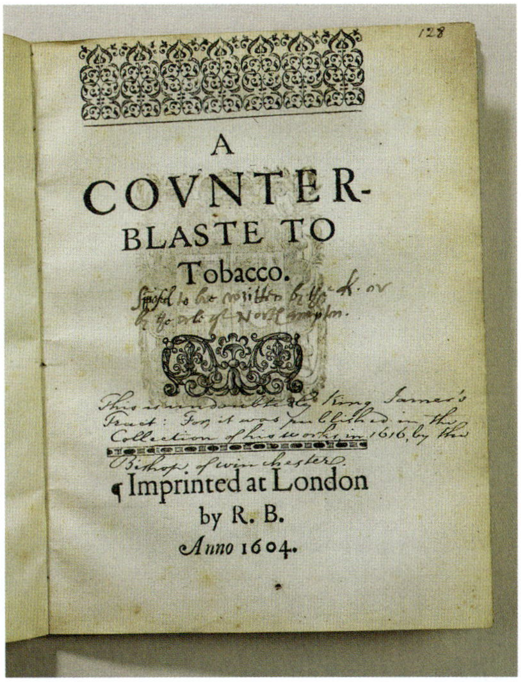

Smoking dramatically increases the risk of disease, including heart disease, throat and lung cancer. One of the earliest anti-smoking articles dates back to 1604. It was written by King James I of Great Britain and Ireland. James described smoking as 'a custome loathsome to the eye, hatefull to the Nose, harmefull to the braine, [and] dangerous to the lungs.' This first edition copy is part of the George Arents Collection, which documents the history, culture and lore of tobacco.

Cis-platin

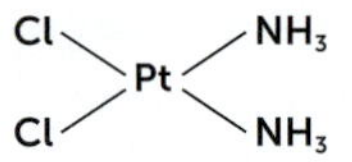

Figure 26.12 *Cis*-platin or *cis*-diamminedichloro-platinum(II). The prefix *cis* is used as in organic molecules since the same ligands are next to each other.

The compound *cis*-platin (Figure 26.12), also known by its brand name Neoplatin, is used in the treatment of many cancers, including those of the lymph glands, testicles, ovaries and brain. It was first synthesized in 1865 but its anti-cancer properties were discovered by accident by American Barnett Rosenberg in 1965 while observing the replication of bacteria in a solution containing platinum electrodes. Although it causes nausea and vomiting, it is particularly effective when combined with surgery and other drugs and is now being used in combination with taxol in the treatment of breast cancer. The way that *cis*-platin destroys cancerous tissue is now understood. First, *cis*-platin reacts with water forming the ion *cis*-$[PtCl(NH_3)_2(H_2O)]^+$. Second, the water molecule is replaced by guanine or adenine bases within DNA; these bases covalently bind to the Pt atom. This prevents DNA from participating in cell division and leading to the death of the cell. Since cancerous cells replicate rapidly, they are disproportionately affected. However, some healthy cells also grow rapidly (including blood cells in bone marrow and cells in the stomach, mouth, skin or hair), and this is the reason that all chemotherapeutic drugs, like *cis*-platin, also produce harmful side effects.

EXERCISE **26H**

Trans-platin

Trans-platin is the *trans*-stereoisomer of *cis*-platin, and does not exhibit a comparably useful pharmacological effect, perhaps because it gets in the way of DNA bases linking to the platinum atom. It is also more toxic, so it is important to ensure that batches of *cis*-platin don't contain the *trans* isomer

Draw the structure of *trans*-platin.

Taxol

Cis-platin is a small molecule, synthesized in factories. On the other hand, paclitaxel (trade name taxol) is a very large molecule (Figure 26.13) that occurs naturally. It is extracted from Pacific yew trees (*Taxus brevifolia*).

Microtubules are protein structures found in the cytoplasm (the fluid in a cell). They help give the cell rigidity but they also disassemble chromosomes during cell division. Taxol stabilizes the microtubes so preventing disassembly, preventing cell formation and causing the cancerous cell to die.

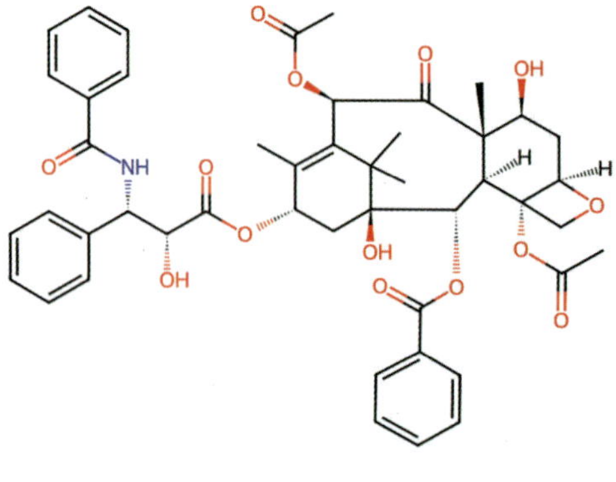

Figure 26.13 Taxol (Paclitaxel).

BOX **26.3**

Paul Ehrlich and chemotherapy

Nobel Prize-winning German Paul Ehrlich (1854–1915) is widely regarded as the father of chemotherapy. Chemotherapy is the treatment of disease using drugs, although the term is commonly given a narrower meaning, that of the specific application of drugs to the treatment of cancer. Using the term more widely, it includes antibiotics and also Ehrlich's personal discovery – the compound arsphenamine – which was used to treat the sexually transmitted disease syphilis until penicillin became available much later.

It was Ehrlich who was credited with the idea of the 'magic bullet' – a drug that targets a bacterium without affecting the host organism (for example, a human being). Many drugs have side effects, and so are not perfect magic bullets. Ehrlich was so famous that Hollywood produced a biographical film in 1940 entitled *Dr. Ehrlich's Magic Bullet*. He was also commemorated on German bank notes.

Paul Ehrlich.

Deutsche Bundesbank, Frankfurt am Main

REVISION QUESTIONS

26.1 Alka-Seltzer is a combination of aspirin, sodium hydrogencarbonate and citric acid, designed as an antacid and pain reliever. When the tablets are dropped into water, the acid reacts with sodium hydrogencarbonate to produce 'fizzing'.

(i) Write a balanced ionic equation for this reaction, bearing in mind that citric acid contains three carboxyl groups.

(ii) The tablets dissolve very quickly, as compared with ordinary aspirin tablets. How do you think this is achieved?

In an experiment to find the mass of citric acid (molar mass 210 g mol^{-1}) in an Alka-Seltzer tablet, two tablets were dissolved in a beaker of water positioned on a balance and the mass of the system was taken every fifteen seconds. It was assumed that the loss in mass of the system was completely due to the evolution of carbon dioxide produced by the reaction. The results are given below:

Loss of mass in system/g	Time/s	Loss of mass in system/g	Time/s
0	0		
0.17	15	0.7	165
0.43	30	0.72	180
0.53	45	0.73	195
0.57	60	0.73	210
0.59	75	0.74	225

Loss of mass in system/g	Time/s	Loss of mass in system/g	Time/s
0.61	90	0.76	240
0.63	105	0.77	255
0.65	120	0.79	270
0.66	135	0.79	285
0.69	150	0.8	300

(iii) Plot a graph of mass of system (y-axis) against time (x-axis). From your graph, estimate the mass of carbon dioxide released during the reaction.

(iv) Using the balanced equation for the reaction, calculate the mass of citric acid in each tablet.

(v) Your answer will be significantly lower than the mass quoted on the packet. Can you suggest deficiencies in the experiment that might account for this?

26.2 The structure of paracetamol (4-hydroxyacetylanilide) and its infrared spectrum are shown below:

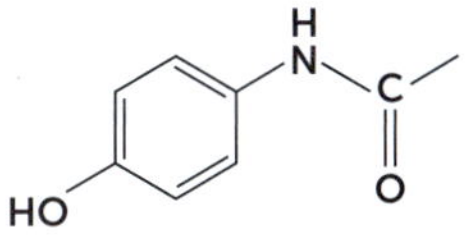

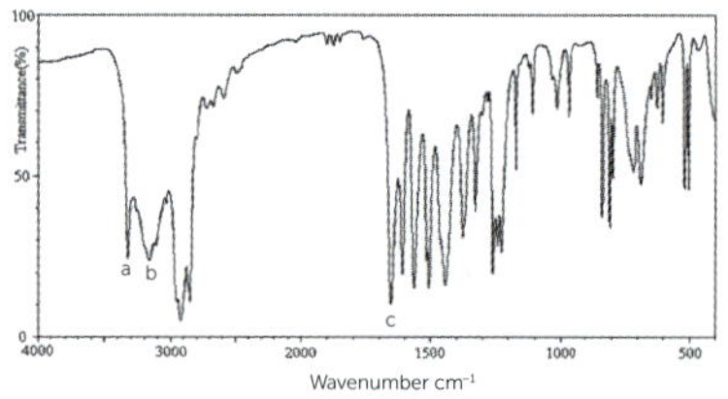

Calculate the molecular mass of paracetamol. Match covalent bonds to the peaks marked a, b and c on the IR spectrum

26.3 The mass spectrum of one of the molecules in Figure 26.5 is shown below. Identify the molecule, state its molecular formula (in the form $C_xH_yN_zO_pS_q$) and give structures for the molecular ions responsible for m/z = 172 and 156.

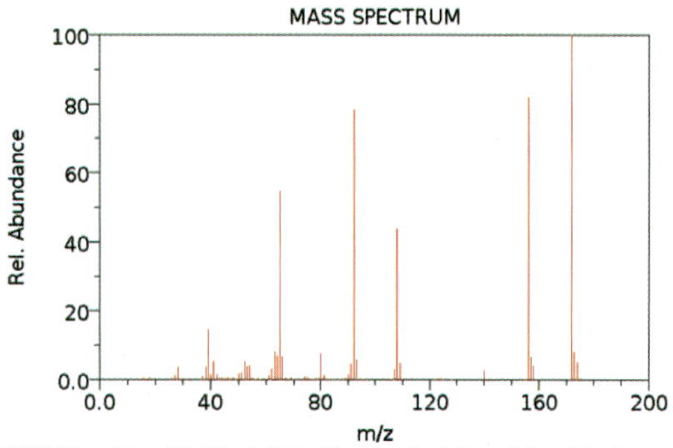

NIST Chemistry WebBook (http://webbook.nist.gov/chemistry)

26.4 When sulfanilamide is added to a petri dish containing streptococcal bacteria in agar (the 'food'), the bacteria cease to grow. If 4-aminobenzoic acid is added to the dish, bacterial growth resumes. Explain this observation.

26.5 *Cis*-platin is synthesized in the following sequence of reactions using reagents [A] to [D]. Identify the reagents in each case.

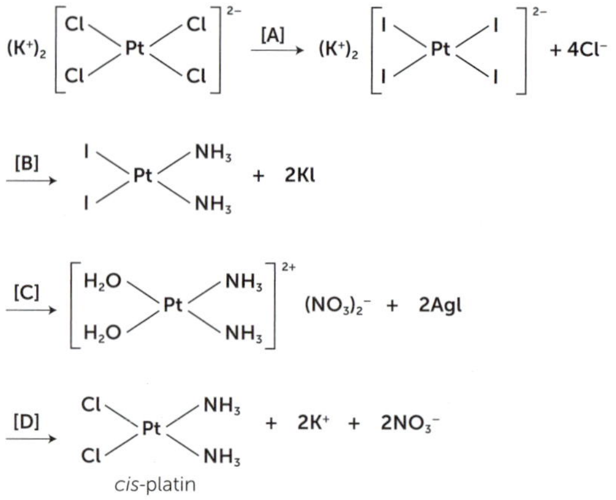

26.6 The neurotransmitter 4-aminobutanoic acid is found mostly as a zwitterion in solution. Can you write an equation for the equilibrium that exists between this molecule and its zwitterion?

26.7 Experimentally, how would you demonstrate the dangerous tars and byproducts from cigarette smoking to school children?

DID YOU KNOW?

In 1941, Albert Alexander, a policeman, was the first person to be treated with penicillin. He had contracted an infection after scratching his face on a rose bush. Despite the efforts of doctors, his head became covered with abscesses, one of which necessitated the removal of an eye. Nothing had helped. As his condition was critical, it was decided to use him for the first human trial of penicillin. He was given an infusion of penicillin and immediately started to improve. Soon, he was well on the way to recovery, but the supply of penicillin began to run out. Despite recycling penicillin recovered from his urine, the supply became exhausted and the bacteria began to regain their hold. Sadly he died soon after.

Extension material for Unit 26, discussing treatment for hayfever, is available on our website at https://www.bloomsburyonlineresources.com/chemistry.

Fire and Flame

27

OBJECTIVES

This unit:

- Summarizes the characteristics of fire and flames
- Defines flammability limits and flash point
- Outlines the reaction mechanism for the combustion of methane
- Explains the principles behind the chimney and the Davy lamp
- Describes the combustion of wood and other materials

CONTENTS

27.1 Fire

A fire is caused by the combustion of a fuel in an oxidant (often oxygen itself). It is well known that the essential preconditions of a fire are fuel, energy ('heat') to raise the temperature of the fuel to its autoignition temperature, an oxidizer (such as oxygen), but we should add a fourth. The fourth precondition is that of an exothermic chain reaction, that part of the chemistry of the reaction between the fuel and oxygen that allows the combustion to keep going at a fast enough rate. See Figure 27.1.

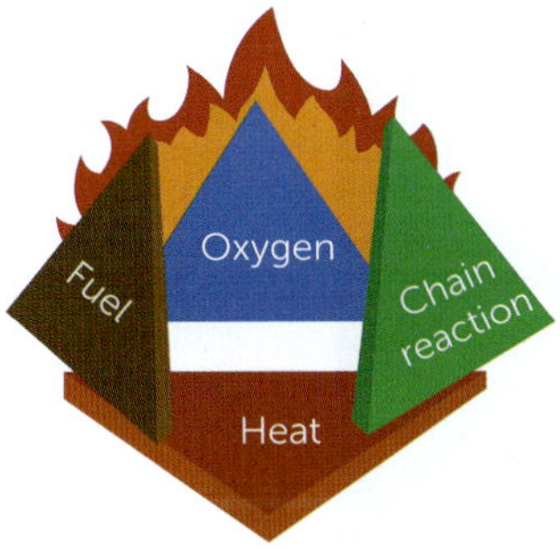

Figure 27.1 The fire tetrahedron, showing the four preconditions for a fire.

The heat is provided by an ignition source: a flaming match, an electric spark or even (for gases as in the diesel engine) by rapid compression. Combustion in a fire produces a flame. If the fuel is very rapidly consumed (e.g. when igniting a test tube of hydrogen with a lighted match), the flame may last for only a fraction of a second and there may also be an additional feature: gases may be heated so rapidly by the combustion that a pressure wave is produced, known as an **explosion**.

Figure 27.2 Fire and flames will continue for as long as the fuel and oxygen are available. Fires in coal-mine seams have been known to rage for hundreds of years.

27.2 Flame

The part of the fire we can see is called the flame. A flame is made from hot gas and particles, some of which emit light. The hot gas consists of the products of combustion and any unused fuel. In a candle flame (Figure 27.3), the fuel is a mixture of long-chain hydrocarbon candle waxes. Hot liquid wax travels along the wick to the flame zone where it is vaporized by heat, so producing a fuel that consists of gaseous hydrocarbons. The candle is an example of a **diffusion flame** because oxygen from the air must diffuse towards the flame, while vaporized hydrocarbon must diffuse outwards from the wick. Once formed, a candle flame is sustained by the heat released during the burning until all the wax is used up, with the oxygen for combustion being continually replenished from cooler heavier air that is drawn from the room and which enters combustion zones from the bottom sides of the flame – this is an example of the 'chimney effect' discussed below.

Figure 27.3 Candles, with many different fuels (paraffin wax, beeswax, sperm whale oil etc), have been used for thousands of years. The purpose of the wick is to draw liquid fuel into the flame by capillary action. One problem with candles was that as the wax was used up and the candle shortened, the top part of the wick would fall outside the flame, so that candle wicks had to trimmed every few hours or the candle would dim. A revolution in candle making came when, about 1825, the braided (plaited) wick was invented. This was self-trimming; it was designed to slowly curl into the flame as the candle burns, so allowing the top of the wick to burn away.

A flame is a burning zone: the frontier at which oxygen and fuel meet and react. The combustion reaction is confined to a surprisingly thin region around the edge of the flame, and this is true of all flames. If you think of the flame as a balloon, the combustion is restricted to the surface of the balloon: in terms of combustion, a flame is hollow. It is possible to place an unused matchstick inside a flame without the match head igniting. (This party trick is harder to do with a candle because it flickers more than (say) the flame of a Bunsen burner.)

The white part of the flame is caused by particles of white hot, unburnt soot. The yellow and red colour is caused by slightly cooler soot. The hottest part of the flame (about 1,400°C), known as the 'veil', is where most of the fuel is burnt because there is plenty of oxygen. The veil is barely visible: it appears as a haze just outside the luminous white part of the flame. The veil is actually blue, but its colour is masked by the intense luminous emission of the rest of the flame and our eyes cannot detect it.

Diffusion flames are inefficient because there is much wasted unburnt fuel. Industrial and domestically useful flames are usually **premixed**, meaning that air or oxygen is mixed with gaseous fuel *before* being ignited in a flame. The Bunsen burner and the domestic cooker fall into this category. The flame is generally bluer, more controllable and hotter (about 1,600°C) than the candle flame, but the exact flame temperature depends on the proportions of fuel and oxygen that are premixed.

The high temperature of a flame provides so much energy on the molecular scale that bonds break, producing low concentrations of free radicals, which are highly reactive molecules because they have odd electrons with which to form covalent bonds. The odd electrons are depicted as 'dots' in molecular formulae. Free radicals include CH and C_2, which are responsible for the blue colour of flames. In the high temperature of the flame, some CH and C_2 move to a high energy state, eventually falling back and emitting the extra energy as blue (and also some green) light.

EXERCISE 27A

(i) Why does a flame point upwards?

(ii) Draw diagrams showing the odd electrons in CH and C_2.

(a)

(b)

Figure 27.4 (a) A roaring Bunsen flame, an example of a premixed flame. (b) Robert Bunsen (1811–1899), father of the Bunsen burner.

BOX **27.1**

Michael Faraday and the 'Chemical History of a Candle'

Michael Faraday was one of the great experimenters of the nineteenth century. He started his scientific career as an assistant to Sir Humphry Davy. Faraday is best known for his pioneering discoveries in electricity and magnetism, but he was also an accomplished chemist. He discovered the important compound benzene (C_6H_6) and was the first to liquefy chlorine. A highly religious and ethical person, Faraday declined a government invitation to develop chemical weapons for the Crimean War of 1853–1856.

Michael Faraday was famous for giving lectures to the public in the Royal Institution, London, a tradition that continues to this day. His lecture series on the 'Chemical History of a Candle' remains in print.

Michael Faraday delivering a Christmas Lecture at the Royal Institution in 1856.

27.3 The chimney effect

Many fossil fires occur under controlled conditions in buildings, where escape channels (which we call chimneys or stacks) allow waste gases to be removed.

If fuel (e.g. coal) is lit in a grate, the flame is seen to flicker in all directions and some smoke might enter the room, but within a few minutes the flame and smoke are seen rushing up the chimney. In wood stoves with a narrow chimney the onset of this stage might involve a sudden 'whoosh' sound. This upward drive is caused by the **chimney** (or stack) **effect**.

The explanation for the chimney effect depends upon the fact that gas particles attempt to spread from regions of higher particle density to regions of lower particle density. When the fire is first lit, the air above the fire warms up. The molecules in the hot air are moving faster than the molecules of colder air, and this causes them to spread out and so become less dense and rise up. In effect, the hot gas is buoyant and it rises just like a cork when released under water. In a fire, the hot gas contains the products of combustion – soot, CO_2, H_2O, tar, unburned fuel and the flame itself (made of burning gas) – and all are drawn up the chimney. Cool air (containing fresh oxygen) then enters the base of the fire from the ventilated room and the process repeats itself. The overall effect is that there is a difference in gas pressure (the 'chimney draught') between the cool air entering the fire and the gas in the chimney, driving waste gases and smoke particles upwards to be vented into the outside air. The importance of the chimney is that it is an insulated physical structure that helps to sustain and contain a high column of hot gas, minimizing the cooling that occurs in fires in the open.

Figure 27.6 A tall chimney dominating the city skyline.

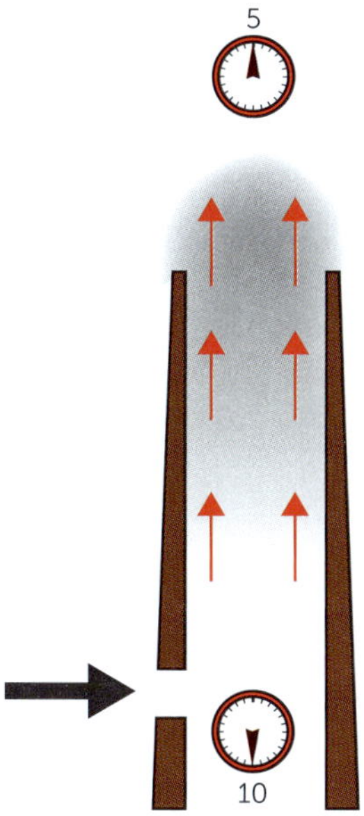

Figure 27.5 The chimney effect. Cold air (black arrow) enters the fire at the base of the chimney and the products of combustion (red arrows) rise. Gas pressure is lowest at the top of the chimney. The chimney is often designed to be narrower at the top as constricted gases move faster, further causing a reduction in gas pressure with height, and making it less likely that smoke will fall downward in the building.

27.4 Mechanism of combustion

The reactions involved in combustion are the same, whether combustion is uncontrolled (as in an explosion) or controlled (as in a continuous flame). The simplest hydrocarbon is methane. The overall equation for the burning of methane in oxygen (Figure 27.7) is deceiving because it suggests that the CO_2 and H_2O are produced in a single collision. In reality, all combustion reactions are complex, with over a 100 reactions in the mechanism, some of which are chain reactions (reactions where reactive products cause further reactions).

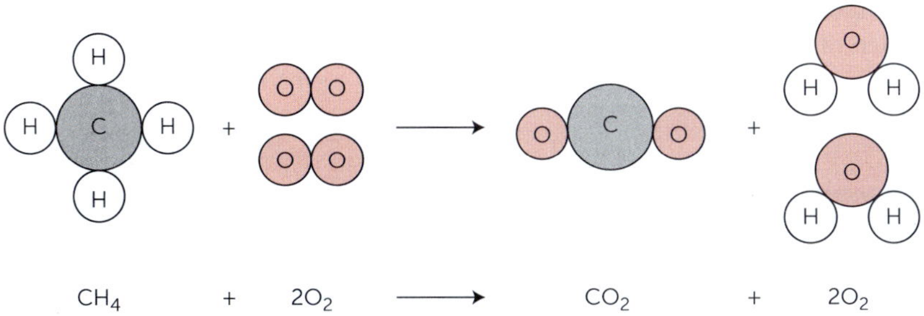

Figure 27.7 Overall equation for the combustion of methane.

EXERCISE 27B

Predict how the gas pressure difference between the bottom and top of a chimney changes with
(i) chimney height; and
(ii) fire temperature.

The complexity of combustion is illustrated by looking at a few of the stages involved when methane is burnt in air. A similar mechanism applies to the combustion of other hydrocarbons. We start with the breakup of the fuel, methane.

- Methane molecules dissociate by absorbing energy from the igniting flame and producing CH_3• (methyl) radicals and H• radicals (atoms):

$$CH_4(g) \rightarrow CH_3\bullet(g) + H\bullet(g) \quad (27.1)$$

- The H• is important because it reacts with oxygen producing oxygen O• and hydroxyl HO• radicals:

$$H\bullet(g) + O_2(g) \rightarrow O\bullet(g) + HO\bullet(g) \quad \textbf{(27.2)}$$

- The $CH_3\bullet$(g), O•(g) and HO•(g) radicals then take part in a number of reactions. In one sequence of reactions, $CH_3\bullet$(g) is converted into the formyl radical •CHO(g) which, in turn, reacts with HO• radicals:

$$\bullet CHO(g) + HO\bullet \rightarrow CO(g) + H_2O(g) \quad \textbf{(27.3)}$$

producing the poisonous gas carbon monoxide and water.

- The hydroxyl radical HO• has a further important role because it converts CO to CO_2, producing more H• that subsequently feeds the reaction (27.2):

$$HO\bullet(g) + CO(g) \rightarrow H\bullet(g) + CO_2(g) \quad \textbf{(27.4)}$$

Once the igniting flame starts reaction (27.1), combustion is self-sustaining, for two reasons. The first reason is a consequence of chemistry. Highly reactive free radicals are produced, consumed and regenerated: it is noteworthy that in reaction (27.2), two radicals are produced from one radical. The second reason is the availability of energy. Although reactions (27.1) and (27.2) are endothermic, reactions (27.3) and (27.4) are strongly exothermic and, overall, combustion generates more heat than it uses. This means that heat is available to perpetuate reactions (27.1) and (27.2) and produce more free radicals. The availability of energy and the chain reaction in which free radicals are regenerated work together, making the subsequent reactions faster and faster until either the fuel or oxygen runs out. This is similar to the chain reaction mechanism for the chlorination of methane (page 256).

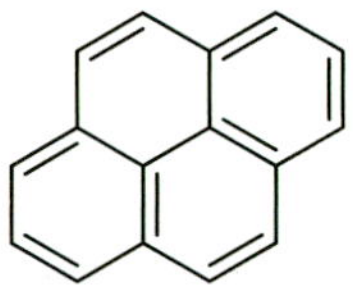

Pyrene, an example of a polycyclic aromatic hydrocarbon.

The only source of the oxygen in the oxygen-containing species (including CO and HO•) is O_2 from the air. If there is insufficient oxygen in the initial fuel mixture, there will be insufficient HO• radicals and the combustion is said to be 'incomplete', meaning that only a small number of the CO molecules produced in reaction (27.3) will be subsequently converted to CO_2 in reaction (27.4). It is for this reason that burning a fuel in the presence of very little air can be hazardous: if the products of combustion are not vented properly, people may be exposed to a dangerously high concentration of CO.

The presence of soot in a flame is an indicator of incomplete combustion. Soot is a mixture of charcoal (carbon) and polycyclic aromatic hydrocarbons (PAHs) – big molecules related to benzene and consisting of rings of carbon atoms bonded together. Both components result from a complex series of reactions that starts with the breakdown of hydrocarbons in the flame. If there is insufficient oxygen for the amount of fuel present, the carbon and PAHs do not burn and form a residue of soot that is deposited on cooler surfaces.

EXERCISE **27C**

(i) Do electric fires produce carbon monoxide?

(ii) Find out why carbon monoxide is so toxic.

(iii) Write a balanced overall equation for the combustion of $C_{25}H_{52}$, one of the hydrocarbons found in candle wax.

(iv) Bromotrifluoromethane (CF_3Br) is known to react with both H• and HO• radicals. Explain why, when a small quantity of CF_3Br is added to a methane/air mixture, it proves very difficult to ignite the mixture.

(v) Explain why the burning of methane in a shortage of oxygen produces less energy than a 2:1 methane/oxygen mixture.

27.5 Energy changes during combustion

The relatively low bond enthalpy of the O=O bond in the oxygen molecule is important in making the combustion of hydrocarbons strongly exothermic. The O=O bond is unusually weak for a double bond (compare with other double bonds in Table 13.4 on page 234) and so when strong bonds (C=O, O–H) are formed in the final products (CO_2, H_2O) of combustion, more energy is released than is needed to break the O=O and C–H bonds. As the calculation on page 234 shows, if the O=O bond were a strong bond, combustion would not be so exothermic and one might have to find another way of keeping warm.

Figure 27.8 shows that the enthalpy of combustion of the alkanes (methane, ethane, propane, butane, pentane, hexane and heptane) varies linearly with the number of carbon atoms.

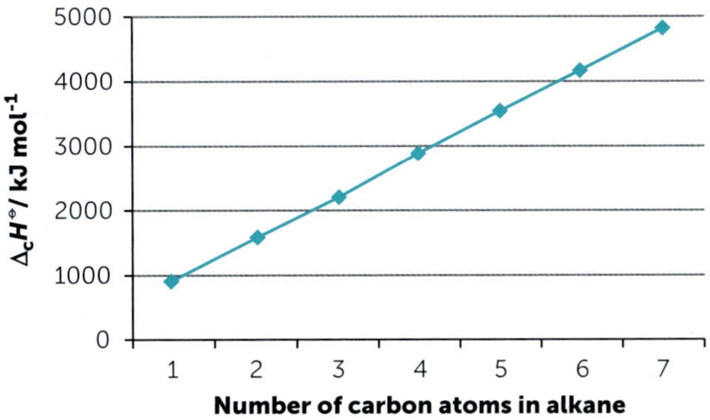

Figure 27.8 Variation of standard enthalpy of combustion of alkanes with the number of carbon atoms in the alkane molecule.

27.6 Autoignition temperature and flammability limits

Think of a fuel (wax, coal, petrol etc.) contained in a metal cup in air. Suppose the cup is placed into an electric oven: the important requirement is that there is no naked flame. If the temperature of the oven is raised, eventually a temperature will be reached when the fuel catches fire. This temperature is called the **autoignition temperature** in air, the 'auto' part of the term reflecting the spontaneous nature of the combustion. Some autoignition temperatures are listed in Table 27.1. Taking a mixture of hydrogen with air as an example, and provided that there is enough hydrogen and air to allow combustion (this will be discussed shortly), the data show that a container containing hydrogen gas and air would have to be heated to 500°C before it ignites.

Table 27.1 Data for flammable materials

Material	State at room temperature[1]	Flash point[2] /°C	Autoignition temperature[3]/°C	Flammability limits (% by volume in air)[4]
Dry pine wood	s	–	427	–
Benzene C_6H_6	l	–11	498	1.2–7.8
Ethyne (acetylene)	g	–	305	2.5–100
Methanol CH_3OH	l	11	464	6–36
Hydrogen H_2	g	–	500	4–74

Table 27.1 (Continued)

Material	State at room temperature[1]	Flash point[2] /°C	Autoignition temperature[3]/°C	Flammability limits (% by volume in air)[4]
Methane CH_4	g	–	537	5–15
Semi anthracite coal	s	–	400	N/A
Sulfur	s	–	243	N/A
White phosphorus	s	–	34	N/A
Paraffin wax	s	235	245–340	N/A

[1] s = solid, l = liquid, g = gas
[2] Applies only to liquids
[3] Also known as the thermal ignition temperature and (for solids) as the kindling temperature
[4] Also known as explosion limits. Data applies to gases or vapours at 25°C in *air* at 1 atm total gas pressure. It is not applicable (N/A) to other states. The lower value (e.g. 4 per cent for H_2) is known as the 'lower flammability limit'. The upper flammability limits are always greater in *pure* O_2.

Heating in a container to 500°C in a flame-free environment is not, of course, the way we would choose to burn hydrogen. We know that a test tube of hydrogen gas at room temperature is easily ignited by a naked flame, and there are two reasons for this. First, the temperature of a flame (even a candle flame) exceeds the autoignition temperature of hydrogen. Second, once ignited, the chain reaction generates heat, which perpetuates the combustion reactions even if the igniting flame has been extinguished.

However, there is always an additional requirement to be met before the ignition of an inflammable gas or vapour becomes possible. Hydrogen is easily ignited in air using a naked flame *provided* that the concentration of hydrogen in the fuel mixture with air is between 4 per cent and 74 per cent by volume, i.e. not too low or too high. Outside these limits, hydrogen does not ignite in a naked flame. Similar observations apply to all flammable fuel/air mixtures, and the ranges of concentration – here 4 per cent to 74 per cent – over which ignition is possible are known as the lower and upper **flammability limits**. They are also known as **explosion limits**, but they apply to combustion in general, irrespective of whether the gas mixture is used to form a steady flame or to take part in an explosion.

It might appear odd that flammable gases cannot be ignited by a naked flame if the concentration of gas is too low or too high and the explanation for flammability limits is to be found in the reaction mechanism of combustion. If the fuel concentration is

The vapour of some alcoholic cocktails exceeds the lower flammability limit of ethanol in air.

too low, the air and any other gases present absorb some of the heat generated in the combustion and prevent the combustion from continuing. If the fuel concentration is too high, there is insufficient oxygen to produce enough free radicals to set up a chain reaction: in this case we need only to allow the fuel (e.g. pure hydrogen in a test tube) to mix with air for a few seconds so as to dilute the fuel until its concentration drops to within the flammability limits.

Flammability limits apply to both gases and the vapours of flammable liquids. We have previously introduced the idea of the **flash point** of a flammable liquid as the lowest temperature at which the vapour above the liquid can be ignited by a naked flame. The connection between flash point and lower flammability limit is provided by the idea of vapour pressure, also introduced in Unit 10. At the temperature of the flash point, the vapour pressure of the flammable liquid is just sufficient for the concentration of vapour to reach its lower flammability limit. For example, the flash point of methanol is 11°C and the lower flammability limit for methanol is 6 per cent in air. 11°C is the minimum temperature at which the methanol concentration in the methanol vapour/air mixture above the liquid methanol surface reaches 6 per cent. Above 11°C, there would be more than 6 per cent methanol in the methanol/air mixture and, provided there is not more than 36 per cent, the methanol vapour will still ignite in a flame. Benzene has a much lower flash point (−11°C) than methanol but also a smaller flammability range: benzene/air mixtures with benzene concentrations above 7.8 per cent will not be ignited by a flame.

EXERCISE **27E**

(i) A sample of white phosphorus is taken from the fridge and placed on a glass dish on a summer's day. At what room temperature would the phosphorus ignite?

(ii) Benzene freezes at 5.5°C. Will solid benzene at its freezing point ignite in the presence of a flame?

(iii) A metal can with a valve is filled with pure methane gas. A lighted match is then held at point A and the valve opened. Immediately, a flame appears at A, followed by an explosion ten seconds later. Explain these observations. Explain why, when the experiment is repeated with pure hydrogen, there is a shorter delay between flame and explosion.

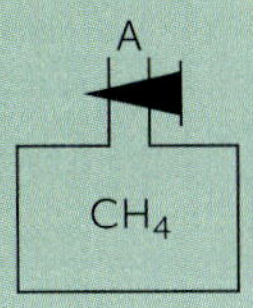

27.7 The burning of wood

There are two main stages in the production of heat from wood in a fire:

- In the first stage, flammable material is produced from the decomposition of the cellulose structure of the wood in the heat of a flame. Decomposition by heat is given the name **pyrolysis**, from the Greek *pyro* (for fire) and *lysis* (break up). The pyrolysis of wood begins at a temperature of 200–300°C and involves a series of reactions that produces wood gas (a mixture of gases including flammable methane, carbon monoxide and hydrogen) and brown tar, and which leaves charcoal as the final product. The pyrolysis reactions are endothermic, i.e. they use up energy, and so pyrolysis can only continue if there is an external source of heat. See Figure 27.9.

The burning of wood in a plentiful supply of oxygen involves many competing reactions. (Image by Matilda Simpson.)

- In the second (or burning) stage the tar, charcoal, CO, H_2 and CH_4 burn (all exothermic reactions) and the heat produced accelerates the pyrolysis of more wood, producing more flammable products. At above 500°C, charcoal is a strong reducing agent and it reacts with steam and also with CO_2:

$$C(s) + H_2O(g) \rightarrow CO(g) + H_2(g) \quad (\textit{exo}\text{thermic})$$

$$C(s) + CO_2(g) \rightarrow 2CO(g) \quad (\textit{endo}\text{thermic})$$

CO and H_2 burn with blue and very faintly blue flames, but sodium salts from the wood ash (the final solid product of wood combustion) colour the flame yellow.

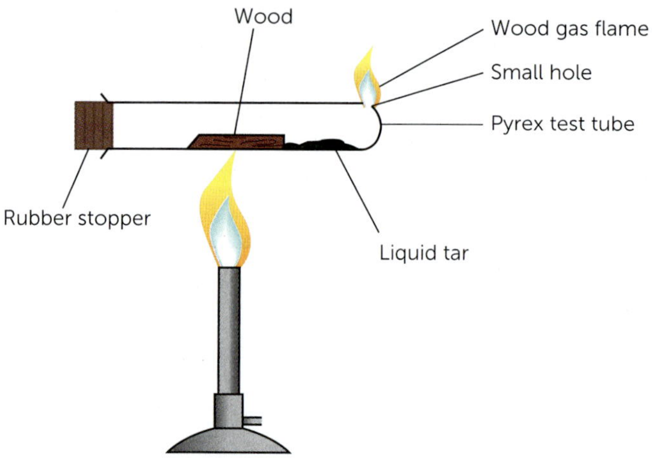

Figure 27.9 Demonstration of wood pyrolysis. The wood (e.g. a broken-up wood splint) is heated in a test tube by the Bunsen flame but there is no direct contact between the Bunsen flame and the wood. Once the wood begins to pyrolyse, a flaming match is used to ignite the wood gas emerging from the small hole in the test tube. Commercially, 1 kg of wood yields roughly 1 m^3 (at room temperature and pressure) of wood gas. Coal may be used instead of wood, producing coal gas and lots of tar.

EXERCISE 27E

(i) Why is it best to dry wood before burning it?

(ii) Why do the chimney emissions from a wood stove smell less as the stove gets hotter?

27.8 The Davy lamp

The autoignition temperature of methane is roughly 500°C, so that even a candle flame is hot enough to ignite methane provided its gaseous concentration is within its flammability limits in air (5–15 per cent by volume). Methane ('fire damp') explosions in coal mines are triply dangerous. First, the methane explosion may cause the collapse of mineshafts. Second, such combustion explosions may produce high concentrations of poisonous carbon monoxide because of a lack of oxygen. Third, a methane explosion could also ignite coal dust in the mine producing an even more powerful explosion.

The Davy safety lamp, designed by Sir Humphry Davy in 1815, depends on the fact that metals are excellent conductors of heat. If a fine metal gauze is placed over a flame and the gauze sealed to a base so that methane gas cannot reach the flame other than through the gauze, then, no matter how much fuel is available, the flame of the Davy lamp never travels through the gauze to the outside. This is because the metal gauze conducts so much heat away that the flammable methane gas does not reach its autoignition temperature and so a flame cannot be supported in the zone on either side of the gauze.[1] While the gauze does not prevent the movement of gases, it does prevent any flammable gases that enter the lamp from being ignited by the flame. It is therefore possible to use the Davy lamp as a light source even in the presence of potentially explosive methane.

Figure 27.10 The Davy lamp.

The Davy lamp also helped miners to detect flammable methane gas because methane causes the flame to burn with a blue tinge. At higher methane concentrations the lamp flame would be extinguished. Carbon dioxide ('black damp' or 'choke damp') does not support combustion and is not flammable, but high concentrations can asphyxiate workers. Knowing that CO_2 is heavier than air, miners suspecting high concentrations of carbon dioxide would place their Davy lamps on the ground. If the flame was extinguished, it was likely that CO_2 was present.

27.9 The Trench Effect and the King's Cross fire

If you light a match and you notice that the match is about to go out, what do you do? The usual remedy is to incline the match at an angle, causing the match to burn more fiercely (Figure 27.11). Fire will always spread more rapidly up a slope than it will on a flat surface. This is a consequence of the **Coanda effect**: if a stream of hot gas strikes an object, it produces a region of lower gas pressure between the gas and the object. If the gas stream is part of a flame and the object is a piece of wood, the flame is pulled towards the wood and it appears to stick to its surface. This is called **flame attachment**. Experiments show that the Coanda effect is particularly important in speeding up fires where the fuel is at an angle of at least 10° to the horizontal.

The Coanda effect is particularly important in a fire if – as in the matchstick – the flame attaches itself to an object that is also made wood. The attachment increases the contact between flame and fuel, causing the wood to decompose producing flammable wood gas and also preheating and drying unburnt wood. For this reason, an industrial or domestic wood fire might appear to be small in scale and not life threatening initially. However, if the concentration of methane in the air has built up so that it exceeds the lower flammability limit, the methane may suddenly ignite, producing an

[1]You can see this yourself. Place a fine *metal* tea strainer over a candle flame: the flame will not pass through the strainer. See also https://www.youtube.com/watch?v=rhavaxv3a40&t=82s.

Figure 27.11 The Coanda effect: both the visible and invisible parts of a fire spread faster up a slope. (Image by Helen Lewis.)

all-consuming fireball of super-heated gas (a **flashover**). This further accelerates the fire and produces dangerous smoke.

These effects were a major contributor to the loss of life in the King's Cross underground station disaster in London in November 1987, when a fire started at the foot of the wooden escalators – perhaps because of a lit cigarette being dropped into some oily workings under the escalator. The escalators were inclined at about 30° to the horizontal and the wood of the escalators acted as the bulk fuel. The escalators and their sides acted like a trench and boxed in the combustion, thus limiting the dispersion of the methane and heat away from the fire (**containment effect**). The fire was further intensified because the underground train tunnel provided fresh air and the escalator tunnel provided an exit upwards, so starting a chimney-like rush of gas (the **chimney effect**) and driving the fire upwards. The combined impact of the Coanda, flashover, containment and chimney effects led to a dangerous set of circumstances collectively known as the '**Trench Effect**'. In King's Cross tube station, the fire suddenly became life threatening when the methane ignited, the explosion heralding the rapid spread of the fire and smoke up the escalator and into the upper areas occupied by more commuters where some people were blown off their feet. Thirty-one people died in the fire and sixty others were injured. (An animation of the trench effect may be found at: http://www.tmba.tv/trial-graphics/arson-investigation/london-underground.)

Devastation caused by the 1987 King's Cross Underground fire.

27.10 'The toxic twins'

Particulates (smoke) are a dangerous feature of many fires. Carbon monoxide is another danger. CO is always produced from carbon-containing fuels in a shortage of oxygen. Since oxygen is rapidly consumed by burning, CO is a by-product of all fires in buildings. An additional poison is hydrogen cyanide gas produced from the combustion of nitrogen-containing fabrics, carpets and foams. HCN is even more poisonous than CO. Together, CO and HCN are known to fire fighters as the 'toxic twins'. (The mechanism by which CO and HCN kills is discussed in Box 12.10, page 210 and page 454.)

'Wildfires' are uncontrolled fires originating from an unplanned ignition source in natural vegetation. (Image by Stefan H. Doerr.)

In 2003 a fire occurred at a night club in Rhode Island, USA. Two pyrotechnic devices were set off during a band performance, igniting a polymer foam board surrounding the stage and producing smoke, CO and HCN. Over 100 people lost their lives.

DID YOU KNOW?

The match girls

Matches used to be made from white phosphorus. Workers at the match factories got 'phossy jaw', a painful, disfiguring and sometimes fatal disease caused by white phosphorus poisoning. In Britain, the use of white phosphorus in matches was prohibited from 1911.

Young women workers from a match factory in 1888. The young person on the right is selling matches in the street.

REVISION QUESTIONS

27.1 Briefly describe the mechanism by which carbon dioxide and water are produced from the combustion of hydrocarbons.

27.2 Ethanol, CH_3CH_2OH, has a flash point of 13°C, an autoignition temperature of 363°C and its flammability limits are 3.3–19 per cent by volume in air.

(i) A beaker of ethanol is cooled by a jacket of ice, taking its temperature to 5°C. A candle flame is held close to the mouth of the beaker. Will the ethanol ignite?

(ii) Diluting ethanol with water increases its flash point. The flash point of a 50 per cent ethanol/water mixture is 24°C. Taking brandy as 50 per cent ethanol by volume, predict whether a lighted match would ignite the vapour from a glass of brandy at 15°C.

(iii) Ethanol is heated until it becomes a gas and the gas pumped into a modified Bunsen burner where it mixes with air forming a mixture containing 19 per cent by volume of ethanol. Will this mixture ignite if exposed to a lighted match? What if the flow of ethanol is increased so that the combustion mixture contains more than 19 per cent of ethanol by volume in air?

(iv) Liquid ethanol and air are sealed into a strong steel container and heated to 400°C. When the container is allowed to cool and opened, black carbon is seen inside the container. What has happened?

27.3 The natural gas used in homes is nearly pure methane, typically containing 94 per cent methane and 5 per cent ethane. Before natural gas was piped long distances to homes, towns used to produce their own gas (known as 'town' gas or 'coal' gas) locally from the pyrolysis of coal. Typically, the coal gas[2] contained 50 per cent hydrogen, 35 per cent methane, 10 per cent carbon monoxide and small amounts of other hydrocarbons and smelly hydrogen sulfide.

(i) Why is the smelly compound t-butyl mercaptan $(CH_3)_3CSH$ added to domestic natural gas?

(ii) There are concerns that $(CH_3)_3CSH$ might react with the rust in gas storage tanks and so become converted into a less smelly compound. Briefly describe how you might investigate this in the laboratory.

(iii) Why is coal gas more poisonous than natural gas?

(iv) Explain why there are still dozens of deaths due to CO poisoning in homes even though natural gas does not contain carbon monoxide.

27.4 Explain why a forest fire spreads more quickly up a slope.

27.5 What part did the wooden escalators play in increasing the intensity and severity of the King's Cross fire of 1987?

27.6 Calculate the standard enthalpy change for the *incomplete* combustion of dicyanoacetylene in *oxygen*:

$$C_4N_2(g) + 2O_2(g) \rightarrow 4CO(g) + N_2(g)$$

given $\Delta_f H^{\ominus}$ (C_4N_2) = +626.2 kJ mol^{-1}. Use Table 13.2 (page 224) for the other data.

27.7 One of the reactions in the combustion of methane is:

$$CH_4(g) \rightarrow CH_3\bullet(g) + H\bullet(g)$$

(i) What is the significance of the 'dots' in the two species on the right-hand side?

(ii) Calculate $\Delta H^{\ominus}$ for this reaction. Use the NIST site (http://webbook.nist.gov/chemistry) for enthalpy of formation data. To use the NIST database, type the formula (e.g. CH3 – ignore subscripts); the data you require is listed under 'gas phase thermochemistry data'.

27.8 Most fatalities in fires arise from CO poisoning. In fires in buildings containing wood, CO is produced as a constituent of wood gas following wood pyrolysis. Very quickly, hot charcoal residues generate more CO by reacting with (a) steam and (b) CO_2. The following graph shows the production of CO in a room fire over the first 270 s after ignition.

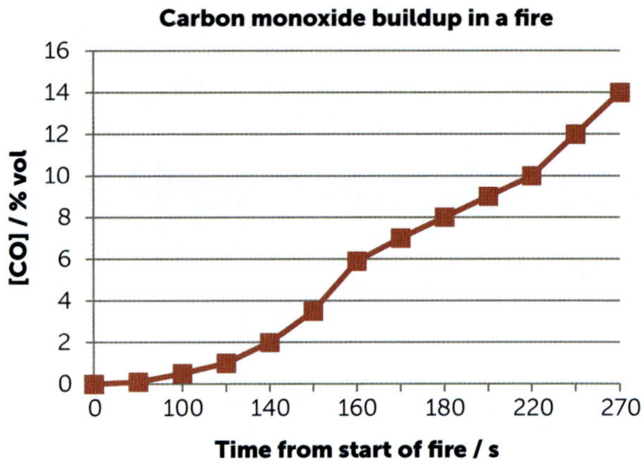

(i) Write equations showing how charcoal produces CO via reactions (a) and (b). Briefly describe an experiment to demonstrate that CO is produced via reaction (b).

[2]Coal gas, produced by the pyrolysis of coal, should not be confused with the gas extracted from underground coal beds which is 90 per cent methane.

(ii) CO is a fuel and readily burns in air making CO_2, yet the concentration of CO increases during the fire. Why?

(iii) The concentration threshold for immediate danger from CO is 1,200 parts per million by volume. Approximately how long did it take to reach this threshold?

(iv) In a fire, how might CO concentrations suddenly increase further?

27.8 What do you predict will happen to the lower flammability limit of methane if the percentage of oxygen in air is artificially increased to above 21 per cent v/v?

Extension 27 covers dangerous gases in coal mines and the use of canary birds as detectors: see https://www.bloomsburyonlineresources.com/chemistry.

28 Climate Change and Decarbonization

CONTENTS

OBJECTIVES

This unit:

- Explains the science of the greenhouse effect
- Introduces the most important greenhouse gases
- Gives examples of decarbonization in transport and industry
- Sets out the stages involved in carbon capture and storage

28.1 Introduction

Although climate change has occurred throughout the life of the Earth, we are specifically interested in climate change brought about by the activities of human beings (anthropogenic changes) that release certain gases ('**greenhouse gases**', GHGs), particularly carbon dioxide, into the atmosphere. Currently, annual global GHG emissions are 50 billion tons of CO_2 equivalent. A major concern is the increase in the Earth's temperature (**global warming)** caused by greenhouse gas emissions, a thermal insulating effect that is usually called the '**greenhouse effect**'. Global warming is believed to cause severe climatic changes including rising sea levels.

Volcanoes are a natural source of the greenhouse gas carbon dioxide but very substantial increases in carbon dioxide have been caused by global industrialization.

28.2 The greenhouse effect

Demonstrating the greenhouse effect in the laboratory

An example of the greenhouse effect is illustrated in Figure 28.1. Two identical glass bottles containing dry nitrogen (N_2) and carbon dioxide (CO_2) respectively, and at equal gas pressure, are exposed to a bright lamp placed at an equal distance from both bottles. Eventually, the gas temperature levels out and reaches a steady-state as the rate of energy losses (by conduction, convection and radiation) equals the rate that energy is being trapped. Temperature probes (electronic thermometers) show that the bottle containing CO_2 is at a higher steady-state temperature. Since the only difference between the two bottles is their content, we conclude that CO_2 is better at retaining energy. This is because CO_2 is a 'stronger' (more effective) greenhouse gas than N_2.

Explanation

CO_2 and N_2 do not absorb *visible* light. CO_2 does absorb (invisible) infrared radiation ('IR light') but N_2 does not. Although glass is a poor transmitter of IR light, some IR from the lamp gets through the walls of the bottle and some of this light is absorbed by the CO_2. The CO_2 then re-emits some of IR light but in all directions, including *back* into the body of the gas. The effect of the CO_2 is to reduce the loss of energy from the bottle of gas *compared* with N_2 and so the CO_2 gets hotter than the N_2.

Many solids absorb *visible* light. If we add charcoal or soil to the bottles, some of the visible light from the lamp is absorbed by the solid, but crucially, some of this energy is subsequently re-emitted as infrared light. Once more, the CO_2 absorbs some of this infrared light and re-emits it in all directions causing an even bigger rise in temperature. This is how a garden greenhouse works.

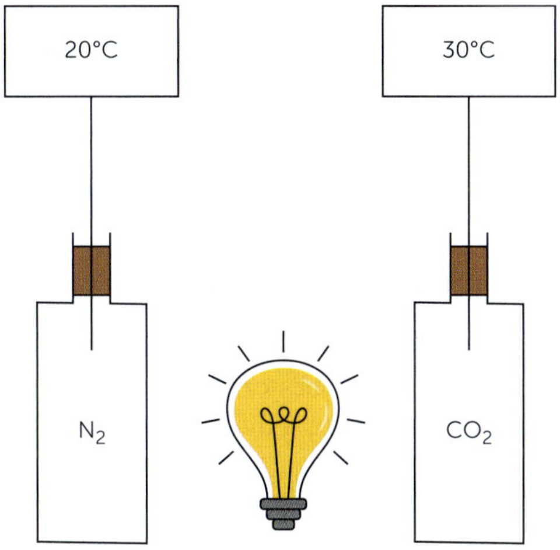

Figure 28.1 Relative greenhouse effect of carbon dioxide and nitrogen gases. Carbon dioxide gas warms up more than the nitrogen gas because CO_2 is a stronger absorber of infrared light.

BOX **28.1**

Eunice Newton Foote and the Greenhouse Effect

Eunice Foote (1819–1888) was an American inventor, scientist and women's rights campaigner.

In 1856 she set up tall glass cylinders and inserted a thermometer in each. She then added different gases to each cylinder and monitored the temperature of the gases when they were all exposed to sunlight at the same time. She found that carbon dioxide got warmer than air. Her publication in the *American Journal of Art and Science* was one of the first to anticipate what we now call the greenhouse effect.

See https://publicdomainreview.org/collection/first-paper-to-link-co2-and-global-warming-by-eunice-foote-1856

The effect of the greenhouse effect in the troposphere and stratosphere

The greenhouse effect has always operated on the Earth and, without it, the temperature on Earth would be a super cold −18°C rather than an average of 15°C. This 'natural' greenhouse effect, essentially a way of thermally insulating the part of the Earth in which we live, is mainly caused by naturally occurring water vapour, carbon dioxide, methane and nitrous oxide. It is not linked with human activity. Our concern is *additional* global warming due to *additional* concentrations of greenhouse gases produced by *human activity* (see Table 28.1); this is sometimes called the **enhanced greenhouse effect**. However, the mechanism of Earth warming is the same whether the greenhouse gases are natural or made by human activity.

In examining the greenhouse effect and its effect on the Earth's temperature, we borrow some of the ideas from Figure 28.1. There is no glass around the Earth as there was in our experiment. and the Sun replaces the lamp.

To start, imagine that there are no greenhouse gases in the atmosphere of the Earth – even naturally occurring ones. *Visible* light from the Sun reaches ground-level where some of it is absorbed and re-emitted by the solid Earth as *infrared* light. With no gases (like CO_2) to absorb the re-emitted IR, the lower part of the atmosphere (the *troposphere*, up to an altitude of 10 km and where we live) remains cold because much of the IR energy radiates upwards and warms the upper atmosphere (called the *stratosphere*, at 10–50 km). It is this scenario (Figure 28.2(a)) that would make the Earth so cold that its average surface temperature would be −18°C.

In fact, we know that CO_2 (and other greenhouse gases) exist in the troposphere that clings around the solid Earth. These greenhouse gases have a substantial impact on the Earth's temperature even though they are (apart from water vapour) present at surprisingly low concentrations (Table 28.1). The mechanism is as follows. Some of the IR emitted from the solid Earth passes upward into the troposphere where it is re-absorbed and re-emitted in all directions by molecules of greenhouse gases (particularly CO_2). The re-emitted IR that is directed *downwards* causes the troposphere to warm up more than it otherwise would, creating an average steady-state temperature that allows us to live on the Earth (Figure 28.2(b)). The mathematical models used by climate scientists show that the higher the concentration of greenhouse gases, particularly CO_2, the hotter the Earth becomes.

The energy trapping in the troposphere caused by greenhouse gases means that there is less IR energy available to radiate into the stratosphere so that the steady-state temperature of the stratosphere is lower than it would be in the absence of greenhouse gases. This is called **stratospheric cooling.** The presence of additional greenhouse gases from human activity increases this effect, bringing about additional stratospheric cooling. Increased stratospheric cooling[1] was observed after it was predicted theoretically and provides evidence of an enhanced natural greenhouse effect, one due to anthropogenic emissions.

Greenhouse gases cause the troposphere to get warmer and the stratosphere to get colder

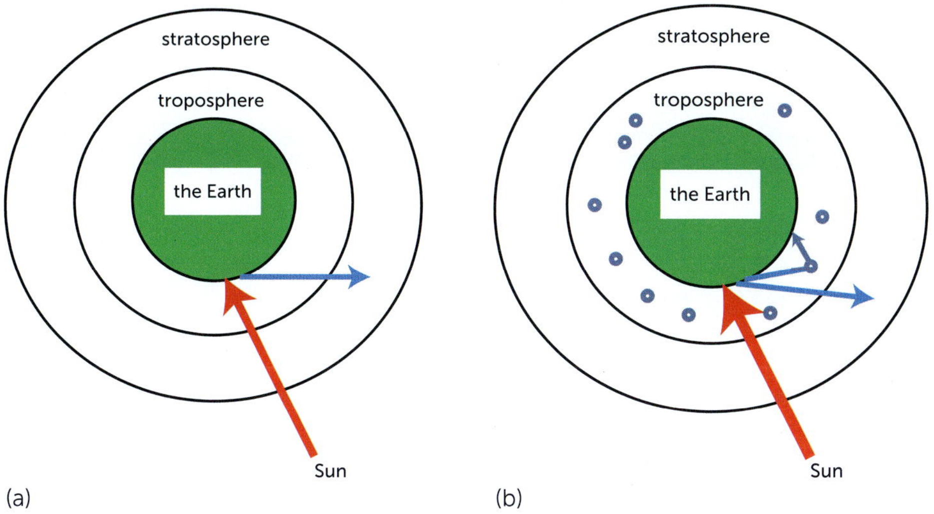

Figure 28.2 The greenhouse effect on Earth

(a) The scenario with no greenhouse gases in the troposphere. Visible light from the Sun (red arrow) is absorbed by the solid Earth which re-emits part of the energy as IR radiation (blue arrow), some of which passes upwards and warms up the stratosphere. The Earth is cold because it is not thermally insulated.
(b) The scenario with greenhouse gases (molecules symbolized by circles) in the troposphere. Here, there is an additional option for the IR re-emitted from the Earth: some of it can be re-absorbed by the greenhouse gases and directed back to Earth. This acts as a kind of thermal insulation and causes the steady-state temperature of the Earth to be higher than it would otherwise be.

---

[1]Reductions in stratospheric ozone are a separate cause of stratospheric cooling, since fewer ozone molecules mean fewer UV photons being absorbed. However, while stratospheric ozone levels are returning to their pre-CFC levels, tropospheric greenhouse gas concentrations are increasing.

Table 28.1 Greenhouse gases. the concentration of CO_2 today is 420,000 ppb or 0.42 per cent

Name	Formula	Source	Global Warming Potential (GWP) relative to CO_2	Per cent in troposphere by parts per billion by volume
Carbon dioxide	CO_2	Burning of fossil fuels and from industry (e.g. cement production)	1	420,000 (It was about 300,000 in 1950).
Methane	CH_4	Burning, production and transport (leakages) of fossil fuels, and from farm animals	30	1,900
(Tropospheric) Ozone	O_3	Produced from the reactions of vehicle generated ground pollutants	1000	Strong regional and local variability
Nitrous oxide	N_2O	Agriculture and wastewater treatment	273	330
Fluorine compounds	e.g. $CHCl_2CF_3$ and SF_6	Includes the solvents CFCs and HCFCs. SF_6 used as electrical insulator	All are > 1,000. SF_6 = 23,000	0.0115 (for SF_6)

DID YOU KNOW?

The temperature of the Earth has increased by an average of 1.1°C since pre-industrial times, alongside increases in atmospheric CO_2. Warming has been greater at the Arctic and Antarctic poles and greater in the northern rather than the southern hemisphere. Water vapour plays an important role in the 'natural' greenhouse effect, but it's not usually listed as a greenhouse gas because although the reservoir of liquid water on Earth is vast, its concentration in the troposphere is controlled by temperature rather than by human activity. Any increase in the Earth's temperature causes more liquid water to evaporate into water vapour, which as a greenhouse gas accelerates temperature change. For this reason water is said to *amplify* global warming.

EXERCISE **28A**

(a) What evidence is there that CO_2 and N_2 do not absorb visible light? What evidence is there that CO_2 absorbs infrared radiation?

(b) Explain why greenhouse gases are always strong absorbers of infrared light.

(c) Eunice Foote discovered that moist CO_2 is a better retainer of energy than dry CO_2. Why?

(d) Chlorofluorocarbons are stronger greenhouse gases than carbon dioxide, but are less important overall. Why?

28.3 Net zero and decarbonization

Under the 2015 Paris climate agreement, about 200 countries agreed to limit global temperature rises to 1.5°C by 2100. This requires greenhouse gas emissions – especially CO_2 – to be massively reduced. '**Net zero**' is a commitment by a government or company to reduce its net greenhouse gas emissions to zero. **Decarbonization** refers to the process of reducing or eliminating carbon dioxide (CO_2) and other global warming gases from the atmosphere, often through the transition to renewable energy sources.

Decarbonization is both revolutionary and highly technological. *Revolutionary*, because it involves altering the way we manufacture, make food and produce energy. Agriculture, for example, is a major producer of carbon dioxide, methane and nitrous oxide. The construction industry is also energy demanding and the materials it uses (including brick, steel and concrete) produce massive amounts of CO_2. *Highly technological* because it is proposed that new technology and improved efficiency will *substitute* for highly carbonized economies and so allow populations to maintain their health services, economies, food supplies and standards of living.

From the chemistry point of view, decarbonization has many benefits. It not only reduces greenhouse gases, it also reduces conventional pollution by gases and particulates. By concentrating on efficiency, and re-using materials, it supports attempts to make industrial chemistry more sustainable.

Decarbonization of energy is different from defossilization. Manufacturing industries produce pharmaceuticals, solvents, paints and fabrics but they usually use oil or methane as the raw materials. In the worldwide chemical and oil industry, decarbonization refers to reducing emissions by electrification and improved process efficiency, but defossilization refers to replacing feedstocks based on carbon (e.g. oil or methane) with alternative, non-fossil sources of carbon such as biomass and plastic waste.

EXERCISE **28B**

Breconshire Foods cooks food within its small factory. It has recently replaced the gas cookers with electrical cookers that produce virtually no greenhouse gas emissions. Can the company claim that it achieves net zero emissions?

Decarbonizing energy

Any industrial process that transports, stores, heats, compresses or pumps materials requires energy. In some countries, greenhouse gas levels have fallen considerably because fossil fuel consumption has fallen, but it is believed that further and drastic reductions are required to avoid the very damaging effects of further global warming. **Renewable energy** is energy produced from natural sources, with those natural sources being replenished at the same rate as they are consumed. Examples include solar energy, wind energy, tidal and sea wind energy, biomass (methane from the digestion of waste), geothermal energy (harnessing the energy deep in the ground) and hydroelectric schemes. Renewable energy is classified as '**low carbon**' because it produces fewer emissions than using fossil fuels. Nuclear energy is also low carbon, as it produces no CO_2 when it is operating, but is not renewable as the nuclear fuel is slowly used up and has to replaced by extracting new uranium from the world's limited supplies of uranium ore.

Decarbonizing energy for transport by using electricity

There are considerable benefits to expanding the uses of electricity ('electrification') on the grounds of efficiency. For example, although the typical range of battery electric vehicles (BEVs), which run solely on electricity, is currently less than diesel or petrol cars, they can travel twice as far *per kilowatt* of energy as a diesel or petrol car. This is because BEVs waste much less energy than fossil fuel engines. BEVs use approximately 90 per cent of the energy in the battery and from regenerative braking to propel the vehicle. Petrol/diesel vehicles only convert 16 to 25 per cent of the energy in the fuel into vehicular motion. Similar benefits apply to the electrification of railways. The extra efficiency reduces overall CO_2 emissions even if the electricity used is obtained from fossil fuels, but to achieve ambitious emission target reductions it is essential that electricity from renewable sources is adopted as quickly as possible.

Battery electric vehicles (like this one) waste less energy than diesel or petrol cars. (Image by Hetty Simpson.)

28.4 Hydrogen as an energy carrier for cars

Hydrogen may be used as an energy store to provide the energy for cars when needed. The most common type of hydrogen vehicle, called a Fuel Cell Electric Vehicle (FCEV), is an electric car that makes its own electricity using an on-board **fuel cell** that breaks down hydrogen gas into electricity and (non-polluting) water:

$$2H_2(g) + O_2(g) \rightarrow 2H_2O(g) + 4e^-$$

Currently, most of the world's hydrogen is made from fossil fuels, mainly by the reaction of steam with methane (see page 327) at high temperature and pressure. This process uses lots of energy and directly produces CO_2. One of the biggest challenges in using hydrogen in cars and buses is to make pure hydrogen without fossil fuels ('green hydrogen') so that CO_2 emissions are reduced.

Hydrogen may be made from the electrolysis of water (see page 119). The thermodynamics of electrolysis may be considered (at constant T and P) using the Gibbs equation (see page 292) which involves thermodynamic quantities of the reaction system only – and not of the surroundings:

$$\Delta G^\circ = \Delta H^\circ - T\Delta S^\circ$$

The decomposition of liquid water:

$$H_2O(l) \rightarrow H_2(g) + 0.5O_2(g) \qquad \Delta G^\circ(298\ K) = +237.13 kJ$$

is not spontaneous at room temperature, as evidenced by a positive value of $\Delta G^\circ(298\ K)$. To force electrolysis to happen, the electric current must provide *more* than 237.13 kJ of energy. In practice, considerably more than 237.13 kJ of energy is required to make electrolysis fast enough. Electrolysis is therefore very energy 'greedy'. The hydrogen gas would also need to be compressed during storage – another source of energy expenditure.

FCEVs have longer ranges than BEVs and shorter filling times. However, for hydrogen to be 'green' the electricity used in electrolysis must have been generated from low-carbon energy, and since the fewer the stages involved in a process the more efficient is that process overall, a more efficient use of that electricity would be to use it directly in a standard electric car (BEV) rather than produce hydrogen by electrolysis for subsequent use in a FCEV hydrogen car. In addition, the basic electric infrastructure (national grid) already exists for use with BEVs, whereas countries wishing to adopt hydrogen for cars on a large scale would require a massive investment in the facilities needed for the transport, storage and re-fuelling of hydrogen.

28.5 Decarbonizing industrial processes

We take the Haber-Bosch process as an example. Commercial hydrogen is produced on a vast scale by the reaction of steam with methane (page 327). The first stage of this process is energy demanding and the second stage directly produces huge amounts of CO_2, which is separated off from the H_2. The CO_2 is usually released into the atmosphere. In the Haber-Bosch process (page 278), which also requires energy to run, hydrogen from the steam/methane reaction is reacted with nitrogen from the air. It is estimated that making hydrogen and making ammonia in this way produces about 1.4 per cent of the world's CO_2 emissions due to human activities, with 90 per cent of this contribution being generated from the manufacture of hydrogen rather than from the Haber-Bosch process itself.

We cannot easily stop producing ammonia because the fertilizers made from ammonia are crucial to increase crop yields to feed a growing world population.

Steam reforming plants are highly efficient at producing hydrogen, but they generate vast quantities of CO_2.

To decarbonize ammonia production, three avenues are being explored: first, and most important, is to produce hydrogen by electrolysing water using electricity generated from renewable sources; second, to use carbon capture and storage (CCS) in both the steam reforming and Haber-Bosch stages; and third, for the Haber-Bosch process itself, to use new catalysts to reduce the temperature and pressure needed in ammonia production; despite some hopeful discoveries a commercially viable new catalyst has yet to be developed.

BOX **28.1**

Using electric arc furnaces in the manufacture of steel

Iron for steel making is produced from iron ore using a blast furnace-basic oxygen furnace. The energy requirements of blast furnaces are huge. Not only does this process produce vast amounts of CO_2 but the method is unsuitable for recycling scrap metal with more than 30 per cent iron or steel. An electric arc furnace (EAF) works by producing a long-lasting high voltage spark ('arc') between two carbon electrodes, so generating the high temperature needed to melt iron. EAF can be used to purify even 100 per cent scrap iron or steel, so reducing the amount of iron needed to be extracted from its ore and causing a net reduction in CO_2 emissions per tonne of steel.

An electric arc furnace

28.6 Carbon capture and storage

A new class of technologies based around the capture and storage (or 'sequestration') of CO_2 is being explored internationally for use with energy intensive industries such as cement making. One approach is to absorb the CO_2 emitted by an industrial process in an amine solution (a base), represented as R-NH_2. The absorption is driven by an exothermic equilibrium reaction producing a protonated amine (RNH_3^+) and a carbamate ion:

$$2R\text{-}NH_2 + CO_2 \rightleftharpoons RNH_3^+ + RNHCOO^-$$

Subsequent heating of the solution reverses the reaction, regenerating the amine (which is recycled and used again) and CO_2 gas; the CO_2 is pumped away and stored in underground geological formations. See Figure 28.3.

CCS is expensive and itself uses energy but overall it is proposed that CCS would move more CO_2 into long-term storage than it generates from its operation. CCS is a major part of plans to reduce CO_2 emissions, but it is currently a long way from achieving the scale needed to make a sizeable impact.

(a)

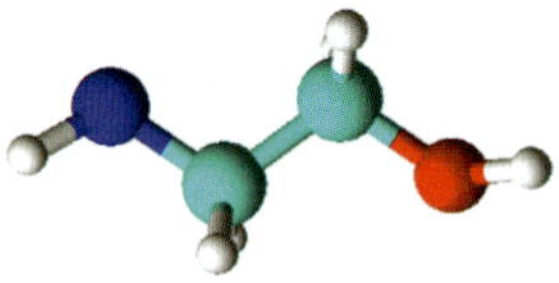

(b)

One of the amines used as a solvent in CCS is monoethanolamine (MEA). (a) structural and (b) ball-and-stick representations of MEA. The second hydrogen atom of the NH_2 group in (b) is not visible because it is pointing into the page.

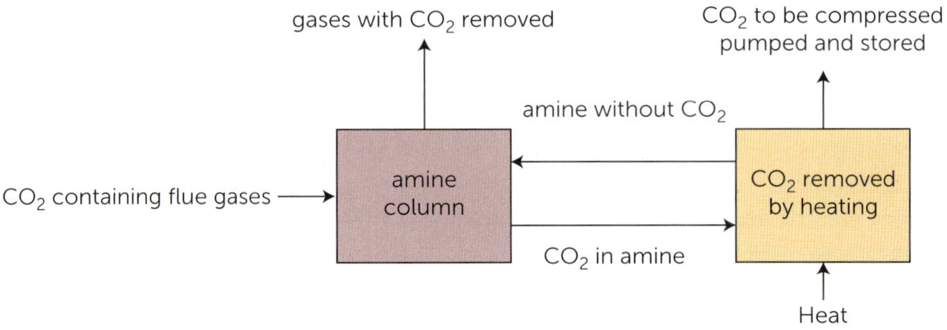

Figure 28.3 Stages in carbon (dioxide) capture and storage using an amine solvent.

DID YOU KNOW?

The North Atlantic region cooled slightly between 1600 to 1850, but it was enough to occasionally freeze the Baltic Sea and to allow fairs to take place on the frozen river Thames. This period, known as the Little Ice Age, was very influential in the novelist Charles Dickens' early life, and although it snows much less frequently at Christmas in London now, the association of snow and Christmas remains. Since then, average temperatures have increased and climatic models indicate that increasing temperatures are linked to elevated levels of CO_2 in the atmosphere.

'Frost fair' on the river Thames, painted by Thomas Wyke (Wijck).

REVISION QUESTIONS

28.1 Comment on whether the following three greenhouse experiments are a 'fair test'.

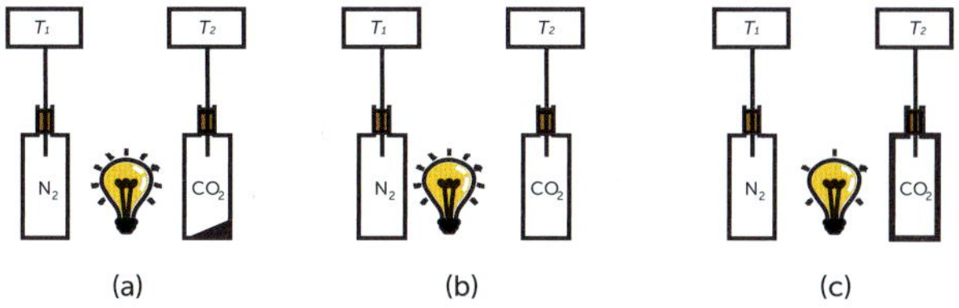

28.2 List some of the factors that must be considered in deciding if cruise ships produce more CO_2 per person than jet flights.

28.3 It is sometimes proposed that global warming can be attributed not to the greenhouse effect but to increasing solar intensity. How does the existence of stratospheric cooling counter this proposal?

28.4 Venus is closer to the Sun than the Earth. Its atmospheric pressure is 100 times greater than the Earth's atmosphere and most of its atmosphere consists of CO_2. Use this information to explain why the average surface temperature of Venus is 464°C.

28.5 It is predicted that the increase in electricity demand worldwide from artificial intelligence (AI) by 2030 could be as high as 9.45×10^{14} watt hours. Assuming a world population of 8 billion, what would be the electricity consumption due to AI per person per hour?

28.6 This question is about the efficiency of electric cars. Suppose that 100 J of electricity is consumed in charging a lithium electric battery. The lithium ion battery in the car absorbs 95 per cent of this electrical energy. Of the energy in the battery, 90 per cent is available to convert to mechanical energy to drive the car forward. What is the efficiency of the electric car?

28.7 In some plans for the capture and the long-term storage of carbon dioxide, it is proposed that the CO_2 would be stored in rock at about a kilometre underground. At this depth, the external pressure is roughly 200 atmospheres and the temperature is above 40°C.

(i) What will be the state of CO_2 under these conditions?

(ii) Will the state of CO_2 make it more or less soluble in rock formations containing oil or salty water?

28.8 Methane, the second most important global warming gas, is destroyed in the atmosphere in about 7–12 years by reaction with hydroxyl radicals:

$$CH_4(g) + {\bullet}OH(g) \rightarrow {\bullet}CH_3(g) + H_2O(g)$$

Hydrogen gas also reacts with hydroxyl radicals (•OH) in the atmosphere:

$$H_2(g) + {\bullet}OH(g) \rightarrow H_2O(g) + {\bullet}H(g)$$

Use this information to explain why hydrogen gas, which does not absorb IR radiation, is an *indirect* global warming gas.

For more about the hydrogen economy, including fuel cells and hydrogen powered vehicles, see our website https://www.bloomsburyonlineresources.com/chemistry.

29 Nanochemistry

CONTENTS

OBJECTIVES

This unit:

- Defines nanoparticles by their size
- Describes several types of nanoparticles
- Discusses some of the uses of nanoparticles
- Highlights concerns about the safety of nanoparticles

29.1 What are nanoparticles?

The word nano is derived from the Greek *nanos*, meaning dwarf. Nanoparticles are defined as particles between 1 to 100 nanometres (nm) in all dimensions. Figure 29.1 illustrates the relative sizes of objects in nanometres (nm).

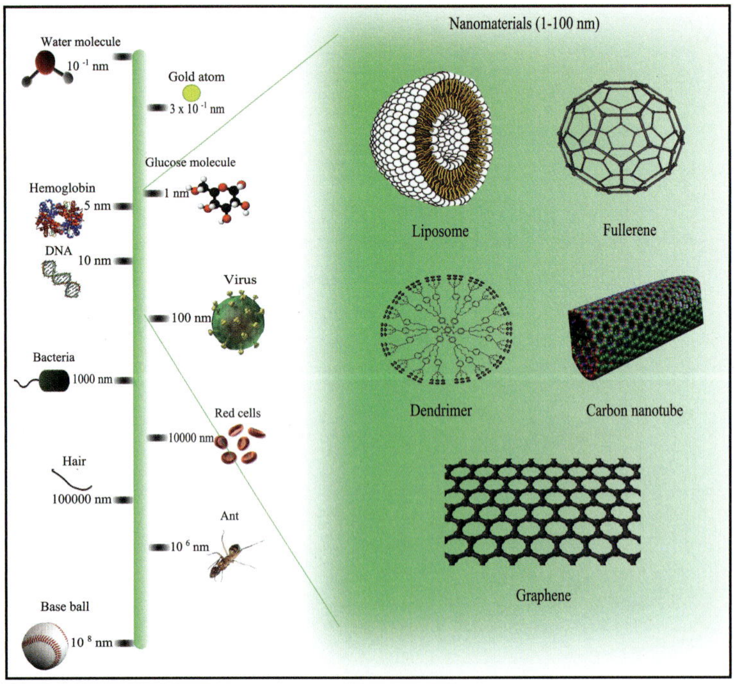

Figure 29.1 Relative sizes (in nanometres, nm) of atoms, molecules, objects and nanoparticles. 1 nm = 10^{-9} m = 1000 pm.

(Image by Sureshbup)

Nanoparticles are not new

Carbon nanotubes have been discovered in ancient pottery in India and iron carbide (FeC_3) nanowires have been found in ancient Damascus steel, an alloy used to make strong swords. In these examples, the tiny **nanoparticles** were produced as part of the manufacturing process even though the ancient artisans had no understanding of their chemistry. Ionic micelles are nanoparticles produced in soap action (Unit 11); soaps have been in use for over 3,500 years. An emulsifier called casein is dispersed as a colloid in milk; it consists of tiny nanoparticle micelles. Liposomes are nanoparticles inside biological cells. Nanoparticles are produced by many natural processes including forest fires; these are termed 'environmental nanoparticles'.

Nanoparticles have different properties from their constituent parts

There are two reasons that nanoparticles behave differently to bulk materials. The first is the enormously increased surface area. The effect of surface area on reaction rate was introduced in Unit 14. The increase in surface area when matter is arranged in tiny nanoparticles is vast. For example, 1 g of gold nanoparticles with a diameter of 1 nm would have a total surface area of over a million times that of a single sphere of gold of 1 g mass. This means that the otherwise very slow reactions of the bulk component become fast enough as a nanoparticle to be useful and commercially viable.

The second reason is much less predictable. Nanoparticles are so small that the electrons are arranged differently to the bulk material. These 'quantum effects' may produce different physical, spectral, magnetic and conduction properties. For example, a sample of copper nanoparticles is harder than bulk copper and gold nanoparticles melt at a higher temperature than a solid lump of gold. The energy levels in nanoparticles are different from that of the constituent atoms in isolation. In a nanoparticle, the energy levels occur in bands of energy that belong to the whole nanoparticle and not to the constituent atoms.

Although the quantum effects observed in nanoparticles might seem strange, we have already met an example where the properties of another structure might have surprised us. Benzene, C_6H_6, consists of hexagonal rings of carbon atoms; it is not a conductor of electricity. Yet when hexagonal carbon rings (minus the H atoms) are arranged in sheets, as in graphite, the resulting structure is electrically conducting (see page 74). Graphite is not a nanoparticle, but if you were presented with it for the first time, its electrical conductivity would be both puzzling and surprising. As scientists study nanoparticles, their properties become less of a surprise but remain remarkable!

While graphite is not a nanoparticle, pieces of a single sheet of graphite, given the name **graphene**, are (Figure 29.2). Unlike graphite, graphene is transparent and virtually colourless and a much better electrical conductor than graphite. Unlike graphite (which is brittle), graphene is 'atom for atom' the strongest material known!

In general, nanoparticles can also be manufactured from several components, as in the biological 'vehicle' to achieve human vaccination for COVID (see later). In general, 'purpose made' nanoparticles confer a wider range of properties to the new material (a *nanomaterial*) with many new potential applications.

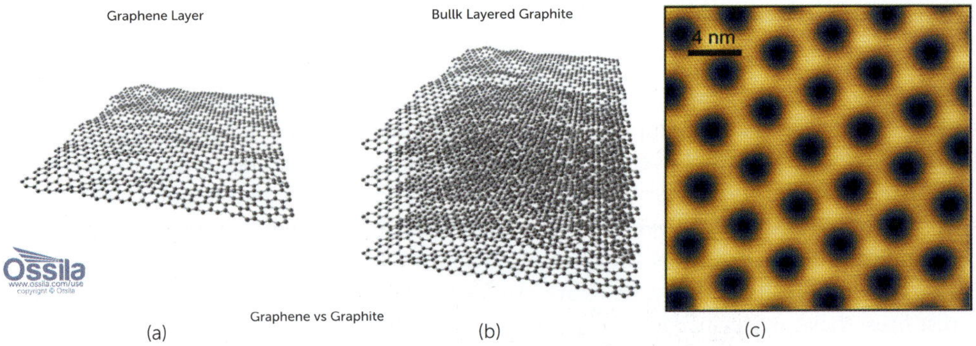

Figure 29.2 (a) Graphene, (b) Graphite, courtesy of Ossila, https://www.ossila.com/pages/graphene-vs-graphite (c) Atomic force microscope image of graphene.

29.2 Gold nanoparticles (AuNPs)

Gold nanoparticles may be prepared in the laboratory by reducing gold ions (with oxidation state +1) to gold (with oxidation state 0) using citric acid:

$$Au^{+1} \rightarrow Au^{0}$$

The Au atoms aggregate together in layers, building up into a nanoparticle with citrate ions attaching (as ligands) on its outermost layer. The size of the nanoparticle is controlled by varying the concentrations of reactants, adding the reactants in a different order and by controlling pH and temperature. For spherical nanoparticles of 15 nm, each 1 gram of AuNP contains a mindboggling 3×10^{16} nanoparticles.

Remarkably, AuNPs can also be produced by some plants after they have been immersed in soluble gold compounds. This offers the possibility of producing AuNPs in more environmentally friendly ways.

Michael Faraday made gold colloids (which we now know consist of AuNPs) in 1856 and one of his colloids is displayed in the Royal Institution in London today (Figure 29.3).

Figure 29.3 Faraday's original gold colloid

The colour of a AuNP solution depends on the particle size (Figure 29.4). From our knowledge of spectroscopy, we can infer that nanoparticles have different colours in solution because different sized nanoparticles have slightly different gaps between the energy levels involved in the transitions associated with light absorption. The size of the nanoparticle controls the wavelengths of white light it absorbs.

Gold nanoparticles are used in electronic chips and sensors, in drug delivery, as catalysts and as biomarkers to detect disease.

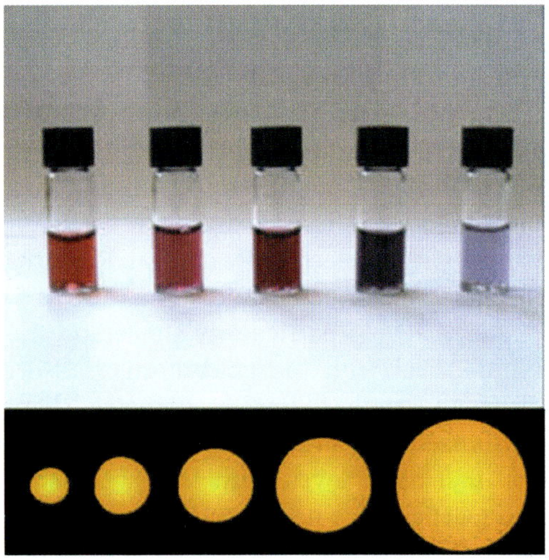

Figure 29.4 Colour changes of gold (AuNP) solutions as nanoparticle size increases from below 20 nm (left) to higher than 100 nm (right). The smaller the nanoparticle, the bigger the energy gap involved in the transition and the stronger is the red colour.

Aleksandar Kondinski

BOX **29.1**

The exquisite Lycurgus Cup

The Lycurgus Cup is red when white light is *transmitted* through it from behind but when white light is shone on the front of the cup the *reflected* light is green. This two-colour phenomenon is called *dichroism*. The cup, displayed in the British Museum, was made by the Romans in the fourth century CE. Its remarkable colours are partly due to the dispersion of tiny amounts of gold and silver nanoparticles in the glass. The effect might have been discovered accidentally and it would not have been understood at the time. (Photo by Chappsnet).

EXERCISE **29A**

Spectrum of gold nanoparticles

The following absorption spectrum is of 20 nm nanoparticles of gold in colloidal solution.

(i) What is the peak absorption in nm?

(ii) Using the 'colour cheese' (page 405), confirm that the gold solution is coloured red.

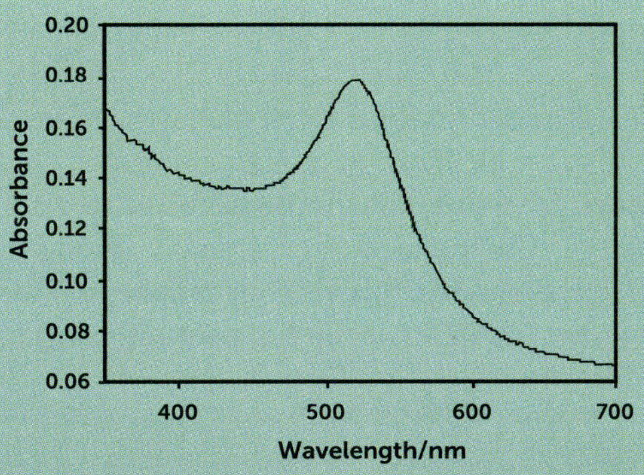

From E. I. Iwuoha et al., Sensors 2008, 8, 8262–74; DOI: 10.3390/s8128262.

$Au_{133}(SR)_{52}$ nanoparticle

An astonishing level of detail is available on a very special gold nanoparticle: $Au_{133}(SR)_{52}$, where SR is shorthand for a thionate ligand – an organic group that contains a sulfur atom. As its formula suggests, the nanoparticle consists of exactly 133 gold atoms and 52 outer SR ligands (Figure 29.5).The diameter of the entire nanoparticle (including the ligand shell) is ~3.0 nm, with the metal (gold) core diameter being ~1.7 nm. The nanoparticle is electrically neutral. Remarkably, and despite the presence of gold atoms, the nanoparticle behaves in some ways as a non-metal!

$Au_{133}(SR)_{52}$ is particularly stable and the way the gold atoms are built up is highly symmetrical with the gold atoms 'self-organizing' themselves into two layers each of which strictly follows a geometric shape called an icosahedron. This arrangement minimizes the internal stress of the nanoparticle. The thionate ligands occupy the final outer layer, with the sulfur atoms binding to the gold atoms. $Au_{133}(SR)_{52}$ nanoparticles are being studied to see if they might be used in new types of catalysis and optoelectronics.

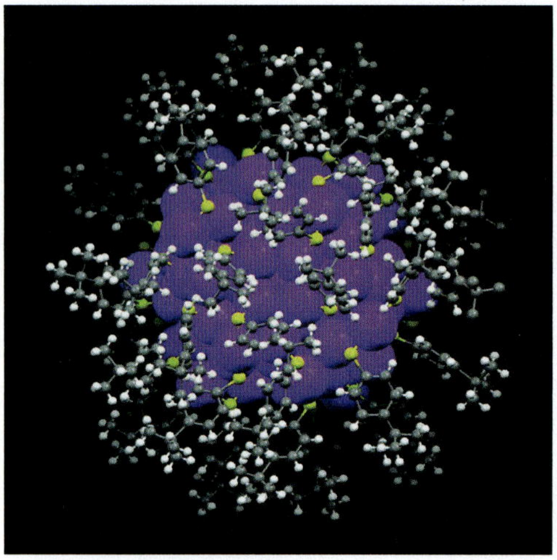

Figure 29.5 The $Au_{133}(SR)_{52}$ nanoparticle. Colour code for atoms: gold = violet; sulfur = yellow; hydrogen atoms = white; carbon atoms = grey. The thionate ligand (SR) consists of a sulfur atom and a butyl group (-C_4H_9) all bonded to a benzene ring. The full formula for the thionate ligand SR is $(CH_3)_3CC_6H_4S$– where – denotes the bond from the sulfur atom to the gold atom.

29.3 Nanoparticles based on carbon

Buckminsterfullerene, C60, is an allotrope of carbon that exists as a nanoparticle. C60 and C70 exist in outer space (p375). We have already mentioned graphene (see Figure 29.2 above). Other carbon nanoparticles are shown in Figure 29.6.

There will likely be many important applications of carbon nanoparticles in the future. They are already being used in water purification, in sensors and in energy storage. Graphene is exceptionally strong and flexible, and an excellent conductor of electricity. For this reason, graphene is being used to make a flexible ('bendable') photovoltaic cell that can be sown into clothing,

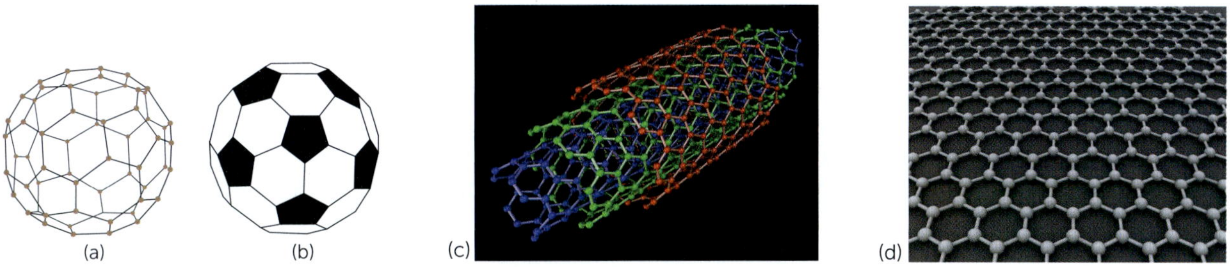

Figure 29.6 Examples of nanoparticles based on carbon.

(a) The arrangement of carbon atoms in buckminsterfullerene
(b) A football for comparison
(c) Three nanotubes made entirely of carbon atoms. The carbon atoms are given different colours for clarity. (Image by Eric Wieser)
(d) Graphene consists of flat hexagons of carbon. It may be thought of as a single layer of graphite.

(Image by AlexanderAlUS).

29.4 Quantum dots

'Quantum dots' are semiconductor nanocrystals between 2 and 10 nm diameter and containing fewer than 100 atoms. When an electric current is passed through the quantum dot nanoparticle, the electrons in the nanoparticle move to a higher energy state; they may then lose energy by emitting light at a single wavelength. The wavelength (colour) of light depends upon the size and shape of the nanoparticle. One type of quantum dot consists of lead sulfide (PbS) and lead selenide (PbSe) molecules in a core, surrounded and stabilized by a layer of oleic acid (an organic acid). The Nobel Prize in Chemistry 2023 was awarded to Louis Brus, Michel Bawendi and Alexei Yekimov for their discovery of quantum dots.

Quantum dots are used in high resolution TV screens, in solar panels and in biomedical imaging.

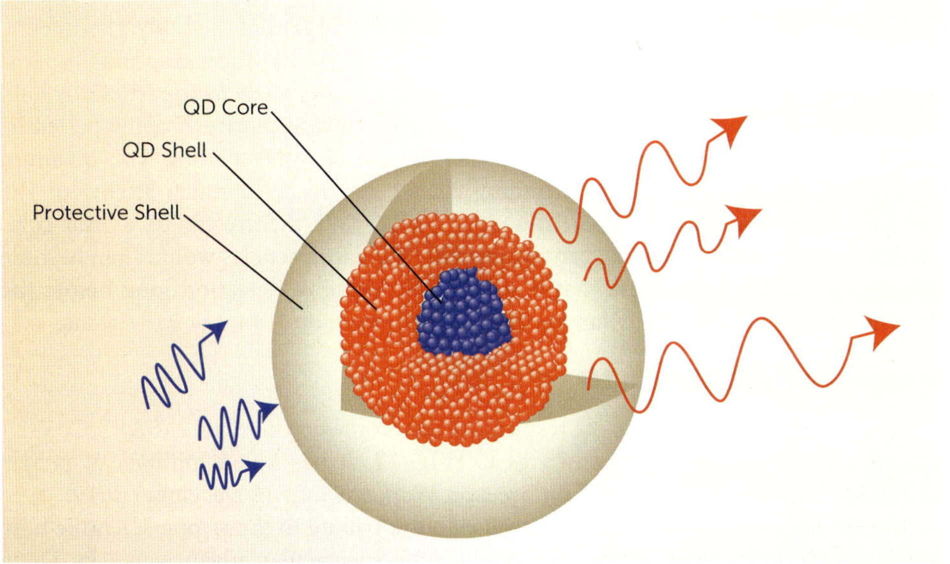

Figure 29.7 A quantum dot nanoparticle. In this example, a blue light from a LED (light-emitting diode) strikes the quantum dot and monochromatic (single-wavelength) red light is emitted.

29.5 Biological nanoparticles

Many naturally occurring biological molecules and structures are between 1 and 100 nm and so qualify as nanoparticles. Examples include lipoproteins (complex structures composed of lipids and proteins that transport water soluble lipids within the body) and exosomes (protein carrying lipid structures that are found in most bodily fluids). See Figure 29.8.

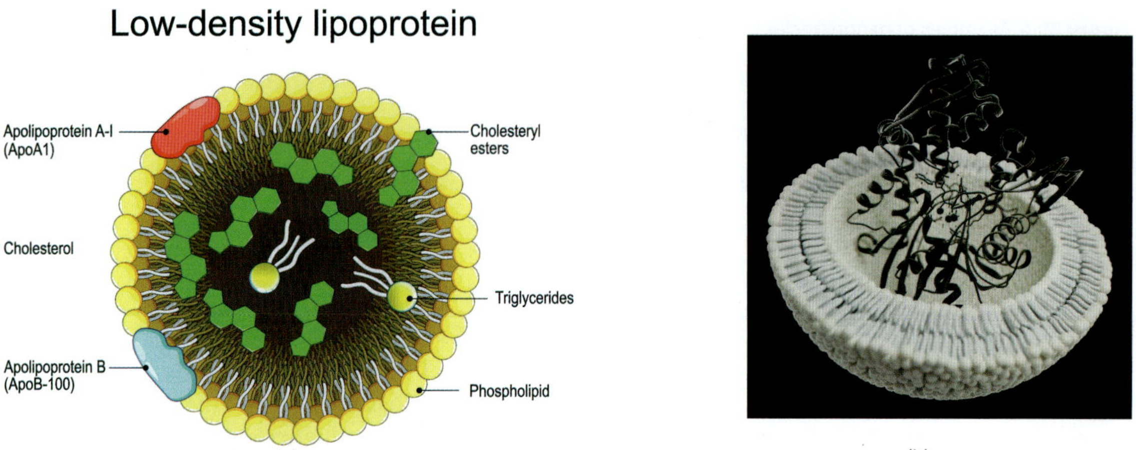

(a) (b)

Figure 29.8 Examples of biological nanoparticles.

(a) Lipoprotein (LDL)
(b) An exosome showing the lipid bi-layer on the outside and a 'cargo' of proteins and nucleic acid (DNA or RNA, shown as black ribbons) inside.

(Drawing by Guillaume Pelletier).

29.6 Nanoparticles in medicine

AuNPs are used in delivering drugs as 'nanocarriers' to organs of the body; therapeutic drugs can be coated onto the surface of gold nanoparticles. The exceptionally large surface area-to-volume ratio of gold nanoparticles allows each nanoparticle to hold hundreds of molecules of drug, so that the effective dosage received by the patient is higher than achieved using standard tablets. The use of AuNPs, along with quantum dots, as biomarkers for disease, including cancer, has received much attention. The requirement is that the nanoparticles specifically bind themselves to a molecular marker (indicator) produced by cancerous cells. The nanoparticles would then become loaded with so many markers that it would be capable of detection long before the disease has spread to produce damage that could be seen on X-rays or MRI scans.

Vaccines for COVID-19

A virus known as the severe acute respiratory syndrome coronavirus 2 (SARS-CoV-2), caused a disease commonly known as COVID-19. It has killed millions of people worldwide since its first identified outbreak in Wuhan, China in December 2019.

The development of the COVID-19 vaccine is a tribute to the strong scientific base built up over many years across the world. One spectacular advance was by Drew Weissman and Katalin Karikó who developed the revolutionary messenger ribonucleic acid (mRNA) technology used in the COVID-19 vaccines. Their work allowed mRNA to enter cells without triggering the body's defences. Traditional vaccines work by injecting an inactive or weakened pathogen (such as a virus) in the human body to trigger an immune response that helps the body defend itself later. mRNA vaccines work differently: they provide the genetic information that the body needs to make antibody proteins to combat the virus without exposing the body to the virus itself. Put another way, the mRNA vaccine 'teaches' the body to protect itself against the COVID-19 virus.

But how would the mRNA be delivered so it could do its work? The solution was to encase the mRNA within a protective bubble called a lipid nanoparticle (Figure 29.9b), which easily enters the cell plasma membrane.

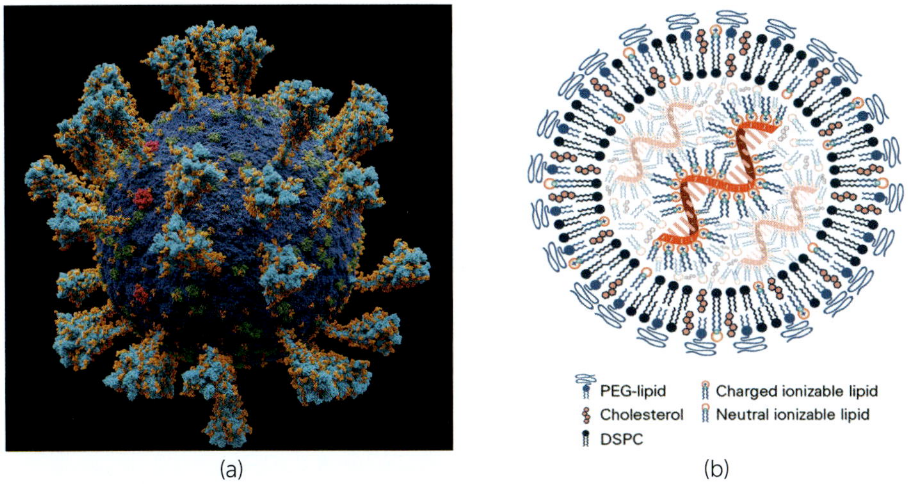

Figure 29.9

(a) Public Enemy Number One: the SARS-CoV-2 virus, a nanoparticle between 80 and 120 nm in size. The protruding 'spikes' are spike glycoproteins that play an important role in penetrating host cells and initiating infection. (Model created by Alexey Solodovnikov.)

(b) mRNA lipid nanoparticle.

(Image by Michael D. Buschmann, Manuel J. Carrasco, Suman Alishetty, Mikell Paige, Mohamad G. Alameh, and Drew Weissman.)

Katalin Karikó and Drew Weissman, who were awarded the Nobel Prize for Physiology/Medicine for their discoveries that led to the development of mRNA vaccines.

29.7 Are nanoparticles safe?

Since nanoparticles have different properties from the bulk material, it is reasonable to assume they may have different levels of toxicity. In many cases, toxicity studies on nanoparticles have yet to be carried out. This, along with worries about the unnoticed intake of nanoparticles through ingestion in food stuffs or through inhalation in their manufacture or use, have prompted calls for greater recognition of the potential impact of their widespread use.

REVISION QUESTIONS

29.1 Show that the ratio (*R*) of the surface area of a sphere to the volume of a sphere is $\frac{3}{r}$ where *r* is the radius of the sphere. Calculate the ratio for a 1 cm sphere of gold and for a 10 nm nanosphere of gold and compare the values.

29.2 Sunscreens are important is reducing the risk of skin cancer. Two substances used in sunscreen are zinc oxide and titanium oxide. List three benefits of using these compounds as nanoparticles in sunscreen mixtures instead of as powdered (bulk) oxides.

29.3 Explain why bulk gold metal (shiny yellow) has a different colour from gold nanoparticles in solution.

29.4 Calculate the *m*/*z* value for the parent molecular ion of $Au_{133}(SR)_{52}$ that would be expected to be produced in a mass spectrometer.

29.5 Briefly discuss how the carbon atoms are bonded in diamond, graphite and graphene. Why is graphite brittle and graphene tough when exposed to external forces? Why doesn't graphene-oxide conduct electricity?

29.6 Look at the mass spectrum of carbon-60 on the NIST website. (To see it, search 'C60' on the NIST Chemistry WebBook, select Buckminsterfullerene and select mass spectrum). Notice the strong peak at $m/z = 360$ (due to two species, $(C_{30})^{+}$ and $(C_{60})^{2+}$. The parent peak (due to $(C_{60})^{+}$) lies at $m/z = 360$.

DID YOU KNOW?

Stephanie Chanteau and James Tour of the Center for Nanoscale Science and Technology at Rice University synthesized a nanoscale human-like molecule using more than a dozen separate chemical reactions. The body of the human (molecule) consists of two benzene rings and methyl groups represent the hands and feet. The whole molecule is about 2 nm tall. Nanoparticles of this type are referred to as 'NanoPutians', after the diminutive Lilliputian people in Jonathan Swift's fictional novel *Gulliver's Travels*. The NanoPutian shown below is called NanoKid. Its systematic name is somewhat longer:

2-[2,5-Bis(3,3-dimethyl-1-butyn-1-yl)-4-{[3,5-di(1-pentyn-1yl)phenyl]ethynyl}phenyl]-1,3-dioxolane

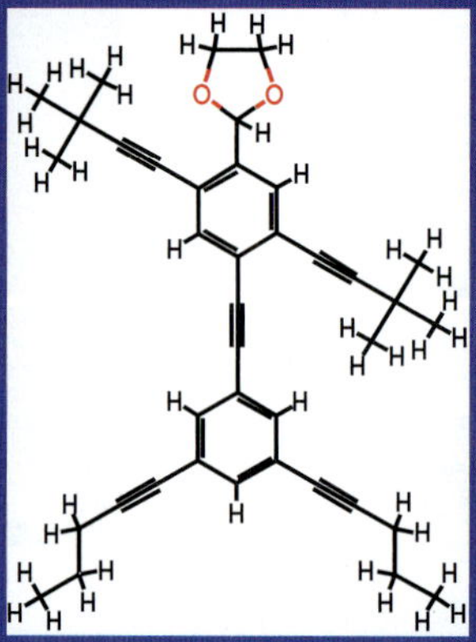

Skeletal representation of the NanoKid nanoparticle/molecule. It has the molecular formula $C_{39}H_{42}O_2$. Other NanoPutians include *NanoMonarch*, *NanoJester*, *NanoTexan* and *NanoChef*.

A short history of the allotropes of carbon is given in Extension 29 at https://www.bloomsbury onlineresources.com/ chemistry.

Electronic Structures

This table gives the electronic configurations of the elements in their ground state

Shell		1	2		3			4				5					6			7
Subshell		1s	2s	2p	3s	3p	3d	4s	4p	4d	4f	5s	5p	5d	5f	5g	6s	6p	6d	7
1	H	1																		
2	He	2																		
3	Li	2	1																	
4	Be	2	2																	
5	B	2	2	1																
6	C	2	2	2																
7	N	2	2	3																
8	O	2	2	4																
9	F	2	2	5																
10	Ne	2	2	6																
11	Na	2	2	6	1															
12	Mg	2	2	6	2															
13	Al	2	2	6	2	1														
14	Si	2	2	6	2	2														
15	P	2	2	6	2	3														
16	S	2	2	6	2	4														
17	Cl	2	2	6	2	5														
18	Ar	2	2	6	2	6														
19	K	2	2	6	2	6		1												
20	Ca	2	2	6	2	6		2												
d-block elements																				
21	Sc	2	2	6	2	6	1	2												
22	Ti	2	2	6	2	6	2	2												
23	V	2	2	6	2	6	3	2												
24	Cr	2	2	6	2	6	5	1												
25	Mn	2	2	6	2	6	5	2												
26	Fe	2	2	6	2	6	6	2												
27	Co	2	2	6	2	6	7	2												
28	Ni	2	2	6	2	6	8	2												
29	Cu	2	2	6	2	6	10	1												
30	Zn	2	2	6	2	6	10	2												
31	Ga	2	2	6	2	6	10	2	1											
32	Ge	2	2	6	2	6	10	2	2											
33	As	2	2	6	2	6	10	2	3											
34	Se	2	2	6	2	6	10	2	4											
35	Br	2	2	6	2	6	10	2	5											
36	Kr	2	2	6	2	6	10	2	6											
37	Rb	2	2	6	2	6	10	2	6			1								
38	Sr	2	2	6	2	6	10	2	6			2								
d-block elements																				
39	Y	2	2	6	2	6	10	2	6	1		2								
40	Zr	2	2	6	2	6	10	2	6	2		2								
41	Nb	2	2	6	2	6	10	2	6	4		1								
42	Mo	2	2	6	2	6	10	2	6	5		1								
43	Tc	2	2	6	2	6	10	2	6	5		2								
44	Ru	2	2	6	2	6	10	2	6	7		1								
45	Rh	2	2	6	2	6	10	2	6	8		1								
46	Pd	2	2	6	2	6	10	2	6	10										
47	Ag	2	2	6	2	6	10	2	6	10		1								
48	Cd	2	2	6	2	6	10	2	6	10		2								
49	In	2	2	6	2	6	10	2	6	10		2	1							
50	Sn	2	2	6	2	6	10	2	6	10		2	2							
51	Sb	2	2	6	2	6	10	2	6	10		2	3							
52	Te	2	2	6	2	6	10	2	6	10		2	4							
53	I	2	2	6	2	6	10	2	6	10		2	5							
54	Xe	2	2	6	2	6	10	2	6	10		2	6							

Shell		1	2	3				4				5					6			7
Subshell								4s	4p	4d	4f	5s	5p	5d	5f	5g	6s	6p	6d	7
55	Cs	2	8	18				2	6	10		2	6				1			
56	Ba	2	8	18				2	6	10		2	6				2			
																				Lanthanoids
57	La	2	8	18				2	6	10		2	6	1			2			
58	Ce	2	8	18				2	6	10	1	2	6	1			2			
59	Pr	2	8	18				2	6	10	3	2	6				2			
60	Nd	2	8	18				2	6	10	4	2	6				2			
61	Pm	2	8	18				2	6	10	5	2	6				2			
62	Sm	2	8	18				2	6	10	6	2	6				2			
63	Eu	2	8	18				2	6	10	7	2	6				2			
64	Gd	2	8	18				2	6	10	7	2	6	1			2			
65	Tb	2	8	18				2	6	10	9	2	6				2			
66	Dy	2	8	18				2	6	10	10	2	6				2			
67	Ho	2	8	18				2	6	10	11	2	6				2			
68	Er	2	8	18				2	6	10	12	2	6				2			
69	Tm	2	8	18				2	6	10	13	2	6				2			
70	Yb	2	8	18				2	6	10	14	2	6				2			
																				d-block elements
71	Lu	2	8	18				2	6	10	14	2	6	1			2			
72	Hf	2	8	18				2	6	10	14	2	6	2			2			
73	Ta	2	8	18				2	6	10	14	2	6	3			2			
74	W	2	8	18				2	6	10	14	2	6	4			2			
75	Re	2	8	18				2	6	10	14	2	6	5			2			
76	Os	2	8	18				2	6	10	14	2	6	6			2			
77	Ir	2	8	18				2	6	10	14	2	6	7			2			
78	Pt	2	8	18				2	6	10	14	2	6	9			1			
79	Au	2	8	18				2	6	10	14	2	6	10			1			
80	Hg	2	8	18				2	6	10	14	2	6	10			2			
81	Tl	2	8	18				2	6	10	14	2	6	10			2	1		
82	Pb	2	8	18				2	6	10	14	2	6	10			2	2		
83	Bi	2	8	18				2	6	10	14	2	6	10			2	3		
84	Po	2	8	18				2	6	10	14	2	6	10			2	4		
85	At	2	8	18				2	6	10	14	2	6	10			2	5		
86	Rn	2	8	18				2	6	10	14	2	6	10			2	6		
87	Fr	2	8	18				2	6	10	14	2	6	10			2	6		1
88	Ra	2	8	18				2	6	10	14	2	6	10			2	6		2
																				Actinoids
89	Ac	2	8	18				2	6	10	14	2	6	10			2	6	1	2
90	Th	2	8	18				2	6	10	14	2	6	10			2	6	2	2
91	Pa	2	8	18				2	6	10	14	2	6	10	2		2	6	1	2
92	U	2	8	18				2	6	10	14	2	6	10	3		2	6	1	2
93	Np	2	8	18				2	6	10	14	2	6	10	4		2	6	1	2
94	Pu	2	8	18				2	6	10	14	2	6	10	6		2	6		2
95	Am	2	8	18				2	6	10	14	2	6	10	7		2	6		2
96	Cm	2	8	18				2	6	10	14	2	6	10	7		2	6	1	2
97	Bk	2	8	18				2	6	10	14	2	6	10	9		2	6		2
98	Cf	2	8	18				2	6	10	14	2	6	10	10		2	6		2
99	Es	2	8	18				2	6	10	14	2	6	10	11		2	6		2
100	Fm	2	8	18				2	6	10	14	2	6	10	12		2	6		2
101	Md	2	8	18				2	6	10	14	2	6	10	13		2	6		2
102	No	2	8	18				2	6	10	14	2	6	10	14		2	6		2
103	Lr	2	8	18				2	6	10	14	2	6	10	14		2	6	1	2
104	Rf	2	8	18				2	6	10	14	2	6	10	14		2	6	2	2
105	Db	2	8	18				2	6	10	14	2	6	10	14		2	6	3	2
106	Sg	2	8	18				2	6	10	14	2	6	10	14		2	6	4	2
107	Bh	2	8	18				2	6	10	14	2	6	10	14		2	6	5	2
108	Hs	2	8	18				2	6	10	14	2	6	10	14		2	6	6	2

Beyond $_{94}$Pu the assignments are conjectural

Answers to Exercises and Revision Questions

Unit 1

Exercises

1A

(i) 3.45×10^{-5} **(ii)** 3×10^{8}
(iii) 8.20575×10^{-2} **(iv)** 3.5×10^{0}
(v) 6.022×10^{23} **(vi)** 1.7×10^{1}

1B

(i) 379.75 (or 380) **(ii)** 17.34 **(iii)** –1.921 (or –1.92)
(iv) 2.75×10^{-5} **(v)** 0.409
(vi) 5.517×10^{-3} (or 5.5×10^{-3})
(vii) 5.5×10^{-6}

1C

(i) 0.040 nm **(ii)** 10 g **(iii)** 6×10^{-7} m

1D

(i) Rearranging, $C = \frac{q}{m \times \Delta T}$ The units of C are

$$\frac{\text{J}}{\text{g} \times \text{K}} = \text{J}\,\text{g}^{-1}\text{K}^{-1}$$

(ii) $M = \frac{m}{n}$ and the units are $\frac{\text{g}}{\text{mol}} = \text{g mol}^{-1}$

1E

Some of the alcohol in the wine has been oxidized.

1F

(i) (a) 1, (b) 4, (c) 2 or 3 (ambiguous), (d) 3
(ii) No, to four significant figures the atomic mass is 15.99 atomic mass units.

Revision questions

1.1 (d) **1.2** (b) **1.3** (d) (information about precision does not allow us to make any comment on the presence or absence of systematic errors).
1.4 Average is 10.5 ppm, error = 10.5 – 10.0 = 0.5 ppm; (b) is true.
1.5 Cr^{3+} 6.9×10^{-11} m, F^{-} 1.36×10^{-8} m, O 1.40×10^{-8} m
1.6 Uncertainty is at least ± 0.1°C
1.7 (i) 0.1235 V, **(ii)** 12.45 m, **(iii)** 0.003558, **(iv)** 1201 K
1.8 (i) 3.29×10^{-4}, **(ii)** 0.449
1.9 ε has units of $\text{mol}^{-1}\,\text{m}^2$
1.10 (i) random **(ii)** systematic (too big to be explained by draughts) **(iii)** systematic (pressure is falling)

Unit 2

Exercises

2A

(i) chemical **(ii)** physical **(iii)** physical

2B

(i) physical **(ii)** physical **(iii)** chemical

2C

(i) extensive **(ii)** intensive **(iii)** extensive
(iv) intensive

2D

(i) caesium (1) bromine (1)
(ii) copper (1) chlorine (2)
(iii) iron (3) oxygen (4)
(iv) carbon (1) hydrogen (4)
(v) lead (1) sulfur (1) oxygen (4)
(vi) potassium (2) chromium (2) oxygen (7)
(vii) sodium (3) phosphorus (1) oxygen (4)
(viii) manganese (1) oxygen (2)

2E

carbon, hydrogen, oxygen, phosphorus, potassium, iodine, nitrogen, sulfur, calcium, iron, sodium, chlorine

2F

1. (i) Natrium (Arabic root)
(ii) *ferrum* (Latin) **(iii)** *plumbum* (Latin)
(iv) *stannum* (Latin) **(v)** *kalium* (Latin)
(vi) *argentum* (Latin)
2. (i) Lise Meitner worked on uranium fission and realized that atomic fission could produce vast quantities of energy. **(ii)** Glenn T. Seaborg, an American nuclear chemist and Nobel Prize winner (1951) for investigating the chemistry of heavy elements. **(iii)** Niels Bohr and **(iv)** Ernest Rutherford made major contributions to our understanding of atomic structure. **(v)** Oganessian is considered the world's leading researcher in superheavy elements. He and his team discovered the heaviest elements in the Periodic Table.

2G

(i) CaO **(ii)** AlF_3 **(iii)** Al_2O_3

2H

(i) Cu_2O **(ii)** CuO **(iii)** PbO_2 **(iv)** Cr_2O_3
(v) V_2O_5 **(vi)** TiF_3 **(vii)** S_2Cl_2 **(viii)** O_2F_2

2I

(i) $CuSO_4$ **(ii)** $Cu(NO_3)_2$ **(iii)** NH_4Cl
(iv) H_2SO_4 **(v)** HNO_3 **(vi)** $Al(NO_3)_3$
(vii) $Ca(OH)_2$ **(viii)** $KHCO_3$ **(ix)** $Ca(HCO_3)_2$

2J

(i) $N_2 + 3H_2 \rightarrow 2NH_3$
(ii) $2Na + Cl_2 \rightarrow 2NaCl$
(iii) $4Al + 3O_2 \rightarrow 2Al_2O_3$
(iv) $Fe_2O_3 + 3CO \rightarrow 2Fe + 3CO_2$

2K

(i) $4K + O_2 \rightarrow 2K_2O$
(ii) $2Fe + 3Cl_2 \rightarrow 2FeCl_3$
(iii) $2Li + 2H_2O \rightarrow 2LiOH + H_2$
(iv) $Cu + 4HNO_3 \rightarrow Cu(NO_3)_2 + 2NO_2 + 2H_2O$

Revision questions

2.1 (i) mixture **(ii)** compound **(iii)** element
(iv) mixture **(v)** compound **(vi)** mixture
(vii) element **(viii)** mixture **(ix)** element
(x) compound.
2.2 m.p. = physical, b.p. = physical, soft = physical, corrodes = chemical, conducts electricity = physical, reacts with water = chemical.
2.3 liquid
2.4 gas
2.5 (i) *cuprum* (Latin for Cyprus, famous for its copper mines)
(ii) *stibium* (Latin for 'mark'; once used for eyebrow pencils)
(iii) *hydrargyrum* (Latin)
2.6 (i) O **(ii)** 2O **(iii)** $2O_2$ **(iv)** $3SO_3$ **(v)** 4Ne
2.7 (i) lithium (2), carbon (1), oxygen (3)
(ii) calcium (1), nitrogen (2), oxygen (6)
(iii) nitrogen (3), hydrogen (12), phosphorus (1), oxygen (4)
2.8 (i) BaF_2 **(ii)** $SnBr_2$ **(iii)** $SnCl_4$ **(iv)** CH_3CO_2Na
(v) Al_2S_3 **(vi)** KOH **(vii)** $NaBrO_3$ **(viii)** $K_2Cr_2O_7$
(ix) $(NH_4)_3PO_4$ **(x)** $Pb(CN)_2$
2.9 As a bleach (hair, textiles, paper, straw, leather, etc.) or as a mild antiseptic.
2.10 $2H_2O_2(l) \rightarrow 2H_2O(l) + O_2(g)$
2.11 (i) $KClO_4 \rightarrow KCl + 2O_2$
(ii) $2S_8 + 3AsF_5 \rightarrow S_{16}(AsF_6)_2 + AsF_3$
(iii) $Hg + 2NH_4I \rightarrow HgI_2 + H_2 + 2NH_3$
(iv) $Al_4C_3 + 12H_2O \rightarrow 4Al(OH)_3 + 3CH_4$
(v) $3Zn + 2H_3PO_4 \rightarrow Zn_3(PO_4)_2 + 3H_2$
(vi) $Fe_2O_3 + Na_2CO_3 \rightarrow 2NaFeO_2 + CO_2$
2.12 (i) $2Zn(s) + O_2(g) \rightarrow 2ZnO(s)$
(ii) $2K(s) + 2H_2O(l) \rightarrow 2KOH(aq) + H_2(g)$
(iii) $CO_2(g) + 2Mg(s) \rightarrow 2MgO(s) + C(s)$
2.13 You could reasonably speculate that aluminium has been oxidized in the dishwasher. Aluminium is protected by a very thin layer of aluminium oxide and the cleaning agent removes it, allowing more aluminium to oxidize to aluminium oxide.

Unit 3

Exercises

3A

Six neutrons and six protons in the nucleus, with six electrons outside.

3B

(i) 1,836 times (using exact masses)
(ii) 3.97 (about 4)

3C

Isotope	Protons	Neutrons	Electrons
U-235	92	143	92
U-238	92	146	92
$^{1}_{1}H$	1	0	1
$^{2}_{1}H$	1	1	1
$^{3}_{1}H$	1	2	1

3D

(i) (a) uranium-238 (b) fluorine
(c) hydrogen-3 (tritium) (d) tritium
(ii) 49.31 per cent

3E

m(O) = [(15.9949 × 99.759) + (16.9991 × 0.0374) + (17.9992 × 0.2039)]/100 = 15.9994 u = 16.00 u (to four significant figures)

3F

The intensities add up to 100, and so correspond to percentage abundances:
m(Ne) = [(91 × 20) + (0.3 × 21) + (8.7 × 22)]/100
= 20.18 u ≈ 20 u

3G

(i) m/z values are 44, 15, 14 (i.e. 28/2)
(ii) Fragmentation follows ionization:

$HF(g) + e^- \rightarrow HF^+(g) + e^- + e^-$
$HF^+(g) \rightarrow F^+(g) + H(g)$

m/z for the F^+ ion is 19, and for HF^+ is 20

(iii) The parent ions $C_6H_6^{\cdot+}$, $C_6H_5NO_2^{\cdot+}$ and $C_6H_5OH^+$ are produced in the ionization stage. They then fragment producing a $C_6H_5^{\cdot+}$ ion, e.g.

$C_6H_5NO_2^{\cdot+}(g) \rightarrow C_6H_5^{\cdot+}(g) + NO_2(g)$
m/z = 77

3H

$Be^{3+}(g) \rightarrow Be^{4+}(g) + e^-$
electrons: 1 0

There are only four electrons in the beryllium atom and so there is no fifth ionization energy.

3I

Element and symbol	Atomic number	Bohr structure	s,p,d,f structure
Hydrogen H	1	1.	$1s^1$
Helium He	2	2	$1s^2$
Lithium Li	3	2.1	$1s^2\ 2s^1$
Beryllium Be	4	2.2	$1s^2\ 2s^2$
Boron B	5	2.3	$1s^2\ 2s^2\ 2p^1$
Carbon C	6	2.4	$1s^2\ 2s^2\ 2p^2$
Nitrogen N	7	2.5	$1s^2\ 2s^2\ 2p^3$
Oxygen O	8	2.6	$1s^2\ 2s^2\ 2p^4$
Fluorine F	9	2.7	$1s^2\ 2s^2\ 2p^5$
Neon Ne	10	2.8	$1s^2\ 2s^2\ 2p^6$
Sodium Na	11	2.8.1	$1s^2\ 2s^2\ 2p^6\ 3s^1$
Magnesium Mg	12	2.8.2	$1s^2\ 2s^2\ 2p^6\ 3s^2$
Aluminium Al	13	2.8.3	$1s^2\ 2s^2\ 2p^6\ 3s^2\ 3p^1$
Silicon Si	14	2.8.4	$1s^2\ 2s^2\ 2p^6\ 3s^2\ 3p^2$
Phosphorus P	15	2.8.5	$1s^2\ 2s^2\ 2p^6\ 3s^2\ 3p^3$
Sulfur S	16	2.8.6	$1s^2\ 2s^2\ 2p^6\ 3s^2\ 3p^4$
Chlorine Cl	17	2.8.7	$1s^2\ 2s^2\ 2p^6\ 3s^2\ 3p^5$
Argon Ar	18	2.8.8	$1s^2\ 2s^2\ 2p^6\ 3s^2\ 3p^6$
Potassium K	19	2.8.8.1	$1s^2\ 2s^2\ 2p^6\ 3s^2\ 3p^6\ 4s^1$
Calcium Ca	20	2.8.8.2	$1s^2\ 2s^2\ 2p^6\ 3s^2\ 3p^6\ 4s^2$

3J

(i)

Element and symbol	1s	2s	$2p_x$	$2p_y$	$2p_z$	s,p,d,f structure
Hydrogen H	↑					$1s^1$
Helium He	↑↓					$1s^2$
Lithium Li	↑↓	↑				$1s^2\ 2s^1$
Beryllium Be	↑↓	↑↓				$1s^2\ 2s^2$
Boron B	↑↓	↑↓	↑			$1s^2\ 2s^2\ 2p^1$
Carbon C	↑↓	↑↓	↑	↑		$1s^2\ 2s^2\ 2p^2$
Nitrogen N	↑↓	↑↓	↑	↑	↑	$1s^2\ 2s^2\ 2p^3$
Oxygen O	↑↓	↑↓	↑↓	↑	↑	$1s^2\ 2s^2\ 2p^4$
Fluorine F	↑↓	↑↓	↑↓	↑↓	↑	$1s^2\ 2s^2\ 2p^5$
Neon Ne	↑↓	↑↓	↑↓	↑↓	↑↓	$1s^2\ 2s^2\ 2p^6$

(ii)

Element and symbol	3d	3d	3d	3d	3d	4s	s,p,d,f structure
Manganese (Mn)	↑	↑	↑	↑	↑	↑↓	$[Ar]\ 3d^5\ 4s^2$
Iron (Fe)	↑↓	↑	↑	↑	↑	↑↓	$[Ar]\ 3d^6\ 4s^2$
Copper (Cu)	↑↓	↑↓	↑↓	↑↓	↑↓	↑	$[Ar]\ 3d^{10}\ 4s^1$

Revision questions

3.1 Following are false:
(ii) should be $^{12}_{6}C$,
(iv) atoms of isotopes are not identical,
(v) a mass spectrometer does not detect neutral atoms.

3.2

Isotope	Number of neutrons	Number of electrons	Number of protons	Abundance/%
$^{32}S_{16}$	16	16	16	95.0
$^{33}S_{16}$	17	16	16	0.76
$^{34}S_{16}$	18	16	16	4.2
$^{36}S_{16}$	20	16	16	0.021

m(S) is given by the equation:

$$\frac{(95\times32)+(0.76\times33)+(4.2\times34)+(0.021\times36)}{100}$$

$$=\frac{3208.6}{100}=32\ \text{u (to two significant figures)}$$

Sources of approximation are **(i)** mass numbers have been used instead of isotopic masses, and **(ii)** some of the abundances are only given to two significant figures. (This explains why the total of the percentage abundances is 99.981 per cent, not 100.000 per cent.)

3.3
$Mg(g) + e^- \rightarrow Mg^+(g) + e^- + e^-$
m/z ≈ 24 (≈ means approximately)
(2.8.1)
$Mg(g) + e^- \rightarrow Mg^{2+}(g) + e^- + e^- + e^-$
m/z ≈ 12
(2.8)

The $Mg^{2+}(g)$ ion, might be formed in one stage (as shown here) or by $Mg^+(g) \rightarrow Mg^{2+}(g) + e^-$.

3.4
(i)
$F_2(g) + e^- \rightarrow F_2^+(g) + e^- + e^-$
m/z = 38
$F_2^+(g) \rightarrow F^+(g) + F(g)$ (fragmentation)
m/z = 19

(ii)
$HCN(g) + e^- \rightarrow HCN^+(g) + e^- + e^-$
m/z = 27
$HCN(g) + e^- \rightarrow HCN^{2+}(g) + e^- + e^- + e^-$
m/z = 13.5
and
$HCN^+(g) \rightarrow CN^+(g) + H(g)$ (fragmentation)
m/z = 26

(iii)
$C_6H_5Cl(g) + e^- \rightarrow C_6H_5Cl(g)^+ + e^- + e^-$
m/z = 114 (due to $C_6H_5^{37}Cl^+$)
and 112 (due to $C_6H_5^{35}Cl^+$)
$C_6H_5Cl(g)^+ \rightarrow C_6H_5^+(g) + Cl(g)$ (fragmentation)
m/z = 77

(iv) $N_2^+(g)$ from air gives a signal at *m/z* = 28

(v)
$Cl_2(g) + e^- \rightarrow Cl_2^+(g) + e^- + e^-$
m/z = 70 (due to $^{35}Cl^{35}Cl^+$)
m/z = 72 ($^{35}Cl^{37}Cl^+$)
m/z = 74 ($^{37}Cl^{37}Cl^+$)
$CH_4(g) + e^- \rightarrow CH_4^+(g) + e^- + e^-$
m/z = 16

If some chlorine and methane react:
$CH_4(g) + Cl_2(g) \rightarrow CH_3Cl(g) + HCl$
$CH_3Cl(g) + e^- \rightarrow CH_3Cl^+(g) + e^- + e^-$
m/z = 50 (for $CH_3^{35}Cl^+$)
m/z = 52 (for $CH_3^{37}Cl^+$)

3.5
$2^1_1H_2(g) + O_2(g) \rightarrow 2^1_1H_2O(l)$
$2^2_1H_2(g) + O_2(g) \rightarrow 2^2_1H_2O(l)$

Heavy water (deuterium oxide) will be more dense than ordinary water.

3.6 $m(^2_1H_2{}^{16}_8O) = (2 \times 2.0140) + (15.9949) = 20.02$ u (to four significant figures).
Since 1 u = 1.66054×10^{-24} g,

$m(^2_1H_2{}^{16}_8O) = 1.66054 \times 10^{-24}$
$\times\ 20.0229 = 3.3249 \times 10^{-23}$ g

or 3.325×10^{-23} g to four significant figures

$m(H_2O) = (2 \times 1.008) + (16.00) = 18.02$ u
$m(H_2O) = 18.02 \times 1.66054 \times 10^{-24}$ g
$= 2.992 \times 10^{-23}$ g to four significant figures

3.7 See text.

3.8 A plot is seen to contain three groups containing eight electrons (the outer electrons are removed first), eight more electrons and two more electrons, respectively.

Sum of ionization energies = 1,389,373 kJ mol^{-1}. This is the energy required to strip all the electrons from the atom.

3.9 For a definition of 'orbital' see the text. The size of a 100 per cent orbital is the size of the universe itself.

3.10 (i) 2.8.7 **(ii)** 2.8.8 **(iii)** 2.8.6 **(iv)** 2
(v) 2.8 **(vi)** [Ar]9
(ii), **(iv)**, and **(v)** are electronic structures in which all the shells are filled.

3.11 (i) $1s^2\ 2s^2\ 2p^6\ 3s^2\ 3p^5$
(ii) $1s^2\ 2s^2\ 2p^6\ 3s^2\ 3p^6$
(iii) $1s^2\ 2s^2\ 2p^6\ 3s^2\ 3p^4$ **(iv)** $1s^2$
(v) $1s^2\ 2s^2\ 2p^6$
(vi) $[Ar]\ 3d^9$. Using the box notation for **(iii)**:

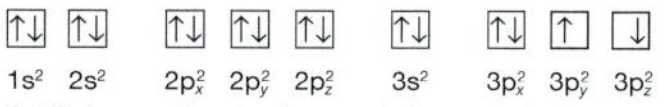

3.12 According to the particle model of the atom (according to which the sub-atomic particles are treated as discrete entities) atoms are mostly empty space, although the electrical charge of electrons and protons will be felt beyond their physical location. But according to the quantum theory of atoms, the electrons are 'smeared out' as clouds. These clouds, or orbitals, extend to infinity, although there are some zones of the orbital of higher electron density.

Unit 4

Exercises

4A

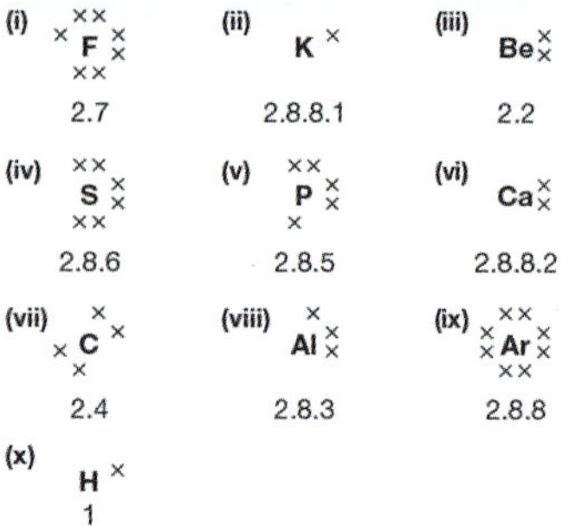

4B

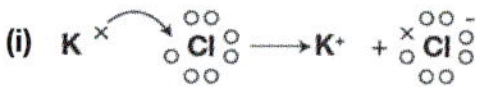

(ii) $2K(s) + Cl_2(g) \rightarrow 2KCl(s)$ (or $2K^+$, $2Cl^-(s)$)

4C

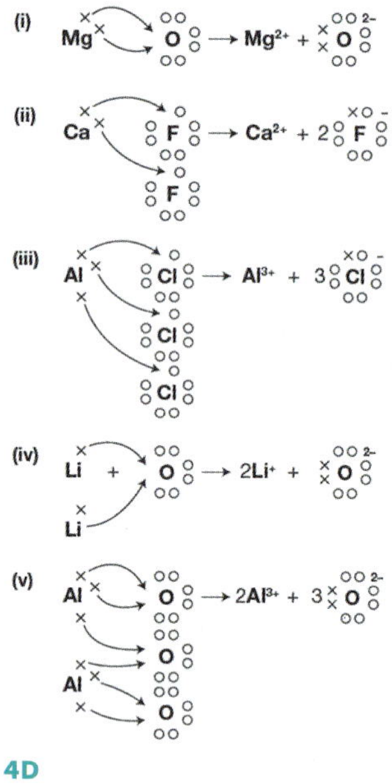

4D

Lewis structures:

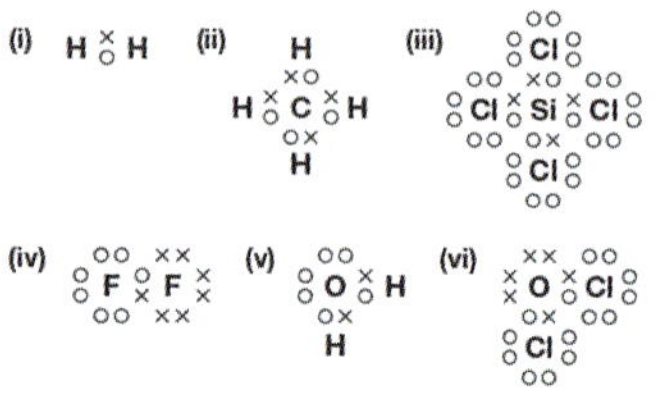

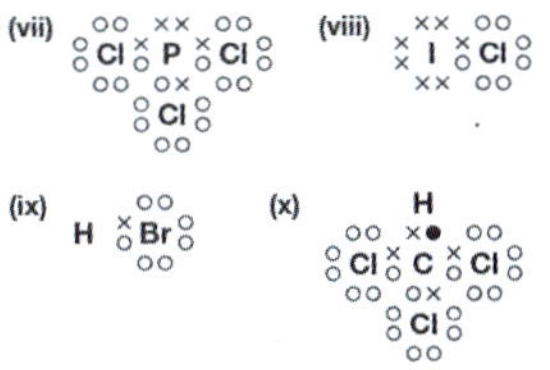

Structural formulae:

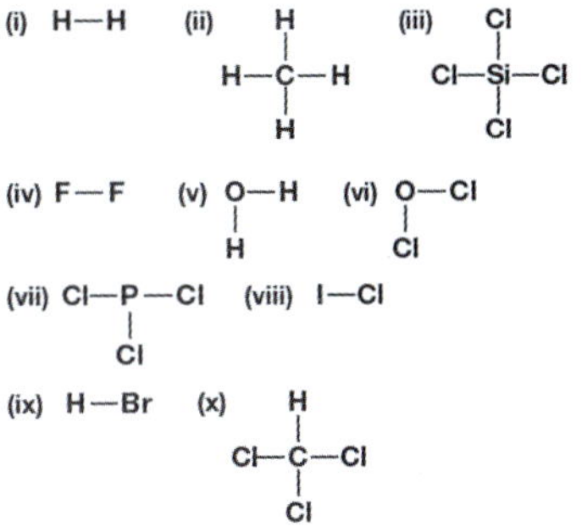

4E

Lewis structures:

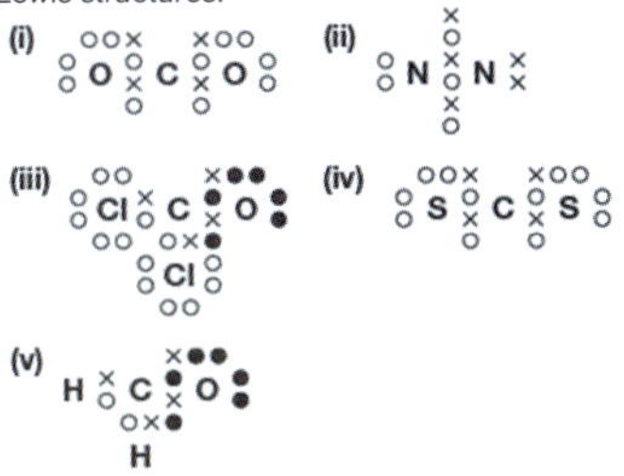

Structural formulae:

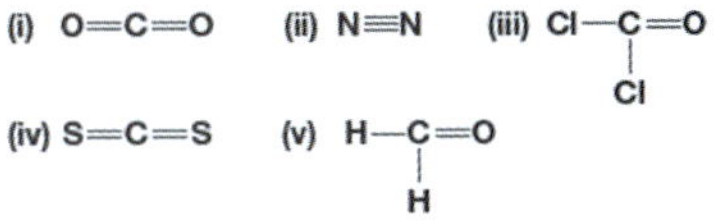

4F

(i) covalent (two non-metals)
(ii) ionic (metal + non-metal)
(iii) covalent (two non-metals)

4G

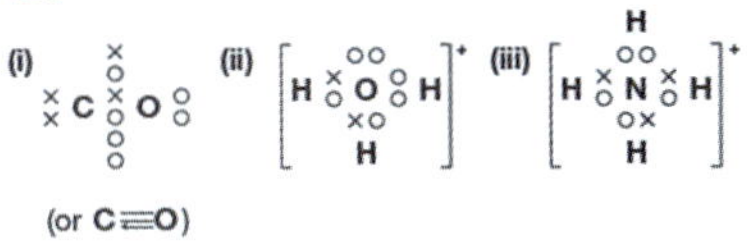

4H

(i) 0.6 **(ii)** 1.4 **(iii)** 0.4 **(iv)** 0.5 **(v)** 1.0
Order: $C^{\delta-}$–$H^{\delta+}$, $C^{\delta+}$–$Cl^{\delta-}$, $Be^{\delta+}$–$H^{\delta-}$, $C^{\delta+}$–$O^{\delta-}$, $O^{\delta-}$–$H^{\delta+}$

4I

(i) LiCl **(ii)** KF **(iii)** Al_2S_3

4J

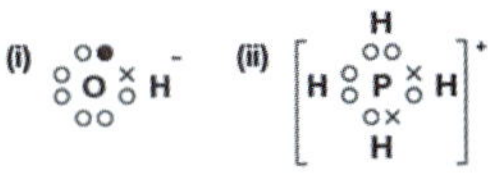

where • is an electron donated from another atom, e.g. Na^+ from Na^+ OH^-.

4K

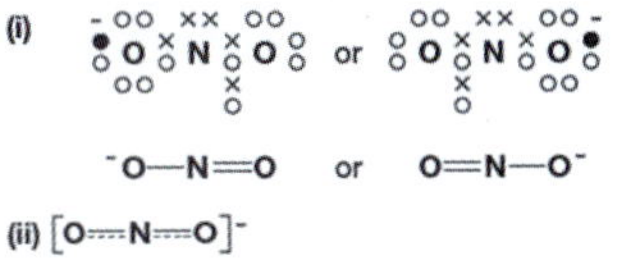

Revision questions

4.1

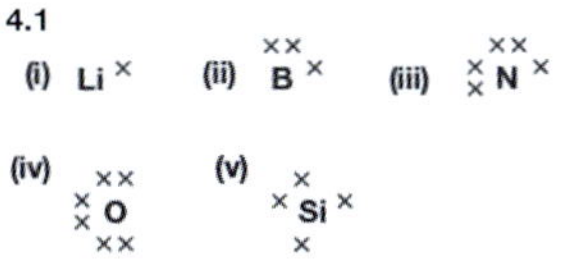

4.2

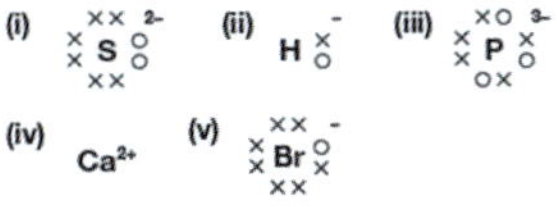

4.3

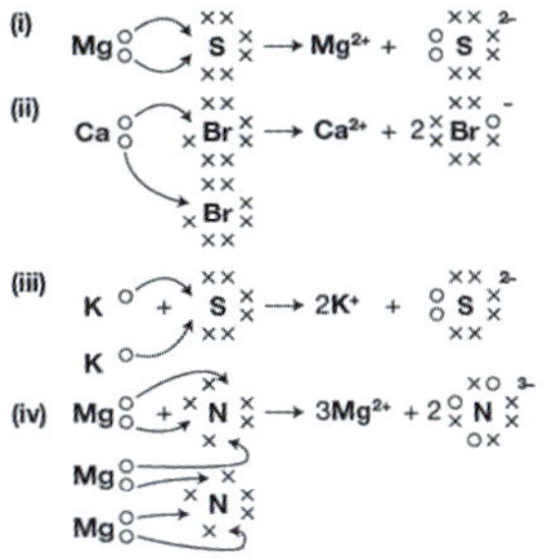

4.4 Because the sodium atom has the electronic structure 2.8.1, it loses only one electron to achieve the stable electronic structure of neon.

4.5

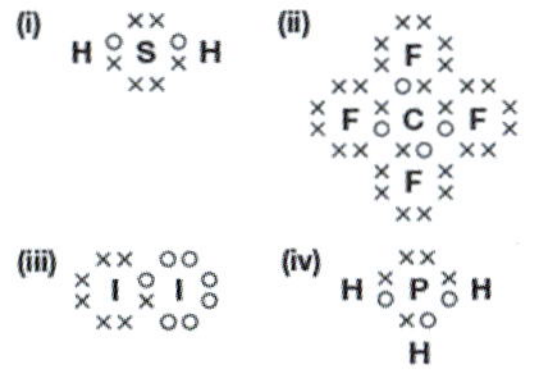

4.6

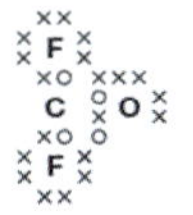

4.7

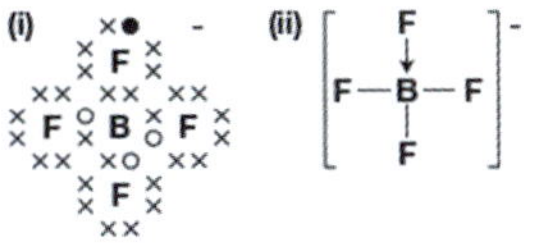

4.8

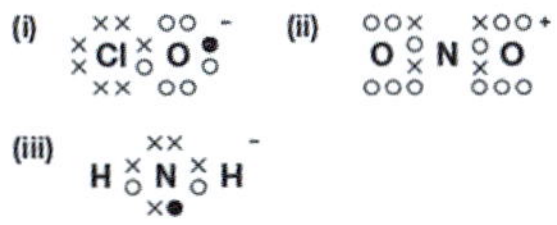

4.9 Si–Br
4.10 1.8, ionic
4.11

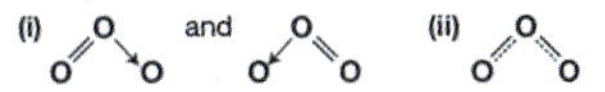

4.12

Na	+	Cl	→	Na^+	+	Cl^-
$2p^6\ 3s^1$		$3s^2\ 3p^5$		$2p^6$		$3s^2\ 3p^6$

The inner orbitals are not shown.

4.13 The cause is a second reaction that competes for the chlorine. The competing reaction is the formation of yellow/brown iron (III) chloride. The iron comes from the metal spoon: $Fe(s) + Cl_2(g) \rightarrow Fe^{3+}, 3Cl^-(s)$

Unit 5

Exercises

5A

Li has to lose one electron only in order to achieve a stable electronic structure. Be and B have to lose two and three electrons, respectively. This takes a great deal more energy than for Li.

5B

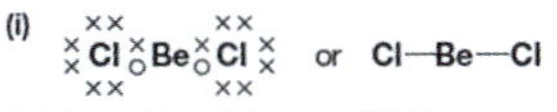

(ii) 4

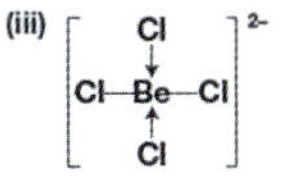

(iv) 8 **(v)** $[BeCl_4]^{2-}$

Beryllium has a more stable electronic configuration in $[BeCl_4]^{2-}$, because the atom is surrounded by eight electrons.

5C

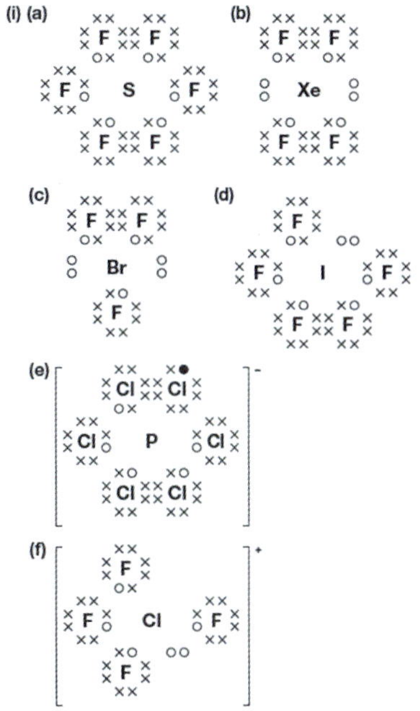

(ii) Nitrogen ($1s^2\ 2s^2\ 2p^3$) cannot expand its octet as there are no available d orbitals to do so.
(iii) Chlorine can expand its octet, but it is too small an atom to accommodate seven atoms of fluorine around it. Iodine is a much bigger atom.

5D

(i) trigonal planar **(ii)** tetrahedral
(iii) roughly tetrahedral (not all the bond lengths are the same)
(iv) tetrahedral **(v)** octahedral

5E

(i) Distorted tetrahedral with respect to the electron pairs; 'bent' with respect to H–S bonds (similar to water).
(ii) Distorted tetrahedral with respect to the electron pairs; trigonal pyramidal with respect to its bonds (similar to ammonia).
(iii) Distorted octahedral with respect to the electron pairs.
(iv) Distorted tetrahedral with respect to the electron pairs; 'bent' with respect to its bonds.
(v) As for **(iv)**.

5F

(i) trigonal planar (remember that all the bond lengths are the same because of resonance)
(ii) trigonal planar **(iii)** linear **(iv)** tetrahedral
(v) trigonal planar

5G

1. (i) no – linear **(ii)** yes **(iii)** no – tetrahedral **(iv)** yes **(v)** yes
2. (i) yes **(ii)** $H^{\delta+}–F^{\delta-}$

5H

(i) two
(ii) As Figure 5.6, with '2+' ions and two free electrons for each ion in the structure.
(iii) three
(iv) Na, Mg, Al (the more valence electrons that are free, the stronger the bonding).
(v) Al (it has the strongest metallic bonding, therefore it requires most energy to break down the structure).

5I

····C≡C–C≡C–C≡C–C····

5J

(i) Kr, Ar, Ne, He (the smaller the atom, the fewer the electrons and the weaker the forces that attract the atoms together).
(ii) Again, I_2 is the largest molecule and has the greatest number of electrons, so the forces that attract I_2 molecules together are the strongest and it has the highest melting and boiling points. The other halogens follow this trend.

Revision questions

5.1

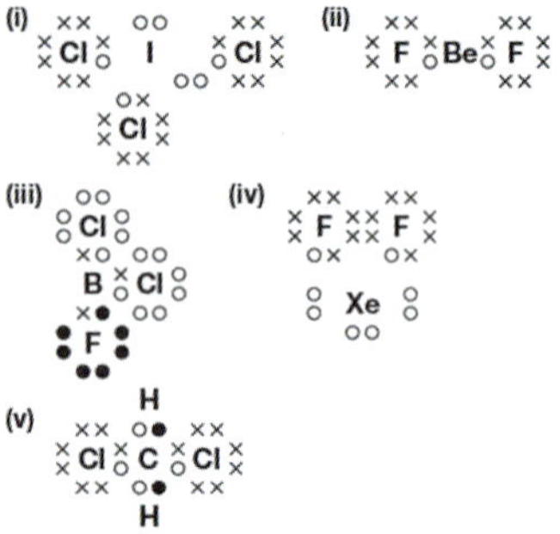

5.2 The shapes with respect to the electron pairs around the central atom are roughly the following:
(i) trigonal bipyramidal **(ii)** linear
(iii) trigonal planar
(iv) trigonal bipyramidal **(v)** tetrahedral

5.3

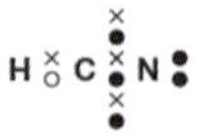

linear

5.4 (i) no **(ii)** no **(iii)** no **(iv)** yes **(v)** yes

5.5 Substances that exist as atoms or non-polar molecules (such as the noble gases or iodine) liquefy or solidify at appropriate temperatures.

5.6 Ca because it has two valence electrons and stronger metallic bonding (more electron density to 'glue' the positive ions together).

5.7 Each carbon atom in graphite has one free, mobile electron. There are no free electrons in diamond, they are all involved in covalent bonding.

5.8 A CO_2 molecule is linear and non-polar. Only London forces attract the molecules together in the solid state.

5.9 The oxygen atom of water has two lone pairs of electrons and can therefore form two hydrogen bonds per molecule. The nitrogen atom in ammonia has only one lone pair and ammonia molecules can only form one hydrogen bond per molecule. In addition, oxygen is more electronegative than nitrogen.

5.10 Argon gas consists of isolated argon atoms and so the only intermolecular forces between argon atoms are the weak London dispersion forces (LDF). This explains low boiling point: argon is a gas at room temperature. LDF is present between propanone molecules and between water molecules, but more substantial attraction is caused by dipole-dipole forces in propanone ($(CH_3)_2C{=}O$, see Figure 5.12) and (even stronger) by hydrogen bonding in water (H_2O, see Figure 5.17). In general, the greater the attraction between molecules, the higher the boiling point.

5.11 If you make a model you will find there are eight hexagons and twelve pentagons.

5.12 The molecular identity of water molecules persists in ice, liquid water or even water vapour; even in ice we can still identify H_2O units, even though they are now linked by intermolecular attractive forces (principally hydrogen bonding but also London dispersion forces). What happens as the ice is heated is a battle between the intermolecular forces holding the ice lattice together and the increasing kinetic energy of the water molecules. The important point here is that the strength of the intermolecular forces does not change as the temperature increases, so that eventually the jostling of the water molecules is so vigorous that the ice lattice breaks down into liquid water where the water molecules slide over each other.

5.13 The table of electronegativities (page 57) allows us to calculate the electronegativity difference of atoms in the Be–F bond as 4.0 – 1.5 = 2.5. This shows that the Be–F bond is polar (it has a non-zero bond dipole). If BeF_2 was bent, like water, the overall (molecular) dipole would be non-zero. A zero molecular dipole can only arise if the BeF_2 is linear, with each Be–F bond dipole pointing in opposite directions and so cancelling out.

Unit 6

Exercises

6A

(i)

$Ca^{2+},2NO_3^-(s) \xrightarrow{H_2O} Ca^{2+}(aq) + 2NO_3^-(aq)$

(ii)

$2K^+,SO_4^{2-}(s) \xrightarrow{H_2O} 2K^+(aq) + SO_4^{2-}(aq)$

(iii)

$2Na^+,CO_3^{2-}\cdot 10H_2O(s) \xrightarrow{H_2O} 2Na^+(aq) + CO_3^{2-}(aq) + 10H_2O(l)$

6B

The nitrate ion is the spectator ion. Overall ionic reaction is:

$Cu^{2+}(aq) + 2NO_3^-(aq) + Zn(s) \rightarrow Zn^{2+}(aq) + 2NO_3^-(aq) + Cu(s)$

or, cancelling spectator ions

$Cu^{2+}(aq) + Zn(s) \rightarrow Zn^{2+}(aq) + Cu(s)$

This is the same ionic equation as for the reaction between copper sulfate solution and zinc solid.

6C

(i) Carbonate, sulfide and hydroxide
(ii) We will need to mix a solution of a soluble barium salt (such as barium ethanoate, $CH_3COO^-)_2$ Ba^{2+}) with a solution of a soluble sulfate (such as magnesium sulfate, Mg^{2+},SO_4^{2-}).
(iii) We will need to mix a solution of a soluble iron salt (such as iron(II) sulfate, Fe^{2+},SO_4^{2-}) with a solution of a soluble hydroxide (such as sodium hydroxide, Na^+,OH^-).

6D

Coordinate bond.

6E

$HBr(g) + H_2O(l) \rightarrow H_3O^+(aq) + Br^-(aq)$
or

$HBr(g) \xrightarrow{H_2O} H^+(aq) + Br^-(aq)$

6F

(i) Calcium chloride will form a white precipitate, whereas magnesium nitrate will not. Potassium sulfate may also give a cloudiness of (slightly soluble) silver sulfate depending upon the concentrations of reactants used.
(ii) The precipitates are silver bromide (AgBr) and silver iodide (AgI):

$Ag^+(aq) + Br^-(aq) \rightarrow AgBr(s)$

$Ag^+(aq) + I^-(aq) \rightarrow AgI(s)$

These precipitates are useful in confirming the presence of bromides and iodides in solution.

6G

(i) The rest do not contain a sulfate ion.

6H

(ii) Ammonium carbonate:
$NH_4^+(aq) + OH^-(aq) \rightarrow NH_3(g) + H_2O(l)$

6I

(i) (a) two **(b)** one
(ii) (a) ethanoates (acetates)
(b) nitrates

6J

$K^+,OH^-(s) \xrightarrow{H_2O} K^+(aq) + OH^-(aq)$

$Ca^{2+},2OH^-(s) \xrightarrow{H_2O} Ca^{2+}(aq) + 2OH^-(aq)$

6K

(i) $Fe(s) + H_2SO_4(aq) \rightarrow FeSO_4(aq) + H_2(g)$
(ii) $Fe(s) + 2H^+(aq) \rightarrow Fe^{2+}(aq) + H_2(g)$
($SO_4^{2-}(aq)$ is the spectator ion)

6L

(i) D is Zn^{2+} (solid ZnO turns yellow upon heating). The white precipitate suggests that C is Cl^-.
(ii) Dilute sulfuric acid. Confirm this by adding $BaCl_2$/HCl (white precipitate observed).
(iii) Mix the solutions together one by one. The copper solution is coloured blue. Both sodium carbonate and sodium hydroxide will form a blue precipitate with copper ions since both copper(II) hydroxide and copper(II) carbonate are insoluble (see Table 6.2). However, only sodium carbonate will 'fizz' with the remaining solution, dilute hydrochloric acid.

Revision questions

6.1 (i) Pure sulfuric 'acid' is a covalent liquid and does not contain any $H^+(aq)$ ions.
(ii) It produces $OH^-(aq)$ ions when it dissolves in water.
(iii) The water present in the ammonia would react with the concentrated acid, producing dilute sulfuric acid, which would neutralize the ammonia.
(iv) When it is hot and concentrated.

6.2 (i) $CuSO_4\cdot 5H_2O(s) \rightarrow Cu^{2+}(aq) + SO_4^{2-}(aq) + 5H_2O(l)$
(ii) $Li^+,OH^-(s) \rightarrow Li^+(aq) + OH^-(aq)$

6.3 (i) $(COOH)_2(aq) + 2KOH(aq) \rightarrow (COOK)_2(aq) + 2H_2O(l)$
or $H^+(aq) + OH^-(aq) \rightarrow H_2O(l)$
(ii) $(COOH)_2(aq) + Zn(s) \rightarrow (COO)_2Zn(aq) + H_2(g)$
or $2H^+(aq) + Zn(s) \rightarrow H_2(g) + Zn^{2+}(aq)$

6.4 (i) $MgCO_3(aq) + H_2SO_4(aq) \rightarrow MgSO_4(aq) + H_2O(l) + CO_2(g)$
or $2H^+(aq) + CO_3^{2-}(aq) \rightarrow CO_2(g) + H_2O(l)$
(ii) $(NH_4)_2SO_4(aq) + 2NaOH(aq) \rightarrow 2NH_3(g) + 2H_2O(l) + Na_2SO_4(aq)$
or $NH_4^+(aq) + OH^-(aq) \rightarrow H_2O(l) + NH_3(g)$
(iii) $Na_2S(aq) + 2HNO_3(aq) \rightarrow 2NaNO_3(aq) + H_2S(g)$
or $S^{2-}(s) + 2H^+(aq) \rightarrow H_2S(g)$
(iv) $Na_2SO_3(aq) + 2CH_3CO_2H(aq) \rightarrow 2CH_3CO_2Na(aq) + SO_2(g) + H_2O(l)$
or $SO_3^{2-}(aq) + 2H^+(aq) \rightarrow SO_2(g) + H_2O(l)$

6.5 (i) Both gases react with water, producing acidic solutions. SO_2 forms a more strongly acidic solution than CO_2, i.e. it produces more $H^+(aq)$ ions per mole of dissolved gas in aqueous solution. (SO_2 is also more soluble than CO_2.) The presence of greater numbers of $H^+(aq)$ ions means that more limestone can be destroyed.
(ii) The $OH^-(aq)$ ions force any dissolved CO_2 to produce carbonate ions, $CO_3^{2-}(aq)$, which react with Ca^{2+} to produce insoluble calcium carbonate:

$Ca^{2+}(aq) + CO_3^{2-}(aq) \rightarrow CaCO_3(s)$

6.6 The blue colour suggests that the substance contains copper(II) ions (K = Cu^{2+}). The fizzing with acid suggests the presence of a carbonate ion (J = CO_3^{2-}). In the fizzing reaction, the carbonate is converted to a salt and to CO_2, while the copper(II) ion is unaffected. The copper(II) ion forms a blue precipitate of copper(II) hydroxide with OH^-(aq).

6.7 G = Fe^{3+}(aq) (confirmed by precipitate of $Fe(OH)_3$); H = NO_3^- (confirmed by brown ring test).

6.8 S = S^{2-} and T = Pb^{2+}:

$S^{2-}(s) + 2H^+(aq) \rightarrow H_2S(g)$

This reaction causes Pb^{2+} ions in the solid PbS to enter aqueous solution where they may now be identified by the addition of sulfate ions, which precipitate white lead(II) sulfate:

$Pb^{2+}(aq) + SO_4^{2-}(aq) \rightarrow PbSO_4(s)$

and by the addition of hydroxide ions:

$Pb^{2+}(aq) + 2OH^-(aq) \rightarrow Pb(OH)_2(s)$

$Pb(OH)_2(s) + 2OH^-(aq) \rightarrow [Pb(OH)_4]^{2-}(aq)$

6.9 Under alkaline conditions, any free aluminium ions are precipitated as $Al(OH)_3$ or as higher hydroxides:

$Al^{3+}(aq) + 3OH^-(aq) \rightarrow Al(OH)_3(s)$

The precipitate 'locks up' Al^{3+}(aq). In acidic conditions, free aluminium ions are produced (see page xx):

$Al(OH)_3(s) + 3H^+(aq) \rightarrow \underset{\text{Toxic}}{Al^{3+}(aq)} + 3H_2O(l)$

6.10

(i) Because as soon as NO meets air (O_2), it forms NO_2.
(ii) X = nitrous oxide; Y = ammonia (the only common alkaline gas); Z = carbon dioxide.

Unit 7

Exercises

7A

(i) $Cu^{2+}(aq) + 2e^- \rightarrow Cu(s)$
(ii) $Zn(s) \rightarrow Zn^{2+}(aq) + 2e^-$
(iii) $Cu^{2+}(aq) + Zn(s) \rightarrow Cu(s) + Zn^{2+}(aq)$
(iv) The sulfate ions remain the same throughout the reaction – they do not take part in the redox reaction. The sulfate ions are **spectator** ions, as discussed in Unit 6.

7B

(i) Zn oxidized (gains O), O_2 reduced (loses O!)
(ii) Mg oxidized (loses electrons), Cu^{2+} reduced (gains electrons)
(iii) Cl_2 reduced (gains electrons), I^- oxidized (loses electrons)
(iv) Since sodium chloride is Na^+Cl^-, Na oxidized (loses electron), Cl_2 reduced (gains electrons)

7C

(i) +4 **(ii)** –3 **(iii)** O **(iv)** +5 **(v)** +5
(vi) +7 **(vii)** +2 **(viii)** +5 (contains VO_3^-)
(ix) +5 **(x)** +6

7D

(i)

SO_4^{2-}, SO_3	+6
SO_3^{2-}, SO_2	+4
S	0
H_2S	–2

(ii) No, the oxidation number of sulfur does not change.
(iii) SO_2 is reduced to S because the oxidation number of sulfur decreases from +4 to 0. H_2S is oxidized to S because the oxidation number of sulfur increases from –2 to 0.

7E

$HClO_4$	+7
ClO_3^-	+5
ClO^-	+1
Cl_2	0
HCl Cl^-	–1

7F

(i) Those with oxidation numbers +4 and 0, because they can be oxidized (increase their oxidation number) and reduced (decrease their oxidation number).
(ii) Those with oxidation number +6, because they can be reduced only.
(iii) H_2S, because it can be oxidized only.

7G

1. (i) $I^- + 3H_2O \rightarrow IO_3^- + 6H^+ + 6e^-$
(ii) $NO_3^- + 4H^+ + 3e^- \rightarrow NO + 2H_2O$
(iii) $Cr_2O_7^{2-} + 14H^+ + 6e^- \rightarrow 2Cr^{3+} + 7H_2O$
(iv) $H_2O_2 + 2H^+ + 2e^- \rightarrow 2H_2O$
(v) To balance $S^{2-} \rightarrow SO_4^{2-}$
Balance the Os with H_2O:
$S^{2-} + 4H_2O \rightarrow SO_4^{2-}$
Balance the hydrogens by adding $8H_2O$ to the opposite side and $8OH^-$ to the same side:
$S^{2-} + 4H_2O + 8OH^- \rightarrow SO_4^{2-} + 8H_2O$
Balance the charges:
$S^{2-} + 4H_2O + 8OH^- \rightarrow SO_4^{2-} + 8H_2O + 8e^-$
Simplify the equation:
$S^{2-} + 8OH^- \rightarrow SO_4^{2-} + 4H_2O + 8e^-$
2. (i) $Cu + 2Ag^+ \rightarrow Cu^{2+} + 2Ag$
(ii) $Cu + 2NO_3^- + 4H^+ \rightarrow Cu^{2+} + 2NO_2 + 2H_2O$
(iii) $I_2 + 2S_2O_3^{2-} \rightarrow S_4O_6^{2-} + 2I^-$
(iv) $2I^- + H_2O_2 + 2H^+ \rightarrow I_2 + 2H_2O$
(v) $2H_2O_2 + ClO_2^- \rightarrow 2O_2 + Cl^- + 2H_2O$

7H

(i)

Pt | H_2(g) | H^+(aq) || Cu^{2+}(aq) | Cu(s)

(ii) +0.34V = $E^\ominus(Cu^{2+}(aq)/Cu(s))$ – 0, therefore $E^\ominus(Cu^{2+}(aq)/Cu(s))$ = + 0.34 V
(iii) $H_2(g) + Cu^{2+}(aq) \rightarrow 2H^+(aq) + Cu(s)$

7I

(i) yes

$I_2(s) + 2e^- \rightleftharpoons 2I^-(aq)$ +0.54
$Br_2(l) + 2e^- \rightleftharpoons 2Br^-(aq)$ +1.09

(ii) no

$2H^+(aq) + 2e^- \rightleftharpoons H_2(g)$ 0.00
$Cu^{2+}(aq) + 2e^- \rightleftharpoons Cu(s)$ +0.34

(iii) yes

$Zn^{2+}(aq) + 2e^- \rightleftharpoons Zn(s)$ –0.76
$Fe^{3+}(aq) + e^- \rightleftharpoons Fe^{2+}(aq)$ +0.77

(iv) yes

$I_2(s) + 2e^- \rightleftharpoons 2I^-(aq)$ +0.54
$Fe^{3+}(aq) + e^- \rightleftharpoons Fe^{2+}(aq)$ +0.77

7J

Work out the reactions by using anticlockwise arrows:
(i) $Mg(s) + 2Ag^+(aq) \rightarrow Mg^{2+}(aq) + 2Ag(s)$
(ii) no reaction
(iii) $Ni(s) + Cu^{2+}(aq) \rightarrow Ni^{2+}(aq) + Cu(s)$
(iv) $Zn(s) + 2H^+(aq) \rightarrow Zn^{2+}(aq) + H_2(g)$
(v) no reaction

7K

(i) There is no dissolved oxygen in boiled water.
(ii) Water is more highly conducting if it carries dissolved salt and rusting occurs more rapidly. A great deal of salt is around at coastal locations.
(iii) Acid rain provides H^+ ions, making rusting occur more readily (see the equations for rusting).

7L

The two main ways are: (1) coating the metal with an unreactive coating (e.g. paint or grease) that acts as a physical barrier to water and air. Electroplating iron with silver falls into this category as the silver is relatively unreactive. (2) Dipping iron in zinc (galvanization). Here, the zinc not only acts a physical barrier, but (as shown by the values of the relevant electrode potentials for the metal/metal ion redox couples) it also more reactive than iron so that it reacts (oxidises) before the iron does. This is a form of 'sacrificial protection'.

7M

NO_3^- HNO_3	+5
NO_2	+4
NO_2^- HNO_2	+3
NO	+2
N_2O	+1
N_2	0
NH_3 NH_4^+	–3

Revision questions

7.1 (i) C +4
(ii) Mn +4
(iii) S +6
(iv) Sn +4
(v) N –3

7.2 yes

$Cu^{2+}(aq) + 2e^- \rightleftharpoons Cu(s)$ +0.34
$Fe^{3+}(aq) + e^- \rightleftharpoons Fe^{2+}(aq)$ +0.77

7.3 (i) $Cu^{2+}(aq) + Fe(s) \rightarrow Fe^{2+}(aq) + Cu(s)$
(ii) no reaction

7.4 (i) Au (oxidation number 0 to +1)
(ii) manganese in MnO_4^- (+7 to +2)
(iii) MnO_4^- it is reduced to Mn^{2+}

7.5 $SO_2(g) + 2H_2O(l) + 2Fe^{3+}(aq) \rightarrow SO_4^{2-}(aq) + 4H^+(aq) + 2Fe^{2+}(aq)$

7.6 Find out the reaction:

$Al^{3+}(aq) + 3e^- \rightleftharpoons Al(s)$ –1.67
$Fe^{2+}(aq) + 2e^- \rightleftharpoons Fe(s)$ –0.44

(i) From the Al electrode to the Fe electrode.
(ii) The reactions occurring are:
$Al(s) \rightarrow Al^{3+}(aq) + 3e^-$
$Fe^{2+}(aq) + 2e^- \rightarrow Fe(s)$
(iii) $Al^{3+}(aq) + 3e^- \rightarrow Al(s)$ $E^\ominus = -1.67$
$Fe^{2+}(aq) + 2e^- \rightarrow Fe(s)$ $E^\ominus = -0.44$
$E^\ominus(cell) = E^\ominus_R - E^\ominus_L$
$= -0.44 - (-1.67)$
$= 1.23V$
(iv) Al(s) | Al^{3+}(aq) || Fe^{2+}(aq) | Fe(s)

7.7 (i) Ag^+, S^{2-}
(ii) Yes, silver ions are reduced to silver by aluminium, which is oxidized to aluminium ions.
(iii) Polish rubs off the sulfide layer, this method 'regenerates' the silver metal and it is not lost.

7.8 $MnO_2(s) + 4H^+(aq) + 2Cl^-(aq) \rightarrow Mn^{2+}(aq) + 2H_2O(l) + Cl_2(g)$

7.9 $Zn + 4OH^- \rightarrow [Zn(OH)_4]^{2-} + 2e^-$ (oxidation)
$NO_3^- + 6H_2O + 8e^- \rightarrow NH_3 + 9OH^-$ (reduction).
Multiply the oxidation reaction by 4 and add both half equations then simplify.
$4Zn(s) + NO_3^-(aq) + 7OH^-(aq) + 6H_2O(l) \rightarrow 4[Zn(OH)_4]^{2-}(aq) + NH_3(g)$

7.10 (i) Ni^{2+} (gains electrons and is reduced)
(ii) Zn (loses electrons and is oxidized).

7.11

(i) A rubber tube is connected to the end of the taps in the Hofmann voltameter and samples of the two causes collected in separate test tubes. Hydrogen would 'pop' in the presence of a flaming splint, and oxygen would re-ignite a glowing splint.
(ii) It is possible that there is a leak in the voltameter, but the most likely explanation is that the teacher used alternating electricity (AC) instead of direct current (DC). This rapidly and continuously shifts the polarity of the electrodes (e.g. fifty times a second) producing equal volumes of a 2:1 mixture of hydrogen and oxygen in *each* arm of the voltameter. Such a mixture is explosive and produces a much louder noise that the ordinary 'hydrogen pop'!

7.12 The greater the difference in $E^\ominus$ values between the couples making up the between the two half-cells in an electrochemical cell, the greater the overall voltage of the cell and the greater the energy transferred per mole of electrons. If sodium and lithium are connected to another metal in a circuit, the lithium couple (with the more negative $E^\ominus$ value) provides more energy per mole of electrons. Lithium is also less dense than sodium, and so batteries are lighter to carry.

Unit 8

Exercises

8A

(i) 63 u **(ii)** 120 u **(iii)** 26 u **(iv)** 256 u
(v) 88 u **(vi)** 404 u **(vii)** 20 u

8B

(i) 20 g (1 mol) **(ii)** 12 × 10^{23} **(iii)** 2.4 g (0.1 mol)
(iv) 159.5 g

8C

1. (i) 14 g **(ii)** 224 g **(iii)** 53.3 g **(iv)** 2520 g
2. (i) 1 mol **(ii)** 0.80 mol **(iii)** 5.1 × 10^{-3} mol
(iv) 0.50 mol
3. Na_2CO_3 (0.01 mol)
4. (i) 3.8 × 10^{22} (0.063 mol)
(ii) 6.6 × 10^{20} (5.5 × 10^{-4} mol Al_2O_3)

8D

(i) 23 per cent **(ii)** 16 per cent **(iii)** 68 per cent
(iv) 80 per cent **(v)** 73 per cent

8E

(i) CH_4 **(ii)** Al_2O_3 **(iii)** Fe_3O_4 **(iv)** Na_2SO_4

8F

(i) C_2H_6 **(ii)** C_5H_7N (empirical formula)
$C_{10}H_{14}N_2$ (molecular formula)

8G

(i) $CuSO_4{\cdot}5H_2O \rightarrow CuSO_4{\cdot}3H_2O + 2H_2O$
(ii) $CuSO_4{\cdot}3H_2O \rightarrow CuSO_4{\cdot}H_2O + 2H_2O$
(iii) $CuSO_4{\cdot}H_2O \rightarrow CuSO_4 + H_2O$
(iv) $CuSO_4 \rightarrow CuO + SO_3$

8H

(i) 2 g **(ii)** 0.32 g **(iii)** 38 g
(iv) $C(s) + O_2(g) \rightarrow CO_2(g)$, 0.60 g
(v) $CuCO_3(s) \rightarrow CuO(s) + CO_2(g)$, 0.82 g

8I

(i) 6 dm^3 **(ii)** 1.6 g **(iii)** 4.8 dm^3 **(iv)** 10 dm^3
(v) 80 cm^3 **(vi)** 7.5 g.

8J

(i) oxygen, 0.1 mol
(ii) magnesium, 2.1 g
(iii) copper(II) oxide, 2.4 g.

8K

56 per cent.

Revision questions

8.1 (i) 28 u **(ii)** 134.5 u **(iii)** 98 u
(iv) 164 u **(v)** 322 u
8.2 (i) 35.5 g **(ii)** 71 g **(iii)** 124 g
(iv) 127 g **(v)** 720 g
8.3 (i) 0.94 mol **(ii)** 0.25 mol **(iii)** 0.25 mol
(iv) 0.051 mol **(v)** 10 mol
8.4 (i) 29.3 g **(ii)** 11 g **(iii)** 4 g
(iv) 624 g **(v)** 6300 g
8.5 7.5 g
There are 2.00/17 mol NH_3. 1 mol NH_3 has the same number of molecules as there are in 1 mol SO_2. 2.00/17 mol NH_3 has the same number of molecules as there are in 2.00/17 mol SO_2 or 2.00/17 × 64 g SO_2 = 7.5 g SO_2.
8.6 6 g
1 mol Mg reacts with 1 mol S; 24g Mg reacts with 32 g S; 6 g Mg reacts with 8 g S.
8.7 (i) 0.8 g:
2 mol $NaNO_3$ produces 1 mol O_2; 170 g $NaNO_3$ produces 32 g O_2; 4.25 g $NaNO_3$ produces 32/170 × 4.25 g O_2.
(ii) 0.6 dm^3:
2 mol $NaNO_3$ produces 24 dm^3 O_2; 170 g $NaNO_3$ produces 24 dm^3 O_2; 4.25 g $NaNO_3$ produces 24/170 × 4.25 dm^3 O_2.
8.8 (i) $2NaHCO_3(s) \rightarrow Na_2CO_3(s) + H_2O(l) + CO_2(g)$
(ii) 168 g **(iii)** 44 g, 24 dm^3
(iv) 0.60 dm^3 (600 cm^3).
8.9 51 per cent:

$$\frac{7\times \text{ molecular mass of water}}{\text{molecular mass of } MgSO_4\cdot 7H_2O}\times 100$$

$$=\frac{7\times 18}{246}\times 100$$

8.10 C_2H_4 (the simplest, or empirical, formula is CH_2).
8.11 Amount of Al in moles = 13.50/27 = 0.50 mol.
Mass of Cl combined with Al = 66.75 − 13.50 = 53.25 g.
Amount of Cl in moles = 53.25/35.5 = 1.50 mol.
Ratio of the number of moles Al atoms to number of moles Cl atoms is 0.50:1.50 or 1:3. Formula is $AlCl_3$.
8.12

	C	H	Cl	N
%	52.3	5.94	14.1	27.7
mol	52.3/12	5.94/1	14.1/35.5	27.7/14
or	4.4	5.9	0.4	2.0
Divide by smallest number (0.4)				
Approx.	11	15	1	5

The molar mass of $C_{11}H_{15}ClN_5$ is 252.5 g mol^{-1} so formula is $C_{22}H_{30}Cl_2N_{10}$.
8.13

	C	H	O
mg	20.5	2.27	27.3
mol	20.5/12	2.27/1	27.3/16
	1.7	2.3	1.7
	1	1.3	1
Approx	3	4	3

The molar mass of $C_3H_4O_3$ is 88 g mol^{-1} so formula is $C_6H_8O_6$.
8.14
(i) $Na_2CO_3.10H_2O$
(ii) Molar mass is 286 g mol^{-1}. Number of mol of salt in 1,000 g is 1,000/286 = 3.50 mol, this is equivalent to 35 mol of water.
(iii) Percentage of water = (35 × 18/1,000) × 100 = 63 per cent.
8.15
(i) 286 g mol^{-1}
(ii) Moles of vitamin A in tablet = 0.015/286, molecules = (0.015/286) × 6 × 10^{23} or 3.1 × 10^{19}

Unit 9

Exercises

9A

(i) 10.0 mol dm^{-3} **(ii)** 0.008 mol dm^{-3}
(iii) 50 dm^3 **(iv)** 28.6 g

9B

(i) K^+ and Cl^- both 0.1 mol dm^{-3}
(ii) Na^+ and CO_3^{2-} are 0.250 mol dm^{-3} and 0.125 mol dm^{-3}, respectively
(iii) NH_4^+ and SO_4^{2-} are 0.50 mol dm^{-3} and 0.25 mol dm^{-3}, respectively
(iv) Cu^{2+} and NO_3^- are 0.15 mol dm^{-3} and 0.30 mol dm^{-3}, respectively

9C

(i) 0.05 mol dm^{-3} **(ii)** 0.10 mol dm^{-3}
(iii) 0.20 mol dm^{-3} **(iv)** 0.40 mol dm^{-3}
(v) 1.00 mol dm^{-3}

9D

(i) 0.53 g
Amount in mol = vol × molar conc = 500/1,000 × 0.01
= 5.0 × 10^{-3} mol
= 5.0 × 10^{-3} × $M(Na_2CO_3)$ g = 0.53 g
(ii) It would be more concentrated than the value calculated.
(iii) 0.02 mol dm^{-3}
Amount in mol present = vol × molar conc = 50/1,000 × 0.1 = 5 × 10^{-3} mol
If this amount is made up to 250 cm^3, then

molar conc = amount in mol/vol
= 5 × 10^{-3}/0.25 = 0.02 mol dm^{-3}
(iv) 125 cm^3
Number of moles needed = vol × conc = 1 × 2 = 2 mol
vol = number mol/conc = 2/16 = 0.125 dm^3
(v) 0.05 dm^3

9E

(i) 50 cm^3
1 mol H_2SO_4 ≡ 2 mol KOH

$$\frac{\text{Vol}\times\text{molar concentration } H_2SO_4}{\text{Vol}\times\text{molar concentration KOH}}=\frac{1}{2}$$

$$\frac{V\times 10}{25/1000\times 40}=\frac{1}{2}$$

(ii) 0.071 mol dm^{-3} **(iii)** 0.263 mol dm^{-3}
(iv) 20 cm^3

9F

(i) $6Fe^{2+}(aq) + Cr_2O_7^{2-}(aq) + 14H^+(aq) \rightarrow 6Fe^{3+}(aq) + 2Cr^{3+}(aq) + 7H_2O(l)$
6 mol Fe^{2+} ≡ 1 mol $Cr_2O_7^{2-}$, so that

$$\frac{\text{vol}\times\text{molar concentration } Fe^{2+}}{\text{vol}\times\text{molar concentration } Cr_2O_7^{2-}}=\frac{6}{1}$$

$$\frac{25/1000\times C}{13.5/1000\times 0.021}=\frac{6}{1}$$

C = 0.065 mol dm^{-3}
(ii) 50 cm^3
(iii) 210 cm^3
The equation is:

$$3MnO_4^- + 5Fe^{2+} + 5C_2O_4^{2-} + 24H^+ \rightarrow 3Mn^{2+} + 5Fe^{3+} + 10CO_2 + 12H_2O$$

$$\frac{\text{vol}\times\text{molar concentration } MnO_4}{\text{amount in moles of } FeC_2O_4}=\frac{3}{5}=\frac{V\times 0.020}{0.007}$$

V = 210 cm^3

9G

(i) 67 cm^3

$$\frac{\text{vol } \times \text{ molar concentration } H_2SO_4}{\text{amount in moles of CuO}}=\frac{1}{1}$$

$$=\frac{V\times 1.5}{7.95/(63.5+16)}$$

V = 0.067 dm^3
(ii) 0.29 g **(iii)** 417 cm^3 **(iv)** 3.1 g

9H

(i) 0.1128 g **(ii)** to ensure they were completely dry
(iii) 3.9043 × 10^{-3} mol dm^{-3} (molar mass of $Ni(DMG)_2$ = 288.914 and amount of $Ni(DMG)_2$ in 100 cm^3 = 0.1128/288.914 mol. Multiply this amount of $Ni(DMG)_2$ by 10 to get the concentration).
(iv) 0.4583 g (mass of Ni^{2+} in original solution of volume 2000 cm^3 = 3.9043 × 10^{-3} × 58.69 × 2).

9I

(i) 1.2 mg **(ii)** 400 ppm

9J

(i) 0.014 mol dm^{-3}
500 mg in 1 dm^{-3} is 500/1,000 × 1/35.5 mol dm^{-3}
(ii) 200 ppm
(the solution has been diluted 250/50 = 5 times)
(iii) 0.5 ppb

9K

(i) 13.4 per cent **(ii)** 42 g **(iii)** 180 cm^3
(iv) 568 × 5/1,000 = 2.8 units

9L

(i)

$$X_R=\frac{n_R}{n_P+n_Q+n_R+n_S}$$

(ii)

$$\frac{n_R}{n_P+n_Q+n_R+n_S}+\frac{n_Q}{n_P+n_Q+m_R+n_S}$$
$$+\frac{n_R}{n_P+n_Q+n_R+n_S}+\frac{n_S}{n_P+n_Q+m_R+n_S}$$
$$+\frac{n_P+n_Q+n_R+n_S}{n_P+n_Q+n_R+n_S}=1$$

(iii) pure P

9M

(i) 3/10 or 0.3
(ii) 0.0089, from

$$\frac{37.25/74.5}{37.25/74.5\ +\ 1000/18}$$

9N

(i) weakly acidic **(ii)** weakly acidic
(iii) strongly basic **(iv)** strongly basic
(v) weakly basic **(vi)** weakly acidic
(vii) strongly basic

Revision questions

9.1 (i) 4.20 g **(ii)** 126 g **(iii)** 4.7 g
9.2 It absorbs water and carbon dioxide from the air and therefore would increase in mass when handled in air.
9.3 (i) 50 cm^3 **(ii)** 100 cm^3
9.4 (i) 100 cm^3 **(ii)** 23.1 g
9.5 (i) 0.35 g

$KOH + HNO_3 \rightarrow KNO_3 + H_2O$

Amount in mol HNO_3 = vol × molar conc, i.e.

$25/1000 \times 0.250 = 6.25 \times 10^{-3}$ mol

The same amount in mol of KOH is needed, i.e.

$6.25 \times 10^{-3} \times (39 + 16 + 1)$ g

(ii) 0.39 g, from

$CuCO_3 + 2HNO_3 \rightarrow Cu(NO_3)_2 + H_2O + CO_2$

9.6 (i) 0.100 mol dm^{-3} **(ii)** approx. 6 g dm^{-3}
9.7 100 cm^3
9.8 20 g
9.9 KBr: 2000 ppm of K^+

0.050 mol dm^{-3} of K^+ is 0.050×39

= approx. 2 g dm^{-3} or 2000 mg dm^{-3}

Na_2SO_4: 460 ppm of Na^+

0.010×2 mol dm^{-3} Na^+ is $0.020 \times 23 = 0.46$ g dm^{-3}

$Pb(NO_3)_2$: approx. 1200 ppm of Pb^{2+}
9.10 0.013 g
9.11 0.12 g
20 mg dm^{-3} of Na^+ is $20/1000 \times 1/23$ mol dm^{-3} Na^+
This corresponds to
2 × the number of mol of $Na_2CO_3 \cdot 10H_2O$
or $20/1{,}000 \times 1/23 \times 1/2$ mol dm^{-3} $Na_2CO_3 \cdot 10H_2O$
or $20/1{,}000 \times 1/23 \times 1/2 \times 286$ g dm^{-3} $Na_2CO_3 \cdot 10H_2O$
9.12 (i) pH = $-\log [H^+(aq)] = -\log [4.48 \times 10^{-9}] = 8.349$
(ii) pH = $-\log [H^+(aq)]$; $5.93 -\log [H^+(aq)]$
$[H^+(aq)] = 10^{-5.93}$
$[H^+(aq)] = 1.2 \times 10^{-6}$ mol dm^{-3}
9.13
(i) 1500 ppm
1 ppm = 1 mg dissolved in 1 dm^3
150 mg in 100 cm^3 is 1,500 mg in 1,000 cm^3 or 1 dm^3
(ii) (mass of solute/mass of solution) × 100 =
(0.150/100) × 100 = 0.150 per cent
9.14
5 μg in 50 cm^3 is equivalent to 100 μg in 1,000 cm^3 or 1 dm^3
This is 100 ppb, well below the limit

Unit 10

Exercises

10A

Changes in phase are reversible. The key to reversing such changes is temperature.

10B

(i) Temperature may be thought of as the average kinetic energy of a large group of molecules. If the same amount of energy is absorbed by fewer water molecules, the resulting temperature change is higher.
(ii) Hot days mean a higher temperature, so that more molecules possess enough energy to break away from the liquid. Windy days push the air (containing water vapour) away so that fewer water molecules can condense back onto the clothes.
(iii) Molecules of a gas are always moving and they spread out to fill any container (diffusion).

10C

No. Cooking is a chemical reaction and the rate of reaction depends upon the temperature. The temperature of boiling water is the same (normally 100°C) whether the water boils slowly or rapidly.

10D

(i) (20/760) × 101,325 = 2,666 Pa ≈ 2,700 Pa
(ii) 10 × 101,325 = 1,013,250 Pa
(iii) 107,000 Pa
(iv) 1,000 Pa

10E

The temperature has increased by a factor of (2,500/500) = 5.00. This means that the volume increases five-fold, from 73.0 cm^3 to 365 cm^3 (370 cm^3 to two significant figures).

10F

Pressure has increased by a factor of 90.0/30.0 = 3.00, causing the volume to reduce by a factor of one-third:

$$\frac{1}{3} \times 200.0 = 66.7\,cm^3$$

10G

The number of moles of Zn = (mass/molar mass) = 0.50/65 = 0.0077 mol. The equation is:

$Zn(s) + H_2SO_4(aq) \rightarrow ZnSO_4(aq) + H_2(g)$

Therefore, 1 mol of Zn produces 1 mol of H_2, so that 0.0077 mol of Zn produces 0.0077 mol of H_2. 1 mol of H_2 occupies 24 dm^3 at room temperature and pressure, so that 0.0077 mol of H_2 occupies 0.0077 × 24 = 0.185 dm^3 (≈ 190 cm^3)

10H

Volume of argon in 1.0 dm^3 = (0.94/100) × 1.00 = 0.0094 dm^3 (9.4 cm^3)

10I

$T = 20.0 + 273.15 = 293.2$ K
$V = 0.0020\ m^3$
$P = nRT/V = (1.0 \times 8.3145 \times 293.2)/0.0020$ =
1,218,698 Pa = 1.2×10^6 Pa (to two significant figures)

10J

The greater the intermolecular forces, the greater the deviations from ideal behaviour. CO_2 is a larger molecule than H_2 or N_2, and so the London dispersion forces between its molecules will be greater. (The vapours of polar molecules, such as ethanol or water, show considerable deviations from ideal behaviour because even stronger intermolecular forces are present.)

10K

(i) A liquid boils when its vapour pressure equals the external (atmospheric) pressure. Atmospheric pressure decreases with increasing altitude, so that the vapour pressure required in order for boiling to occur is lower than is needed at sea level. The (reduced) vapour pressure is therefore achieved at a lower liquid temperature, i.e. the boiling point is lower.
(ii) Salty water contains ions from the salt at the surface of the solution. The ions occupy sites that would be taken by water molecules in pure water, and so the number of water molecules in the surface that can escape is reduced in comparison with pure water at the same temperature. A higher temperature than 100°C is needed in order for the water molecules in the surface of the solution to provide a vapour pressure of 1 atm (boiling point).

10L

Total gas pressure = air pressure + vapour pressure of ethanol, so

101 = ? + 7.9

and air pressure = 101 − 7.9 ≈ 93 kPa.

10M

(i) Treat the gases separately. −119°C is at the critical temperature of oxygen so that liquid oxygen will condense if subjected to its critical pressure. However, −119°C is above the critical temperature of hydrogen; hydrogen will not condense out, whatever the applied pressure.
(ii) The problem with liquefying hydrogen gas is getting its temperature at (or below) the critical temperature of −240°C. This is well below the normal boiling point of liquid oxygen or liquid nitrogen. Liquid helium (bp −269°C) could be used as coolant.

Revision questions

10.1 (i) (a) (0.124/760) = 1.63×10^{-4} atm
(b) 1.5 × 101,325 = 151,988 Pa
(c) 89 × 1,000 = 89,000 N m^{-2}
(d) 100.0 kPa is 1 bar so that 101.325 kPa is 1.01325 bar
(e) −100 + 273.15 = 173.15 K
(ii) $V = nRT/P$ where $T = 273 + 20 = 293$ K and P is in Pa. This gives molar volumes of 24.0, 23.4, 25.0 dm^3
10.2 Avogadro's law states that equal volumes of gases (at the same pressure and temperature) contain the same number of molecules. Therefore, if (within experimental error) equal volumes of H_2 and Cl_2 react together, producing twice that volume of HCl, the ratio in which the molecules react must be 1:1 and the ratio of HCl product molecules to H_2 or Cl_2 molecules must be 2:1. This confirms the ratio contained in the equation.
10.3 $M(KClO_3)$ = 122.5 g mol^{-1}, so that the number of moles of $KClO_3$ = 10.0/122.5 = 0.0816.
According to the chemical equation, 2 mol of $KClO_3$ gives 3 mol of O_2. Therefore, 0.0816 mol of $KClO_3$ gives (3/2) × 0.0816 = 0.122 mol of O_2.
1 mol of O_2 occupies 24 dm^3 at room temperature and pressure. Therefore, 0.122 mol of O_2 occupies 0.122 × 24 = 2.9 dm^3. Summarizing, 10.0 g of potassium chlorate decomposes to make about 2.9 dm^3 of oxygen gas.
10.4 (i) P_T = 25 + 50 = 75 kPa = 75,000 Pa.
(ii) Rearrangement of the equation $PV = nRT$ gives

$$n = \frac{PV}{RT}$$

To work out the number of moles of chlorine gas and oxygen gas, substitute into this equation using the partial pressure of each gas for P:

number of moles of Cl_2 =
(25 000 × 0.0100)/8.3145 × 293 = 0.103

number of moles of O_2 =
(50 000 × 0.0100)/8.3145 × 293 = 0.205

Total number of moles of gas = 0.103 + 0.205
= 0.308, so
mole fraction of $O_2 = \dfrac{0.205}{0.308} = 0.666$

(Quicker way! The mole fraction equals the ratio p_{O_2}/P_T = 50.0 kPa/75.0 kPa = 0.666.)
10.5 Number of moles of N_2 = 2.0/28 = 0.071
number of moles of H_2 = 6.0/2.0 = 3.0
total number of moles of gas = 3.071

$$P_T = \frac{nRT}{V} = \frac{3.071 \times 8.3145 \times 473}{0.0020}$$

$P_T = 6.0 \times 10^6$ Pa = 60 atm
10.6 Start by calculating the number of moles of He in the balloon:

$$n = \frac{PV}{RT} = \frac{1.00 \times 105 \times 100}{8.3145 \times 293} = 41.05$$

Then apply the equation

$$T = \frac{PV}{nR}$$

At the new altitude

$$T = \frac{PV}{nR} = \frac{5.00 \times 10^4 \times 166}{41.05 \times 8.3145} = 243K$$

We conclude that the temperature at the new altitude was 243 K.
10.7 First, calculate the molar volume of an ideal gas at 273 K and 1 atm pressure:

$$V = \frac{nRT}{P}$$

$$= \frac{1.000 \times 8.3145 \times 273}{101325} \approx 0.0224\ m^3 \text{ or } 22.4\,dm^3$$

The closer the experimentally determined molar volume of a real gas is to 22.4 dm^3, the more ideally behaved is the gas. The experimental molar volumes suggest that H_2 behaves most ideally, followed by CO_2 and (least ideally) NH_3. This reflects the greater degree of intermolecular force between NH_3 molecules.

10.8 At 40°C, the vapour pressure of water = 7,370 Pa. The vapour occupies half the flask so that $V = 1.0\,dm^3 = 0.0010\,m^3$.

$$n = \frac{PV}{RT} = \frac{7370 \times 0.0010}{8.3145 \times 313} = 2.8 \times 10^{-3}\,\text{mol}$$

To convert the number of moles of water to molecules of water, we multiply by N_A:

number of molecules = $2.8 \times 10^{-3} \times 6.022 \times 10^{23} \approx 1.7 \times 10^{21}$

10.9 If a solvent has a high vapour pressure at room temperature, it is likely it has a sufficiently high vapour pressure at lower temperatures to be ignited by a naked flame.

10.10

(i) Equation is $2NaN_3(s) \rightarrow 2Na(s) + 3N_2(g)$. One mole of azide has a mass of 65.01 g and generates 3 moles of N_2. One mole of any gas has a volume of about 24 dm^3 at room conditions so the N_2 occupies 72 dm^3. In a crash, the N_2 is produced at higher temperature and so the pressure of the N_2 in the bag would be initially greater than one atmosphere, so allowing the airbag to retain its shape for longer.

(ii) The airbag mechanism may be initiated during crushing. If not, unignited and toxic sodium azide may be spread into the atmosphere during crushing. Airbag mechanisms must be taken out before disposing of the rest of the car.

10.11

(i) They are logarithmic scales, which exaggerates the diagram at lower pressures and temperatures.

(ii) see text. **(iii)** (a) ice; (b) liquid water; (c) This point is on the curve, so the equilibrium is a mixture of water vapour and liquid water; (d) supercritical state (red zone).

Unit 11

Exercises

11A

Oil floats on water – this was concealed by the opaque earthenware jug. The captain poured the oil layer into his opponent's tankard.

11B

(i) Cyclohexane is less dense than water (see Table 11.1) and floats to the top.

(ii) Table 11.1 shows the density of cyclohexane as 0.77 g cm^{-3}, so that 130 cm^3 of cyclohexane has a mass of 100 g. Table 11.3 shows that 100 g of cyclohexane saturated with water contains 0.01 g of water. Therefore, 130 cm^3 of cyclohexane contains $0.01 \times 130/100 = 0.013$ g of water.

11C

All the molecules are very similar and differ only in their chain length. Butan-1-ol, possessing the shortest chain, is predicted to be the most soluble in water.

11D

Organic molecules, including benzene, are discussed in later units. The hydrochloride form is:

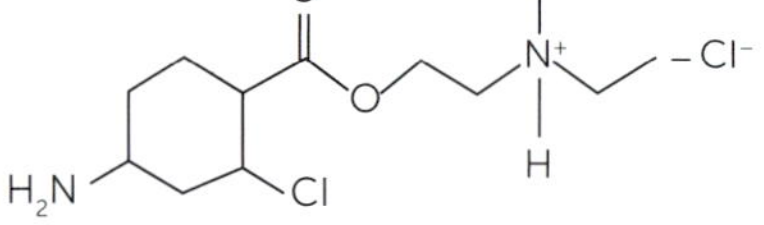

11E

The concentration of SO_2 in the water layer is $0.90 \times 0.10 = 0.090$ mol dm^{-3}

Revision questions

11.1 The only true answer is **(i)**.

11.2 (i) Miscible, presumably because strong hydrogen bonding occurs between the water and propan-2-ol molecules.

(ii) Miscible, presumably because the intermolecular forces between propanone and methyl benzene (a big molecule) are at least as strong as those between pairs of molecules of methyl benzene and between pairs of propanone molecules.

11.3 C is the least polar solvent. Ionic solids are at best only slightly soluble in non-polar solvents.

11.4

[drug in octanol] = [drug in water] × 1000
= 0.01 × 1000 = 10 mg dm^{-3} of wall

(We are assuming that octanol behaves similarly to the lipids in the stomach wall.)

11.5 4.6 m^3 = 4,600 dm^3. This volume of gas contains 4,600/22.4 = 205 mol of Cl_2. Since 1 m^3 = 1,000 dm^3, the concentration of gas is

moles of gas/volume of solution
= 205/1000 ≈ 0.21 mol dm^{-3}.

11.6 The molar concentration of O_2 at 0°C is

$$\frac{14.6}{1000 \times 32} = 4.6 \times 10^{-4}\,\text{mol dm}^{-3}\,O_2$$

The volume of O_2 at 0°C is $22.4 \times 4.6 \times 10^{-4}$ = 0.0100 dm^3 or 10.0 cm^3 O_2.

Similarly at 30°C, the molar concentration of O_2 is

$$\frac{7.6}{1000 \times 32} = 2.4 \times 10^{-4}\,\text{mol dm}^{-3}$$

Therefore, the volume of O_2 at 30°C is

$24.9 \times 2.4 \times 10^{-4} = 0.0060$ dm^3 or 6.0 cm^3 O_2

11.7 Tyndall effect (see text). In soap, colloidal particles are ionic micelles. Such particles cannot be separated using filter paper.

11.8 The vinegar produces an egg without its shell. The outer part of the remaining egg, which looks like a thin opaque white skin, is a semipermeable membrane. The inner part of the egg contains protein, and water travels across the membrane (driven by osmosis) in an attempt to equalize concentrations. This causes the egg to swell. In syrup, the reverse occurs – water passes out of the egg and the egg shrinks.

11.9 This is an example of the common ion effect. Flooding the solution with Na^+ by adding Na^+Cl^- saturates the solution with sodium ions. Excess sodium ions are removed by precipitating the soap $RCOO^-\,Na^+$.

Unit 12

Exercises

12A

(i) Na 2.8.1 or $1s^22s^22p^63s^1$
K 2.8.8.1 or $1s^22s^22p^63s^23p^64s^1$
Rb 2.8.18.8.1 or $1s^22s^22p^63s^23p^63d^{10}\,4s^24p^65s^1$
[inert gas]ns^1 where n is a whole number greater than 1

(ii) F 2.7 or $1s^22s^22p^5$
Cl 2.8.7 or $1s^22s^22p^63s^23p^5$
Br 2.8.18.7 or $1s^22s^22p^63s^23p^63d^{10}4s^24p^5$
I 2.8.18.18.7 or $1s^22s^22p^63s^23p^63d^{10}4s^24p^64d^{10}5s^25p^5$
outer electronic structure ns^2np^5 where n is a whole number greater than 1

12B

(i) No – the second ionization energy is too high.

(ii) K^+ has the electron arrangement of an inert gas, argon.

(iii) The second electron is nearer the nucleus and less shielded from the nucleus by electrons in shells between it and the nucleus. To take away a second electron would involve 'disruption' of a stable octet of electrons.

12C

(i) The boiling points decrease as the group is descended. The metallic bonds become longer and weaker as the atoms get larger.

(ii) About 960 K.

(iii) The alkali metals have relatively large atomic radii and their outer electrons are weakly held – the outer electron is well shielded from the nucleus by the full inner shells. The large atomic radii of the atoms result in longer, weaker metallic bonding as compared with the bonding in other metals, making them relatively 'soft'.

(iv) The large atomic radii of the elements result in them taking up more 'room' (volume), per unit mass, than other metals. They therefore have lower densities.

12D

(i) Extremely reactive:

$Fr(s) + O_2(g) \rightarrow FrO_2(s)$
$2Fr(s) + 2H_2O(l) \rightarrow 2FrOH(aq) + H_2(g)$

(ii) Fr_2CO_3, very soluble, very thermally stable.

(iii) Thermally the most stable. Likely to decompose to the nitrite with strong heating:

$2FrNO_3(s) \rightarrow 2FrNO_2(s) + O_2(g)$

12E

The metallic bonds are shorter and stronger because the atomic radii of the Group 2 metals are smaller than those of the corresponding alkali metals. Also two valence electrons are released per ion, to 'glue' the cations together in the structure.

12F

(i) The atom has lost its outer shell electrons and the overall positive charge 'pulls' the remaining electrons closer to the nucleus.

(ii) Although there is an increase in the positive charge on the nucleus, the outer electrons are farther away from the nucleus and shielded by more full shells, as the group is descended.

(iii) The atomic and ionic radii are similar.

12G

(i) $SiH_3–SiH_2–SiH_2–SiH_2–SiH_2–SiH_3$

(ii) Covalent, similar shape to ethane; each Ge atom has four bonds arranged tetrahedrally around it.

12H

(i) Covalent Ionic

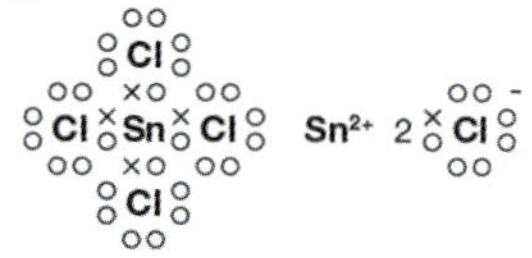

(ii) Pb(II) is the more stable oxidation state of lead.

(iii) Pb–H bonds are longer and weaker than C–H bonds, because Pb is a much larger atom than C.

(iv) CO.

(v) GeO and GeO_2. You would expect GeO_2 to be the more stable, because Ge(IV) is the more stable oxidation state (GeO is readily oxidized to GeO_2 in air).

12I

(i) $SiCl_4(l) + 2H_2O(l) \rightarrow SiO_2(s) + 4HCl(aq)$

(ii) It is longer and weaker. Si is a larger atom than C.

12J

(i) Si–Si bonds are longer and weaker than C–C bonds. The shared electrons can 'jump out' of the Si–Si bonds more easily.

(ii) Only a small proportion of the bonding electrons are set free; each silicon atom still has three covalent bonds to keep it in place if one bonding electron is lost.

12K

(i) Black/dark.

(ii) Solid.

(iii) The least stable hydride with respect to decomposition into its elements.

(iv) Would be the halogen with the least oxidizing power.

(v) As the largest halide ion, At^- would be the most highly polarized – you would expect the salt to have a great deal of covalent character.

(vi) Dark yellow or a darker colour, such as orange.

12L

(i) $Xe(g) + 2F_2(g) \rightarrow XeF_4(s)$

(ii)

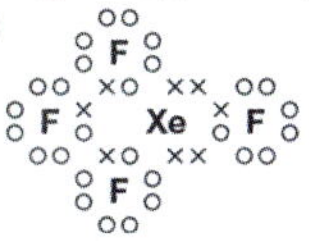

Octahedral with respect to the electron pairs around Xe

(iii) In XeF_4, Xe has oxidation state +4. It would 'prefer' to be reduced to oxidation state 0, i.e. atoms of the element Xe, which have a stable octet.

12M

[Ar] ↑ ↑ ↑ _ _ (3d) ↑↓ (4s) Paramagnetic

[Ar] ↑ ↑ ↑ ↑ ↑ (3d) ↑ (4s) Paramagnetic

[Ar] ↑↓ ↑↓ ↑↓ ↑↓ ↑↓ (3d) ↑↓ (4s) Diamagnetic

Metal (ii) has the most unpaired electrons.

12N

(i) $1s^2 2s^2 2p^6 3s^2 3p^6 3d^5$ **(ii)** $1s^2 2s^2 2p^6 3s^2 3p^6 3d^4$

(iii) +7, MnO_4^-

12O

(i) Sc^{3+} has no d electrons.

(ii) Zn^{2+} has a full subshell of d electrons – there are no 'gaps' for d electrons to be promoted.

12P

(i) 2– **(ii)** $[CuCl_4]^{2-}$

(iii) 4 **(iv)** tetrachlorocuprate(II)

12Q

1. **(i)** hexacyanoferrate(III)
   (ii) tetraamminezinc(II)
   (iii) hexaaquatitanium(III)
   (iv) tetraamminediaquacopper(II)
   (v) pentaamminebromocobalt(III)
   (vi) diamminetetrachloroplatinum(IV)
2. **(i)** $[PtCl_6]^{2-}$ **(ii)** $[CrCl_2(H_2O)_4]^+$
   (iii) $[Pd(CN)_6]^{2-}$ **(iv)** $[CoBr_4]^{2-}$
   (v) $[Fe(H_2O)_6]^{3+}$

12R

(i) 6 **(ii)** 2–

12S

(i) A sodium atom loses its outer shell electron when it becomes an ion. The positive charge on the ion draws the remaining electrons closer into the nucleus. A chlorine atom gains an electron to its outer shell when it becomes an ion. The negative charge on the ion makes the electrons 'spread out'.

(ii) The 2+ charge on the magnesium ion draws the electrons closer to the nucleus.

(iii) The greater the negative charge on the ion, the more the electrons repel each other and 'spread out'.

Revision questions

12.1 (i) H **(ii)** X, L or Z **(iii)** Y or G
(iv) X or D **(v)** Z, M or K **(vi)** I **(vii)** J
(viii) H **(ix)** D **(x)** L

12.2 (i) Violently, $Ra(s) + 2H_2O(l) \rightarrow Ra(OH)_2(aq) + H_2(g)$

(ii) The most thermally stable and insoluble carbonate of Group 2. The carbonate should decompose to the oxide and carbon dioxide, on strong heating.

12.3 Tin reverts to the crumbly α allotrope at low temperatures.

12.4 CS

12.5 (i) Li_2O

$Li_2O(s) + H_2O(l) \rightarrow 2LiOH(aq)$

(ii) NO_2 **(iii)** BeO

12.6 (i) ionic **(ii)** covalent

12.7 (ii) C (network solid) and Al (strong metallic bonding).

(iii) Any of H, He, N, O, F, Ne, Ar; London dispersion forces between simple molecules or atoms.

(iv) Li, Na, Mg or Al. See Unit 5.

(v) Yes. There is a pattern of low boiling points then a gradual rise to very high boiling point(s), then a drop.

12.8 (i) 6 **(ii)** +2 **(iii)** tetraaquadichloronickel(II)

12.9 (i) $[Cu(NH_3)_4(H_2O)_2]^{2+}$ **(ii)** $[PtBr_6]^{2-}$

12.10 (i) $[H_3N–Cu–NH_3]^+$ linear

(ii)

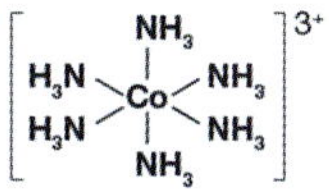

octahedral.

Unit 13

Exercises

13A

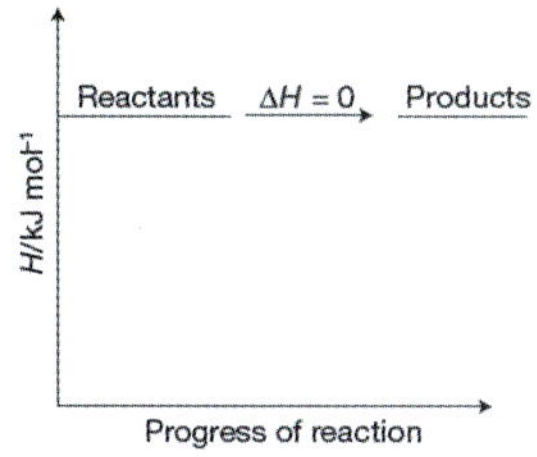

13B

(i) A balanced equation should be also written. Only then do we know how much sodium and chlorine have reacted.

(ii) Number of moles of carbon
$= 120\,g/12\,g\,mol^{-1} = 10.$
The equation shows that complete combustion of 1 mol of carbon produces 393.15 kJ of heat. Therefore, the complete burning of 10 mol of carbon produces 3931.5 kJ of heat energy.

13C

601.7 kJ is required to decompose 2 mol of MgO, so 300.9 kJ is required to decompose 1 mol of MgO.

13D

Eqn (13.14) is obtained by adding eqns (13.12) and (13.13). By Hess's law,

$$\Delta H^\ominus_{(13.14)} = \Delta H^\ominus_{(13.12)} + \Delta H^\ominus_{(13.13)} = -296.8 + (-98.9) = -395.7\,kJ\,mol^{-1}$$

13E

To complete the equations, add the elements in their standard states. The physical states of the elements and compounds at 298 K should also be included because we are assuming that $\Delta H^\ominus_f$ refers to this temperature.

$H_2(g) + S(s) + 2O_2(g) \rightarrow H_2SO_4(l)$

$6C(s) + 6H_2(g) + 3O_2(g) \rightarrow C_6H_{12}O_6(s)$

13F

(i) $2CO(g) + O_2(g) \rightarrow 2CO_2(g)$
Using the standard enthalpy data in Table 13.2,
$\Delta H^\ominus_f(CO_2(g)) = -393.51\,kJ\,mol^{-1}$,
$\Delta H^\ominus_f(CO(g)) = -110.53\,kJ\,mol^{-1}$ and
$\Delta H^\ominus_f(O_2)(g)) = 0\,kJ\,mol^{-1}$. Therefore,

$$\Delta H^\ominus = [2\Delta H^\ominus_f(CO_2(g))] - [2\Delta H^\ominus_f(CO(g)) + \Delta H^\ominus_f(O_2(g))] = [2(-393.51)] - [2(-110.53) + 0] = -565.96\,kJ\,mol^{-1}$$

(ii) $Fe_2O_3(s) + 3CO(g) \rightarrow 2Fe(s) + 3CO_2(g)$
$\Delta H^\ominus_f(CO_2(g)) = -393.51\,kJ\,mol^{-1}$;
$\Delta H^\ominus_f(CO(g)) = -110.53\,kJ\,mol^{-1}$;
$\Delta H^\ominus_f(Fe_2O_3)(s) = -824.2\,kJ\,mol^{-1}$;
remember that
$\Delta H^\ominus_f(Fe(s)) = 0\,kJ\,mol^{-1}$

$$\Delta H^\ominus = [3\Delta H^\ominus_f(CO_2(g)) + 2\Delta H^\ominus_f(Fe(s))] - [3\Delta H^\ominus_f(CO(g)) + \Delta H^\ominus_f(Fe_2O_3)] = -1180.5 - (-1155.8) = -24.74\,kJ\,mol^{-1}$$

(iii) $NH_3(g) \rightarrow \frac{1}{2}N_2(g) + \frac{3}{2}H_2(g)$
$\Delta H^\ominus_f(NH_3(g)) = -46.11\,kJ\,mol^{-1}$, and
$\Delta H^\ominus_f(H_2(g)) = 0\,kJ\,mol^{-1}$
$\Delta H^\ominus_f(N_2(g)) = 0\,kJ\,mol^{-1}$

$$\Delta H^\ominus = [\tfrac{1}{2}\Delta H^\ominus_f(N_2(g)) + \tfrac{3}{2}\Delta H^\ominus_f(H_2(g))] - [\Delta H^\ominus_f(NH_3(g))] = -(-46.11) = +46.11\,kJ\,mol^{-1}$$

The positive sign shows that energy is absorbed in order to decompose the molecule.

13G

(i) $M(C_{12}H_{22}O_{11}) = 342\,g\,mol^{-1}$. 5,644 kJ of heat is given out when 342 g of sucrose are completely burned in oxygen.
Therefore, 1 g of sucrose gives out 5,644/342 = 16.5 kJ; energy value = 16.5 kJ g^{-1}.

(ii) 1 g of H_2 produces 142 kJ of heat. Therefore, the mass of H_2 that produces 100 kJ of heat is

$$\frac{100}{142} \times 1 = 0.704\,g$$

0.704 g of H_2 contains (0.704/2.00) = 0.352 mol of H_2; 1 mol of H_2 occupies 24 dm^3 at room temperature and pressure. Therefore, 0.352 mol of H_2 occupies

$0.352 \times 24 = 8.5\,dm^3$

i.e. 8.5 dm^3 of gas burns to give 100 kJ of heat.

13H

20 g produces 120 Cal (= 120 kcal) when burned.
The energy value of the chocolate bar is
$4.18 \times 120 \approx 500$ kJ. The energy value of 1 g of chocolate is
$(1/20) \times 500 \approx 25\,kJ\,g^{-1}$ (compare with the materials in Table 13.3).

13I

Consumption of $O_2 = 0.27 \times 60 \times 24 = 390\,dm^3\,day^{-1}$.
Energy requirement per day = 390 × 20 = 7,800 kJ (1,870 Cal).

13J

2237 × 0.1 = 223.7 kJ (endothermic)

13K

(i) The process $Na(s) \rightarrow Na^+(g) + e^-$ is the sum of the atomization and ionization steps in the Born–Haber cycle. Therefore, the energy requirement is

$\Delta H^\ominus_1 + \Delta H^\ominus_2 = 109 + 494 = +603\,kJ\,mol^{-1}$

(ii) The process $Na^+,Cl^-(s) \rightarrow Na(g) + {}^1/_2Cl_2(g)$ involves the sum of $(-\Delta H^\ominus_5) + (-\Delta H^\ominus_4) + (-\Delta H^\ominus_3) + (-\Delta H^\ominus_2)$
$= 771 + 364 - 121 - 494 = 520\,kJ\,mol^{-1}$

13L

The formation of magnesium fluoride is

$Mg(s) + F_2(g) \rightarrow Mg^{2+}, F_2^-(s)\ \Delta H^\ominus_f = -1121\,kJ\,mol^{-1}$

This may also be split up into steps, as with NaCl in the text:

$\Delta H^\ominus_5 = \Delta H^\ominus_f - \Delta H^\ominus_1 - \Delta H^\ominus_2 - \Delta H^\ominus_3 - \Delta H^\ominus_4$ (A)

Note that:
(a) $\Delta H^\ominus_2$ involves two ionization steps:

$Mg(g) \rightarrow Mg^+(g) + e^-$ $\Delta H^\ominus = 744\,kJ\,mol^{-1}$

$Mg^+(g) \rightarrow Mg^{2+}(g) + e^-$ $\Delta H^\ominus = 1456\,kJ\,mol^{-1}$

so that $\Delta H^\ominus_2 = 744 + 1456 = 2{,}200\,kJ\,mol^{-1}$.
(b) Step 3 involves the production of two fluorine atoms:
$F_2(g) \rightarrow 2F(g)$ $\Delta H^\ominus_3 = 158\,kJ\,mol^{-1}$

(c) Two fluorine atoms accept electrons:

$2F(g) + 2e^- \rightarrow 2F^-(g)$ $\Delta H^\ominus_4 = (2 \times -339) = -678\,kJ\,mol^{-1}$

Substituting into equation (A),

$$\Delta H^\ominus_5 = -1121 - 148 - 2200 - 158 + 678 = -2{,}949\,kJ\,mol^{-1}$$

or, $\Delta H^\ominus_L = -\Delta H^\ominus_5 = +2{,}949\,kJ\,mol^{-1}$.

13M

Average = (531 + 1075)/2 = 803 kJ mol^{-1}.

13N

(i) The standard enthalpy change for the process $CH_4(g) \rightarrow CH_3(g) + H(g)$ will approximately equal $\Delta H^\ominus_{C-H}$. Therefore the enthalpy change for the reverse process:

$CH_3(g) + H(g) \rightarrow CH_4(g)$

will be $-\Delta H^\ominus_{C-H}$, i.e. $-412\,kJ\,mol^{-1}$.

(ii) Comparing the bond lengths with the bond enthalpy data in Table 13.4, the trend supports the

proposition that long bonds are weak bonds (or, conversely, that short bonds are strong bonds).
(iii) Double and triple bonds are stronger than single bonds. The general order of bond strength is triple bonds > double bonds > single bonds.
(iv) N≡N has a large bond energy (Table 13.4) so that when N_2 is formed, large amounts of energy are given out. This helps to make the explosion reaction highly exothermic and thermodynamically spontaneous. It also causes rapid gas expansion. Of course, N_2 can only be formed if N is present in the explosive!

13O

(i) $CH_2 = CH_2(g) + Cl_2(g) \rightarrow CH_2Cl\text{–}CH_2Cl(g)$
To completely break up the molecules on the left-hand side of this equation we need to dissociate: one C=C bond, four C–H bonds and one Cl–Cl bond. From Table 13.4,

energy absorbed = 612 + 4(412) + 242 = 2,502 kJ mol^{-1}

i.e. the energy change due to bond destruction is + 2,502 kJ.

To form the reactant molecules, we need to make four C–H bonds, two C–Cl bonds and one C–C bond. From Table 13.4,

energy given out = 4(412) + 2(338) + 348 = 2,672 kJ mol^{-1}

i.e. the energy change due to bond formation = –2,672 kJ. Therefore, the net change in energy is

$\Delta H^{\circ}_{f}(298\text{ K}) = 2{,}502 - 2{,}672 = -170\text{ kJ mol}^{-1}$

(ii) Bond energies can be used directly only if all the reactants and products are gases. If some reactants or products are not gases then the enthalpy change for the phase change (solid → gas or liquid → gas) must be allowed for.

We start by working out the enthalpy change for the reaction in which all the reactants and products are gaseous:

$2H_2O(g) \rightarrow 2H_2(g) + O_2(g)$ (A)

We now apply the average bond energies in Table 13.4 for the heat absorbed in breaking bonds in reactant molecules, this involves breaking four O–H bonds:

4 × 463 = 1,852 kJ mol^{-1}

From Table 13.4 the heat released when bonds in product molecules are assembled, which involves making two H–H bonds and one O=O bond, is

(2 × 436) + 497 = 1,369 kJ mol^{-1}

The overall change in enthalpy is therefore

$\Delta H^{\circ}_{A} = 1{,}852 - 1{,}369 = +483\text{ kJ mol}^{-1}$

However, we want the enthalpy change for the reaction

$2H_2O(l) \rightarrow 2H_2(g) + O_2(g)$ (B)

and we are given

$H_2O(l) \rightarrow H_2O(g)\ \Delta H^{\circ} = +44\text{ kJ mol}^{-1}$

or

$2H_2O(l) \rightarrow 2H_2O(g)\ \Delta H^{\circ} = +88\text{ kJ mol}^{-1}$

$H_2O(g)$ possesses a higher enthalpy than $H_2O(l)$: this means that the enthalpy change for reaction (B) will be larger than that of reaction (A) by 88 kJ (if you can't see this, draw an energy diagram for the processes involved):

$\Delta H^{\circ}_{B}(298\text{K}) = 483 + 88 = +571\text{ kJ mol}^{-1}$

Revision questions

13.1 Following are false: **(i)**, **(ii)**, **(iv)** (2 mol $SO_3(g)$ made) and **(vi)**.
13.2 (i) See text.
(ii) Combustion of methyl benzene:

$C_7H_8(l) + 9O_2(g) \rightarrow 7CO_2(g) + 4H_2O(l)$

$\Delta H^{\circ}_{c} = [4\Delta H^{\circ}_{f}(H_2O(l)) + 7\Delta H^{\circ}_{f}(CO_2(g))] - [\Delta H^{\circ}_{f}(C_7H_8(l)) + 9\Delta H^{\circ}_{f}(O_2(g))]$

From Table 13.2:

$= [(-285.83 \times 4) + (-393.51 \times 7)] - [12.0 + 0]$
$= (-3897.89) - (12.0) = -3909.9$ kJ per mole of methyl benzene

(iii) 1 mol of methyl benzene has a molar mass of approximately 92 g mol^{-1}. Energy value of methyl benzene is 3909.9/92 ≈ 42 kJ g^{-1}.1 cm^3 of methyl benzene has a mass of 0.86 g. The heat given out in the complete combustion of 1 cm^3 of methyl benzene is approximately
42 × 0.86 = 36 kJ cm^{-3}.
13.3 (i) $\Delta H^{\circ} = [\Delta H^{\circ}_{f}(P(s)\text{ (red)})] - [\Delta H^{\circ}_{f}(P(s)\text{ (white)})]$
$= -17.6 - 0 = -17.6\text{ kJ mol}^{-1}$
i.e. red phosphorus is the more energetically stable form of phosphorus.
(ii)

$\Delta H^{\circ} = [\Delta H^{\circ}_{f}(H_2O(l)) + \Delta H^{\circ}_{f}(CH_3CO_2C_2H_5(l))]$
$\quad - [\Delta H^{\circ}_{f}(C_2H_5OH(l)) + \Delta H^{\circ}_{f}(CH_3CO_2H)(l)]$
$= (-285.83 - 481) - (-277.69 - 485)$
$= (-766.83) - (-762.69) \approx -4\text{ kJ mol}^{-1}$

(iii)

$\Delta H^{\circ} = [3\Delta H^{\circ}_{f}((N_2(g)) + 9\Delta H^{\circ}_{f}(H_2O(g))$
$\quad + 5\Delta H^{\circ}_{f}(Al_2O_3(s)) + 6\Delta H^{\circ}_{f}(HCl(g))]$
$\quad - [6\Delta H^{\circ}_{f}(NH_4ClO_4(s)) + 10\Delta H^{\circ}_{f}(Al(s))]$

The ΔH°_{f} of elements in their reference states is zero. $\Delta H^{\circ}_{f}(NH_4ClO_4(s))$ is obtained from Table 13.2.

$\Delta H^{\circ} = [(-241.82 \times 9) + (-1675.7 \times 5)$
$\quad + (-92.31 \times 6)] - [-295.31 \times 6]$
$= [-2176.38 - 8378.5 - 553.86] + [1771.86]$
$\approx -9337\text{ kJ mol}^{-1}$

(Notice that the effect of producing ($Al_2O_3(s)$) is to make the overall reaction highly exothermic.)
13.4 $HCOOH(l) + {}^1/_2O_2(g) \rightarrow CO_2(g) + H_2O(l)$
$\Delta H^{\circ}_{c} = -277.2\text{ kJ mol}^{-1}$
$\Delta H^{\circ}_{c} = [\Delta H^{\circ}_{f}(CO_2(g)) + \Delta H^{\circ}_{f}(H_2O(l))]$
$- [\Delta H^{\circ}_{f}(HCOOH(l)) + {}^1/_2\Delta H^{\circ}_{f}(O_2(g))]$

where $\Delta H^{\circ}_{f}(HCOOH(l))$ is the only unknown, $\Delta H^{\circ}_{f}(CO_2(g)) = -393.51\text{ kJ mol}^{-1}$ and $\Delta H^{\circ}_{f}(H_2O(l)) = -285.83\text{kJmol}^{-1}$ (from Table 13.2) and $\Delta H^{\circ}_{f}(O_2(g)) = 0$. Substitution and rearrangement gives:

$\Delta H^{\circ}_{f}(HCOOH(l) = (-393.51 - 285.83) - (-254.6)$
$= -424.7\text{ kJ mol}^{-1}$

13.5 (i) $C_6H_{12}O_6(s) + 6O_2(g) \rightarrow 6CO_2(g) + 6H_2O(l)$
(ii) $M(C_6H_{12}O_6) = 180\text{ g mol}^{-1}$. The complete combustion of 180 g of glucose releases 2,816 kJ of energy. The complete combustion of 10 g of glucose would release (10/180) × 2,816 = 160 kJ (two significant figures). If the combustion of glucose within the body were only 40% efficient, only 64 kJ of energy would be available for work for every 10 g of glucose consumed.
13.6 (i) The third standard ionization energy of aluminium is the standard enthalpy change for the process
$Al^{2+}(g) \rightarrow Al^{3+}(g) + e^-$

$\Delta H^{\circ}_{ion(3)} = 2{,}751\text{ kJ mol}^{-1}$

(ii) The equation

$Al(s) \rightarrow Al^{3+}(g) + 3e^-$ (A)

is the sum of the equations:

$Al(s) \rightarrow Al(g)$ (B)

$Al(g) \rightarrow Al^{+}(g) + e^-$ (C)

$Al^{+}(g) \rightarrow Al^{2+}(g) + e^-$ (D)

$Al^{2+}(g) \rightarrow Al^{3+}(g) + 3e^-$ (E)

The enthalpy change for reaction (A) is the sum of the enthalpy changes for reactions (B)–(E) (if you are unsure about this, sketch an energy diagram):
$\Delta H^{\circ}_{A} = \Delta H^{\circ}_{B} + \Delta H^{\circ}_{C} + \Delta H^{\circ}_{D} + \Delta H^{\circ}_{E}$
$= 326 + 584 + 1{,}823 + 2{,}751 = 5{,}484\text{ kJ mol}^{-1}$

13.7

(a) $Ca(s) \rightarrow Ca(g)$ ΔH°_{1}

(b) $Ca(g) \rightarrow Ca^{+}(g) + e^-$ ΔH°_{2A}

(c) $Ca^{+}(g) \rightarrow Ca^{2+}(g) + e^-$ ΔH°_{2B}

(d) $O_2(g) \rightarrow O(g) + O(g)$ ΔH°_{3}

(e) $O(g) + 2e^- \rightarrow O^{2-}(g)$ ΔH°_{4}
(f) $Ca^{2+},O^{-}(s) \rightarrow Ca^{2+}(g) + O^{2-}(g)$ $\Delta_{latt}H^{\circ}$

(ii) The formation of calcium oxide,

$Ca(s) + {}^1/_2O_2(g) \rightarrow Ca^{2+},O^{-}(s)$ ΔH°_{f}

may also be split up into steps, as for NaCl. We follow the pattern for NaCl in the text, with

$\Delta H^{\circ}_{2} = \Delta H^{\circ}_{2A} + \Delta H^{\circ}_{2B}$. Therefore,

$\Delta H^{\circ}_{f} = \Delta H^{\circ}_{1} + \Delta H^{\circ}_{2} + {}^1/_2\,\Delta H^{\circ}_{3} + \Delta H^{\circ}_{4} + (-\Delta_{latt}H^{\circ})$

Notice that only half of ΔH°_{3} is required because only one atom of O is required to combine with one atom of Ca.

$\Delta H^{\circ}_{f} = 193 + 1{,}740 + 248.5 + 703 - 3513 = -629\text{ kJ mol}^{-1}$
13.8

$2F_2(g) + S(g) \rightarrow SF_4(g)$ $\Delta H^{\circ} = -763\text{ kJ mol}^{-1}$

In breaking bonds in the reactant molecules, we break two F–F bonds:

heat absorbed = 2 × 158 = 316 kJ

The heat released when bonds in product molecules are assembled involves making four S–F bonds, each of bond enthalpy x. The total enthalpy change is $-4x$ (which is negative because heat is given out when bonds are formed). Therefore, the overall change in enthalpy is

$-763 = -4x + 316$

which gives

$x = \Delta H^{\circ}_{S\text{–}F} = (763 + 316)/4$
$x = 270$ kJ per mole of S–F bonds.

Unit 14

Exercises

14A

(i) NO disappears at twice the rate at which H_2 disappears = 0.02 mol dm^{-3} s^{-1}.
(ii) Nitrous oxide appears at the same rate at which H_2 disappears = 0.01 mol dm^{-3} s^{-1}.
(iii) Water appears at the same rate at which H_2 disappears = 0.01 mol dm^{-3} s^{-1}.

14B

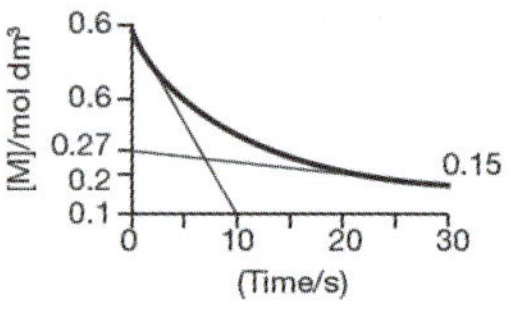

$$\text{initial rate} = \frac{0.60 - 0.10}{10} = 0.050\text{ mol dm}^{-3}\text{ s}^{-1}$$

$$\text{rate at }25\text{s} = \frac{0.27 - 0.15}{30} = 4.0 \times 10^{-3}\text{ mol dm}^{-3}\text{ s}^{-1}$$

(Your answers might be slightly different due to the difficulty of drawing tangents.)

14C

(i) The cut divides the marble into two rectangular blocks of dimensions 0.5 × 1 × 1 cm. The surface area of each block is (0.5 + 1 + 1 + 0.5 + 0.5 + 0.5) = 4 cm^3.
(ii) The total surface area is the sum of the surface area for both blocks = 8 cm^3. This is 33 per cent greater than the surface area of the *undivided* block (6 cm^3). Dividing the block into even smaller pieces greatly increases the total surface area available for attack by the acid.

14D

(i) 100 – 20°C = 80°C so that the increase in rate is 2 × 2 × 2 × 2 × 2 × 2 × 2 × 2 (i.e. 2^8) = 256 times.
(ii) The heat generated by the exothermic reaction raises the temperature of the reaction mixture, so increasing its rate.

14E

(i) Nails have a lower surface area than powder.
(ii) (a) Water dilutes the concentration of acid, so reducing the number of acid–zinc collisions per second. **(b)** Ice cools the mixture, thereby reducing reaction rate. **(c)** Sodium carbonate reacts with the acid, so drastically reducing the concentration of hydrogen ions in solution.

14F

(i) 1 **(ii)** 2 **(iii)** 1 **(iv)** 3

14G

(i) Third order **(ii)** Rate = 7000 × $(0.060)^3$ = 1.5 mol dm^{-3}s^{-1}

14H

(i) Second order.

(ii) $k = \dfrac{\text{initial rate}}{[C_{12}H_{22}O_{11}][H^+]} = \dfrac{4.0\times10^{-6}}{(0.20)(0.10)}$

$= 2.0 \times 10^{-4}\ \text{mol}^{-1}\,\text{dm}^3\,\text{s}^{-1}$

14I

Rate of reaction = $k[M][Y]^0 = k[M]$

14J

rate of reaction = $k[Br^-(aq)]^x[BrO_3^-(aq)]^y[H^+(aq)]^z$
Comparison of experiments 1 and 2 shows that doubling $[Br^-(aq)]$ doubles the initial rate; therefore, $x = 1$. Comparing experiments 1 and 3 shows that increasing $[BrO_3^-(aq)]$ by a factor of four increases the initial rate by four; therefore, $y = 1$. Comparing experiments 1 and 4 shows that doubling $[H^+(aq)]$ increases the rate by four; therefore, $z = 2$ and the rate expression is

rate = $k[Br^-(aq)]\,[BrO_3^-(aq)]\,[H^+(aq)]^2$

14K

$t_{1/2} = \dfrac{0.693}{k} = \dfrac{0.693}{9.6\times10^{-3}} = 72\,\text{s}$

14L

$10\,\text{h} = 10 \times 60 \times 60 = 36{,}000\,\text{s}$

$[A]_t = [A]_0\,e^{-kt} = 0.01 \times e^{-(6.5\times10^{-6}\times36000)}$
$= 0.01 \times 0.791 = 0.0079\,\text{mol}\,\text{dm}^{-3}$

14M

$t_{1/2} = \dfrac{0.693}{5.73\times10^{-7}} = 1209000\,\text{s}$ (14.0 days)

Forty-two days is three half-lives. During this period the concentration of pesticide will have fallen to one-eighth of its initial concentration, i.e. from $0.10\,\text{mg}\,\text{dm}^{-3}$ to $0.0125\,\text{mg}\,\text{dm}^{-3}$.

Revision questions

14.1 See text.
14.2 (i), (vi)
14.3 (i) II **(ii)** I
14.4 (i) 1
(ii) initial rate
$= 2.00 \times 10^{-5} \times 0.100 = 2.00 \times 10^{-6}\,\text{mol}\,\text{dm}^{-3}\,\text{s}^{-1}$
(iii) rate $= 2.00 \times 10^{-5} \times 0.0523 = 1.05 \times 10^{-6}\,\text{mol}\,\text{dm}^{-3}\,\text{s}^{-1}$
(iv) half-life $= 0.693/2.00\times10^{-5} = 3465\,\text{s} \approx 9.6\,\text{h}$

14.5
(i) 1
(ii) Plot as follows:

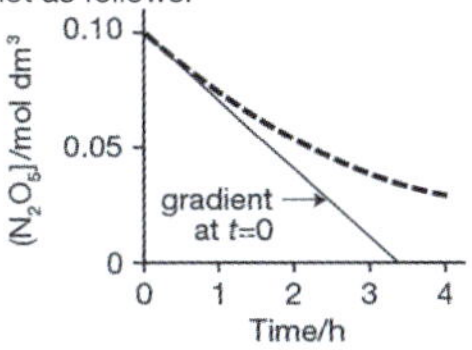

(iii) initial rate $= \dfrac{0.10-0.0}{3.3\times60\times60} = 8.4\times10^{-6}\ \text{mol dm}^{-3}\,\text{s}^{-1}$

(iv) initial rate = $k\,[N_2O_5(g)]_0$
$k\,(65°C) = 8.4 \times 10^{-6}/0.10 = 8.4 \times 10^{-5}\,\text{s}^{-1}$

(v) $t_{1/2} = \dfrac{0.693}{8.4\times10^{-5}} = 8300\,\text{s}\ (2.3\,\text{h})$

(Inspection of the graph and table confirms that the reaction half-life is roughly 2 h.)

14.6 (i) Consumption of hydroxide ion is negligible and rate of reaction depends upon the concentration of the bromomethane only:
rate of reaction = $k'[CH_3Br(l)]$

with $k' = k[OH^-(aq)]$. If the bromomethane were in excess, the reaction would also be pseudo first order and the rate expression would be

rate of reaction = $k''[OH^-(aq)]$

with $k'' = k[CH_3Br(l)]$

(ii) (a) Initial rate of reaction = $k\,[CH_3Br(l)][OH^-(aq)]$
$= 3.0 \times 10^{-4} \times 1.0$
$\times\ 10^{-2} \times 1.0 \times 10^{-4}$
$= 3.0 \times 10^{-10}\,\text{mol}\,\text{dm}^{-3}\,\text{s}^{-1}$
(b) pseudo rate constant = $k' = k[OH^-(aq)]$
$= 3.0 \times 10^{-4} \times 1.0 \times 10^{-2} = 3.0 \times 10^{-6}\,\text{s}^{-1}$
(c) pseudo half-life of reaction = $0.693 / 3.0 \times 10^{-6}$
$= 2.3 \times 10^5\,\text{s}$.

14.7 (i)

rate = $k[ClO]^x[NO_2]^y[N_2]^z$

Comparison of experiments 1 and 2 shows that halving the concentration of ClO halves the initial rate; therefore, $x = 1$. Experiments 1 and 3 show that doubling $[NO_2]$ doubles the rate; therefore, $y = 1$. Similarly, comparison of experiments 2 and 4 shows that doubling $[N_2]$ doubles the initial rate; therefore, $z = 1$. The rate expression becomes

rate = $k[ClO][NO_2][N_2]$

(ii) From the data of experiment 1,

$k = \dfrac{\text{initial rate}}{[ClO][NO_2][N_2]} = \dfrac{6}{1}$

$= \dfrac{3.5\times10^{-4}}{[1.0\times10^{-5}][2.0\times10^{-5}][3.0\times10^{-5}]}$

$= 5.8 \times 10^{10}\,\text{mol}^{-2}\,\text{dm}^6\,\text{s}^{-1}$

From the data of the remaining experiments:

experiment 2: $k = 6.0 \times 10^{10}$
experiment 3: $k = 5.9 \times 10^{10}$
experiment 4: $k = 6.0 \times 10^{10}$

Average k (298 K) $= 6.0 \times 10^{10}\,\text{mol}^{-2}\,\text{dm}^6\,\text{s}^{-1}$.
(iii) If the reaction between ClO, NO_2 and N_2 were elementary, we would expect the rate expression to be $K[ClO][NO_2][N_2]$, as above. The rate expression is therefore *consistent* with a one-step mechanism.

14.8

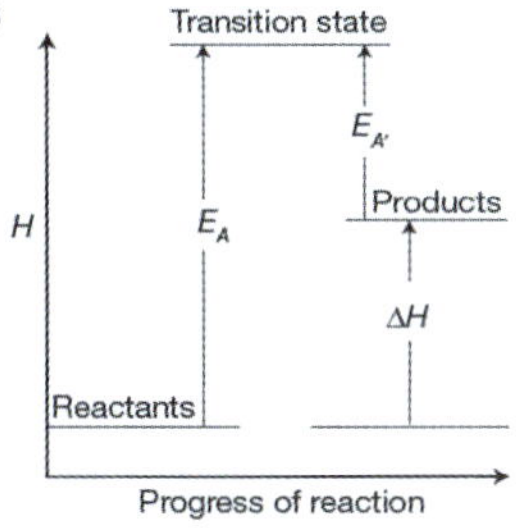

14.9 First derive the expression

$\ln\left(\dfrac{[A]_0}{[A]_t}\right) = kt$

as in the text. Symbolize the time at which 99 per cent of the concentration of A has disappeared as $t_{0.99}$. After this time, the concentration of A will have fallen to one-hundredth of its initial value:

$[A]_{0.99} = \dfrac{[A]_0}{100}$

Substitution gives

$\ln\left(\dfrac{[A]_0}{[A]_0/100}\right) = \ln(100) = k \times t_{0.99}$

since $\ln(100) = 4.605$,
$4.605 = k \times t_{0.99}$

or $t_{0.99} = \dfrac{4.605}{k}$

14.10
(a) Starting materials

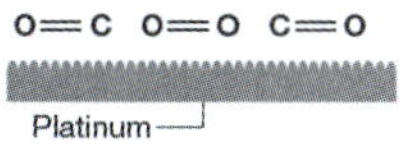

(b) Adsorption

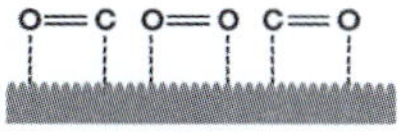

(c) Rearrangement

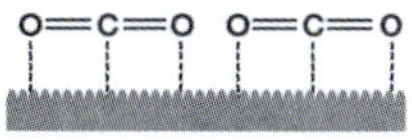

(d) Desorption

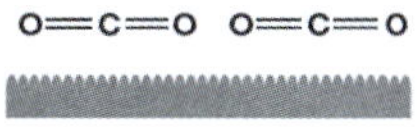

14.11
Box 14.1 introduces the idea of a transition state. We cannot isolate transition states to 'see them', but we know that their structure will be somewhere between the starting and product molecule(s). One key advantage of the catalysed route to make ethane is that the H–H bond has already been broken by the time the ethane-producing reaction takes place. This is shown in the following diagrams of the likely molecular arrangements in the transition states, the dashed lines show bonds that are *either* weakening or strengthening. The C=C bond is weakening in both cases; the H–H bond in (a) is weakening; the desorption 'bond' to the Ni is weakening in (b); and the C–H bond is strengthening in both cases.

(a)

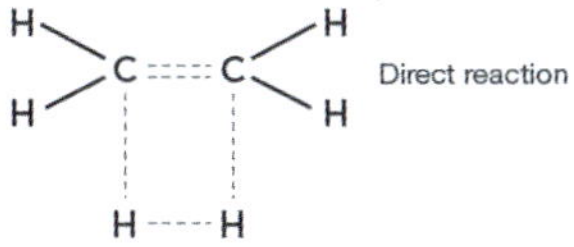

(b)

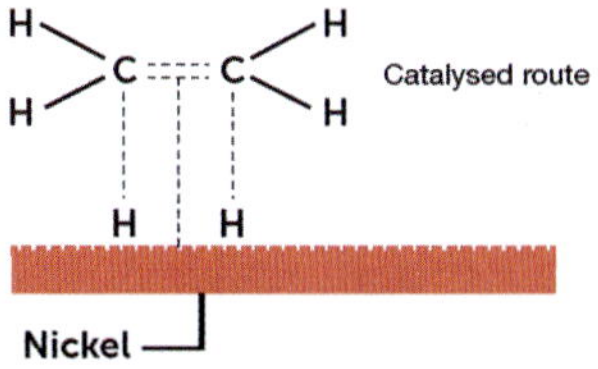

Unit 15

Exercises

15A

Statement (b) is correct. Statement (a) is incorrect because the double arrows show that the reaction is an equilibrium reaction and complete conversion of reactants to products does not occur at the temperature of the experiment.

15B

(i) $K = \dfrac{[Fe^{3+}(aq)]^2[Cl^-(aq)]^2}{[Fe^{2+}(aq)]^2[Cl_2(g)]}$ the units are $\text{mol}\,\text{dm}^{-3}$

(ii) $K = \dfrac{[ClONO_2(g)]}{[NO_2(g)][ClO(g)]}$ the units are $\text{mol}^{-1}\,\text{dm}^3$

(iii) $K = \dfrac{[NO_2(g)]^2}{[N_2O_4(g)]}$ the units are $\text{mol}\,\text{dm}^{-3}$

15C

(i) Reactions in Table 15.1 where $K > 10^6$ are 2, 7, 8 (at 298 K) and 10.
(ii) The reaction between sodium ions and EDTA

$Na^+(aq) + EDTA^{4-}(aq) \rightleftharpoons NaEDTA^{3-}(aq)$

does not go to completion. This means that the $EDTA^{4-}$ will not complex all the sodium ions in a mixture and the $Na^+(aq)$–$EDTA^{4-}$ reaction is therefore worthless in quantitative analysis. (See also Exercise 12R.)

15D

Yes. The equilibrium constant for the reverse reaction at 1,100 K is $1/(7 \times 10^{-9}) = 1.4 \times 10^8$. We conclude that decomposition goes to completion.

15E

(i) K (average) = 45 at 490°C.

$K = \dfrac{[HI(g)]^2}{[H_2(g)][I_2(g)]}$

or
$[HI(g)] = \sqrt{(K[H_2(g)][I_2(g)])}$
$= \sqrt{(45 \times 3.00 \times 3.00)} = 20\,\text{mol}\,\text{dm}^{-3}$

In the reaction between hydrogen and iodine, $1H_2(g) \equiv 1I_2(g) \equiv 2HI(g)$. Therefore, if 20 mol dm^{-3} of HI is produced, (20/2) = 10 mol dm^{-3} of each reactant is used up. This means that initially there must have been 10 + 3.00 = 13 mol dm^{-3} of $H_2(g)$ and $I_2(g)$.

(ii)

$$K(800\text{ k}) = \frac{[HI(g)]^2}{[H_2(g)][I_2(g)]} = \frac{[19.1]}{[3.50][2.50]} = 41.7$$

This value is lower than the value at 490°C (=45), suggesting that the equilibrium constant decreases with increasing temperature.

15F

$$\frac{[N_2(g)][H_2(g)]^3}{[NH_3]^2} = \frac{0.04 \times (0.03)^3}{(0.02)^2} \approx 3 \times 10^{-3}\text{ mol}^2\text{ dm}^6$$

This is not equal to the equilibrium constant at this temperature. Therefore, the reaction is not at equilibrium.

15G

One of the products, water vapour, is driven away (by heating and by the flow of unused hydrogen gas). This prevents the reaction reaching equilibrium. In an attempt to restore the concentration of water, more iron(III) oxide reacts with hydrogen. Eventually, all the Fe_2O_3 is consumed.

15H

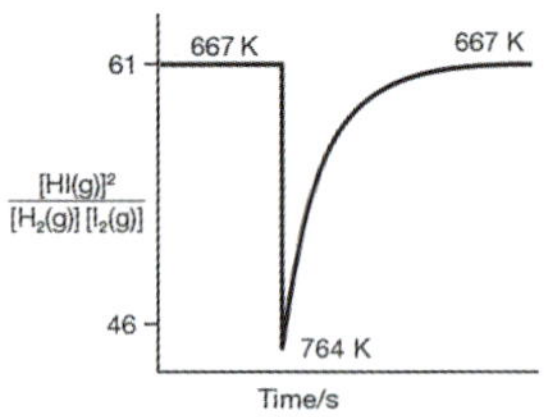

15I

Contents becomes less coloured (more colourless) at lower temperatures (colourless compounds are often white when frozen), indicating a higher $[N_2O_4]$ at lower temperatures. This suggests that reaction (A) is exothermic.

15J

(i) Equilibrium concentration of ethane, CH_3CH_3, decreases. K is unaffected.
(ii) This increases the concentration of hydrogen reactant. Therefore the equilibrium concentration of ethane increases; K is unaffected.
(iii) No effect on composition or upon K.
(iv) Exothermic reaction, so K increases. Thus the concentration of ethane increases.

15K

(i) Huge energy consumption – about 1 per cent of all manmade energy is used in the Haber–Bosch process.
(ii) Excess fertilizer runs off into rivers and lakes, causing algae bloom (and using up dissolved oxygen) which kills some animals.

15L

Applying the equilibrium law,

$$K = \frac{[H_2O(g)]}{[H_2O(l)]}$$

However, the concentration of a pure liquid is fixed at that temperature. We can now write

$K[H_2O(l)] = [H_2O(g)]$

By the ideal gas equation,

$P = cRT$

where c is the concentration of gas (= n/V) and P is its partial pressure. Therefore,

$p_{(H2O)} = K[H_2O(l)] \times RT$

where R is the universal gas constant. All the quantities on the right-hand side of this equation are fixed at a particular temperature. We conclude that the partial pressure of water vapour in equilibrium with its liquid is also fixed at a particular temperature.

Revision questions

15.1 The chemical equation is missing.
15.2 (i) The completed table is

Time(s)	$[H_2(g)]$ (Concentration/mol dm^{-3})/10^{-3}	$[I_2(g)]$	$[HI(g)]$	Q
0	1.000	1.000	0.000	0.000
50	0.897	0.897	0.206	0.053
100	0.813	0.813	0.374	0.211
200	0.685	0.685	0.630	0.846
300	0.592	0.592	0.817	1.905
400	0.521	0.521	1.040	3.985
500	0.465	0.465	1.069	5.285
600	0.420	0.420	1.160	7.628
700	0.383	0.383	1.234	10.381
800	0.352	0.352	1.296	13.556
850	0.333	0.333	1.334	16.048
900	0.332	0.332	1.336	16.193
950	0.331	0.331	1.338	16.340

(ii) A plot of $[H_2(g)]$ and $[HI(g)]$ against time:

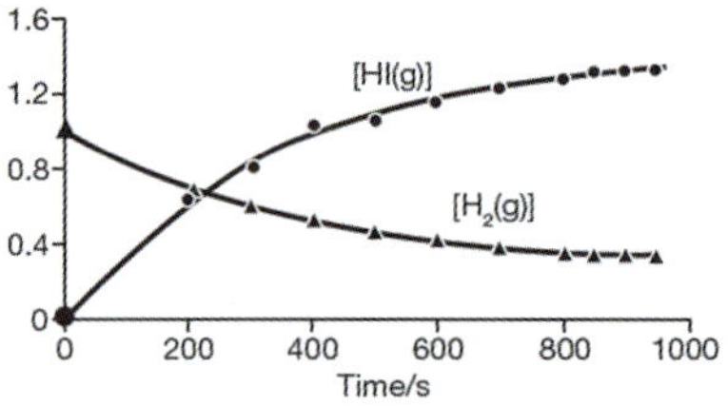

The initial rate of consumption of hydrogen is (1.000 – 0.897) × 10^{-3}/50 = 2.06 × 10^{-6} mol dm^{-3} s^{-1}. The initial rate of HI production is 0.206 × 10^{-3}/50 = 4.12 × 10^{-6} mol dm^{-3} s^{-1}. This is double the rate at which $H_2(g)$ is consumed, as expected by the reaction equation $H_2(g) + I_2(g) \rightleftharpoons 2HI(g)$
(iii) A plot of Q against time reveals that equilibrium has been reached at about 900 s. K is the value of Q when Q has levelled off. K is roughly 17.

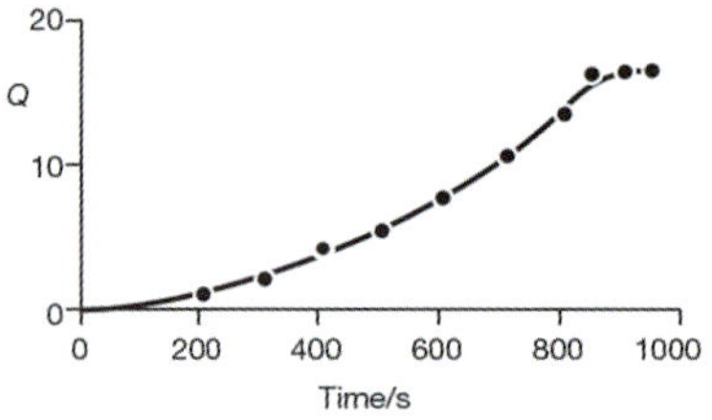

(iv) Figure 15.6 suggests a crude estimate of the temperature as being at least 1200 K.
15.3 (i) The presence of water in the initial mixture produces a lower yield of ester in the equilibrium mixture in order to maintain K at that temperature. (Le Chatelier's principle allows us to come to the same conclusion.)
(ii) (a)

$$K = \frac{[CH_3COOC_2H_5(l)][H_2O(l)]}{[CH_3COOH(l)][C_2H_5OH(l)]} = \frac{(4.48)(4.48)}{(2.67)(1.93)} = 3.89$$

(b) According to the chemical equation for this reaction, 1 mol of ester (and 1 mol of water) are produced using 1 mol of acid and 1 mol of ethanol. Therefore, the initial concentration of ethanol was 2.67 + 4.48 = 7.15 mol dm^{-3}. Similarly, the initial concentration of ethanoic acid was 4.48 + 1.93 = 6.41 mol dm^{-3}.
(iii)

$$K = \frac{k_f}{k_b}$$

Here,

$$3.89 = \frac{2.4 \times 10^{-4}}{k_b}$$

giving,

$$k_b = \frac{2.4 \times 10^{-4}}{3.89} = 6.2 \times 10^{-5}\text{ mol}^{-1}\text{ dm}^{-3}\text{ s}^{-1}$$

15.4

$$k_p = \frac{(p_{NOCl})^2}{(p_{NO})^2 \times p_{Cl2}} = \frac{(0.03)^2}{(0.02)^2(0.11)} = 19\text{ bar}^{-1}$$

15.5

$$K = \frac{[Sn^{4+}(aq)][Fe^{2+}(aq)]^2}{[Sn^{2+}(aq)][Fe^{3+}(aq)]^2}$$

The equilibrium concentration of iron(II) ions will be double that of tin(IV), i.e. $[Fe^{2+}(aq)]$ = 0.080 mol dm^{-3}. The following equilibrium concentrations are also known: $[Sn^{4+}(aq)]$ = 0.040 mol dm^{-3}; $[Sn^{2+}(aq)]$ = 0.050 mol dm^{-3}. We also know that $K = 1.0 \times 10^{10}$. Substituting into the equilibrium constant expression,

$$K = \frac{0.040 \times (0.080)^2}{0.050 \times [Fe^{3+}(aq)]^2}$$

giving

$$[Fe^{3+}(aq)] = \sqrt{\left(\frac{0.040 \times (0.080)^2}{0.050 \times 1.0 \times 10^{10}}\right)}$$
$$= 7.2 \times 10^{-7}\text{ mol dm}^{-3}$$

15.6 No, K is so small that virtually no phosgene has decomposed at this temperature.
15.7 We make two observations before tackling this problem:
(a) We take 'yield' to mean the gain in concentration of methanol at equilibrium.
(b) This reaction involves a reduction in the number of molecules, going from left to right.
This allows us to make the following conclusions:
(i) Reduced gain in $[CH_3OH]$ at equilibrium.
(ii) The equilibrium concentration of CH_3OH increases.
(iii) This decreases the equilibrium concentration of CH_3OH.
(iv) No effect.
(v) Equilibrium composition contains more product (CH_3OH) at lower temperatures.
K is only affected in (v) – K increases with decreasing temperature for an exothermic reaction.
15.8 To see the relationship, substitute the following equilibrium expressions into the equation $K(3) = K(1) \times K(2)$:

$$K(1) = \frac{[NO(g)]^2}{[O_2(g)][N_2(g)]};\ K(2) = \frac{[NO_2(g)]^2}{[O_2(g)][NO(g)]^2}$$

$$K(3) = \frac{[NO_2(g)]^2}{[N_2(g)][O_2(g)]^2}$$

15.9
(i) Applying the equilibrium law:

$$K = \frac{[O_2(g)]_{water}}{[O_2(g)_{air}]}$$

or

$$[O_2(g)]_{air} = \frac{[O_2(g)]_{water}}{K_{c(T)}}$$

By the ideal gas equation ($P=(n/v)RT$), the partial pressure of oxygen, p_{O_2}, in the air is proportional to its concentration:
$p_{O_2} = [O_2(g)]_{air}RT$
so that

$$p_{O_2} = \frac{[O_2(g)]_{water} \times RT}{K}$$

or $[O_2(g)]_{water} = \dfrac{p_{O_2} \times K}{RT}$

which may be expressed in the form of Henry's law (see Unit 11):

$c = K_H \times p$

where $p = p_{O_2}$, $c = [O_2(g)]_{water}$ and $K_H = \dfrac{K}{RT}$
(ii) Applying the equilibrium law

$K = \dfrac{[I_2]_{CCl_4}}{[I_2]_{H_2O}}$

leads directly to a constant distribution ratio K, also symbolised K_d.

15.10
No. K is controlled by the ratio of forward and back reactions:

$$K = \frac{k_f}{k_b}$$

so that even if k_f is large, K will not be large unless k_b is relatively small *compared* with k_f.

Unit 16

16A

(i) The order is: SO_3(s), SO_3(l) and SO_3(g). Data tables show $S^{\ominus}$ (J K^{-1} mol^{-1}) as 70.7, 113.8 and 256.8, respectively. (These values are theoretically corrected to 298 K, as is the standard molar entropy data for ice, liquid water and water vapour in Figure 16.1.)
(ii) Yes. The molecules – and therefore their energy – are dispersed over a reduced volume and so S falls (i.e. the change in entropy, ΔS, is negative).

16B

(i) $\Delta S^{\ominus}$ for reaction (16.6) is strongly negative, reflecting the fact that after the reaction, all of the energy previously spread out in six species is now concentrated in one species, $CuSO_4.5H_2O$(s). The entropy change for the back reaction is equal in size but of opposite sign:

$CuSO_4.5H_2O(s) \rightarrow CuSO_4(s) + 5H_2O(l)$ $\Delta S^{\ominus}$ = + 158.6 J K^{-1} mol^{-1} (–16.6)

with the positive value reflecting the greater dispersal of energy associated with the reaction. (Note that $\Delta S^{\ominus}$ is given in units of J K^{-1} mol^{-1} not J K^{-1}; this because we are using standard molar entropies to calculate $\Delta S^{\ominus}$.)
(ii) $\Delta S^{\ominus}$ is positive because (a) a greater number of species is formed, dispersing energy more widely (as equation (16.7) shows, we started with three and end up with thirteen species); and (b), one of the products (NH_3) will exist as a gas and the water will exist in the liquid state, so dispersing energy relatively more widely than the two solid reactants. **Comment**: this reaction is fairly unusual in that it is a spontaneous endothermic reaction. Most (but by no means all) spontaneous reactions observed in the lab are exothermic (see explanation on page 294) and the large positive increase in entropy in reaction (16.7) is significant here.
(iii) Two molecules of liquid water are produced per molecule of methane: liquid molecules disperse energy less than gases and so $\Delta S^{\ominus}$ is predicted to be negative. If the water is produced as a water vapour rather than a liquid, as in reaction (16.9), three gaseous reactant molecules produce three gaseous product molecules and it is difficult to make a prediction of sign other than that $\Delta S^{\ominus}$ will be less negative than for reaction (16.8). That $\Delta S^{\ominus}$(16.9) is fairly close to zero suggests that in this case it is the *number* rather the nature of the gaseous molecules on each side of the equation that is important in determining $\Delta S^{\ominus}$.

16C

(i) (a) The shape of the curve and the equilibrium composition (point X) would change since Gibbs energies (and equilibrium constants) change with temperature.
(b) A catalyst does not change the curve at all. A catalyst would speed up the attainment of the reaction compositions shown on the x-axis, but it would not affect the Gibbs energy at each composition.
(ii) Assuming that the volume remains at 1.0 litre, then K = [molar conc of B at equil]/[molar conc of A at equil] = [(0.7/1.0)/(0.3/1.0)] = 2.3.
(iii) Your sketch should show that the equilibrium lies on the far left:

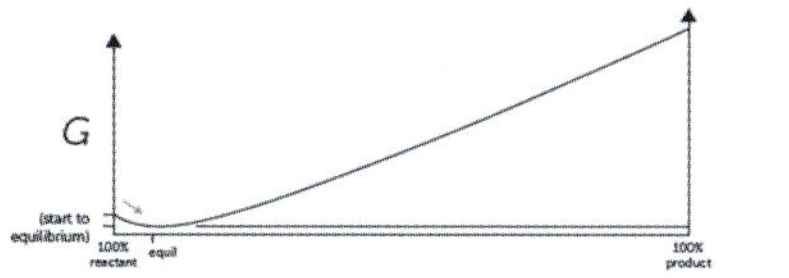

16D

Addition of the ethyl ethanoate increases the Gibbs energy of the reaction system. By forcing some of the ethyl ethanoate to react with water, the reacting system achieves a lower G.

16E

$$\left(-\frac{\Delta G^{\ominus}}{RT}\right) = -(-30{,}500/(8.3145\times298) = 12.31;$$

$$K = \exp(12.31) = 2.2\times10^5$$

16F

$\Delta S^{\ominus}$ = +591 J K^{-1} mol^{-1} = +0.591 kJ mol^{-1}

Converting T°C to K gives (a) 273 K, (b) 278 K, (c) 373 K

$\Delta G^{\ominus}$ (273 K) = $\Delta H^{\ominus} - T\Delta S^{\ominus}$ = (164) – (273 × 0.591) = +2.66 kJ mol^{-1}. Following a similar pattern, $\Delta G^{\ominus}$ (278 K) = –0.30 kJ mol^{-1} and $\Delta G^{\ominus}$ (373 K) = –56.4 kJ mol^{-1}.

The reaction is not 'allowed to go' at 273 K (0°C) but it is 'allowed' at the other two temperatures, although marginally so at 278 K (5°C). In this reaction, the positive enthalpy change means that even a relatively large positive entropy change can only compensate and make $\Delta G^{\ominus}$ negative at temperatures above 4°C.

16G

(a) This is an example of thermodynamic V kinetic control. The relative ΔG values show that **B** is more stable than **A**. To explain the observations, we suppose that at 25°C, the rate of reaction (1) > rate of reaction (2) because it has the lowest activation energy, E_A. Reaction (2) does not reach equilibrium at 25°C so **A** dominates the product mixture. At 80°C, rate (1) is faster than it was at 25°C, but reaction (2) is now sufficiently fast enough to cause the reaction to reach equilibrium in a reasonable time, producing an equilibrium in which the more stable isomer (**B**) dominates.
(b) Even at 25°C equilibrium will eventually be achieved and so much of A will break down into B, but it will take very much longer at the lower temperature.

Revision questions

16.1 (T = true; F = false) (a) F, reaction rate is not part of the definition of a spontaneous reaction; (b) F; (c) T; (d) T (the energy change between reactants and products is unaffected by a catalyst. K is also unaffected by a catalyst); (e) T

16.2

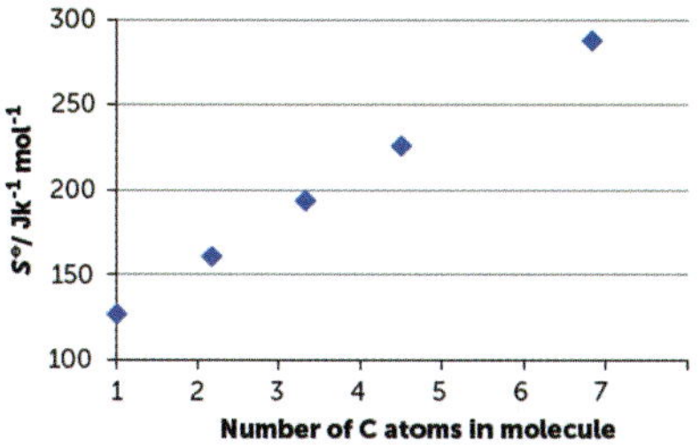

It would be expected that there is a connection between molecular size and entropy and this is most obvious when the series of molecules being considered has a common molecular structure. Ignoring the point for 2-propanol, the graph is almost perfectly linear, with $S^{\ominus}$ = [(32n) + 96] J K^{-1} mol^{-1}, where n is the number of carbon atoms. $S^{\ominus}$ for 2-propanol is slightly smaller than that of 1-propanol, presumably because its different shape allows for energy to be dispersed differently.

16.3
$\Delta S^{\ominus}$ = $S^{\ominus}$(CH_4(g)) – [$S^{\ominus}$(•CH_3(g)) + [$S^{\ominus}$(H•(g))] = 186.13 – (194.2 + 114.7) = –122.8 J K^{-1} mol^{-1}. The negative entropy change is rationalized because we are going from two gaseous molecules to one, with the corresponding concentration of energy in only one species. Back reactions have a change in entropy that is equal but opposite to that of the forward reaction. Therefore, $\Delta S^{\ominus}$ for $CH_4(g) \rightarrow$ •CH_3(g) + •H(g) is +122.8 J K^{-1} mol^{-1}.

16.4
(i) The breakup of the crystalline lattice in solution suggests that $\Delta S^{\ominus}$ would be strongly positive, which is the case. $\Delta H^{\ominus}$ is positive. **Comment**: $\Delta S^{\ominus}$ is so positive that despite the enthalpy change being positive, $\Delta G^{\ominus}$ is negative and so the dissolution is 'allowed to go'.
(ii) Since $\Delta H^{\ominus}$ is close to zero, we must look at $\Delta S^{\ominus}$, the standard change in entropy for the reacting system, to explain why $\Delta G^{\ominus}$ is negative (hence explaining why K >1, the criterion for an 'allowed' reaction). The production of more species (and gaseous species at that) allows us to predict that $\Delta S^{\ominus}$ is strongly positive. **Comment**: Since for this reaction, $\Delta H^{\ominus}$ may be taken to be zero, the equation, $\Delta G^{\ominus} = \Delta H^{\ominus} - T\Delta S^{\ominus}$ reduces to $\Delta G^{\ominus} = -T\Delta S^{\ominus}$.

16.5
$\Delta G^{\ominus}$ (298 K) = $\Delta H^{\ominus} - T\Delta S^{\ominus}$ = –180.58 – (298 × 0.020) = –186.54 kJ mol^{-1}

$\Delta G^{\ominus}$ (600 K) = $\Delta H^{\ominus} - T\Delta S^{\ominus}$ = –180.58 – (600 × 0.020) = –192.58.6 kJ mol^{-1}
The reaction is 'allowed to go' at both temperatures. (Note that $\Delta S^{\ominus}$ is very small and $T\Delta S$ has only a minor influence on $\Delta G^{\ominus}$ even at 600 K.)
K(298 K) = exp(–$\Delta G^{\ominus}/RT$) = exp[186 540/(8.3145 × 298)] = exp(+75.29) = 5.0 × 10^{32}
K(600 K) = exp(–$\Delta G^{\ominus}/RT$) = exp[192 580/(8.3145 × 600)] = exp(–38.60) = 5.80 × 10^{16}. Note that K falls with increasing T for an exothermic reaction.

16.6
(i) $\Delta S^{\ominus}$(6.3) = –96 J K^{-1} mol^{-1} = –0.096 kJ mol^{-1}.

$\Delta G^{\ominus}$ (298 K) = $\Delta H^{\ominus} - T\Delta S^{\circ}$ = (–483.64) – (300 × – 0.096) = –454.8 kJ mol^{-1}.

$\Delta G^{\ominus}$ (600 K) = $\Delta H^{\ominus} - T\Delta S^{\ominus}$ = (–483.64) – (600 × –0.096) = –426.0 kJ mol^{-1}. This confirms that the reaction is 'allowed to go' at both temperatures. A doubling of temperature makes only a minor difference to $\Delta G^{\ominus}$ because in this reaction the $\Delta H^{\ominus}$ component (which we have assumed does not vary with T) is so large and negative. The $\Delta G^{\ominus}$ for the back reaction, (–16.3), is simply the reverse, i.e. +454.8 kJ mol^{-1} at 300 K and +426.0 kJ mol^{-1} at 600 K.
$\Delta G^{\ominus}$(298 K) = –454,800 J mol^{-1}

$$\left(-\frac{\Delta G}{RT}\right) = -(-454{,}800/(8.3145\times300) = 182.3$$

$$K(298\,\text{K}) = \exp\left(-\frac{\Delta G^{\ominus}}{RT}\right) = \exp(183.3)$$

$$= K(298\,\text{K}) = 1.5\times10^{79}$$

If $\Delta G^{\ominus}$(600 K) = –426,000 J mol^{-1}, and = K (600 K) = 1.2 × 10^{37}
(ii) This confirms the prediction of Le Chatelier's principle: for an exothermic reaction K falls with increasing temperature.

16.7
$H(g) + H(g) \rightarrow H_2(g)$
$\Delta H^{\ominus}$ = –435.94 kJ mol^{-1} and $\Delta S^{\ominus}$ = –0.0987 kJ mol^{-1}
$\Delta G^{\ominus} = \Delta H^{\ominus} - T.\,\Delta S^{\ominus}$ = –435.94 – (298 × (–)0.0987) = –483.64 + 29.4 = –406.5 kJ mol $^{-1}$
This substantial negative free energy change reflects the strong tendency of free H(g) radicals (better represented as H•(g) to combine together by forming a covalent bond.

16.8
(i) $\Delta H^{\ominus}$ = –234 – (52.3 – 242.8) = –43.5 kJ mol^{-1}

$\Delta S^{\ominus}$ = 282.6 – (219.5 + 188.8) = –125.7 J K^{-1} mol^{-1} = –0.1257 kJ K^{-1} mol^{-1}

Rearranging, $\Delta G^{\ominus}$ (298 K) = $\Delta H^{\ominus} - T\Delta S^{\ominus}$:

$T = (\Delta H^{\ominus} - \Delta G^{\ominus})/\,\Delta S^{\circ}$; we now substitute for $\Delta G^{\ominus}$ = –1.0 and the same values for $\Delta H^{\ominus}$ and $\Delta S^{\ominus}$:

T= (–43.5–(–1.0))/(–0.1257) = 354 K

Conclusion: $\Delta G^{\ominus}$ for the reaction is –1.0 kJ mol^{-1} at 354 K (81°C). ΔG becomes less negative, and therefore K becomes smaller, as T increases, as predicted by Le Chatelier's principle for an exothermic reaction.

(ii) One tactic is to try and increase the reaction rate while still operating at a lower temperature, which would favour a higher concentration (pressure) of ethanol in the equilibrium mixture, e.g. by increasing the pressure of ethene and steam, or using a catalyst.

16.9 (i) shown in the following figures. The equilibrium mixture is labelled 'equil'.

(ii) By measuring how far the equilibrium point is along the axis and expressing this as a percent of the axis length (use a ruler!), we arrive at equilibrium compositions of about 25 per cent K and 75 per cent J for the reaction J → K; and 95 per cent N and 5 per cent M for M → N.

(iii) $K_{(M \rightarrow N)}$ would fall with increasing temperature.

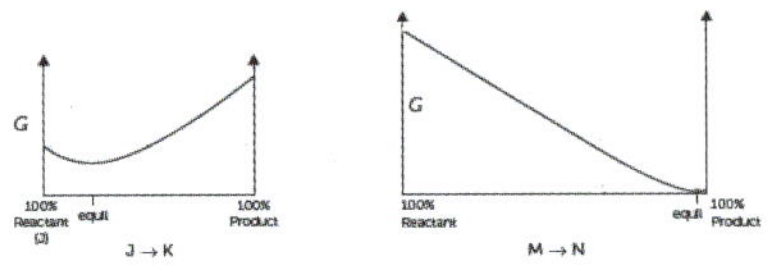

16.10 $K = \exp(+62{,}000)/(8.3145 \times 298) = 7.4 \times 10^{10}$. ΔG°(298 K) per residue is about –0.62 kJ mol^{-1}. This is very small. ΔG°(298 K) for unfolding (denaturing the protein) will be +62 kJ mol^{-1}, but per residue this is only about +0.62 kJ mol^{-1} – less than the kinetic energy of a molecular group at room temperature, so that the question is 'why don't proteins just shake apart?' The answer seems to be that just like a 3-D puzzle, unfolding is an 'all or nothing' process and the integrated structure makes it hard to break off one part of it. Since proteins are part of the engine of living things, this is just as well!

16.11

(i) We write equations for both forward and back equilibrium constants with the corresponding standard Gibbs energy changes (F forward, B for back).

$$K_b = \exp\left(-\frac{\Delta G^\circ(B)}{RT}\right) \qquad K_f = \exp\left(-\frac{\Delta G^\circ(F)}{RT}\right)$$

We want to prove that:

$$\exp\left(-\frac{\Delta G^\circ(B)}{RT}\right) = \frac{1}{\exp\left(-\frac{\Delta G^\circ(F)}{RT}\right)}$$

but we know that $G^\circ(B) = -\Delta G^\circ(F)$ and so:

$$\exp\left(-\frac{-\Delta G^\circ(F)}{RT}\right) = \frac{1}{\exp\left(-\frac{\Delta G^\circ(F)}{RT}\right)}$$

On the left-hand side, (–) × (–) = + , giving exp

$$\exp\left(+\frac{\Delta G^\circ(F)}{RT}\right) = \frac{1}{\exp\left(-\frac{\Delta G^\circ(F)}{RT}\right)}$$

This is now of the form:

$y^x = 1/y^{-x}$ and mathematically both the right- and left-hand sides are equivalent so the relationship is proved.

(ii) For the reaction $2HI(g) \rightarrow H_2(g) + I_2(g)$, $K = 1/45.6 = 0.0219$ at 764 K and ΔG°(764 K) = + 24.26 kJ mol^{-1}

16.12

Decomposition is the reverse process for which ΔG° (298 K) = +80.1 kJ. This is non-spontaneous and in water at room temperature, hardly any water molecules will be broken up, i.e. K is exceedingly large.

16.13

Any reaction that is exothermic in the forward direction is endothermic in the reverse direction. There are therefore equal numbers of both. The reason is it appears that exothermic reactions are more common, is that exothermic reactions are more likely to be 'allowed to go' (be spontaneous) and so we encounter them more often in the laboratory.

Unit 17

Exercises

17A

K_w increases as the temperature is increased.

17B

(i) $[H_3O^+(aq)] = 0.10$ mol dm^{-3}

$[OH^-(aq)] = 1.0 \times 10^{-14}/[H_3O^+(aq)]$
$= 1.0 \times 10^{-14}/0.10$
$= 1.0 \times 10^{-13}$ mol dm^{-3}

pH = –log (0.10) = 1.0

(ii) $[H_3O^+(aq)] = 10^{-2.20} = 6.3 \times 10^{-3}$ mol dm^{-3}

17C

(i) The number of moles of $Ca(OH)_2$ is

$$\frac{0.10 \text{ g}}{(40+17+17) \text{ g mol}^{-1}} = 0.0014 \text{ mol}$$

The concentration of calcium hydroxide is

$$\frac{0.0014}{0.200} = 0.0070 \text{ mol dm}^{-3}$$

One $Ca(OH)_2$ provides two hydroxide ions in solution. Thus, $[OH^-(aq)] = 0.014$ mol dm^{-3}.

At room temperature,

$1.0 \times 10^{-14} = [0.014]\,[H_3O^+(aq)]$

Rearranging;

$[H_3O^+(aq)] = 1.0 \times 10^{-14}/0.014$
$= 7.1 \times 10^{-13}$ mol dm^{-3}

pH = –log $[H_3O^+(aq)]$ = 12.15

(ii) We answer (b) first:

(b)

$[H_3O^+(aq)] = 10^{-11.10} = 7.9 \times 10^{-12}$ mol dm^{-3}

(a)

$[OH^-(aq)] = 1.0 \times 10^{-14}/[H_3O^+(aq)]$
$= 1.0 \times 10^{-14}/7.94 \times 10^{-12}$
$= 1.3 \times 10^{-3}$ mol dm^{-3}

17D

pOH = 14 – pH = 14 – 5.80 = 8.20;

$[OH^-(aq)] = 10^{-8.20} = 6.3 \times 10^{-9}$ mol dm^{-3}

17E

$[H_3O^+(aq)]$	$[OH^-(aq)]$	pH	pOH
10^{-14}	1	14	0
10^{-13}	10^{-1}	13	1
10^{-12}	10^{-2}	12	2
10^{-11}	10^{-3}	11	3
10^{-10}	10^{-4}	10	4
10^{-9}	10^{-5}	9	5
10^{-8}	10^{-6}	8	6
10^{-7}	10^{-7}	7	7
10^{-6}	10^{-8}	6	8
10^{-5}	10^{-9}	5	9
10^{-4}	10^{-10}	4	10
10^{-3}	10^{-11}	3	11
10^{-2}	10^{-12}	2	12
10^{-1}	10^{-13}	1	13

Note the pattern: the pH decreases steadily and the pOH increases steadily. A high pOH value indicates a strongly acidic solution. A high pH indicates a strongly basic solution.

17F

(i)

$[H_3O^+(aq)] \approx \sqrt{(K_a \times C_A)}$

$C_A = 0.020$ mol dm^{-3}

$K_a = 4.9 \times 10^{-10}$ mol dm^{-3} (from Table 16.1)

$[H_3O^+(aq)] \approx \sqrt{(4.9 \times 10^{-10} \times 0.020)}$
$= 3.1 \times 10^{-6}$ mol dm^{-3}

The percentage of HCN molecules that are ionized is calculated as follows:

$$\text{per cent}_{\text{ionized}} = \frac{[H_3O^+(aq)]}{C_A} \times 100$$

$$= \frac{3.1 \times 10^{-6} \times 100}{0.020} = 0.016 \text{ per cent}$$

pH = –log $[H_3O^+(aq)]$ = –log (3.1×10^{-6}) = 5.5

(ii) Mass of benzoic acid in solution = 4,000 × 0.050/100 = 2.0 g.

The formula of benzoic acid is obtained from Table 17.2; $M(C_6H_5COOH) = 122$ g mol^{-1}.

Number of moles of benzoic acid = $2.0/122 = 1.6 \times 10^{-2}$

Concentration of benzoic acid = $1.6 \times 10^{-2} / 1.0 = 1.6 \times 10^{-2}$ mol dm^{-3}

$[H_3O^+(aq)] \approx \sqrt{(K_a \times C_A)}$

$C_A = 1.6 \times 10^{-2}$ mol dm^{-3};

$K_a = 6.5 \times 10^{-5}$ mol dm^{-3} (from Table 17.2)

$[H_3O^+(aq)] \approx \sqrt{(6.5 \times 10^{-5} \times 1.60 \times 10^{-2})}$
$= 1.0 \times 10^{-3}$ mol dm^{-3}

pH = –log $[H_3O^+(aq)]$ = –log (1.0×10^{-3}) = 3.00

The percentage of benzoic acid molecules ionized is $[1.0 \times 10^{-3}/1.6 \times 10^{-2}] \times 100 \approx 6$ per cent.

Rearrangement of the equation

$$K_a = \frac{[H_3O^+(aq)]^2}{C_A - [H_3O^+(aq)]} = 0.0070 \text{ mol dm}^{-3}$$

gives an accurate value of $[H_3O^+(aq)]$ as 9.85×10^{-4} mol dm^{-3}, close to the approximate value obtained above.

17G

(i) $CH_3COOH(aq) + H_2O(l) \rightleftharpoons CH_3COO^-(aq) + H_3O^+(aq)$

As more hydronium ions are mopped up by $OH^-(aq)$ ions in the neutralization reaction, more ethanoic acid molecules ionize. The titration volumes are then identical with those of a strong acid.

(ii) The members of the first two pairs are strong acids and strong bases. The CH_3COOH–NaOH pair contains a weak acid and most of its molecules must ionize before neutralization can take place. The ionization process is endothermic and reduces the overall amount of heat given out in the reaction between ethanoic acid and sodium hydroxide.

17H

(i) $K_b = 10^{-5.98} = 1.1 \times 10^{-6}$ mol dm^{-3}

(ii)

$$K_b = \frac{[OH^-(aq)]^2}{C_B}$$

Rearranging;

$[OH^-(aq)] = \sqrt{(K_b C_B)} = \sqrt{(1.8 \times 10^{-5} \times 0.200)}$
$= 1.9 \times 10^{-3}$ mol dm^{-3}

$$[H_3O^+(aq)] = \frac{1.0 \times 1.0^{-14}}{1.90 \times 10^{-3}} = 5.3 \times 10^{-12} \text{ mol dm}^{-3}$$

pH = –log (5.3×10^{-12}) = 11.28

(iii) $C_6H_5NH_2(aq) + H_2O(l) \rightleftharpoons C_6H_5NH_3^+(aq) + OH^-(aq)$ (conjugate acid)

$[H_3O^+(aq)] = 10^{-7.80} = 1.6 \times 10^{-8}$ mol dm^{-3}

$$[OH^-(aq)] \approx \frac{1.0 \times 1.0^{-14}}{1.6 \times 10^{-8}} = 6.3 \times 10^{-7} \text{ mol dm}^{-3}$$

$$C_B = \text{initial conc} \approx \frac{[OH^-(aq)]^2}{K_b} \quad K_b \text{ listed in Table 17.2}$$

$$\approx \frac{[6.3 \times 10^{-7}]^2}{4.3 \times 10^{-10}} = 9.2 \times 10^{-4} \text{ mol dm}^{-3} = 9.2 \times 10^{-4} \text{ mol dm}^{-3}$$

17I

Salt	Formula	Parent acid and base	Acidic	Basic	Neutral
Iron(III) nitrate	$Fe(NO_3)_3$	HNO_3 (SA) $Fe(OH)_3$ (WB)	✓		
Calcium chloride	$CaCl_2$	HCl (SA) $Ca(OH)_2$ (SB)			✓
Sodium sulfate	Na_2SO_4	H_2SO_4 (SA) NaOH(SB)			✓
Ammonium benzoate	$C_6H_5CO_2NH_4$	$C_6H_5CO_2H$ (WA) NH_3(aq) (WB)	✓		

$(K_{a\text{ (benzoic acid)}} > K_{b\text{ (ammonia)}})$.

17J

Measure pH at different acid and salt concentrations. Plot $[H_3O^+(aq)]$ against the ratio C_A/C_S. The slope of the graph equals K_a.

17K

(i) Number of moles of ethanoic acid = $0.10 \times 25/1000 = 2.5 \times 10^{-3}$

Number of moles of sodium ethanoate = $0.10 \times 50/1000 = 5.0 \times 10^{-3}$

Concentration of ethanoic acid = $(2.5 \times 10^{-3})/(75/1000)$ = 3.3×10^{-2} mol dm^{-3} = C_A.
Concentration of sodium ethanoate = $(5.0 \times 10^{-3})/(75/1000)$ = 6.6×10^{-2} mol dm^{-3} = C_S.
$C_A/C_S = 0.50$

$$[H_3O^+(aq)] \approx \frac{C_A \times K_a}{C_S}$$

= $0.50 \times 1.8 \times 10^{-5} = 9.0 \times 10^{-6}$ mol dm^{-3}
pH = $-\log(9.0 \times 10^{-6}) = 5.05$
(ii) Similar calculations show that if $C_A/C_S = 2.0$,

$$[H_3O^+(aq)] \approx \frac{C_A \times K_a}{C_S}$$

= $2.0 \times 1.8 \times 10^{-5} = 3.6 \times 10^{-5}$ mol dm^{-3}
pH = $-\log(3.6 \times 10^{-5}) = 4.44$

17L

$C_S/C_A = 1$; and $\log(C_S/C_A) = 0$

$$pH = pK_a + \log\frac{C_S}{C_A} = 4.18 + 0, \text{ so } pK_a = 4.18$$

$K_a = 10^{-4.18} = 6.6 \times 10^{-5}$ mol dm^{-3}. This suggests that the acid is benzoic acid (see Table 17.2).

17M

pH	Phenolphthalein	Bromocresol green
2	colourless	yellow
3.7	colourless	yellow
5	colourless	blue/yellow
6	colourless	blue
8	colourless	blue
11	pink	blue

17N

(i) Alizarin yellow.
(ii) Pairs (a) and (b); (c) involves a weak base.

17O

(i) Bromocresol green changes over the pH range 4.0–5.6. This is too soon, and its use in the titration of a weak acid and a strong base will lead to an underestimation of the equivalence point.
(ii) For the mixture in question, 12.50 cm^3 represents half-equivalence, i.e. the volume of NaOH required to neutralize half the ethanoic acid. At half-equivalence the concentration of ethanoic acid equals the concentration of sodium ethanoate, i.e. $C_S = C_A$ and $\log C_S/C_A = 0$. The Henderson–Hasselbalch equation then simplifies to pH of solution = pK_a. This gives us an approximate method (in addition to the method outlined in Exercise 17J) of finding the acidity constant of a weak acid. For example, Figure 17.3(b) shows that after the addition of 12.5 cm^3 of NaOH, pH ≈ 4.7 ≈ $pK_a(CH_3CO_2H)$. Therefore, $K_a \approx 2 \times 10^{-5}$ mol dm^{-3}.

17P

(i) From Table 17.2, $K_a = 4.3 \times 10^{-7}$ mol dm^{-3}. Applying the equation

$$K_a \approx \frac{[H_3O^+(aq)]^2}{[CO_2(aq)]}$$

(you may recognize this as equation (17.5)).
$[H_3O^+(aq)]^2 \approx 4.3 \times 10^{-7} \times 1.2 \times 10^{-5}$
= 5.2×10^{-12} mol^2 dm^{-6}
$[H_3O^+(aq)] \approx \sqrt{(5.2 \times 10^{-12})} \approx 2.3 \times 10^{-6}$ mol dm^{-3}
pH = $-\log(2.3 \times 10^{-6}) = 5.62$
(ii) Salt occupies some of the space that would otherwise be taken up by dissolved carbon dioxide. Therefore, salty water is less acidic than deionized water.

Revision questions

17.1 (i) F
(ii) T
(iii) T **(iv)** T **(v)** T
17.2 (i) HNO_3 provides 0.080 mol dm^{-3} of $H^+(aq)$
pH = $-\log(0.080) = 1.10$
(ii) KOH provides 0.080 mol dm^{-3} of $OH^-(aq)$ in solution.

$$[H^+(aq)] = \frac{1.0 \times 1.0^{-14}}{0.080} = 1.3 \times 10^{-13} \text{ mol dm}^{-3}$$

pH = $-\log[1.3 \times 10^{-13}] = 12.89$
17.3 (i) Table 17.1 shows that $K_w = 51.3 \times 10^{-14}$ mol^2 dm^{-6} at 100°C. In pure water $[OH^-(aq)] = [H_3O^+(aq)]$; therefore $51.3 \times 10^{-14} = [H_3O^+(aq)]^2$ or $[H_3O^+(aq)] = 7.16 \times 10^{-7}$ mol dm^{-3}.
pH (100°C) = $-\log[7.16 \times 10^{-7}] = 6.145$
(ii) Number of moles of NaOH = $(11.0/1000) \times 0.040$ = 4.4×10^{-4}.
Number of moles of HCl = $(9.0/1000) \times 0.040$ = 3.6×10^{-4}.
Since 1 HCl ≡ 1 NaOH, after neutralization we have 0.80×10^{-4} mol of $OH^-(aq)$ spread out over 20 cm^3 of solution. This is a concentration of
$(0.80 \times 10^{-4})/20/1000 = 4.0 \times 10^{-3}$ mol dm^{-3} of $[OH^-(aq)]$
At 25°C,

$$[H^+(aq)] = \frac{1.0 \times 1.0^{-14}}{4.0 \times 10^{-3}} = 2.5 \times 10^{-12} \text{ mol dm}^{-3}$$

pH = $-\log[2.5 \times 10^{-12}] = 11.60$
(iii)

$$C_{17}H_{19}O_3N(aq) + H_2O(l) \rightleftharpoons C_{17}H_{19}O_3NH^+(aq) + OH^-(aq) \quad (A)$$

Since:
$K_b = 1.6 \times 10^{-6}$ mol dm^{-3};
$M(C_{17}H_{19}O_3N) = 285$ g mol^{-1};
1.0 mg = 0.0010 g
Number of mols of morphine = 0.0010 g/285 g mol^{-1} = 3.5×10^{-6} mol.
Concentration of morphine = $C_B = 3.5 \times 10^{-6}$ mol per dm^3
We now use the equation

$$K_b \approx \frac{[OH^-(aq)]^2}{C_B}$$

Rearranging and substituting values,
$[OH^-(aq)] \approx \sqrt{(K_b \times C_B)}$
= $\sqrt{(3.5 \times 10^{-6} \times 1.6 \times 10^{-6})}$
= 2.4×10^{-6} mol dm^{-3}
pOH = $-\log(2.4 \times 10^{-6}) = 5.62$
Also, since pH + pOH = 14,
pH = 14 – 5.62 = 8.38
(iv) The $OH^-(aq)$ produced in reaction (A) reacts with the $H_3O^+(aq)$ from HCl(aq):

$$H_3O^+(aq) + OH^-(aq) \rightarrow 2H_2O(l) \quad (B)$$

Addition of equations (A) and (B) gives the overall reaction

$$C_{17}H_{19}O_3N(aq) + H_3O^+(aq) \rightleftharpoons C_{17}H_{19}O_3NH^+(aq) + H_2O(l)$$

The number of moles of morphine in 100 cm^3 of solution is 3.5×10^{-7}. This requires an equal number of moles of HCl(aq) for neutralization. We calculate the volume of 0.0010 mol dm^{-3} HCl(aq) required as follows:

$$\text{volume of HCl (aq)} = \frac{\text{moles}}{\text{concentration}} = \frac{3.5 \times 10^{-7}}{0.0010}$$

= 3.5×10^{-4} dm^3 or 0.35 cm^3
17.4 The ionization of phenol is represented as

$$C_6H_5OH(aq) + H_2O(l) \rightleftharpoons C_6H_5O^-(aq) + H_3O^+(aq)$$

using $[H_3O^+(aq)] \approx \sqrt{(K_a \times C_A)}$, $C_A = 5.00 \times 10^{-3}$ mol dm^{-3} and $K_a = 1.3 \times 10^{-10}$ mol dm^{-3} (from Table 17.2),
$[H_3O^+(aq)] \approx \sqrt{(1.3 \times 10^{-10} \times 5.00 \times 10^{-3})}$
≈ 8.1×10^{-7} mol dm^{-3}
pH = $-\log[H_3O^+(aq)] = -\log(8.06 \times 10^{-7}) = 6.094$
17.5 The acidity constant for $NH_4^+(aq)$ is the equilibrium constant that applies to the equation

$$NH_4^+(aq) + H_2O(l) \rightleftharpoons NH_3(aq) + H_3O^+(aq)$$

that is,

$$K_a(NH_4^+(aq)) = \frac{[NH_3(aq)][H_3O^+(aq)]}{[NH_4^+(aq)]}$$

For the equation

$$NH_3(aq) + H_2O(l) \rightleftharpoons NH_4^+(aq) + OH^-(aq)$$

the following is true:

$$K_b(NH_3(aq)) = \frac{[NH_4^+(aq)][OH^-(aq)]}{[NH_3(aq)]}$$

We also know that
$K_w = [H_3O^+(aq)][OH^-(aq)]$
The relationship
$K_a \times K_b = K_w$ (A)
is now proved by substitution and cancellation of terms:

$$\frac{\cancel{[NH_3(aq)]}[H_3O^+(aq)]}{\cancel{[NH_4^+(aq)]}} \times \frac{\cancel{[NH_3(aq)]}[H_3O^+(aq)]}{\cancel{[NH_3(aq)]}} = [H_3O^+(aq)][OH^-(aq)]$$

$$K_a \quad \times \quad K_b \quad = \quad K_w$$

Equation (A) applies for all pairs of conjugate acids and bases. Here,
$K_a(NH_4^+(aq)) = K_w/K_b(NH_3(aq))$
= $1.0 \times 10^{-14}/(1.8 \times 10^{-5})$
= 5.6×10^{-10} mol dm^{-3}
17.6 (i) Assuming that the dissolved CO_2 concentration equals its initial concentration, and applying the equilibrium law,

$$K_a \approx \frac{[HCO_3^-(aq)][H_3O^+(aq)]}{[CO_2(aq)]}$$

If the only acidity is due to dissolved carbon dioxide, $[HCO_3^-(aq)] = [H_3O^+(aq)]$, giving

$$K_a \approx \frac{[H_3O^+(aq)]^2}{[CO_2(aq)]}$$

(ii) $[H_3O^+(aq)] \approx \sqrt{(4.3 \times 10^{-7} \times 4.0 \times 10^{-2})}$
= 1.3×10^{-4} mol dm^{-3}
(corresponding to 0.3 per cent ionization).
pH = $-\log(1.3 \times 10^{-4}) = 3.89$
17.7 (i) WB–SA, therefore acidic **(ii)** WA–WB, $K_a = K_b$ so approximately neutral **(iii)** WB–SA, therefore acidic **(iv)** SA–SB, therefore neutral **(v)** SB–WA, so basic.
17.8 $M(C_6H_5COOH) = 122$ g mol^{-1},
$M(C_6H_5COONa) = 144$ g mol^{-1}
Number of moles of C_6H_5COOH = 0.100/122
= 8.20×10^{-4}
Number of moles of C_6H_5COONa = 0.080/144
= 5.6×10^{-4}
Concentration of (C_6H_5COOH) = 8.2×10^{-4} / 0.100
= 8.2×10^{-3} mol dm^{-3}
Concentration of (C_6H_5COONa) = 5.6×10^{-4} / 0.100
= 5.6×10^{-3} mol dm^{-3}

$$[H_3O^+(aq)] \approx \frac{C_A \times K_a}{C_S} \approx \frac{8.20 \times 10^{-3} \times 6.5 \times 10^{-5}}{5.6 \times 10^{-3}}$$

= 9.5×10^{-5} mol dm^{-3}
pH = $-\log(9.5 \times 10^{-5}) = 4.02$

$$C_6H_5COOH(aq) + H_2O(l) \rightleftharpoons C_6H_5COO^-(aq) + H_3O^+(aq)$$

Explanation of buffer action is as follows. (a) Addition of $H_3O^+(aq)$ causes association of the $C_6H_5COO^-(aq)$ and $H_3O^+(aq)$, so restoring $[H_3O^+(aq)]$. (b) Addition of $OH^-(aq)$ neutralizes $H_3O^+(aq)$, and more benzoic acid ionizes to compensate.
17.9 Choice of indicators from Table 17.4 for a weak base–strong acid titration is bromocresol green and methyl red. The remaining indicators begin to change colour before or after the pH at the equivalence point (≈ pH 4–8).
17.10 The ingested hydrogencarbonate increases the buffer capacity of the athlete to combat lactic acid production.
17.11 Firstly, this is quite different from asking whether or not water is acidic, basic or neutral. As we have seen, water is neutral because when pure it contains equal numbers of H_3O^+ (or $H^+(aq)$) and $OH^-(aq)$ ions. This question is about the reactions of the H_2O *molecule*. The relevant part of Unit 17 is page 304, where an acid is defined as a proton donor and a base as a proton acceptor. We know that water autoionizes: $H_2O(l) + H_2O(l) \rightleftharpoons H_3O^+(aq) + OH^-(aq)$. In this reaction, one water molecule is donating a proton (and so behaving as an acid) and another is accepting a proton (and so behaving as base). Water acts as both acid and base. In other words, water is amphoteric (see Unit 6, Box 6.2).

Unit 18

Exercises

18A

(i)

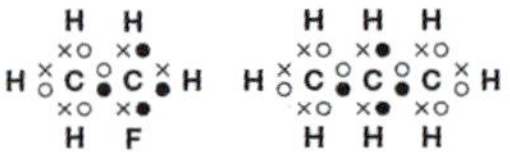

(ii) C_8H_{18}

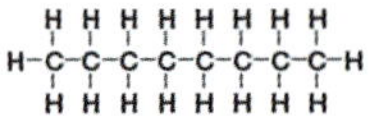

(iii) $C_{12}H_{26}$

18B

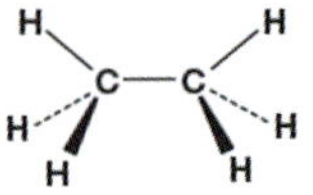

18C

(i)

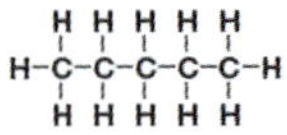

(ii)

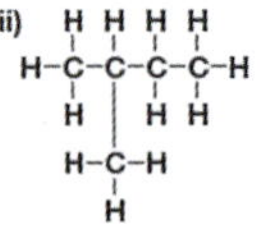

(iii)

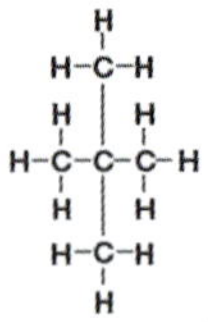

18D

(i) 2-methylpropane or, better, methylpropane (the 2 isn't necessary in this case)
(ii) 4-ethylheptane
(iii) 2,2-dimethylpropane
(iv) 4-ethyl-3,5-dimethyloctane

18E

(i)

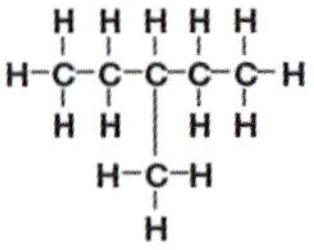

(ii)

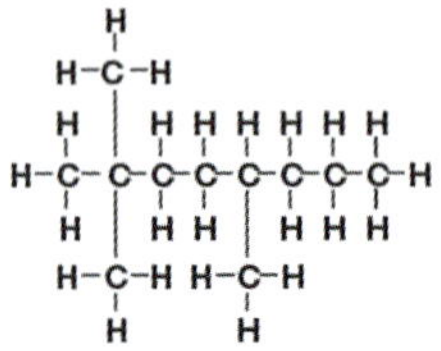

(iii)

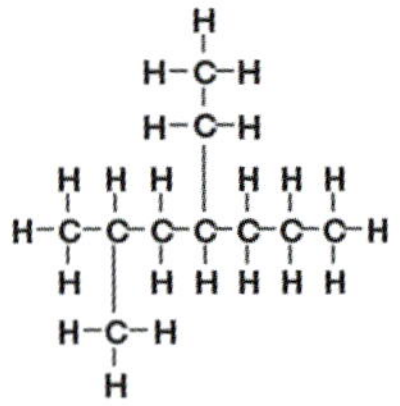

(iv)

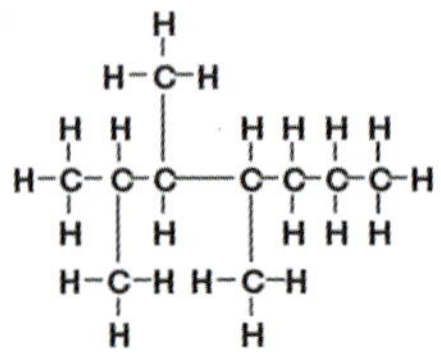

18F

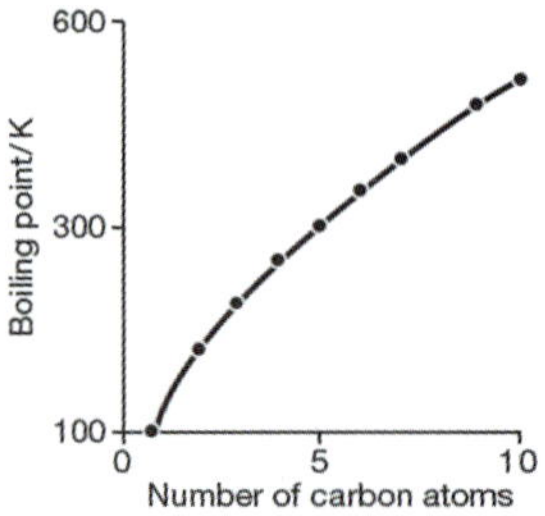

(i) butane **(ii)** 399 K

18G

(i) C_nH_{2n}
(ii) The ring is far less 'strained' – the bond angle is larger than in cyclopropane.
(iii) Hexane

C_6H_{14}

$CH_3CH_2CH_2CH_2CH_2CH_3$

Cyclohexane

C_6H_{12}

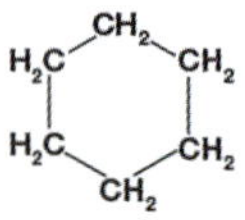

18H

(i) Copper metal; water vapour and carbon dioxide gas.
(ii) $4CuO(s) + CH_4(g) \rightarrow 4Cu(s) + 2H_2O(g) + CO_2(g)$
(iii) $CuO(s) + H_2(g) \rightarrow Cu(s) + H_2O(g)$

18I

(i) See Box 18.6
(ii) No, unlike ethane.

18J

(i) hex-1-ene
(ii) but-2-ene
(iii) 4,4-dimethylpent-2-ene
(iv) 3-bromopropene

18K

(i) $CH_3CH_2CH{=}CHCH_2CH_3$

(ii)

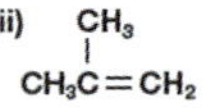

(iii)

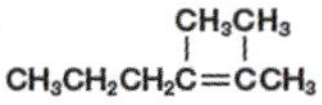

(iv)

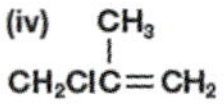

(v) $CH_3CH_2CH_2CHBrCH{=}CH_2$

18L

(i) 3 =

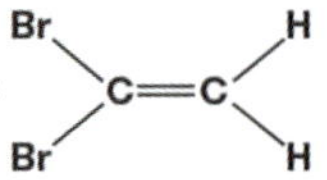

(ii) geometric
(iii) structural
(iv) 1 is *trans*-1,2-dibromoethene, **2** is *cis*-1,2-dibromoethene, **3** is 1,1-dibromoethene.

18M

(i) $CH_2{=}CH_2 + HBr \rightarrow CH_3{-}CH_2Br$ (bromoethane)
(ii) $CH_2{=}CH_2 + H_2SO_4 \rightarrow CH_3{-}CH_2{-}OSO_3H$ (ethyl hydrogensulfate)
(iii) $CH_3{-}CH{=}CH_2 + Br_2 \rightarrow CH_3{-}CHBr{-}CH_2Br$ (1,2-dibromopropane)
(iv) $CH_3{-}CHOH{-}CH_2OH$
(v) $n(CH_2{=}CHCN) \rightarrow \cdots CH_2{-}CHCN{-}CH_2{-}CHCN{-}CH_2{-}CHCN{-}CH_2{-}CHCN{-}CH_2{-}CHCN\cdots$

18N

(i) $CH_3{-}CH_2C{\equiv}CH$
(ii) propene, $CH_3{-}CH{=}CH_2$ and propane, $CH_3CH_2CH_3$
(iii) $CHBr_2{-}CHBr_2$ 1,1,2,2-tetrabromoethane.

18O

(i)

mass of C present = $21.99 \times \frac{12}{44} = 6.00$ mg

percentage C = $\frac{6}{7.02} \times 100 = 85.5$ per cent

mass of H present = $8.95 \times \frac{2}{18} = 1$ mg

percentage H = $\frac{1}{7.02} \times 100 = 14.3$ per cent

Empirical formula:

	C	H
no of moles	$\frac{85.5}{12}$	$\frac{14.3}{1}$
ratio	7.1 : 14.3	
	1 : 2	

Therefore, the empirical formula is CH_2.
(ii) molecular formula X = C_3H_6, structure $CH_3{-}CH{=}CH_2$
Y is $CH_3CH_2CH_3$
Z is $CH_3CHBrCH_2Br$

18P

(i)

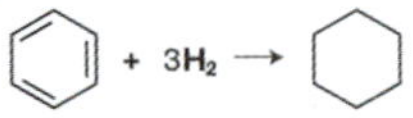

(ii) $3 \times -119 = -357$ kJ mol^{-1}
(iii) Benzene has a lower enthalpy of formation than that suggested by the Kekulé structure.

18Q

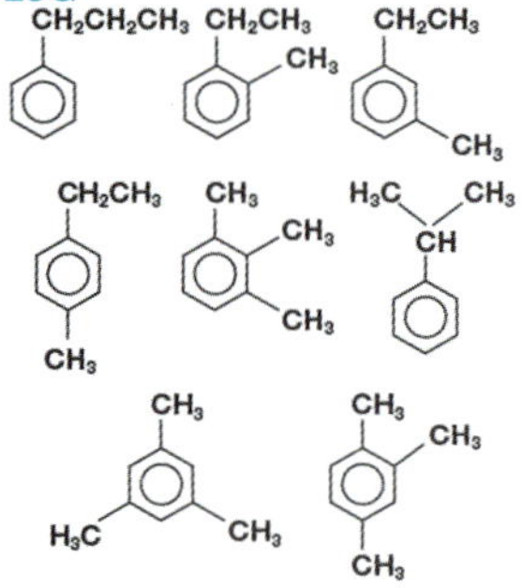

Revision questions

18.1 C_6H_{14}, $CH_3CH_2CH_2CH_2CH_2CH_3$, $CH_3(CH_2)_4CH_3$

18.2
$CH_3CH_2CH_2CH_2CH_2CH_3$ hexane

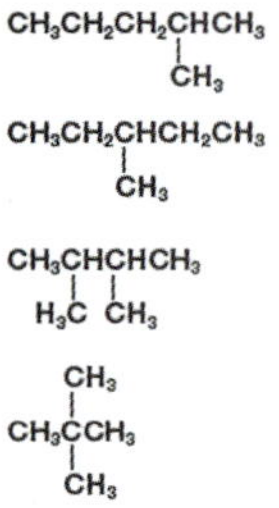

2-methylpentane
3-methylpentane
2,3-dimethylbutane
2,2-dimethylpropane

18.3 (i) pent-2-ene **(ii)** 2,5-dimethylhex-2-ene

18.4 (i) $CH_2{=}CHCH_2CH_2CH_3$

(ii)

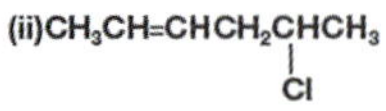

(iii)

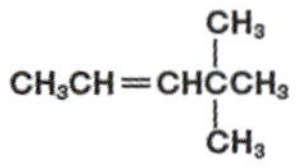

(iv) $CH_2{=}CHCH{=}CH_2$

(v)

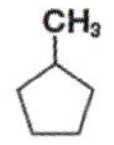

(vi) $CH_3CH_2CH{\equiv}CHCH(CH_3)CH_2CH_2CH_3$

18.5 (ii) and **(iii)**

18.6 (i)

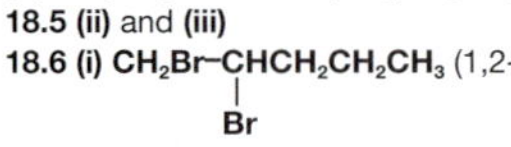

(1,2-dibromopentane)

(ii)

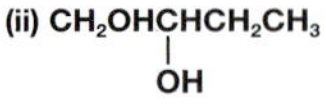

(butan-1,2-diol)

(iii) $CH_3CH_2CH_2CH_3$ (butane)

18.7

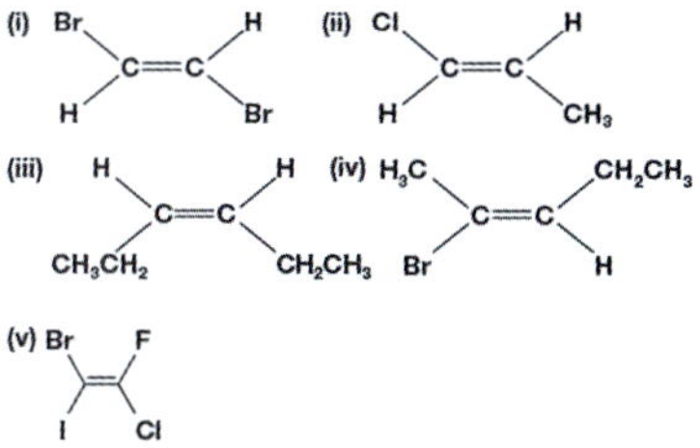

18.8 tetrafluoroethene: $CF_2{=}CF_2$

PTFE: $\cdots CF_2\text{–}CF_2\text{–}CF_2\text{–}CF_2\text{–}CF_2\text{–}CF_2\text{–}CF_2\text{–}CF_2\text{–}CF_2\text{–}CF_2 \cdots$

a non-stick coating for frying pans.

18.9 (i)

	C	H
no of moles	$\frac{88.9}{12}$	$\frac{11.1}{1}$
ratio	7.4 : 11.1 2 : 3	

The empirical formula is C_2H_3.

(ii) C_4H_6

(iii) $CH_3CH_2C{\equiv}CH$, but-1-yne

$CH_3C{\equiv}CCH_3$, but-2-yne

18.10

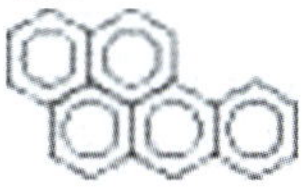

18.11 We used bond energies to find the standard enthalpy change for the reaction:

$6C(s) + 3H_2(g) \rightarrow C_6H_6(l) \quad \Delta H^\ominus_1 = ?$ (equation 1)

We start by calculating the standard enthalpy change for the reaction in which gaseous carbon makes gaseous benzene:

$6C(g) + 3H_2(g) \rightarrow C_6H_6(g) \quad \Delta H^\ominus_2 = ?$ (equation 2)

From Table 13.4,

$\Delta H^\ominus(\text{H–H}) = 436\ \text{kJ mol}^{-1}$
$\Delta H^\ominus(\text{C–H}) = 412\ \text{kJ mol}^{-1}$
$\Delta H^\ominus(\text{C=C}) = 612\ \text{kJ mol}^{-1}$
$\Delta H^\ominus(\text{C–C}) = 348\ \text{kJ mol}^{-1}$

Assuming that benzene has the Keluké molecular structure, the heat absorbed in breaking bonds in reactant molecules, i.e. three H–H bonds, is

heat absorbed = 3 × 436 = 1308 kJ

The heat released when bonds in product molecules are assembled, which involves making three C=C bonds, three C–C bonds and six C–H bonds, is

heat given out = (3 × 612) + (6 × 412) + (3 × 348)
= 5352 kJ

The overall change in enthalpy is

$\Delta H^\ominus_2 = -5352 + 1308 = -4044\ \text{kJ mol}^{-1}$

The difference between equations (2) and (1) is that we have to allow for the production of gaseous carbon atoms and formation of gaseous benzene being endothermic:

$6C(s) \rightarrow 6C(g)\ \Delta H^\ominus_3(298\ \text{K}) = (717\times6) = 4302\ \text{kJ mol}^{-1}$

$C_6H_6(l) \rightarrow C_6H_6(g)\ \Delta H^\ominus_4(298\ \text{K}) = 31\ \text{kJ mol}^{-1}$

Use of the enthalpy diagram confirms that:

$\Delta H^\ominus_1 + \Delta H^\ominus_4 - \Delta H^\ominus_2 = \Delta H^\ominus_3$ or

$\Delta H^\ominus_1 = \Delta H^\ominus_3 - \Delta H^\ominus_4 + \Delta H^\ominus_2$

$\Delta H^\ominus_1 = 4302 - 31 + (-4044) = +227\ \text{kJ mol}^{-1}$

Therefore, the difference between the theoretical and experimental values is (227 – 50) = 177 kJ mol^{-1}. We explain this difference by assuming that the Keluké structure is not the true structure of benzene – benzene has a lower enthalpy than this structure suggests.

18.12 (a)

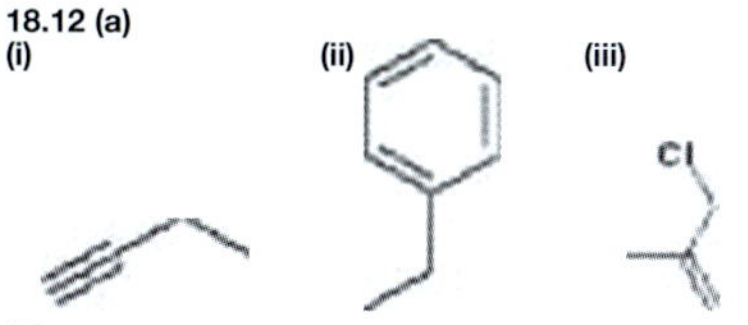

(b)

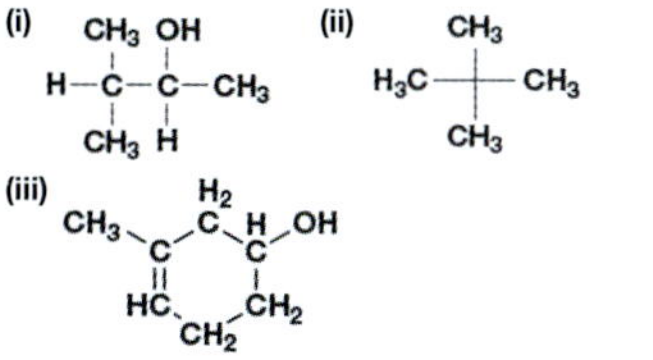

Unit 19

Exercises

19A

(i) (a) CH_3CH_2Cl (b) $CH_3CHBrCH_3$

(c) $CH_2ClCHClCHFCH_3$

(d)

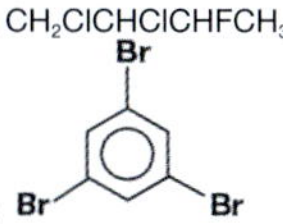

(ii) (a) 1-bromo-1-chloroethane (b) 3-iodopentane

(c) trichlorofluoromethane

(d) 3-chloromethylbenzene (or 3-chlorotoluene)

19B

(i)

I: pentan-1-ol
$CH_3CH_2CH_2CH_2CH_2OH$

II: pentan-2-ol
$CH_3CH_2CH_2CHOHCH_3$

III: pentan-3-ol
$CH_3CH_2CHOHCH_2CH_3$

IV: 2,2-dimethylpropan-1

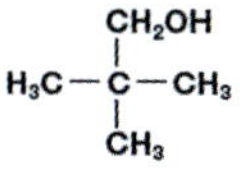

V: 2-methylbutan-2-ol
$CH_3CH_2\underset{|\ CH_3}{C}OHCH_3$

VI: 3-methylbutan-2-ol
$CH_3CHOH\underset{|\ CH_3}{C}HCH_3$

VII: 2-methylbutan-1-ol
$CH_3CH_2\underset{|\ CH_3}{C}HCH_2OH$

VIII: 3-methylbutan-1-ol
$CH_3\underset{|\ CH_3}{C}HCH_2CH_2OH$

(ii) $CH_2OHCHOHCH_2OH$

19C

(i) approx: 64.5, 58, 60

(ii) propan-1-ol is a liquid, the others are gases

(iii) hydrogen bonding between the alcohol molecules results in stronger intermolecular attractions than in the other compounds.

19D

I, IV, VII, VIII are primary; II, III, VI are secondary; V is tertiary

19E

(i) aldehyde **(ii)** ketone **(iii)** aldehyde

(iv) ketone **(v)** ketone.

19F

(i) 'eth' is a prefix that refers to two carbon atoms. The simplest ketone, propanone, contains three.

(ii)

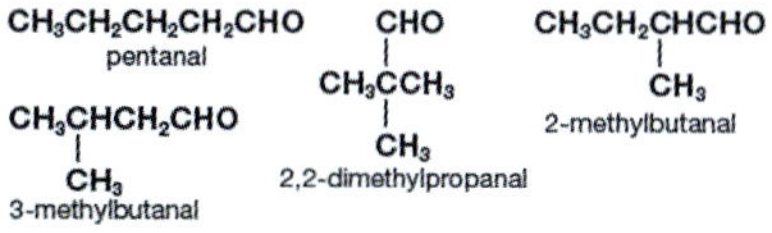

(iii)

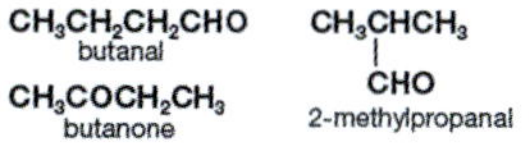

19G

(i) butanone $CH_3COCH_2CH_3$

(ii) propanoic acid C_2H_5COOH

(iii) octan-2-one $C_6H_{13}COCH_3$.

19H

(i) $CH_3CH_2CH_2COOH$

(ii)

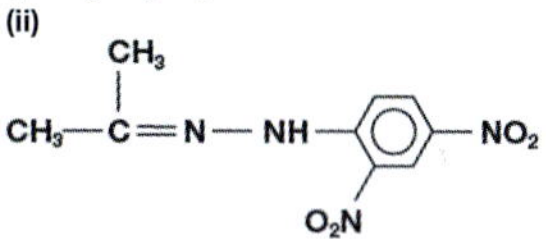

(iii) $CH_3\text{–}\underset{|\ H}{C}{=}NNH_2$ **(iv)** CH_3CH_2COOH

19I

(i) butanoic acid **(ii)** heptanoic acid

(iii) butane-1,4-dioic acid.

19J

CH_3COOH, $CH_2BrCOOH$, $CH_2ClCOOH$, $CHCl_2COOH$, CCl_3COOH, CF_3COOH

(chlorine is more electronegative than bromine).

19K

(i) (a) methyl ethanoate CH_3COOCH_3

(b) ethyl octanoate $C_7H_{15}COOCH_2CH_3$

(c) methyl methanoate $HCOOCH_3$

(d) diethyl oxalate (diethyl ethanedioate) $(COOCH_2CH_3)_2$.

(ii) (a) propanoic acid, methanol

(b) methanoic acid, ethanol

(c) ethanoic acid, propan-1-ol

(d) chloroethanoic acid, methanol.

19L

NH_3, CH_3NH_2, CH_3NHCH_3.

19M

(i) Middle C has an asterisk.

(ii) 2 methylbutan-1-ol, 2-bromo-1-chlorobutane.

19N

(i)

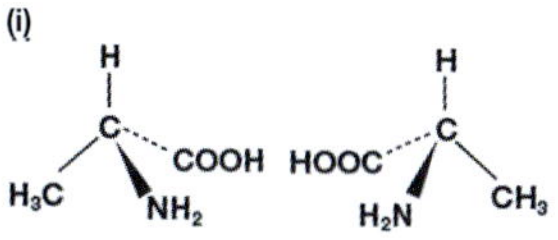

(ii) $CH_3\underset{|\ NH_3^+}{C}HCOO^-$

19O

(i)

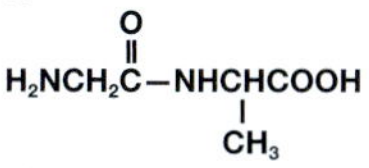

(ii) Gly-Ala-Cys, Gly-Cys-Ala, Cys-Gly-Ala, Cys-Ala-Gly, Ala-Cys-Gly, Ala-Gly-Cys (six in all)

Revision questions

19.1 (i) $CH_3CH_2CH_2CH_2CHOHCH_3$

(ii) $CH_3\underset{|\ CH_3}{C}OHCH_3$

(iii) $CH_2OHCH_2CHOHCH_3$

(iv) 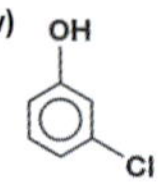

19.2 ethanol, water, phenol, ethanoic acid.

19.3

(i) 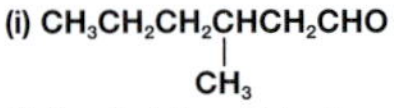

Notice that the aldehyde group contains the carbon atom numbered 1.

(ii) $CH_3COCH_2COCH_3$ **(iii)** 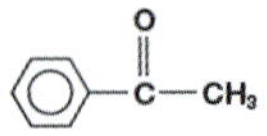

(iv) 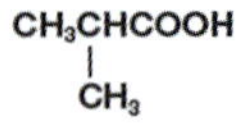

(v) 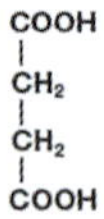

19.4 oxalic acid

19.5 (i) $CH_3CH_2COOCH_2CH_3$ **(ii)** $CH_3CH_2CH_2OH$ **(iii)** $CH_3(CH_2)_3COOH$ **(iv)** CH_3COOH

19.6 (i) $HCOOCH_2CH_2CH_3$ **(ii)** $HCONH_2$ **(iii)** HCOCl

19.7

(i) 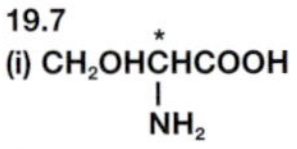

(ii) Yes, it has a carbon atom attached to four different groups, marked '*'.

(iii) **(iv)** 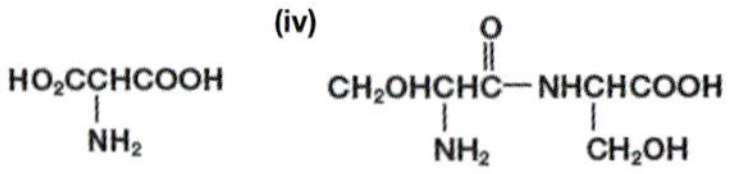

19.8 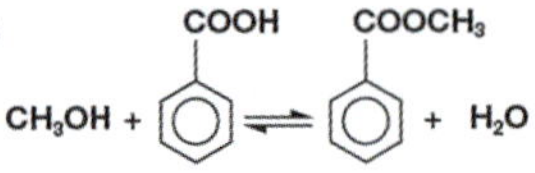

19.9

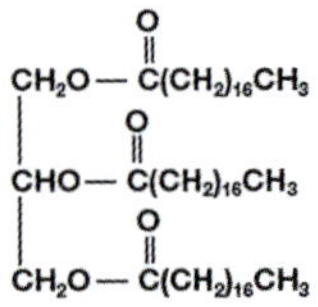

19.10 (i) ester **(ii)** amino group **(iii)** ester and carboxylic acid **(iv)** amide **(v)** phenol, alcohol, amino **(vi)** halogen.

19.11 (i) X would form an orange precipitate with 2,4-dinitrophenylhydrazine.

(ii) X would reduce Tollens' reagent or Fehling's solution.

(iii) CH_3CH_2CHO; to confirm this, check the melting point of the DNP derivative against that in a data book.

19.12

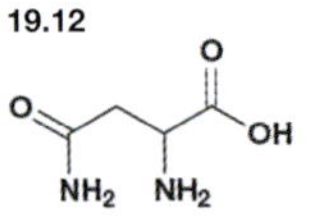

19.13 It is an alcohol (phenols occur when the –OH group is bonded to a benzene ring).

19.14

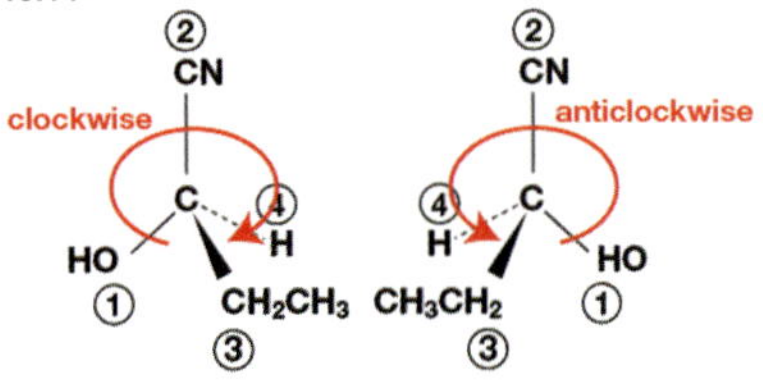

***R* isomer** ***S* isomer**

Unit 20

Exercises

20A

(i) The bridging bromine atom is in the way of the carbon atoms on one side – there is steric hindrance by the bridging atom.

(ii) *trans*-1,2-dibromoethene

20B

The possible carbocations are:

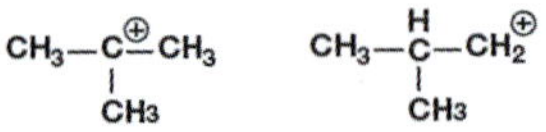

The left-hand structure is the more stable because three alkyl (methyl) groups are pushing electron density on to the positive carbon atom. The major product would therefore be:

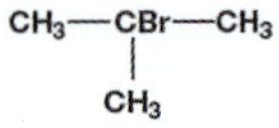

20C

The overall reaction is:

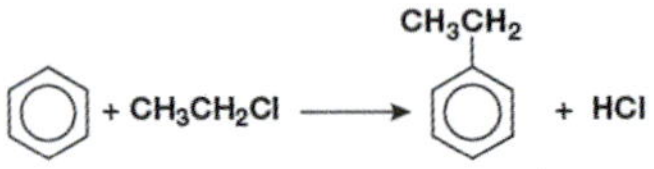

The mechanism is as follows:

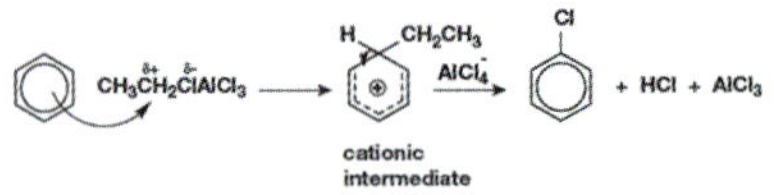

20D

It would be expected to be S_N2, because the carbon attacked is primary.

Revision questions

20.1

(i) $C_{10}H_{22}(l) + 15½O_2(g) \rightarrow 10\,CO_2(g) + 11H_2O(l)$

(ii) Incomplete combustion would yield C and CO – in addition to CO_2 and H_2O.

20.2

(i) $CH_2{=}CH{-}CH_2CH_3 + Br_2 \rightarrow CH_2Br{-}CH\ Br{-}CH_2CH_3$
1,2-dibromobutane

(ii) 1 mol of $CH_2{=}CHCH_2\ CH_3$ produces 1 mol of $CH_2BrCHBrCH_2CH_3$; 56 g of $CH_2{=}CHCH_2CH_3$ produces 216 g $CH_2\ BrCHBrCH_2CH_3$; and 15 g of $CH_2{=}CHCH_2\ CH_3$ produces $216 \times \frac{15}{56}$ $CH_2BrCHBrCH_2CH_3$ = 58 g

This is the theoretical yield.

Therefore the percentage yield obtained =

$$\frac{\text{actual yield}}{\text{theoretical yield}} \times 100 = \frac{35}{58} \times 100 = 60 \text{ per cent}$$

20.3

(i) $CH_2{=}CHCH_3$

(ii) 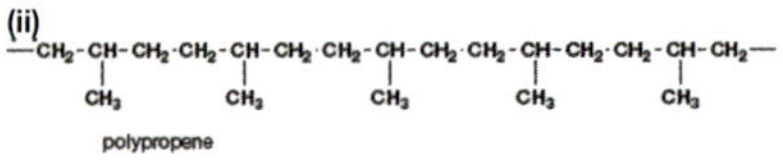

20.4

Ethene reacts rapidly in an addition reaction:

$CH_2{=}CH_2 + Br_2 \rightarrow CH_2Br{-}CH_2\ Br$

The electron density in benzene is delocalized, therefore it does not undergo addition reactions so readily. In the presence of $AlBr_3$ benzene undergoes substitution:

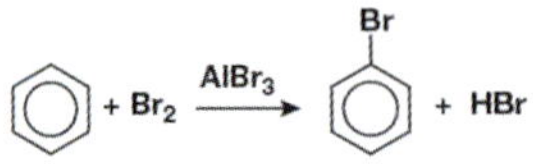

20.5

(i) No, there are eight electrons (one on each carbon) available for delocalization. This does not conform to Hückel's Rule (4*n* + 2). The shape of cyclooctatatraene is puckered and non-planar.

(ii) Yes, each carbon atom donates one electron and there is an extra negative charge making six electrons in all.

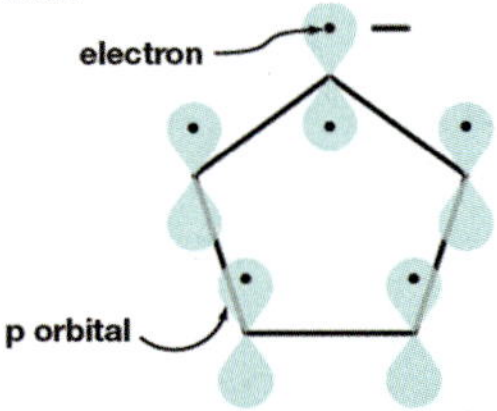

(iii) Yes each carbon provides one electron making fourteen – this is 4*n* + 2 if *n* = 3.

(iv) Yes, the positive charge indicates an electron has been lost – there are therefore two electrons, one on each of the carbons that are not charged. This conforms to Hückel's Rule if *n* = 0.

20.6

(i) Yes, pyridine is aromatic. Nitrogen has a lone pair of electrons, but they do not get involved in the delocalized ring of electrons – each carbon contributes one electron, and nitrogen, which has five valence electrons, also donates one electron making six. The lone pair on nitrogen is available for bonding and this makes pyridine basic.

(ii) Yes, sulfur donates two electrons (as oxygen does in furan) and there is one electron per carbon atom, making six in all.

(iii) Yes, nitrogen has a lone pair of electrons available, which, together with one from each carbon atom, makes six.

(iv) No, the carbons are bonded to two hydrogens and two carbons so cannot contribute any electrons.

20.7

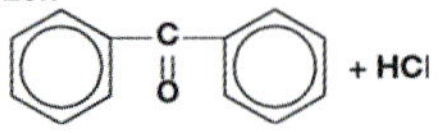

20.8

B > D > A > C

(see Table 20.1)

20.9

The alkyl group directs the bromine atom to the *ortho* and *para* positions. It is a bulky group and bromine is quite a large atom, so you would expect the *para* isomer to be the predominant product.

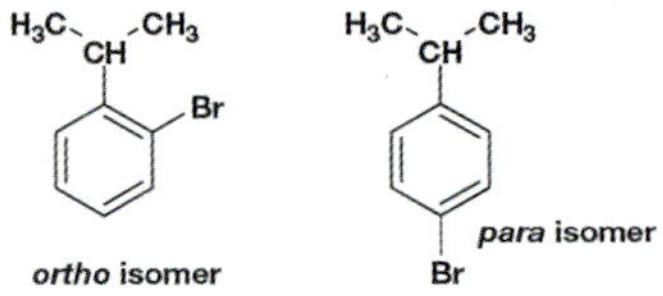

20.10

(Introduction of the methyl group first, would result in the *ortho* and *para* isomers)

20.11

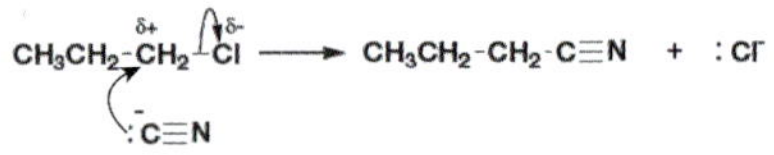

20.12

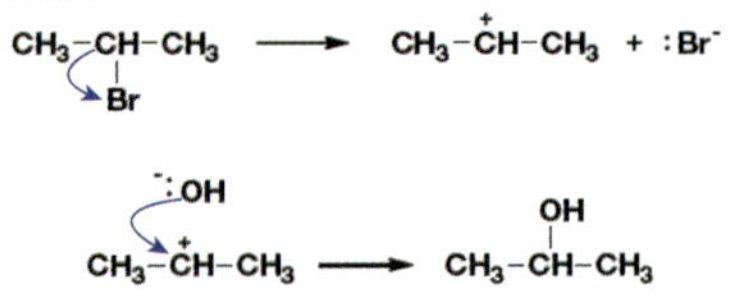

Unit 21

Exercises

21A

Heat the sample of sea water to dryness, so that all the water is evaporated off. The solid left would not be pure sodium chloride, because sea water contains other dissolved salts such as magnesium chloride and potassium chloride.

21B

(i) A separating funnel.
(ii) Add mixture to water and stir → separate sand off by gravity filtration → wash sand well and dry in an oven → filtrate contains dissolved sugar → evaporate water from filtrate to obtain concentrated solution (solid sugar decomposes when heated, so this is 'safer' than evaporating solution to dryness) → allow concentrated solution to cool and sugar crystals should separate out → filter solution to remove sugar crystals.
(iii) Dissolve the mixture in water at a temperature just less than 100°C → cool solution to 20°C → crystalline potassium chlorate separates out, which can be removed by filtering (potassium chloride remains in solution).
(iv) Fractional distillation will result in a mixture that is much richer in the volatile component (the one that boils at 80°C) collecting in the receiving container.
(v) Simple distillation.

21C

(i) Let the amount of X in the organic layer be x g, then

$$K_d = 7 = \frac{x/200}{(1-x)/100}$$

or $x = 0.93$ g.
(ii) First extraction: let the amount of X in the organic layer be z g, then

$$K_d = 7 = \frac{z/100}{(1-z)/100}$$

Therefore $z = 0.875$ g, i.e. 0.125 g left in the aqueous layer.
Second extraction: let amount of X in organic layer be y g, then

$$K_d = 7 = \frac{y/100}{(0.125-y)/100}$$

Therefore, $y = 0.109$ g and the total amount extracted is 0.875 + 0.109 = 0.98 g
that is, 0.05 g more than for one extraction.

21D

(i)

$$R_f \text{ for A} = \frac{y+z}{x+y+z}$$

(ii) R_f for B $= \dfrac{z}{x+y+z}$

Revision questions

21.1 (i) evaporation
(ii) simple distillation
(iii) fractional distillation
(iv) fractional distillation
(v) column chromatography
(vi) separation using a separating funnel
(vii) stir mixture in water → filter to remove chalk → evaporate off water from filtrate to give salt.
21.2 Paper chromatography using propanone as solvent. If substances are not all coloured, use a suitable method to develop the spots.
21.3 Sublimation – ammonium chloride 'sublimes', potassium chloride does not.
21.4 (i) Paper chromatography (or TLC)
(ii) Column chromatography. Ethanol can be used as solvent in both techniques.
21.5 Paper chromatography → spot urine and spots of amino acids it is thought to contain on the paper → run chromatogram using a suitable solvent (water) → develop colourless spots by spraying with ninhydrin.
21.6 Gas chromatography → compare retention times of components of mixture with the retention times of known alcohols (or use GC–MS!).
21.7 Paper chromatography → compare pattern of spots of suspect ink with a sample of ink obtained from the forged cheque.
21.8 Let mass extracted be x g, then

$$3.5 = \frac{x/50}{(0.5-x)/150}$$

i.e. $x = 0.27$ g
21.9
(i) Area ABC = 1/2 × 16 × 40 = 320 mm²
Area PQR = 1/2 × 20 × 32 = 320 mm²
(ii) The same area under each of the peaks suggests there are equal amounts of each constituent in the mixture.
(iii) You could cut out each peak from the chart and weigh them, then compare masses.
21.10 Dissolve 'dyes' from smarties by stirring them in a *little* water. Use paper chromatography to see if more than one spot occurs.

Unit 22

Exercises

22A

Rearranging,
$\nu = \frac{c}{\lambda}$ gives $\lambda = \frac{c}{\nu}$
ν = 98 MHz = 98,000,000 Hz, therefore

$$\nu = \frac{3.00 \times 10^8}{9.8 \times 10^7} = 3.1 \text{ m}$$

22B

(i)

$$E = \frac{3.00 \times 10^8 \times 3.99 \times 10^{-13}}{700 \times 10^{-9}} = 171 \text{ kJ mol}^{-1}$$

(ii)

$$E = \frac{3.00 \times 10^8 \times 3.99 \times 10^{-13}}{500 \times 10^{-9}} = 214 \text{ kJ mol}^{-1}$$

22C

$\Delta E = hN_A\nu$, therefore

$$\nu = \frac{\Delta E}{hN_A} = 100 \text{ kJ mol}^{-1} / 3.99 \times 10^{-13} \text{ kJ s mol}^{-1}$$

= 2.51 × 10¹⁴ Hz (i.e. s⁻¹)

22D

(i) Follow Example 22.2. If 10 per cent of the light is absorbed, 90 per cent of the light is transmitted, i.e.

$$\frac{I_0}{I} = \frac{100}{90} = 1.11$$

absorbance = $A = \log\left(\frac{I}{I_0}\right) = \log(1.11) = 0.0453$
(ii) We start by expressing equation (22.2) on page 398, in the alternative form $I_0/I = 10^A$. Therefore:

$$\frac{I_0}{I} = 10^{3.00} = 1.0 \times 10^3 \therefore \frac{I_0}{I} = 0.0010$$

therefore, the percentage of light transmitted is

$$\frac{100 \times 1}{I_0} = 100 \times 0.0010 = 0.10 \text{ per cent}$$

22E

The higher temperatures mean that many hydrogen atoms are present in the $n = 2$ level.

22F

The energy gap between the $n = 4$ and $n = 1$ level is 1,228 kJ mol⁻¹. Using the equation $\Delta E = hN_A\nu$

$$\nu = \frac{\Delta E}{hN_A} = 1228/3.99 \times 10^{-13} = 3.08 \times 10^{15} \text{ Hz}$$

22G

The rate at which the excited states are produced can be controlled like any other chemical reaction.
(i) Lowering the temperature of the reaction mixture will slow down the reaction and decrease the intensity of the emitted light.
(ii) Increasing temperature increases the emission intensity.

22H

(i) red
(ii) blue green
(iii) colourless (absorptions are outside the visible region)

22I

Only absorptions of visible light (400-720 nm) are relevant. The absorptions between 400 nm and 450 nm suggest that the solution would be yellow (this is the observed colour).

22J

(i) propionic acid and ethanal
(ii) propionic acid and butanol
(iii) all three.

22K

Its simplest formula is $C_4H_{11}N$; infrared suggests $-NH_2$ present; $m(C_4H_{11}N)$ = 73 u (as confirmed by mass spectrum). Possible structures include $(CH_3)_2CHCH_2NH_2$ and $CH_3(CH_2)_3NH_2$.

22L

(i) $A_\lambda = \varepsilon_\lambda cb$, so that $c = A_\lambda/\varepsilon_\lambda b = 1.1/(1.1 \times 10^4 \times 0.00010) = 1.0$ mol m⁻³, i.e. 0.0010 mol dm⁻³.
(ii) (a) $c = A_\lambda/\varepsilon_\lambda b = 0.010/(1000 \times 0.010)$ = 0.0010 mol m⁻³
(b) 0.00 025 mol m⁻³.
(iii) c = 2.0 mol m⁻³; $\varepsilon_\lambda = A_\lambda/cb = 0.40/(2 \times 1.0 \times 10^{-2})$ = 20 mol⁻¹ m².

22M

For 2-nitrophenol, $A_2 = \varepsilon_2 c_2 b$ (where the subscript '2' stands for 2-nitrophenol).
For 4-nitrophenol, $A_4 = \varepsilon_4 c_4 b$ (where the subscript '4' stands for 4-nitrophenol).
Therefore, the total absorbance at 313 nm is

$$\varepsilon_2 c_2 b + \varepsilon_4 c_4 b = (1.2 \times 10^2 \times 10^{-3} \times 10^{-2}) + (9.6 \times 10^2 \times 10^{-3} \times 10^{-2}) = 0.0012 + 0.0096 = 0.0108.$$

22N

(i) The concentrations are converted to mol m⁻³ by multiplying by 1000. The plot is:

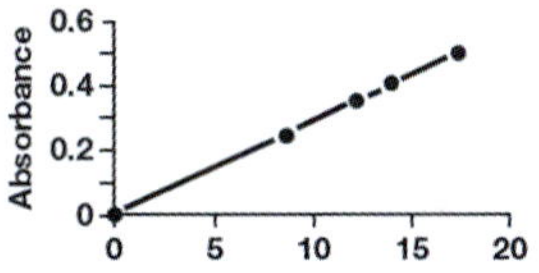

(ii) Slope = $\varepsilon \times b$ = (0.50 – 0.00)/(17.5 × 10⁻² – 0.0) = 2.9 mol⁻¹ m³. Given that b = 10⁻² m,
$\varepsilon_{526 \text{ nm}}$ = slope/b = 2.9 × 10² mol⁻¹ m².

22O

(i) Three (protons from CH_3, CH_2 and CHO).
(ii) Two (protons from CH_3 and CH).

22P

(i)

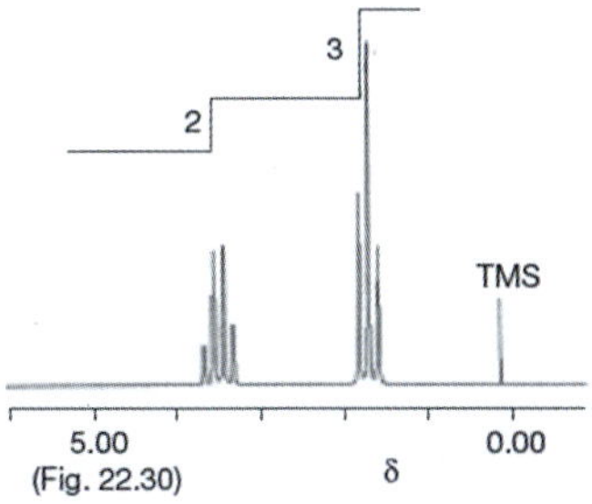

(Fig. 22.30)

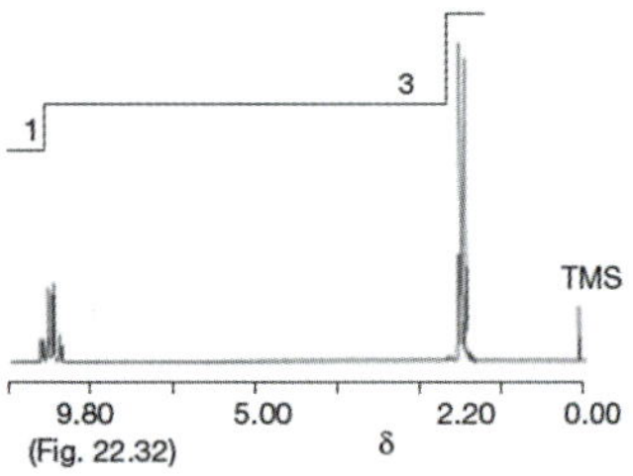

(Fig. 22.32)

(ii) 1:6:15:20:15:6:1

22Q

(i) (b) **(ii)** (a) **(iii)** (c)

22R

Count the number of carbon nuclei in different environments and compare the chemical shifts with those in Table 22.2.

(a) ii **(b)** iii **(c)** iv **(d)** i

Revision questions

22.1 (i) $\nu = c/\lambda = 3.00 \times 10^8/2 \times 10^{-7} = 1.5 \times 10^{15}$ Hz,
(ii) 6.0×10^{14} Hz,
(iii) 3.00×10^{13} Hz.

22.2 The missing words are emission (or fluorescence), absorbance, chemiluminescence and molar absorption coefficient.

22.3 $\nu = c/\lambda = 3.00 \times 10^8/1000 \times 10^{-9} = 3.00 \times 10^{14}$ Hz

$E = hN_A\nu = 3.99 \times 10^{-13} \times 3.00 \times 10^{14} = 120\ \text{kJ mol}^{-1}$.

22.4 $\Delta E = hN_A\nu$; $\nu = \Delta E/hN_A = 30/3.99 \times 10^{-13} = 7.5 \times 10^{13}$ Hz.

22.5 (i) Percentage transmitted light = 100 – 84 = 16 per cent, therefore

$$16\ \text{per cent} = \frac{100 \times I}{I_0}$$

By rearrangement,

$$\frac{I_o}{I} = \frac{100}{16} = 6.25$$

$A = \log(6.25) = 0.796$

(ii) $1.00 \times 10^{-5}\ \text{mol dm}^{-3} \equiv 1.00 \times 10^{-2}\ \text{mol m}^{-3}$

$A = \varepsilon cb = 1000 \times 1.00 \times 10^{-2} \times 10^{-2} = 0.100$

$$\frac{I_o}{I} = 10^{0.100} = 1.26$$

$$\frac{I}{I_o} = 0.79,\quad 100 \times \frac{I}{I_o} = 79\ \text{per cent}$$

To give $A = 0.010$, dilute by a factor of 10.

22.6 $\nu = \Delta E/hN_A = 494/3.99 \times 10^{-13} = 1.24 \times 10^{15}$ Hz

$\lambda = c/\nu = 3.0 \times 10^8/1.24 \times 10^{15} = 2.4 \times 10^{-7}$ m (240 nm)

22.7 (i) $E = -1310/50^2 = -0.524\ \text{kJ mol}^{-1}$.

(ii) E for $n = 3$ level is $-1310/3^2 = -145.6\ \text{kJ mol}^{-1}$. Therefore,

$\Delta E = -0.524 - (-)\ 145.6 = 146.1\ \text{kJ mol}^{-1}$

$\nu = \Delta E/hN_A = 146.1/3.99 \times 10^{-13} = 3.66 \times 10^{14}$ Hz

22.8 (i) Dark lines show wavelengths at which elements in the sun's cooler outer region absorb light. (Some lines are also produced by absorbing elements in the Earth's atmosphere.)

(ii)

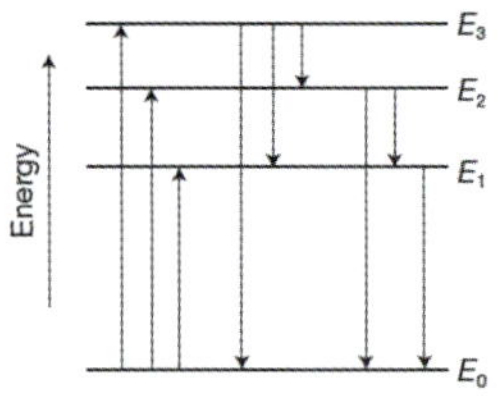

There are three possible transitions in absorption, and six in emission. (The smaller number in absorption follows from the fact that all molecules start at E_0.) Generalizing, an emission spectrum contains more lines than the corresponding absorption spectrum because the excited state does not have to fall to the ground state in 'one go'.

22.9 (i) It is likely that the absorption spectrum will contain a single strong peak between 490 and 530 nm. **(ii)** The absorptions below 400 nm may be ignored. The colour cheese allows us to predict that the solution will absorb light at about 430 nm in wavelength and that it will therefore appear yellow. **(iii)** The mixture will absorb virtually all the colours of white light. Very little light will be reflected and the paint will appear black.

22.10 The plot is very similar to Fig. 24.2 (see Unit 24, page 439). Concentration of ozone c is

$$c = \frac{P}{RT} = \frac{100.0}{8.314 \times 298} = 4.04 \times 10^{-2}\ \text{mol m}^{-3}$$

Therefore,

$A(254\ \text{nm}) = \varepsilon cb = 369 \times 4.04 \times 10^{-2} \times 0.100 = 1.49$.

22.11

37.5 per cent C and 12.5 per cent H leads to the ratio of H to C as 4:1.

Infrared spectrum contains a broad band at about 3,400 cm^{-1}, which is recognizable as characteristic of a hydrogen-bonded –OH group. There is no carbonyl absorption present, suggesting that the compound is an alcohol or phenol. The parent molecular ion possesses a molecular mass of 32 u (too low for a phenol). This suggests that the alcohol is CH_3OH. This contains only two types of protons (as indicated by the NMR spectrum) and the H:C ratio is 4:1 as required.

22.12 (i) Chloroform contains C–H bonds that would absorb in the same region as the hydrocarbons. CF_3CCl_3 contains no C–H bonds.

(ii) $b = 20\ \text{mm} = 2.0 \times 10^{-2}$ m. $c = A/\varepsilon b = 0.050/(5.0 \times 2.0 \times 10^{-2}) = 0.50\ \text{mol m}^{-3}$ or $5.0 \times 10^{-4}\ \text{mol dm}^{-3}$. $M(C_8H_{18}) = 114\ \text{g mol}^{-1}$. $c = 114 \times 5.0 \times 10^{-4} \times 10^3 = 57\ \text{mg dm}^{-3}$.

22.13 (i)

$CrO_4^{2-}(aq) \rightarrow Cr^{3+}(aq)$
+6 +3 (reduced)

(ii) $5.0 \times 10^{-4}\ \text{mol dm}^{-3} = 5.0 \times 10^{-1}\ \text{mol m}^{-3}$. Initial absorbance = $\varepsilon cb = 200 \times 5.0 \times 10^{-1} \times 1.0 \times 10^{-2} = 1.0$. After reaction, the concentration of $Cr^{3+}(aq)$ is also $5.0 \times 10^{-1}\ \text{mol m}^{-3}$; no $CrO_4^{2-}(aq)$ remains. Final absorbance = $\varepsilon cb = 1.0 \times 5.0 \times 10^{-1} \times 1.0 \times 10^{-2} = 5.0 \times 10^{-3}$.

22.14 The flame atomizes the sample so that atoms of the elements are produced. The flame does not produce the excited states of atoms, and this function is carried out by the light sources, which are known as hollow cathode lamps. A different lamp is required for each element.

22.15 1,2-dibromoethane, CH_2BrCH_2Br (all protons in the same chemical environment, so only one resonance in the NMR spectrum).

22.16 Butanone, $CH_3COCH_2CH_3$.

The mass spectrum shows the molecular ion at 72, indicating a molecular formula of C_4H_8O.

The infrared spectrum shows a sharp peak at about 1700 cm^{-1}, indicating a carbonyl group.

The ^{1}H-NMR has a quartet at δ2.5 (CH_2 protons split into a quartet by CH_3), a singlet at δ2.0 (CH_3 attached to CO and too far away from the other protons for them to split the resonance) and a triplet at δ1.0 (CH_3 split by CH_2 next to it).

22.17

(i)

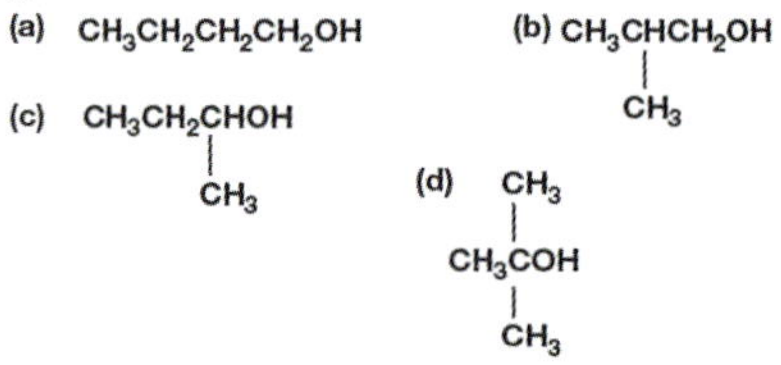

(ii) In a and c the carbon nuclei are in four different environments, whilst in d there are only two. Only in b are there three different carbon environments, leading to three different peaks in the ^{13}C-NMR spectrum. So c is the answer.

22.18

You might take the view that it doesn't matter! UV, visible, IR and NMR spectroscopy involve the absorption of electromagnetic radiation (light) of varying frequencies, whereas mass spectra involves the production of ions by ionization using electric charges. Viewed like this, MS is not strictly a form of spectroscopy.

22.19

The rate of the chemical reaction producing the excited state increases at higher temperature, producing more transitions to the lower energy level per second. This causes an increase in observed emission intensity. However, if the temperature is too high, the enzyme involved in the reaction denatures and/or the insect dies.

Unit 23

Exercises

23A

$^{238}_{92}U \rightarrow {}^{234}_{90}X + {}^{4}_{2}He^{2+}$

Examination of the Periodic Table shows that X = Th.

23B

$^{215}_{83}Bi \rightarrow {}^{215}_{84}X^{+} + {}^{0}_{-1}e^{-}$

Inspection of a Periodic Table shows that X = Po (polonium).

23C

P and S are isotopes of the same element.

23D

From Unit 22,

$\nu = c/\lambda = 3.00 \times 10^8/10^{-10} = 3.0 \times 10^{18}$ Hz

$E = hN_A\nu = 3.99 \times 10^{-13} \times 3.0 \times 10^{18}$
$= 1.2 \times 10^6\ \text{kJ mol}^{-1} \approx 10^6\ \text{kJ mol}^{-1}$.

23E

^{238}U has an extremely long half-life (Table 23.2). Its decay would be represented by a horizontal straight line as for ^{137}Cs in Figure 23.3.

23F

(i) ^{235}U decays faster than ^{238}U, so that more ^{235}U would have been present millions of years ago.

(ii) (a)

$^{201}_{87}Fr \rightarrow {}^{197}_{85}At^{2-} + {}^{4}_{2}He^{2+}$
astatine

$^{231}_{90}Th \rightarrow {}^{231}_{91}Pa^{+} + {}^{0}_{-1}e^{-}$
protactinium

(b)

It may be shown that after ten half-lives there is only a tiny fraction of radionuclide left. The reduction in mass after ten half-lives is actually $1/(2 \times 2 \times 2 \times 2 \times 2 \times 2 \times 2 \times 2 \times 2 \times 2) = 1/2^{10} = 1/1024$. If 0.0100 g is reduced by 1,024, only about 1×10^{-5} g of Fr would be left. In fact, 100 h represents 7.5×10^7 half-lives of Fr! We can safely assume that after 100 h, all the Fr would be gone.

For thorium, 100 h is four half-lives so that the original mass will fall by a factor of $2 \times 2 \times 2 \times 2 = 16$. One-sixteenth of 0.0100 g is 0.000 625 g.

These calculations may also be made using the formula $m_t = m_0 \times e^{-kt}$, with k being calculated from the equation

$$t_{1/2} = \frac{0.693}{k}$$

For example, for thorium,

$$k = \frac{0.693}{t_{1/2}} = 0.693/(25 \times 60 \times 60) = 7.7 \times 10^{-6}\ \text{s}^{-1}$$

$$m_t = m_0 \times e^{-kt} = 0.0100 \times e^{-(7.7 \times 10^{-6} \times 3.6 \times 10^5)}$$

(where 100 h ≡ 3.6×10^5 s) = 0.0100 × 0.0625 = 0.000 625 g (as above).

Revision questions

23.1 ${}^{14}_{7}N + {}^{4}_{2}He^{2+} \rightarrow {}^{17}_{8}O^{2+} + {}^{1}_{1}H$

23.2 (i) ${}^{3}_{1}H \rightarrow {}^{3}_{2}He^{+} + {}^{0}_{-1}e^{-}$

(ii) ${}^{6}_{3}Li + {}^{1}_{0}n \rightarrow {}^{3}_{1}H + {}^{4}_{2}He$

23.3 From the half-life of ${}^{210}_{84}Po$ (140 days) we calculate that $k = 5.7 \times 10^{-8}\,s^{-1}$.

(a) $t_{0.99} = 4.605/5.7 \times 10^{-8} \approx 8.0 \times 10^{7}$ s or ≈ 926 days.

(b) $t_{0.999} = 6.908/5.7 \times 10^{-8} \approx 1.21 \times 10^{8}$ s or 1,403 days.

23.4

(a)

${}^{222}_{86}Rn \rightarrow {}^{218}_{84}Po + {}^{4}_{2}He^{2+}$

Don't forget to check that the mass number on the LHS equals the combined mass numbers on the right. Similarly, the number of protons on the LHS equals the combined number of protons on the right.

(b) 7.6 days is two half-lives so 1.0 mg remains.

23.5 (i) Radon gas.

(ii) The harm caused depends upon the activity of the source (i.e. the half-life and mass of the radionuclide), the type of nuclear radiation, the distance of the individual from the source, the health and age of the individual, the area or part of body exposed, the duration of exposure, the degree of protection provided etc.

23.6 m = 1 g = 0.001 kg. $E = mc^2 = 0.001 \times (3.0 \times 10^8)^2 = 9 \times 10^{13}$ J. **Comment**: This is enough energy to keep 1 million 1 kW electric fires operating for a day. Table 13.3 (page 227) shows that 1 g of coal produces about 30 kJ of energy = 3.0×10^4 J. The ratio = $(9 \times 10^{13})/3.0 \times 10^4 = 3 \times 10^9$, so that the complete conversion of 1 g of nuclear fuel produces three billion times as much energy as the complete combustion of 1 g of coal. The secret to the huge amount of energy lies in the equation $E = mc^2$, since it contains the speed of light (itself a big number) squared. The huge energy output is a reflection of the extremely strong forces that hold the particles in the nucleus together, forces which are billions of times stronger than those in chemical bonds.

Unit 24

Exercises

24A

$CO(g) + O_2(g) \rightarrow CO_2(g)$ – oxidation of CO

$2NO(g) \rightarrow N_2(g) + O_2(g)$ – reduction of NO to N_2

24B

(i) $C_6H_{12}O_6(aq) + 6O_2(g) \rightarrow 6CO_2(g) + 6H_2O(l)$

(ii) Waste contains 600×10^5 mg glucose or $600 \times 10^2/180$ mol glucose. This needs

$$\frac{600 \times 10^2 \times 6}{180}\ \text{mol } O_2$$

or

$$\frac{600 \times 10^2}{180} \times 6 \times 32\ g O_2 \ \text{ in } 10^5\ dm^3$$

which is 0.64 g in 1 dm³, or 640 mg dm⁻³ (or ppm).

Revision questions

24.1

(a) iv **(b)** v **(c)** iii **(d)** ii **(e)** i

24.2

(a) Cl• radicals are regenerated in reaction (24.3), and are free to react with more ozone in (24.1).

(b) If the reaction was very fast, the depletion of ozone would be small because chlorine oxide (ClO) would be 'trapped' as chlorine nitrate and the regeneration of Cl• radicals are (equation 24.3 in book) would cease. Unfortunately, the rate of chlorine nitrate formation is not fast enough to do this!

24.3

(a) Good ozone is ozone in the stratosphere. It screens out harmful UVB light (see text). 'Bad ozone' is tropospheric ozone, which is found closer to the Earth's surface. It is harmful to health.

(b) 30 (NO^+), 42 ($C_3H_6^+$), 46 (NO_2^+), 48 (O_3^+) and less obviously, 44 (CH_3CHO^+). The production of ozone, O_3, is discussed in the text. Experiments show that methanal (HCHO) is produced as well as ethanal (CH_3CHO), but m/z for ($HCHO^+$) is the same as for NO^+, so mass spectrometry is not ideal in identifying methanal in this case.

24.4

(i) 0.012 g in 250 cm³ water is 0.048 g in 1 dm³ (or 48 ppm (mg dm⁻³).

0.048 g in 1 dm³ is $0.048/62 = 7.7 \times 10^{-4}$ mol dm⁻³

(ii) No. 48 ppm is above the maximum safe limit of 45 ppm.

(iii) %N = (2 × 14 × 100)/80 = 35 per cent

24.5

$[Pb^{2+}]$ = 10 mg dm⁻³

$$= \frac{10}{1000}\ g\ dm^{-3} = \frac{10 \times 1}{1000 \times 207}\ mol\ dm^{-3}$$

Rate = 4.8×10^{-5} mol dm⁻³ min⁻¹

24.6 (i) $C_8H_{18}(l) + 12^1/_2O_2(g) \rightarrow 8CO_2(g) + 9H_2O(l)$

(ii) 200 g octane is 200/114 mol.

This would produce 200/114 × 8 mol CO_2 at 100 per cent efficiency. At 95 per cent efficiency, 200/114 × 8 × 95/100 = 13.3 mol CO_2 (about 585 g).

24.7 Limestone is calcium carbonate, which reacts with the acid and neutralizes it.

24.8 (i) Amount in mol of CO = PV/RT

$= 1.5 \times 10^5 \times 10^{-3}/8.3145 \times 298$

$= 0.0606$

The flask is released into a sealed room of volume 100 m³. The concentration of gas (in mol m⁻³) is $n/v = 0.0606/100 = 0.000606$ mol m⁻³

$= 6.06 \times 10^{-4}$ mol m⁻³

$= 17$ mg m⁻³

(ii) This is well below the dangerous level. However, higher concentration will be experienced near the bottle.

24.9

$$\frac{100 \times 10}{1000} = 1\ g$$

24.10

(i) $$\frac{15 \times 1.25\ g}{1000} = 1.88 \times 10^2\ g$$

(ii) $$\frac{85}{1000 \times 96} = 8.9 \times 10^{-4}\ mol\ dm^{-3}$$

24.11 pH of lake before pollution is $10^{-8.40} = 4.0 \times 10^{-9}$ mol dm⁻³, pH of lake after pollution is $10^{-8.10} = 7.9 \times 10^{-9}$ mol dm⁻³. So the hydrogen ion concentration has *doubled*.

Unit 25

Exercises

25A

The time interval, 8 h, is double the half-life, so that the calculation is straightforward since eight hours earlier the concentration would have been four times greater, i.e. 200 mg dm⁻³. Alternatively, you could rearrange equation 25.1, page 450 to calculate $[A]_t/e^{-kt}$: the answer should be the same.

25B

We want $[A]_0$. Rearranging equation 25.2 gives $[A]_0 = [A]_t + kt$

Since $[A]_t = 100$

$[A]_0 = 100 + (6 \times 15) = 190$ mg of ethanol per 100 cm³ of blood. This is dangerously high (see Table 25.2).

Revision questions

25.1 (i) The data are plotted as follows:

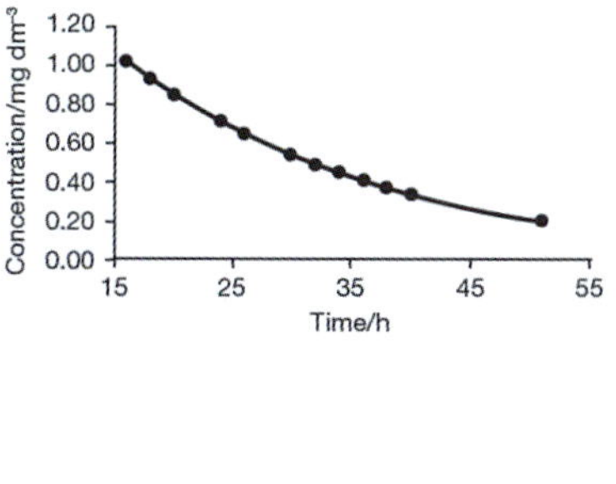

From the graph, the time it takes for the initial concentration of poison to fall by one half is about 15 h. This is true whatever the starting time (for a first-order decay, the half-life is independent of concentration). Conclusion: $t_{1/2} \approx 15$ h.

(ii) $k = 0.693/15 = 0.0462$ h.

$[A]_0 = [A]_t/e^{-kt} = 1.02/e^{(-0.0462 \times 16)} = 1.02/0.4752$, or approximately 2.1 mg dm⁻³.

25.2 We will work out the concentrations in mol per dm⁻³. Molar mass of ethanol = 46 u.

One microgram (μg) is one millionth of a gram. The amount of ethanol in 35 μg of ethanol is $(35 \times 10^{-6}/46)$ mol. The amount of ethanol per dm³ of *breath* is $(35 \times 10^{-6}/46) \times 10$, i.e. $[\text{ethanol}]_{breath} = 7.60 \times 10^{-6}$ mol dm⁻³.

One milligram (mg) is one-thousandth of a gram. The amount of ethanol in 80 mg of ethanol is $(80 \times 10^{-3}/46)$ mol. The amount of ethanol per dm³ of *blood* is $(80 \times 10^{-3}/46) \times 10$, i.e. $[\text{ethanol}]_{blood} = 1.74 \times 10^{-2}$ mol dm⁻³.

Concentration ratio = $1.74 \times 10^{-2}/7.60 \times 10^{-6} \approx 2{,}300$ (to three significant figures). Put another way, there are about 2,300 more molecules of ethanol in blood than in the same volume of breath.

25.3 (i) Reduction is 15 mg per 100 ml of blood per hour. Therefore two hours earlier the blood alcohol concentration is estimated to have been 79 + 15 + 15 = 129 mg/100 ml.

(ii) Urine: 1.34 × 80 = 107 mg alcohol per 100 ml of urine. Saliva: (109/100) × 80 = 87 mg per 100 ml of saliva

25.4

(i) Two, one in each of the O-CO units.

(ii) $C_{17}H_{21}NO_4$

(iii) 303 u

(iv) C=O

(v) parent molecule ion (M^+)

25.5 At first sight we could choose any two peaks. There is, however, a subtle reason why we should not choose the peaks due to the hydroxyl group, OH. Such peaks are greatly affected by hydrogen bonding. The extent of hydrogen bonding turns out to be dependent on the concentration of ethanol, and this interferes with a simple analysis of peak heights based on concentration. For the other peaks, such as those due to the C–H stretch and C–O bend, we would expect the absorbance and not the transmittance to be proportional to ethanol concentration, in accordance with the Beer–Lambert law.

25.6 (a) A generous allowance of 6 mg/100 ml is subtracted from the measurement. The statement would be that 'the sample contained not less than 94 mg of ethanol per 100 ml of blood'.

(b) 100 – (3 × 15) = 100 – 45 = 55 mg/100 ml.

25.7 Workers who make hats are called hatters. In the nineteenth century, hatters used mercury nitrate to help prepare the felt and this poison caused neurological symptoms that were misdiagnosed as madness. The phrase 'Mad as a Hatter' first appeared in print in 1829, long before Lewis Carroll introduced the Mad Hatter in *Alice's Adventures in Wonderland*.

Unit 26

Exercises

26A

(i) Ibuprofen

(ii) 2-(4-(2-methylpropyl)phenyl)propanoic acid

26B

(a) The hydrogen bonds between the water molecules are much stronger than the forces between the water and drug molecules (see page 180).

(b) By making the hydrochloride of the active ingredient (see page 181).

(c) The key here is to cover the active ingredient with a material that slowly allows water to enter and dissolve the drug, releasing the drug at a pre-determined rate. That is the answer in a nutshell, but for information I now outline how this may be done: (i) by mixing and compressing the drug with a harmless polymer; or (ii)

by putting the drug inside a polymer case that may be permeated by water after a set time. In the second method, you can sometimes hear the drug rattling around in the polymer capsule. In *bi-layer tablets*, the outer layer is designed to release the drug faster than the inner layer, so that the drug is released in the bloodstream for immediate and longer-term relief. Common polymers used as a coating are based on cellulose: the release of drug is triggered by the slow penetration of water through the polymer.
(d) The most important feature of experimental design is to control the variables. We assume that the mass of drug in each capsule is the same. To measure the rate at which the capsule releases drug X we first identify an analytical method to determine the concentration of X in water: this might be a titration or analysis by chromatography or UV spectroscopy. Experiments might be carried out as follows.
A capsule is gently lowered into (say) 250 cm^3 of water at temperature *T*. After (say) one minute the capsule is withdrawn and the solution stirred and analysed for X. The experiment is repeated with a new capsule with the measurement taken at (say) five minutes. Repeating the experiment at further intervals will show how quickly the capsule dissolves. The experiment may be repeated at different temperatures and at different pHs.

26C

The oral lethal dose for a 43 kg person is 143 × 43 = 6149 mg or 6.149 g. To achieve this dose would require 6.149/0.5 or about twelve tablets.

26D

The compound slowly hydrolyses in a damp atmosphere and ethanoic acid (the main ingredient of vinegar) is formed:

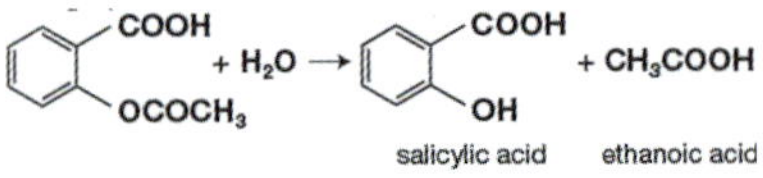

26E

There might be several possible explanations (!), but the true explanation is that while bacteria produce their own 4-aminobenzoic acid, human beings obtain it through their diet.

26F

The chiral centre is marked with a star. The two enantiomers (S and R) are:

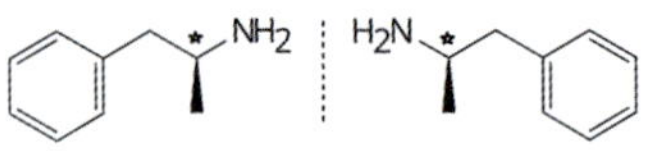

Note that we are not showing the CH_3 group coming out of the plane of the page. The black triangular bond implies that the methyl group is there.

26G

$2HCl(aq) + Mg(OH)_2(s) \rightarrow MgCl_2(aq) + 2H_2O(l)$

26H

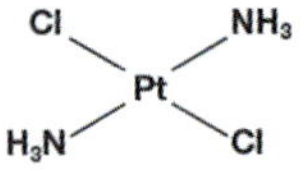

Revision questions

26.1
(i) $3H^+ + 3HCO_3^- \rightarrow 3H_2O + 3CO_2$
(ii) Production of CO_2 agitates the system, which helps dissolve the aspirin.
(iii) The graph levels out at around 0.81 g (your answer may be slightly different). This is 0.81/44 = 0.018 mol CO_2.
(iv) From **(i)** this is equivalent to 0.018/3 = 0.006 mol citric acid, which is 0.006 × 210 = 1.260 g citric acid in two tablets or 0.630 g per tablet.
(v) Carbon dioxide dissolves in water; vigorous effervescence results in some water being lost.
26.2 m = 151 u; a: N–H stretch, b: O–H stretch, c: C=O stretch
26.3 Identified by parent ion as sulfanilamide. $C_6H_8N_2O_2S$

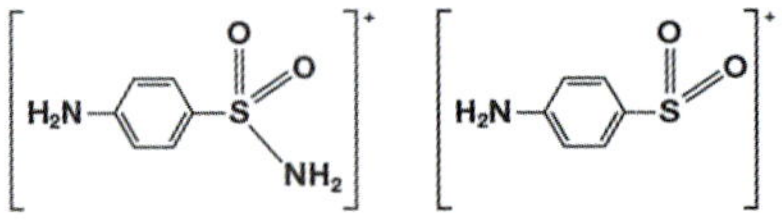

Parent ion (172) and (156).

26.4 Sulfanilamide prevents bacteria growing by stopping 4-aminobenzene being used in cell replication. Addition of 4-aminobenzoic acid compensates for this.
26.5 [A] = KI (potassium iodide), [B] = NH_3 (ammonia), [C] = $AgNO_3$ (silver nitrate), [D] = KCl (potassium chloride).
26.6

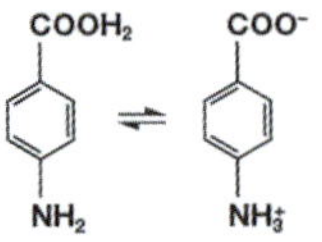

26.7
There are a number of ways of drawing smoke from a lit cigarette and condensing the tar etc. See https://science.cleapss.org.uk/resource-info/the-smoking-machines-demonstration.vid

Unit 27

Exercises

27A

(i) Hot gases are less dense than cold gases, so the hot air rises upwards. Gravity opposes this upward motion and drags the flame down, causing the characteristic tear shape. Experiments show that in a gravity free environment such as the Space Station, a flame is much more spherical.
(ii) HĊ· :C=C:

27B

(i) Pressure difference (and chimney draught) increases with chimney height because the taller the chimney the greater the column of hot air it sustains before venting to the outside.
(ii) A hotter fire makes the gases hotter, so increasing the pressure difference.

27C

(i) No. CO is only produced by incomplete combustion of hydrocarbon fuels. Electricity is made from fossil fuels but the combustion occurs in the power station and not at the point of use of the electricity.
(ii) CO reacts with the body's haemoglobin, converting it to the stable compound carboxyhaemoglobin which, since it is unable to carry oxygen, causes oxygen starvation and even death.
(iii) $C_{25}H_{52}(g) + 38O_2(g) \rightarrow 26H_2O(g) + 25CO_2(g)$ + heat. The probability of thirty-eight O_2 molecules successfully colliding with one hydrocarbon molecule is zero, so it is clear that the reaction cannot proceed in one collision. As we have seen, the mechanism of all burning reactions is complicated, consisting of many elementary steps and involving many reaction intermediates.
Balancing combustion reactions for big hydrocarbons can be tedious and there is a general formula to help which works for all compounds that contain carbon, hydrogen and oxygen – but no other elements:

$$C_aH_bO_c + \left[a + \frac{1}{4}b - \frac{1}{2}c\right]O_2 \rightarrow aCO_2 + \frac{1}{2}b\,H_2O$$

For $C_{25}H_{52}$, *a* = 25, *b* = 52 and *c* = 0 and so

$$\left[a + \frac{1}{4}b - \frac{1}{2}c\right]O_2 = 25 + 13 + 0 = 38O_2$$

(iv) This confirms the important role of the free radicals in combustion. The reaction of CF_3Br with both H• and HO• radicals stops the combustion chain reaction in its tracks. CF_3Br makes a very effective fire extinguishing agent, but like CFCs it is a potent destroyer of the ozone layer and its use has been banned.
(v) The shortage of O_2 produces a less hot flame, since the exothermic reaction of HO•(g) + CO(g) occurs on a smaller scale. In fact, the flame temperature also falls if there is too much O_2 present: a 2:1 ratio of CH_4 to O_2 is ideal.

27D

(i) 34°C, although white P smoulders (without a flame) at lower temperatures.
(ii) Yes, since 5.5°C is higher than its flashpoint (–11°C).
(iii) The gas is initially 100 per cent CH_4. This is above its flammability limit and so the gas in the can cannot be ignited. However, as soon as the valve is opened the CH_4 gas escapes from the can and quickly mixes with outside air creating a flammable CH_4/air mixture which burns with a flame outside the can at point A. Seconds later, the CH_4 pressure has fallen to atmospheric pressure and so some air enters the can: at soon as the mixture composition in the can reaches 15 per cent (Table 27.1) the mixture explodes. If the demonstration is repeated with pure hydrogen, the delay between the flame and explosion is shorter because the mixture has only to reach 74 per cent H_2 before exploding and also because H_2 escapes at a faster rate than heavier CH_4 molecules.

27E

(i) Burning wet wood wastes some of the heat energy of a fire, as some of the energy is used to vaporize the water in the wood or the wood sap in the wood. Dry wood also occupies a smaller volume than wet wood, which on an industrial scale means that dry wood produces more energy per kg than wet wood.
(ii) The smell is caused by unburnt fuel (organic compounds, including those in tar). As the temperature of the stove rises, more of organic compounds are burned in the fire itself, thereby improving its efficiency and generating more heat. It is for this reason that wood stoves should be made very hot as quickly as possible after ignition of the fuel. Thereafter, the room temperature should be controlled by the amount of wood in the stove rather than throttling back the air too much or the efficiency of the stove will be reduced.

Revision questions

27.1
CH_3•(g), O•(g) and HO•(g) radicals are produced during combustion, which then take part in a number of reactions. In one sequence of reactions, CH_3•(g) is converted into the formyl radical •CHO(g) which, in turn, reacts with HO• radicals producing water. The hydroxyl radical, HO• has a further important role because it converts CO to CO_2 via reaction (27.4).
27.2
(i) and **(ii)** No, in both cases the temperature is below the flash point.
(iii) The 19 per cent mixture will ignite because it is (just) within its flammability limits, but if the percentage of ethanol in the fuel mixture exceeds 19 per cent, the flame will go out.
(iv) 400°C is above its autoignition temperature so it will ignite inside the container. The container is likely to have contained insufficient air for complete combustion of the ethanol as soot has been formed.
27.3
(i) To make the detection of natural leaks easier. The odour of $(CH_3)_3CSH$ can be detected even if its concentration is only 1 part per million by volume (1 ppmv).
(ii) The principle of such an investigation is that the effect of rust upon t-butyl mercaptan is compared with a 'control'. Dry air containing a known concentration of t-butyl mercaptan is left in contact with a known mass of fine powdered rust (of a particular particle size) in a sealed and gas-tight container for (say) one month. The gas above the rust is then analysed by passing it into a gas chromatograph (GC) to

measure the concentration of t-butyl mercaptan. If the concentration has fallen, this suggests that chemical reaction or physical adsorption has taken place. Finally, a repeat experiment is carried out with mercaptan-free dry air: this is the control.

(iii) Because it contains CO. All gases have the potential to cause asphyxiation – death due to lack of oxygen – if their presence in air is accompanied by lower oxygen concentrations, but coal gas has an additional risk due to the presence of toxic CO.

(iv) Incomplete combustion of any fuel, including coal and natural gases, produces CO even if there was none in the original fuel. Poor ventilation and blocked chimneys raise the risk of poisoning.

27.4 The Coanda effect causes flame attachment.

27.5 The escalators operated at a slope, so causing flame attachment. They were also wooden, a source of fuel which led to flashover from methane produced from the decomposing wood. Finally, the escalator walls (sides) helped to contain the fire, so building up heat and methane concentrations which made flashover even more likely.

27.6

$\Delta H^{\ominus}$ for reaction = $\Delta_c H^{\ominus}$ (C_4N_2 in O_2) = [(0) + (4 × −110.5) − [(626.2) + 0] = −1068.2 kJ mol^{-1}.

27.7

(i) The dots signify spare electrons, i.e. that both species are free radicals.

(ii) $\Delta_f H^{\ominus}$ (CH_4(g)) = −74.87 kJ mol^{-1}; $\Delta_f H^{\ominus}$ (CH_3•(g)) = 145.69 kJ mol^{-1}; $\Delta_f H^{\ominus}$ (H•(g)) = 217.998; $\Delta H^{\ominus}$ for reaction = (217.998 + 145.69) − (−74.87) = 438.6 kJ mol^{-1} (At data from NIST)

27.8

(i)

(a) C(s) + H_2O(g) → CO(g) + H_2(g) (Note that CO and H_2 are both fuels)

(b) C(s) + CO_2(g) → 2CO(g)

To demonstrate that reaction **(b)** takes place, pass CO_2 slowly over very hot charcoal in a heat resistant pyrex tube:

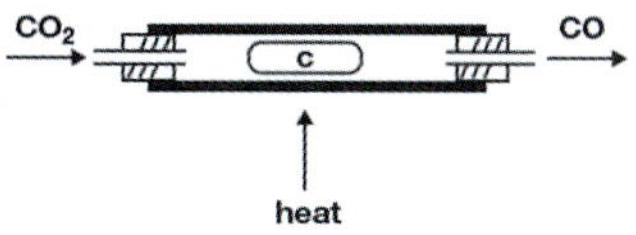

Collect the emerging gas by passing it through a rubber tube into water and allow the bubbles to collect in a bottle full of water. The CO displaces the water in the bottle. Ignore the first collection (which will contain mostly CO_2). CO can be detected using a flaming splint (see identification of gases, Unit 6). It burns with a blue flame.

(ii) The fire quickly consumes most the oxygen (air) in the room. Subsequent combustion is incomplete, and CO levels will be high.

(iii) 1200 parts per million by volume is (1200/1,000,000) × 100 = 0.12 per cent by volume. The graph shows that this threshold is reached in a few seconds. To put this into perspective, a wood burner's exhaust gases contain about 5,000 ppmv or 0.5 per cent CO by volume. In a well serviced burner, nearly all of that goes up the chimney, but the graph shows that the CO concentrations in a room fire are twenty times that level only two minutes after ignition.

(iv) A flashover e.g due to a LPG cylinder rupturing.

27.9

If the percentage of O_2 was zero, the fuel cannot catch fire. Simplistically then, we might predict that increasing the percentage of O_2 would *lower* the lower flammability limit so that burning is capable of being sustained at lower fuel concentrations. Experiments show that this is the case. In fact, not only does the lower limit goes lower but the upper limit goes higher, i.e. the flammability *range* increases with O_2.

Unit 28

28A

(a) CO_2 and N_2 are colourless so they do not absorb visible light. CO_2 but not N_2 absorbs IR radiation, as demonstrated by the fact that CO_2 has an IR spectrum (see Figure 24.3, page 440) but N_2 does not.

(b) The greenhouse effect in the atmosphere involves the absorption re-emission of IR. Therefore, strong absorbers of IR are always powerful greenhouse gases.

(c) H_2O is a strong absorber of IR light.

(d) The concentration of CO_2 is massively greater than that of CFCs. Note that thanks to the Montreal Protocol, the concentration of CFCs in the atmosphere is falling.

28B

Probably not. When considering the CO_2 emissions produced by a company or process, the emissions connected with the whole chain of operations must be counted. Breconshire foods might cook using electric ovens that produce no CO_2 when they are being used, but if the electricity that passes through the national grid for the ovens was generated using fossil fuels, the company's 'carbon footprint' – a count of all emissions associated with the company's work – shows that it is not overall at net-zero. Similarly, if the food ingredients supplied to the company are transported to the factory by diesel vehicles, the associated production of CO_2 needs to be counted.

Revision questions

28.1

All three are unfair. In (a) one bottle has a solid in it. In (b) the lamp is closer to one bottle. In (c) the N_2 bottle has slightly thinner glass and the light intensity getting through the CO_2 bottle is correspondingly reduced. Note that the light source in (c), although dimmer than in the other experiments, affects both bottles equally.

28.2

The environmental impact of cruise ship emissions is considerable but there are plans for some cruise ships to use sails and solar panels in the future. A cruise ship carries its own hotel, so that the comparison should be with a flight, and a hotel-based vacation. To start your study see https://www.energymonitor.ai/sectors/transport/weekly-data-is-it-better-to-take-a-cruise-than-fly/?cf-view&cf-closed

28.3

See details of stratospheric cooling in book.

28.4

The greater proximity of Venus to the Sun increases the intensity of sunlight reaching the planet's surface, so causing a higher surface temperature. This increase is amplified by the enhanced greenhouse effect caused by very high atmospheric CO_2 concentrations.

28.5

9.45 × 10^{14} / 8 × 10^{9} = 1.25 × 10^{5} watt hours, i.e. roughly 100 *kilo*watts per person per hour. That's the equivalent of the electricity consumption of 50–100 small electric heaters per hour!

28.6

We start with 100 J. The energy available to propel the car is 100 J × 0.95 × 0.90 = 85.5 J, an efficiency of 85.5 per cent. The efficiency of a petrol car is about 20 per cent. Hydrogen cars, using fuel cells, will be less efficient, as they involve an additional stages, including expenditure of energy in the compression of hydrogen for storage in the car. For more details, see our website: https://www.bloomsburyonlineresources.com/chemistry.

28.7

T > 31°C (the critical temperature of CO_2), so the CO_2 will be a supercritical fluid and therefore more soluble.

28.8

The H_2 mops up the OH radicals that would otherwise destroy methane. This extends the lifetime of methane so that the impact of the global warming of CH_4 increases. Studies indicate that the equivalent GWP of H_2 is about 10. In other words, the effect of H_2 is equivalent to a gas that directly absorbs IR with a GWP of 10. However, this would only be important if the production of H_2 were vastly scaled up as part of a drive towards a 'hydrogen economy', when any leaks of H_2, although difficult to stop, would assume a much greater importance.

Unit 29

29A

(i) 525 nm.

(ii) The colour cheese shows that the peak corresponds to green light. The opposite segment indicates the transmitted light, which is red.

Revision questions

29.1

Surface area of a sphere = $4\pi r^2$; volume of sphere $= \frac{4}{3}\pi r^3$

$$R = \frac{\text{Surface area}}{\text{Volume}} = \frac{4\pi r^2}{4/3\pi r^3} = \frac{3}{r}$$

In other words, the ratio is inversely proportional to r. To compare the two cases we express the radii in the same units of metres (m) and R then has the units of m^{-1}.

r = 1 cm = 0.01 m so R = 3/(0.01) = 300 = 3 × 10^{2} m^{-1}
r = 10 nm = 10^{-8} m so R = 3/(10^{-8}) = 3 × 10^{8} m^{-1}

The nanoparticle has a 1 million greater ratio of surface area to volume.

29.2

The nanoparticles have a huge surface area to volume, exposing more of an area to incoming UV light; therefore, they absorb more UV light than bulk material. They will also cause more scattering of UV light, again reducing the intensity of UV reaching the skin. The use of nanoparticles also creates a sunscreen mixture that is less thick and so it is easier to apply to the skin. (It is also largely colourless, whereas sunscreen containing bulk ingredients is often white).

29.3

The different colours show that the energy transitions involved for atoms in bulk gold and for gold nanoparticles are different.

29.4

The expected parent ion = $Au_{133}(SR)_{52}^+$. The formula of the thionate ligand SR = $(CH_3)_3CC_6H_4S$. Using the atomic masses in the Periodic Table at the front of the book gives $m((CH_3)_3CC_6H_4S)$ = 34791.171u. The atomic mass of Au = 196.967 u. Therefore, the molecular mass of $Au_{133}(SR)_{52}$ = 34791.171 u and m/z of $Au_{133}(SR)_{52}^+$ ion = 34791.171u.

29.5

See p.**75**. In graphene, the delocalized fourth valence electron of the carbon atoms (it is formally described as a π electron) sits on the surface of the honeycomb structure. Graphite is brittle because its layers are relatively weakly held and can be sheared away upon the application of a horizontal force. Graphene bends but since all the C-C hexagons are strongly held together, it flexes but does not break easily. When graphene oxide is formed, the C–O bonds use up the delocalized electrons that allow the graphene to conduct electricity.

Glossary

absolute error The difference between the measured value and the true value.

absolute zero (0 K or −273.15°C) The lowest possible temperature.

absorbance A measure of the extent to which a sample in a cell absorbs light at a particular wavelength. Defined as log (I_0/I).

absorption (of light) The taking in of a photon of light by an atom, molecule or ion.

abundance of an isotope The percentage (in terms of numbers of atoms) of an isotope in a sample of an element.

accuracy The closeness of measurement(s) to the true value.

acid A substance that produces hydronium ions (H_3O^+ or $H^+(aq)$) in solution.

acidic solution A solution with a pH below 7 at 25°C.

acidity constant (K_a) The equilibrium constant for the ionization of an acid in water.

activation energy The minimum energy required by colliding reactant molecules so that chemical change occurs.

addition reaction A reaction in which a carbon–carbon multiple bond is broken and parts of another molecule are added to each carbon atom of the multiple bond.

alcohol A series of organic compounds in which an alkyl group is bonded to −OH, e.g. CH_3OH.

aldehyde One of a series of organic compounds that contain a −CHO group, e.g. CH_3CHO.

alkali A water-soluble base.

alkali metal A Group 1 element.

alkaline earth metal A Group 2 element.

alkaline solution A solution of a base – a solution with a pH above 7 at 25°C.

alkane A hydrocarbon of general formula C_nH_{2n+2}, e.g. CH_4.

alkene A hydrocarbon of general formula C_nH_{2n}, e.g. C_2H_4.

alkyne A hydrocarbon of general formula C_nH_{2n-2}, e.g. C_2H_2.

allotropes Alternative forms of an element in which the atoms are joined together in different ways, e.g. diamond, graphite and buckminsterfullerene.

alpha particle A helium ion, produced in some types of radioactive decay.

amine Compounds in which organic groups replace the hydrogen atoms in ammonia, e.g. CH_3NH_2, $(CH_3)_2NH$ and $(CH_3)_3N$.

amphoteric substance A substance is amphoteric when it shows reactions that are characteristic of both acids and bases.

antibacterials Substances that stop bacteria from growing.

antibiotic A substance or organism that attacks and kills another organism.

antibiotics A substance or organism that attacks and kills bacterial infections.

anode In an electrochemical cell (or battery), the electrode at which oxidation takes place.

aromatic compound An organic compound that contains a benzene ring, e.g. C_6H_5Cl.

aryl group An aromatic group, e.g. $-C_6H_5$.

atmosphere (1) The gaseous envelope around a planet. (2) A unit of pressure, equal to 101 kPa.

atom An atom is the smallest part of an element that chemically reacts like a bulk sample of the element.

atomic mass (of an element) The average mass of one atom of that element in atomic mass units; symbolized *m*(element).

atomic mass unit (u) A very small unit of mass ($1u \approx 1.66 \times 10^{-24}$ g).

atomic number (*Z*) The number of protons in an atom.

autoignition temperature The minimum temperature, in the absence of a flame, at which a fuel ignites spontaneously in air.

Avogadro's constant (N_A) The number of objects per mole of objects ($N_A = 6.022 \times 10^{23}$ mol^{-1}).

Avogadro's law (or principle) Equal volumes of all gases at the same pressure and temperature contain the same number of molecules (or moles).

base A substance that accepts a proton from an acid, forming a salt and water only.

basicity constant (K_b) The equilibrium constant for the ionization of a base in water.

Beer–Lambert equation (law) An equation (applying at one wavelength) that relates the absorbance A of a substance to the concentration of the absorber *c*, the cell path length *b* and the molar absorption coefficient ε of the substance. Formally expressed as $A = \varepsilon cb$.

beta particles Fast moving electrons produced within the atomic nucleus during radioactive decay.

blood alcohol concentration (BAC) The concentration of ethanol in blood, usually expressed in units of mg of alcohol per 100 ml of blood.

boiling point Defined as the temperature at which the vapour above a liquid equals the external atmospheric pressure.

bond dipole The dipole belonging to a particular bond in a molecule. See molecular dipole.

bond enthalpy ($\Delta H^\ominus$A–B) A measure of the strength of a bond. The standard enthalpy change when 1 mol of bonds are broken, all species being in the gas phase.

buffer (solution) A solution that resists changes in pH upon the addition of acid, base, or upon dilution.

Carbon capture and storage (CCS) Any technology that allows the capture and storage of greenhouse gases (particularly CO_2).

carboxylic acid One of a series of organic compounds that contain a –COOH group, e.g. CH_3COOH.

catalyst A substance which, although it is not consumed, speeds up a chemical reaction.

cathode In a battery or electrochemical cell, the electrode at which reduction takes place.

chain reaction A sequence of self-sustaining chemical reactions or physical processes.

chemiluminescence The emission of light that accompanies a chemical reaction.

chiral centre (in optical Isomerism) A key stereo molecular feature in a molecule that creates optical isomers that are non-superimposable mirror images of each other. A common chiral centre is a carbon atom (C*) bonded to four different molecules or groups.

chromatography A technique used to separate mixtures.

colloid Particles (1–1000 nm) of substances (size range 1–1000 nm) in solution which are smaller than those in suspensions but larger than individual molecules or ions.

colour A physical property of a material that is controlled by the wavelengths of light that are reflected or absorbed by that material.

common ion effect The precipitation of a species 'AB' in solution, due to the swamping of the mixture with A or B ions from another source.

competing reactions Reactions that consume the same reactant but which produce different products.

complex (inorganic chemistry) Species consisting of ligands bonded to a central metal ion, e.g. $[Ag(NH_3)_2]^+$.

compound A substance consisting of atoms of different elements joined together in a definite ratio.

conjugate acid Species made when a base gains a proton.

conjugate base Species made when an acid loses a proton.

coordination number The number of nearest neighbours. These are ions of opposite charge in an ionic crystal, or the number of atoms that are attached to the central metal ion in a complex.

critical point In a phase diagram, the critical point lies at the intersection of the critical temperature and critical pressure. The liquid phase does not exist above the critical point.

critical pressure The pressure required to liquefy a substance at its critical temperature.

critical temperature The temperature above which a substance cannot be liquefied, whatever the applied pressure.

decarbonization refers to activities that reduce carbon dioxide (CO_2) and other global warming gases from the atmosphere

decay constant (*k*) The first-order rate constant for the decay of a radionuclide.

deionized water Pure water obtained by passing tap water through an ion-exchange resin.

delocalization The 'spreading' out of electron density over a molecule or ion.

density Mass per unit volume of substance; common unit g cm^{-3}.

diatomic molecule A molecule consisting of two atoms, e.g. HCl and O_2.

dipole A partial positive and equal partial negative charge separated by a bond.

dissociation Another word for 'break-up'.

distillation The separation of a mixture, making use of the fact that its components have different boiling points.

electrochemical cell Two electrodes in electrical contact with each other and with a solution (the electrolyte).

electrode An experimental arrangement at which oxidation or reduction occurs at the surface of a conductor.

electrolysis The decomposition of a dissolved or molten substance using electricity.

electron A negatively charged sub-atomic particle.

electronegativity The measure of the ability of an atom to attract bonding electrons to itself when it is joined to another atom by a covalent bond.

element A substance in which all the atoms have the same atomic number. Elements cannot be broken down into anything simpler by chemical reactions.

elementary reaction An individual reaction step; part of an overall reaction mechanism.

emission (of light) The giving out of a photon of light by an atom, molecule or ion.

empirical formula A chemical formula that uses the smallest whole number ratio of the atoms present in a compound.

emulsifying agent A substance that stabilizes an emulsion so allowing emulsification.

emulsion A colloidal solution consisting of one liquid dispersed in another.

endpoint In a titration, the point at which enough solution has been added that the indicator begins to change colour. Ideally, the endpoint should equal the equivalence point.

endothermic reaction A reaction in which thermal energy ('heat') is taken in.

energy value (of a fuel) The amount of heat produced when 1 g of fuel is completely burned. Units are $kJ\ g^{-1}$.

enthalpy The energy (or 'heat content') of a substance at constant pressure.

enthalpy of electron gain ($\Delta_{eg}H^{\ominus}$) The standard enthalpy change when one mole of anions is formed, all species being in the gas phase.

enthalpy of fusion ($\Delta_{fus}H^{\ominus}$) The standard enthalpy change when 1 g (or 1 mol) of a liquid is formed from 1 g (or 1 mol) of solid without a change in temperature; units are $J\ g^{-1}$ or $J\ mol^{-1}$.

enthalpy of vaporization ($\Delta_{vap}H^{\ominus}$) The standard enthalpy change when 1 g (or 1 mol) of a vapour is formed from 1 g (or 1 mol) of liquid without a change in temperature. Units are $J\ g^{-1}$ or $J\ mol^{-1}$.

entropy (S) A thermodynamic quantity which is a measure of the degree to which energy spreads out during a physical or chemical change.

equilibrium A state of dynamic balance that is produced by two opposing processes (e.g. forward and back reactions) that are occurring at the same rate.

equilibrium composition A particular set of concentrations of reactants and products at equilibrium at that temperature.

equilibrium constant (K) A ratio, symbolised K (or K_p) that contains the equilibrium molar concentrations (or equilibrium partial pressures) of the reactants and products raised to the coefficients shown in the chemical equation for the reaction.

equivalence point In a titration, the point when exactly enough solution has been added from a burette to completely react with the solution in the flask.

ester One of a series of organic compounds which contain the –COOR group, e.g. $CH_3COOC_2H_5$.

excited state An energy state of a molecule, atom or ion other than the ground state.

exothermic reaction A reaction in which thermal energy ('heat') is given out.

explosion A violent increase in gas pressure in a confined space.

extensive property A property of a substance that depends on the amount of substance present, e.g. volume.

first-order reaction (decay) A reaction which follows first-order kinetics.

flammability limits The range of concentration of fuel in air over which ignition is possible.

flash point The lowest temperature at which vapour can be ignited by a naked flame.

fluorescence A type of light emission.

freezing point The temperature at which a liquid turns into a solid.

frequency The number of waves of light passing a fixed point per second.

gas chromatography Mass Spectrometry (GC-MS) An analytical technique used to separate and then identify the constituents in a mixture.

gamma radiation Short wavelength electromagnetic radiation emitted during nuclear change, e.g. during radioactive decay.

Gibbs energy (G) A hybrid physical quantity that is related to the enthalpy, entropy and temperature of a system by the relationship $G = H - TS$.

greenhouse effect A process where certain gases in the Earth's atmosphere trap thermal energy ('heat') from the sun, warming the planet.

greenhouse gas A gas, such as CO_2 or CH_4, that (via the Greenhouse Effect), thermally insulates the system (the earth, a car or a garden greenhouse), increasing steady-state temperatures.

ground state The lowest energy state of an atom, molecule or ion.

half-life ($t_{1/2}$) The time taken for the concentration of a reactant to fall to half its original value.

halogen A Group 17 element.

hard water Water that contains dissolved magnesium or calcium salts, and which does not lather well with soap.

Henry's law This states that 'the solubility of a gas is proportional to the pressure of the gas above the liquid'.

hydration of ions The attachment of water molecules to a central ion.

hydrogen bond A hydrogen atom bridging two strongly electronegative atoms (N, F, Cl or O).

hydrolysis The reaction of a substance with water.

hydrophilic Water-attracting.

hydrophobic Water-repelling.

ideal gas An ideal gas is one whose pressure, volume and temperature may be predicted with perfect accuracy by the ideal gas equation ($PV = nRT$).

immiscible liquids Liquids that do not dissolve in each other (i.e. they do not mix).

inert pair A pair of electrons that form part of the outer (or valence) shell of electrons in an atom, but do not take part in bond formation.

inert pair effect This refers to the observation that the heavier elements of Groups 13–17 have oxidation states that are lower by two than the maximum predicted for their group.

intensive property A property of a substance that does not depend on the amount of substance present, e.g. temperature.

intermolecular forces Forces between molecules.

ion An atom or group of atoms that possesses a positive or negative charge.

ion exchange The exchange of one type of ion in solution with another.

ionic product constant (K_w) The product of the hydronium and hydroxide ion concentrations for any aqueous solution; the product varies with temperature.

ionization The formation of an ion.

ionization energy (or enthalpy) The standard energy change when one electron is removed from a gaseous atom, molecule or ion.

isomers Compounds that have the same molecular formula but different structures.

isotopes Atoms with the same number of protons but different numbers of neutrons.

isotopic mass The mass of one atom of an isotope in atomic mass units; symbolized m(isotope).

kelvin (K) The SI unit of temperature. 0°C is equivalent to 273.15 K.

ketone Compounds that contain the carbonyl group (–C=O) between organic groups, e.g. CH_3COCH_3.

lattice enthalpy ($\Delta_{latt}H^{\circ}$) The standard enthalpy change when 1 mol of a crystal lattice is broken into isolated gaseous particles.

Le Chatelier's principle States that the concentrations of reactants and products in an equilibrium mixture will alter so as to counteract any changes in pressure, temperature or concentration.

ligand A molecule or anion that attaches itself to the central metal ion in a complex.

light A electromagnetic wave that travels at 3.00×10^8 metres per second.

limiting reagent The reactant that, not being in excess, controls the amount of products that are made.

London dispersion forces Weak attractive intermolecular or interatomic forces.

lone pair A pair of electrons in the outer (valence) shell of an atom not shared with another atom.

mass number (A) The sum of the number of protons and neutrons in an atomic nucleus.

mass spectrometer An instrument that is used to find the isotopic masses and abundance of isotopes in a sample of an element. Also used to provide information about the molecules present in a mixture.

mass spectrum When a molecule or atom is ionized, a characteristic group of ions with different masses is made. A plot of the relative abundance of these ions against their mass by charge ratio (m/e) is called a mass spectrum.

mean bond enthalpy (ΔH°(A–B)) An average bond dissociation energy for a bond A–B, averaged over several different types of molecules.

melting point The temperature at which a solid melts (see normal melting point).

metalloid Elements with properties that lie between those typical of metals and non-metals. Also known as semimetals.

miscible Liquids that dissolve in each other (i.e. mix).

molar absorption coefficient Symbolized ε; a measure of the intensity of absorption of a substance at a particular wavelength. See Beer–Lambert equation.

molar solubility The concentration of a substance (in units of moles per cubic decimetre, i.e. mol dm^{-3}) in a saturated solution of that substance at that temperature.

mole (mol) The SI unit of amount of substance.

molecular dipole The overall (net) dipole of the molecule, obtained by a vector sum of the individual bond dipoles. In a diatomic molecule, the bond dipole and the molecular dipole are the same thing. See also bond dipole.

mole fraction The number of moles of a substance in a mixture divided by the total number of moles of all the components of the mixture.

molecule A neutral particle consisting of two or more atoms combined in a definite ratio.

nanoparticles Particles between 1 to 100 nm in all dimensions.

nanoPutian A nanoputian is a nanoparticle molecule that resembles the shape of a person.

'Net zero' A commitment of a company, household or government to reduce its net greenhouse gas emissions to zero.

neutral solution A solution with equal numbers of hydronium and hydroxide ions, with a pH equal to 7 at 25°C.

neutralization reaction The chemical reaction in which a base 'cancels out' an acid, producing a salt and water only.

neutron A neutral particle found in atoms of all elements (except for protium $^{1}_{1}H$).

Nitrile (or cyanide) Nitriles are organic molecules containing the cyanide group, C≡N.

NMR Nuclear magnetic resonance – a type of spectroscopy.

noble gas A Group 18 element.

normal boiling point Also known as normal boiling temperature – the temperature at which the vapour pressure of a liquid equals 1 atm.

normal freezing point Also called normal freezing temperature – the temperature at which a liquid freezes under a pressure of 1 atm.

normal melting point Also called melting temperature – the temperature at which a solid melts under a pressure of 1 atm. Equals the normal freezing point.

nuclear fission is a type of nuclear reaction where a large, unstable atomic nucleus breaks up into two or more smaller nuclei, releasing energy, neutrons and gamma rays.

nuclear fusion is a type of nuclear reaction where two light atomic nuclei combine to form a single heavier one, releasing energy.

nucleon A name for the particles in the nucleus (i.e. for neutrons and protons).

nuclide An *atom* of a particular isotope.

optical isomerism See Chiral Centre.

orbital The volume in which an electron is to be found with a certain probability (e.g. 90 per cent). Each orbital holds up to two electrons, which may be thought of as spinning in opposite directions.

osmosis The flow of solvent (usually water) through a semipermeable membrane.

osmotic pressure The pressure which must be applied across a semipermeable membrane in order to stop osmosis.

oxidizing agent A substance which causes another substance to be oxidized in a chemical reaction and is itself reduced.

pascal (Pa) SI unit of pressure, 101,325 Pa = 1 atm.

per cent transmittance The percentage of light at a particular wavelength that is allowed through a sample in a spectrometer; defined as $(I/I_0) \times 100$.

pH The negative logarithm of the hydronium ion concentration in an aqueous solution. At 25°C, pure water is neutral with a pH of 7.0.

phase diagram A plot of the equilibrium states of substances or mixtures at different pressures and temperatures.

phenol An organic compound in which an –OH group is directly attached to an aromatic ring, e.g. C_6H_5OH.

photolysis The decomposition of a substance using light as the energy source.

photon A 'particle' of light; it has no mass and consists of pure energy.

pK_a The negative logarithm of an acidity constant.

pK_b The negative logarithm of a basicity constant.

polar molecule A molecule having one end that is positively charged while the other is equally negatively charged.

polyatomic molecule A molecule that contains more than two atoms, e.g. O_3, H_2O_2.

precipitation The formation of an insoluble substance in solution.

precision The closeness of repeat measurements.

proton A positively charged particle in the nucleus of all atoms; also an isolated hydrogen ion.

pyrolysis The decomposition of a substance using heat as the energy source.

radicals (free radicals) Atoms or molecules that possess unpaired electrons e.g. •OH

radioactivity The spontaneous disintegration of nuclei.

radioisotope An isotope that is radioactive.

radionuclide An *atom* of a radioisotope.

random errors Errors that cause repeat measurements to be scattered (up and down) around the true value.

rate constant (k) The rate of a reaction (at a particular temperature) when the reacting substances are at a concentration of 1 mol dm^{-3}; k values measure the relative speeds of different reactions.

rate expression An expression, found by experiment, that relates the reaction rate to the rate constant and to the concentrations of the reactants.

rate of reaction The speed of a reaction at an instant; usual units are mol $dm^{-3}\ s^{-1}$.

reaction 'allowed to go' A reaction, starting from pure reactants and at temperature T, is thermodynamically allowed to go if the standard Gibbs energy change for the reaction at temperature T is negative. A negative standard Gibbs energy change also means the equilibrium constant for the reaction (at T) is greater than one.

reaction intermediate A species that is produced and consumed during an overall reaction.

reaction mechanism The collection of elementary reactions making up an overall reaction.

Reaction that goes to completion A reaction goes to completion when the limiting reagent is entirely converted to the products shown in the chemical equation for that reaction

redox couple The oxidized and reduced species in a half-reaction.

Redox reaction In a redox reaction, one reactant is oxidized, so causing the simultaneous reduction of another reactant.

reducing agent A substance that causes another substance to be reduced in a chemical reaction and is itself oxidized.

relative atomic mass (A_r) The average mass of one atom of an element relative to one atom of carbon-12.

relative molecular mass (M_r) The mass of a molecule relative to one atom of carbon-12.

resonance Mixing Lewis structures to attempt to obtain the actual structure of a species.

room temperature and pressure (RTP) In this book, taken as 20°C and 1 atm pressure.

salt A substance made by the neutralization of an acid with a base.

saturated hydrocarbon A hydrocarbon that contains no carbon–carbon multiple bonds and cannot react with hydrogen, e.g. an alkane.

semiconductor A material that is not a good conductor of electricity at room temperature, but becomes conducting if certain other elements are added to it in small amounts.

spectrometer An instrument that measures the frequencies and intensities of emitted or absorbed light. (A simpler hand-held instrument is called a *spectroscope*.)

spectrum The pattern of frequencies (and the intensities) of light that is emitted (*emission spectrum*) or absorbed (*absorption spectrum*) by a species.

spontaneous A process is thermodynamically spontaneous when, once started, it continues without assistance until equilibrium is achieved.

standard ambient temperature and pressure (SATP) 25°C (298 K) and 100 kPa.

standard analytical method A sequence of experimental operations agreed by professional bodies, designed to optimize the accuracy and precision of analytical measurements.

standard enthalpy of formation ($\Delta_f H^{\circ}$) The standard enthalpy change that accompanies the formation of 1 mol of a substance from its constituent elements in their standard states.

Standard solution A solution of known concentration, which is used to calibrate the analytical method by which the concentration of unknown solutions is determined.

standard state A pure form of a substance at 1 bar pressure.

standard temperature and pressure (STP) 0°C and 1 atm pressure.

stoichiometric point Another name for *equivalence point*.

strong acid An acid that is completely ionized in solution.

strong base A base that is completely ionized in solution.

strongly acidic solution A solution with a pH below 3.

strongly basic solution A solution with a pH above 12.

sublimation Changing directly from a solid to a vapour.

substitution reaction A reaction in which an atom or group of atoms replaces another atom or group of atoms in a molecule.

Supercritical Fluid The physical state of a substance above its critical temperature and critical pressure.

supercritical fluid extraction (SFE) A type of compound extraction that uses a supercritical fluid as the solvent.

systematic errors Errors that cause measurements to be either higher or lower than the true value, and cannot be averaged out.

titration A technique in which a solution of known concentration (a standard solution) is reacted with a solution of unknown concentration, until equivalent quantities have reacted.

transition element (metal) A group of elements found between Groups 2 and 13 of the Periodic Table.

The Trench Effect A particular set of conditions that dangerously accelerates fire. The effect came to the public notice because of the 1987 King's Cross Fire.

triple point The particular temperature and external pressure at which the three phases (gas, liquid and solid) of that substance co-exist in stable thermodynamic equilibrium.

uncertainty (in a measurement) A quantity that indicates the degree of doubt in a measurement. It is the part of a measurement given after the ± sign. For example, if a temperature is recorded as 25.5 ± 0.4°C, the uncertainty is ± 0.4°C.

unsaturated hydrocarbon A hydrocarbon that contains carbon–carbon multiple bonds and can react with hydrogen, e.g. an alkene or alkyne.

van der Waals forces A collective term for the weakest attractions between molecules or atoms. Two major van der Waals forces are London dispersion forces and dipole–dipole interactions.

vaporization The conversion of a liquid to a gas.

vapour Commonly used to mean a gas above a liquid. Formally defined as a gas below its critical temperature.

vapour pressure The pressure of a vapour in equilibrium with a liquid.

VOCs (volatile organic compounds) are organic compounds of environmental importance that exists as gases at room temperature.

wavelength The length of a single wave of light of a particular frequency.

wavenumber (cm^{-1}) The number of wavelengths of light per centimetre.

weak acids/weak bases Acids or bases that are incompletely ionized in solution.

Index

Where a definition also appears in the glossary, the relevant book pages are given first, followed by the letter G.

A

B

C

P

Q

R

S

T

U

V

W

Y

Z

Valencies of Common Ions

Name*	Valency	Symbol
Ammonium	1	NH_4^+
Copper(I) or *cuprous*	1	Cu^+
Caesium	1	Cs^+
Potassium	1	K^+
Silver	1	Ag^+
Sodium	1	Na^+
Lithium	1	Li^+
Hydrogen	1	H^+
Barium	2	Ba^{2+}
Calcium	2	Ca^{2+}
Cobalt(II)	2	Co^{2+}
Copper(II) or *cupric*	2	Cu^{2+}
Iron(II) or *ferrous*	2	Fe^{2+}
Lead(II)	2	Pb^{2+}
Magnesium	2	Mg^{2+}
Mercury(II) or *mercuric*	2	Hg^{2+}
Nickel	2	Ni^{2+}
Strontium	2	Sr^{2+}
Tin(II) or *stannous*	2	Sn^{2+}
Zinc	2	Zn^{2+}
Aluminium	3	Al^{3+}
Chromium (III)	3	Cr^{3+}
Iron(III) or *ferric*	3	Fe^{3+}
Tin(IV) or *stannic*†	4	Sn^{4+}

*Common or alternative names are shown in italics.
†Does not exist as an ion, but exhibits the valency tabulated.

Name*	Valency	Symbol
Ethanoate or *acetate*	1	CH_3COO^-
Hydrogencarbonate or *bicarbonate*	1	HCO_3^-
Hydrogensulfate (*hydrogensulphate*)	1	HSO_4^-
Hydrogensulfite (*hydrogensulphite* or *bisulphite*)	1	HSO_3^-
Bromide	1	Br^-
Bromate or *bromate(V)*	1	BrO_3^-
Chloride	1	Cl^-
Chlorate or *chlorate(V)*	1	ClO_3^-
Cyanide	1	CN^-
Fluoride	1	F^-
Hydride	1	H^-
Hydroxide	1	OH^-
Iodide	1	I^-
Nitrate or *nitrate(V)*	1	NO_3^-
Nitrite or *nitrate(III)*	1	NO_2^-
Manganate(VII) or *permanganate*	1	MnO_4^-
Carbonate	2	CO_3^{2-}
Chromate or *chromate(VI)*	2	CrO_4^{2-}
Dichromate or *dichromate(VI)*	2	$Cr_2O_7^{2-}$
Oxide	2	O^{2-}
Sulfide or *sulphide*	2	S^{2-}
Sulfite, *sulfate(IV)* or *sulphite*	2	SO_3^{2-}
Sulfate, *sulfate(VI)* or *sulphate*	2	SO_4^{2-}
Nitride	3	N^{3-}
Phosphate	3	PO_4^{3-}